Systems Biology Approaches: Prevention, Diagnosis, and Understanding Mechanisms of Complex Diseases

Sanket Joshi • Rina Rani Ray
Moupriya Nag • Dibyajit Lahiri
Editors

Systems Biology Approaches: Prevention, Diagnosis, and Understanding Mechanisms of Complex Diseases

Springer

Editors
Sanket Joshi
Amity Institute of Microbial Technology
Amity University Rajasthan
Jaipur, Rajasthan, India

Rina Rani Ray
Department of Biotechnology
Maulana Abul Kalam Azad University
Haringhata, West Bengal, India

Moupriya Nag
University of Engineering and
Management, Kolkata
Kolkata, West Bengal, India

Dibyajit Lahiri
University of Engineering and
Management, Kolkata
Kolkata, West Bengal, India

ISBN 978-981-99-9461-8 ISBN 978-981-99-9462-5 (eBook)
https://doi.org/10.1007/978-981-99-9462-5

This Springer imprint is published by the registered company Springer Nature Singapore Pte Ltd.
The registered company address is: 152 Beach Road, #21-01/04 Gateway East, Singapore 189721, Singapore

Preface

At present, Systems Biology is considered an interdisciplinary approach and an important analytical tool, providing powerful and novel insights into various aspects of human health and disease. Acute and chronic disorders can be easily analyzed using this tool in a more vivid manner. Various types of complex disorders are considered to be a combination of immunological, genetic, environmental, and neurological factors.

Classical biological research has focused on isolated components rather than complex systems. This mechanism has helped in the characterization and identification of organism parts. However, this approach has not provided a clear view of the interaction of various components. Hence, Systems Biology emerges as a new field of study that helps in understanding organisms at the systems level, thereby unraveling the dynamics and structure of cells and organisms. This helps comprehend the entire organism and the interaction of its components.

This book aims to highlight the use of Systems Biology in understanding various diseases associated with organisms. This book provides a clear idea of how Systems Biology techniques can be used for studying both acute and chronic disorders.

Jaipur, Rajasthan, India	Sanket Joshi
Haringhata, West Bengal, India	Rina Rani Ray
Kolkata, West Bengal, India	Moupriya Nag
Kolkata, West Bengal, India	Dibyajit Lahiri

Contents

1 **Introduction to Systems Biology** 1
Nitish Rai, Namita Ashish Singh, Vivek Jain, Preet Jain, Alexander Choi, and Saurabh K. Sinha

2 **Systems Biology and Human Diseases** 27
R. Jasmine, Sowmiya Soundararajan, A. Sherlin Rosita, and H. Deena Priscilla

3 **Medical Applications of Systems Biology** 55
Devendra Singh, Perugu Shyam, Sunil Kumar Verma, and Anjali

4 **Systems Biology for Metabolic Disorder and Disease** 71
Mukta Sharma and Atul Singh

5 **Systems Biology Consortium for Infectious Diseases** 93
Ashwini Sunil Pisey and Gokul Shankar Sabesan

6 **Case Study-Based Approaches of Systems Biology in Addressing Infectious Diseases** 115
Sourav Dutta and Avijit Ghosh

7 **Systems Biology and Hospital-Associated Infections**............... 145
Sreejita Ghosh, Dibyajit Lahiri, Moupriya Nag, Rina Rani Ray, and Debasmita Bhattacharya

8 **Systems Biology Approach in Understanding Mitochondrial Disease** 173
Moupriya Nag, Debarati Chatterjee, Sumana Paul, Arina Guha, Debasmita Bhattacharya, and Dibyajit Lahiri

9 **Systems Approaches in Identifying Disease-Related Genes and Drug Targets**.. 195
Sohini Chakraborty and Satarupa Banerjee

10 **Systems Approach for Identifying Drug Targets by Computational Approaches** 257
Sonali S. Shinde, Komal B. Padule, Swapna L. Sawant, and Aniket P. Sarkate

11 **Systems Biology Approaches for Precision Medicine** 271
Subrahmanya Hegde, S. M. Andrian, Harshith Sajal, Anil K. K. Mallikarjunappa, Angel Benny, and Sumitha Elayaperumal

12 **Systems Biology Approaches for Autoimmune Diseases** 299
Sonali S. Shinde, Gresi D. Mate, Sneha S. Mali, Neha D. Naikwadi, Mohammad Abuzar, and Aniket P. Sarkate

13 **Systems Biology: A New Era for Vaccine Development; from Deductive Toward Inductive** 321
Usnik Debnath and Sahana Ghosh

14 **The Important Role of Systems Biology in Neuroscience and Neurology and Its Associated Diseases** 343
M. Haripriyaa and K. Suthindhiran

15 **Computational and Experimental Approaches to Decipher the Complexity of Diseases** 393
Sanket S. Rathod, Sonali S. Shinde, Prafulla B. Choudhari, Rakesh P. Dhavale, and Aniket P. Sarkate

16 **Complex Disease Prediction Using Systems Biology Approach** 415
Meghashree H. R., S. Yuvaraj, Sunitha C. Mesta, J. Satish Kumar, and E. Sumitha

17 **Systems Biology in Understanding the Human Gut Microbiome and Related Diseases Highlighting Metabolic Modeling and Analysis** 437
Priyanka Narad, Romasha Gupta, Sudeepti Kulshrestha, and Abhishek Sengupta

18 **Understanding Colon Cancer: Causes, Prevention, and Complementary Therapies Including Therapeutic Foods, Beverages, and Meditation** 467
Mahalakshmi S. Biradar, Shankar Thapa, Sonali S. Shinde, Aniket P. Sarkate, and Kalusing S. Padvi

19 **The Potential of Systems Biology to Understand the Tenets of Host-pathogen Interaction, Toxicology, and Aging** 489
Koushik Nayak, Sumanta Daw, and Pratick Singha

20 **Systems Biology Approaches as a Tool in Understanding Neurodevelopmental Disorders: Some Case Studies** 511
Anuvab Dey, Subhrojyoti Ghosh, Sourav Majhi, Aishee Nishat Hoque, and Aneshwa Chakrabarti

21 **Systems Biology Approaches for Cancer Biology** 537
Bandita Dutta, Dibyajit Lahiri, Moupriya Nag, Niladri Sarkar, Rina Rani Ray, and Debasmita Bhattacharya

22 **Prevention, Diagnosis, and Understanding Mechanisms of Metabolic Diseases Using Systems Biology** 561
Bhavani P. V, B. Vinutha, Renuka B, and B. B. Borse

Editors and Contributors

About the Editors

Sanket Joshi is currently serving as a Professor and Deputy Director at Amity Institute of Microbial Technology, Amity University Rajasthan, Jaipur, India. He served as Deputy Director, Oil & Gas Research Center (2018–2023), and an Application Specialist, at Central Analytical and Applied Research Unit, Sultan Qaboos University, Oman (2013–2023). Prof. Joshi has 19 years of academic teaching and research experience and 4 years of industrial R&D experience in India and Oman. While working in Indian pharma companies, he undertook several turnkey projects (both upstream and downstream) on antibiotics (β-lactams, macrolides, glycopeptides), antidiabetic drug (for type II diabetes), and Co-enzyme Q10. His current research interests encompass energy, microbial products, and environmental bioremediation. He serves as an Academic/Associate Editor for some of the highly reputed journals and was awarded the "NRI Senior Scientist Award" from the Microbiologists Society, India, during the academic year 2019–2020. Prof. Joshi was listed among the top 2% of highly cited scientists worldwide in the "Stanford University–Elsevier" list for excellence in scientific research in the disciplines of "Energy" and "Biotechnology" for 2022 and 2023.

Rina Rani Ray joined the Department of Biotechnology as associate professor in 2017. She has done her doctoral research in the Department of Microbiology, Bose Institute, Kolkata, India, and was awarded a Ph.D. degree from the University of Calcutta in 1996. Since then, she is working on different branches of microbiology, including enzyme biotechnology, metal bioremediation, and fungal biotechnology. She has published more than 75 research papers in various national and international journals. She serves on the editorial board of many reputed journals. She is the member of various scientific associations and was nominated fellows of Zoological Society, India, and Academy for Environment and Life Sciences, India.

Moupriya Nag is presently working as associate professor in the Department of Biotechnology, University of Engineering & Management, Kolkata, India. Her postdoctoral work was focused on exploring the aggregation kinetics and mechanism of action of peptides associated with neurodegenerative diseases in the presence of various membrane surfaces with solution NMR. Her research interests include single molecule biophysics of protein unfolding/refolding and protein aggregation kinetics and its mechanism of action, especially the amyloid proteins in neurodegenerative disorders. Recently, her research focuses on functional amyloids in bacteria and their relation in forming biofilms. Her work deals with developing novel antimicrobial and anti-biofilm compounds from natural sources, including plants, microbes, and green synthesized nanoparticles. She has about 5 years of teaching experience in biophysics, bioinformatics, bioprocess and biotechnology, enzyme technology, computing, molecular modeling, and drug design. She has also published many research articles in peer-reviewed international journals and authored or coauthored numerous book chapters. She is also a member of many scientific societies like Indian Biophysical Society and Indian Science Congress Association.

Dibyajit Lahiri is working as associate professor in the Department of Biotechnology, University of Engineering & Management, Kolkata, India. He has experience of working with various chromatographic techniques and isolation of bioactive compounds from different plant samples. His current research interest is precisely on biofilm isolated from human prosthesis and its inhibition by various novel "phyto and nano" compounds. He is keen to explore the molecular mechanism behind the removal of biofilm by natural compounds. His research arena also encompasses computational drug development by using bioinformatics. Dr. Lahiri was awarded the best teacher award and best paper awards from UEM, Kolkata. He is a life member of Indian Science Congress Association and Institution of Engineers (India).

Contributors

Mohammad Abuzar Department of Pharmaceutical Quality Assurance, Dr. D. Y. Patil Institute of Pharmaceutical Sciences and Research, Pune, Maharashtra, India

S. M. Andrian Department of Biotechnology and Bioinformatics JSS Academy of Higher Education and Research, Mysore, Karnataka, India

Anjali CSIR-Central Drug Research Institute, Lucknow, Uttar Pradesh, India

Satarupa Banerjee Department of Biotechnology, School of Biosciences and Technology, Vellore Institute of Technology, Vellore, Tamil Nadu, India

Angel Benny Department of Biotechnology and Bioinformatics JSS Academy of Higher Education and Research, Mysore, Karnataka, India

Debasmita Bhattacharya Department of Basic Science and Humanities, Institute of Engineering and Management, Kolkata, West Bengal, India

University of Engineering and Management, Kolkata, West Bengal, India

Department of Basic Science and Humanities, Institute of Engineering and Management, Salt Lake, University of Engineering and Management, Kolkata, West Bengal, India

P. V. Bhavani Pentavalent Bio Sciences Private Limited, Bangalore, India

Mahalakshmi S. Biradar Department of Pharmaceutical Chemistry, Hillside College of Pharmacy, Rajiv Gandhi University of Health Science, Bengaluru, India

B. B. Borse PPSFT/SFS Department, Central Food Technological Research Institute, Mysore, Karnataka, India

Aneshwa Chakrabarti Department of Chemistry and Chemical Biology, IIT ISM Dhanbad, Dhanbad, India

Sohini Chakraborty Department of Biotechnology, School of Biosciences and Technology, Vellore Institute of Technology, Vellore, Tamil Nadu, India

Debarati Chatterjee Department of Biotechnology, University of Engineering and Management, Kolkata, West Bengal, India

Alexander Choi Department of Neurology, University of Arizona-Phoenix, Phoenix, AZ, USA

Prafulla B. Choudhari Department of Pharmaceutical Chemistry, Bharati Vidyapeeth College of Pharmacy, Kolhapur, Maharashtra, India

Sumanta Daw Department of Microbiology, Maulana Abul Kalam Azad University of Technology, Nadia, West Bengal, India

Usnik Debnath Department of Microbiology, VijaygarhJyotish Ray College, Kolkata, India

H. Deena Priscilla Department of Biotechnology and Bioinformatics, Bishop Heber College (Autonomous), Trichy, Tamil Nadu, India

Affiliated to Bharathidasan University, Trichy, Tamil Nadu, India

Anuvab Dey Department of Biological Sciences and Biological Engineering, Indian Institute of Technology Guwahati, North Guwahati, Assam, India

Rakesh P. Dhavale Department of Pharmaceutics, Bharati Vidyapeeth College of Pharmacy, Kolhapur, India

Bandita Dutta Microbiology Research Laboratory, Department of Biotechnology, Maulana Abul Kalam Azad University of Technology, Kolkata, West Bengal, India

Sourav Dutta GCC Biotech India Pvt Ltd, Kolkata, West Bengal, India

Sumitha Elayaperumal Department of Biotechnology and Bioinformatics JSS Academy of Higher Education and Research, Mysore, Karnataka, India

Avijit Ghosh GCC Biotech India Pvt Ltd, Kolkata, West Bengal, India

Sahana Ghosh Faculty, Department of Microbiology, VijaygarhJyotish Ray College, Kolkata, India

Sreejita Ghosh Microbiology Research Laboratory, Department of Biotechnology, Maulana Abul Kalam Azad University of Technology, Haringhata, West Bengal, India

Subhrojyoti Ghosh Department of Biotechnology, IIT Madras, Chennai, Tamil Nadu, India

Arina Guha Department of Biotechnology, University of Engineering and Management, Kolkata, West Bengal, India

Romasha Gupta Amity Institute of Biotechnology, Amity University, Noida, Uttar Pradesh, India

M. Haripriyaa Marine Biotechnology and Bioproducts Lab, Department of Biomedical Sciences, School of Biosciences and Technology, Vellore Institute of Technology, Vellore, India

Subrahmanya Hegde Department of Biotechnology and Bioinformatics JSS Academy of Higher Education and Research, Mysore, Karnataka, India

Aishee Nishat Hoque Department of Biological Sciences and Biological Engineering, Indian Institute of Technology Guwahati, North Guwahati, Assam, India

Preet Jain Department of Biotechnology, Mohanlal Sukhadia University, Udaipur, Rajasthan, India

Vivek Jain Department of Pharmaceutical Sciences, Mohanlal Sukhadia University, Udaipur, Rajasthan, India

R. Jasmine Department of Biotechnology and Bioinformatics, Bishop Heber College (Autonomous), Trichy, Tamil Nadu, India

Affiliated to Bharathidasan University, Trichy, Tamil Nadu, India

Sudeepti Kulshrestha Amity Institute of Biotechnology, Amity University, Noida, Uttar Pradesh, India

Dibyajit Lahiri Department of Biotechnology, University of Engineering and Management, Kolkata, West Bengal, India

Sourav Majhi Department of Biological Sciences and Biological Engineering, Indian Institute of Technology Guwahati, North Guwahati, Assam, India

Sneha S. Mali Department of Pharmaceutical Quality Assurance, Dr. D. Y. Patil Institute of Pharmaceutical Sciences and Research, Pune, Maharashtra, India

Anil K. K. Mallikarjunappa Department of Environmental Science JSS Academy of Higher Education and Research, Mysore, Karnataka, India

Gresi D. Mate Department of Pharmaceutical Quality Assurance, Dr. D. Y. Patil Institute of Pharmaceutical Sciences and Research, Pune, Maharashtra, India

Meghashree H. R. Department of Biotechnology and Bioinformatics JSS Academy of Higher Education and Research, Mysore, Karnataka, India

Sunitha C. Mesta Department of Microbiology and Tissue Culture, JSS Academy of Higher Education and Research, Mysore, Karnataka, India

Moupriya Nag Department of Biotechnology, University of Engineering and Management, Kolkata, West Bengal, India

Neha D. Naikwadi Department of Pharmaceutical Quality Assurance, Dr. D. Y. Patil Institute of Pharmaceutical Sciences and Research, Pune, Maharashtra, India

Priyanka Narad Amity Institute of Biotechnology, Amity University, Noida, Uttar Pradesh, India

Koushik Nayak Department of Medical Biotechnology, Ramakrishna Mission Vivekananda Educational and Research Institute, Narendrapur, Kolkata, West Bengal, India

Komal B. Padule Department of Quality Assurance Technique, Dr. D. Y. Patil Institute of Pharmaceutical Sciences and Research, Pune, Maharashtra, India

Kalusing S. Padvi Department of Chemical Technology, Dr. Babasaheb Ambedkar Marathwada University, Aurangabad, Maharashtra, India

Sumana Paul Department of Biotechnology, University of Engineering and Management, Kolkata, West Bengal, India

Ashwini Sunil Pisey Department of Microbiology, Faculty of Medicine, Manipal University College Malaysia, Melaka, Malaysia

Nitish Rai Department of Biotechnology, Mohanlal Sukhadia University, Udaipur, Rajasthan, India

Sanket S. Rathod Department of Pharmaceutical Chemistry, Bharati Vidyapeeth College of Pharmacy, Kolhapur, Maharashtra, India

Rina Rani Ray Microbiology Research Laboratory, Department of Biotechnology, Maulana Abul Kalam Azad University of Technology, Kolkata, West Bengal, India

B. Renuka Promic Svasthya Private Limited, Mysore, India

Gokul Shankar Sabesan Department of Microbiology, Faculty of Medicine, Manipal University College Malaysia, Melaka, Malaysia

Harshith Sajal Department of Biotechnology and Bioinformatics JSS Academy of Higher Education and Research, Mysore, Karnataka, India

Niladri Sarkar Microbiology Research Laboratory, Department of Biotechnology, Maulana Abul Kalam Azad University of Technology, Kolkata, West Bengal, India

Aniket P. Sarkate Department of Chemical Technology, Dr. Babasaheb Ambedkar Marathwada University, Aurangabad, Maharashtra, India

J. Satish Kumar Department of Environmental Science, JSS Academy of Higher Education and Research, Mysore, Karnataka, India

Swapna L. Sawant Department of Pharmaceutics, Dr. D. Y. Patil Institute of Pharmaceutical Sciences and Research, Pune, Maharashtra, India

Abhishek Sengupta Amity Institute of Biotechnology, Amity University, Noida, Uttar Pradesh, India

Mukta Sharma School of Life Sciences and Technology, IIMT University, Meerut, Uttar Pradesh, India

A. Sherlin Rosita Department of Biotechnology and Bioinformatics, Bishop Heber College (Autonomous), Trichy, Tamil Nadu, India

Affiliated to Bharathidasan University, Trichy, Tamil Nadu, India

Sonali S. Shinde Department of Chemical Technology, Dr. Babasaheb Ambedkar Marathwada University, Aurangabad, Maharashtra, India

Perugu Shyam National Institute of Technology, Warangal, Telangana, India

Pratick Singha Department of Marine Biotechnology, Bharathidasan University, Tiruchirappalli, Tamil Nadu, India

Atul Singh School of Life Sciences and Technology, IIMT University, Meerut, Uttar Pradesh, India

Devendra Singh National Institute of Technology, Warangal, Telangana, India

Namita Ashish Singh Department of Microbiology, Mohanlal Sukhadia University, Udaipur, Rajasthan, India

Saurabh K. Sinha Department of Pharmaceutical Sciences, Mohanlal Sukhadia University, Udaipur, Rajasthan, India

Sowmiya Soundararajan Department of Biotechnology and Bioinformatics, Bishop Heber College (Autonomous), Trichy, Tamil Nadu, India

Affiliated to Bharathidasan University, Trichy, Tamil Nadu, India

E. Sumitha Department of Biotechnology and Bioinformatics JSS Academy of Higher Education and Research, Mysore, Karnataka, India

K. Suthindhiran Marine Biotechnology and Bioproducts Lab, Department of Biomedical Sciences, School of Biosciences and Technology, Vellore Institute of Technology, Vellore, India

Shankar Thapa Department of Pharmacy, Universal College of Medical Sciences, Bhairahawa, Nepal

Sunil Kumar Verma BN College of Engineering and Technology, Lucknow, Uttar Pradesh, India

B. Vinutha G. R Medical College, Hospital and Research Centre, Mangalore, India

S. Yuvaraj Department of Pharmaceutical Chemistry, Cauvery College of Pharmacy, Mysore, Karnataka, India

Introduction to Systems Biology 1

Nitish Rai, Namita Ashish Singh, Vivek Jain, Preet Jain, Alexander Choi, and Saurabh K. Sinha

Abstract

The area of system biology is a combination of experimental large-scale measurement techniques and computational approaches to interrogate, model and understand life at the level of organismal system itself. The applications of system-wide approaches in understanding the factors that affect human health and diseases is opening a novel area with great prospects. In this chapter, the area of system biology has been described with the current status, and its impact on the individual and population. The application of system biology in developing a framework for modelling living systems and diseases is discussed. Further the utility of systems biology in molecular diagnostics, genetic techniques, deletion/duplication analysis, targeted variant analysis, DNA methylation, pathway-based biomarker analysis, gene interaction maps and disease genes identification etc. is described. These current knowledge of significant contribution in the field may stimulate further discussion and debates in other areas in addition to systems biology. In this way, the systems medicine will realize its full potential and promises within societal standards.

N. Rai (✉) · P. Jain
Department of Biotechnology, Mohanlal Sukhadia University, Udaipur, Rajasthan, India
e-mail: nitish.rai@mlsu.ac.in

N. A. Singh
Department of Microbiology, Mohanlal Sukhadia University, Udaipur, Rajasthan, India

V. Jain · S. K. Sinha
Department of Pharmaceutical Sciences, Mohanlal Sukhadia University, Udaipur, Rajasthan, India

A. Choi
Department of Neurology, University of Arizona-Phoenix, Phoenix, AZ, USA

S. Joshi et al. (eds.), *Systems Biology Approaches: Prevention, Diagnosis, and Understanding Mechanisms of Complex Diseases*,
https://doi.org/10.1007/978-981-99-9462-5_1

Keywords

System biology · Computational approaches · System medicine · Pathophysiology · Disease model

1.1 Introduction

It has been close to a decade since the term 'systems biology' was first utilized in the context of contemporary biological nomenclature. It currently encompasses a significant portion of the biological sciences and is utilized to address a wide variety of research challenges involving both humans and a diverse range of model organisms (Chuang et al. 2010).

While interactions between multiple cells contribute to determine a tissue response, different organs made up of different tissue types further coordinate to generate a physiological response. These connections serve as the foundation for a network structure that encodes the core design principle directing the physiological response dynamics (Somvanshi and Venkatesh 2014). In essence, the field of systems biology revolves around the principle that the phenotype of any individual living organism is a reflection of the simultaneous multitude of molecular interactions from various levels occurring at any one time, combined in a holistic manner to produce such a phenotype (Ayers and Day 2015; Friboulet and Thomas 2005).

A network is formed by the interactions of DNA, RNA, transcription factors, enzymes, proteins and metabolites. The network-based approach is widely utilized to find the potential genes causing various diseases and syndromes (Chen et al. 2012). The authors of an excellent article by Lage et al. (2007) used this type of analysis to build the network of interactions between many proteins and genes for human disorders. They discovered a total of 669 links using network analysis and Bayesian predictor, of which 298 properly identified the disease-causing proteins as leading possibilities. A total of 870 links without disease-causing genes were discovered, and these can serve as new candidates for the examination of diseases such as type II diabetes, cancer, inflammatory bowel disease, Alzheimer's, retinal pigmentosa, coronary heart disease and amyotrophic lateral sclerosis. To uncover the genes and interaction network for various uncommon diseases as obesity, metabolic syndrome, ageing and cancer, similar investigations were conducted by numerous researchers (Wang et al. 2007; Ehrhart et al. 2021). This method of network reconstruction and analysis has excellent potential to pinpoint numerous genes and proteins that cause diseases and to aid in the selection of more effective medication targets (Cuccato et al. 2009; Zhang et al. 2014; Kovács et al. 2019).

This chapter provides an overview about various aspects of system biology along with their relevance in human health and disease. Further, the application of system biology in modelling living system is also discussed.

1.2 Systems Biology: Current Status

Disease can be caused by bio-molecular interactions. As a result, it is critical to investigate interacting partners and examine the network for accurate illness diagnosis and processes (Houtman et al. 2005; Terentiev et al. 2009). Intrinsic and extrinsic factors, like gene mutations and environmental stressors, influence network behaviour, resulting in illness (Del Sol et al. 2010; Stelling et al. 2004). The natural network's robustness is weakened as a result of disturbances, resulting in pathological conditions (Kitano et al. 2004). Understanding the dynamic regulatory aspects of the integrated network is critical for disease state perturbation analysis. Some disease symptoms are monogenic, produced by mutations in a single gene, whereas the vast majority are multigenic, caused by mutations in multiple genes. Understanding protein-protein interaction networks is critical for dealing with multigenic disease (Zhu et al. 2007). Protein interaction networks are used to discover new genes, investigate network characteristics, find sickness sub-networks and classify network-based disorders (Dunker et al. 2001). This method aids in the understanding of complicated systems underlying systemic illnesses such as tumour growth, neurological, cardiovascular, autoimmune and metabolic syndromes (Somvanshi and Venkatesh 2014) (Fig. 1.1).

Disease states could be viewed as a network property rather than a single protein or gene malfunction (Jordán et al. 2012). Defects in regulatory interaction networks collectively define a disease phenotype. The majority of multi-genetic disorders are caused by interactions and feedback loops that span several temporal and

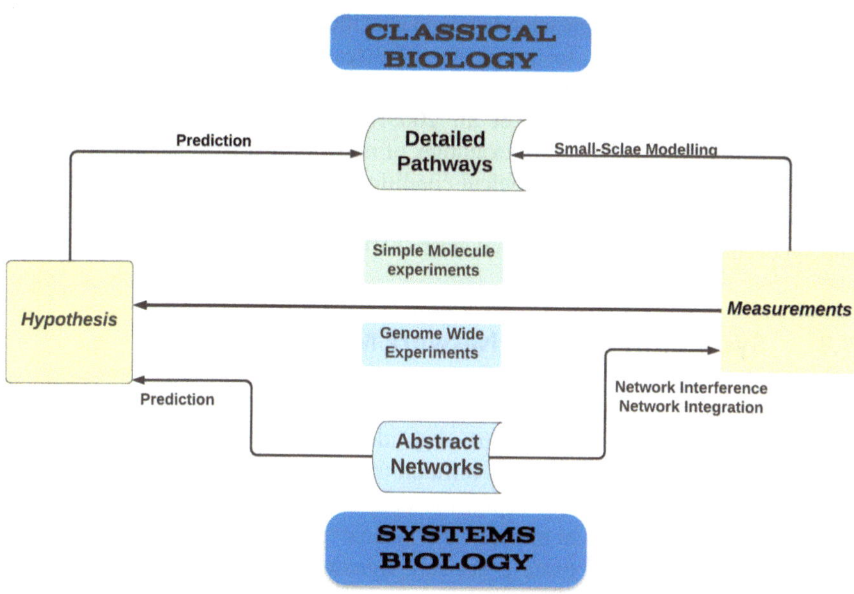

Fig. 1.1 System biology

geographical scales, from the cellular to the organism level (Klipp et al. 2010). The organism and phenotypes are thus shown to be emergent features of the interaction networks of all individual cells, without which processes such as tissue repair, immunity and homeostasis would be impossible. Cancer, autoimmunity and diabetes are consequently caused by errors in cellular information processing (Klipp et al. 2010). Recent research on a variety of diseases and their corresponding physiological abnormalities has yielded voluminous information on protein and gene deficiencies associated with a specific function. Attempts have been made to link a disease's phenome to its associated genome. The Online Mendelian Inheritance in Man (OMIM) (Hamosh et al. 2005) repository contains 1284 diseases and 1777 disease genes, with new gene—disease links being discovered on a regular basis. The diseasome is the accumulation of knowledge about the majority of disease diseases and their associated mutations in genes or disease genes (Goh et al. 2007; Somvanshi and Venkatesh 2014).

Furthermore, disease information is represented (Goh et al. 2007) by establishing two types of networks: the 'Human disease Network' (HDN) and the 'Disease gene Networks' (DGN). Nodes in the HDN represent disorders and are connected by two disorder nodes if they share a common genetic flaw. In DGN, nodes represent genes, and two genes are linked if they cause the same condition. According to the HDN research, 867 of the 1284 disorders have at least one link to other disorders, while 516 belong to the large component, indicating that the majority of diseases are caused by other diseases. In DGN, however, of the 1777 illness genes, 1377 are linked to other disease genes, and 903 are part of a large component. It implies that the protein products of genes involved in common diseases are likely to interact with other proteins via the interactome. Furthermore, many genes have synchronized and elevated expression levels in specific tissues, indicating an increased interconnection that leads to the cluster of disease hubs. This demonstrates the global functional interdependence of multiple disorders, resulting in a network arrangement of diseases (Park et al. 2009; Somvanshi and Venkatesh 2014). Diabetes, obesity and cancer are examples of network diseases that are precursors to a wide range of other physiological conditions. The network-based technique is being extensively used to identify candidate genes responsible for various diseases and syndromes (Somvanshi and Venkatesh 2014).

1.3 Systems Biology for Network Modelling

Despite the fact that network generation and graph-based analysis are critical for understanding disease relationships and identifying disease possibilities, the dynamic nature of these networks is difficult to grasp. Networks are static representations of systems that cannot depict the time evolution of a state in response to mutations and environmental impacts. It is critical to model interactions and mechanisms in order to extract disease network dynamics (Somvanshi and Venkatesh 2014). A study define disease modelling as the mathematical representation of a disease's molecular to physiological mechanisms (Tegnér et al. 2009).

Modelling disease states are required for computer studies of disease processes, which include quantification of biological data (Cho et al. 2006). Kinetic equations are founded on first principles, such as the rule of mass action, the Michalis-Menten equation and the Hill equation. Because these mechanisms are stochastic, stochastic modelling tools such as Monte-Carlo simulations, Foker-Plank formulations and rule-based approaches can be used to investigate disease pathway dynamics. Feedback effects are included in mathematical representations of dynamics. These models are required to gain mechanistic insights and predict parameter outcomes in a variety of situations or perturbations (Del Sol et al. 2010; Somvanshi and Venkatesh 2014). Kinetic modelling and flux balance analysis have been used to simulate the dynamical and steady-state responses of network systems (Chen et al. 2010; Moreno-Sánchez et al. 2008). Modelling software tools and rule-based and compartmental modelling techniques reduce computing work. The modelling approach chosen is determined by the available data, model scale and illness perturbation representation (Chandra 2009; Karp et al. 2005; Klipp et al. 2010) (Fig. 1.2).

1.3.1 Disease Modelling and Model Integration

It is challenging to intuitively grasp the dynamic nature of these networks, despite the fact that network creation and graph-based analysis are effective for comprehending the relationships between various diseases and identifying disease

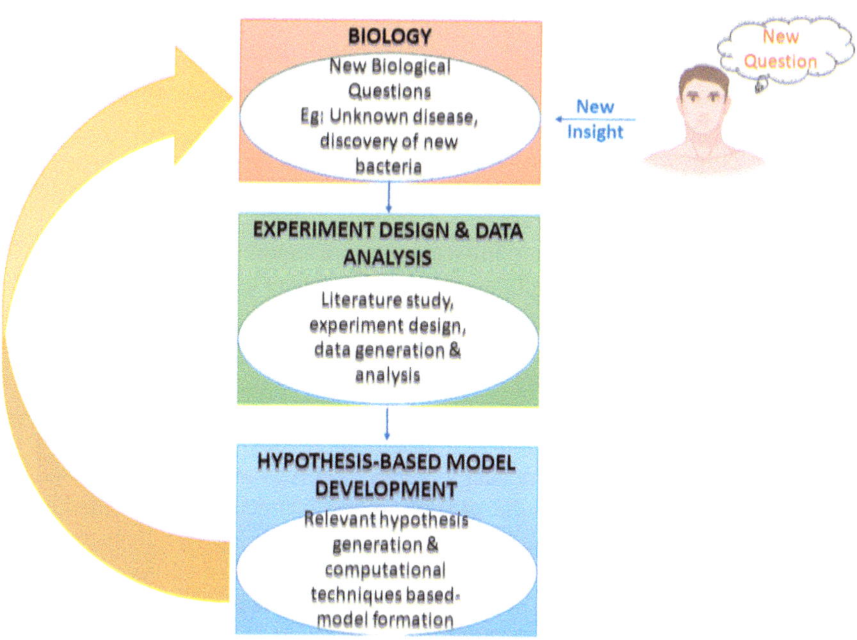

Fig. 1.2 System biology and its utility in model development

candidates (Kann 2007; Younesi and Hofmann-Apitius 2013). Networks represent systems in a static manner and are unable to show how the status of the system has changed over time due to changes in the many governing factors, such as mutations and environmental influences (Ma'ayan 2011). In order to extract dynamic behaviour of networks describing diseases, understanding the relationships and mechanisms of these networks is therefore crucial (Liu and Thiagarajan 2012; Rintala et al. 2022).

Disease modelling can be defined as the mathematical depiction of a disease mechanism from the molecular to the physiological level (Tegnér et al. 2009; Huppert and Katriel 2013). For computational study of the disease process, which essentially entails converting the biological data into quantitative terms, modelling of disease states is crucial (Cho et al. 2006). Based on the kinds of reactions and interactions present in the routes, kinetic equations are constructed using fundamental principles, the law of mass action, Michalis-Menten equations and Hill equations. The dynamics of the disease pathways can be studied using a variety of stochastic modelling techniques, including Monte Carlo simulations, Foker-Plank formulations and rule-based approaches, because these mechanisms are likewise stochastic in nature. The underpinning dynamics are mathematically represented by incorporating the impacts of feedback systems.

In order to forecast the results for the set of parameters under various situations or perturbations, these models are necessary (Vicini 2008; Del Sol et al. 2010). As of now, the dynamical and steady-state responses of the network systems have been modelled, using kinetic modelling and flux balance analysis methodologies (Moreno-Sánchez et al. 2008; Sahu et al. 2021). To further lessen the computing cost, rule-based and compartmental modelling methodologies, as well as a number of modelling software tools, are now accessible. The quantity of data provided, the model's scale and the way the perturbation causing the disease is represented in the modelling strategy.

For the purposes of biological modelling and drug development, a variety of network modelling tools and databases are listed in Table 1.1 (Klipp et al. 2010; Shaukat et al. 2021).

In fact, building the intricate multi-scale network requires extensive data mining. For system parameters, different databases and pertinent experimental data can be sorted. These models will be improved by growing research, the availability of more trustworthy data and knowledge of interaction networks. The databases for pathways and protein interactions are listed in Table 1.2.

Another crucial step in effective disease modelling is the building of hierarchical models at different network levels (Kirouac et al. 2010; Heiner and Gilbert 2013; Zhang et al. 2017). One can conduct a hierarchical regulation study based on the illness pathway inferred from the protein-protein interaction and metabolic networks to comprehend the transcriptional, signalling and metabolic contributions to the disease phenotype (Ideker and Sharan 2008; Handorf and Klipp 2012). The concept is to build models for specific networks and examine dynamics at corresponding scales (Terentiev et al. 2009; Koschorreck et al. 2007). Analysis of the dynamics of a particular phenotype is made more difficult by the integration of these models of sub-networks with different timelines and heterogeneity in the cellular environment (Caetano-Anollés et al. 2019; Przytycka and Kim 2010; Wilkinson

Table 1.1 Pathway and protein interaction databases

S. No.	Pathway	Web address
1	aMAZE (molecular function, interaction and biochemical process database)	http://www.amaze.ulb.ac.be/
2	BIND (biomolecular interaction database)	http://binddb.org
3	BioCarta (pathway database)	https://maayanlab.cloud/Harmonizome/dataset/Biocarta+Pathways
4	BioCyc pathway/genome databases	https://biocyc.org/
5	BioGrid	https://thebiogrid.org/
6	BRENDA (comprehensive enzyme information system)	https://www.brenda-enzymes.org/
7	DIP (database of interacting proteins)	https://dip.doe-mbi.ucla.edu/dip/Main.cgi
8	Metabolic pathways from all domains of life	https://metacyc.org/
9	HPIDB	https://hpidb.igbb.msstate.edu/
10	HPRD (human protein reference database)	https://www.hsls.pitt.edu/obrc/index.php?page=URL1055173331
11	IntAct	https://www.ebi.ac.uk/intact/
12	KEGG (Kyoto encyclopedia of genes and genomes)	https://www.genome.jp/kegg/
13	MINT (molecular INTeraction database)	https://mint.bio.uniroma2.it/
14	NCI/nature pathway interaction database	http://www.ndexbio.org/
15	Panther (genes, proteins and pathway database)	http://www.pantherdb.org/pathway/
16	Pathway commons	https://www.pathwaycommons.org/
17	Pathway Hunter tool	https://www.hsls.pitt.edu/obrc/index.php?page=URL1251224885
18	Reactome	https://reactome.org/
19	STRING (functional protein–protein interaction database)	https://string-db.org/
20	WikiPathways	https://www.wikipathways.org/index.php/WikiPathways

2009; Chandra 2009). The capacity to identify targets is improved by integrating various modelling techniques into a unified framework since they each offer a unique perspective.

1.3.2 Disease Dynamics Analysis: Network Perturbations

It is possible to understand the dynamics of these networks using disease models based on the quantitative analysis of network systems and the integration of various types of data (Buschur et al. 2020). This will allow for the generation of various predictions about the physiological systemic response. One can investigate various network dynamics possibilities by performing a perturbation analysis (Ma'ayan 2011). To track the system's response in different settings, perturbation analysis can

Table 1.2 Database and tools for network modelling and target identification

S.No.	Databases	Weblink
1	BATMAN-TCM	http://bionet.ncpsb.org.cn/batman-tcm/
2	Biomodels.net (biological modelling tool)	https://www.ebi.ac.uk/biomodels/
3	Cell designer	https://www.celldesigner.org/help/CDH_Database_T.html
4	CellML (XML markup language and model repository)	https://www.cellml.org/
5	COBRA (constraint-based reconstruction and analysis)	https://opencobra.github.io/
6	COPASI—COmplex PAthway SImulator	https://copasi.org/
7	Cytoscape (software tool for analysing networks)	https://cytoscape.org/
8	DIP (database of interacting proteins)	https://dip.doe-mbi.ucla.edu/dip/Main.cgi
9	Drugbank (drug and drug target database)	https://go.drugbank.com/
10	iPATH—Interactive pathways explorer	https://pathways.embl.de/
11	JDesigner (graphical modelling for biochemical networks)	https://www.sys-bio.org/
12	Matador (drug target database)	https://www.hsls.pitt.edu/obrc/index.php?page=URL1209757429
13	PDTD (potential drug target database)	http://allie.dbcls.jp/pair/PDTD;potential+drug+target+database.html
14	SABIO-RK (relational database for biochemical kinetics)	http://sabio.h-its.org/
15	STITCH (chemical association networks)	http://stitch.embl.de/
16	SuperTarget (web source for analysing drug-target interaction)	https://ngdc.cncb.ac.cn/databasecommons/database/id/564
17	SYCAMORE (computational modelling and simulation tool)	http://sycamore.h-its.org/sycamore/
18	TDR: Targets database	https://www.tdrtargets.org/
19	TIde: Target identification (scanning drug targets in network models)	http://sysbio.molgen.mpg.de/tide
20	TTD: Therapeutic targets database	http://db.idrblab.net/ttd/

be used to change system parameters, protein concentration, node and edge diameters, interaction intensity and nodal knockouts. This can result in complex computations where a wide range of parameter space must be compared to the phenotypic response of the system. However, the parameter space should fall within medically reasonable limits. The system's response to a single parameter alteration at a time, while maintaining other system-native parameters at their original values, can also provide more clarity on the dynamics of the disease. A matrix of sensitive characteristics that characterize the phenotypes of health and sickness can be produced by such an approach.

The nodes that are crucial for the structural and dynamic integrity of the network can also be identified using this research (Koster et al. 2009; Wang et al. 2014). By adjusting the network parameters and determining the exact objective, one can improve the network performance once the crucial and dominant interactions have been identified. A fundamental foundation for synthetic biology treatments to regulate diseases is also provided by this kind of analysis. Another potential method for quantifying the impact of system parameter governance on network performance is flux control analysis (Moreno-Sánchez et al. 2008; Basler et al. 2016). Recently, these techniques were used to discover deregulation of lung cell metabolism on SARS-CoV-2 infection (Nanda and Ghosh 2021).

Hit and lead optimization studies are used to further screen the list of potential pharmacological targets. Following the identification of the targets, their biological characteristics are investigated and the appropriate (antagonist/or modulating) molecules are developed to regulate the target and subsequent network dynamics (Heifetz et al. 2018). The substance perturbs the system in an experimental way. In clinical trials, the effectiveness of the medications is mostly assessed by seeing how the disease develops at various levels of pathophysiology (Post et al. 2005). For the purpose of establishing a correlation between a drug's effectiveness and a disease's response, it is crucial to analyse these models for illness progression with respect to medications. Additionally, data gathered from the evaluation of drug targets in clinical studies can be found at various nodes of the network and given back into the model to improve predictability. These models aid in anticipating the illness state's time course progression in terms of how well a medicine would work. Furthermore, by analysing the morbidity and mortality linked to diseases, these models can be used to forecast the time courses of allied problems in particular diseases like diabetes (De Gaetano et al. 2008) and cardiovascular events. This forecast will aid in developing the therapy regimens that can be used beforehand to reduce the danger.

Furthermore, the analysis of dynamic biological networks may benefit from using DynOVis. It enables the user to look into long-term changes in biological networks and visualizes dynamic perturbations in those networks. Identification of the network nodes' biological relevance is made simple by the integration of data from numerous internet databases (Kuijpers et al. 2019).

1.3.3 Physiological and Immunological Models

The network models' ability to accurately translate their reaction into physiologically quantifiable output has its limitations. The synchronization of molecular models with the tissue and organ level dynamics suffers from a significant gap as a result of the different scales of the models (Kumar et al. 2006). This gap is being actively closed by research in physiological engineering and tissue engineering. One such project is the Physiome Project (Hunter et al. 2005). The thorough comprehension of the system with a high level of abstraction from the molecular to physiological level is crucial for the development of physiological models. It involves comprehending the physiological system and all of its components, including signal processing, cellular and tissue architecture, biophysical, chemical and electro-mechanical properties, as well as many layers of system information processing (Butcher et al. 2004; Dhillon et al. 2020). If these models are effectively developed, they can accurately imitate physiological behaviour and have a high level of predictability. Numerous diseases, including the growth of tumours, Parkinson's disease, Alzheimer's disease, respiratory disorders, diabetic nephropathy and osteoporosis, have been studied using physiological models, with remarkable results.

In addition, immunological models that can depict host-pathogen interactions are in high demand, particularly for lethal infectious illnesses like AIDS (Bertoletti et al. 2010). Since the pathogens, such as bacteria, fungi and viruses, are foreign entities to the native physiology, the features of host-pathogen interactions have their own unique peculiarities that must be considered during model construction. With the aid of these immunological models, it is possible to find the best therapeutic targets that are selective antagonists of infections without upsetting the immune system. These models may be used to characterize the dynamics of the infection inside the body and develop more effective anti-infection treatments.

The ideas of virtual patients and virtual humans are growing with the development of sophisticated disease network modelling, illness progression modelling and physiological modelling (Hunter et al. 2013; Zhang and Tyson 2022). Virtual patients are made up of a variety of these mathematical representations of the disease state's structural and dynamic features (Klinke 2008). This makes an effort to accurately depict human physiology and illness. The platform for conducting investigations like perturbation analysis, drug trials, clinical trials and simulating the associated physiological states might be the virtual human models. Recently, scientist develop virtual patient cohorts for cardiac models. Although it presents many technical difficulties, the creation of comprehensive biophysical virtual patient cohorts for cardiac models has a lot of potential. The conversion of virtual patient cohorts into industrial and clinical tools will require interaction with industry and regulators. Applications of modelling, such as physiologically based pharmacokinetic models, offer an example of how to create and evaluate models for usage in clinical settings. Similar to this, the industry's employment of models during the early, low-regulatory stages of product, medicine or device development will offer a real-world setting for creating and using virtual patient cohorts. A cheap method for calculating the effect of patient variability on physiology, pathology and therapies is offered by virtual cohorts of cardiac models. An essential step in converting computational models of the heart into practical applications in industry and medicine is the capacity to run low-cost simulations across a useful representation of a patient population (Niederer et al. 2020).

1.3.4 Networks as Drug Target

The illness network is an interaction network made up of disease-marker biomolecules that affect how the network functions in relation to the phenotype of the disease (Yıldırım et al. 2007; Butcher et al. 2004). Due to the interaction networks' inherent complexity, an emergent quality that is necessary for smooth operation is produced by a number of different components. To find targets, it is crucial to identify the elements needed for these particular features, whose disturbance alters the phenotypic response and causes diseases. This suggests that various component nodes may be engaged in a disease state, which further affects the multigenic functional modules. Cancer and the metabolic syndrome are examples of diseases that have several players at play and result in phenotypic problems rather than being single target diseases.

These disorders have scarcely been cured by conventional therapy. In order to reduce the impact of the disease, a network-based strategy can target network subcircuits rather than a single node (Zanzoni et al. 2009). Network dynamics can be controlled by one or more targets (Neduva and Russell 2006). Complex disorders have been found to impact hub genes and proteins and targeting them can result in a monotonous response that might be unwelcome and cause a number of side effects. Therefore, minimizing side effects should be the target selection method.

By connecting pharmaceuticals with their particular protein targets, Yıldırım et al. (2007) created the Drug-Target network based on US-FDA authorized medications. Drug targets are shown to have distinct topological signatures from critical proteins and to inhabit particular areas in interactome networks (Yıldırım et al. 2007). Illness genes are located on the peripheral of the disease network, which can be efficiently targeted without suffering many side effects, according to topological study of the disease network (Hase et al. 2009). Network dynamic analysis can be used to find several targets at various network levels, such as those involved in diabetes and cancer (Moller 2001). As a result, choosing a target and developing pharmacological molecules that can affect network response may be a better alternative than choosing medications that only change how disease genes function (Pujol et al. 2010).

Targeting entire networks of genes or proteins becomes more challenging (Barabási 2007). This calls for a thorough understanding of the network's regulatory dynamics and the availability of the precise system characteristics. By performing perturbation analysis as previously mentioned, it is possible to carry out dynamic simulations of network reaction. The benchmark for target identification for the network can be established as the ideal set of parameters that produces the healthy (non-diseased) system response.

The system responsiveness will essentially be tracked in relation to the network's node and edge strengths. It is crucial to identify the target such that redundant pathways are blocked or restricted because illness states are robust in nature and can even use redundant pathways to maintain their state (Kitano 2007). These methods, which are still developing, have a great deal of potential for finding more sensible pharmacological targets. For target identification, Table 1.2 offers a variety of databases and technologies.

1.4 Application of System Biology in Pathophysiology

1.4.1 Molecular Diagnostics

The value of systems biology comes to light readily through its application in everyday clinical biochemistry. In defining homeostatic range and population database for cellular and extracellular milieu, emergency states of sepsis, cancer, unknown or previously undescribed genetic disorders are accurately identified in time for the appropriate medical care. In diseases of development and pathologic aging, the carefully sequestered and delivered neurotransmitters, the construction and repair of the cell membrane phospholipid bilayer and the motility and fission of mitochondria may become degraded in feed forward cascades of susceptibility to stressors and cellular work. The technologies allowing for a molecular diagnostic to define health and disease, including ELISA, Western Blot with SDS-PAGE, confocal immunofluorescent microscopy and immunohistochemistry, ionization, separation by chemical chromatography, NMR mass spectrometry and the ability to perform massively parallel tandem single cell DNA and RNA sequencing and proteomics, necessitate a systems biology approach to the organism in a population as a reference standard. This is readily illustrated in the case of epigenetic histone methylation patterns related to stimuli. Of pathological importance, childhood abuse leads not only to direct physical harm but lasting psychological damage that may manifest in an overt post-traumatic stress disorder characterized by flashbacks, avoidance, and emotional disorder and it may manifest even years after a conscious reconciliation with a covert somatic disorder, termed conversion disorder or functional neurologic disorder. The same trauma may also be shared via a mass calamity, such as war or epidemic, and studied as a uniform independent variable. However, the impact of the stressor is anything but uniform, with some individuals escaping unscathed and others with lasting harm. The epigenetic effects of trauma, e.g. in serotonin or noradrenaline signalling (see pathway-based biomarker analysis) accordingly vary and may be studied as a proxy of trauma response. To interpret the effects, the data must be handled statistically. According to the framework of systems biology, the responses will follow patterns of regulation/dysregulation and resilience/susceptibility. In other words, effects that are balanced may be termed adaptive or 'healthy', including at the cellular level as per the term 'physiological hormesis,' and those that are unbalanced leading to processes of cellular toxicity and death.

The sequencing of the 3.1 billion DNA bases comprising the human genome in 2001 as global collaboration and the subsequent advancements in the speed and throughput of genetic sequencing from traditional Sanger method to massively parallel tandem sequencing ushered in a new era of molecular diagnostics where the myriad arrays of quaternary amino acid structures, modifications and functions could now be correlated to processes in gene expression, with both constitutively active and inducible expression, with modifiers such as nuclear actin in chromatin modifying the level of activity of RNA polymerase. Thus, biologic function could now be identified based on structural coding elements in their entirety, leading to concepts of transcriptome—and proteome—based analysis of function and disease, and application to cellular biology such as the lineage differentiation of stem cells for cell models and identification of pathogens. There remains a great deal to be understood about the human genome as to the variability and unpredictability of phenotypic expression, including the significance of reduced penetrance of mutant genes, the principle of anticipation and effect of the sex chromosome on differential pathogenesis and mosaic expression of mitochondrial DNA.

1.4.2 Genetic Techniques

Harnessing the leverage of systems biology informed by accurate molecular diagnostics may be aided by appropriate use of genetic and ancillary techniques, which are outlined below. Target polymerase chain reaction (PCR) provides high accuracy with both sensitivity and specificity to the variants tested, and at practically low turnaround time and cost. It thus remains ideal for a wide variety of clinical applications including genetic syndromes, cancer and infectious disease. Other techniques include single nucleotide polymorphism (SNP) microarrays, array comparative genomic hybridization (CGH)/virtual karyotyping, Sanger sequencing, Southern blot, panel/pathway sequencing and whole exome or genome sequencing (WES or WGS). Many of these have longer turnaround and higher expertise or operating cost, so are reserved for special applications (Katsanis and Katsanis 2013). This review focuses on gene sequencing. Less complete analysis such as Fluorescence In Situ Hybridization (FISH) and comparative genomic hybridization (CGH) may find use in specific applications including developmental disorders and cancer classification/monitoring. Cytogenetic applications have been mostly replaced by SNP-array CGH (aCGH) that uses probes of varying size to detect both chromosomal and genomic rearrangements and deletions/duplications to a resolution of a few kilobases in size.

1.4.3 Deletion/Duplication Analysis

An older, however, still evolving genetic method is fluorescent in situ hybridization with whole-chromosome paints of peripheral blood lymphocytes. This may be applied to genetic screening of the high-risk newborn, developmental delay in paediatric diagnostic assessment, and effects of radiation exposure (Kodama et al. 2001).

Deletions below the limit of detection by FISH or aCGH may be detected on a Southern blot, which uses restriction enzymes to cleave DNA to fragments that are matched with a cDNA probe, and mutation detected due to absence of the hybridization band (classify electrophoresis-separated DNA fragments depending on sequence differences). It remains the gold standard for immunoglobulin and T cell rearrangements. Southern blot may be helpful for detecting insertion of a transposable element such as an Alu or LINE repeat that may cause disease.

1.4.4 Targeted Variant Analysis

Gene expression microarray techniques were developed in the 1990s including DNA/RNA and BioChip, involving fixation of nucleic acids with known sequences onto a support structure with labelled sample molecules, with both quantity and sequence of the sample DNA or RNA identified by hybridization to the probe molecule (Schena et al. 1995). The techniques include phenotype-specific SNP panels, copy number/structural homology via array CGH, expressed RNA detection of multiple types and methylation detection. Multiplex PCR sequencing provides another solution that may be faster and accurate to the level of point mutation that may change protein function or cause truncation of the protein with loss of function or toxic gain of function. A variety of solutions offer genetic panels based on systems of disease, such as movement disorders, epilepsy, mitochondrial disorders. Companies including Invitae, Blueprint Genetics and Athena Diagnostics provide mostly phenotype or disease-specific panels generally with the traditional health care provider-directed model, whereas direct-to-consumer commercial services tend to offer genome-wide SNP tests, which are generally broader but may be less disease-focused.

1.4.5 Sequence Analysis: Nucleic Acid Amplification of Target Sequence

Synthesis of DNA using the polymerase chain reaction (PCR) in vitro via a heat-stable polymerase (helicobacter) was first described by Mullis and Faloona in 1987. Traditional Sanger method of binding each nucleotide to a separate polymerase is time intensive, giving way in common use to faster tandem, massively parallel sequence methods. Models of embryonic development using comparative sequence analysis have shown areas of gene regulation that are highly conserved across species, including non-coding sequences that modify or enhance sequencing (Woolfe et al. 2005). Using chromatin immunoprecipitation with massively parallel sequencing (ChIP-seq) method, 55,000 tissue-specific enhancers have been identified in the human genome, in which around 50,000 are cell-type specific (Heitzmann et al. 2009). A complementary approach is defined through computational identification of tissue specificity in transcription factor binding sites based on sequence homology (Hallikas et al. 2006), which has been adapted to embryonic human heart development validated with an in vivo zebrafish reporter assay.

Analysing splicing mutations may be inferred from mRNA for a gene of interest by reverse transcription PCR (RT-PCR), where a cDNA is made from RNA by reverse transcription which may be primed to be a substrate for PCR, and amplified products may then be analysed by electrophoresis, which may then infer a splicing mutation by abnormal product size. RT-PCR may be used more broadly when amplification of RNA is desired, e.g. to detect gene expression at low levels in certain cell populations.

Real-time PCR provides accuracy to point mutations and small insertions/deletions, with monitoring provided by hybridization and hydrolysis (aka TaqMan) probes. This newer technique developed in the 1990s does not target an endpoint like traditional PCR, but rather using the Thermus aquaticus (Taq) polymerase, the fluorescent emission of a dual-labelled fluorescent probe is captured using the 5′-3′ exonuclease activity of Taq polymerase. The target DNA is quantified based on incremental fluorescent emission per cycle of the PCR reaction, and has found common use in combination with reverse transcription PCR to provide a reliable quantification of mRNA expression. Real-time PCR has undergone exponential growth in use to fields of infectious disease, oncology, autoimmune disease and gene therapy. However, a main limitation for RT-qPCR is its ability only to detect known sequences by predesigned probes, and this has fuelled more recent studies using unbiased RNA sequencing methods, with capture including non-coding and novel RNA molecules.

Single nucleotide polymorphism (SNP) consists of variability on the single nucleotide level. Minisatellites (variable number tandem repeats) consisting of around 30 base pair sequences that are repeated may be detected by Southern analysis and are useful for identity, as are linkage analysis, consisting of analysis of polymorphic markers close to gene of interest in a family, and to test whether family members are likely to have inherited the mutant gene (Harada and Korf 2013).

1.4.6 DNA Methylation: Luminometric Methylation Assay (LUMA) and Other Techniques

Genomic DNA methylation varies from clear, reproducible biological effects such as gender (e.g. male vs. female people with Parkinson disease) and in less patterned ways, for instance obesity with or without inflammatory phenotype. Capturing genomic level DNA methylation has been defined by a luminometric methylation assay (LUMA) based on composite DNA cleavage by methylation-sensitive restriction enzymes and polymerase assay by Pyrosequencing, allowing for high throughput of up to 48 samples in 20 min using only 200–500 ng of DNA per sample (Karimi et al. 2006). The DNA methylation data capture genome-wide markers of methylation, CpG (cytosine-guanine) islands, which can be divided into functionally correlated genes such as systemic immune response, mitochondrial function and cytoskeletal organization, with differences found in disease state, e.g. Parkinson disease compared to controls (Chuang et al. 2017). However, subsequent studies have ranked other methods as providing more accurate determination of genomic methylation by quantifying deoxycytidine (dC) and methylated cytosine (5mC) through liquid chromatography coupled with tandem mass spectrometry (LC-MS/MS), which requires a smaller DNA sample and is more sensitive to small differences in global DNA methylation to 5% (Song et al. 2005), with issues of less availability and higher cost of testing. Another method involves bisulphite conversion of DNA and PCR amplification of long interspersed nuclear element 1 (LINE-1) conservative sequences, which also can be more sensitive to methylation changes, and has been adapted to high throughput analysis of cancer samples for prognostic information to guide treatment.

1.4.7 RNA Single Cell Sequencing and Array-Based Testing

RNA sequencing (RNA-Seq) allows for detection of novel transcripts that microarray does not detect (Garber et al. 2011; Zhao et al. 2014). Accordingly, RNA-Seq publications and budgets have continued to increase over the past decade. Pipelines differ and typically include trimming, alignment to a reference genome or transcriptome, quantification by assignment to gene or transcript and normalization to reduce sequencing bias, followed by differential expression analysis across conditions (Corchete et al. 2020). Like other next-generation sequence methods there are advantages in RNA-Seq of unbiased genome coverage, small amount of sample required for PCR amplification, and high resolution to the base-pair level. Special applications include single-cell RNA-Seq, spatial RNA-Seq, circulating cell-free RNA (cfRNA), and simultaneous protein detection via antibodies tagged with oligonucleotide barcodes. These methods may be combined with epigenetic mapping of open chromatin and cis-regulatory elements using ATAC-Seq, probe-based enrichment of transcripts of interest or 3′ polyA capture of messenger RNA. A complementary method, ChIP-Seq (chromatin immunoprecipitation sequencing) identifies DNA binding sites of transcription factors and other chromatin-associated proteins across the genome that is combined with RNA-Seq to examine how DNA-associated proteins influence gene expression (Mimura et al. 2014).

1.4.8 Non-coding RNAs: Biomarker of Disease

Non-coding RNAs, which are not translated to protein, include transfer RNA and ribosomal RNA, as well as small RNAs including micro-RNA (miRNA) and small interfering RNA (siRNA). MiRNA particularly is found to be stable in blood and regulate multiple genes per molecule. These properties lend to possibility of biomarker of disease, and this has been explored in neurodegenerative diseases such as Alzheimer's disease (Wang et al. 2012) and Parkinson's disease (Goh et al. 2019). MiRNAs, particularly those serving a tumour suppressor role, have also been targeted for cancer therapy (Mollaei et al. 2019).

1.4.9 DNA: Whole Exome and Whole Genome Sequencing

Whole exome sequencing (WES) provides clear advantages over SNP-array CGH in greater sensitivity to genetic disorders. This has been demonstrated particularly in developmental disorders, i.e. failure to meet growth, motor, speech and learning or behavioural-social milestones or childhood onset of motor, cognitive or behavioural disorder, with an increase in sensitivity from 5.7% aCGH to 20% with WES (Martinez-Granero et al. 2021). However, with WES providing the analysis of only the 1.5% exome portion of the whole genome, there are clear limitations including the absence of analysing introns or other important non-coding regions for gene regulation. For instance, splice sites for gene transcription may be located in either exons or introns and WES may not detect the intronic mutation. The Exome Sequencing, like Whole Genome Sequencing may be combined with a machine learning algorithm such as Invitae's Moon, to match known pathogenic variants with relevant clinical features, including age of onset, phenomenology, and classification of deficits according to the genetic database. These curated databases feature daily updates using automated literature searches with weighted phenotypic features according to degree of overlap with the disease (Apollo database). This then allows for future report to update a variant of unknown significance into either a likely pathogenic or pathogenic variant or a likely benign/benign variant.

Next-generation sequencing (NGS) involves sequencing of target nucleic acid fragments and analysing the data using bioinformatic tools, which captures nearly all of the euchromatic genome. It is useful when searching broadly for pathogen outside of a panel. It also has been shown to provide higher accuracy of mutation detection than either Sanger sequencing or quantitative PCR in non-small cell lung cancer, including 'non-hotspot' mutations (Gao et al. 2016). However, the use in clinical practice remains limited due to longer turnaround time, requirement for higher number of samples, higher cost of running the test and bioinformatics requirement (Souf 2016). Due to the complexity of the whole genome, there is a solution of pairing massive data with a machine algorithm. This cuts down the amount of data by around 1/1000 (3 B nucleotide bases to 4 M SNPs of established pathologic significance listed on gnomAD and dbSNP genetic libraries), as a sampling of the data can provide accuracy. Companies such as Invitae (Diploid) provide a phenotype-genotype correlation based on an AI-driven decision-support algorithm (Moon), matching submitted key clinical features (e.g. spasticity, ataxia, cognitive impairment), gender, age of onset through search of available genetic libraries. The clinician/geneticist uploads NGS data containing features of single nucleotide or copy number variation (SNV/CNV). As the genetic databases are growing, harmonization becomes necessary to ensure accurate classification of a variant such as by the American College of Medical Genetics and Genomics (ACMG) classification. Databases with centralized harmonization focused on genetic disease include the National Center for Biotechnology Information's ClinVar database, (http://www.ncbi.nlm.nih.gov/clinvar) and OMIM, http://www.omim.org. Other databases directed at genetic loci include www.hgvs.org/dblist/dblist.html, https://www.narcis.nl/, National Human Genome Research Institute ClinGen, http://www.clinicalgenome.org, MitoMap, http://www.mitomap.org/MITOMAP/HumanMitoSeq, DECIPHER, http://decipher.sanger.ac.uk.

1.4.10 Detection of Nucleotide Expansion Disorders

Nucleotide expansion disorders are implicated in a large array of neurologic disorders including myotonic dystrophy type 1, a number of spinocerebellar ataxias, Friedrich's ataxia, Huntington Disease and CANVAS. These disorders may be detected by specific PCR probe at the genome locus of expansion, or in the case of CANVAS requires a bioinformatics approach due to the expansion location in intronic DNA. Importantly, these disorders are not detected by standard whole exome sequencing, and specified algorithms have been developed for WGS data such as ExpansionHunter, STRetch and TREDPARSE. However, a method termed exSTRa (expanded short tandem repeat algorithm) has been developed to use detect repeat expansion disorders using both WES and WGS data (Tankard et al. 2018).

1.4.11 Pathway-Based Biomarker Analysis

One of the key initial applications of molecular diagnostics is to define normative values, health changes related to age and variability not directly linked to health. For example, in the gut microbiome there is a predominance of *Bacteroides, Firmicutes* and *Actinobacteria*, etc. The gut microbiome is comprised of 10/13 microorganisms, including bacteria, protozoans, metazoans (parasites) and viruses. Systems biology directs the work of putative disease gene identification beyond building population level references to use for Control versus Disease (e.g. Genome-Wide Association Studies (GWAS)), to build a methodology that is practical and replicable of precision medicine and personalized therapy, such as nutrition. The diet determines dominant enterotype, or diet-related microbiome such as increase in *Prevotella* with carbohydrate-rich diet and increase in *Bacteroides* in animal fat and protein-rich diet. With age, there is increased individual variability, reduction in butyrate/SCFA producing species, and increased prevalence of hypochloraemia and bacterial overgrowth linked to H pylori infection. Diet, e.g. ketogenic diet × 16 weeks has shown major effects on the microbiome in mice, with increase in SCFA-producing microbiota, and related increase in paramedian cerebral blood flow including appetite regulation centre in hypothalamus (Ma et al. 2018). Strains linked with anti-inflammation include Bacteroides and Firmicutes (SCFA-producing) and Lactobacillus (GABA, serotonin synthesis) and modulation of microbiome and monitoring as a biomarker of disease have been of growing interest for clinical trials. Microbiome testing via 16S rRNA gene sequencing has become increasingly utilized, and Illumina Miseq sequencing provides high throughput of stool microbiota and metabolite profiling.

Leading from established libraries, the genomic, transcriptomic and proteomic analytic techniques are the mainstay of modern biomarker development. A biomarker is a biologically meaningful proxy of organism state or function most commonly applied to the diagnosis and prognosis of disease, but it may also be defined in terms of normal or healthy functioning. Molecular diagnostics has shown greatest inroads to date in informing biomarkers of therapeutic agents in cancer (Chen et al. 2004; Liang et al. 2019). Other applications are in profiling tissue response to standard of care therapy such as muscle mitochondrial profiling pre versus post strength training in young versus old healthy subjects, and emerging fields of personalized medicine involving pharmacogenomics screening, or markers of disease in response to therapy, such as FDG-PET (Fluorodeoxyglucose-positron emission tomography) in early Alzheimer's Disease (AD) brain pre/post aerobic exercise. Rather than an isolated biomarker, pathway-based biomarker development locates panels of biomarkers for diagnosis and therapy. One important methodology of pathway-based biomarkers through molecular diagnostics is the development of library screening, which can accelerate the development and increase efficacy of medical or surgical therapy. For instance, a number of clinical therapeutics that have advanced to market approval have originated from library screening (Beutler 2009). More than 1/3 of new drugs on the market between 1981 and 2014 were derived from natural products, and both drug discovery and the understanding of pharmacodynamic effects have been furthered by genetic sequencing technology applied to gene expression microarray and RNA-seq methods. These elucidate the molecular mechanism of the drug and its differential effects (pharmacogenomics). The development of RNA-Seq high-throughput sequencing has brought personalized medicine to reality, for instance in cancer profiling (Liu et al. 2021).

1.4.12 Gene Interaction Maps

Gene interaction maps appeared in the early 2000s with advancements in gene sequencing and the advent of genomic libraries. These maps consist of pairwise measures of the degree to which a given gene influences the expression or function of other genes in the organism, thereby producing a map of systems of gene function that cluster and are interactive. This allows for building an 'interactome' based on available cluster analysis to inform cell and disease specific systems biology, [with imputation modelling]. Using publicly available databases allows for wide scope meta-analysis of normal and pathologic cell lines and tissues for gene expression, which may be applied to interactive mapping. One solution to gene interaction is de-cycling of the pathway derived gene maps (e.g. KEGG) and performing an enrichment analysis (Subramanian et al. 2005), such as by using GSEA software (GSEA (gsea-msigdb.org)) and population comparison statistics to generate association/enrichment scores such as mixed effects linear models. However, limitations include lack of generalizability outside the data modelled, and many pathways are highly overlapping such that interaction may not be detected, rendering the defined interactome incomplete. A complementary approach has been to define disease modules based on gene concordance and protein interaction validation across studies, e.g. from GWAS to disease-associated SNPs, including most recently use of artificial learning neural networks (Dwivedi et al. 2020).

1.4.13 Disease Genes Identification

One of the primary sources of molecular diagnostics knowledge has been gene expression RNA-Seq experiments. These discoveries from exome and genome-wide association studies have included thousands of disease-associated genes as curated in the Online Mendelian Inheritance in Man (OMIM, URL <https://omim.org/statistics/entry>), which are linked to external genome databases such as ExAC and gnomAD and catalogued in ClinVar (Amberger et al. 2019). Another open-source gene disease association database, DisGeNET (<https://www.disgenet.org>) integrates human data with inferences from mouse and rat models with integration of CTD, UniProt, Rat Genome Database and Mouse Genome Database, and data from Genome-wide association studies (GWAS), Genetic Association Database (GAD), Literature-derived Human Gene-Disease Network (LHGDN), GeneRIF and BeFree. Putative disease-associated gene expression may be examined gene-wise to determine *p*-value for differential expression of a gene across phenotypes, which works well for highly penetrant, causative genes following the

Mendelian law of inheritance. Database of mRNA and protein from different species such as RefSeq may be searched against a sample and match to target sequence for gene mapping such as using BLAST (basic local alignment search tool). Given that variability of expression is the mode of most genes in the human genome with significant effect of the environment, an alternative method using gene expression matrices becomes required to account for the prevalent interactive effects. Gene set enrichment analysis (GSEA) as described above allows for a transcriptome or given set of expressed genes to be assessed across phenotypes to look for differential expression (Shi and Walker 2007). The statistic method to generate association score may include linear regression such as t-test, signal-to-noise (SNR), non-parametric (e.g. Wilcoxon) or non-linear, e.g. artificial neural network and subsequently generates enrichment score (ES) based on permutation *p*-value.

1.5 Conclusions

Network modelling and analysis are required to treat network disorders (systemic diseases). This approach has several advantages, including the identification of rational drug targets, effective drug design with few side effects, effective therapeutic strategies, diagnosis of disease source rather than symptoms, early and reliable diagnosis of diseases using predictive models and rational toxicological and drug safety assessments leading to improved healthcare. This strategy has enormous potential in biomedical research and pharmaceutical firms, as it can cut clinical trial efforts. Thus, illness systems engineering will revolutionize human health with modern treatments and personalized medicine. Global research and academic collaboration are required to address these difficulties and achieve speedier results. With developments in computational systems biology and other interdisciplinary disciplines and technology, a 'in silico Clinic' is on the horizon.

Acknowledgement The authors are grateful to the authorities of Mohanlal Sukhadia University, Udaipur and the University of Arizona-Phoenix, United States for supporting this work.

Conflict of InterestThe authors declare no conflict of interest.

References

Amberger JS, Bocchini CA, Scott AF, Hamosh A (2019) OMIM.org: leveraging knowledge across phenotype–gene relationships. Nucleic Acids Res 47(D1):D1038–D1043

Ayers D, Day PJ (2015) Systems medicine: the application of systems biology approaches for modern medical research and drug development. Mol Biol Int 2015:1

Barabási AL (2007) Network medicine—from obesity to the "diseasome". N Engl J Med 357(4):404–407

Basler G, Nikoloski Z, Larhlimi A, Barabási AL, Liu YY (2016) Control of fluxes in metabolic networks. Genome Res 26(7):956–968

Bertoletti A, Maini MK, Ferrari C (2010) The host–pathogen interaction during HBV infection: immunological controversies. Antivir Ther 15(3_suppl):15–24

Beutler JA (2009) Natural products as a foundation for drug discovery. Curr Protoc Pharmacol 46(1):9–11
Buschur KL, Chikina M, Benos PV (2020) Causal network perturbations for instance-specific analysis of single cell and disease samples. Bioinformatics 36(8):2515–2521
Butcher EC, Berg EL, Kunkel EJ (2004) Systems biology in drug discovery. Nat Biotechnol 22(10):1253–1259
Caetano-Anollés G, Aziz MF, Mughal F, Gräter F, Koç I, Caetano-Anollés K, Caetano-Anollés D (2019) Emergence of hierarchical modularity in evolving networks uncovered by phylogenomic analysis. Evol Bioinform Online 15:1176934319872980
Chandra N (2009) Computational systems approach for drug target discovery. Expert Opin Drug Discovery 4(12):1221–1236
Chen HW, Yu SL, Chen JJ, Li HN, Lin YC, Yao PL, Chou HY, Chien CT, Chen WJ, Lee YT, Yang PC (2004) Anti-invasive gene expression profile of curcumin in lung adenocarcinoma based on a high throughput microarray analysis. Mol Pharmacol 65(1):99–110
Chen Q, Wang Z, Wei D (2010) Progress in the applications of flux analysis of metabolic networks. Chin Sci Bull 55:2315–2322
Chen Y, Gan M, Jiang R, Zhang W (2012) Constructing human phenome-interactome networks for the prioritization of candidate genes. Stat Interface 5(1):137–148
Cho CR, Labow M, Reinhardt M, van Oostrum J, Peitsch MC (2006) The application of systems biology to drug discovery. Curr Opin Chem Biol 10(4):294–302
Chuang HY, Hofree M, Ideker T (2010) A decade of systems biology. Annu Rev Cell Dev Biol 26:721–744
Chuang YH, Paul KC, Bronstein JM, Bordelon Y, Horvath S, Ritz B (2017) Parkinson's disease is associated with DNA methylation levels in human blood and saliva. Genome Med 9:1–2
Corchete LA, Rojas EA, Alonso-López D, De Las RJ, Gutiérrez NC, Burguillo FJ (2020) Systematic comparison and assessment of RNA-seq procedures for gene expression quantitative analysis. Sci Rep 10(1):19737
Cuccato G, Della Gatta G, di Bernardo D (2009) Systems and synthetic biology: tackling genetic networks and complex diseases. Heredity 102(6):527–532
De Gaetano A, Hardy T, Beck B, Abu-Raddad E, Palumbo P, Bue-Valleskey J, Pørksen N (2008) Mathematical models of diabetes progression. Am J Physiol Endocrinol Metab 295(6):E1462–E1479
Del Sol A, Balling R, Hood L, Galas D (2010) Diseases as network perturbations. Curr Opin Biotechnol 21(4):566–571
Dhillon BK, Smith M, Baghela A, Lee AH, Hancock RE (2020) Systems biology approaches to understanding the human immune system. Front Immunol 11:1683
Dunker AK, Lawson JD, Brown CJ, Williams RM, Romero P, Oh JS, Oldfield CJ, Campen AM, Ratliff CM, Hipps KW, Ausio J, Nissen MS, Reeves R, Kang C, Kissinger CR, Bailey RW, Griswold MD, Chiu W, Garner EC, Obradovic Z (2001) Intrinsically disordered protein. J Mol Graph Model 19(1):26–59. https://doi.org/10.1016/s1093-3263(00)00138-8. PMID: 11381529
Dwivedi SK, Tjärnberg A, Tegnér J, Gustafsson M (2020) Deriving disease modules from the compressed transcriptional space embedded in a deep autoencoder. Nat Commun 11(1):856
Ehrhart F, Willighagen EL, Kutmon M, van Hoften M, Curfs LM, Evelo CT (2021) A resource to explore the discovery of rare diseases and their causative genes. Sci Data 8(1):124
Friboulet A, Thomas D (2005) Systems biology—an interdisciplinary approach. Biosens Bioelectron 20(12):2404–2407
Gao J, Wu H, Shi X, Huo Z, Zhang J, Liang Z (2016) Comparison of next-generation sequencing, quantitative PCR, and sanger sequencing for mutation profiling of EGFR, KRAS, PIK3CA and BRAF in clinical lung tumors. Clin Lab 62(4):689–696
Garber M, Grabherr MG, Guttman M, Trapnell C (2011) Computational methods for transcriptome annotation and quantification using RNA-seq. Nat Methods 8(6):469–477
Goh KI, Cusick ME, Valle D, Childs B, Vidal M, Barabási AL (2007) The human disease network. Proc Natl Acad Sci 104(21):8685–8690

Goh SY, Chao YX, Dheen ST, Tan EK, Tay SS (2019) Role of MicroRNAs in Parkinson's disease. Int J Mol Sci 20(22):5649
Hallikas O, Palin K, Sinjushina N, Rautiainen R, Partanen J, Ukkonen E, Taipale J (2006) Genome-wide prediction of mammalian enhancers based on analysis of transcription-factor binding affinity. Cell 124(1):47–59
Hamosh A, Scott AF, Amberger JS, Bocchini CA, McKusick VA (2005) Online Mendelian Inheritance in Man (OMIM), a knowledgebase of human genes and genetic disorders. Nucleic Acids Res 33(suppl_1):D514–D517
Handorf T, Klipp E (2012) Modeling mechanistic biological networks: an advanced Boolean approach. Bioinformatics 28(4):557–563
Harada S, Korf BR (2013) Overview of molecular genetic diagnosis. Curr Protoc Hum Genet 76(1):9–1
Hase T, Tanaka H, Suzuki Y, Nakagawa S, Kitano H (2009) Structure of protein interaction networks and their implications on drug design. PLoS Comput Biol 5(10):e1000550
Heifetz A, Southey M, Morao I, Townsend-Nicholson A, Bodkin MJ (2018) Computational methods used in hit-to-lead and lead optimization stages of structure-based drug discovery. Methods Mol Biol 1705:375–394
Heiner M, Gilbert D (2013) Biomodel engineering for multiscale systems biology. Prog Biophys Mol Biol 111(2–3):119–128
Heitzmann N, Seidel T, Opitz A, Hetmanek A, Wecker C, Fischer M, Ufer S, Schmidmaier R, Neuhaus B, Siebeck M, Stürmer K (2009) Facilitating diagnostic competences in simulations in higher education. Front Learn Res 7(4):1–24
Houtman JC, Barda-Saad M, Samelson LE (2005) Examining multiprotein signaling complexes from all angles: the use of complementary techniques to characterize complex formation at the adapter protein, linker for activation of T cells. FEBS J 272(21):5426–5435
Hunter P, Smith N, Fernandez J, Tawhai M (2005) Integration from proteins to organs: the IUPS physiome project. Mech Ageing Dev 126(1):187–192
Hunter P, Chapman T, Coveney PV, De Bono B, Diaz V, Fenner J, Frangi AF, Harris P, Hose R, Kohl P, Lawford P (2013) A vision and strategy for the virtual physiological human: 2012 update. Interface Focus 3(2):20130004
Huppert A, Katriel G (2013) Mathematical modelling and prediction in infectious disease epidemiology. Clin Microbiol Infect 19(11):999–1005
Ideker T, Sharan R (2008) Protein networks in disease. Genome Res 18(4):644–652
Jordán F, Nguyen TP, Liu WC (2012) Studying protein–protein interaction networks: a systems view on diseases. Brief Funct Genomics 11(6):497–504
Kann MG (2007) Protein interactions and disease: computational approaches to uncover the etiology of diseases. Brief Bioinform 8(5):333–346
Karimi M, Johansson S, Stach D, Corcoran M, Grandér D, Schalling M, Bakalkin G, Lyko F, Larsson C, Ekström TJ (2006) LUMA (LUminometric methylation assay)—a high throughput method to the analysis of genomic DNA methylation. Exp Cell Res 312(11):1989–1995
Karp PD, Ouzounis CA, Moore-Kochlacs C, Goldovsky L, Kaipa P, Ahrén D, Tsoka S, Darzentas N, Kunin V, López-Bigas N (2005) Expansion of the BioCyc collection of pathway/genome databases to 160 genomes. Nucleic Acids Res 33(19):6083–6089
Katsanis SH, Katsanis N (2013) Molecular genetic testing and the future of clinical genomics. Nat Rev Genet 14(6):415–426
Kirouac DC, Ito C, Csaszar E, Roch A, Yu M, Sykes EA, Bader GD, Zandstra PW (2010) Dynamic interaction networks in a hierarchically organized tissue. Mol Syst Biol 6(1):417
Kitano H (2007) Towards a theory of biological robustness. Mol Syst Biol 3(1):137
Kitano H, Oda K, Kimura T, Matsuoka Y, Csete M, Doyle J, Muramatsu M (2004) Metabolic syndrome and robustness tradeoffs. Diabetes 53(suppl_3):S6–S15
Klinke DJ (2008) Integrating epidemiological data into a mechanistic model of type 2 diabetes: validating the prevalence of virtual patients. Ann Biomed Eng 36:321–334
Klipp E, Wade RC, Kummer U (2010) Biochemical network-based drug-target prediction. Curr Opin Biotechnol 21(4):511–516

Kodama Y, Pawel D, Nakamura N, Preston D, Honda T, Itoh M, Nakano M, Ohtaki K, Funamoto S, Awa AA (2001) Stable chromosome aberrations in atomic bomb survivors: results from 25 years of investigation. Radiat Res 156(4):337–346

Koschorreck M, Conzelmann H, Ebert S, Ederer M, Gilles ED (2007) Reduced modeling of signal transduction–a modular approach. BMC Bioinformatics 8(1):1–24

Koster ES, Rodin AS, Raaijmakers JA, Maitland-vander Zee AH (2009) Systems biology in pharmacogenomic research: the way to personalized prescribing? Pharmacogenomics 10(6):971–981

Kovács IA, Luck K, Spirohn K, Wang Y, Pollis C, Schlabach S, Bian W, Kim DK, Kishore N, Hao T, Calderwood MA (2019) Network-based prediction of protein interactions. Nat Commun 10(1):1240

Kuijpers TJ, Wolters JE, Kleinjans JC, Jennen DG (2019) DynOVis: a web tool to study dynamic perturbations for capturing dose-over-time effects in biological networks. BMC Bioinformatics 20(1):1

Kumar N, Hendriks BS, Janes KA, de Graaf D, Lauffenburger DA (2006) Applying computational modeling to drug discovery and development. Drug Discov Today 11(17–18):806–811

Lage K, Karlberg EO, Størling ZM, Olason PI, Pedersen AG, Rigina O, Hinsby AM, Tümer Z, Pociot F, Tommerup N, Moreau Y (2007) A human phenome-interactome network of protein complexes implicated in genetic disorders. Nat Biotechnol 25(3):309–316

Liang Y, Feng G, Wu L, Zhong S, Gao X, Tong Y, Cui W, Qin Y, Xu W, Xiao X, Zhang Z (2019) Caffeic acid phenethyl ester suppressed growth and metastasis of nasopharyngeal carcinoma cells by inactivating the NF-κB pathway. Drug Des Devel Ther 13:1335–1345

Liu B, Thiagarajan PS (2012) Modeling and analysis of biopathways dynamics. J Bioinforma Comput Biol 10(04):1231001

Liu X, Su L, Li J, Ou G (2021) Identification of pathway-based biomarkers with crosstalk analysis for overall survival risk prediction in breast cancer. Front Genet 12:689715

Ma D, Wang AC, Parikh I, Green SJ, Hoffman JD, Chlipala G, Murphy MP, Sokola BS, Bauer B, Hartz AM, Lin AL (2018) Ketogenic diet enhances neurovascular function with altered gut microbiome in young healthy mice. Sci Rep 8(1):6670

Ma'ayan A (2011) Introduction to network analysis in systems biology. Sci Signal 4(190):tr5

Martinez-Granero F, Blanco-Kelly F, Sanchez-Jimeno C, Avila-Fernandez A, Arteche A, Bustamante-Aragones A, Rodilla C, Rodríguez-Pinilla E, Riveiro-Alvarez R, Tahsin-Swafiri S, Trujillo-Tiebas MJ (2021) Comparison of the diagnostic yield of aCGH and genome-wide sequencing across different neurodevelopmental disorders. NPJ Genom Med 6(1):25

Mimura I, Kanki Y, Kodama T, Nangaku M (2014) Revolution of nephrology research by deep sequencing: ChIP-seq and RNA-seq. Kidney Int 85(1):31–38

Mollaei H, Safaralizadeh R, Rostami Z (2019) MicroRNA replacement therapy in cancer. J Cell Physiol 234(8):12369–12384

Moller, D.E. (2001) New Drug Targets for Type 2 Diabetes and the Metabolic Syndrome. Nature, 414, 821-827. https://doi.org/10.1038/414821a

Moreno-Sánchez R, Saavedra E, Rodríguez-Enríquez S, Olín-Sandoval V (2008) Metabolic control analysis: a tool for designing strategies to manipulate metabolic pathways. Biomed Res Int 2008:597913

Nanda P, Ghosh A (2021) Genome scale-differential flux analysis reveals deregulation of lung cell metabolism on SARS-CoV-2 infection. PLoS Comput Biol 17(4):e1008860

Neduva V, Russell RB (2006) Peptides mediating interaction networks: new leads at last. Curr Opin Biotechnol 17(5):465–471

Niederer SA, Aboelkassem Y, Cantwell CD, Corrado C, Coveney S, Cherry EM, Delhaas T, Fenton FH, Panfilov AV, Pathmanathan P, Plank G (2020) Creation and application of virtual patient cohorts of heart models. Phil Trans R Soc A 378(2173):20190558

Park JH, Kim YK, Yun PY, Yi YJ, Yeo IS, Lee HJ, Park JY (2009) Analysis of factors affecting crestal bone loss around the implants. J Korean Dent Sci 2(2):12–17

Post TM, Freijer JI, DeJongh J, Danhof M (2005) Disease system analysis: basic disease progression models in degenerative disease. Pharm Res 22:1038–1049

Przytycka TM, Kim YA (2010) Network integration meets network dynamics. BMC Biol 8(1):1–3

Pujol A, Mosca R, Farrés J, Aloy P (2010) Unveiling the role of network and systems biology in drug discovery. Trends Pharmacol Sci 31(3):115–123
Rintala TJ, Ghosh A, Fortino V (2022) Network approaches for modeling the effect of drugs and diseases. Brief Bioinform 23(4):bbac229
Sahu A, Blätke MA, Szymański JJ, Töpfer N (2021) Advances in flux balance analysis by integrating machine learning and mechanism-based models. Comput Struct Biotechnol J 19:4626–4640
Schena M, Shalon D, Davis RW, Brown PO (1995) Quantitative monitoring of gene expression patterns with a complementary DNA microarray. Science 270(5235):467–470
Shaukat Z, Aiman S, Li CH (2021) Protein-protein interactions: methods, databases, and applications in virus-host study. World J Virol 10(6):288
Shi J, Walker MG (2007) Gene set enrichment analysis (GSEA) for interpreting gene expression profiles. Curr Bioinforma 2(2):133–137
Somvanshi PR, Venkatesh KV (2014) A conceptual review on systems biology in health and diseases: from biological networks to modern therapeutics. Syst Synth Biol 8:99–116
Song L, James SR, Kazim L, Karpf AR (2005) Specific method for the determination of genomic DNA methylation by liquid chromatography-electrospray ionization tandem mass spectrometry. Anal Chem 77(2):504–510
Souf S (2016) Recent advances in diagnostic testing for viral infections. Biosci Horiz 9:hzw010
Stelling J, Sauer U, Szallasi Z, Doyle FJ, Doyle J (2004) Robustness of cellular functions. Cell 118(6):675–685
Subramanian A, Tamayo P, Mootha VK, Mukherjee S, Ebert BL, Gillette MA, Paulovich A, Pomeroy SL, Golub TR, Lander ES, Mesirov JP (2005) Gene set enrichment analysis: a knowledge-based approach for interpreting genome-wide expression profiles. Proc Natl Acad Sci 102(43):15545–15550
Tankard RM, Bennett MF, Degorski P, Delatycki MB, Lockhart PJ, Bahlo M (2018) Detecting expansions of tandem repeats in cohorts sequenced with short-read sequencing data. Am J Hum Genet 103(6):858–873
Tegnér JN, Compte A, Auffray C, An G, Cedersund G, Clermont G, Gutkin B, Oltvai ZN, Stephan KE, Thomas R, Villoslada P (2009) Computational disease modeling–fact or fiction? BMC Syst Biol 3(1):1–3
Terentiev AA, Moldogazieva NT, Shaitan KV (2009) Dynamic proteomics in modeling of the living cell. Protein-protein interactions. Biochem Mosc 74:1586–1607
Vicini P (2008) Kinetic models at the intersection of biology, engineering and medicine. Drug Discov Today Dis Model 4(5):271–272
Wang E, Lenferink A, O'Connor-McCourt M (2007) Genetic studies of diseases: cancer systems biology: exploring cancer-associated genes on cellular networks. Cell Mol Life Sci 64:1752–1762
Wang S, Song J, Tan M, Albers KM, Jia J (2012) Mitochondrial fission proteins in peripheral blood lymphocytes are potential biomarkers for Alzheimer's disease. Eur J Neurol 19(7):1015–1022
Wang P, Lü J, Yu X (2014) Identification of important nodes in directed biological networks: a network motif approach. PLoS One 9(8):e106132
Wilkinson DJ (2009) Stochastic modelling for quantitative description of heterogeneous biological systems. Nat Rev Genet 10(2):122–133
Woolfe A, Goodson M, Goode DK, Snell P, McEwen GK, Vavouri T, Smith SF, North P, Callaway H, Kelly K, Walter K (2005) Highly conserved non-coding sequences are associated with vertebrate development. PLoS Biol 3(1):e7
Yıldırım MA, Goh KI, Cusick ME, Barabási AL, Vidal M (2007) Drug—target network. Nat Biotechnol 25(10):1119–1126
Younesi E, Hofmann-Apitius M (2013) From integrative disease modeling to predictive, preventive, personalized and participatory (P4) medicine. EPMA J 4:1–2
Zanzoni A, Soler-López M, Aloy P (2009) A network medicine approach to human disease. FEBS Lett 583(11):1759–1765
Zhang T, Tyson JJ (2022) Understanding virtual patients efficiently and rigorously by combining machine learning with dynamical modelling. J Pharmacokinet Pharmacodyn 49(1):117–131

Zhang B, Tian Y, Zhang Z (2014) Network biology in medicine and beyond. Circ Cardiovasc Genet 7(4):536–547

Zhang Y, Wang Z, Wang Y (2017) Multi-hierarchical profiling: an emerging and quantitative approach to characterizing diverse biological networks. Brief Bioinform 18(1):57–68

Zhao S, Fung-Leung WP, Bittner A, Ngo K, Liu X (2014) Comparison of RNA-Seq and microarray in transcriptome profiling of activated T cells. PLoS One 9(1):e78644

Zhu X, Gerstein M, Snyder M (2007) Getting connected: analysis and principles of biological networks. Genes Dev 21(9):1010–1024

Systems Biology and Human Diseases

2

R. Jasmine, Sowmiya Soundararajan, A. Sherlin Rosita, and H. Deena Priscilla

Abstract

Systems biology is concerned with the analysis and modelling of complex biological systems using computational and mathematical tools. This deals with the mechanism of functioning and signalling of metabolic networks. It helps to decipher the transcription and translation products that are involved in the regulation of any physiological response. The process includes data mining analysis, networking modelling and experimental validation for understanding and providing possible solutions to diseases by studying human physiology with detailed knowledge on the mode of action. This would be of immense significance in determining suitable drug targets and developing of effective therapies against various diseases associated with human and understanding the disease mechanism collaborated with drug target identification and development associated with systems biology. Among the several diseases and disorders, we would like to highlight and conceive an affirmative notion that forms the framework linking system biology with colorectal cancer.

R. Jasmine · A. Sherlin Rosita · H. Deena Priscilla
Department of Biotechnology and Bioinformatics, Bishop Heber College (Autonomous), Trichy, Tamil Nadu, India

Affiliated to Bharathidasan University, Trichy, Tamil Nadu, India

S. Soundararajan (✉)
Department of Biotechnology and Bioinformatics, Bishop Heber College (Autonomous), Trichy, Tamil Nadu, India

Affiliated to Bharathidasan University, Trichy, Tamil Nadu, India

PG and Research Department of Biotechnology, Vivekanandha College of Arts and Sciences for Women (Autonomous), Elayampalayam, Tiruchengode, Namakkal, Tamil Nadu, India

S. Joshi et al. (eds.), *Systems Biology Approaches: Prevention, Diagnosis, and Understanding Mechanisms of Complex Diseases*,
https://doi.org/10.1007/978-981-99-9462-5_2

Keywords

System biology · Disease · Drug therapy · Colorectal cancer · Network modelling

2.1 Introduction

'Wellness is the finest potential condition of health for individuals and organizations'. There seem to be two main concerns: realizing an individual's full potential physically, mentally, spiritually, socially, economically and fulfilling one's expectations and responsibilities in the family, society, house of worship, as well as other situations (Kühn and Rieger 2017). Several more common ailments such as respiratory disorders, diabetes, obesity and cancer involve altered interactions between thousands of genes. Such genes must be identified and high-throughput techniques (omics) are now available to identify such genetic traits as well as their own products and though understanding and analysing their functions, which is still a challenge (Wörheide et al. 2021). Network-based analyses of disease-associated genes facilitate understanding a disease in molecular level. In colorectal cancer (CRC), for example, a module was employed to start investigating candidate genes that had been confirmed by functional and pharmacological studies. In systems medicine such analyses are critical. It is a new field that seeks to gain a translational understanding of the complex mechanisms that underpin common infections (Beklen et al. 2021).

'A Healthy mind is a healthy body'. But the well-being of women especially Indian women is increasingly under threat. Health is complex and is dependent on a variety of factors (Doan et al. 2022). Even social and environmental factors have a great impact on one's health (Ungar and Theron 2020). It is also said, 'Nutrition is a determinant of health'. One report states that it is the quantitative examination of energetic interactions between various components of a biological system with the aim of understanding the system behaviour as a whole rather than the behaviour of its individual constituents. It is interdisciplinary in nature dealing with engineering, medicine, pharmacology, physiology, biology, chemistry, computer, mathematics and physics. In short, they are a combination of bioinformatics, mathematical biology and omics sciences (Picó et al. 2019). Systems biology is a multidisciplinary field of study that examines the human body's systems as a whole, including biochemical, physiological and environmental interactions. Systems medicine is based on systems science and systems biology, and it takes into account complex interactions within the human body in relation to a patient's genomics, behaviour and environment (Revo 2018). Thus, this emerging discipline aims to address the multiple genes or their products that interact causing a malfunction or disease in a complex network (Austin 2021).

Systems biology is concerned with the analysis and modelling of complex biological systems using computational and mathematical tools. This deals with the mechanism of functioning and signalling of metabolic networks. It helps to decipher the transcription and translation of products that are involved in the regulation of physiological response (Pathak and Singh 2020). The process includes data mining analysis, networking modelling experimental validation for understanding and providing possible solutions to disease by way of studying the human physiology with detailed knowledge about the mode of action of our entire system (Joshi et al. 2021). This would be of immense significance in determining suitable drug targets and developing effective therapies against various diseases occurring within the human body. It also helps in understanding the disease mechanism collaborated with drug target identification with systems biology (Roessler et al. 2021). We would like to highlight and conceive an affirmative notion that forms the framework linking the system biology and human diseases.

2.2 Human Microbiome and Health

The microbiome of human has a notable impact on health (Brial et al. 2018). It is essential to a variety of host-related activities. Although comparative assessments of microbiome have shown significant variation in species and gene composition linked to a number of disease states, they may not fully explain how this variation affects the community and the host (Gilbert et al. 2018). Here, we present a computational framework for metagenomic systems biology that combines an in silico analysis of metabolic networks with metagenomic data (Greenblum et al. 2012). The gut microbiome serves the host in a variety of vital biochemical capacities, and microbiome abnormalities are linked to a wide range of human disease processes (Harris et al. 2012). Systems biology perspectives focusing on next-generation 'omics' technologies can now characterize the gut microbiome in precise manner at the genetic, functional and structural levels in order to map microbiome variability between species, persons and communities (transcriptomic, proteomic and metabolic), and also provide new insights about the significance of the gut microbiome in human health (Heinken et al. 2021). This has proven the significance of the gut microbiome in the disease pathogenesis for several systemic disease such as overweight, heart and respiratory disease, as well as for intestinal disorders, such inflammatory bowel disease (Tang et al. 2017). The host and the microbiota make together a sophisticated 'super-organism' that offers the host advantages in numerous vital areas of life (Bigeh et al. 2020). However, disruptions in the host's regulatory circuits that regulate bacterial sensing and homeostasis or changes to the microbiome brought on by external factors (infection, nutrition, or way of living culture) may interfere with this symbiotic relationship and encourage disease (Freitas 2022). It has not been well investigated how genetic background influences the development of an effective immune response to illness or immunization (Rowland et al. 2018). However, it is well known that both in humans and other animals, susceptibility (Kobayashi et al. 2019) and resistance to a wide range of infectious agents as well as on the ensuing host inflammatory and immunological responses are related to genes (Bigeh et al. 2020). The 35,000 genes in the human genome, 1–10 million proteins (including antibodies) and 2000–3000 metabolites in the human metabolome all contribute to the measurement of human complexity (García 2020). The assessment of human diversity is even more challenging because many of the factors, including the notion of race, are not even universally accepted (Wishart et al. 2018).

2.3 Systems Biology: A Tool for Disease Analysis

A systematic approach must be taken in order to account for human complexity, variability and its impact on health and disease. Systems biology has grown in popularity over the past 10 years as a result of the development of analytical platforms and bioinformatics tools (Fig. 2.1). Such a method has a lot of potential for identifying important networks and pathways for achieving optimal human health (Dash et al. 2019). Biology has made immensive spectacular progress by way of technological and innovative approaches that have improved our understanding of living organisms providing us detailed information at the system level (Bongard and Levin 2021). The field of system biology is at its budding stage, but it has got the huge potential to create extraordinary opportunities and overcome the challenges attributed towards medicinal research enabling in-depth knowledge about the interrelationship between genotype and phenotype (El Karoui et al. 2019). This could offer valuable insights on the mechanism and functional role which in turn may help to achieve predictive models that might improve our understanding of the fundamentals of human biology. In the wake of this conceptual maturity, a possible blooming new discipline that might act as interface between human health and disease, in short 'Molecular Systems Biology', could gain momentum (Berg et al. 2020).

In any area of research, inclusive of cancer biology, it is essential to unravel the mechanism underlying the diseased condition. This focus leads us to the need to gain data on the modus operandi of signal transduction pathways. Also, sometimes multiple signalling molecules tend to operate activating the cellular cascade.

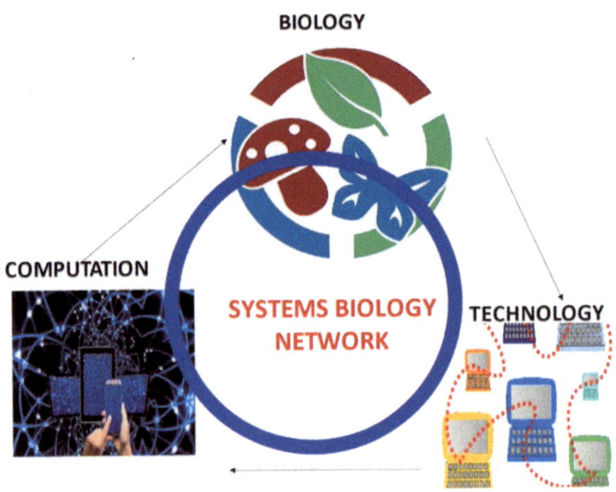

Fig. 2.1 Network flow of systems biology

Researches all over the world have unified in one concept that states the fact suggesting metabolic pathways are interconnected; hence, system-level approach to biology is inadvertently the main motivating factor behind our understanding of human diseases (Cryan et al. 2019).

Another study undertaken by few scientists shows the biological outcome of applied drugs tends to exert certain synergistic and some non-intuitive correlation which impacts the cause of drug design as well as the field of pharmacology at large (Kuznetsov et al. 2021). In this context, there is a need to combine and organize the various diseases along with their classification and treatment into a framework that is network oriented. The next challenging part of the work piece is the identification of novel strategies to prepare a structure based on biological process and disruption in diseased condition (Jarada et al. 2020).

2.4 Origin of Diseases

It is our greatest sense of belief that diseases arise from exposure to toxic factors present on the environment. A different focus has been made to ensure the truth behind this notion. At the end, it could be concluded that disease is the response and inability to deal with imbalance changes within its internal environment. This effect is caused by one or more undesirable changes exerted by the external environment coupled with unfortunate evolutionary legacies (Chu and Karr 2017).

Rudolf Virchow has stated that 'Diseases are not independent and isolated in existence' (Marya and Patel 2021). Exchange of energy and matter with the surrounding environment is generally consider advantageous but problems persist or initiate when this exchange is unbalanced; it could be either insufficient or excessive (Wiedmann et al. 2020).

Infectious disease is considered to be a natural component of ecosystem as microbial pathogens form a part of this biodiversity. Disease outbreaks and dramatic epidemics are said to be a normal phenomenon in the functioning of the ecosystem. The greatest threat to human well-being arises when humans interfere and degrade the ecosystem. So, maintaining the degradation of nature may have a higher influence on the individuals human health (Schmid and Schöb 2022).

2.5 Network Flow of System Biology

The visible signs and symptoms are seen on a phenotypic level, and this data could be linked to genes at molecular level that indicates the disease's cause. Other layers, such as socioeconomic and environmental factors, may need to be considered by the physician (Struy et al. 2021). However, many diseases are more difficult to diagnose. External causes, disease mechanisms, or cell, tissue, or organ involvement, may be highly complex or only partially understood. In these types of cases, a standard diagnostic methodology that provided decision support by presenting the variables would be beneficial to the physician, allowing contributory disease mechanisms

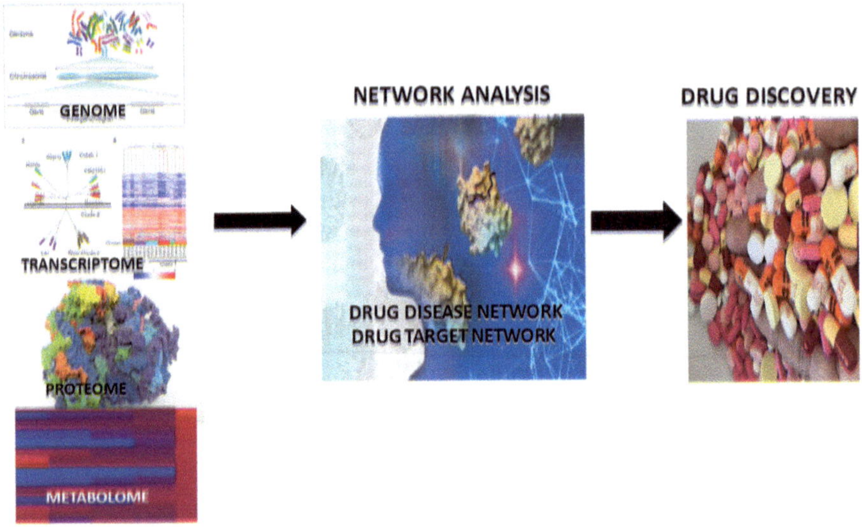

Fig. 2.2 Network analysis flow chart

to be elucidated and diagnostic predictions to be computed (Coffin and Duboc 2022). Different clinical variables, such as symptoms and laboratory variables, could be described in different network layers using this diagnostic model (Borzouei et al. 2020).

2.6 Networks and Interactions

Networks are used to create graphical presentations using graphs and charts of complex systems. Molecules such as genes and proteins are represented as nodes in cellular networks, and their interactions as links (Fig. 2.2). Disease-associated genes discovered through omics research can be computationally mapped onto models of the human protein–protein interaction (PPI) network (Capriotti et al. 2019). In other words, each disease-related gene is linked to its corresponding protein product. The characteristics of the resulting maps are similar to those of other types of networks. One of the most important features is that functionally-related genes tend to cluster together and form disease modules.

2.7 Analysis of Disease Modules Using Network

Disease modules could indeed aid in the organization and prioritization of disease-associated genes discovered through high-throughput sequencing. The analysis of disease modules makes use of general network principles, such as the fact that changes to hub genes are likely to have large effects, whereas changes to many

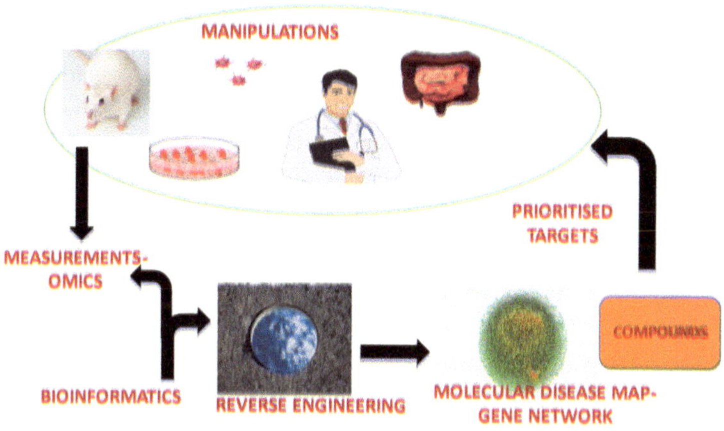

Fig. 2.3 Introduction to networks

genes with few links are likely to correspond to small effect genes (Benson 2016). A significant study of network design principles in model organisms such as yeast an important observation is that nodes in a network that are highly interconnected are likely to be functionally related. Novel candidate genes could therefore be discovered among the interactors of known disease genes. Disease modules can also aid in the discovery of new disease genes, biomarkers and therapeutic targets (Fig. 2.3). Surprisingly, one seminal study in systems medicine was initiated by researchers with no clinical training who had studied network design principles in model organisms such as yeast cells or worms (Gupta et al. 2022).

2.8 Colorectal Cancer

Despite the advances made each year, colorectal cancer (CRC) is now the third most frequent malignant cancer in both men and women globally. As a result, ongoing efforts must be undertaken to comprehend the molecular pathways underlying this disease and engage new potential treatments (Xi and Xu 2021). Stepping back from the perspective of prevention (i.e., personal screening), one of the major challenges to comprehending cancer biology is its inherent heterogeneity, as has been observed in numerous cancer forms. As a result, it becomes challenging to pinpoint the main molecular triggers of the tumoral process and comprehend the variances between various cancer kinds without comparing their expression profiles to look at their most prominent variations (Narrandes and Xu 2018).

2.8.1 Gene Expression-Methodology

Gene expression data can now be obtained from a variety of methods, including microarray and RNA-seq analyses, all of which offer useful data that can be used to paint a more realistic picture of these molecular changes (Feltes et al. 2020). However, it is neither intuitive nor clear due to the enormous volume of papers that are released every year finding tumoral expression patterns and biomarkers that may be biologically relevant in those contexts. Therefore, combining both forms of data and designing a multi-approach, computational strategy to extract pertinent information would be the best way to generate this expressive panorama because different computational techniques produce varied outcomes and interpretations. The utilization of various types of data and computational techniques to understanding the biological pathways underlying the tumoral process and comprehend what could lie beyond the heterogeneity in cancer is already broadly accepted (Feltes et al. 2020).

Furthermore, a distinct dataset-mining procedure would be essential to guarantee correct comparison and accurate results, a topic that, however, is commonly neglected and insufficiently covered in most computational research. CRC investigations are currently one of the promising methods for identifying novel genes that may serve as therapeutic targets or prognostic biomarkers (Singh et al. 2021). When performing a cancer study, there are many options. For instance, they could be accompanied by network approaches (Cava and Castiglioni 2019), signalling pathway analysis (Neapolitan et al. 2015), or networks associated with functional pathways analysis (González-Reymúndez and Vázquez 2020), or they could take a pure machine-learning approach, exploring the combination of omics-data (Cava et al. 2021).

2.8.2 Multi-Approach Omics Analysis for Colorectal Cancer

To access the key expression similarities and variances of different cancer types, we describe in this chapter a multi-approach omics analysis of both microarray and RNA-seq data. This analysis is integrated with a machine learning (ML) approach and systems biology tools (Lacalamita et al. 2021). We selected only the most trustworthy and homogeneous datasets for future analysis by manually curating the Gene Expression Omnibus (GEO) [https://www.ncbi.nlm.nih.gov/geo/] and individually reviewing the available cancer-related microarray and RNA-seq experiments (Fasterius et al. 2017). We were able to create a picture that adequately explains the various gene expression patterns observed in various tumoral kinds using data from microarray experiments and RNA-seq investigations. In addition to evaluating a potential common molecular route that may oversee the tumoral process, we have also supplied an overall survival analysis for further debate. We have emphasized that this work aims to create and report results of a new, rigorous and reproducible multi-omics in in silco approach (Subramanian et al. 2020).

To the best of our knowledge, there are currently no precise tumour markers for colorectal cancer; nevertheless, the discovery of such biomarkers may aid in the early detection and screening of diseases. Deep sequencing procedures that are complicated and expensive are needed for the validation of the identification of mRNAs as possible biomarkers. In order to detect CRC and predict patient survival, the current study set out to uncover genetic biomarkers and therapeutic targets (Lin et al. 2021).

- Retrieving omics data.
- Analysis of differentially expressed genes (DEGs).
- Identification of potent pathways.
- Protein–protein interaction network (Eddy et al. 2020).

2.8.2.1 Retrieving Omics Data

The publicly accessible GEO database was manually mined for obtaining its numerous omics datasets (GSEs) of microarray and RNA-seq data (Comte et al. 2020). For colorectal tumours. The following criteria were used to choose the most reliable and homogeneous datasets:

1. Elimination of datasets that applied knockdown cultures or artificially induced mutations;
2. Exclusion of studies that used any type of pharmacological manipulation;
3. Exclusion of studies that used interfering molecules, such as miRNAs or siRNAs;
4. Selection of studies with at least three control and three experimental samples and.
5. Selection of studies conducted solely in Homo sapiens. In the present analysis the NCBI-GEO was used to obtain the GSE95513 gene expression profile. Based on GPL96 platforms (HG-U133A Affymetrix Human Genome U133A), the microarray data of GSE95513 (Aziz et al. 2020).

2.8.2.2 Analysis of Differentially Expressed Genes

The DEGs (whose *p*-values <0.05 and logFC >1) were obtained from the standard dataset on CRC tumours. Out of the total genes, 2363 genes were found to be upregulated in GSE95513 dataset respectively. Further to confirm their states of gene expression, the upregulated and downregulated genes mentioned in Fig. 2.4 were chosen to be the high-confidence DEGs. The volcano plots illustrate the dataset of genes that have been either upregulated or downregulated.

The selected microarray studies' raw data were downloaded using the GEO query programme (for the R platform). Samples were carefully excluded if they showed any errors or file corruption during preparation. Each dataset is provided separately for background adjustment and normalization. The data processing and analysis involved the use of two packages: (1) affy (Gautier et al. 2004) for Affymetrix-derived datasets, and (2) lumi for raw output files from Illumina Bead Studio toolkit. After preprocessing, the array Quality Metrics program (Kauffmann et al. 2009) was employed to evaluate sample quality data for all microarray datasets.

Prior to differential gene expression analysis, samples that showed poor quality in at least half of the array QualityMetrics parameters were always eliminated. Finally, for the examination of differential gene expression, the programmes limma and Biobase (Huber et al. 2015) were used. By using a filter of |log2FC|1 with the Benjamini Hochberg for FDR correction of p 0.05, differentially expressed genes

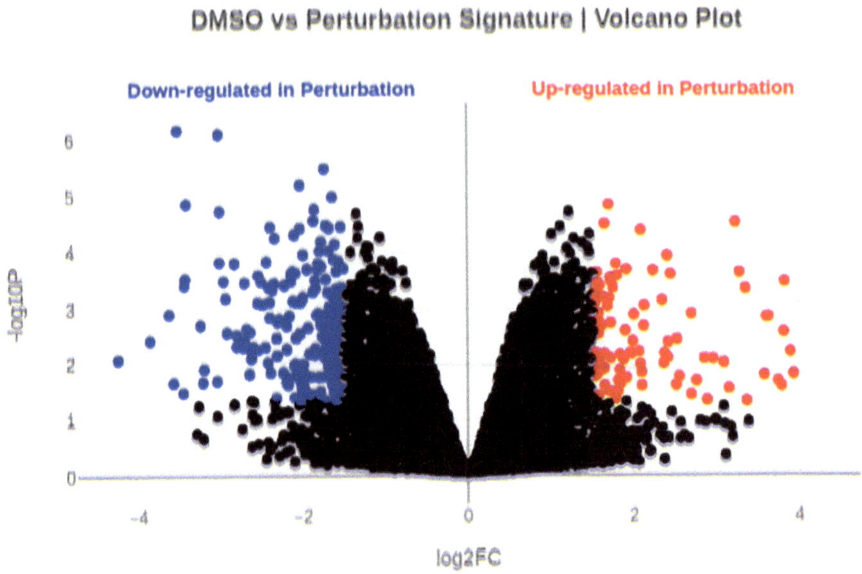

Fig. 2.4 Volcano plot of the dataset (GSE95513). Red plots represent genes with [logFC] > 1 and $p < 0.05$ (upregulation). Blue plots represent the remaining genes (downregulation). Black spots show no significant change

(DEGs) were found. Datasets were examined separately. The GeneOverlap R programme was used to retrieve Jaccard indexes (Shen and Sinai 2020). Results from the standard dataset (GSE95513) on CRC tumours with|log2FoldChange|(|log2FC|) >1.585, also known as|FoldChange| > 3 and adjusted *p*-value, were obtained (Table 2.1).

2.8.2.3 Protein–Protein Interaction Network

We used a systems biology approach to further evaluate the major molecular pathways connected to the ultimately chosen targets. We used the STRING 11 metasearch tool to create the primary protein–protein interaction (PPI) networks (Szklarczyk et al. 2019). The networks were saturated after simultaneously searching all the targets—which will be discussed further—until all the proteins that had been input were connected, or almost connected (Fig. 2.5). The following STRING parameters were used: (1) a degree of confidence of 0.400; (2) the use of just experiments and coexpression search parameters; and (3) a limit of 200 interactors on the first shell and none on the second. Networks were then imported into Cytoscape 3.7.1 to be analysed and worked with (Shannon et al. 2003). Based on the STRING online database, a DEG PPI network was constructed containing 20 DEGs (Fig. 2.6). The genes CTNNB1, TNIK, TCF4, LRP6 were found to be closely interacting partners (Cava et al. 2018; Cava et al. 2019).

Table 2.1 GSE95513 important Wnt genes and their expression levels

Gene symbol	Log FC	*p*-value	Regulation
MIDN	1.697320871	0.024767356	Upregulated
TUFT1	1.218203785	0.024767356	Upregulated
HIST2H2AA4	3.227040929	0.024767356	Upregulated
OGDH	1.647932894	0.024767356	Upregulated
INPP5K	1.113281107	0.024767356	Upregulated
DLL4	2.087652385	0.024767356	Upregulated
ULK1	1.466115268	0.029177963	Upregulated
PYGB	1.011579955	0.029177963	Upregulated
DGCR2	1.279628712	0.029420895	Upregulated
FAM234A	1.385196663	0.033146334	Upregulated
SLFN5	1.446990498	0.035147163	Upregulated
TMEM175	1.193889298	0.035147163	Upregulated
PLXNA3	1.460839736	0.036903904	Upregulated
FAM222A	2.40414412	0.037359993	Upregulated
PKD1	1.235070621	0.038010902	Upregulated
TUBA1A	1.784313768	0.040596604	Upregulated
GPS2	1.045528364	0.040596604	Upregulated
POLG	1.296551682	0.040596604	Upregulated
HEATR5A	1.260106289	0.040596604	Upregulated
TLN2	1.912349844	0.040596604	Upregulated
EIF4EBP2	1.285233209	0.040596604	Upregulated
ERCC6	2.23286636	0.040596604	Upregulated
CLTCL1	1.183645388	0.040596604	Upregulated
RSRP1	3.274478021	0.040596604	Upregulated
FAM89B	1.551309146	0.040596604	Upregulated
PNPLA2	1.511039021	0.040596604	Upregulated
MYO9A	2.442494092	0.040639474	Upregulated
CKS2	1.799716865	0.040789481	Upregulated
DOPEY2	1.383555363	0.040789481	Upregulated
TNIK	1.356666624	0.041560243	Upregulated
CTNNB1	1.226596098	0.041560243	Upregulated
LRP6	1.495820274	0.042370302	Upregulated
TCF4	1.710171135	0.042370302	Upregulated
GAR1	−1.638462166	2.00674826	Downregulated
TRIB3	−1.597915158	2.359766007	Downregulated
DUS2	−1.694069962	2.380868724	Downregulated
LRFN4	−1.871061508	2.030960099	Downregulated
ACVR2B	−1.301499566	2.539804141	Downregulated
PCGF6	−1.350546884	2.75544099	Downregulated
TSKU	−1.295626878	2.611193605	Downregulated
ROBO3	−1.264929848	2.571547609	Downregulated
COA7	−1.353222496	2.846470675	Downregulated
SLC16A9	−2.35654393	2.596605291	Downregulated

(continued)

Table 2.1 (continued)

Gene symbol	Log FC	*p*-value	Regulation
DNAJC17	−1.417652597	2.844578975	Downregulated
CTPS2	−1.581770944	2.77168993	Downregulated
RP13-726E6.2	−1.424844326	3.125306764	Downregulated
RPF2	−1.356692216	3.2610781	Downregulated
PLD6	−1.838190273	2.648910476	Downregulated
THRA	−1.014582285	3.434420072	Downregulated
PALM3	−1.52743528	3.714826311	Downregulated
ADAP1	−1.089448766	4.045761659	Downregulated
FAM19A3	−2.403687552	4.450845354	Downregulated
MDFI	−3.433217225	15.92081875	Downregulated
PRDM1	−2.211687561	11.14752001	Downregulated
TIFA	−1.258655125	15.37778598	Downregulated
KLHL23	−1.186996603	28.00656377	Downregulated
TMEM74B	−2.338177738	9.233587153	Downregulated
HRK	−2.69633262	30.43651891	Downregulated
WDR12	−1.07603123	26.59687948	Downregulated

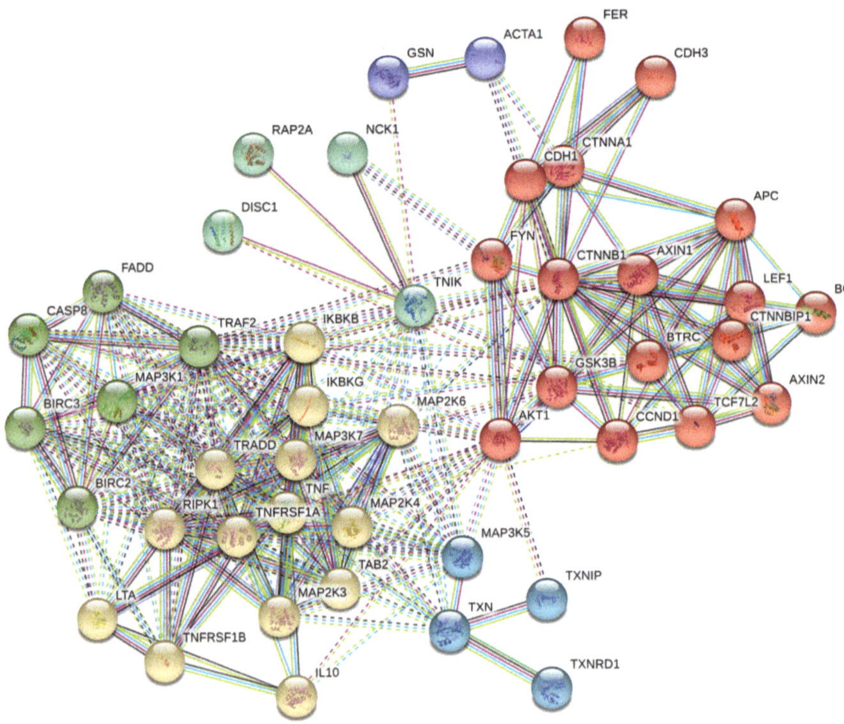

Fig. 2.5 Protein interaction network from STRING database

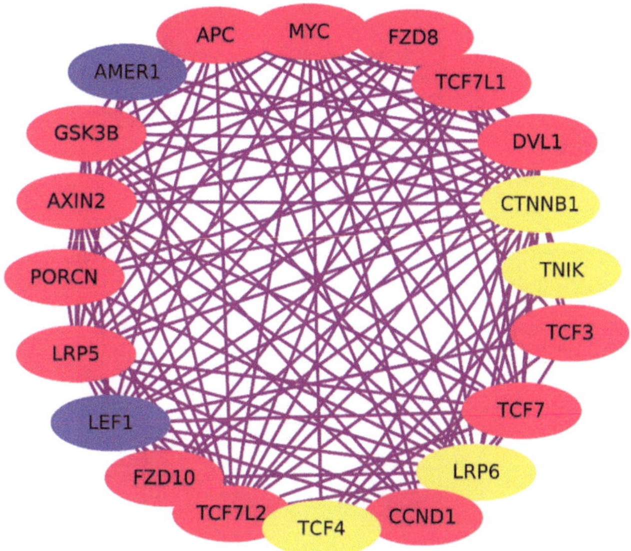

Fig. 2.6 The first-level interacting partners obtained from STRING database

2.8.2.4 Identification of Potent Pathways: Key for Early Treatment

To identify the nodes and biochemical pathways most significant to each PPI network's topology, each network's topology was examined in terms of clustering, centralities and gene ontology. The following centralities were examined using the Cytoscape plug-in CentiScaPe 2.2 (Scardoni et al. 2009): node degree, betweenness and eigenvector. Node degree calculates the number of direct connections of a given node; betweenness calculates the number of shortest paths that pass through each node; and eigenvector calculates how regulated a node is based on the node's number of connections and how PathLinker was used to determine potential signalling pathways (Huang et al. 2018) (Fig. 2.5). The following were the PathLinker parameters: (1) treat network as undirected option disabled; (2) allow sources and targets in path option enabled; (3) connect sources to each other option enabled; and (4) k: 500; (3) edge penalty: 1; (3) edge weight: unweighted. The canonical Wnt

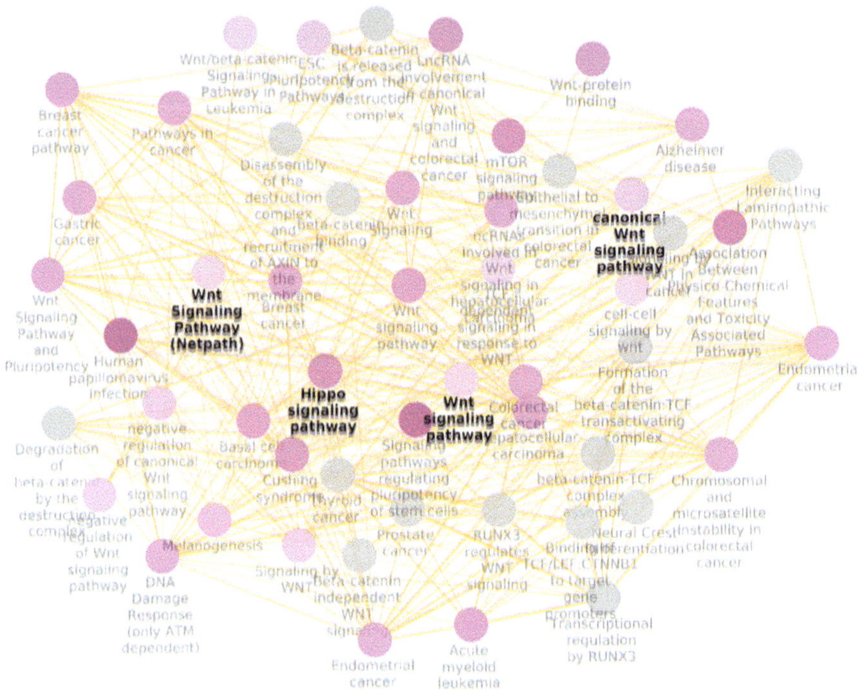

Fig. 2.7 Interrelation between biological process pathways from the identified GSE95513

signalling is the most significant pathway observed from the interaction in screening the CRC dataset GSE95513 (Fig. 2.7).

It becomes evident (through annotation based on biological pathway) that most of the DEGs fall under Wnt signalling cascade inferring that in abnormalities like CRC, this signalling pathway is one of the critical pathways involved in the survival and development of cancer cells. Hence, the systems biological approach of discovering biomarkers for CRC treatment is validated through the gene expression analysis for DEGs in the Wnt pathway, network analysis and annotation of the identified DEGs (Rahman et al. 2019).

2.9 Alzheimer's Disease

Alzheimer's disease (AD) is a neurodegenerative disorder with no effective cure wherein age is considered as the major risk factor contributing to this disease (Yadav et al. 2022). Therapeutic focus has been made by targeting two different groups labelled early-onset AD (EOAD) and late onset (AD) (LOAD). EOAD has been associated with mutation in anyone of the genes namely PSEN-1 or presenilin—1, APP or the Amyloid Protein Precursor and PSEN—2 or Presenilin—2, which is found to be a rare symptomatic case prior to age 65 (Rujeedawa et al. 2021). LOAD is a common form arising in persons above 65 years ruling out sole genetic basis or familial occurrence but a pathobiologic profile caused by environmental or epigenetic sources (Lee at al. 2022). Though we have produced enough surrogates for genetic forms of EOAD, we have not developed suitable models to study and analyse LOAD (Berge-Seidl 2022).

Studies suggest that amyloid beta and tau protein overproduction are visible end-stage biomarkers present in the brain of persons affected with EOAD and LOAD. Researchers and scientists feel bringing about a controlled regulation of these proteins might help in the treatment of AD on a more effective manner (Sarkar and Biswas 2022).

A systems biological approach to AD requires sufficient integration of varied data sets in order to gain detailed understanding of the complexity involved in this disorder. Nowadays, use of high-throughput DNA sequencing in order to analyse patient-derived DNA for observing disease-oriented alleles and drawing conclusions for genetic risk in genes have already found its way in the market supported by medical inference (Suhail et al. 2019). In-depth pedigree analysis has helped us to confirm the role of three autosomal dominant genes as the major causative agent of EOAD (A1, A2). The Alzheimer's disease genetics consortium has assessed various evaluation and derived datasets showing enormous SNP–SNP and gene–gene interactions (50, 51) in the LOAD genetic loci (Sutherland et al. 2019). Evaluation obtained from systems analyses LOAD affected individuals their specific genetic nodes, edge interactions and perturbations or disturbances in networks have been able to unravel the complexity associated with pathobiological process (Lagisetty et al. 2022). The categorical simultaneous fusion of network biology and gene modules attributed to function could pave way for developing therapeutic milestones pertaining to LOAD as well as cardiovascular diseases (Lee et al. 2021).

Metabolomic analyses help to reveal the downstream plight of disease relevant to phenotypic clinical implications. In-depth analysis of brain tissues showed signs of metabolomic dysregulations in case of persons affected by AD with quantitative decrease in sphingolipids (Hajjar et al. 2020). Further studies have led to confirmation of dysregulation in lipids and some metabolites inside the brain as well as the periphery that denotes the clinical significance of the pathology behind LOAD and also helps to bridge the link between health and disease state (Kao et al. 2022).

A key to understanding the basics of health and disease at large by means of systems biology approach has thrown light on the lifestyle choices of individuals as a major factor contributing to the risk of AD. There runs a thread of joint between metabolism and genome by means of modifications created epigenetically such as DNA methylation (Lefèvre-Arbogast et al. 2019).

Without the help of medicine, behavioural modifications seem to offer best disease prevention options if adopted as a long-time perspective. Studies have narrowed their results onto diet and exercise as a major effective component for the prevention of AD. Moderate exercise has been found to lower the risk of dementia (Sullivan and Lachman 2017). High fat diet seems to interfere with the metabolism and associated with AD by disrupting circadian clocks, which alters the microbiome of the gut. Nutritional support provided by means of specific fatty acids and other essential supplements has proved beneficial in showing down the progression of AD (Ding et al. 2022). From the findings, we believe systems biology approach considering the diversity of internal and external sources responsible for the prevalence of disease could offer valuable insights into identifying therapeutic targets effective in the prevention and treatment of AD.

Systems biology has been engaged in the transformation of modern health care from disease diagnosis and treatment to personalized medicine. The success of this upcoming future scenario has been made possible by the use of high-throughput technologies and mass spectrometry analysis to decipher the role of 'omics' in the interpretation of information pertaining to health and disease (Goetz and Schork 2018). This revolution has helped to gain significant knowledge on genome and exosome digital data so as to infer their role in disease and health. There has been tremendous development in the field of cancer research owing to whole-genome sequencing and whole exome sequencing. Sequenced cancer genomes that could act as effective bio markers have been identified in breast cancer, ovarian cancer, small cell lung cancer, melanoma, chromic lymphatic leukaemia, Sonic-Hedgehog medulloblastoma, paediatric glioblastoma and hepatocellular carcinoma (Yadav et al. 2020).

This detection has amplified our understanding on mutations present in somatic cells when compared to normal tissues. Also, it has helped to identify potential markers for subtyping thereby focusing on suitable targets and providing personalized treatment to cancer patients (Tran et al. 2021). The ability of whole genome sequencing (was) has been so immense that Link et al. were able to locate a germline de novo p53detection that affected a female to be a victim of 3 types of cancer in a span of 5 years. It has also proven to help identify spontaneous mutations in normal genome of patients suffering from cancer (Zhao et al. 2019).

2.10 Systems Biology–Predictor of Disease Risk and Provider of Personalized Health Care

One cannot underestimate the power of systems biology as it not only offers remedies to the onset era of personalized medicine but also predicts the degree of disease risk (Cohen et al. 2021) along with providing personalized health care coupled with preventive medicine. In present scenario, pathologists and physicians have started to rely on systems approach to search higher standards in case of disease diagnosis at the molecular level (Flores et al. 2013). A case study undertaken to analyse the genome of a patient with family history has been identified and linked to rare variants on three genes with post—test probability risk causing myocardial infarction and coronary artery disease (CAD) (Dai et al. 2016). The scientists have stressed the need to analyse the genome of patients to know about the genetic risk and possible variants attributed to pharmacogenomics so as to provide personalized care in a patient specific manner (Crews et al. 2012).

Always a need to assess the transcriptome, proteome and metabolomics has been stresses as they act as definitive indicators of phenotypes rather than genomes (Shen et al. 2022). This would help to analyse individual's physiological profile. Researchers across the globe have reported that systems biology should be combined with ecological model and clinical implication should be in line with validated data on selection and screening of population subject to strong evidence on this information. This kind of personalized medicine may increase cost of drugs to individuals. It can be overcome by combinatorial therapy of developing network-oriented target-specific compounds that have been clinically validated (Vineis et al. 2020). Systems biology works in such a way that it helps to interpret multiple biological networks and offers ideal skills to infer relationship between genotype and phenotype (Fiocchi and Iliopoulos 2021).

Human physiology encompasses numerous biological processes that include molecular interactions occurring intracellularly to obvious phenotypic response taking place in the body. The phenotype of a typical cell is governed by structural and dynamic properties controlled by complex biological network (Joyce et al. 2021). An organ-level response is exhibited by the coordinated action of various cells and tissues. The above-mentioned network acts in a regulated way controlled by sequence in events in particular order. This network pathway when altered or disturbed leads to undesirable physiological state called disease (Magnani et al. 2020). In order to analyse the problems encountered in human physiology and the development of disease, it has been studied by means of five modules that include data mining, networking, modelling, experiment and finally validation of results (Silverman et al. 2020).

It is difficult to characterize a disease owing to the complexity of mammalian systems thereby creating a challenge to study them in detail. System-level analysis of biological framework by combining the regulatory mechanisms to physiological output requires the properties such as adaptation, dual stability, degree of robustness and oscillation that help a cell to respond in a controlled synchronized fashion with great specificity (Papin et al. 2005; Kholodenko 2007).

2.11 Network Modelling System Analysis

The greatest saviour to characterize and treat diseases such as diabetes and cancer comes in the form of network modelling system analysis, their further target and therapy identification (Mardinoglu and Nielsen 2012). In order to address the problems of human disease, cell to human approach has been employed due to advancements in the field of systems biology (Brooks and Alper 2021).

This framework consists of five modules,

- Data mining refers to derivation of hypothesis or taking into consideration a predefined problem and collecting relevant information at the multiple levels of omics such as genomes, proteome, metabolome and transcriptome.

Fig. 2.8 Modules of network modelling framework

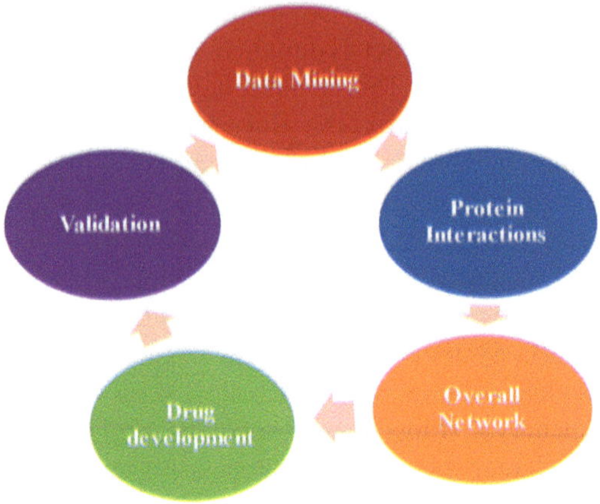

- Networking module deals with protein–protein interactions embedded within the genetic and signalling framework (Langley et al. 2013).
- Modelling and analysis module comprises overall network that forms the initial layer for kinetics designing and interactions occurring within the network. These modules are constructed based on experimental and literature information. The proposed models have been optimized and parameterized to project expected phenotypic response by means of flux balance and pathway methods of analysis (Walpole et al. 2013).
- Experimental cum drug development module helps to identify specific targets-based theoretical knowledge, their selection and screening along with assessment for viability and sustainability in the pharmacological setting (Gustafsson et al. 2014). The strategy for screening of potential targets has been addressed by way of HITS and lead optimization studies (Renner et al. 2011). This is followed by a drug discovery process wherein potential drugs/ligands are formulated to influence the target site and bring about desired physiological response (Fig. 2.8). Further, investigation on animal models has helped to study impact of response therapy paving way for knowledge about drug–target interaction with minimal or least side effects but elevated immune response (Hauser et al. 2017).
- The validation process ensures the facts proposed by experimental results that have confronted the truth behind the drug likeness of selected compound and authenticity of selected/chosen target in the treatment of particular disease. Also, drugs are assessed for their safety (ethical and legal), toxicity and possible side effects. It then transverses through preclinical and clinical trials in order to be approved by FDA (Food & Drug Administration) and enters into the pharmaceutical market coupled with biomedical sciences (Lu et al. 2020).

2.12 Predicting Therapeutic Measures Through System Biology

The goal of cancer research has been to distinguish between healthy and malignant cells molecularly. There are molecules involved in numerous parallel signal transduction pathways. Their behaviour seems to be influenced by a number of variables (Imran et al. 2017). Understanding and predicting the results of intracellular signalling are difficult due to the interaction of regulatory circuits, cross-talk between pathways and the non-linear reaction kinetics of biochemical processes. A sophisticated supra-cellular communication network is also created as a result of interactions between tumour and other cell types (Nair et al. 2019). By transforming the impacts of targeted medicines into long-lasting responses, systems biology can help to improve the efficacy of treatment. Future tailored cancer therapies may benefit clinically from the application of systems biology methods (Werner et al. 2014). Pharmacology has long employed systems techniques to comprehend how drugs can effect organs and organisms. We can broaden the scope of systems pharmacology to include network analysis at many scales of biological organization and to explain both the beneficial and harmful effects of medications by applying computational and experimental systems biology techniques to pharmacology (Zhao and Iyengar 2012) Genomic changes increase the complexity of cancer and are a significant barrier to a thorough understanding of the molecular processes underpinning oncogenesis (Fares et al. 2020). It is necessary to evaluate the intricate web of pathways involving gene regulation, signalling and cell metabolism (Naylor and Chen 2010), as well as the changes in its dynamics brought on by the various types of mutations leading to malignancy, in order to comprehend how a cancer cell differs functionally from a normal cell (Sever and Brugge 2015).

2.13 Challenges in Application of Systems Biology

The greatest challenge underlying systems biology approach is to combine genetic, signalling transduction pathways and metabolic mesh network to develop a detailed in silico cell for understanding the physiological changes brought about by a disease. An interaction network comprises of the functional attributes of biomolecules such as DNA, RNA transcription factors, proteins enzymes and metabolites. The relevant data and information pertaining to this interaction and their subsequent regulation is difficult to obtain and hence it remains a challenge (Mace et al. 2005; Sam et al. 2007). It has been reported that disturbances in these interaction give rise to the onset of innumerable diseases. Both intrinsic and extrinsic factors like mutations and environmental stress contribute to alterations in the network causing disease (Del Sol et al. 2010).

A disease phenotype is collectively attributed to defects in regulatory network interaction (Jordán et al. 2012). It is stated that interaction network leads to tissue repair, immunity and maintenance of equilibrium throughout the human body. Thereby, any changes or modification in the information processing across cells have been responsible for diseases such as cancer, autoimmunity and diabetes (Schulz and Klipp 2010). The data comprising of collection of information of various disease or disorders and their respective mutations in gene responsible for a disease are called a diseasome (Madhavan and Mustafa 2022). There has been great demand to study the mechanism of host–pathogen interaction by means of immunological model on deadly diseases that underlie infections such as AIDS and SARS. The concept of virtual humans and virtual patients has started to evolve owing to advancements in disease network modelling, progression modelling and physiological model systems (Shah et al. 2020).

Systemic studies help in the reconstruction of altered network occurring between genes, signalling proteins and metabolites as nodes and their subsequent interaction taking place mutually as edges in a system-oriented disease (Becker et al. 2007; Bajikar and Janes 2012). This metabolic syndrome arises from complex interplay of several factors such as hormones of the gastrointestinal tract, pancreas, stress and sex hormones, few stations, sterols, adipokins, certain metabolites, few molecules involved in signalling cascade, mutations occurring in genes, inflammation, changes in lifestyle such as diet, exercise, sleep, digestive disorders and behaviour. These levels are intra/interconnected with numerous feedback and crosstalk mechanism that are important to regulate/mediate a coordinated response on the system. The lack of proper signalling and action on the metabolic regulation has been observed in insulin resistance, which has attributed to a variety of factors claiming responsibility leading to defective homeostasis. That multifunctional genetic and metabolomic level impairment gives rise to cardiovascular diseases, hyperglycaemia, dyslipidaemia and complications arising from diabetes such as diabetic nephropathy, retinopathy and neuropathy (Pignatti et al. 2020).

2.14 Future of Systems Biology

'Prediction is difficult, especially about the future' (Mark Twain)—humorist and writer (1835–1910). System biology could be the developed field by 2030 that relies on technological advancements. The realization of modelling of human disease will be approved rather than rejection. Synthetic biology would have gained momentum. Multidisciplinary science will be the normal routine with researchers spanning different disciplines such as biology, maths, computers and omics science being able to communicate in a common language (Lockie et al. 2020). The present role of molecular biology that is used to understand and study different model will be additionally supported by system biology tools for experimentation (Zou et al. 2020).

Nowadays, computational techniques in combination with genome profiling methodologies have benefitted us to identify and locate mediators involved in diseased state. Added to this, the successful completion of human genome sequence has helped us to transverse to the next level of identifying potential genetic determinants responsible for some of the most common diseases with the help of Hap Map project (Course et al. 2020). The most crucial part or outcome of disease not just arise from genetic level but has close proximity to metabolites produced from physiological output. This has led to the growth and development of yet another field of research termed metabolomics steered by Jeremy Nicholson (Ala-Korpela and Davey Smith 2016). A study undertaken by Martin and colleagues (2007) states that significant changes are brought about by gut bacterial flora on the host creating an error in metabolic equilibrium as a result affecting the health of the individual (Schwabe and Jobin 2013). This work highlights the fact that genotype–phenotype relationship is not just the factor responsible for a disease but also physiological facets incorporating the microbiome and the environments seem to be crucial factor underlying a particular disease.

Evaluating on the importance of risk imposed on the health of human, the Director of National Institute of Health has suggested possible investigation to be focused on 4 P's—**Predictive, Personalized, Preventive** medicine along with P**articipation** by the people at large (public) (Pinu et al. 2019). Development of personalized health profiles is a herculean task. The combinatorial impact of genetic and metabolic activities contributes to the changes as defined by systems biology framework concept (Irvine and Nguyen 2019). It comprises of three sub-domains.

1. high-throughput screening by means of bioinformatics tools.
2. technological development leading to experimentation involving in vivo analysis and finally.
3. mathematical calculations and computational modelling coupled with simulations. From the earlier researches, we have been well informed about the genomic, proteomic and metabolic factors responsible for alteration in an individual's health. Presently, much of the research is focused on computational system's biology which allows for prediction and personalization of health risk in humans.

Earlier, traditional system of medicine greatly depended on treatment based on diagnosis which strives to reduce the problem with optimal result. This practice tends to ignore interactions among different organs, extracellular molecules, endocrine and immune functions as well as impact of the environment. Based on many studies, we now know diseases such as Alzheimer's and cardiovascular diseases (CVD) arise due to interactions between intrinsic and extrinsic modulations -3-(4,5) (Mordeniz 2019).

The advent of 'omics' technology comprising of genomics, epigenomics, transcriptomics, proteomics, metabolomics along with analytical tools has helped us to improve our understanding on the role of individual elements and their relevance to a disease. In some diseases, the specific gene or the gene encoded protein or any element responsible has been studied and identified (Hasin et al. 2017). The identification of specific biomarkers like breast cancer type—2 (BRCA2), prostate specific antigen (PSA) and their mutated erroneous gene has helped to study the health in the initial stages of disease. This has increased our understanding of the complexity occurring in biological systems (Boussios et al. 2022).

2.15 Summary and Conclusion

To visualize and assist the concept of system biology, we need to focus on multiomics methods along with linking of biological molecules database encompassing gene transcription, protein translation, protein–protein interaction, enzymes and immune function metabolites and facilitating a common ID to create a map of this cluster (Pinu et al. 2019). Systems biology may be regarded as the promise for refining our knowledge while dealing with diseases and have great impact in the field of medicine. A quantitative, holistic systems biology approach might be applied in the diagnosis and treatment of human disease with opening up new avenues of developing personalized medicine for individuals. A comprehensive network-based systems approach can help in designing suitable drug—targets with minimal side effects and lesser toxicity (Rai et al. 2018).

The concept of networking and engineering to deal with problems/diseases/disorders arising in biological system required the help of systems biology approach to minimize and revolutionize the complications involved in present day therapeutics. This has opened up a new era of pharmacology wherein personalized medicine seems to rule the roost. To realize this dream innovation, we need to bridge the gap between researchers and academicians across the globe. This might help to provide fruitful enduring outcomes in limited span of time. The intricate closely knitted framework of computational system biology and the combined effect of knowledge gained from other relevant interdisciplinary sciences with due efforts in technological and statistical advancements have created a new platform and has opened up a twilight of space on the horizon encompassing the concept of 'in silico clinic'. The field of systems biology is said to revolutionize the science and practice of medicine.

References

Ala-Korpela M, Davey Smith G (2016) Metabolic profiling-multitude of technologies with great research potential, but (when) will translation emerge? Int J Epidemiol 45(5):1311–1318. https://doi.org/10.1093/ije/dyw305

Austin CP (2021) Opportunities and challenges in translational science. Clin Transl Sci 14(5):1629–1647

Aziz F, Gul H, Muhammad I, Uddin I (2020) Link prediction using node information on local paths. Physica A: Stat Mech Appl 557:124980

Bajikar SS, Janes KA (2012) Multiscale models of cell signaling. Ann Biomed Eng 40:2319–2327

Becker SA, Duarte NC, Jamshidi N, Thiele I, Mo ML, Vo TD et al (2007) Global reconstruction of the human metabolic network based on genomic and bibliomic data. Proc Natl Acad Sci 104(6):1777–1782

Beklen H, Yildirim E, Kori M, Turanli B, Arga KY (2021) Systems-level biomarkers identification and drug repositioning in colorectal cancer. World J Gastrointest Oncol 13(7):638

Benson M (2016) Clinical implications of omics and systems medicine: focus on predictive and individualized treatment. J Intern Med 279(3):229–240

Berg G, Rybakova D, Fischer D et al (2020) Microbiome definition re-visited: old concepts and new challenges. Microbiome 8:103. https://doi.org/10.1186/s40168-020-00875-0

Berge-Seidl V (2022) Genetic studies of sporadic Parkinson's disease: on the identification of genetic risk factors and the path towards a better understanding of underlying pathogenic mechanisms
Bigeh A, Sanchez A, Maestas C, Gulati M (2020) Inflammatory bowel disease and the risk for cardiovascular disease: does all inflammation lead to heart disease? Trends Cardiovasc Med 30(8):463–469
Bongard J, Levin M (2021) Living things are not (20th century) machines: updating mechanism metaphors in light of the modern science of machine behavior. Front Ecol Evol 9:650726
Borzouei S, Mahjub H, Sajadi NA, Farhadian M (2020) Diagnosing thyroid disorders: comparison of logistic regression and neural network models. J Family Med Prim Care 9(3):1470
Boussios S, Rassy E, Moschetta M, Ghose A, Adeleke S, Sanchez E, Pavlidis N (2022) BRCA mutations in ovarian and prostate cancer: bench to bedside. Cancers 14(16):3888
Brial F, Le Lay A, Dumas ME, Gauguier D (2018) Implication of gut microbiota metabolites in cardiovascular and metabolic diseases. Cell Mol Life Sci 75(21):3977–3990
Brooks SM, Alper HS (2021) Applications, challenges, and needs for employing synthetic biology beyond the lab. Nat Commun 12(1):1–16
Capriotti E, Ozturk K, Carter H (2019) Integrating molecular networks with genetic variant interpretation for precision medicine. Wiley Interdiscip Rev Syst Biol Med 11(3):e1443
Cava C, Castiglioni I (2019) In silico perturbation of drug targets in pan-cancer analysis combining multiple networks and pathways. Gene 698:100–106
Cava C, Bertoli G, Colaprico A, Olsen C, Bontempi G, Castiglioni I (2018) Integration of multiple networks and pathways identifies cancer driver genes in pan-cancer analysis. BMC Genomics 19(1):1–16
Cava C, Bertoli G, Castiglioni I (2019) Portrait of tissue-specific coexpression networks of non-coding RNAs (miRNA and lncRNA) and mRNAs in normal tissues. Comput Math Methods Med 2019:1
Cava C, Sabetian S, Castiglioni I (2021) Patient-specific network for personalized breast cancer therapy with multi-omics data. Entropy 23(2):225
Crews KR, Hicks JK, Pui CH, Relling MV, Evans WE (2012) Pharmacogenomics and individualized medicine: translating science into practice. Clin Pharmacol Ther 92(4):467–475. https://doi.org/10.1038/clpt.2012.120
Chu EW, Karr JR (2017) Environmental impact: concept, consequences, measurement. In: Reference module in life sciences. https://doi.org/10.1016/B978-0-12-809633-8.02380-3
Coffin B, Duboc H (2022) Diagnostic and therapeutic approach to persistent abdominal pain beyond irritable bowel syndrome. Aliment Pharmacol Ther 56(3):419–435
Cohen SP, Vase L, Hooten WM (2021) Chronic pain: an update on burden, best practices, and new advances. Lancet 397(10289):2082–2097
Comte B, Baumbach J, Benis A, Basílio J, Debeljak N, Flobak Å, Franken C, Harel N, He F, Kuiper M, Méndez Pérez JA, Pujos-Guillot E, Režen T, Rozman D, Schmid JA, Scerri J, Tieri P, Van Steen K, Vasudevan S, Watterson S, Schmidt H (2020) Network and systems medicine: position paper of the European collaboration on science and technology action on open multiscale systems medicine. Netw Syst Med 3(1):67–90. https://doi.org/10.1089/nsm.2020.0004
Course MM, Gudsnuk K, Smukowski SN, Winston K, Desai N, Ross JP, Valdmanis PN (2020) Evolution of a human-specific tandem repeat associated with ALS. Am J Hum Genet 107(3):445–460
Cryan JF, O'Riordan KJ, Cowan CS, Sandhu KV, Bastiaanssen TF, Boehme M, Dinan TG (2019) The microbiota-gut-brain axis. Physiol Rev 99:1877
Dai X, Wiernek S, Evans JP, Runge MS (2016) Genetics of coronary artery disease and myocardial infarction. World J Cardiol 8(1):1–23. https://doi.org/10.4330/wjc.v8.i1.1
Dash S, Shakyawar SK, Sharma M (2019) Big data in healthcare: management, analysis and future prospects. J Big Data 6:54. https://doi.org/10.1186/s40537-019-0217-0
Del Sol A, Balling R, Hood L, Galas D (2010) Diseases as network perturbations. Curr Opin Biotechnol 21(4):566–571

Ding L, Liu J, Zhou L, Jia X, Li S, Zhang Q, Yu M, Xiao X (2022) A high-fat diet disrupts the hepatic and adipose circadian rhythms and modulates the diurnal rhythm of gut microbiota-derived short-chain fatty acids in gestational mice. Front Nutr 9:925390

Doan T, Ha V, Strazdins L, Chateau D (2022) Healthy minds live in healthy bodies–effect of physical health on mental health: evidence from Australian longitudinal data. Curr Psychol 1:1–12

Eddy S, Mariani LH, Kretzler M (2020) Integrated multi-omics approaches to improve classification of chronic kidney disease. Nat Rev Nephrol 16(11):657–668

El Karoui M, Hoyos-Flight M, Fletcher L (2019) Future trends in synthetic biology—a report. Front Bioeng Biotechnol 7:175

Fares J, Fares MY, Khachfe HH, Salhab HA, Fares Y (2020) Molecular principles of metastasis: a hallmark of cancer revisited. Signal Transduct Target Ther 5(1):1–17

Fasterius E, Raso C, Kennedy S, Rauch N, Lundin P, Kolch W, Al-Khalili Szigyarto C (2017) A novel RNA sequencing data analysis method for cell line authentication. PLoS One 12(2):e0171435

Feltes BC, Poloni JDF, Nunes IJG, Faria SS, Dorn M (2020) Multi-approach bioinformatics analysis of curated omics data provides a gene expression panorama for multiple cancer types. Front Genet 11:586602

Fiocchi C, Iliopoulos D (2021) IBD systems biology is here to stay. Inflamm Bowel Dis 27(6):760–770

Flores M, Glusman G, Brogaard K, Price ND, Hood L (2013) P4 medicine: how systems medicine will transform the healthcare sector and society. Per Med 10(6):565–576. https://doi.org/10.2217/pme.13.57

Freitas GA (2022) The coming singularity: the rapid evolution of human identity. Austin Macauley Publishers

García LF (2020) Immune response, inflammation, and the clinical spectrum of COVID-19. Front Immunol 11:1441

Gautier L, Cope L, Bolstad BM, Irizarry RA (2004) Affy—analysis of affymetrix GeneChip data at the probe level. Bioinformatics 20(3):307–315

Gilbert JA, Blaser MJ, Caporaso JG, Jansson JK, Lynch SV, Knight R (2018) Current understanding of the human microbiome. Nat Med 24(4):392–400. https://doi.org/10.1038/nm.4517

Goetz LH, Schork NJ (2018) Personalized medicine: motivation, challenges, and progress. Fertil Steril 109(6):952–963

González-Reymúndez A, Vázquez AI (2020) Multi-omic signatures identify pan-cancer classes of tumors beyond tissue of origin. Sci Rep 10(1):1–13

Greenblum S, Turnbaugh PJ, Borenstein E (2012) Metagenomic systems biology of the human gut microbiome reveals topological shifts associated with obesity and inflammatory bowel disease. Proc Natl Acad Sci 109(2):594–599

Gupta M, Choudhary M, Singh A, Sheoran S, Singla D, Rakshit S (2022) Meta-QTL analysis for mining of candidate genes and constitutive gene network development for fungal disease resistance in maize (Zea mays L.). Crop J

Gustafsson M, Nestor CE, Zhang H, Barabási AL, Baranzini S, Brunak S, Benson M (2014) Modules, networks and systems medicine for understanding disease and aiding diagnosis. Genome Med 6(10):1–11

Hajjar I, Liu C, Jones DP, Uppal K (2020) Untargeted metabolomics reveal dysregulations in sugar, methionine, and tyrosine pathways in the prodromal state of AD. Alzheimers Dement (Amst) 12(1):e12064. https://doi.org/10.1002/dad2.12064

Harris K, Kassis A, Major G, Chou CJ (2012) Is the gut microbiota a new factor contributing to obesity and its metabolic disorders? J Obes 2012:879151

Hasin Y, Seldin M, Lusis A (2017) Multi-omics approaches to disease. Genome Biol 18(1):1–15

Hauser AS, Attwood MM, Rask-Andersen M, Schiöth HB, Gloriam DE (2017) Trends in GPCR drug discovery: new agents, targets and indications. Nat Rev Drug Discov 16(12):829–842. https://doi.org/10.1038/nrd.2017.178

Heinken A, Basile A, Thiele I (2021) Advances in constraint-based modelling of microbial communities. Curr Opin Syst Biol 27:100346

Huang LJ, Law JN, Murali TM (2018) Automating the PathLinker app for cytoscape. F1000Res 7:727

Huber W, Carey VJ, Gentleman R, Anders S, Carlson M, Carvalho BS, Morgan M (2015) Orchestrating high-throughput genomic analysis with bioconductor. Nat Methods 12(2):115–121

Imran A, Qamar HY, Ali Q, Naeem H, Riaz M, Amin S, Kanwal N, Ali F, Sabar MF, Nasir IA (2017) Role of molecular biology in cancer treatment: a review article. Iran J Public Health 46(11):1475–1485

Irvine GW, Nguyen S (2019) An overview of the "-omics" fields at the forefront of next-generation personalized medicine and fundamental systems biology studies. Biomed Genet Genom 4(2):1

Jarada TN, Rokne JG, Alhajj R (2020) A review of computational drug repositioning: strategies, approaches, opportunities, challenges, and directions. J Chem 12(1):1–23

Joshi A, Rienks M, Theofilatos K, Mayr M (2021) Systems biology in cardiovascular disease: a multiomics approach. Nat Rev Cardiol 18(5):313–330

Joyce K, Fabra GT, Bozkurt Y, Pandit A (2021) Bioactive potential of natural biomaterials: identification, retention and assessment of biological properties. Signal Transduct Target Ther 6(1):1–28

Jordán F, Nguyen TP, Liu WC (2012) Studying protein–protein interaction networks: a systems view on diseases. Brief Funct Genomic 11(6):497–504

Kao PH, Baiya S, Lai ZY, Huang CM, Jhan LH, Lin CJ et al (2022) An advanced systems biology framework of feature engineering for cold tolerance genes discovery from integrated omics and non-omics data in soybean. Front Plant Sci 13:1019709

Kauffmann A, Gentleman R, Huber W (2009) Array Quality Metrics—A bioconductor package for quality assessment of microarray data. Bioinformatics 25(3):415–416

Kholodenko BN (2007) Employing systems biology to quantify receptor tyrosine kinase signaling in time and space. In: Introduction to systems biology, pp 300–318

Kobayashi N, Takahashi D, Takano S, Kimura S, Hase K (2019) The roles of Peyer's patches and microfold cells in the gut immune system: relevance to autoimmune diseases. Front Immunol 10:2345

Kühn S, Rieger UM (2017) Health is a state of complete physical, mental and social well-being and not merely absence of disease or infirmity. Surg Obes Relat Dis 13(5):887

Kuznetsov M, Clairambault J, Volpert V (2021) Improving cancer treatments via dynamical biophysical models. Phys Life Rev 39:1–48

Langley SR, Dwyer J, Drozdov I, Yin X, Mayr M (2013) Proteomics: from single molecules to biological pathways. Cardiovasc Res 97(4):612–622

Lacalamita A, Piccinno E, Scalavino V, Bellotti R, Giannelli G, Serino G (2021) A gene-based machine learning classifier associated to the colorectal adenoma—carcinoma sequence. Biomedicine 9(12):1937

Lagisetty Y, Bourquard T, Al-Ramahi I, Mangleburg CG, Mota S, Soleimani S, Shulman JM, Botas J, Lee K, Lichtarge O (2022) Identification of risk genes for Alzheimer's disease by gene embedding. Cell Genom 2(9):100162. https://doi.org/10.1016/j.xgen.2022.100162

Lee LY, Pandey AK, Maron BA, Loscalzo J (2021) Network medicine in cardiovascular research. Cardiovasc Res 117(10):2186–2202. https://doi.org/10.1093/cvr/cvaa321

Lee WJ, Cho H, Baek MS, Kim HK, Lee JH, Ryu YH, Seong JK (2022) Dynamic network model reveals distinct tau spreading patterns in early-and late-onset Alzheimer disease. Alzheimers Res Ther 14(1):1–13

Lefèvre-Arbogast S, Wagner M, Proust-Lima C, Samieri C (2019) Nutrition and metabolic profiles in the natural history of dementia: recent insights from systems biology and life course epidemiology. Curr Nutr Rep 8(3):256–269. https://doi.org/10.1007/s13668-019-00285-1

Lockie S, Fairley-Grenot K, Ankeny R, Botterill L, Howlett B, Mcbratney A, Woodhead I (2020) The future of agricultural technologies. Australian Council of Learned Academies (ACOLA)

Lu RM, Hwang YC, Liu IJ et al (2020) Development of therapeutic antibodies for the treatment of diseases. J Biomed Sci 27:1. https://doi.org/10.1186/s12929-019-0592-z

Lin JS, Perdue LA, Henrikson NB, Bean SI, Blasi PR (2021) Screening for colorectal cancer: updated evidence report and systematic review for the us preventive services task force. JAMA 325(19):1978–1998. https://doi.org/10.1001/jama.2021.4417
Madhavan M, Mustafa S (2022) Systems biology–the transformative approach to integrate sciences across disciplines. Phys Sci Rev 1:1
Magnani ND, Marchini T, Calabró V, Alvarez S, Evelson P (2020) Role of mitochondria in the redox signaling network and its outcomes in high impact inflammatory syndromes. Front Endocrinol 11:568305
Mardinoglu A, Nielsen J (2012) Systems medicine and metabolic modelling. J Intern Med 271(2):142–154
Marya R, Patel R (2021) Inflamed: deep medicine and the anatomy of injustice. Penguin
Mordeniz C (2019) Introductory chapter: traditional and complementary medicine. In: Traditional and complementary medicine. IntechOpen. https://doi.org/10.5772/intechopen.86373
Nair A, Chauhan P, Saha B, Kubatzky KF (2019) Conceptual evolution of cell signaling. Int J Mol Sci 20(13):3292. https://doi.org/10.3390/ijms20133292
Narrandes S, Xu W (2018) Gene expression detection assay for cancer clinical use. J Cancer 9(13):2249–2265. https://doi.org/10.7150/jca.24744
Naylor S, Chen JY (2010) Unraveling human complexity and disease with systems biology and personalized medicine. Pers Med 7(3):275–289
Neapolitan R, Horvath CM, Jiang X (2015) Pan-cancer analysis of TCGA data reveals notable signaling pathways. BMC Cancer 15(1):1–12
Papin JA, Hunter T, Palsson BO, Subramaniam S (2005) Reconstruction of cellular signalling networks and analysis of their properties. Nat Rev Mol Cell Biol 6(2):99–111. https://doi.org/10.1038/nrm1570
Pathak RK, Singh DB (2020) Systems biology approaches for food and health. In: Advances in agri-food biotechnology. Springer, Singapore, pp 409–426
Picó C, Serra F, Rodríguez AM, Keijer J, Palou A (2019) Biomarkers of nutrition and health: new tools for new approaches. Nutrients 11(5):1092. https://doi.org/10.3390/nu11051092
Pignatti C, D'Adamo S, Stefanelli C, Flamigni F, Cetrullo S (2020) Nutrients and pathways that regulate health span and life span. Geriatrics 5(4):95
Pinu FR, Beale DJ, Paten AM, Kouremenos K, Swarup S, Schirra HJ, Wishart D (2019) Systems biology and multi-omics integration: viewpoints from the metabolomics research community. Meta 9(4):76. https://doi.org/10.3390/metabo9040076
Rahman MR, Islam T, Gov E, Turanli B, Gulfidan G, Shahjaman M, Moni MA (2019) Identification of prognostic biomarker signatures and candidate drugs in colorectal cancer: insights from systems biology analysis. Medicina 55(1):20
Rai S, Raj U, Varadwaj PK (2018) Systems biology: a powerful tool for drug development. Curr Top Med Chem 18(20):1745–1754. https://doi.org/10.2174/1568026618666181025113226
Renner S, Popov M, Schuffenhauer A, Roth HJ, Breitenstein W, Marzinzik A et al (2011) Recent trends and observations in the design of high-quality screening collections. Future Med Chem 3(6):751–766
Revo V (2018) Systemic and semantic features of two concepts: "systemic medicine" and "systems medicine"
Roessler HI, Knoers NV, van Haelst MM, van Haaften G (2021) Drug repurposing for rare diseases. Trends Pharmacol Sci 42(4):255–267
Rowland I, Gibson G, Heinken A, Scott K, Swann J, Thiele I, Tuohy K (2018) Gut microbiota functions: metabolism of nutrients and other food components. Eur J Nutr 57(1):1–24. https://doi.org/10.1007/s00394-017-1445-8
Rujeedawa T, Carrillo Félez E, Clare IC, Fortea J, Strydom A, Rebillat AS, Zaman SH (2021) The clinical and neuropathological features of sporadic (late-onset) and genetic forms of Alzheimer's disease. J Clin Med 10(19):4582
Sam L, Liu Y, Li J, Friedman C, Lussier YA (2007) Discovery of protein interaction networks shared by diseases. In: Biocomputing 2007, pp 76–87

Sarkar S, Biswas SC (2022) Studies on contribution of astrocyte secreted proteins on neuronal health in models of Alzheimer's disease (Doctoral dissertation, UNIVERSITY OF CALCUTTA)
Scardoni G, Petterlini M, Laudanna C (2009) Analyzing biological network parameters with CentiScaPe. Bioinformatics 25(21):2857–2859
Schmid B, Schöb C (2022) Biodiversity and ecosystem services in managed ecosystems. In: The ecological and societal consequences of biodiversity loss, pp 211–231
Schulz M, Klipp E (2010) Introduction to systems biology. In: Systems biology in psychiatric research: from high-throughput data to mathematical modeling, pp 81–95
Schwabe RF, Jobin C (2013) The microbiome and cancer. Nat Rev Cancer 13(11):800–808
Sever R, Brugge JS (2015) Signal transduction in cancer. Cold Spring Harb Perspect Med 5(4):a006098. https://doi.org/10.1101/cshperspect.a006098
Shannon P, Markiel A, Ozier O, Baliga NS, Wang JT, Ramage D et al (2003) Cytoscape: a software environment for integrated models of biomolecular interaction networks. Genome Res 13(11):2498–2504
Shah VK, Firmal P, Alam A, Ganguly D, Chattopadhyay S (2020) Overview of immune response during SARS-CoV-2 infection: lessons from the past. Front Immunol 11:1949
Shen L, Sinai M (2020) GeneOverlap: test and visualize gene overlaps. R package version 1.26.0; http://shenlab-sinai.github.io/shenlab-sinai/
Shen S, Zhan C, Yang C, Fernie AR, Luo J (2022) Metabolomics-centered mining of plant metabolic diversity and function: past decade and future perspectives. Mol Plant 1:1
Silverman EK, Schmidt HH, Anastasiadou E, Altucci L, Angelini M, Badimon L, Baumbach J (2020) Molecular networks in network medicine: development and applications. Wiley Interdiscip Rev Syst Biol Med 12(6):e1489
Singh MP, Rai S, Singh NK, Srivastava S (2021) Transcriptomic landscape of early age onset of colorectal cancer identifies novel genes and pathways in Indian CRC patients. Sci Rep 11(1):1–11
Struyf T, Deeks JJ, Dinnes J, Takwoingi Y, Davenport C, Leeflang MM, Spijker R, Hooft L, Emperador D, Domen J, Horn S, Van den Bruel A, Cochrane COVID-19 Diagnostic Test Accuracy Group (2021) Signs and symptoms to determine if a patient presenting in primary care or hospital outpatient settings has COVID-19. Cochrane Database Syst Rev 2(2):CD013665. https://doi.org/10.1002/14651858.CD013665.pub2
Subramanian I, Verma S, Kumar S, Jere A, Anamika K (2020) Multi-omics data integration, interpretation, and its application. Bioinfo Biol Insights 14:1177932219899051
Suhail Y, Cain MP, Vanaja K, Kurywchak PA, Levchenko A, Kalluri R (2019) Systems biology of cancer metastasis. Cell Syst 9(2):109–127
Sullivan AN, Lachman ME (2017) Behavior change with fitness technology in sedentary adults: a review of the evidence for increasing physical activity. Front Public Health 4:289. https://doi.org/10.3389/fpubh.2016.0028
Sutherland HG, Albury CL, Griffiths LR (2019) Advances in genetics of migraine. J Headache Pain 20(1):72. https://doi.org/10.1186/s10194-019-1017-9
Szklarczyk D, Gable AL, Lyon D, Junge A, Wyder S, Huerta-Cepas J, Mering CV (2019) STRING v11: protein–protein association networks with increased coverage, supporting functional discovery in genome-wide experimental datasets. Nucleic Acids Res 47(D1):D607–D613
Tang WW, Kitai T, Hazen SL (2017) Gut microbiota in cardiovascular health and disease. Circ Res 120(7):1183–1196
Tran L, Xiao JF, Agarwal N, Duex JE, Theodorescu D (2021) Advances in bladder cancer biology and therapy. Nat Rev Cancer 21(2):104–121
Ungar M, Theron L (2020) Resilience and mental health: how multisystemic processes contribute to positive outcomes. Lancet Psychiatry 7(5):441–448
Vineis P, Robinson O, Chadeau-Hyam M, Dehghan A, Mudway I, Dagnino S (2020) What is new in the exposome? Environ Int 143:105887
Walpole J, Papin JA, Peirce SM (2013) Multiscale computational models of complex biological systems. Annu Rev Biomed Eng 15:137–154

Werner HM, Mills GB, Ram PT (2014) Cancer systems biology: a peek into the future of patient care? Nat Rev Clin Oncol 11(3):167–176
Wiedmann T, Lenzen M, Keyßer LT et al (2020) Scientists' warning on affluence. Nat Commun 11:3107. https://doi.org/10.1038/s41467-020-16941-y
Wishart DS, Feunang YD, Marcu A, Guo AC, Liang K, Vázquez-Fresno R, Sajed T, Johnson D, Li C, Karu N, Sayeeda Z, Lo E, Assempour N, Berjanskii M, Singhal S, Arndt D, Liang Y, Badran H, Grant J, Serra-Cayuela A, Scalbert A (2018) HMDB 4.0: the human metabolome database for 2018. Nucleic Acids Res 46(D1):D608–D617. https://doi.org/10.1093/nar/gkx1089
Wörheide MA, Krumsiek J, Kastenmüller G, Arnold M (2021) Multi-omics integration in biomedical research–a metabolomics-centric review. Anal Chim Acta 1141:144–162
Xi Y, Xu P (2021) Global colorectal cancer burden in 2020 and projections to 2040. Transl Oncol 14(10):101174
Yadav G, Vashisht M, Yadav V, Shyam R (2020) Molecular biomarkers for early detection and prevention of ovarian cancer-a gateway for good prognosis: a narrative review. Int J Prev Med 11:135. https://doi.org/10.4103/ijpvm.IJPVM_75_19
Yadav M, Pandey P, Sharma P (2022) Understanding the genetic, molecular, and cellular basis of ageing as the biggest risk factor of Alzheimer's disease. Eur J Biol Res 12(1):62–76
Zhao S, Iyengar R (2012) Systems pharmacology: network analysis to identify multiscale mechanisms of drug action. Annu Rev Pharmacol Toxicol 52:505
Zhao EY, Jones M, Jones S (2019) Whole-genome sequencing in cancer. Cold Spring Harb Perspect Med 9(3):a034579. https://doi.org/10.1101/cshperspect.a034579
Zou X, Kim DW, Gotoh T, Liu J, Kim JK, Finkielstein CV (2020) A systems biology approach identifies hidden regulatory connections between the circadian and cell-cycle checkpoints. Front Physiol 11:532355

Medical Applications of Systems Biology

3

Devendra Singh, Perugu Shyam, Sunil Kumar Verma, and Anjali

Abstract

The vast figure of described molecular agents connected to pathogenesis could not be integrated easily or handled by traditional analytical procedures due to the exponential growth of highly sophisticated medical and scientific analysis tools over the past 30 years. In fact, the increased number of complicated diseases being defined has, in part, been caused by the recognition that various moieties represent disease markers. Thanks to currently accessible cutting-edge technology, scientists and doctors may now look into and evaluate any particular dysregulations occurring at the genomic, microRNA (miRNA), transcriptome, and protein levels. The fact that just isolated molecular levels are separately examined for its influence on each specific health condition presents a limitation in this scientific brave new world. Later its inception in the 1992 year, systems biology and remedy have mostly focused on the changes in complete route kinetics that lead to the beginning and/or worsening of the condition(s) under investigation. In order to shed light on a variety of study settings, systems medicine methodologies can be used. This has the practical benefit of revealing unique dynamic binding interactions that are crucial for affecting the progression of medical disorders. As a result, systems medicine helps to pinpoint clinically significant drug targets for preventative and curative actions against this illness.

D. Singh (✉) · P. Shyam
National Institute of Technology, Warangal, Telangana, India

S. K. Verma
BN College of Engineering and Technology, Lucknow, Uttar Pradesh, India

Anjali
CSIR-Central Drug Research Institute, Lucknow, Uttar Pradesh, India

S. Joshi et al. (eds.), *Systems Biology Approaches: Prevention, Diagnosis, and Understanding Mechanisms of Complex Diseases*,
https://doi.org/10.1007/978-981-99-9462-5_3

Keywords
System biology · Application · Human disease · Diagnostic

3.1 Introduction

Over the past 30 years, very sophisticated scientific and medical research analytical methods have grown exponentially where the majority of the major molecular factors thought to influence human states, and diseases can be examined. Credit to developing wet lab technologies like Mass spectrometry (MS), Next-generation sequencing (NGS), Quantitative polymerase chain reaction (qPCR), and comprehensive bio-informatics sets, clinicians and researchers could start to explore whichever distinct dysregulations are befalling at the transcriptomic, genomic, metabolomics, proteomic, and miRnomic levels. All of these technologies have the ability to extract data from large databases, enabling the creation of disease models for wet-lab testing. In addition to being a key element of translational medicine at the moment, the interaction among the dry and wet lab and particular clinical skills is made possible by systems medicine. The fact that most scientific studies only individually examine certain molecular levels for their influence on a given health problem highlights one of the downsides of this brave new scientific era. Any type of therapeutic research with the goal of identifying molecular pathway connections which are dysregulated and need to ideally concentrate on examining all facets of the complicated and multi-factorial medical ailment or conditions. This entails a thorough evaluation of all concurrent molecular interactions taking place in stages (i.e., transcriptomic). Such "bigger-picture" research viewpoints help researchers gain a deeper understanding of multifaceted and multi-factorial disease ailments and, in the end, "fast-track" the discovery as well as a medical analysis of a particular dysregulated pathway (molecular) in conjunction with pathogenesis significance, as well as the joint novel medicine targets identification for the creation of efficient translational therapeutic toward the ailment (Singh et al. 2022). As a result, the innovative research field of systems biology has emerged in the past 10 years in response to the urgent need to address these research flaws (Bruggeman and Westerhoff 2007; Keene et al. 2004).

3.2 Principles of Systems Biology Research Methods

The fundamental tenet of systems biology is that whichever exclusive living organism's phenotype is a rumination of the numerous molecular interactions transpiring simultaneously at different levels and combining in a holistic way to form that phenotype (Friboulet and Thomas 2005; Singh et al. 2023). In order to identify discrete variations in the configuration of inter-molecular interactions in relation to the organism's examined phenotypes, information from the dysregulations of several crucial molecular targets (Tillmann et al. 2015). Consequently, today's maximum self-reliant System Biology (SB) research factions are made up of investigators with a discernible understanding of research work for the majority of molecular genetics. The primary research techniques differ according to the molecular level being studied, the amount of data generated, and other factors. As a result, SB is a highly multidisciplinary area of study which requires people with a variety of technological backgrounds and research specialties (Friboulet and Thomas 2005; Tillmann et al. 2015). Although it is far from usually possible to measure every molecular component of an organ or even a biological pathway, significant efforts are being made to increase analytical sensitivity and give the results a quantitative relevance through the use of field standards (Brazma 2009; Bustin et al. 2009). Boolean methods are helping to produce first-generation systems biology models given the limitations at hand (Kaushik and Sahi 2015). The assumption that the data are accurate and usable is one of the key differences among systems medicine and systems biology. While the data generation skill of wet lab is frequently not the foremost objective, now it is considered to be accurate and practical in system biology. In order to produce exquisite data sets used to create pathway models and treatments, systems medicine (SM) guarantees in the way toward taking the spearhead with molecular and clinical expertise. This, in turn, should unswervingly add to stratified medicine on the way to personalized healthcare (Cardinal-Fernández et al. 2014; Kamada 1992; Karnes et al. 2014; Tillmann et al. 2015; Zou et al. 2013; Singh et al. 2022a, b).

Systems biology research methodologies mainly trust on mathematical modeling and computational/bioinformatics mechanisms to provide solutions to the targeted research topics in toting toward serving as an interdisciplinary field of study (Araúzo-Bravo 2016; Balazki et al. 2015; Cromar et al. 2015; de Graaf et al. 2009; Karp et al. 2010; Wolstencroft et al. 2015; Zhang et al. 2010; Singh et al. 2022c). In system biology, such informatics technology applications might take either an experimental data-driven "bottom-up" method or a hypothesis-driven "top-down" method (Logan et al. 2010; Zou et al. 2013). The bottom-up, data-driven method begins with the assortment of massive data sets obtained from a variety of omics-centered experimental techniques. Next, careful mathematical modeling evaluations are conducted to conglomerate the correlations amongst important molecular players from various omics data results achieved (Lehnert et al. 2015; Logan et al. 2010; Rolfsson and Palsson 2015). Network modeling is the main approach used by the bottom-up systems biology theoretical method (Breitling et al. 2008; Edwards and Thiele 2013; Fouladiha et al. 2015; Stevens et al. 2014). A standard biological network model is made up of many nodes networking with one another across edges. Nodes are characterized as distinct significant molecular actors from any omics level, in addition to edges that characterize experimentally verified molecular interactions (Logan et al. 2010). Any given biological network may contain nodes and edges that vary in both kind and level of detail. Additionally, hubs are regarded as extremely dynamic nodes engaging in a close-knit network (Andorf et al. 2013). The two categories of "party" hubs and "dating" hubs are further divisions of hubs. Party hubs are dynamic nodes which can frequently connect with several other partners (molecular) at once, while date hubs are significantly active as they act together with other molecular partners over a wider range of durations and locations (Andorf et al. 2013; Singh and Agarwal 2022).

The hypothesis-driven top-down technique, in contrast to bottom-up experimental methodologies, mainly depends on mathematical modeling to study interactions (molecular) at a small scale for a particular phenotype or biological situation (Çakır and Khatibipour 2014; Logan et al. 2010). The dynamical modeling used for this includes converting the pathway (molecular) interactions existing in the examined entity into predefined mathematical designs, like partial as well as ordinary differential equations, that can be investigated and queried in a "dry lab" environment (Barbolosi et al. 2009; Schramm and Dimitrov 2014). Since the majority of intermolecular activities have particular kinetics that may be replicated by suitable mathematical derivations (e.g., Michaelis–Menten kinetics), this technique can be used (Sobie 2011). Dynamic modeling, however, is only useful if certain presumptions are made about the biomolecular interactions that are occurring, for example, the choice of predetermined reaction rate kinetics happening inside the investigated bio-molecular collaborations (Raue et al. 2015; Sobie 2011). The middle-out (rational) approach is an intriguing third methodology for systems biology research models, which implements the bottom-up approaches as well as top-down approaches (Giuliani et al. 2014; Majumder and Mukherjee 2011).

3.3 Computational Methods and Biological Systems in Systems Biology

A network of biological organisms that perform a particular function is referred to as a biological system. Research has lately concentrated on examining the dynamics and structure of a complete biological system, even if understanding the genome and how it is expressed is still crucial. The interactions between a system's various components are determined by how well it functions. The following four fundamental properties must be understood in order to comprehend the interaction and functions of a system:

1. the system's structural components, such as the metabolic pathway and its gene relationships, over and above the controls over the aforementioned processes,
2. system dynamics that includes arrays of system behavior through identification and time of the key processes that regulate particular behaviors under different circumstances,
3. the control strategy that involves setting up cellular systems to reduce malfunctions,
4. the design process that includes methods for altering and creating biological systems to cut down on errors.

There is no agreed-upon meaning of SB; there are many distinct definitions, dependent on the investigator. SB is a multidisciplinary branch of biology that focuses on simulating complex biological systems through computation and mathematics. Understanding complicated connections and activities at the organism, tissue, and cell levels is the goal of this subject, and it has immediate applications in biomedical research. Systems biology, therefore, includes a methodology that departs from the previous reductive biological study.

In order to analyze the phenotype produced by the components of the system, systems biology looks at the relations amid numerous modules more willingly than the individual properties of the molecules. Systems biology uses computational methods to accomplish this by building potential in silico models which can correspondingly have tested experimentally in in vitro or in vivo, enabling the study of a lot of data (Bartocci and Lió 2016; Breitling 2010; Hillmer 2015; Kitano 2002; Zou et al. 2013; Singh et al. 2021). Numerous computational tools are employed in the examination of biological systems, including methods for determining or predicting the molecular structure, and molecular interactions, for capturing molecular dynamics and for sequence alignment (Nobile et al. 2017).

The two sorts of computational procedures are those conducted by data and those conducted by hypotheses. Data are used in network modeling to show how various molecular components interact with one another. Dynamic modeling, on the other hand, is predicated on a theory to describe the interactions and quantitative relationships among the molecular components (Fig. 3.1).

This group of computational methods comprises methods involving more compact biological systems with a small number of molecular components (Faratian et al. 2009). There is a connection between the creation of a computer program and the design of a biological model from the perspective of computational model use in SB (Barnat et al. 2009; Müssel et al. 2010). As a result, the researcher is able to

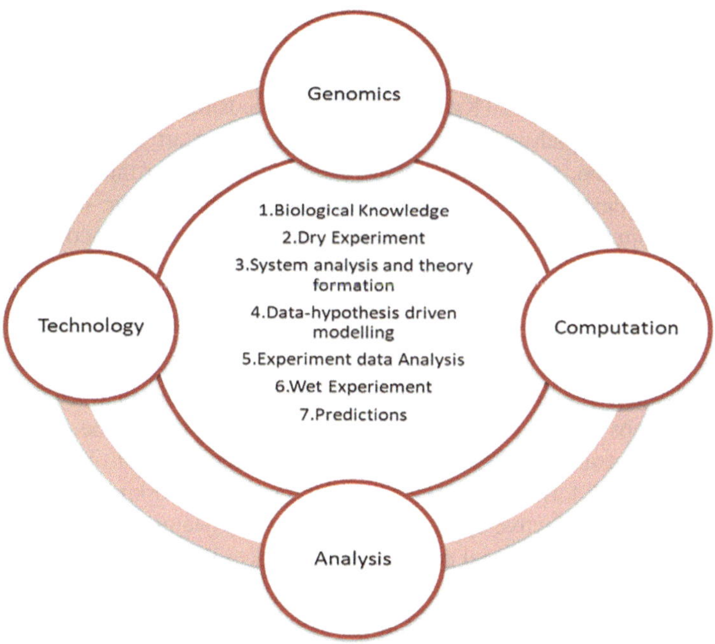

Fig. 3.1 System biology research with a hypothesis-driven approach

see how the reactions in the system flow inside and examine how these reactions alter based on the system's environment (Gay et al. 2010).

Systems biology refers to the molecular components as "nodes" in a biological network, and their interactions are referred to as "edges." Genes, metabolites, proteins, medications, illnesses, as well as phenotypes, are examples of molecular components (Sobie et al. 2011). In contrast, known molecular routes are represented mathematically as part of the dynamical modeling of a biological system. Established on the physico-chemical characteristics that govern reaction rate and the contact's affinity, mathematical calculations are used to describe biological processes (Benque et al. 2012). It is necessary to deliver the preliminary concentrations of cellular components, plus their connectivity must subsequently be built sequentially to obtain the utmost effective implementation and functioning of a dynamic model. Thus, by dynamic modeling, the component concentrations at later times can be computed and compared to those that take place during the experimental time (Zou et al. 2013). When the starting data and parameters are established and point in a specific direction, a dynamic model is said to be definitive. However, it can also be stochastic when it can move to multiple states with varying probabilities based on the initial data (Hoops et al. 2006).

3.4 Systems Biology Applied to Human Illness

The classic mechanistic method of therapeutic examination has been explored and may be limited to the examination of the biological impact of a minute or individual quantities of crucial molecules in complex, multi-factorial human illnesses, such as cancer. SM is the use of SB contained by the scope of current medical research. Its concept dates back to 1992 (Kamada 1992). By focusing primarily on the disconcertion of general pathway kinetics for the ensuing start or worsening of the researched situation, such a larger perception makes way for innovative and newfangled understanding and perception keen on the all-inclusive temperament of these illness situations. When using such an approach, systems medicine necessitates the use of certain essential components in order to achieve its therapeutic theranostic objectives (Costa 2008). In order to effectively and quickly execute any unique research methodology targeted at achieving the intended outcome for systems medicine-based initiatives, it is ideal to have established the fundamental aspects of systems medicine (Singh and Agarwal 2022d; Kalia et al. 2020). Without a doubt, the labs working on systems medicine projects need to be equipped with the appropriate research protocol and infrastructure modifications toward intensive inter-disciplinary information flow, data handling, and networking which are essential for facilitating effective approaches in a way to systems medicine. The use of computing from computational and modeling sciences is a crucial aspect of systems medicine. Such knowledge is a requirement for processing "large" data sets effectively as well as for interpreting wet lab data in terminologies leading to the emergence of intricate interactions amid many major molecular actors. The first human condition to be recently studied from a systems-level approach is neuroblastoma (Logan et al. 2010).

In order to identify fundamental molecular agents influencing the severity and pathogenesis of the infection, in the human immunodeficiency virus (HIV) case, the study of Sarmady et al. is an excellent example of the resourcefulness of computer-aided testing on heretofore available data (experimental) from specific interactions at the molecular level (Sarmady et al. 2011; Sharma et al. 2019). A motif discovery technique was used in the study to analyze a certain subset of human immunodeficiency virus viral protein sequences as well as binding immediate protein sequences of associates discovered on the host entity (Sarmady et al. 2011). A motif discovery technique was used in the study to analyze certain groups of human immunodeficiency virus viral protein sequences as well as sequences of immediate binding protein partners discovered on the host organism (Sarmady et al. 2011). Such a sequence-based prediction and interactome techniques enabled the identification of human immunodeficiency virus Nef protein as the primary binding site for numerous host proteins, including HCK, VAV1, CD4, LCK, FYN, HLA-A, MAPK1, and GNB2L1 (Sarmady et al. 2011). A systematic representation was shown in Fig. 3.2.

A review by Verma et al. that created a protein regulatory network built on systems that considers the effects of microRNAs (miRNA) affecting breakpoint cluster region (BCR) would be an additional case in point of the application of modeling science in system medicine. Chronic myeloid leukemia's ABL phosphorylation and oncoprotein expression levels cell lines with imatinib resistance (Verma et al. 2013). BCR.ABL-specific microRNAs and tyrosine kinase inhibitors were shown to have diverse impacts on cell lines with varying chemo-resistance qualities along with expression profiles, and this network of protein regulation was found to be consistent in determining these differences (Verma et al. 2013). Additionally, high-throughput quantitative polymerase chain reaction-based microRNAs expression profiles were created for this study's purposes, demonstrating how a systems-based

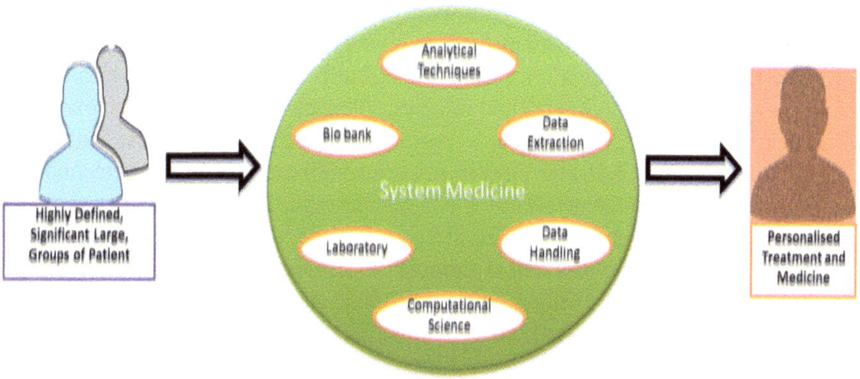

Fig. 3.2 A systematic representation of the elements needed to deploy systems medicine techniques in contemporary medical research

methodology may be used to create a network for protein regulation with the help of massive experimental data sets (Verma et al. 2013). This research can be used to demonstrate the value of quantitative analytical tools for encouraging innovative data gathering in systems medicine research.

The subject of biomarker discovery is an alternate study area where systems medicine approaches are particularly beneficial (Cardinal-Fernández et al. 2014). In a recent work, Zhang et al. used high-throughput mRNA and microRNA expression profiling studies to in silico analyze the expression data for equally metastatic and primary prostate cancer (Zhang et al. 2012). The analysis's findings showed that primary and metastatic prostate cancers have different microRNAs–mRNA correlation and regulatory modular networks (Zhang et al. 2012). This research serves as a classic illustration of the value of systems-level investigation for the discovery of an extremely interactive bio-marker defining various severity and classes for a particular disease circumstance which can eventually be used to examine (or forecast) particular specific patient response to treatment. The availability of relatively large but well-defined patient groups is another essential condition for the up-and-coming application of systems medicine. If curated biobanks are used to source patient samples, such patient groupings can be structured very well.

As the results of High-throughput (HT) profiling of metabolites are being examined for the control of urate crystals, that unswervingly causes gout in the body of humans, as well as is also linked to type II diabetes and cardiovascular disease, Albrecht et al.'s study examined the pathophysiology of hyperuricemia (Mani et al. 2008). This study used a hypothesis-free approach using Gaussian Graphical Modeling to analyze 355 metabolites from 1764 patients in order to build a set of connections of metabolites that regulate the generation of serum urate (Albrecht et al. 2014). The findings of this investigation revealed a unique serum urate regulatory mechanism that involved 38 important metabolite components, a large percentage of which had a gender-specific characteristic (Albrecht et al. 2014).

A study supervised by Mani et al. adds to the growing body of evidence supporting the need of using a systems-level approach to forecast oncogenes in cancer circumstances (Mani et al. 2008). This work analyzed the B-cell interactome plus microarray-based data sets to identify molecular dysregulation interactions in 3 distinct non-lymphomas Hodgkin's (Mani et al. 2008). The development of quantitative data extraction technologies and imaging sciences has been of enormous value in achieving good outcomes using systems medicine methodologies. Single-cell sequencing, laser-assisted microdissection, and high content imaging technology are a few examples of such technologies (Benner et al. 2015; Faruque et al. 2015; Navin 2015). Breast cancer and Dupuytren's disease, a proliferative fibromatosis illness, are among the two medical illnesses examined with the help of systems-level research approaches. Immunology research is another area in which these conditions are examined (Bolt et al. 2015; Kidd et al. 2014; Rehman et al. 2008, 2011, 2012). Essentially this shift in study focus, which focuses on general molecular network interactions more willingly than specific molecules, enables additional productive and clinically useful research conclusions.

3.5 System Medicine Implications in the Development of Novel Medicines

Various areas of the drug R&D processes contained in the context of the pharmaceutical business are currently being expedited by the ever-expanding influence of systems medicine. In the initial stages of drug development, among the primary research issues where systems medicine's point of view can significantly differ is in the prediction of pharmacological side effects (Zou et al. 2013). In the past, pharmacogenetics research for drug development nearly exclusively concentrated on the impact of changes in specific genes for creating a particularly bad outcome (Zou et al. 2013). But given that these negative effects are likely the result of multiple factors, a systems-based inquiry can be much more successful in identifying and/or anticipating negative side effects before any unique lead molecule advances further along the drug development process (Zou et al. 2013). In their latest findings, Zhao et al. looked into the prospect of finding a secondary medication that might be used in addition to the diabetes medication rosiglitazone (Zhao et al. 2013). Because rosiglitazone has a significant risk of an adverse effect called myocardial infarction, the researchers looked for a second medication that could lower this risk in those who were also on rosiglitazone (Zhao et al. 2013). Through its capacity to control blood clotting procedures, the results of this systems-based methodology revealed that exenatide is an appropriate medicine to administer for reducing the risks of rosiglitazone cardiac side effects (Zhao et al. 2013). This study showed that systems medicine techniques can be used to anticipate the best drug combination therapy to be used for particular illness situations, in addition to having a significant role in forecasting drug adverse effects.

Systems techniques are important for predicting target/drug interactions, which is another field of drug research. Investigations of gene expression profiling of drug perturbations sought to identify new inhibitors of the human ethera-go-go related (hER) gene potassium channel, that is crucial for controlling tumor cell growth and death (Babcock et al. 2013; Pakladok et al. 2014). The Connectivity Map (CMap) was used in the study to choose potential inhibitors of human ether-a-go-go related gene (hERG) with like induction of pattern of gene expression, along with the techniques (analytical) from databases (Babcock et al. 2013). For the determination of conducting target/drug interaction prediction studies, additional systems-led approaches can be used, especially when dealing with large and diverse molecular interaction networks. The probabilistic matrix factorization and the NRWRH are two examples of such techniques (Chen et al. 2012; Cobanoglu et al. 2013).

Similar to how the use of multidrug therapy avenues to maximize the efforts involved in the development of new drugs via improved therapy success is a recent trend, the significance of the network pharmacology approach for identifying the new target/drug for single as well as multiple illness situations is becoming much more vital (Tang and Aittokallio 2014). This strategy is extremely valuable, especially for complicated and multivariate clinical disorders like cancer pathways, which have a large range of druggable targets (Tang and Aittokallio 2014). In the developing field of drug repositioning, where treatments that are thought to be outdated or ineffectual for a specific human disease may instead turn out to be very successful for a completely different ailment, SM techniques also play a key role (Jin and Wong 2014). The exploratory investigation by Jin et al. concentrated on the examination and minimization of off-target drug effects for important signaling (cancer) pathways through the investigation of transcriptome expression patterns occurring prior to as well as after giving the drug (Jin et al. 2012). This study combined the Bayesian factor regression model (BFRM), an established systems-based analytical approach, with the innovative network of CSB components, resulting in an approach known as CSB-BFRM (Jin et al. 2012). Concrete evidence research in leukemia, prostate, as well as breast cancer models demonstrated the validity of the unique systems medicine approach. For the bulk majority of the medications that the FDA has approved, the CSB-BFRM were effective in predicting the clinical response results (Jin et al. 2012).

Other bioinformatics software is also being created to help investigators conduct medication repositioning studies successfully. In order to improve quick access to this information, one such application is Drugmap Central, which enables users to download any multilayer data for recognized medications and compounds within a single framework (Fu et al. 2013). Finally, systems medicine approaches, where the critical examination emphasizes interactions of molecules influencing pathogenesis which were frequently active as well as dysregulated in numerous sickness situations, can also have a significant influence on the opportunities of identifying a new infection network. miRNA research is a typical illustration of how systems medicine approaches are applied in this research environment. It is not surprising that miRNAs can be linked to numerous clinical diseases, possibly simultaneously, given that miRNAs regulate transcripts and can each downregulate the expression of hundreds of downstream target genes (Mestdagh et al. 2011). The creation of bioinformatics web tools like miRNA Body Map makes it easier for miRNA scientists to identify the nodes as well as hubs for their network modeling tactics relevant to the particular circumstances they are studying by providing a perfect visual of the degree of molecular interactions which were straightly affected by the members of miRNA of the miRNome (Mestdagh et al. 2011).

3.6 Analysis and Future Prospects

There is no doubt that systems medicine is changing how researchers, clinicians, and academics approach medical research experiments. A greater understanding of the intricate and entangled, complicated molecular interactions is possible when multi-level data from both computational in-silico and real experimental sources are examined together. Otherwise, the interactome would not be visible because the connections between regulatory processes are not immediately apparent. This results in the discovery of novel dynamic interactions that play a crucial role in determining how diseases develop and, as a result, serve as therapeutically significant key molecules for the creation of new diagnostic and therapeutic agents. However, there are still significant obstacles because new technologies like next-generation sequencing and mass spectrometry (MS) produce enormous amounts of information, in addition to the computational approaches currently in use, which have only recently been able to handle sorting across such astronomical volumes of data to find important conjectures for the modeled research uncertainties. The choice of the best systems-led approaches is subject to a great level of subjectivity because there is not a single particular instrument that can be used to assist in the amalgamation of multi-omics information.

Systems medicine is refocusing on molecular hubs that have a significant and varied impact on the outcomes of poly-pharmacology treatments (Kell 2013; Kell et al. 2013). Additionally, systems medicine and systems biology techniques ought to be used more extensively in order to construct the "virtual human," an extremely ambitious aim (Kell 2007).

The traditional "symptom" for the development of the new drug and the use of Lipinski's Rule for the assortment of new drugs for the development are two conceptual barriers that must be completely removed in order for systems-led research to fully realize its potential (Chen et al. 2012; Cobanoglu et al. 2013; Fu et al. 2013; Jin et al. 2012; Jin and Wong 2014; Kell 2013; Mestdagh et al. 2011; Tang and Aittokallio 2014). The need for efficient systems medicine models that combine various data masses for the precise discovery of therapies, novel or new drug targets, as well as improved classifying the suffers risk category, are further issues. Barely proper handling of quantitative data, achieved by means of standardized and approved approaches, with efficient data transmission abilities amongst various software programs as well as information handling platforms in a seamless way, can lead to such effectiveness (Lefever et al. 2009). To enable more rapid theranostic breakthroughs, in the end, SM will prove to be one of the cornerstones of how future research will be conducted by adopting these novel research perceptions.

References

Albrecht E, Waldenberger M, Krumsiek J, Evans AM, Jeratsch U, Breier M, Adamski J, Koenig W, Zeilinger S, Fuchs C, Klopp N, Theis FJ, Wichmann H-E, Suhre K, Illig T, Strauch K, Peters A, Gieger C, Kastenmüller G, Meisinger C (2014) Metabolite profiling reveals new insights into the regulation of serum urate in humans. Metabolomics 10(1):141–151. https://doi.org/10.1007/s11306-013-0565-2

Andorf CM, Honavar V, Sen TZ (2013) Predicting the binding patterns of hub proteins: a study using yeast protein interaction networks. PLoS One 8(2):e56833. https://doi.org/10.1371/journal.pone.0056833

Araúzo-Bravo MJ (2016) Computational biology methods for characterization of pluripotent cells. Methods Mol Biol 1357:195–220. https://doi.org/10.1007/7651_2015_279

Babcock JJ, Du F, Xu K, Wheelan SJ, Li M (2013) Integrated analysis of drug-induced gene expression profiles predicts novel hERG inhibitors. PLoS One 8(7):e69513. https://doi.org/10.1371/journal.pone.0069513

Balazki P, Lindauer K, Einloft J, Ackermann J, Koch I (2015) MONALISA for stochastic simulations of petri net models of biochemical systems. BMC Bioinformatics 16(1):215. https://doi.org/10.1186/s12859-015-0596-y

Barbolosi D, Benabdallah A, Hubert F, Verga F (2009) Mathematical and numerical analysis for a model of growing metastatic tumors. Math Biosci 218(1):1–14. https://doi.org/10.1016/j.mbs.2008.11.008

Barnat J, Brim L, Černá I, Dražan S, Fabriková J, Láník J, Šafránek D, Ma H (2009) Bio DiVinE: a framework for parallel analysis of biological models. Electron Proceed Theor Comp Sci 6:31–45. https://doi.org/10.4204/EPTCS.6.3

Bartocci E, Lió P (2016) Computational modeling, formal analysis, and tools for systems biology. PLoS Comput Biol 12(1):e1004591. https://doi.org/10.1371/journal.pcbi.1004591

Benner S, Kakeyama M, Endo T, Yoshioka W, Tohyama C (2015) Application of neuro trace staining in the fresh frozen brain samples to laser microdissection combined with quantitative RT-PCR analysis. BMC Res Notes 8(1):252. https://doi.org/10.1186/s13104-015-1222-9

Benque D, Bourton S, Cockerton C, Cook B, Fisher J, Ishtiaq S, Piterman N, Taylor A, Vardi MY (2012) Bma: visual tool for modeling and analyzing biological networks. In: Madhusudan P, Seshia SA (eds) Computer aided verification, vol 7358. Springer, Berlin Heidelberg, pp 686–692. https://doi.org/10.1007/978-3-642-31424-7_50

Bolt MJ, Stossi F, Callison AM, Mancini MG, Dandekar R, Mancini MA (2015) Systems level-based RNAi screening by high content analysis identifies UBR5 as a regulator of estrogen receptor-α protein levels and activity. Oncogene 34(2):154–164. https://doi.org/10.1038/onc.2013.550

Brazma A (2009) Minimum information about a microarray experiment (MIAME)–successes, failures, challenges. Sci World J 9:420–423. https://doi.org/10.1100/tsw.2009.57

Breitling R (2010) What is systems biology? Front Physiol 1:9. https://doi.org/10.3389/fphys.2010.00009

Breitling R, Gilbert D, Heiner M, Orton R (2008) A structured approach for the engineering of biochemical network models, illustrated for signalling pathways. Brief Bioinform 9(5):404–421. https://doi.org/10.1093/bib/bbn026

Bruggeman FJ, Westerhoff HV (2007) The nature of systems biology. Trends Microbiol 15(1):45–50. https://doi.org/10.1016/j.tim.2006.11.003

Bustin SA, Benes V, Garson JA, Hellemans J, Huggett J, Kubista M, Mueller R, Nolan T, Pfaffl MW, Shipley GL, Vandesompele J, Wittwer CT (2009) The MIQE guidelines: minimum information for publication of quantitative real-time PCR experiments. Clin Chem 55(4):611–622. https://doi.org/10.1373/clinchem.2008.112797

Çakır T, Khatibipour MJ (2014) Metabolic network discovery by top-down and bottom-up approaches and paths for reconciliation. Front Bioeng Biotechnol 2:1. https://doi.org/10.3389/fbioe.2014.00062

Cardinal-Fernández P, Nin N, Ruíz-Cabello J, Lorente JA (2014) Medicina de sistemas: Una nueva visión de la práctica clínica. Arch Bronconeumol 50(10):444–451. https://doi.org/10.1016/j.arbres.2013.10.010

Chen X, Liu M-X, Yan G-Y (2012) Drug–target interaction prediction by random walk on the heterogeneous network. Mol BioSyst 8(7):1970–1978. https://doi.org/10.1039/C2MB00002D

Cobanoglu MC, Liu C, Hu F, Oltvai ZN, Bahar I (2013) Predicting drug–target interactions using probabilistic matrix factorization. J Chem Inf Model 53(12):3399–3409. https://doi.org/10.1021/ci400219z

Costa J (2008) Systems medicine in oncology. Nat Clin Pract Oncol 5(3):117–117. https://doi.org/10.1038/ncponc1070

Cromar GL, Zhao A, Yang A, Parkinson J (2015) Hyperscape: visualization for complex biological networks. Bioinformatics 31(20):3390–3391. https://doi.org/10.1093/bioinformatics/btv385

de Graaf AA, Freidig AP, Roos BD, Jamshidi N, Heinemann M, Rullmann JAC, Hall KD, Adiels M, van Ommen B (2009) Nutritional systems biology modeling: from molecular mechanisms to physiology. PLoS Comput Biol 5(11):e1000554. https://doi.org/10.1371/journal.pcbi.1000554

Edwards LM, Thiele I (2013) Applying systems biology methods to the study of human physiology in extreme environments. Extrem Physiol Med 2(1):8. https://doi.org/10.1186/2046-7648-2-8

Faratian D, Clyde RG, Crawford JW, Harrison DJ (2009) Systems pathology—taking molecular pathology into a new dimension. Nat Rev Clin Oncol 6(8):455–464. https://doi.org/10.1038/nrclinonc.2009.102

Faruque J, Beaulieu CF, Rosenberg J, Rubin D, Yao D, Napel S (2015) Content-based image retrieval in radiology: analysis of variability in human perception of similarity. J Med Imaging (Bellingham) 2(2):025501. https://doi.org/10.1117/1.JMI.2.2.025501

Fouladiha H, Marashi S-A, Shokrgozar MA (2015) Reconstruction and validation of a constraint-based metabolic network model for bone marrow-derived mesenchymal stem cells. Cell Prolif 48(4):475–485. https://doi.org/10.1111/cpr.12197

Friboulet A, Thomas D (2005) Systems biology—an interdisciplinary approach. Biosens Bioelectron 20(12):2404–2407. https://doi.org/10.1016/j.bios.2004.11.014

Fu C, Jin G, Gao J, Zhu R, Ballesteros-villagrana E, Wong STC (2013) DrugMap central: an on-line query and visualization tool to facilitate drug repositioning studies. Bioinformatics 29(14):1834–1836. https://doi.org/10.1093/bioinformatics/btt279

Gay S, Soliman S, Fages F (2010) A graphical method for reducing and relating models in systems biology. Bioinformatics 26(18):i575–i581. https://doi.org/10.1093/bioinformatics/btq388

Giuliani A, Filippi S, Bertolaso M (2014) Why network approach can promote a new way of thinking in biology. Front Genet 5:83. https://doi.org/10.3389/fgene.2014.00083

Hillmer RA (2015) Systems biology for biologists. PLOS Pathog 11(5):e1004786. https://doi.org/10.1371/journal.ppat.1004786

Hoops S, Sahle S, Gauges R, Lee C, Pahle J, Simus N, Singhal M, Xu L, Mendes P, Kummer U (2006) COPASI—a COmplex PAthway SImulator. Bioinformatics 22(24):3067–3074. https://doi.org/10.1093/bioinformatics/btl485

Jin G, Wong STC (2014) Toward better drug repositioning: prioritizing and integrating existing methods into efficient pipelines. Drug Discov Today 19(5):637–644. https://doi.org/10.1016/j.drudis.2013.11.005

Jin G, Fu C, Zhao H, Cui K, Chang J, Wong STC (2012) A novel method of transcriptional response analysis to facilitate drug repositioning for cancer therapy. Cancer Res 72(1):33–44. https://doi.org/10.1158/0008-5472.CAN-11-2333

Kalia M, Singh D, Sharma D, Narvi S, Agarwal V (2020) Senna alexandriana mill as a potential inhibitor for quorum sensing-controlled virulence factors and biofilm formation in Pseudomonas aeruginosa PAO1. Pharmacogn Mag 16(72):797. https://doi.org/10.4103/pm.pm_315_20

Kamada T (1992) System biomedicine: a new paradigm in biomedical engineering. Front Med Biol Eng 4(1):1–2

Karnes JH, Van Driest S, Bowton EA, Weeke PE, Mosley JD, Peterson JF, Denny JC, Roden DM (2014) Using systems approaches to address challenges for clinical implementation of pharmacogenomics. WIREs Syst Biol Med 6(2):125–135. https://doi.org/10.1002/wsbm.1255

Karp PD, Paley SM, Krummenacker M, Latendresse M, Dale JM, Lee TJ, Kaipa P, Gilham F, Spaulding A, Popescu L, Altman T, Paulsen I, Keseler IM, Caspi R (2010) Pathway tools version 13.0: integrated software for pathway/genome informatics and systems biology. Brief Bioinform 11(1):40–79. https://doi.org/10.1093/bib/bbp043

Kaushik AC, Sahi S (2015) Boolean network model for GPR142 against type 2 diabetes and relative dynamic change ratio analysis using systems and biological circuits approach. Syst Synth Biol 9(1):45–54. https://doi.org/10.1007/s11693-015-9163-0

Keene JD, Mesarovic MD, Sreenath SN (2004) Search for organising principles: understanding in systems biology. Syst Biol 1(1):19–27. https://doi.org/10.1049/sb:20045010

Kell DB (2007) The virtual human: towards a global systems biology of multiscale, distributed biochemical network models. IUBMB Life 59(11):689–695. https://doi.org/10.1080/15216540701694252

Kell DB (2013) Finding novel pharmaceuticals in the systems biology era using multiple effective drug targets, phenotypic screening and knowledge of transporters: where drug discovery went wrong and how to fix it. FEBS J 280(23):5957–5980. https://doi.org/10.1111/febs.12268

Kell DB, Dobson PD, Bilsland E, Oliver SG (2013) The promiscuous binding of pharmaceutical drugs and their transporter-mediated uptake into cells: what we (need to) know and how we can do so. Drug Discov Today 18(5):218–239. https://doi.org/10.1016/j.drudis.2012.11.008

Kidd BA, Peters LA, Schadt EE, Dudley JT (2014) Unifying immunology with informatics and multiscale biology. Nat Immunol 15(2):118–127. https://doi.org/10.1038/ni.2787

Kitano H (2002) Systems biology: a brief overview. Science 295(5560):1662–1664. https://doi.org/10.1126/science.1069492

Lefever S, Hellemans J, Pattyn F, Przybylski DR, Taylor C, Geurts R, Untergasser A, Vandesompele J, RDML Consortium (2009) RDML: structured language and reporting guidelines for real-time quantitative PCR data. Nucleic Acids Res 37(7):2065–2069. https://doi.org/10.1093/nar/gkp056

Lehnert T, Timme S, Pollmächer J, Hünniger K, Kurzai O, Figge MT (2015) Bottom-up modeling approach for the quantitative estimation of parameters in pathogen-host interactions. Front Microbiol 6:608. https://doi.org/10.3389/fmicb.2015.00608

Logan JA, Kelly ME, Ayers D, Shipillis N, Baier G, Day PJ (2010) Systems biology and modeling in neuroblastoma: practicalities and perspectives. Expert Rev Mol Diagn 10(2):131–145. https://doi.org/10.1586/erm.10.4

Majumder D, Mukherjee A (2011) A passage through systems biology to systems medicine: adoption of middle-out rational approaches towards the understanding of therapeutic outcomes in cancer. Analyst 136(4):663–678. https://doi.org/10.1039/C0AN00746C

Mani KM, Lefebvre C, Wang K, Lim WK, Basso K, Dalla-Favera R, Califano A (2008) A systems biology approach to prediction of oncogenes and molecular perturbation targets in B-cell lymphomas. Mol Syst Biol 4(1):169. https://doi.org/10.1038/msb.2008.2

Mestdagh P, Lefever S, Pattyn F, Ridzon D, Fredlund E, Fieuw A, Ongenaert M, Vermeulen J, De Paepe A, Wong L, Speleman F, Chen C, Vandesompele J (2011) The microRNA body map: dissecting microRNA function through integrative genomics. Nucleic Acids Res 39(20):e136. https://doi.org/10.1093/nar/gkr646

Müssel C, Hopfensitz M, Kestler HA (2010) BoolNet—an R package for generation, reconstruction and analysis of Boolean networks. Bioinformatics 26(10):1378–1380. https://doi.org/10.1093/bioinformatics/btq124

Navin NE (2015) Delineating cancer evolution with single-cell sequencing. Sci Transl Med 7(296):296fs29. https://doi.org/10.1126/scitranslmed.aac8319

Nobile MS, Cazzaniga P, Tangherloni A, Besozzi D (2017) Graphics processing units in bioinformatics, computational biology and systems biology. Brief Bioinform 18:870–885

Pakladok T, Hosseinzadeh Z, Almilaji A, Lebedeva A, Shumilina E, Alesutan I, Lang F (2014) Up-regulation of hERG K+ channels by B-RAF. PLoS One 9(1):e87457. https://doi.org/10.1371/journal.pone.0087457

Raue A, Steiert B, Schelker M, Kreutz C, Maiwald T, Hass H, Vanlier J, Tönsing C, Adlung L, Engesser R, Mader W, Heinemann T, Hasenauer J, Schilling M, Höfer T, Klipp E, Theis F, Klingmüller U, Schöberl B, Timmer J (2015) Data2Dynamics: a modeling environment tailored to parameter estimation in dynamical systems. Bioinformatics 31(21):3558–3560. https://doi.org/10.1093/bioinformatics/btv405

Rehman S, Salway F, Stanley JK, Ollier WER, Day P, Bayat A (2008) Molecular phenotypic descriptors of Dupuytren's disease defined using informatics analysis of the transcriptome. J Hand Surg Am 33(3):359–372. https://doi.org/10.1016/j.jhsa.2007.11.010

Rehman S, Goodacre R, Day PJ, Bayat A, Westerhoff HV (2011) Dupuytren's: a systems biology disease. Arthritis Res Ther 13(5):238. https://doi.org/10.1186/ar3438

Rehman S, Day P, Bayat A, Westerhoff H (2012) Understanding Dupuytren's disease using systems biology: a move away from reductionism. Front Physiol 3:316. https://doi.org/10.3389/fphys.2012.00316

Rolfsson Ó, Palsson BO (2015) Decoding the jargon of bottom-up metabolic systems biology. BioEssays 37(6):588–591. https://doi.org/10.1002/bies.201400187

Sarmady M, Dampier W, Tozeren A (2011) Sequence- and interactome-based prediction of viral protein hotspots targeting host proteins: a case study for HIV Nef. PLoS One 6(6):e20735. https://doi.org/10.1371/journal.pone.0020735

Schramm HC, Dimitrov NB (2014) Differential equation models for sharp threshold dynamics. Math Biosci 247:27–37. https://doi.org/10.1016/j.mbs.2013.10.009
Sharma N, Singh D, Rani R, Sharma D, Pandey H, Agarwal V (2019) Chitosan and its nanocarriers: applications and opportunities. In: Nanomaterials in plants, algae and microorganisms. Academic Press, pp 267–286. https://doi.org/10.1016/B978-0-12-811488-9.00013-5
Singh D, Agarwal V (2022) An in vivo acute toxicity and anti-shigellosis effect of designed formulation on rat. J Ayurveda Integr Med 14:100536. https://doi.org/10.1016/j.jaim.2021.100536; Advance online publication
Singh D, Agarwal V (2022d) Herbal antibacterial remedy against upper respiratory infection causing bacteria and in vivo safety analysis. Vegetos 35(1):264–268. https://doi.org/10.1007/s42535-021-00281-3
Singh D, Sharma D, Agarwal V (2021) Screening of anti-microbial, anti-biofilm activity, and cytotoxicity analysis of a designed polyherbal formulation against shigellosis. J Ayurveda Integr Med 12(4):601–606. https://doi.org/10.1016/j.jaim.2021.06.007
Singh D, Pandey H, Shrivastava NK, Das R, Singh V (2022) Green synthesized gold and silver nanoparticles for antimicrobial applications. In: Baskar C, Ramakrishna S, Daniela La Rosa A (eds) Encyclopedia of green materials. Springer, Singapore. https://doi.org/10.1007/978-981-16-4921-9_254-1
Singh V, Pandey H, Misra V, Tiwari V, Srivastava P, Singh D (2022a) Hypolipidemic effect of [6]-Gingerol-loaded Eudragit polymeric nanoparticles in high-fat diet-induced rats and gamma scintigraphy evaluation of gastric-retention time. J Appl Pharm Sci 12(6):156–163. https://doi.org/10.7324/JAPS.2022.120615
Singh D, Singh V, Bhushan Mishra S, Sharma D, Agarwal V (2022b) Evaluation of anti-biofilm, anti-quorum, anti-dysenteric potential of designed polyherbal formulation: in vitro and in vivo study. J Appl Biomed 20(1):7–14. https://doi.org/10.32725/jab.2022.005
Singh D, Pandey H, Singh V (2022c) Natural products that target cancer stem cells. In: Handbook of research on natural products and their bioactive compounds as cancer therapeutics. IGI Global, pp 169–186. https://doi.org/10.4018/978-1-7998-9258-8.ch008
Singh D, Pandey H, Singh V, Pandey D (2023) The role of IoT in healthcare services: an extensive review. In: Internet of things and data mining for modern engineering and healthcare applications, pp 1–14. https://doi.org/10.1201/9781003217398
Sobie EA (2011) An introduction to dynamical systems. Sci Signal 4(191):tr6. https://doi.org/10.1126/scisignal.2001982
Sobie EA, Lee Y-S, Jenkins SL, Iyengar R (2011) Systems biology—biomedical modeling. Sci Signal 4(190):tr2. https://doi.org/10.1126/scisignal.2001989
Stevens A, Leonibus CD, Hanson D, Dowsey AW, Whatmore A, Meyer S, Donn RP, Chatelain P, Banerjee I, Cosgrove KE, Clayton PE, Dunne MJ (2014) Network analysis: a new approach to study endocrine disorders. J Mol Endocrinol 52(1):R79–R93. https://doi.org/10.1530/JME-13-0112
Tang J, Aittokallio T (2014) Network pharmacology strategies toward multi-target anticancer therapies: from computational models to experimental design principles. Curr Pharm Des 20(1):23–36
Tillmann T, Gibson AR, Scott G, Harrison O, Dominiczak A, Hanlon P (2015) Systems medicine 2.0: potential benefits of combining electronic health care records with systems science models. J Med Internet Res 17(3):e3082. https://doi.org/10.2196/jmir.3082
Verma M, Karimiani EG, Byers RJ, Rehman S, Westerhoff HV, Day PJR (2013) Mathematical modelling of miRNA mediated BCR.ABL protein regulation in chronic myeloid leukaemia Vis-a-Vis therapeutic strategies. Integr Biol 5(3):543–554. https://doi.org/10.1039/c3ib20230e
Wolstencroft K, Owen S, Krebs O, Nguyen Q, Stanford NJ, Golebiewski M, Weidemann A, Bittkowski M, An L, Shockley D, Snoep JL, Mueller W, Goble C (2015) SEEK: a systems biology data and model management platform. BMC Syst Biol 9(1):33. https://doi.org/10.1186/s12918-015-0174-y
Zhang S-Q, Ching W-K, Tsing N-K, Leung H-Y, Guo D (2010) A new multiple regression approach for the construction of genetic regulatory networks. Artif Intell Med 48(2):153–160. https://doi.org/10.1016/j.artmed.2009.11.001

Zhang W, Edwards A, Fan W, Flemington EK, Zhang K (2012) MiRNA-mRNA correlation-network modules in human prostate cancer and the differences between primary and metastatic tumor subtypes. PLoS One 7(6):e40130. https://doi.org/10.1371/journal.pone.0040130

Zhao S, Nishimura T, Chen Y, Azeloglu EU, Gottesman O, Giannarelli C, Zafar MU, Benard L, Badimon JJ, Hajjar RJ, Goldfarb J, Iyengar R (2013) Systems pharmacology of adverse event mitigation by drug combinations. Sci Transl Med 5(206):206ra140. https://doi.org/10.1126/scitranslmed.3006548

Zou J, Zheng M-W, Li G, Su Z-G (2013) Advanced systems biology methods in drug discovery and translational biomedicine. Biomed Res Int 2013:e742835. https://doi.org/10.1155/2013/742835

Systems Biology for Metabolic Disorder and Disease

4

Mukta Sharma and Atul Singh

Abstract

The simplistic approach frequently used in biological sciences is no longer sufficient if we want to research complicated multi-factorial illnesses like Metabolic Syndrome (MetS) by utilizing cutting-edge 'omics' techniques. A thorough understanding of the system is required to comprehend the adaptive changes in molecular mechanisms at various phases of pathogenicity since activating various pathways may still result in the same functionality but at multiple metabolic costs. Systems biology is an interdisciplinary branch of research that employs a much more thorough overview to tackle biomedical and biologic research. It focuses on complex interactions within and across biological systems. Whereas the significance of the systems biology approach has long being understood, experimental and simulation methods have progressed to the point where thorough molecular characterization of biological systems is now attainable. The term metabolic syndrome describes a collection of illnesses that include dyslipidaemia, fatty liver, diabetes, obesity, resistance to insulin and other cardiovascular issues. Metabolic syndrome is significant due to its frequency, possible severity and expense. The control of the metabolic pathways that control the overall balance of the body's systems depends heavily on the liver. An intricate web of hormones, transcription factors, and signalling pathways regulates how much glucose and lipids are produced in the liver. A prevalent understanding nowadays is that the metabolic syndrome's hepatic manifestation is fatty liver that is associated with dysfunctional lipoprotein, fatty acid and glucose metabolism. In order to replicate the recognized metabolic activities in a cell, metabolic systems biology provides significant abstracted techniques, which results in a picture which is near to the observed phenotype. This will enable identifying the

M. Sharma (✉) · A. Singh
School of Life Sciences and Technology, IIMT University, Meerut, Uttar Pradesh, India

S. Joshi et al. (eds.), *Systems Biology Approaches: Prevention, Diagnosis, and Understanding Mechanisms of Complex Diseases*,
https://doi.org/10.1007/978-981-99-9462-5_4

patient's illness pattern and giving accurate medical solutions, leading to prophylactic medicine, less treatment and in silico clinical studies when combined with cutting-edge machine learning techniques.

Keywords

System biology · Metabolic syndrome · Insulin resistance · Lipid metabolism · Glucose metabolism · Gut microbiota · Diabetes

4.1 Introduction

System biology is a method for studying the complexities of biological systems. The term system biology has recently emerged to describe the frontier of cross-disciplinary research in biology. It is an interdisciplinary branch of study with a biological foundation that concentrates on intricate connections within biological systems. Nearly 10 years ago, the Human Genome Project (HGP) was finished, and 'omics' technologies—in particular, transcriptomics, proteomics and metabolomics—were simultaneously launched. These events marked the widespread introduction of this term. Computational and mathematical modelling is integrated with biological data and data science. Datasets are often extensive and contain information on numerous biological characteristics and variables, such as genes, proteins and metabolites. Different mathematical and computational techniques are utilized and developed depending on the study subject or application. Our mix of machine learning and dynamic modelling is a key strength. Data originates from experimental and clinical studies conducted in close collaboration with clinical research institutes and hospitals. A prevalent understanding nowadays is that the metabolic syndrome's hepatic manifestation is fatty liver that is associated with dysfunctional lipoprotein, fatty acid and glucose metabolism. Adipocytes are highly specialized cells that are crucial to the energy equilibrium of vertebrate beings. In addition to being the major illness of fat cells, obesity significantly raises the possibility of type-two diabetes, a cardiovascular illness, hypertension and metabolic disorder. Numerous changes in skeletal muscle metabolism brought on by the metabolic syndrome are unquestionably concerning the emergence of insulin resistance. Patients with type-two diabetes and obesity have reduced rates of glycogen synthesis, muscular glucose uptake and mitochondrial oxidation. Oxidative stress has been connected to MetS muscle insulin resistance, increased lipid availability and inflammation. Changes in their rates of synthesis, conversion and catabolism in the intricate system known as lipoprotein metabolism might result in abnormal amounts of certain lipoprotein particles. Understanding lipid problems in people is crucial for creating novel preventative and therapeutic approaches for dyslipidaemia. Trillions of bacteria inhabit a human intestine, 150 times more microbial genes being added to our own human genomes.

4.2 The Metabolic Syndrome and System Biology

The same patient will frequently have diagnosis of fatty liver, dyslipidaemia, insulin resistance, obesity and diabetes. Since this link occurs frequently, it cannot be a coincidence. It follows that some aetiopathogenic pathways may be shared by these seemingly unrelated disorders. Because of this, it has been determined that it is feasible to include these manifestations according to the idea of the metabolic syndrome from a diagnostic perspective, risk assessment, education and global therapy reasons despite the fact that their pathogenic relationship has not yet been established. The main international risk factors for heart disease is the main reason why morbidity in people with obesity and type two diabetes mellitus are pathogenic entities covered by the MetS concept. The fact that the MetS concept is regarded to be sophisticated and heterogeneous yet appearing to be simple is another advantageous feature of it. The MetS pathological range can be expanded to include more pathologies, ranging from psychiatric illnesses to certain cancer kinds, thanks to this flexibility, which will be especially helpful as information technology develops and patient epidemiological data is better integrated. Since each of the MetS's component characteristics has a varied predictive strength and relevance and separate features are used for Type 2 diabetes mellitus (T2DM) and Cardio Vascular Diseases (CVD) risk prediction, the model's usefulness as a risk prediction tool is highly contested due to the model's diversity and complexity (Eckel et al. 2010). Although the MetS has significant limitations, it might be claimed that these limitations are outweighed by the diagnostic and therapeutic awareness it offers, which surpasses the limitations brought on by its fuzzy aetiology and murky link to its constituent parts. In any case, it is anticipated that further MetS research will shed light on both the specific nosological entities that make up the syndrome as well as the underlying causes of a wide spectrum of metabolic disorders. The two main objectives of MetS research in this respect are: (1) Accurately forecasting the risk of metabolic disorders and, consequently, implementing preventive measures in the best way possible. (2) To uncover fresh methods for both therapy and prevention by figuring out the pathophysiology of these related illnesses. MetS is no exception, and several aetiologies or entrances corresponding to discrete pathogenic groups with varying early disease histories are likely. Most likely, an individual's phenotype will be determined by the equilibrium between adaptive allostatic responses and organ-specific pathogenic effector mechanisms, which will ultimately result in global homeostatic failure. A few of the flaws that could lead to the onset of MetS include those in the beta cells, which result in primary hyperinsulinaemia, the liver; this results in fatty liver and insulin resistance, the skeletal muscle; the skeletal muscle; that causes insulin resistance and the primary hypothalamus; which affects peripheral energy homeostasis.

4.3 Metabolic Syndrome and Liver Disease

4.3.1 Lack of Fatty Liver Can Cause Metabolic Syndrome

Since the metabolic syndrome and non-alcoholic fatty liver disease both have a pathogenic aetiology and increase the risk of diabetes and heart disease. These are clinical signs of the identical metabolic dysfunction that raises the risk of diabetes and heart disease.

If three or more people exhibit any one of the following signs: visceral obesity, elevated triglycerides, elevated blood pressure, glucose intolerance or type two diabetes and low high-density-lipoprotein (HDL) cholesterol, MetS is thought to be present. According to the third report of the National Cholesterol Education Program's Expert Panel on the detection, evaluation and treatment of high blood cholesterol in adults, this is the case. All MetS traits are linked by insulin resistance (IR), which is a necessary element and possible pathogen (Balkau and Charles 1999; Olufadi and Byrne 2008). Non-alcoholic fatty liver disease, i.e. NAFLD refers to a collection of ailments that includes hepatic steatosis alone, nonalcoholic steatohepatitis (NASH), fibrosis and cirrhosis. When there is of macro-vesicular hepatic steatosis, which develops when triglycerides (TGs) build up inside hepatocytes without a lot of alcohol use, is the common denominator. However, steatosis is not a human physiological finding. It is a characteristic of the livers of migratory birds, who must store energy to sustain lengthy fasts. Therefore, in contrast to earlier theories, NAFLD is a chronic condition that progresses over time and results in NASH, fibrosis, cryptogenic cirrhosis and an elevated risk of hepatocellular cancer (Bugianesi et al. 2005; Caldwell et al. 1999). Hepatic steatosis and IR's cause and temporal correlations, as well as whether IR or steatosis develops first, are uncertain because both of these conditions have the ability to worsen one another. Hepatic IR may occur independently of peripheral IR if IR and higher plasma levels of glucose and fatty acids lead to enhanced hepatic lipogenesis, reduced -oxidation and hepatic lipid storage. As evidence, peripheral IR can exist without hepatic IR or excessive liver lipid accumulation. Short-term high-fat diet exposure in rats results in changes to hepatic fat accumulation and IR but not to total body IR (Samuel et al. 2004). Contrarily, the development of metabolic syndrome and NAFLD is linked to hyperglycaemia, hyperinsulinaemia and reduced peripheral insulin sensitivity because these conditions encourage hepatic lipogenesis and the release of TGs into the bloodstream as well as intrahepatic, peripheral (obesity and atherosclerosis) and hepatic lipid accumulation which aggravates IR (Unger 2003).

4.3.2 Hepatic Disease's Clinical Presentation in the Metabolic Syndrome

For those suffering from MetS, NAFLD is the most common hepatic disease and the most common cause of abnormal liver examinations. Blood indicators for liver damage are often within normal levels in NAFLD, a silent illness with a broad clinical manifestation. It is possible to evaluate steatosis, but NAFLD is not diagnosed until all other potential reasons of fatty liver have been eliminated. Most instances of NAFLD are asymptomatic; uncommon symptoms include persistent tiredness, discomfort in the right upper abdomen region and unexplained hepatomegaly. When abnormal blood liver function tests are observed, imaging may mistakenly identify fatty liver (hyperechogenic liver during an ultrasound scan). Strong suspicion exists that this case involves NASH. A patient's albumin levels and bilirubinaemia should be taken into account as well if the AST: ALT ratio is more than one, which is a sign that fibrosis and cirrhosis may be advancing. GGT is frequently increased and has been associated with a rise in mortality (Ghouri et al. 2010; Haring et al. 2009). To rule out metabolic syndrome and a heightened potential for cardiovascular disease, a thorough metabolic examination of the patient is indicated when NAFLD and NASH are found. Because of this, NAFLD/NASH patients usually have elevated levels of cholesterol, triglyceridaemia, insulin and fasting blood glucose. The liver enzymes and ultrasonography (US) should always be checked for NAFLD/NASH in individuals with obesity, diabetes, hypertriglyceridaemia and MetS. A variety of imaging approaches, such as CT, MRI, ultrasonography (US), and other non-invasive grading techniques for histological prediction, can be used to detect NAFLD (Joy et al. 2003; Saadeh et al. 2002). Ultrasound is the most simple and affordable diagnostic method to demonstrate increased liver echodensity (e.g. 'brilliant liver'). In contrast, the sensitivity of ultrasound is constrained since it cannot identify the existence and/or degree of fibrosis and inflammation. When lipid build-up covers more than 30% of the hepatic tissue, hepatic steatosis can only be diagnosed (Harrison et al. 2003). However, due to the high cost of CT and MRI scans as well as how difficult it is to differentiate between basic steatosis and steatohepatitis and/or fibrosis, their widespread usage is not supported. Numerous non-invasive diagnostic approaches and there are now scoring systems available, in an effort to improve the non-invasive evaluation of liver fibrosis or the existence of NASH. This holds true for numerous quantitative techniques that are suggested to be substitute markers of the development of NASH, such as elastography, which by analysing the strength (elasticity) of the liver may be able to better detect vascular blurring and parenchymal echo texture. Despite the fact that each of these replacement signs may support clinical evaluations, a liver biopsy is still required. The ideal method of diagnosis is a liver biopsy since it can determine the level of steatosis, the presence of necroinflammatory lesions and the presence of fibrosis, which distinguishes NASH from regular steatosis. The most popular method for judging NASH histological severity is the Brunt grading system. It is supported by the four fundamental histological abnormalities of steatosis, hepatocyte ballooning, inflammation and fibrosis (Brunt et al. 1999).

4.4 The Function of Adipose Tissue in the Pathogenesis and Treatment of Metabolic Syndrome

Standard definitions of the term 'MetS' include elevated cardiovascular disease, type two diabetes, insulin resistance and obesity as distinguishing signs. Along with more standard risk factors like dyslipidaemia and hypertension, this condition is connected to non-traditional risk factors such coagulopathy and inflammation (Isomaa et al. 2001; Grundy 2007; Lorenzo et al. 2007).

It also goes by the terms X syndrome, cardio metabolic syndrome, insulin resistance syndrome and pre-diabetes. Whatever the name, MetS incidence has been steadily rising due to the alarming rise in obesity and T2DM prevalence, which have already reached epidemic levels. The MetS is currently one of the most serious public health issues in the globe, and it always will be.

4.4.1 Adipocytes

Adipocytes, highly specialized cells, are important for maintaining the energy homeostasis of invertebrate species. Obesity is currently a global pandemic and is characterized by excessive adipocyte size or quantity. In addition to being the main cause of fat cell disease, the development of the metabolic syndrome is greatly influenced by obesity, cardiovascular disease, hypertension and type two diabetes. Because of obesity and conditions that are linked to it, the systems that control how metabolic genes are expressed in adipocytes are dysregulated. Therefore, understanding the pathophysiology of metabolic illnesses as well as finding proteins or pathways that may make attractive targets for pharmaceutical treatments requires an understanding of adipocyte differentiation. These insights have an influence on our knowledge of the aetiology and management of metabolic syndrome, a significant category of metabolic illnesses that reduce life expectancy and constitute a hazard on a worldwide level owing to an epidemic.

4.4.2 Adipokinase Secreted by Adipose Tissue

Adipokines are generally any hormone/cytokine released by one of the cells in adipose tissue. The largest kind of adipocyte in adipose tissue, the adipocyte, produces leptin and adiponectin, two hormones that can only be produced and released by fat cells (Table 4.1). When these hormones have a bearing on distant target tissues including the brain, liver and skeletal muscle, they appear to have a main endocrine-based function. There are numerous additional cell kinds, such as immune system cells like T cells and macrophages, are also present in adipose tissue. Additionally, these immune cells participate in the cellular hormone mediator process. Particularly, cytokines that macrophages secrete appear to have a paracrine influence on neighbouring adipocytes. The three most frequent adipocytokines—or, as they are more commonly called these inflammatory mediators—are TNF, IL-6 and IFN (Table 4.1).

4.4.3 Leptin

The fat gene was positionally cloned, which resulted in the discovery of the first hormone made only from adipocytes. Leptin has garnered a great deal of attention since it was discovered in 1994 as one of the most crucial cues for

Table 4.1 The function and regulation of adipokines in metabolic syndrome

Adipokine	Function	Regulation
Adiponectin	Enhances insulin sensitivity, anti-inflammatory, anti-atherogenic, regulates glucose and lipid metabolism	Decreased in obesity, insulin resistance and metabolic syndrome
Leptin	Regulates energy intake and expenditure, enhances insulin sensitivity, influences glucose and lipid metabolism	Increased in obesity, insulin resistance and metabolic syndrome
Resistin	Impairs insulin sensitivity, promotes inflammation, influence glucose and lipid metabolism	Increased in obesity, insulin resistance and metabolic syndrome
Visfatin	Regulates insulin sensitivity, promotes inflammation	Increased in obesity, insulin resistance and metabolic syndrome
TNF-α	Promotes inflammation, impairs insulin signalling, influences glucose and lipid metabolism	Increased in obesity, insulin resistance and metabolic syndrome
IL-6	Promotes inflammation, impairs insulin signalling, influence glucose and lipid metabolism	Increased in obesity, insulin resistance and metabolic syndrome

TNF-α tumour necrosis factor alpha, *IL-6* interleukin-6

controlling calorie intake and energy balance. Leptin alters glucose homeostasis, improves fatty acid oxidation, reduces body weight by reducing fat mass and decreases hunger, all of which help maintain a healthy energy balance (Ahima 2006). Leptin receptors may be found in the brainstem and hypothalamus, two central nervous system components. Reduced hunger is an outcome of active brain pathways, as opposed to greater sympathetic nervous system activity and higher energy expenditure (Hall et al. 2007). Blood leptin levels and adiposity are connected; however, this is only one part of how adiposity is managed. When you fast, leptin drops immediately, followed by a rise in glucocorticoids, a fall in T4 (thyroid hormone), sex hormones and a rise in growth hormone. Liver, pancreatic and skeletal muscle cells respond to insulin more favourably and have lower intracellular lipid concentrations when leptin is present. Leptin treatment, which improves glucose tolerance and decreases triglyceride levels, is actually beneficial for individuals with lipodystrophy and leptin insufficiency (Ronti et al. 2006).

4.4.4 Interlukin-6 (IL-6)

Adipocytes in adipose tissue, along with fibroblasts, preadipocytes, endothelial cells and monocytes, generate interleukin-6 (IL-6). White adipose tissue contributes around 30% of the circulating IL-6, with visceral fat producing more of it than subcutaneous fat (Dube et al. 2008). According to studies, human obesity, insulin resistance, hypertension and hyperlipidaemia all have favourable correlations with plasma IL-6 concentrations (Bastard et al. 2006; Bruun et al. 2003). It has been demonstrated that elevated IL-6 levels can limit adiponectin secretion and may be a predictor of T2DM. IL-6 affects metabolism through decreasing hepatic insulin sensitivity and adipocyte glucose absorption, among other things. Furthermore, it has been proven that IL-6 promotes lipolysis and fat oxidation. IL-6 significantly increases the synthesis of reactive oxygen and nitrogen by macrophages and monocytes, this can cause oxidative stress to increase. As a result, this cytokine seems to be crucial in the inflammation and modifications in glucose and lipid metabolism linked to metabolic syndrome).

4.5 Metabolic Syndrome and Skeletal Muscle

4.5.1 Skeletal Muscle Insulin Resistance in the Metabolic Syndrome

To maintain the body's overall glucose balance, skeletal muscle is crucial. Skeletal muscle absorbs between 70 and 80% of the glucose that is consumed, which is then converted to glycogen used as energy to a lesser amount, stored as fat (DeFronzo et al. 1985; Nuutila et al. 1992). Studies utilizing euglycemic-insulin clamps suggest that the predominant pathway for glucose elimination in healthy individuals is non-oxidative glucose metabolism. Skeletal muscle also absorbs the bulk of the glucose during a hyperglycaemic-hyperinsulinaemic clamp in healthy adults, with muscle glycogen making up about 80% of this polysaccharide. But after a typical meal, about 30% of the glucose is used for muscle glycogen production (Woerle et al. 2003). Insulin resistance has an impact on skeletal muscle in people with type 2 diabetes and obesity, according to research using in vitro human skeletal muscle preparations. As a result, diabetes type 2 and obesity both result in reduced insulin-stimulated skeletal muscle glucose transfer (Dohm et al. 1988; Goodyear et al. 1995). In type-2 diabetics, there is a reduction in the recruitment of GLUT-4 glucose transporters that reach muscle fibres' cell surfaces, which results in this impairment (Ryder et al. 2000).

4.5.2 Metabolic Syndrome-Related Change in Insulin Signalling in Muscle

The phosphorylation of IRSs, notably IRS1 and IRS2, is correlated with insulin signalling via the cell-surface receptor as shown in Fig. 4.1. Mitogen-activated protein kinases and phosphatidylinositol 3 kinase are two metabolic pathways that can carry insulin signals (MAPKs). Each of these procedures performs as a node that is crucial to the control of the metabolic and genetic events regulating insulin sensitivity. At each of these nodes, a large number of substrate isoforms and protein kinases are expressed, which adds to the complexity. As a result, IRS1 signals to Akt in

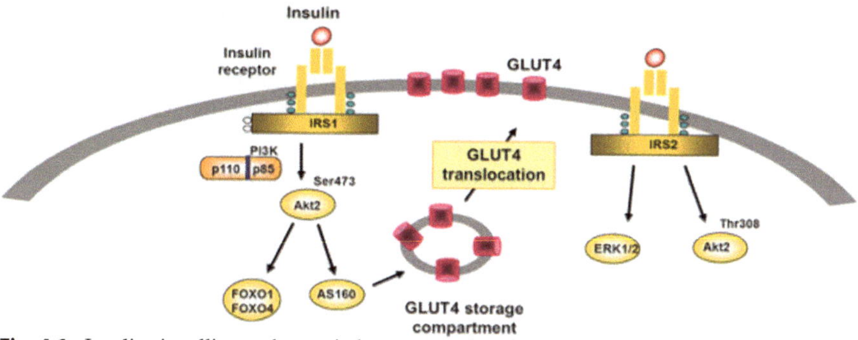

Fig. 4.1 Insulin signalling pathways in human muscle cells

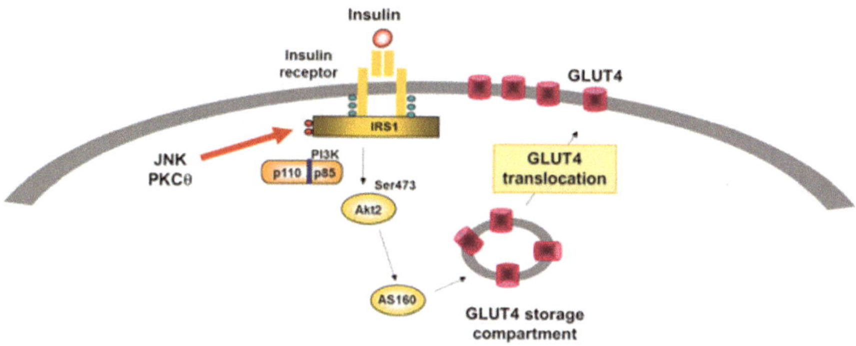

Fig. 4.2 Major metabolic alterations detected in insulin resistant muscles

human myotubes while IRS-2 signals to MAPK. These findings show that isoform specificity exists in node-to-node transmission (Bouzakri et al. 2006).

When macrophages invade adipose tissue during obesity, they release cytokines like TNF- or IL-6, which can activate cell-surface receptor signalling pathways. An oversupply of glucose or lipids can also negatively influence insulin signalling (Nuutila et al. 1992; Chibalin et al. 2008; Taube et al. 2009). The IRS isoforms and perhaps the insulin receptor are involved in serine phosphorylation activities that negatively control the insulin cascade as shown in Fig. 4.2.

Numerous lines of study indicate that patients with type 2 diabetes and obesity have altered insulin signalling in their muscle. Studies on the insulin receptor's phosphorylation in skeletal muscle of type 2 diabetes patients have produced conflicting results. In type-2 diabetics who are slim, insulin receptor phosphorylation appears to be reduced or normal. Similar to this, it has been observed that in obese non-diabetic individuals, insulin receptor phosphorylation is either normal or decreased. Skeletal muscle insulin-induced IRS1 tyrosine phosphorylation is diminished in type 2 diabetics. The cause of this impairment is not low IRS-1 protein expression (Bjornholm et al. 1997; Krook et al. 2000). In type 2 diabetic muscle, a similar impairment is observed at the PI3K-level. Additionally, it has been discovered that obese individuals have less insulin-induced tyrosine phosphorylation of IRS-1 and PI3K in their skeletal muscle.

One theory holds that inhibiting the IRS1/PI3K pathway by phosphorylating IRS proteins at specific serine sites inhibits insulin signalling. This theory is validated by the findings that (a) serine phosphorylation of IRS proteins decreases their ability to bind PI3K and (b) serine phosphorylation of IRS1 speeds up protein degradation.

4.5.3 Metabolic Syndrome and Glucose Metabolism in Muscle

By stimulating GLUT-4, the plasma membrane to reach and then delivering glucose to the cell, insulin increases skeletal muscle's ability to absorb glucose. The glycolytic process uses glucose after it is converted into glycogen by glycogen synthase after being phosphorylated by hexokinase and glucose-6-phosphate.

According to studies by (Carey et al. 2003; Shulman et al. 1990), type 2 diabetes patients have baseline muscle glycogen levels that are 30% lower than those of matched controls, and they also synthesise glycogen at a pace that is 50% slower.

4.5.3.1 Glucose Transport

Skeletal muscle takes in more glucose during the presence of the insulin signalling pathway because GLUT4 is transported to the plasma membrane from intracellular vesicles. Insulin-resistant people with type 2 diabetes or obesity who have normal or elevated levels of GLUT4 m-RNA or protein expression have significantly decreased glucose transport. According to these findings, the underlying cause of the deficit in muscle glucose transport is either decreased GLUT4 translocation or increased glucose transporter intrinsic activity. An issue with GLUT4 recruitment to the cell surface is also suggested by type 2 diabetes patients' muscle, which has decreased AS160 phosphorylation (Karlsson et al. 2005).

4.5.3.2 Glucose Phosphorylation

After entering the muscle cell, hexokinase-II uses glucose to phosphorylate it to G-6-P (HK-II). The processes of phosphorylation and glucose transport are antagonistic. Skeletal muscle in humans, insulin elevates HK-II mRNA, protein levels and cytosolic activity are connected to the translocation of the enzyme to the mitochondria. Data on whether type 2 diabetics' glucose phosphorylation is impacted are contradictory, which is probably due to the various methodologies utilized in the investigations. Due to the fact that impaired glucose phosphorylation is more serious than impaired glucose transport, several articles contend that decreased the rate-limiting step in type 2 diabetes patients for insulin action is HK-II activation. According to this theory, type 2 diabetics have decreased HK-II activity in their muscles (Kruszynska et al. 1998).

4.5.3.3 Glycogen Synthesis

G-6-P can be utilized to produce or oxidize glycogen. A crucial insulin-regulated enzyme called GS regulates the production of glycogen (Dent et al. 1990;

Yki-Jarvinen et al. 1987). GS may exist in both active (dephosphorylated) and inactive (phosphorylated) states. Insulin produces GS by inactivating glycogen synthesis kinase and activating glycogen synthesis phosphatase or PP-1. Muscle GS protein and mRNA levels are decreased in people with type 2 diabetes. Furthermore, basal GS activity is generally lower in type 2 diabetics, and insulin activation of glycogen synthesis is markedly impaired (Damsbo et al. 1991). However, a basic abnormality in the regulation of glycogen formation occurs in type 2 diabetes due to the malfunction of the insulin signaling system, which results in lack of insulin activation.

A recent rat study found that one of the main ways insulin boosts glycogen's metabolic activity, synthesis in skeletal muscle is through activation brought on by G-6-P. Therefore, in compared to controls, type 2 diabetes patients or their first-degree relative's muscles had lower levels of G-6-P during a hyperglycaemic hyperinsulinaemic clamp (Rothman et al. 1995). This may have a substantial influence on inadequate glycogen synthesis.

4.5.3.4 Oxidation of Glucose

Human muscle uses aerobic glycolysis to break down 90% of the glucose that enters the glycolytic pathways, whereas anaerobic glycolysis uses the remaining 10% to produce lactate. Phosphofructokinase (PFK) and pyruvate dehydrogenase (PDH) are the main enzymes that control glycolysis and the oxidation of glucose, respectively. PFK is not insulin controlled, despite the fact that PDH is an essential insulin-controlled enzyme whose activity is quickly increased by insulin in muscle. Type 2 diabetics have anomalies in insulin-stimulated PDH activity but normal PFK activity, demonstrating a disruption of the glycolytic/glucose oxidative pathway (Mandarino et al. 1986).

4.5.4 Metabolism of Muscle Lipids in Metabolic Syndrome

Metabolizing lipids requires skeletal muscle. For instance, most of the energy required by skeletal muscle during rest is produced by the oxidation of fatty acids. Furthermore, lipid metabolism plays a crucial role in the physiology of skeletal muscles, including metabolic switching when going from a feeding to a fasting state and adjusting to the energy requirements of exercise (Achten et al. 2002). The pathogenic elements of the condition, such as insulin resistance, are aided by the aberrant lipid metabolism in skeletal muscles that characterizes the MetS.

At both the organ and whole-body levels, numerous factors control the complex process of fatty acid metabolism in skeletal muscle. Although lipolysis in adipose tissue, which is a major determinant of fatty acid availability, plays a significant role in skeletal muscle's usage of fatty acids, the control of lipid metabolism at the muscular level is equally crucial. In muscles, fatty acid transport into cells is regulated, and intramuscular lipids are activated, oxidized, esterified or stored, as well as subjected to lipolysis. We'll focus on the mechanisms that the metabolic syndrome is said to have altered, though.

4.5.4.1 Transport of Fatty Acids

Proteins that are circulating plasma free fatty acid that are associated to albumin can diffuse or transport those fatty acids to a muscle cell. Transporters of fatty acids linked to membranes, including FAT/CD-36, Membrane Bound Fatty-Acid Binding Protein (MBFABP) and a family of fatty-acid transport protein, transfer fatty acids into the myocyte (FATP-1 and 4). Co-expressed in skeletal muscle are the membrane transporters FAT/CD-36, FABPpm, FATP1 and FATP4, with current studies indicating that FAT/CD-36 and FATP4 are the most significant membrane transporters in terms of quantity (Nickerson et al. 2009).

4.5.4.2 Oxidation of Fatty Acids

The mitochondrial matrix oxidizes fatty acids through a combination of the Krebs cycle, the electron transport chain and -oxidation (ETC). The carnitine-palmitoyltransferase (CPT) system that is made up of CPT-1, carnitine acylcarnitine translocase (CACT) and CPT-2 transports long-chain fatty-acyl-CoAs into mitochondria (Kerner and Hoppel 2000; Mcgarry and Brown 1997). Malonyl-CoA exerts significant allosteric control over the CPT-1-catalysed phase; it is believed to be the step that limits the rate of fatty acid oxidation. Two carbon atoms are formed through the breakdown of fatty acids when they enter the mitochondria during -oxidation (acetyl-CoA). An essential enzyme in -oxidation that affects the rate of fatty acid oxidation is called -hydroxy acyl-CoA dehydrogenase (-HAD). The Krebs-cycle and the electron-transport-chain finally use the acetyl-CoA produced during this process to produce ATP.

4.5.4.3 Fatty Acid Storage

Triacylglycerides (TAG) and other glycerolipids, phospholipids, sphingolipids and cholesterol esters are examples of lipid species; can be made from unoxidized fatty acids as well as other lipid species. The amount of fatty acids that enters the body, their type (saturated/unsaturated), the kind of muscle fibre, the hormonal milieu and the amount of energy required by the muscles, all affect how much fat is stored (Watt and Hoy 2012).

4.6 Lipid Metabolism in Metabolic Syndrome

4.6.1 Lipoprotein Metabolism

A monolayer of proteins, unesterified cholesterol and surface phospholipids surround the triglyceride and/or cholesterol ester core of lipoproteins. For the production of steroid hormones, cholesterol is an essential part of cellular membranes.

Triglycerides enable adipose tissue to store energy while simultaneously providing energy to tissues. Apolipoproteins are the protein components of lipoproteins that serve as ligands for receptors and cofactors for enzymes (Mahley et al. 1984).

Chylomicrons: When triglycerides from the diet are absorbed, the gut produces chylomicrons, which are triglyceride-rich lipoprotein particles. Chylomicrons are injected into the bloodstream alongside lymph vessels. The lipoprotein lipase (LPL) enzyme breaks down triglycerides in the blood, releasing fatty acids that can either be used as a source of energy by cells or stored as triglycerides in adipose tissue. The remaining chylomicron is eventually absorbed and broken down by the liver. Apolipoprotein (apo) B48 is the major protein found in chylomicrons.

An endogenous metabolic cascade of particles for the transportation of triglycerides is formed by the lipoproteins VLDL-IDL-LDL, which also include the apoB100 protein. The liver produces the lipoprotein particle known as Very-Low-Density-Lipoprotein (VLDL), which is high in triglycerides. The amount of triglycerides decreases as a result of the enzyme LPL's breakdown of VLDL, which results in the particle getting denser and smaller. It first changes into Intermediate-Density-Lipoprotein (IDL), which is further broken down into low density lipoprotein, which is then partially absorbed by the liver. IDL is changed into LDL by the hepatic lipase and LPL enzymes. LDL is a cholesterol-rich particle because it carries roughly 80% of the plasma cholesterol. The majority of LDL is absorbed by the liver, which then excretes it through the bile either as cholesterol or after being changed into bile acids.

Both the larger, higher-triglyceride VLDL1 and the smaller, higher-cholesterol VLDL2 particles are secreted by the liver. These lipoproteins undergo intravascular lipolysis which results in normal-sized LDL that stays in the bloodstream for around 2 days. When triglyceride levels are high, on the other hand, VLDL1 particles initially transform into large low-density-lipoprotein particles. These massive LDL particles are changed into small, dense LDL particles with a five-day retention period by the cholesterol ester transport protein (CETP), which is increased in type 2 diabetes, and hepatic lipase. Additionally, larger levels of very-low-density-lipoprotein-1 change the properties of HDL (high-density-lipoprotein), causing these particles to be catabolized more quickly (Rashid et al. 2003). As a result, when VLDL-1 is raised in diabetic dyslipidaemia, sdLDL is formed and the risk of CVD increases.

HDL: These lipoproteins are the tiniest and most compact, and they have a higher proportion of phospholipids than other varieties. The HDL performs a complex function. An apoAI particle that lacks lipids and binds to phospholipid to produce a disc-shaped particle is the secreted particle. A core of cholesterol ester is created when cholesterol is removed from cell and tissue. The cholesterol is absorbed by the lipid-filled macrophages and then is deposited in the atherosclerotic plaque. One of two ways—direct liver absorption or transfer by the transport protein CETP to LDL particles—allows cholesterol from HDL to reach the liver.

4.6.2 Studies in Living Systems Provide Insight into the Effects of Type 2 Diabetes and the Metabolic Syndrome on Lipid Metabolism

Our knowledge of the defective lipoprotein metabolism in diabetic dyslipidaemia has been greatly enhanced by the model. The apoB100 triglyceride and apoB100 production, for instance, were shown to have a linear relationship by the model. Having a high total triglyceride level is primarily obtained by an expanded VLDL1 triglyceride pool; however, this is owing to an increase in particle number rather than particle size, according to the plasma's VLDL1 and VLDL2 particles' triglyceride to apoB100 ratios. In comparison to VLDL1 triglyceride pools, VLDL2 triglyceride pools generally made less of an overall contribution to plasma triglyceride levels. Our hypothesis states that whilst VLDL2 secretion represents basal triglyceride production, VLDL1 secretion is controlled by dynamic variation in substrate-fluxes. Studies show that insulin has no effect on the lipid makeup of VLDL2 particles or the direct production of VLDL2 apoB100 (Malmstrom et al. 1997).

Furthermore, the simulation revealed a significant variation in the duration of glycerol and leucine delays (i.e., VLDL triglycerides and VLDL apoB100AQ7). The completed lipoprotein has already entered the bloodstream after glycerol has been converted into triglycerides for 21 min. However, apoB100 molecules have a 33-min delay time. These findings support the successive assembly theory of VLDL and are likely the first to demonstrate the addition of triglyceride to an initial apoB100-containing particle in a human.

We have examined the features of type 2 diabetes and insulin resistance that are related to the emergence of VLDL-1 and identified significant relationships with plasma glucose. We found that markers of the production of VLDL1 apo-B100 and VLDL1-triglycerides included fasting insulin, intra-abdominal, plasma glucose and liver-fat, as well as the homeostasis model of evaluating insulin resistance. By extending our research to include adiponectin, intra-abdominal and subcutaneous fat, we were able to corroborate this. Only liver fat and plasma glucose, however, remained significant after a multiple regression analysis. Additionally, plasma glucose, adiponectin and intra-abdominal fat are the primary markers of liver fat (Adiels et al. 2006).

Non-alcoholic fatty-liver-disease (NAFLD), which affects people who do not regularly consume alcohol, is characterized by liver fat accumulation that surpasses 5–10% of the liver's weight. Type 2 diabetes, obesity and hyperlipidaemia are all closely linked to NAFLD, according to new research. We studied the relationship between liver fat and inhibition of VLDL-1 in people with varying levels of liver fat. This study demonstrated a relationship between liver fat and both basal VLDL-1 production and a deficiency in insulin-induced VLDL-1 suppression. In fact, people with low amounts of liver fat see a reduction in VLDL-1 secretion from insulin, but participants with large levels of liver fat do not. Since the amount of liver fat has no effect on how much the non-esterified fatty acid (NEFA) pool is suppressed by insulin, a higher proportion of systemic NEFA or the use of other sources of triglycerides, such as hepatic reserves, must be used to facilitate high levels of VLDL-1 synthesis in those with high levels of liver fat. The study found that visceral obesity may not be the best indicator of metabolic disruption and CVD risk in obese individuals.

4.7 Gut Microbiota in Metabolic Syndrome

4.7.1 Adipose Tissue Microbial Energy Extraction and Storage

Short chain fatty acids (SCFA) must be produced in the gut microbiota from a wide range of glycosidases and lyases that would otherwise be indigestible the diet. It turns out that the microbiota of obese mice is more adept at digesting complex polysaccharide and storing the extra calories in adipose tissue (Turnbaugh et al. 2006). The ability of germ-free mice to break down polysaccharides is similar to that of germ-free mice, and the reduced amounts of short chain fatty acids that these mice's liver obtains through portal circulation are associated to decreased lipogenesis. The gut bacteria enhances glucose absorption from the small intestine through a currently accepted mechanism and replenishes energy by turning carbs into short chain fatty acids. Sterol regulatory element binding protein-1c (SREBP-1c) and carbohydrate element binding protein (CEBP), which are the principal substrates for de-novo-lipogenesis in the liver, are activated by high amounts of circulating SCFA and glucose. These transcription factors subsequently control the rate-limiting lipogenic enzymes acetyl-CoA carboxylase and fatty acid synthase.

The gut microbiome enhances lipogenesis, VLDL synthesis and fat accumulation in adipose tissue through elevating lipoprotein lipase activity. The gut microbiota controls intestinal Angptl4 expression via promoting LPL activity by inhibiting Angptl4 and is linked to increased adiposity, according to a research of germ-free Angptl4/– mice (Bäckhed et al. 2007).

4.7.2 Host Metabolism Modulated by Microbes

The gut flora affects the host's metabolism in addition to how energy is derived from meals. Diet-induced obesity does not appear to be a concern in germ-free mice. This finding suggests a significant function for the gut bacteria in both peripheral and central metabolism. The protection against diet-induced obesity in germ-free mice has recently been validated by several independent labs (Ding et al. 2010; Fleissner et al. 2010; Rabot et al. 2010), and additional molecular insights are now being developed. Because complex polysaccharides are absent from the Western diet and other high fat-diet, the gut bacteria encourage obesity through methods other than carbohydrate fermentation.

Acetyl-coA carboxylase, carnitine palmitoyl-transferase and other downstream molecular targets that are all involved in fatty acid oxidation are all linked to increased AMP-activated protein kinase (AMPK) phosphorylation, which is connected with the thin phenotype of germfree mice. Enzyme AMPK acts as a 'fuel gauge', monitoring the level of cellular energy. It could be the reason why germ-free mice have better skeletal muscle fatty acid oxidation promotion abilities. Angptl4 has a significant effect on LPL regulation in addition to acting as a potent-regulator for oxidation of fatty acid. As was earlier documented, the intestinal tract's gut bacteria prevent this gene from being expressed (Mandard et al. 2006). These results are supported by the fact that germ free mice missing Angptl4 lost their defence against 'diet induced obesity'. Additional phenotypic research showed that in germ-free environments, Angptl4 regulated the genes involved in fatty acid oxidation. The absence of communication between Angptl4 and AMPK is crucial because it demonstrates that two complementary yet different pathways that lead to improved fatty acid metabolism protect germ-free mice from diet-induced obesity. The gut microbiota may also affect other hormones that regulate the host metabolism in terms of expression and function.

4.7.3 Gut Microbiota and the Obesity

The gut-microbiota is different in obese persons, despite the fact that some evidence suggests it may have an effect on obesity. An ingenious research method that involves transplanting intestinal microbiota of obese and lean mice into recipient devoid of infections was used to find this connection. Unexpectedly, mice given an obese microbiota experienced a considerable increase in adipose mass, demonstrating that the fat and lean phenotypes were passed on to the recipients. A recent study found that the pattern-recognition receptor 'Toll-like receptor-5 (Tlr-5)' is a key modulator of gut microbial ecology and recognizes the bacterial protein flagellin. Surprisingly, Tlr-5-deficient animals have been related to obesity, the metabolic syndrome and inflammatory bowel illness (Vijay-Kumar et al. 2010). The transmission of the obese phenotype to germ-free mice was significant since it demonstrated that the microbiota was the root source of the condition. The gut microbiota may have an impact on the central nervous system, as evidenced by subsequent research that found higher food consumption in Tlr-5-deficient mice was associated with changed gut microbiota. It is also becoming more and more obvious that the gut microbiota affects animal physiology and behaviour. Germ-free mice, for example, exhibit altered anxiety behaviour and increased locomotor activity. As a result, the gut microbiota can alter dietary and behavioural traits as well as peripheral metabolism.

4.7.4 Gut Microbiota as a Modulator of Glucose Metabolism

The gut microbiota not only influences host obesity through its microbial effects but also promotes low-grade metabolic inflammation, which aids in the development of metabolic disorders. Endotoxin exposure in germ-free mice is very low, but it is still quite low when compared to levels reported in colonized animals. Compared to animals that had been colonized, mice that were germ-free showed decreased adipose inflammation and greater insulin sensitivity. It's noteworthy to notice that antibiotic treatment improves host glucose metabolism, decreases liver triglyceride levels, lowers plasma-endotoxin levels, reduces obesity in obese mice and lowers adipose-inflammation (Cani et al. 2008; Membrez et al. 2008).

4.7.5 Metabolomics for the Identification of the Microbial Metabolites in Disease

The examples above show how gut–bacteria influence human physiology by controlling dietary component metabolism. Researchers used metabolomic analysis of urine from mice resistant or predisposed to developing non-alcoholic fatty-liver disease to discover that vulnerable animals had larger levels of the bacteria that break down phosphatidylcholine into trimethylamines (TMA), which are excreted in the urine. Because phosphatidylcholines are essential for VLDL secretion and insufficient VLDL secretion can lead to the triglyceride accumulation in liver, phosphatidylcholine deficiency can cause steatosis and insulin resistance. However, current study has connected the conversion of dietary choline to trimethylamines to cardiovascular illness and plasma concentrations of choline derived metabolites to highly reliable predictor of cardiovascular disease (Wang et al. 2011). As a result, altered gut microbial structure and function that result in dysbalanced choline metabolism may have an effect on the host's metabolism and physiology.

4.8 Future Directions

A better understanding of how the disease progresses and how its constituent parts interact may result from the identification of the metabolic pathways that are disrupted in metabolic system and of the genetic networks that control them. This could afterwards offer the most logical targets for the creation of efficient and secure therapeutic approaches.

Recent significant technological developments in systems biology and metabolic phenotyping are diverse. In the future, it will be crucial to:

- Identify the individual contributions of disorders (e.g. metabolic syndrome, obesity and diabetes), genetics and the environment to the metabolic phenotype.
- Describe the impact of disease susceptibility genes on metabotype.
- Find out how metabolic syndrome impacts or is influenced by the microbiota.
- Translate cultured metabotypes.

References

Achten J, Gleeson M, Jeukendrup AE (2002) Determination of the exercise intensity that elicits maximal fat oxidation. Med Sci Sports Exerc 34:92–97

Adiels M, Taskinen MR, Packard C, Caslake MJ, Soro-Paavonen A, Westerbacka J, Vehkavaara S, Hakkinen A, Olofsson SO, YKI-Jarvinen H, Boren J (2006) Overproduction of large VLDL particles is driven by increased liver fat content in man. Diabetologia 49:755–765

Ahima RS (2006) Adipose tissue as an endocrine organ. Obesity 14(5):242–249

Balkau B, Charles MA (1999) Comment on the provisional report from the WHO consultation. European Group for the Study of insulin resistance (EGIR). Diabet Med 16(5):442–443; (PM: 10342346)

Bäckhed F, Manchester JK, Semenkovich CF, Gordon JI (2007) Mechanisms underlying the resistance to diet-induced obesity in germ-free mice. Proc Natl Acad Sci U S A 104:979–984

Bastard JP, Maachi M, Lagathu C, Kim MJ, Caron M, Vidal H, Capeau J, Feve B (2006) Recent advances in the relationship between obesity, inflammation and insulin resistance. Eur Cytokine Netw 17:4–12

Bjornholm M, Kawano Y, Lehtihet M, Zierath JR (1997) Insulin receptor substrate-1 phosphorylation and phosphatidylinositol 3-kinase activity in skeletal muscle from NIDDM subjects after in vivo insulin stimulation. Diabetes 46:524–527

Bouzakri K, Zachrisson A, Al-Khalili L, Zhang BB, Koistinen HA, Krook A, Zierath JR (2006) siRNA-based gene silencing reveals specialized roles of IRS-1/Akt2 and IRS-2/Akt1 in glucose and lipid metabolism in human skeletal muscle. Cell Metab 4:89–96

Bruun JM, Verdich C, Toubro S, Astrup A, Richelsen B (2003) Association between measures of insulin sensitivity and circulating levels of interleukin-8, interleukin-6 and tumor necrosis factor-α. effect of weight loss in obese men. Eur J Endocrinol 148:535–542

Brunt EM, Janney CG, Di-Bisceglie AM, Neuschwander-Tetri BA, Bacon BR (1999) Nonalcoholic steatohepatitis: a proposal for grading and staging the histological lesions. Am J Gastroenterol 94(9):2467–2474; (PM: 10484010)

Bugianesi E, Gastaldelli A, Vanni E, Gambino R, Cassader M, Baldi S, Ponti V, Pagano G, Ferrannini E, Rizzetto M (2005) Insulin resistance in non-diabetic patients with non-alcoholic fatty liver disease: sites and mechanisms. Diabetologia 48(4):634–642; (PM: 15747110)

Caldwell SH, Oelsner DH, Iezzoni JC, Hespenheide EE, Battle EH, Driscoll CJ (1999) Cryptogenic cirrhosis: clinical characterization and risk factors for underlying disease. Hepatology 29(3):664–669; (PM: 10051466)

Cani PD, Bibiloni R, Knauf C, Waget A, Neyrinck AM, Delzenne NM, Burcelin R (2008) Changes in gut microbiota control metabolic endotoxemia-induced inflammation in high-fat diet-induced obesity and diabetes in mice. Diabetes 57:1470–1481

Carey PE, Halliday J, Snaar JE, Morris PG, Taylor R (2003) Direct assessment of muscle glycogen storage after mixed meals in normal and type 2 diabetic subjects. Am J Physiol Endocrinol Metab 284:E688–E694

Chibalin AV, Leng Y, Vieira E, Krook A, Bjornholm M, Long YC, Kotova O, Zhong Z, Sakane F, Steiler T, Nylen C, Wang J, Laakso M, Topham MK, Gilbert M, Wallberg-Henriksson H, Zierath JR (2008) Downregulation of diacylglycerol kinase delta contributes to hyperglycemia-induced insulin resistance. Cell 132:375–386

Damsbo P, Vaag A, Hother-Nielsen O, Beck-Nielsen H (1991) Reduced glycogen synthase activity in skeletal muscle from obese patients with and without type 2 (non-insulin-dependent) diabetes mellitus. Diabetologia 34:239–245

DeFronzo RA, Gunnarsson R, Bjorkman O, Olsson M, Wahren J (1985) Effects of insulin on peripheral and splanchnic glucose metabolism in noninsulin-dependent (type II) diabetes mellitus. J Clin Invest 76:149–155

Dent P, Lavoinne A, Nakielny S, Caudwell FB, Watt P, Cohen P (1990) The molecular mechanism by which insulin stimulates glycogen synthesis in mammalian skeletal muscle. Nature 348:302–308

Ding S, Chi MM, Scull BP, Rigby R, Schwerbrock NM, Magness S, Jobin C, Lund PK (2010) Highfat diet: bacteria interactions promote intestinal inflammation which precedes and correlates with obesity and insulin resistance in mouse. PLoS One 5:e12191

Dohm GL, Tapscott EB, Pories WJ, Dabbs DJ, Flickinger EG, Meelheim D, Fushiki T, Atkinson SM, Elton CW, Caro JF (1988) An in vitro human muscle preparation suitable for metabolic studies. Decreased insulin stimulation of glucose transport in muscle from morbidly obese and diabetic subjects. J Clin Invest 82:486–494

Dube MG, Torto R, Kalra SP (2008) Increased leptin expression selectively in the hypothalamus suppresses inflammatory markers CRP and IL-6 in leptin-deficient diabetic obese mice. Peptides 29:593–598

Eckel RH, Alberti KG, Grundy SM, Zimmet PZ (2010) The metabolic syndrome. Lancet 375:181–183

Fleissner CK, Huebel N, AbdEl-Bary MM, Loh G, Klaus S, Blaut M (2010) Absence of intestinal microbiota does not protect mice from diet-induced obesity. Br J Nutr 104:919–929

Ghouri N, Preiss D, Sattar N (2010) Liver enzymes, nonalcoholic fatty liver disease, and incident cardiovascular disease: a narrative review and clinical perspective of prospective data. Hepatology 52(3):1156–1161; (PM: 20658466)

Goodyear LJ, Giorgino F, Sherman LA, Carey J, Smith RJ, Dohm GL (1995) Insulin receptor phosphorylation, insulin receptor substrate-1 phosphorylation, and phosphatidylinositol 3-kinase activity are decreased in intact skeletal muscle strips from obese subjects. J Clin Invest 95:2195–2204

Grundy SM (2007) Metabolic syndrome: a multiplex cardiovascular risk factor. J Clin Endocrinol Metab 92(2):399–404

Hall JE, da Silva AA, Brandon E, Stec DE, Ying Z, Jones DW (2007) Pathophysiology of obesity hypertension and target organ injury. In: Lip GYP, Hall JE (eds) Comprehensive hypertension. Elsevier, New York, NY, pp 447–468

Harrison SA, Torgerson S, Hayashi PH (2003) The natural history of nonalcoholic fatty liver disease: a clinical histopathological study. Am J Gastroenterol 98(9):2042–2047; (PM: 14499785)

Haring R, Wallaschofski H, Nauck M, Dorr M, Baumeister SE, Volzke H (2009) Ultrasonographic hepatic steatosis increases prediction of mortality risk from elevated serum gamma-glutamyl transpeptidase levels. Hepatology 50(5):1403–1411; (PM: 19670414)

Isomaa B, Almgren P, Tuomi T, Forsén B, Lahti K, Nissén M, Taskinen MR, Groop L (2001) Cardiovascular morbidity and mortality associated with the metabolic syndrome. Diabetes Care 24:683–689

Joy D, Thava VR, Scott BB (2003) Diagnosis of fatty liver disease: is biopsy necessary? Eur J Gastroenterol Hepatol 15(5):539–543; (PM: 12702913)

Karlsson HK, Zierath JR, Kane S, Krook A, Lienhard GE, Wallberg-Henriksson H (2005) Insulinstimulated phosphorylation of the Akt substrate AS160 is impaired in skeletal muscle of type-2 diabetic subjects. Diabetes 54:1692–1697

Kerner J, Hoppel C (2000) Fatty acid import into mitochondria. Biochim Biophys Acta 1486:1–17

Krook A, Bjornholm M, Galuska D, Jiang XJ, Fahlman R, Myers MG Jr, Wallberg-Henriksson H, Zierath JR (2000) Characterization of signal transduction and glucose transport in skeletal muscle from type-2 diabetic patients. Diabetes 49:284–292

Kruszynska YT, Mulford MI, Baloga J, Yu JG, Olefsky JM (1998) Regulation of skeletal muscle hexokinase II by insulin in nondiabetic and NIDDM subjects. Diabetes 47:1107–1113

Lorenzo C, Williams K, Hunt KJ, Haffner SM (2007) The National Cholesterol Education Program Adult treatment panel III, international diabetes federation, and World Health Organization definitions of the metabolic syndrome as predictors of incident cardiovascular disease and diabetes. Diabetes Care 30(1):8–13

Mandarino LJ, Madar Z, Kolterman OG, Bell JM, Olefsky JM (1986) Adipocyte glycogen synthase and pyruvate dehydrogenase in obese and type II diabetic subjects. Am J Phys 251:E489–E496

Mandard S, Zandbergen F, van Straten E, Wahli W, Kuipers F, Muller M, Kersten S (2006) The fasting-induced adipose factor/angiopoietin-like protein 4 is physically associated with lipoproteins and governs plasma lipid levels and adiposity. J Biol Chem 281:934–944

Mahley RW, Innerarity TL, Rall SC Jr, Weisgraber KH (1984) Plasma lipoproteins: apolipoprotein structure and function. J Lipid Res 25:1277–1294

Malmstrom R, Packard CJ, Caslake M, Bedford D, Stewart P, Yki-Jarvinen H, Shepherd J, Taskinen MR (1997) Defective regulation of triglyceride metabolism by insulin in the liver in NIDDM. Diabetologia 40:454–462

Mcgarry JD, Brown NF (1997) The mitochondrial carnitine palmitoyltransferase system. From concept to molecular analysis. Eur J Biochem 244:1–14

Membrez M, Blancher F, Jaquet M, Bibiloni R, Cani PD, Burcelin RG, Corthesy I, Mace K, Chou CJ (2008) Gut microbiota modulation with norfloxacin and ampicillin enhances glucose tolerance in mice. FASEB J 22:2416–2426

Nickerson JG, Alkhateeb H, Benton CR, Lally J, Nickerson J, Han XX, Wilson MH, Jain SS, Snook LA, Glatz JF, Chabowski A, Luiken JJ, Bonen A (2009) Greater transport efficiencies of the membrane fatty acid transporters FAT/CD36 and FATP4 compared with FABPpm and FATP1 and differential effects on fatty acid esterification and oxidation in rat skeletal muscle. J Biol Chem 284:16522–16530

Nuutila P, Koivisto VA, Knuuti J, Ruotsalainen U, Teras M, Haaparanta M, Bergman J, Solin O, Voipio-Pulkki LM, Wegelius U et al (1992) Glucose-free fatty acid cycle operates in human heart and skeletal muscle in vivo. J Clin Invest 89:1767–1774

Olufadi R, Byrne CD (2008) Clinical and laboratory diagnosis of the metabolic syndrome. J Clin Pathol 61(6):697–706; (PM: 18505888)

Rabot S, Membrez M, Bruneau A, Gerard P, Harach T, Moser M, Raymond F, Mansourian R, Chou CJ (2010) Germ-free C57BL/6J mice are resistant to high-fat-diet-induced insulin resistance and have altered cholesterol metabolism. FASEB J 24:4948

Rashid S, Watanabe T, Sakaue T, Lewis GF (2003) Mechanisms of HDL lowering in insulin resistant, hypertriglyceridemic states: the combined effect of HDL triglyceride enrichment and elevated hepatic lipase activity. Clin Biochem 36:421–429

Ronti T, Lupattelli G, Mannarino E (2006) The endocrine function of adipose tissue: an update. Clin Endocrinol 64:355–365

Rothman DL, Magnusson I, Cline G, Gerard D, Kahn CR, Shulman RG, Shulman GI (1995) Decreased muscle glucose transport/phosphorylation is an early defect in the pathogenesis of non-insulin-dependent diabetes mellitus. Proc Natl Acad Sci U S A 92:983–987

Ryder JW, Yang J, Galuska D, Rincon J, Bjornholm M, Krook A, Lund S, Pedersen O, Wallberg-Henriksson H, Zierath JR, Holman GD (2000) Use of a novel impermeable biotinylated photolabeling reagent to assess insulin- and hypoxia-stimulated cell surface GLUT4 content in skeletal muscle from type 2 diabetic patients. Diabetes 49:647–654

Saadeh S, Younossi ZM, Remer EM, Gramlich T, Ong JP, Hurley M, Mullen KD, Cooper JN, Sheridan MJ (2002) The utility of radiological imaging in nonalcoholic fatty liver disease. Gastroenterology 123(3):745–750; (PM: 12198701)

Samuel VT, Liu ZX, Qu X, Elder BD, Bilz S, Befroy D, Romanelli AJ, Shulman GI (2004) Mechanism of hepatic insulin resistance in non-alcoholic fatty liver disease. J Biol Chem 279(31):32345–32353; (PM: 15166226)

Shulman GI, Rothman DL, Jue T, Stein P, Defronzo RA, Shulman RG (1990) Quantitation of muscle glycogen synthesis in normal subjects and subjects with non-insulin-dependent diabetes by 13 C nuclear magnetic resonance spectroscopy. N Engl J Med 322:223–228

Taube A, Eckardt K, Eckel J (2009) Role of lipid-derived mediators in skeletal muscle insulin resistance. Am J Physiol Endocrinol Metab 297(5):E1004–E1012

Turnbaugh PJ, Ley RE, Mahowald MA, Magrini V, Mardis ER, Gordon JI (2006) An obesity associated gut microbiome with increased capacity for energy harvests. Nature 44:1027

Unger RH (2003) Lipid overloads and overflows: metabolic trauma and the metabolic syndrome. Trends Endocrinol Metab 14(9):398–403; (PM: 14580758)

Vijay-Kumar M, Aitken JD, Carvalho FA, Cullender TC, Mwangi S, Srinivasan S, Sitaraman SV, Knight R, Ley RE, Gewirtz AT (2010) Metabolic syndrome and altered gut microbiota in mice lacking toll-like receptor 5. Science 328:228–231

Wang Z, Klipfell E, Bennett BJ, Koeth R, Levison BS, Dugar B, Feldstein AE, Britt EB, Fu X, Chung YM, Wu Y, Schauer P, Smith JD, Allayee H, Tang WHW, Didonato JA, Lusis AJ, Hazen SL (2011) Gut flora metabolism of phosphatidylcholine promotes cardiovascular disease. Nature 472:57–63

Watt MJ, Hoy AJ (2012) Lipid metabolism in skeletal muscle: generation of adaptive and maladaptive intracellular signals for cellular function. Am J Physiol Endocrinol Metab 302(11):E1315–E1328

Woerle HJ, Meyer C, Dostou JM, Gosmanov NR, Islam N, Popa E, Wittlin SD, Welle SL, Gerich JE (2003) Pathways for glucose disposal after meal ingestion in humans. Am J Physiol Endocrinol Metab 284:E716–E725

Yki-Jarvinen H, Mott D, Young AA, Stone K, Bogardus C (1987) Regulation of glycogen synthase and phosphorylase activities by glucose and insulin in human skeletal muscle. J Clin Invest 80:95–100

Systems Biology Consortium for Infectious Diseases

5

Ashwini Sunil Pisey and Gokul Shankar Sabesan

Abstract

Infectious diseases continue to be major health concerns worldwide. In the twenty-first century, more people are at risk of life-threatening acute and chronic infections than ever before due to the ongoing tremendous social and ecological changes. The need of the time is to develop new diagnostic, therapeutic, and curative, prophylactic strategies, which should be based on detailed understanding of the immensely complex relationship between pathogens and their hosts. Here, systems biology plays an important role. But currently the traditional, reductionist approaches to investigate this dynamic interaction often lack the scale and/or scope to faithfully model the dual and codependent nature of this relationship. This is limiting the success of translational efforts. This chapter highlights the global trends in infectious diseases until the pandemic happened in 2019–2020. With different aspects of systems biology in relation to the host–pathogen relationships for the infections that have become a burden to mankind, the new analytical strategies adapted to model host–pathogen relationships for translational applications have been discussed. Understanding this concept in detail can drive innovation and discovery. The goal of discovering the predictive signs of disease and developing novel diagnostic and therapeutic strategies to alleviate this burden can be achieved. It would be able to provide solutions to these complex public health challenges and disease outbreaks.

Keywords

Infectious diseases · Systems biology · Gene · Consortia

A. S. Pisey · G. S. Sabesan (✉)
Department of Microbiology, Faculty of Medicine, Manipal University College Malaysia, Melaka, Malaysia
e-mail: gokul.shankar@manipal.edu.my

S. Joshi et al. (eds.), *Systems Biology Approaches: Prevention, Diagnosis, and Understanding Mechanisms of Complex Diseases*,
https://doi.org/10.1007/978-981-99-9462-5_5

5.1 Introduction

The twentieth century was marked by extraordinary advances in our understanding of microbes and infectious diseases, but pandemics remain, food and waterborne illnesses are frequent, multidrug-resistant microbes are on the rise, and the needed drugs and vaccines have not been developed. The scientific approaches of the past including the intense focus on individual genes and proteins typical of molecular biology have not been sufficient to address these challenges. The first decade of the twenty-first century reported remarkable innovations in technology and computational methods. These new tools have seemed to be providing comprehensive views of complex biological systems and correspondingly deeper understanding of pathogen–host interactions.

5.1.1 Infectious Diseases "A Global Burden"

Infectious diseases are being considered a burden to mankind as they contribute to heavy disease load and mortality annually for decades. There have been changing trends in the prevalence of these diseases. Few of them have been observed to be persistently present as chronic infections. Let us have a look at Table 5.1 that presents the global trend observed from 1990 until 2019 for all the age groups (Roser et al. 2021).

Max Roser and colleagues have also reported age-wise incidences of these diseases. Among children, lower respiratory infections mainly tuberculosis, diarrheal diseases, malaria, meningitis, whooping cough, and sexually transmitted infections are among the top ten causes, which are adding up to the global stress. Overall, due to advancement in science and technologies, a decline has been reported for these infections in children younger than 5 years while rate of incidence remains high in the youngest and oldest population of the community (Roser et al. 2021). In the adult age group, HIV/AIDS and diarrheal diseases are recorded to have the highest incidence (Roser et al. 2021).

According to the data provided by the World Health Organization (WHO) for the incidence of infectious diseases globally, 5.7 million deaths have been reported due

Table 5.1 Comparison of the global trend for infectious diseases—just before the COVID pandemic (2019) and three decades before (1990)

Year 1990	Year 2019
Neonatal diseases	Neonatal diseases
Lower respiratory infections	Lower respiratory infections
Diarrheal diseases	Diarrheal diseases
Tuberculosis	HIV/AIDS
Measles	Tuberculosis
Malaria	Malaria
Meningitis	Meningitis
Whooping cough	Whooping cough
HIV/AIDS	Measles

to lower respiratory infections, diarrheal diseases, and tuberculosis until the year 2016. The epidemiological data remain the same for the year 2020 causing the majority of the deaths. High incidences of other diseases like malaria, HIV/AIDS, hepatitis, and tuberculosis are seen in low-income countries (Miao et al. 2019).

The major concern today is these disease-causing morbidities in the population all around the world. Disability-adjusted life year (DALY) is the measure of total burden of disease—both from years of life lost due to premature death and years lived with a disability. One DALY equals one lost year of healthy life. Disability-adjusted life year (DALY) percentages from all causes per 100,000 individuals, contributed by these infectious diseases from years 1990 and 2019, are shown in Figs. 5.1 and 5.2, respectively (Roser et al. 2021).

Currently, one third of the population is infected with the tuberculosis bacillus; millions of people are infected by influenza virus annually; and even though therapeutic drugs have slowed the threat of HIV-1, there is still no definitive cure or viable vaccine in sight. Although influenza virus vaccines provide protection, they must be reformulated annually, and the time lag from the spread of a pandemic virus to the availability of a vaccine is far too long. Newly emerging infectious diseases like human H3N2, swine H1N1 influenza, avian H5N1 influenza, dengue fever, and now coronavirus disease 2019 (COVID-19) have become a constant threat, bacterial infections, such as the recurrent *Salmonella* outbreaks in the food industry, are costly to society (Aderem et al. 2011).

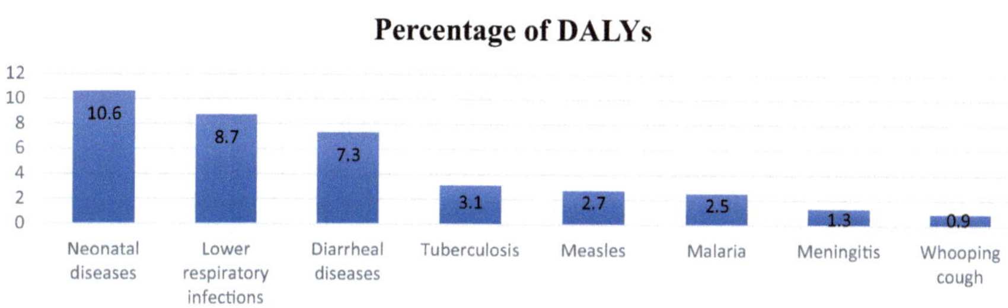

Fig. 5.1 Top ten causes of morbidity in the year 1990

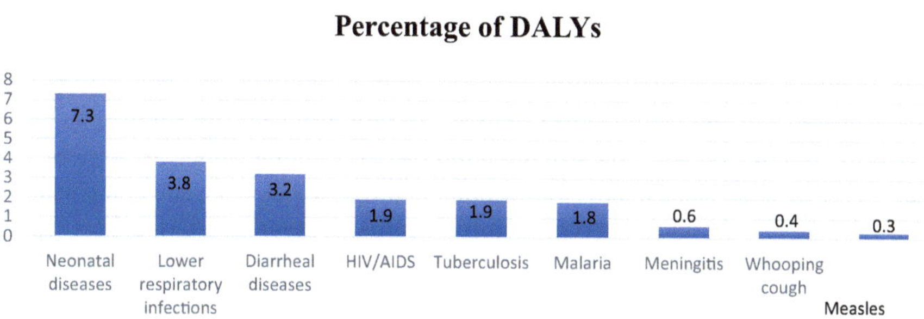

Fig. 5.2 Top ten causes of morbidity in the year 2019

There have been drastic changes in the climatic conditions due to flourishing globalization and industrialization. Researchers believe these changes are to some extent responsible for the emergence of the superbugs or the multidrug-resistant organisms. Due to this, people have been exposed to life-threatening risk of acute and chronic infections. Thus, it becomes the need of this era to further understand these diseases and their pathogens in order to succeed to develop more effective drugs and vaccines to check the spread of these diseases and lessen the global burden.

To meet these challenges, infectious disease researchers are increasingly turning toward systems biology approaches, which enable high-throughput, quantitative descriptions of molecular networks for the underlying infection (Eckhardt et al. 2020). Thus, the ideas and techniques of systems biology are becoming popular day by day.

This chapter focuses on the host–pathogen relationships in an infection, different systems biology approaches used for the understanding these interactions. Along with it new analytical strategies discovered to model host–pathogen relationships and the information acquired through systems biology about translational applications, which are implemented to overcome the obstacles for these infectious diseases, are also discussed.

5.1.2 Host–Pathogen Relationship

All pathogens share the ability to interact with host cells in diverse ways that promote the replication and spread of the pathogen. Pathogens often colonize the host by adhering to or invading the epithelial surfaces that line the respiratory, gastrointestinal, and urinary tracts, as well as the other body surfaces in direct contact with the environment. Intracellular pathogens, including all viruses and many bacteria and protozoa, invade host cells by one of several mechanisms (Bruce et al. 2015). All these organisms either inherently possess special mechanisms to escape the immune response inside the host cell and prevent phagocytosis or possess some virulence factors that cause destruction of the host cells. Virulence factors are proteins specifically present in bacteria, which lead to behavioral changes in the host cells in such a way it promotes replication and spread of the bacteria inside the host. Viruses rely largely on receptor-mediated endocytosis, whereas bacteria exploit cell adhesion and phagocytic pathways; in both cases, the host cell provides the machinery and energy for the invasion (Bruce et al. 2015). Organisms like fungi and protozoan parasites exist in different morphological life forms. They pass through several life stages to become infective and produce disease. Protozoa, by contrast, employ unique invasion strategies that usually require significant metabolic expense on the part of the invader. Once inside, intracellular pathogens seek out a cell compartment that is favorable for their survival and replication, frequently altering host membrane traffic and exploiting the host cell cytoskeleton for intracellular movement (Bruce et al. 2015). Pathogens evolve rapidly, so that new infectious diseases frequently emerge, and old pathogens acquire new ways to evade our attempts at treatment, prevention, and eradication (Bruce et al. 2015).

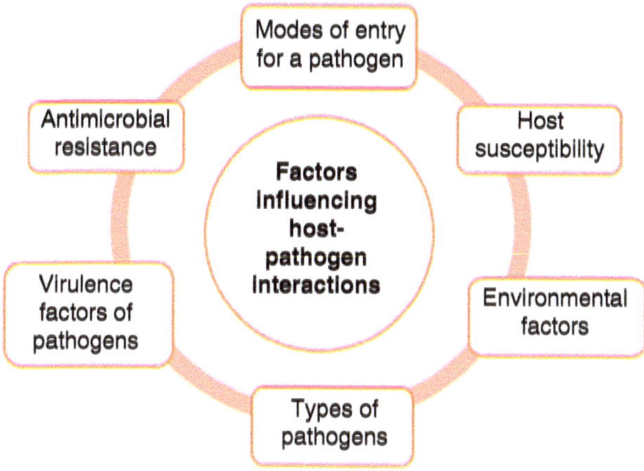

Fig. 5.3 Factors influencing host–pathogen interactions

Host–pathogen interactions (HPI) can be explained as the processes involved during the stages of any pathogenic infection between the microbial pathogens and hosts. These mechanisms are highly dynamic in nature. The outcome of any disease is mainly dependent on HPI. It can either be development of immunity or infection in an individual. An interesting fact has been reported about the same pathogen showing up different outcomes in different patients.

For example, COVID-19 patients were seen to present with different clinical symptoms, scenarios, and complications. This explains the complexity and variability of HPI processes.

This variation in the outcome and the dynamic nature may be due to several influencing factors of the pathogen, host, or environment. Few of these factors affecting HPI are listed in Fig. 5.3.

5.2 Understanding HPI

Researchers were curious since beginning to study HPI. Several attempts are still being made to gather more information in this area. One of the researchers, Yu and colleagues has implicated these postulates in an attempt to understand the host and pathogen-centered pathogenic mechanism at a molecular level. Along with the understanding of the disease processes, these postulates also help us understand the outcome of any disease. These postulates are interlinked and are stated under four headings as shown in Fig. 5.4.

(a) Host–pathogen evolutionary dispositions.
(b) HPI dynamic outcomes.
(c) The role of HPI components.
(d) The checkpoints.

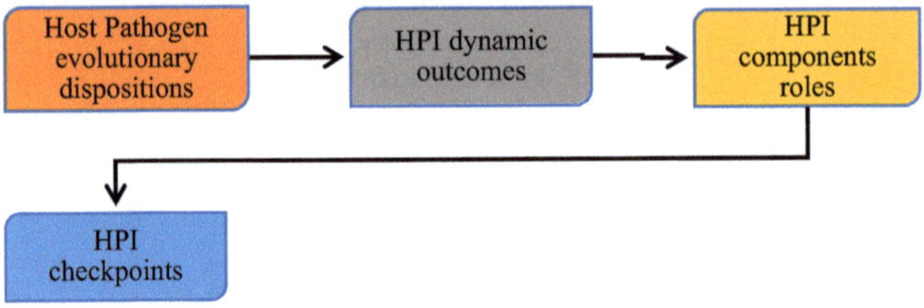

Fig. 5.4 Important components of host–pathogen interactions

Table 5.2 Four postulates implicated to describe host–pathogen interactions

Postulate 1: Host–pathogen evolutionary dispositions	– This part of the postulate characterizes the root cause of HPI dynamics on the part of both the host and the pathogen disposition that is selected for by evolution – Explains about the emergence of HPI owing to evolutionary pressures
Postulate 2: HPI dynamic outcomes	– This part of the postulate describes the relationship between dynamic HPIs and disease outcomes – States possibility of other factors like comorbidities, age, and sex affecting disease outcomes
Postulate 3: HPI roles	– This part of the postulate states that different components(molecules) during the HPI dynamics have differential evolving roles, which should be studied precisely as any change in it will contribute to disease outcomes
Postulate 4: HPI checkpoints	– This part of the postulate is an extension of postulate 3 that highlights how disruption of HPI components may affect HPI outcome networks

Description of the postulates is given in Table 5.2 (Yu et al. 2022):

These postulates form the basis for understanding molecular and cellular host–pathogen interactions and their relations to disease outcomes. Thus, they form a foundation for effectively studying host–pathogen interaction processes. Using the information provided under these postulates while studying systems biology for any infection, complete understanding of its molecular components is acquired. Being able to understand the necessary components of HPI gaps can be filled and thus with new insights novel targets for both host effectors and microbial factors can be identified. These studies and knowledge will provide a better opportunity to effectively check the spread of infectious diseases.

5.3 What Is Systems Biology?

In systems biology, researchers aim to understand complex biological systems, which is often achieved by mathematical modeling and the analyses of high-throughput data (Dix et al. 2016). Thus, it is complex. There have been several attempts to define systems biology. Few of the research scientists defined it as "holistic approach, future of biology; we all will have to do it" or simply "the future of biology." One of the scientists working in this field, Rainer Breitling defined it as "a new kind of biology" or "the successor of molecular biology." Since decades systems biology has been used as a scientific foundation for the success of molecular biology. It is neither a magic bullet nor a fishing expedition, but a rapidly evolving science that presents many challenges and even more opportunities to revolutionize our understanding about host–pathogen relationships, thus can be implicated for the betterment of human health (Breitling 2010; Westerhoff et al. 2009). It is found that there is a vast diversity in thoughts and ideas about systems biology among different biologists all over the world. There have even been interdisciplinary diversities in understanding systems biology. But this is a blessing in disguise—these diverse thoughts can be generously welcomed and accepted. It will provide us with new ideas about this field to sustain research for molecular biology.

To conclude, systems biology can be considered as a comprehensive study of the molecular diversity of living systems, both natural and synthetic, the identification of simplifying general principles and patterns that are recurring features in living and engineered systems, and the integration of our biological knowledge in complex models of the regulatory networks that characterize life.

5.3.1 Systems Biology of Host–Pathogen Interactions

The study of systems biology of host cells, pathogen cells, and their interaction with each other from a clinical perspective is becoming popular recently few years back. Pathogens manipulate the cellular mechanisms of host organisms via host–pathogen interactions (HPIs) in order to take advantage of the capabilities of host cells, leading to infections. The crucial role of these interspecies molecular interactions in initiating and sustaining infections necessitates a thorough understanding of the corresponding mechanisms. Unlike the traditional approach of considering the host or pathogen separately, a systems-level approach, considering the HPI system is indispensable to elucidate the mechanisms of infection (Durmuş et al. 2015).

The following mentioned are the events of HPI and the systems biology-based methods used to analyze and record these significant events (regulatory, metabolic, and protein–protein networks).

- Any interaction between the host and the pathogen triggers **signaling cascades**, that is understood to change the activity of proteins in the cells. This results in emergence of phenotypic response or changes in the host (Dix et al. 2016). These changes happen at different molecular levels. These are **quantified using omics and image data**, for which computational analysis and modeling are used, which is quite challenging (Dix et al. 2016; Durmuş et al. 2015).
- Often **gene expression** data are also used as an alternative, which gives a representative picture of the response of an organism (host or pathogen) to any change occurring after the interaction (Dix et al. 2016).
- **Single-species transcriptomes** *(study of full range of messenger RNA or molecules expressed by any organism) and **dual transcriptomics** are available,

which determine single species-specific expression, or simultaneously the gene expression is measured for both the host and the pathogen during an infection, respectively (Dix et al. 2016).

- There are few examples where **dual microarray analysis** has been used in dual transcriptomics to generate the data and allow researchers to investigate the expression state of a large number of genes, i.e., in many cases, an organism's entire genome (Dix et al. 2016; Ithal et al. 2007).
- In addition, the advent of species-independent RNA-sequence (RNA-seq) platforms allows for high-quality dual RNA-seq (Dix et al. 2016). These are *RNA sequencing* that uses the capabilities of high-throughput sequencing methods to provide insight into the *transcriptome* of a cell.
- Besides analyses restricted to the transcriptome, **systems-level approaches** have been used to elucidate the virulence of microorganisms. These procedures also take metabolic pathways and gene protein interactions into account. Multiple databases have been set up to support the analysis of host–pathogen interactions (Dix et al. 2016; Durmuş et al. 2015).
- Based on **genome-scale metabolic modeling**, analyses of the gut microbiome have been performed, revealing interaction of the microbial ecosystems, their interactions with the host and associations with disorders (Shoaie et al. 2013; Shoaie and Nielsen 2014; Gao et al. 2014; Sadhukhan and Raghunathan 2014; Ji and Nielsen 2015) also addressing symbiotic relationships (Dix et al. 2016; Heinken et al. 2013).
- Recently, the tool **CASINO (Community and Systems Level Interactive Optimization)** has been developed, which allows examination of microbe–microbe, microbe–host, and diet–microbe interactions based on metabolic modeling (Ithal et al. 2007).

5.3.2 Elements of Systems Biology

There are three main elements of systems biology, which were adapted to explain the role of important molecules (biomarkers) and their interactions during a disease process.

1. **Identification of key molecules, i.e., biomarkers:** Identification of biomarkers will help to monitor a disease or predict its progression. It gives us an idea about the severity of the disease and the condition of the host. Thus, it ultimately supports therapeutic decision-making.
2. **Network inference:** This step involves identification of relationships between these involved components of the network to uncover potential drug targets.
3. **Identification of disease modules**: This step involves uncovering the associated disease phenotype underlying the molecular interaction networks.

These three approaches provide excellent results in the research area. Let us discuss each one of them in detail with their application in the clinical field.

5.3.2.1 Biomarker Discovery

The use of biomarkers in basic and clinical research as well as in clinical practice has become so commonplace that their presence as primary endpoints in clinical trials is now accepted almost without question. In the case of specific biomarkers that have been well characterized and repeatedly shown to correctly predict relevant clinical outcomes across a variety of treatments and populations, this use is entirely justified and appropriate (Strimbu and Tavel 2010). The term "biomarker," a portmanteau of "biological marker," refers to a broad subcategory of medical signs—that is, objective indications of medical state observed from outside the patient—which can be measured accurately and reproducibly (Strimbu and Tavel 2010). A biomarker can be considered as an objectively measurable medical sign that indicates a condition or state. Biomarkers can be of any kind, including omics data, like DNA/RNA sequences, proteins, mutations (single-nucleotide polymorphisms [SNPs]), and other attributes of an organism, e.g., an increased heart rate or body temperature (Dix et al. 2016).

Biomarkers can originate from both players, i.e., host and pathogen. For example, the detection of the *Aspergillus* cell wall component galactomannan in the serum of the host has emerged as an important biomarker for supporting the diagnosis of invasive aspergillosis (Dix et al. 2016; Rogers et al. 2013). An example for a host-specific marker is an increased procalcitonin level, which is used as an indicator for bacterial infections, including sepsis. Furthermore, procalcitonin correlates with the severity of the infection and thus also serves as a prognostic biomarker for severe sepsis and septic shock (Dix et al. 2016; Lee 2013). Table 5.3 shows different methods for the identification of biomarkers.

Clinical significance of proteomic methods for large-scale analysis of proteins:

- **Transcriptional biomarker identification through gene expression:** These biomarkers can improve the diagnosis and prognosis of infectious diseases. Proteins represent the functional level of gene expression, and thus they may better reflect cellular responses. This has been applied to identify gene signatures that distinguish between latent and active tuberculosis (Lu et al. 2011) and to study regulatory patterns to distinguish bacterial and fungal whole blood infections (Dix et al. 2015). There are successful advances in infection biology through the identification of transcriptional biomarkers. In tuberculosis research, protein biomarkers were recently found that may allow tuberculosis diagnosis even in immunocompromised patients (Achkar et al. 2015).
- **Noncoding RNAs (ncRNAs), such as microRNAs (miRNAs) or piwi-interacting RNAs (piRNAs):** miRNAs are the best investigated ncRNAs. They have been successfully linked to infectious diseases such as active tuberculosis (Ling et al. 2013), hepatitis B virus infections (Qi et al. 2012), and the discrimination of the systemic inflammatory response syndrome with sepsis (Li et al. 2010).

Table 5.3 Methods utilized for biomarker discovery

Method/technique	Advantages	Drawbacks
(a) Analysis of differential regulation of biomarkers based on statistical significance from research studies (Maertzdorf et al. 2011; Shahabi et al. 2013)	– Fast and well-established method	– Large-scale quantifications (omics) become difficult, the resulting set of potential biomarkers may contain large redundancies and thus giving false-positive findings – Results must be cross-checked or validated by the other method
(b) Incorporation of machine learning techniques to extract data (Tsalik et al. 2016; Dix et al. 2015; Reif et al. 2009; Yousef et al. 2009)	– Combinatorial effects of biomarker candidates can be considered through this method	– Requires a higher computational effort – Often reduces the set of potential biomarkers to the most relevant candidates only
(c) Identification of transcriptional biomarkers through gene expression (Lu et al. 2011; Dix et al. 2015)	– It reflects the cellular response during a disease process – Gene expression is a well-established, fast, and cost-effective method	– Complex – Proteins have larger diversity because of posttranslational modifications. This can form multiprotein complexes, which may alternate the function of the individual proteins
(d) Biomarker discovery through noncoding RNAs (ncRNAs), such as microRNAs (miRNAs) or piwi-interacting RNAs (piRNAs) (Achkar et al. 2015; Ling et al. 2013; Qi et al. 2012; Li et al. 2010)	– It contributes to a broader knowledge about a disease or condition.	
(e) Next-generation genome sequencing for SNP—single-nucleotide polymorphisms (biomarker) identification (Erratum 2008; Werner 2010; Zuniga et al. 2012; Abu-Maziad et al. 2010)	– High precision – Aims to find connections between SNPs and disease phenotypes	–

- **Next-generation genome sequencing for SNP identification (with or without DNA microarray):** Many SNPs in protein-coding regions are linked to severe health risks. SNPs also have been associated with the susceptibility to infections and diseases, like pneumonia in H1N1 influenza infection (Zuniga et al. 2012) or sepsis (Abu-Maziad et al. 2010).

5.3.2.2 Network Inference

Network inference involves modeling and analysis of a network depicting activities of the tiny micromolecules involved during any pathogenic process. Therefore, it is basically about **generation and structural analysis of interaction networks** and then **integrating this information into discovery of disease-associated modules** supported and based on omics data supported by prior knowledge.

A. **Molecular network modeling**: In the field of systems biology, the ultimate focus is always toward understanding the molecular mechanisms underlying all the biological processes on a systems level by means of mathematical modeling. These mechanisms are commonly represented in a network, as shown in Fig. 5.5. (Dix et al. 2016).
B. **Automated inference of gene regulatory networks:** In gene expression using microarray or RNA-seq data, these interaction networks are also termed gene-regulatory networks (GRNs) (Dix et al. 2016). After generation of these interaction networks, data reasoning is conducted using the concept of automated reverse engineering. The working process of an automated gene regulatory network is mentioned in the flowchart below.

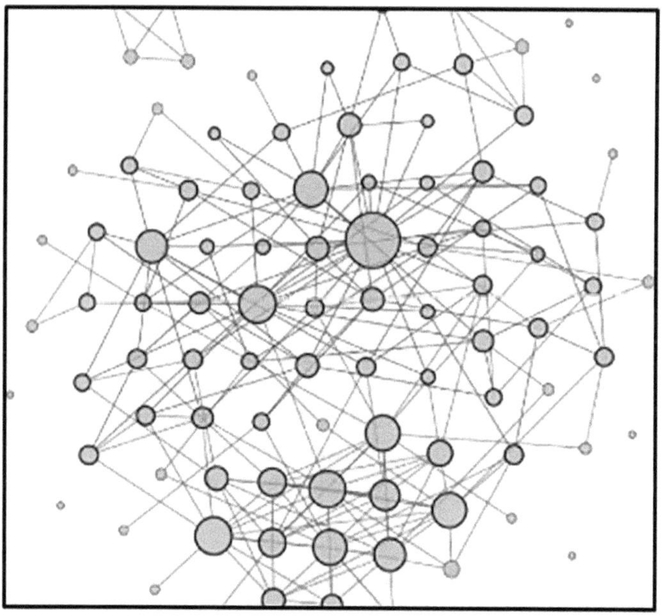

Fig. 5.5 A network, composed of nodes denoting molecules (e.g., proteins, DNA, RNA, or metabolites) and edges representing an interaction between the connected nodes (e.g., protein–protein and protein–DNA)

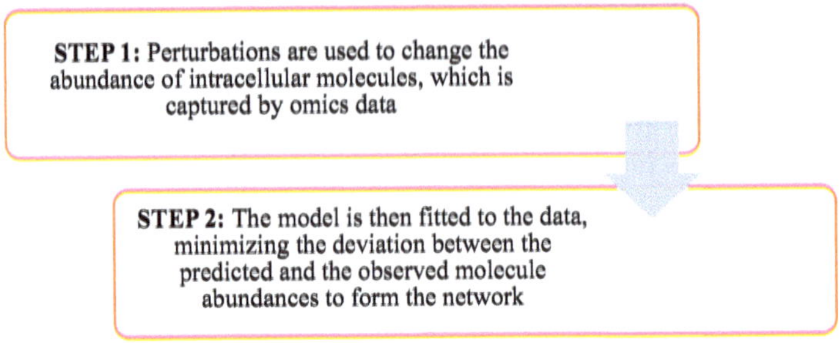

Recent updates in GRNs: Initially this was an expensive setup, difficult to afford, but now the availability of omics data is increased by decreasing costs and standard operating procedures. The reliability of the measurements in conjunction with growing public data repositories is also being updated periodically (Dix et al. 2016).

Drawbacks of GRN: Usually the number of available measurements is much lower compared to the number of genes in the mathematical model underlying the GRN. This results in an underdetermined system (Dix et al. 2016).

To conclude, the following components are related to each other and should be integrated correctly for the development of a successful understanding about how these molecule interactions work:

1. The complexity of the model in terms of the number of free parameters per gene.
2. The data required to explain the observed systems behavior.
3. The quality of the inferred GRN (Hecker et al. 2009b).

Several approaches have been proposed for the inference of gene-regulatory networks (GRNs): Based on the number of genes integrated or involved in the final network, these approaches can be roughly grouped into two classes, as shown in Table 5.4.

Applications:

- Network inference has been mostly applied to predict GRNs of either the host or the pathogen. However, the advent of dual transcriptomics provides the possibility to infer molecular host–pathogen interaction networks (Schulze et al. 2015).
- In a pioneering study, inference of the network underlying *Candida albicans* infection of murine dendritic cells revealed two molecular host–pathogen interactions, indicating an interaction between the inflammatory receptor PTX3 and the fungal transcription factor HAP3, which were experimentally validated (Tierney et al. 2012).

5.3.2.3 Identification of Disease Modules

Network analysis uncovering disease modules: Recent advances in the field of network science suggest the existence of disease modules in molecular networks, i.e., a group of proteins residing in the same neighborhood of the underlying interaction network (Ghiassian et al. 2015).

Table 5.4 Inference of gene-regulatory networks (GRNs)

Approaches used for large-scale or genome-wide networks	Approaches used for small-scale networks
(a) Computationally fast	(a) More sophisticated
(b) Use simplified static mathematical models that display either correlation between genes predicting undirected edges, e.g., correlation-based algorithms (Stuart 2003) and information theory-based algorithms (Margolin et al. 2006; Meyer et al. 2007; Faith et al. 2007; Zoppoli et al. 2010)	(b) Use dynamic mathematical models introducing biologically relevant aspects, such as nonlinearity, in the relation between genes or the activation of genes, e.g., by detailed kinetic knowledge (Mjolsness 2007)
(c) Or causal relations among genes using weighted, directed edges, e.g., regression-based algorithms (Tibshirani 1996; Efron et al. 2004; Hecker et al. 2009a) can also be displayed	(c) Use models based on Boolean logic (Müssel et al. 2010), Bayesian networks (Hartemink et al. 2001), difference equation systems (Vlaic et al. 2012), and linear (Weber et al. 2013) or nonlinear differential equation systems (Yang et al. 2012)
(d) **Clinical significance:** Network inference based on expression data of the human pathogen *Candida albicans* was successfully applied for the identification of hub genes as new potential drug targets (Altwasser et al. 2012)	(d) **Clinical significance:** In the human pathogenic fungus *Aspergillus fumigatus*, the inference of a dynamic GRN was used to predict novel interactions between transcription factors and target genes relevant for the regulation of iron homeostasis (Linde et al. 2012)

Several studies that identify modules underlying diseases such as obesity and type 2 diabetes (Barrenas et al. 2012), asthma (Sharma et al. 2015), and inflammatory and malignant diseases (Gustafsson et al. 2014a) demonstrated that the identified disease modules partially overlap, sharing molecular mechanisms as well as proteins. It was further shown that the degree of overlap correlates with biological similarity, disease symptoms, and increased evidence of comorbidity (Menche et al. 2015).

The analysis of disease modules, in conjunction with diverse data types (e.g., genomics, genome-wide association studies, gene expression, gene disease association, and clinically relevant information), has already proven a successful strategy to the discovery of affected molecular mechanisms, biomarkers, and new drug targets as well as the repositioning of drugs.

Clinical research findings:

- Analysis of the disease module underlying asthma revealed **GAB1 signaling pathway as a novel modulator of asthma** and successfully linked the pathway to glucocorticoids as a therapeutic agent (Ghiassian et al. 2015).
- In a study of breast cancer, the combined use of an interaction network, disease-specific gene expression data, and patient survival data revealed **a module enriched in Aurora B signaling and kinetochore-associated genes**, whose

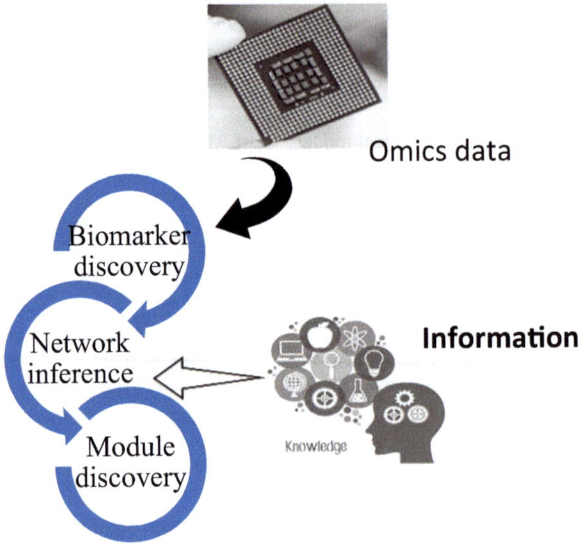

Fig. 5.6 Overview of the relations of systems biology approaches for biomarker discovery, network inference, and module discovery

expression was successfully validated to anticorrelate with patient survival (Wu and Stein 2012).

Translation of the concept of disease modules on the host–pathogen relationship remains problematic due to a lack of genome-wide interaction networks comprising both host and pathogen. Currently, such models are restricted to gene subsets (Remmele et al. 2015) or describe gene islands (Smeekens et al. 2016; Yoon et al. 2015) (Fig. 5.6).

5.3.3 Contributions of Systems Biology in the Field of Systemic Medicine & Personalized Medicine

5.3.3.1 Systemic Medicine

- The analysis of HPI studies has culminated in medical applications to improve diagnostics and therapies, thereby increasing survival chances and reducing adverse effects has to be evaluated in clinical trials (Zhang et al. 2014; Caie et al. 2013; Auffray et al. 2009).
- The creation of associations between gene/protein behavior to diseases and disorders is the topic of systems medicine (Zhang et al. 2014; Caie et al. 2013; Auffray et al. 2009).
- This interdisciplinary approach aims to extend the spectrum of available therapeutic and diagnostic molecules by systems-wide analysis.

- Reyes-Palomares in his study further incorporated a new component, i.e., study of clinical features to understand the disease (Reyes-Palomares et al. 2013).
- They combined disease network models with phenotype data and thereby found new connections between genes and diseases. This approach has improved the understanding of the pathologic process and the molecular mechanisms behind the host response.
- Comprehensive analysis of already known biomarkers has helped to identify associations with conditions that contributed to having a broader knowledge and an extended range of applications. For example, increased levels of glycoprotein acetylation have been determined as biomarkers for incidence and mortality of cardiovascular disease (Fischer et al. 2014; Akinkuolie et al. 2014).
- In addition to this, Ritchie et al. 2015 discovered that these elevated levels of glycoprotein acetylation serve as markers for increased long-term risk of infection, especially septicemia and pneumonia.

Therefore, the identification of biomarkers to develop patient- and tissue-specific therapeutic solutions is one of the key aspects in systems medicine (Capobianco E. Ten 2012). The elucidation of interactions of genes and/or proteins by network analysis is crucial to reveal disease processing (Gustafsson et al. 2014b).

Overview of the relations of systems biology approaches for biomarker discovery, network inference, and module discovery

- With the help of omics data, biomarkers can be identified that indicate patient conditions and may serve as diagnostic or prognostic markers.
- Omics data are used for network inference, where, with support of prior knowledge, regulatory networks are reconstructed. These networks predict novel molecular interactions as well as important key regulatory molecules, which may become novel therapeutic targets.
- The interaction network, thus derived, forms the basis for the discovery of disease modules.
- Creation of hypotheses in this field is supported by omics data and knowledge about the associations between diseases and molecules.
- All these three approaches require experimental validation of the generated hypotheses, which then serve as new prior knowledge for future experiments and analyses.

5.3.3.2 Personalized Medicine

- The aim of personalized medicine is to improve individual treatment and therapy to maximize benefit and minimize harm to patients (Pearson 2016).
- The focus here is to account for patient individual characteristics, including molecular markers (e.g., genetic markers) (Hu et al. 2011).
- Personalized medicine is strongly connected to systems medicine research as biomarkers are needed to monitor the influence of a treatment to an individual (Hu et al. 2011).

- Genomics, transcriptomics, proteomics, and metabolomics approaches are of indispensable need to identify such biomarkers using unbiased approaches (Hu et al. 2011). As a result of rapidly decreasing costs of omics technologies, especially in the field of next-generation sequencing, there is no long way to gain personal omics data of individuals at different molecular levels (Mardis 2006).
- Personalized medicine heavily relies on quantitative proteomics. The identification and utilization of proteomics marker is most common in clinical settings as most drug targets are proteins (Guest et al. 2013).
- Biomarker prediction is initiated by a genome-wide transcriptomic screening based on microarray or RNA-seq. This screening is followed by validation of selected genes using quantitative real-time polymerase chain reaction (PCR) and quantification of the proteins coded by these genes (e.g., by the Western blot test or ELISA) (Schwanhausser et al. 2011; Vogel and Marcotte 2012). Sometimes this validation fails due to some reasons; therefore, generation of proteomic data at large scale is of particular importance.
- Individual omics data allow two ways of application to personalized medicine. Static personal and clinical information (e.g., previous diseases) in combination with static biomarkers (e.g., genetic markers) (Cunha et al. 2014; Lu et al. 2014) will help to stratify the personal risk of an individual gaining infection, suffering from a disease or responding to a therapy.
- Furthermore, dynamic clinical data in combination with dynamically changing biomarkers detected from omics data will help to support therapeutic decisions such as the end of a therapy or when to change drug treatment. Also, it can predict the outcome of an infection.
- One major research topic in systems medicine is the combination of biomarkers from multiple omics data together with clinical biomarkers to ultimately improve predictions and diagnosis (Hu et al. 2011; Liu et al. 2013).

5.3.4 Translational Systems Biology

- Translational systems biology applies modeling and engineering approaches with the aim of optimizing clinical practice (Vodovotz et al. 2008).
- Dynamic biomarkers are a starting point for "translational systems biology." This phrase was coined by Vodovotz et al. (2010) and Vodovotz and An (2015) to describe the utilization of computer simulation to optimize the design of drugs, as well as to simulate clinical trials and the results of a therapy on individuals.
- The extensive application of detailed mathematical modeling makes translational systems biology a promising research area since they not only predict a certain result but also contain information about the reason of the result (Cedersund and Roll 2009). Thus, these models will give clinicians a starting point for the change of therapy.
- One particularly important direction of research is to set up models, which are not only built upon data from one (molecular) level but models including data

from different levels (i.e., molecular, tissue, organ, and whole body) (Schleicher et al. 2016).

- Examples of success stories including multiscale models are models which allow modulation of acute inflammation (Vodovotz et al. 2010), computer-optimized antibiotic treatment for tuberculosis (Pienaar et al. 2015), and a novel mechanism for insulin resistance (Nyman et al. 2014).
- Also, in this field, the future task is to incorporate not only data from different molecular or macromolecular levels in the model but also personal data, which may affect the system's response to stimuli via differences in the initial conditions (Dix et al. 2016).

5.4 Conclusion

Despite the changing trends in infectious diseases, especially after the pandemic that happened recently and other technical challenges involved, the contribution of systems biology today has been enormous. Substantial progress and a lot of information related to the pathogens has been acquired through it. A number of extensive studies are being carried out to decode the genetic information about the medically important bacteria, their virulence factors, and antimicrobial immune mechanisms using transcriptomics and proteomics. Efficacious vaccines and new antimicrobial drugs are on the decline; therefore, more research in areas with a systems biology approach would be the way forward in the future management of infectious diseases. Analysis and translation of this acquired knowledge into effective new control strategies is the target to be achieved, followed by the integration of this extensive information into the health sciences, which will help to acquire fruitful solutions.

It is always not just enough to develop a vaccine or new antimicrobial for future use but also to study the impact of these new inventions on the physiological, molecular, and immunological changes in the host body. In the postpandemic era, as we proudly boast about the lives that were saved with the help of vaccines, we should also acknowledge the deaths and other complications associated with COVID vaccines. A systems biology-based approach to comprehend such faulty host and vaccine encounters is the need of the hour to prevent inadvertent errors. In short, the better understanding of HPI paves the way for individualized therapies, optimal strategies of delivery, and manipulation of outcomes with timely interventions.

References

Abu-Maziad A, Schaa K, Bell EF, Dagle JM, Cooper M, Marazita ML et al (2010) Role of polymorphic variants as genetic modulators of infection in neonatal sepsis. Pediatr Res 68:323e9

Achkar JM, Cortes L, Croteau P, Yanofsky C, Mentinova M, Rajotte I et al (2015) Host protein biomarkers identify active tuberculosis in HIV uninfected and coinfected individuals. EBioMedicine 2:1160e8

Aderem A, Adkins JN, Ansong C, Galagan J, Kaiser S, Korth MJ, Law GL, McDermott JG, Proll SC, Rosenberger C, Schoolnik G, Katze MG (2011) A systems biology approach to infectious disease research: innovating the pathogen-host research paradigm. MBio 2(1):e00325–e00310. PMID: 21285433; PMCID: PMC3034460. https://doi.org/10.1128/mBio.00325-10

Akinkuolie AO, Buring JE, Ridker PM, Mora S (2014) A novel protein glycan biomarker and future cardiovascular disease events. J Am Heart Assoc 3:e001221

Altwasser R, Linde J, Buyko E, Hahn U, Guthke R (2012) Genome-wide scale-free network inference for Candida albicans. Front Microbiol 3:1–10

Auffray C, Chen Z, Hood L (2009) Systems medicine: the future of medical genomics and healthcare. Genome Med 1:2

Barrenas F, Chavali S, Alves A, Coin L, Jarvelin MR, Jornsten R et al (2012) Highly interconnected genes in disease-specific networks are enriched for disease associated polymorphisms. Genome Biol 13:R46

Breitling R (2010) What is system biology? Front Physiol 1:9. https://doi.org/10.3389/fphys.2010.00009

Bruce A, Johnson A, Lewis J, Morgan D, Raff M, Roberts K, Walter P (2015) Molecular biology of the cell, 6th edn

Caie PD, Schuur K, Oniscu A, Mullen P, Reynolds PA, Harrison DJ (2013) Human tissue in systems medicine. FEBS J 280:5949e56

Cedersund G, Roll J (2009) Systems biology: model based evaluation and comparison of potential explanations for given biological data. FEBS J 276:903–922

Cunha C, Aversa F, Lacerda JF, Busca A, Kurzai O, Grube M et al (2014) Genetic PTX3 deficiency and aspergillosis in stem-cell transplantation. N Engl J Med 370:421–432

Dix A, Hünniger K, Weber M, Guthke R, Kurzai O, Linde J (2015) Biomarker-based classification of bacterial and fungal whole-blood infections in a genomewide expression study. Front Microbiol 6:171

Dix A, Vlaic S, Guthke R, Linde J (2016) Use of systems biology to decipher host-pathogen interaction networks and predict biomarkers. Clin Microbiol Infect 22(7):600–606. Epub 2016 Apr 22. https://doi.org/10.1016/j.cmi.2016.04.014

Durmuş S, Çakır T, Özgür A, Guthke R (2015) A review on computational systems biology of pathogen–host interactions. Front Microbiol 6:235. https://doi.org/10.3389/fmicb.2015.00235

Eckhardt M, Hultquist JF, Kaake RM, Hüttenhain R, Krogan NJ (2020) A systems approach to infectious disease. Nat Rev Genet 21(6):339–354. Epub 2020 Feb 14. PMID: 32060427; PMCID: PMC7839161. https://doi.org/10.1038/s41576-020-0212-5

Efron B, Hastie T, Johnstone I, Tibshirani R (2004) Least angle regression. Ann Stat 32:407–499

Erratum PJ (2008) SNP genotyping: six technologies that keyed a revolution. Nat Methods 5:575

Faith JJ, Hayete B, Thaden JT, Mogno I, Wierzbowski J, Cottarel G et al (2007) Largescale mapping and validation of Escherichia coli transcriptional regulation from a compendium of expression profiles. PLoS Biol 5:e8

Fischer K, Kettunen J, Würtz P, Haller T, Havulinna AS, Kangas AJ et al (2014) Biomarker profiling by nuclear magnetic resonance spectroscopy for the prediction of all-cause mortality: an observational study of 17,345 persons. PLoS Med 11:e1001606

Gao YD, Zhao Y, Huang J (2014) Metabolic modeling of common Escherichia coli strains in human gut microbiome. Biomed Res Int 2014:1–11

Ghiassian SD, Menche J, Barabasi AL (2015) A DIseAse MOdule detection (DIAMOnD) algorithm derived from a systematic analysis of connectivity patterns of disease proteins in the human interactome. PLoS Comput Biol 11:e1004120

Guest PC, Gottschalk MG, Bahn S (2013) Proteomics: improving biomarker translation to modern medicine? Genome Med 5:17

Gustafsson M, Edstrom M, Gawel D, Nestor CE, Wang H, Zhang H et al (2014a) Integrated genomic and prospective clinical studies show the importance of modular pleiotropy for disease susceptibility, diagnosis and treatment. Genome Med 6:17

Gustafsson M, Nestor CE, Zhang H, Barabasi AL, Baranzini S, Brunak S et al (2014b) Modules, networks and systems medicine for understanding disease and aiding diagnosis. Genome Med 6:82

Hartemink AJ, Gifford DK, Jaakkola TS, Young RA (2001) Using graphical models and genomic expression data to statistically validate models of genetic regulatory networks. Pac Symp Biocomput:422–433

Hecker M, Goertsches R, Engelmann R, Thiesen HJ, Guthke R (2009a) Integrative modeling of transcriptional regulation in response to antirheumatic therapy. BMC Bioinformatics 10:262

Hecker M, Lambeck S, Toepfer S, van Someren E, Guthke R (2009b) Gene regulatory network inference: data integration in dynamic models review. Biosystems 96:86–103

Heinken A, Sahoo S, Fleming RMT, Thiele I (2013) Systems-level characterization of a host microbe metabolic symbiosis in the mammalian gut. Gut Microb 4:28–40

Hu ZZ, Huang H, Wu CH, Jung M, Dritschilo A, Riegel AT et al (2011) Omics-based molecular target and biomarker identification. Methods Mol Biol 719:547–571

Ithal N, Recknor J, Nettleton D, Hearne L, Maier T, Baum TJ, Mitchum MG (2007) Parallel genome-wide expression profiling of host and pathogen during soybean cyst nematode infection of soybean. Mol Plant Microbe Interact 20(3):293–305. https://doi.org/10.1094/MPMI-20-3-0293

Ji B, Nielsen J (2015) From next-generation sequencing to systematic modeling of the gut microbiome. Front Genet 6:219

Lee H (2013) Procalcitonin as a biomarker of infectious diseases. Korean J Intern Med 28:285

Li LM, Hu ZB, Zhou ZX, Chen X, Liu FY, Zhang JF et al (2010) Serum microRNA profiles serve as novel biomarkers for HBV infection and diagnosis of HBV positive hepatocarcinoma. Cancer Res 70:9798–9807

Linde J, Hortschansky P, Fazius E, Brakhage AA, Guthke R, Haas H (2012) Regulatory interactions for iron homeostasis in aspergillus fumigatus inferred by a systems biology approach. BMC Syst Biol 6:6

Ling H, Fabbri M, Calin GA (2013) MicroRNAs and other non-coding RNAs as targets for anticancer drug development. Nat Rev Drug Discov 12:847–865

Liu LY, Yang T, Ji J, Wen Q, Morgan AA, Jin B et al (2013) Integrating multiple 'omics' analyses identifies serological protein biomarkers for preeclampsia. BMC Med 11:236

Lu C, Wu J, Wang H, Wang S, Diao N, Wang F et al (2011) Novel biomarkers distinguishing active tuberculosis from latent infection identified by gene expression profile of peripheral blood mononuclear cells. PLoS One 6:e24290

Lu Y, Goldstein DB, Angrist M, Cavalleri G (2014) Personalized medicine and human genetic diversity, vol 4. Cold Spring Harb Perspect Med, p a008581

Maertzdorf J, Repsilber D, Parida SK, Stanley K, Roberts T, Black G et al (2011) Human gene expression profiles of susceptibility and resistance in tuberculosis. Genes Immun 12:15–22

Mardis ER (2006) Anticipating the 1,000 dollar genome. Genome Biol 7:112

Margolin AA, Nemenman I, Basso K, Wiggins C, Stolovitzky G, Favera R et al (2006) ARACNE: an algorithm for the reconstruction of gene regulatory networks in a mammalian cellular context. BMC Bioinformatics 7:S7

Menche J, Sharma A, Kitsak M, Ghiassian SD, Vidal M, Loscalzo J et al (2015) Disease networks. Uncovering disease disease relationships through the incomplete interactome. Science 347:1257601

Meyer PE, Kontos K, Lafitte F, Bontempi G (2007) Information-theoretic inference of large transcriptional regulatory networks. EURASIP J Bioinform Syst Biol 2007:1e9

Miao J, Chard LS, Wang Z, Wang Y (2019) Syrian hamster as an animal model for the study on infectious diseases. Front Immunol 10:2329. https://doi.org/10.3389/fimmu.2019.02329

Mjolsness E (2007) On cooperative quasi-equilibrium models of transcriptional regulation. J Bioinform Comput Biol 5:467–490

Müssel C, Hopfensitz M, Kestler HA (2010) BoolNetdan R package for generation, reconstruction and analysis of Boolean networks. Bioinformatics 26:1378–1380

Nyman E, Rajan MR, Fagerholm S, Brannmark C, Cedersund G, Stralfors P (2014) A single mechanism can explain network-wide insulin resistance in adipocytes from obese patients with type 2 diabetes. J Biol Chem 289:33215–33230
Pearson E (2016) Personalized medicine in diabetes: the role of 'omics' and biomarkers. Diabet Med 33:712
Pienaar E, Dartois V, Linderman JJ, Kirschner DE (2015) In silico evaluation and exploration of antibiotic tuberculosis treatment regimens. BMC Syst Biol 9:79
Qi Y, Cui L, Ge Y, Shi Z, Zhao K, Guo X et al (2012) Altered serum microRNAs as biomarkers for the early diagnosis of pulmonary tuberculosis infection. BMC Infect Dis 12:384
Reif DM, Motsinger-Reif AA, McKinney BA, Rock MT, Crowe JE, Moore JH (2009) Integrated analysis of genetic and proteomic data identifies biomarkers associated with adverse events following smallpox vaccination. Genes Immun 10:112–119
Remmele CW, Luther CH, Balkenhol J, Dandekar T, Müller T, Dittrich MT (2015) Integrated inference and evaluation of hostefungi interaction networks. Front Microbiol 6:764
Reyes-Palomares A, Rodríguez-Lopez R, JAG R, Jimenez FS, Medina MA (2013) Global analysis of the human pathophenotypic similarity gene network merges disease module components. PLoS One 8:e56653
Ritchie SC, Würtz P, Nath AP, Abraham G, Havulinna AS, Fearnley LG et al (2015) The biomarker GlycA is associated with chronic inflammation and predicts long-term risk of severe infection. Cell Syst 1:293–301
Rogers TR, Morton CO, Springer J, Conneally E, Heinz W, Kenny C et al (2013) Combined real-time PCR and galactomannan surveillance improves diagnosis of invasive aspergillosis in high risk patients with haematological malignancies. Br J Haematol 161:517–524
Roser M, Ritchie H, Spooner F (2021) "Burden of disease". Published online at OurWorldInData.org. Accessed from https://ourworldindata.org/burden-of-disease [Online Resource]
Sadhukhan PP, Raghunathan A (2014) Investigating hostepathogen behavior and their interaction using genome-scale metabolic network models. Methods Mol Biol 1184:523–562
Schleicher J, Conrad T, Gustafsson M, Cedersund G, Guthke R, Linde J (2016) Facing the challenges of multiscale modelling of bacterial and fungal pathogene host interactions. Brief Funct Genomics 16:57–69
Schulze S, Henkel SG, Driesch D, Guthke R, Linde J (2015) Computational prediction of molecular pathogenehost interactions based on dual transcriptome data. Front Microbiol 6:65
Schwanhausser B, Busse D, Li N, Dittmar G, Schuchhardt J, Wolf J et al (2011) Global quantification of mammalian gene expression control. Nature 473:337–342
Shahabi V, Berman D, Chasalow SD, Wang L, Tsuchihashi Z, Hu B et al (2013) Gene expression profiling of whole blood in ipilimumab-treated patients for identification of potential biomarkers of immune-related gastrointestinal adverse events. J Transl Med 11:75
Sharma A, Menche J, Huang CC, Ort T, Zhou X, Kitsak M et al (2015) A disease module in the interactome explains disease heterogeneity, drug response and captures novel pathways and genes in asthma. Hum Mol Genet 24:3005–3020
Shoaie S, Nielsen J (2014) Elucidating the interactions between the human gut microbiota and its host through metabolic modeling. Front Genet 5:86
Shoaie S, Karlsson F, Mardinoglu A, Nookaew I, Bordel S, Nielsen J (2013) Understanding the interactions between bacteria in the human gut through metabolic modeling. Sci Rep 3:2532
Smeekens SP, van de Veerdonk FL, Netea MG (2016) An omics perspective on Candida infections: towards next-generation diagnosis and therapy. Front Microbiol 7:154
Strimbu K, Tavel JA (2010) What are biomarkers? Curr Opin HIV AIDS 5(6):463–466. PMID: 20978388; PMCID: PMC3078627. https://doi.org/10.1097/coh.0b013e32833ed177
Stuart JM (2003) A gene-coexpression network for global discovery of conserved genetic modules. Science 302:249–255
Ten CE (2012) Challenges for systems medicine. Front Genet 3:193
Tibshirani R (1996) Regression shrinkage and selection via the lasso. J R Stat Soc Ser B 58:267–288

Tierney L, Linde J, Müller S, Brunke S, Molina JC, Hube B et al (2012) An interspecies regulatory network inferred from simultaneous RNA-seq of Candida albicans invading innate immune cells. Front Microbiol 3:85

Tsalik EL, Henao R, Nichols M, Burke T, Ko ER, McClain MT et al (2016) Host gene expression classifiers diagnose acute respiratory illness etiology. Sci Transl Med 8:322ra11

Vlaic S, Schmidt-Heck W, Matz-Soja M, Marbach E, Linde J, Meyer-Baese A et al (2012) The extended TILAR approach: a novel tool for dynamic modeling of the transcription factor network regulating the adaption to in vitro cultivation of murine hepatocytes. BMC Syst Biol 6:147

Vodovotz Y, An G (2015) Translational systems biology. Elsevier, Amsterdam

Vodovotz Y, Csete M, Bartels J, Chang S, An G (2008) Translational systems biology of inflammation. PLoS Comput Biol 4:e1000014

Vodovotz Y, Constantine G, Faeder J, Mi Q, Rubin J, Bartels J et al (2010) Translational systems approaches to the biology of inflammation and healing. Immunopharmacol Immunotoxicol 32:181–195

Vogel C, Marcotte EM (2012) Insights into the regulation of protein abundance from proteomic and transcriptomic analyses. Nat Rev Genet 13:227–232

Weber M, Henkel SG, Vlaic S, Guthke R, van Zoelen EJ, Driesch D (2013) Inference of dynamical gene-regulatory networks based on time-resolved multi-stimuli multi-experiment data applying NetGenerator V2.0. BMC Syst Biol 7:1

Werner T (2010) Next generation sequencing in functional genomics. Brief Bioinform 11:499–511

Westerhoff HV, Winder C, Messiha H, Simeonidis E, Adamczyk M, Verma M, Bruggeman FJ, Dunn W (2009) Systems biology: the elements and principles of life. FEBS Lett 583(24):3882–3890. https://doi.org/10.1016/j.febslet.2009.11.018

Wu G, Stein L (2012) A network module based method for identifying cancer prognostic signatures. Genome Biol 13:R112

Yang X, Dent JE, Nardini C (2012) An SSystem parameter estimation method (SPEM) for biological networks. J Comput Biol 19:175–187

Yoon SH, Park YK, Kim JF (2015) PAIDB v2.0: exploration and analysis of pathogenicity and resistance islands. Nucleic Acids Res 43:D624–D630

Yousef M, Ketany M, Manevitz L, Showe LC, Showe MK (2009) Classification and biomarker identification using gene network modules and support vector machines. BMC Bioinformatics 10:337

Yu H, Li L, Huffman A, Beverley J, Hur J, Merrell E, Huang HH, Wang Y, Liu Y, Ong E, Cheng L, Zeng T, Zhang J, Li P, Liu Z, Wang Z, Zhang X, Ye X, Handelman SK, Sexton J, Eaton K, Higgins G, Omenn GS, Athey B, Smith B, Chen L, He Y (2022) A new framework for host-pathogen interaction research. Front Immunol 13:1066733. PMID: 36591248; PMCID: PMC9797517. https://doi.org/10.3389/fimmu.2022.1066733

Zhang H, Gustafsson M, Nestor C, Chung KF, Benson M (2014) Targeted omics and systems medicine: personalising care. Lancet Respir Med 2:785–787

Zoppoli P, Morganella S, Ceccarelli M (2010) TimeDelay-ARACNE: reverse engineering of gene networks from time-course data by an information theoretic approach. BMC Bioinformatics 11:154

Zuniga J, Buendia-Roldan I, Zhao Y, Jimenez L, Torres D, Romo J et al (2012) Genetic variants associated with severe pneumonia in a/H1N1 influenza infection. Eur Respir J 39:604–610

Case Study-Based Approaches of Systems Biology in Addressing Infectious Diseases

6

Sourav Dutta and Avijit Ghosh

Abstract

The Systems Biology Consortium for Infectious Diseases presents a novel approach to understanding and combating infectious diseases by integrating the principles and techniques of systems biology. Systems biology is an approach that studies biological systems as a whole and their interactions with the environment, using experimental and computational methods to generate and analyze multi-omics data. The chapter explains the definition, classification, and global impact of infectious diseases, which are caused by various pathogens and pose significant threats to human health. It also discusses the challenges in diagnosing, treating, and preventing infectious diseases, such as complexity, heterogeneity, variability, and drug resistance. The chapter shows how systems biology can help identify biomarkers and drug targets, predict disease outcomes and responses, and accelerate drug development and repurposing. The chapter discusses the rationale, objectives, structure, and governance of the consortium, such as data integration and harmonization, standardization of methods and protocols, establishment of databases and resources, education and training in systems biology, leadership and management, membership and collaborations, funding, and sustainability. The chapter provides case studies on viral (HIV), bacterial (tuberculosis), and parasitic (malaria) infections to illustrate the successful application of systems biology approaches. The chapter also addresses the challenges and future perspectives in the field, such as ethical and privacy issues, regulatory and policy implications, emerging technological advances, and translational impact. The chapter concludes by highlighting the potential impact

S. Dutta (✉) · A. Ghosh
GCC Biotech India Pvt. Ltd., Joychandipur Bakrahat, West Bengal, India
e-mail: sourav.dutta@gccbiotech.co.in; avijit.ghosh@gccbiotech.co.in

S. Joshi et al. (eds.), *Systems Biology Approaches: Prevention, Diagnosis, and Understanding Mechanisms of Complex Diseases*,
https://doi.org/10.1007/978-981-99-9462-5_6

and benefits of the consortium in revolutionizing infectious disease research and improving global health outcomes.

Keywords

Sytems Biology · Diseases · Mechanism · HIV · Tuberculosis

6.1 Introduction

Infectious diseases pose a significant threat to global health, causing immense morbidity and mortality worldwide. The emergence of new pathogens, the re-emergence of existing ones, and the increasing prevalence of drug resistance underscore the urgent need for innovative approaches to combat these diseases. In recent years, systems biology has emerged as a powerful framework that integrates biological data, computational modeling, and advanced analytical techniques to gain a comprehensive understanding of complex biological systems. Harnessing the potential of systems biology in the context of infectious diseases is the concept of a systems biology consortium (Oh et al. 2018).

The systems biology consortium represents a collaborative and integrative approach to address the complex challenges posed by infectious diseases. It brings together researchers, clinicians, industry partners, and policymakers, with the aim of applying systems biology approaches to unravel the intricate molecular mechanisms of infections, identify key biomarkers and drug targets, and accelerate the development of interventions (Aderem et al. 2011). By fostering collaboration, data sharing, and standardization, the consortium aims to overcome the limitations of individual studies, facilitate comprehensive analyses, and optimize translational outcomes.

The first part of this chapter provides an overview of infectious diseases, including their definition, classification, and global impact. It highlights the challenges posed by infectious diseases, such as the emergence of drug resistance, the high variability and adaptability of pathogens, and the complexity of host–pathogen interactions. These challenges underscore the need for innovative and integrative approaches, such as systems biology, to enhance our understanding and management of infectious diseases (Eckhardt et al. 2020).

The second part explores the principles and concepts of systems biology, providing an overview of experimental techniques and computational approaches used in this field. It highlights the importance of multi-omics data integration, computational modeling, and advanced analytical tools in unraveling complex biological systems (Germain et al. 2011). Furthermore, it discusses the specific application of systems biology approaches in the context of infectious diseases, emphasizing their potential in understanding the dynamics of host–pathogen interactions, immune responses, and the identification of novel therapeutic strategies (Dix et al. 2016).

Building upon this foundation, the third part delves into the rationale for establishing a systems biology consortium for infectious diseases. It discusses the unique advantages offered by systems biology in addressing the complexity and heterogeneity of infectious diseases. The integration of multi-omics data, the identification of key biomarkers and drug targets, and the use of predictive modeling and simulation techniques are highlighted as essential components of the consortium's approach. Moreover, the chapter explores how a systems biology consortium can accelerate drug development and repurposing, leading to more effective interventions against infectious diseases (Stumpf et al. 2007).

The subsequent parts of the chapter focus on the objectives and scope of a systems biology consortium. They discuss the importance of facilitating collaboration and data sharing among researchers, standardizing experimental and computational methods, and establishing integrated databases and resources (Dräger and Palsson 2014). Additionally, the chapter emphasizes the significance of education and training in systems biology to foster the growth of expertise in this field.

The structure and governance of a systems biology consortium are discussed in detail, including aspects such as leadership and management, membership and collaborations, and funding and sustainability. These elements are essential for establishing a well-coordinated and sustainable consortium that can effectively drive systems biology research for infectious diseases (Kitano 2002).

The chapter then presents case studies that demonstrate the application of systems biology approaches in viral, bacterial, and parasitic infections (Medyukhina et al. 2015). These case studies highlight the potential of systems biology in unraveling the complex mechanisms of infection, identifying novel therapeutic targets, and guiding the development of personalized treatment strategies.

Finally, the chapter explores the challenges and future perspectives of a systems biology consortium. It addresses issues related to data integration and harmonization, ethics and privacy concerns, regulatory and policy implications, and technological advances. It emphasizes the need for collaborative efforts, robust ethical frameworks, supportive regulatory policies, and ongoing technological advancements to maximize the potential impact of systems biology in combating infectious diseases (Davidson et al. 1995).

In conclusion, this chapter provides a comprehensive overview of the concept of a systems biology consortium for infectious diseases. It highlights the potential impact and benefits of such a consortium, outlines its objectives and scope, discusses its structure and governance, presents case studies, and explores the challenges and future perspectives. By embracing the collaborative and integrative nature of systems biology, we can revolutionize our approach to infectious diseases and pave the way for improved diagnostics, therapeutics, and preventive strategies.

6.2 Overview of Infectious Diseases

6.2.1 Definition and Classification

Infectious diseases are illnesses caused by pathogenic microorganisms, including bacteria, viruses, fungi, and parasites. These microorganisms invade and multiply within the body, leading to a wide range of symptoms and clinical manifestations.

Infectious diseases can be transmitted from person to person, through contaminated water or food, or via vectors such as mosquitoes or ticks.

Infectious diseases can be classified based on the type of pathogen involved and the mode of transmission. Common classifications include:

6.2.1.1 Bacterial Infections

Bacterial infections result from the invasion and colonization of the body by bacteria. Examples include tuberculosis (Russell 2011), pneumonia (Jung et al. 2017), urinary tract infections (Olin and Bartges 2015), and strep throat (Jarmusch et al. 2014). Bacterial infections can be treated with antibiotics, although the emergence of antibiotic-resistant strains poses a significant challenge (Vivas et al. 2019).

6.2.1.2 Viral Infections

Viral infections are caused by viruses, which are smaller than bacteria and depend on host cells for replication. Common viral infections include influenza (Dangi and Jain 2012), HIV/AIDS (Anne Crock 2017), hepatitis (Pallavi et al. 2018), and COVID-19 (Schröder 2020). Antiviral medications are available for some viral infections, but many viral diseases are managed through supportive care and vaccination (Ellebedy and Ahmed 2016).

6.2.1.3 Fungal Infections

Fungal infections occur when pathogenic fungi invade the body. Examples include candidiasis (Hani et al. 2015), aspergillosis (Gonty and Page 1977), and ringworm (Anosike et al. 2005). Fungal infections can affect various parts of the body, including the skin, nails, respiratory system, and internal organs. Antifungal medications are used to treat fungal infections (Pound et al. 2011).

6.2.1.4 Parasitic Infections

Parasitic infections are caused by parasites that live in or on the host organism. Malaria (Talapko et al. 2019), leishmaniasis (Burza et al. 2018), schistosomiasis (Gryseels et al. 2006), and intestinal parasites such as giardiasis and hookworm infection (Escobedo and Cimerman 2007; Hotez et al. 2004) are common examples. Treatment for parasitic infections may involve antiparasitic drugs and measures to prevent reinfection.

6.2.2 Global Impact and Challenges of Infectious Diseases

Infectious diseases have a significant impact on global public health, economies, and societal well-being (Morens et al. 2004). They pose several challenges that require effective prevention, control, and treatment strategies.

6.2.2.1 Public Health Impact

Infectious diseases are a leading cause of morbidity and mortality worldwide. They can cause widespread outbreaks and epidemics, putting a strain on healthcare systems, overwhelming healthcare facilities, and leading to increased healthcare costs. Control measures such as vaccination, hygiene practices, and vector control are essential for preventing the spread of infectious diseases.

6.2.2.2 Economic Burden

Infectious diseases impose substantial economic burdens on individuals, communities, and countries. The direct costs include medical expenses, hospitalizations, and treatments. Indirect costs result from productivity losses due to illness, disability, and premature death. Additionally, infectious diseases can negatively impact tourism, trade, and economies, especially in regions heavily affected by epidemics.

6.2.2.3 Emergence of Drug Resistance

The emergence and spread of drug-resistant strains of infectious pathogens, such as antibiotic-resistant bacteria and drug-resistant malaria parasites, pose a significant challenge to treatment and control efforts. The development of new antimicrobial agents and strategies to prevent the further spread of drug resistance is crucial to combat these infectious diseases effectively (Gold and Moellering 1996).

6.2.2.4 Complex Interactions and Heterogeneity

Infectious diseases involve complex interactions between pathogens, hosts, and the environment. Pathogens can evolve rapidly, leading to the emergence of new strains and the ability to evade host immune responses. Additionally, host factors, including genetics, immune status, and comorbidities, influence susceptibility, disease progression, and treatment outcomes. The heterogeneity of pathogens and host populations adds to the complexity of understanding and managing infectious diseases.

Addressing the global impact and challenges of infectious diseases requires a comprehensive and interdisciplinary approach. The field of systems biology offers a promising framework to gain a deeper understanding of the complex mechanisms underlying infectious diseases and develop innovative strategies for prevention, diagnosis, and treatment.

6.3 Systems Biology and its Application in Infectious Diseases

6.3.1 Principles and Concepts of Systems Biology

Systems biology is an interdisciplinary field that aims to understand the behavior and function of complex biological systems as a whole, rather than focusing on individual components in isolation. It integrates experimental techniques, computational modeling, and advanced data analysis to gain a comprehensive understanding of the dynamic interactions and regulatory networks within living organisms. The principles and concepts underlying systems biology provide a foundation for studying the complexity of biological systems and are key to its success in addressing intricate biological questions (Kelley 2007).

6.3.1.1 Holistic Approach

Systems biology takes a holistic approach by considering the system as a whole, acknowledging the interconnectedness of its components, and recognizing the emergent properties that arise from their interactions. This approach considers biological systems as integrated networks of molecules, cells, tissues, and organisms, where the behavior of the system emerges from the collective actions of its components.

6.3.1.2 Integration of Multi-Omics Data

Systems biology integrates data from various omics disciplines, including genomics, transcriptomics, proteomics, and metabolomics. By analyzing these multi-omics datasets collectively, researchers can uncover the relationships and dependencies between different molecular components, identify regulatory mechanisms, and gain insights into the functional dynamics of biological systems (Kitano 2002).

6.3.1.3 Quantitative and Dynamic Analysis

Systems biology emphasizes quantitative analysis and dynamic modeling of biological systems. By quantifying molecular components, their interactions, and their changes over time, researchers can develop mathematical and computational models that capture the behavior of the system. These models help in understanding the system's response to perturbations, predicting its behavior under different conditions, and generating testable hypotheses.

6.3.1.4 Feedback and Feedforward Loops

Biological systems often exhibit feedback and feedforward loops, where the output of a process or component influences its own regulation or the regulation of other processes. Systems biology aims to identify and characterize these regulatory loops, as they play a crucial role in maintaining system stability, resilience, and responsiveness. Understanding the dynamics of feedback and feedforward loops provides insights into the control mechanisms that govern biological systems.

6.3.1.5 Emergent Properties and Systems Robustness

Biological systems exhibit emergent properties that arise from the interactions of their components, rather than being properties of individual components alone. Systems biology seeks to understand and predict these emergent properties, such as self-organization, robustness, and adaptability. Robustness, in particular, refers to the system's ability to maintain its functions and stability even in the face of internal or external perturbations.

6.3.1.6 Network Analysis and Topological Properties

Systems biology employs network analysis to understand the structure and connectivity of biological systems. Network representations enable the identification of key components, hubs, modules, and pathways within the system. Analyzing the topological properties of these networks provides insights into their organization, information flow, and resilience to perturbations (Albert 2007).

6.3.1.7 Iterative Experimental and Computational Cycles

Systems biology adopts an iterative cycle of experimental data generation and computational modeling. Experimental data guide the development and refinement of computational models, which, in turn, generate predictions that can be tested experimentally. This iterative process allows researchers to refine their understanding of the system, validate hypotheses, and uncover new insights.

6.3.1.8 Data Integration and Model Integration

Integrating diverse datasets and models is a fundamental principle of systems biology. By integrating data from multiple sources and incorporating different modeling approaches, researchers can build more comprehensive and accurate models of biological systems. This integration allows for a more holistic understanding of system behavior and can uncover hidden relationships and mechanisms (Ng et al. 2006).

Overall, the principles and concepts of systems biology provide a framework for understanding and studying the complexity of biological systems. By adopting an integrative and holistic approach, incorporating quantitative and dynamic analysis, and leveraging the power of computational modeling and data integration, systems biology enables researchers to unravel the intricacies of biological systems and make significant contributions to our understanding of living organisms.

6.3.2 Experimental Techniques in Systems Biology

Various experimental techniques are employed in systems biology to generate quantitative data on molecular components, their interactions, and system dynamics. These techniques include:

6.3.2.1 High-Throughput Omics Technologies

Technologies such as next-generation sequencing, microarray analysis, and mass spectrometry enable the simultaneous measurement of thousands to millions of molecules (genes, transcripts, proteins, and metabolites) in a biological sample. These data provide insights into the global changes and dynamics of biological systems (Karahalil 2016).

6.3.2.2 Molecular Interaction Analysis

Methods such as yeast two-hybrid assays, co-immunoprecipitation, and protein–protein interaction networks (e.g., protein microarrays) help identify and characterize physical interactions between proteins. They provide insights into the architecture and organization of cellular networks (Durek and Walther 2008).

6.3.2.3 Imaging and Single-Cell Analysis

Advanced imaging techniques, such as confocal microscopy and single-molecule imaging, allow the visualization and quantification of cellular and subcellular processes in real time. Single-cell analysis techniques enable the study of cellular heterogeneity and the identification of rare cell populations (Megason and Fraser 2007).

6.3.2.4 Perturbation Experiments

Perturbation experiments involve manipulating biological systems using genetic, pharmacological, or environmental interventions. Techniques such as gene knockouts, RNA interference, and drug treatments help elucidate the functional roles of specific genes and pathways and their effects on system behavior.

6.3.3 Computational Approaches in Systems Biology

Computational approaches play a vital role in systems biology by integrating and analyzing complex datasets, constructing mathematical models, and simulating system behavior (Meisner and Reif 2015). Key computational techniques used in systems biology include:

6.3.3.1 Network Analysis

Network analysis involves the construction and analysis of biological networks, such as protein–protein interaction networks, gene regulatory networks, and metabolic networks. Network-based approaches help identify key components, detect functional modules, and understand the flow of information within the system.

6.3.3.2 Data Integration and Harmonization

Computational methods are used to integrate diverse omics data from different experimental platforms and sources. Data harmonization techniques standardize and normalize datasets to enable meaningful comparisons and integrative analyses.

6.3.3.3 Mathematical Modeling and Simulation

Mathematical modeling enables the formal representation of biological processes using equations that describe the relationships between components. Computational simulations of these models can predict system behavior under different conditions, test hypotheses, and guide experimental design.

6.3.3.4 Machine Learning and Data Mining

Machine learning algorithms are applied to analyze large-scale omics datasets and identify patterns, correlations, and predictive models. Data mining techniques help uncover hidden relationships and extract meaningful information from complex biological data.

6.3.4 Applications of Systems Biology in Infectious Diseases

Systems biology has emerged as a powerful approach to study infectious diseases, offering insights into pathogen–host interactions, disease progression, and potential therapeutic interventions. Key applications include:

6.3.4.1 Host–Pathogen Interactions

Systems biology approaches enable the comprehensive characterization of host–pathogen interactions at the molecular level. They help identify host factors that contribute to susceptibility or resistance to infections, elucidate pathogen virulence mechanisms, and uncover host immune responses (Dix et al. 2016).

6.3.4.2 Disease Progression and Biomarker Discovery

By integrating omics data with clinical information, systems biology can identify molecular signatures and biomarkers associated with disease progression, severity, and treatment response. This knowledge aids in early diagnosis, personalized medicine approaches, and the development of new therapeutic targets (Wang et al. 2010).

6.3.4.3 Drug Discovery and Repurposing

Systems biology approaches can accelerate drug discovery by identifying potential drug targets and predicting the effects of drug candidates on the system. Computational models and simulations help optimize drug dosing regimens and facilitate drug repurposing for infectious diseases (Peyvandipour et al. 2018).

6.3.4.4 Vaccine Development

Systems biology provides insights into host immune responses to vaccination and helps identify correlates of protection. This knowledge aids in vaccine design, formulation, and optimization, leading to the development of more effective vaccines against infectious diseases (Oberg et al. 2011).

6.3.4.5 Predictive Modeling and Outbreak Control

By integrating epidemiological data, environmental factors, and host–pathogen interactions, systems biology can develop predictive models to understand the spread of infectious diseases and evaluate control strategies. These models aid in outbreak prediction, resource allocation, and public health decision-making (Scarpino and Petri 2019).

In conclusion, systems biology offers a comprehensive and integrative approach to studying infectious diseases. By combining experimental techniques, computational modeling, and data analysis, systems biology provides insights into the complexity of biological systems and aids in the development of novel strategies for prevention, diagnosis, and treatment.

6.4 Rationale for a Systems Biology Consortium

6.4.1 Addressing Complexity and Heterogeneity

Infectious diseases are complex and exhibit significant heterogeneity in terms of pathogen variability, host responses, and environmental factors. Traditional reductionist approaches often fail to capture this complexity. A systems biology consortium provides a rationale for addressing the intricate nature of infectious diseases by considering them as dynamic systems. This approach acknowledges the interconnectedness of molecular components and the need to study their collective behavior (Wright and Wagner 2008).

6.4.2 Integration of Multi-Omics Data

Multi-omics data, encompassing genomics, transcriptomics, proteomics, metabolomics, and other omics disciplines, offer a comprehensive view of biological systems. However, the integration and analysis of these diverse datasets pose significant challenges. A systems biology consortium can leverage expertise in data integration and harmonization, enabling a holistic understanding of infectious diseases by unraveling intricate molecular interactions across different omics layers.

6.4.3 Identification of Key Biomarkers and Drug Targets

Systems biology approaches facilitate the identification of key biomarkers and drug targets associated with infectious diseases. By combining multi-omics data, computational modeling, and machine learning techniques, consortium members can identify molecular signatures that differentiate disease states, predict treatment responses, and discover potential therapeutic targets. This knowledge can accelerate the development of precise diagnostic tools and targeted therapies.

6.4.4 Predictive Modeling and Simulation

Predictive modeling and simulation play a crucial role in systems biology by generating hypotheses, guiding experimental design, and providing insights into disease progression and treatment outcomes. A systems biology consortium can develop and refine mathematical models that simulate the dynamics of infectious diseases. By incorporating experimental data and clinical parameters, these models enable the prediction of disease outcomes, assessment of intervention strategies, and optimization of public health measures.

6.4.5 Accelerating Drug Development and Repurposing

The traditional drug discovery and development process is time-consuming and costly. Systems biology offers a more efficient approach by facilitating the identification of potential drug candidates and repurposing existing drugs for infectious diseases. Through the integration of omics data, computational modeling, and network analysis, a systems biology consortium can identify key molecular targets, predict drug efficacy, optimize dosing regimens, and expedite the translation of promising candidates into clinical trials.

By establishing a systems biology consortium for infectious diseases, the scientific community can collectively address the complexity of these diseases, leverage the power of multi-omics data, identify biomarkers and drug targets, develop predictive models, and accelerate drug development. This collaborative effort will enable researchers, clinicians, and public health professionals to gain a deeper understanding of infectious diseases, improve diagnostic and therapeutic approaches, and ultimately enhance global efforts to prevent and control infectious diseases.

6.5 Objectives and Scope of a Systems Biology Consortium

6.5.1 Facilitating Collaboration and Data Sharing

One of the primary objectives of a systems biology consortium for infectious diseases is to facilitate collaboration among researchers, institutions, and organizations. By fostering a collaborative environment, the consortium can promote the sharing of data, resources, and expertise. This collaboration enables the integration of diverse datasets, promotes interdisciplinary research, and accelerates scientific discoveries in the field of infectious diseases (Simpson et al. 2010).

6.5.2 Standardization of Experimental and Computational Methods

To ensure the reproducibility and comparability of research findings, a systems biology consortium can establish standardized protocols for experimental and computational methods. This standardization includes sample collection and processing, data generation and analysis, and model development and simulation. By adhering to common standards, researchers can effectively share and compare data, enabling more robust and reliable scientific conclusions.

6.5.3 Establishing Integrated Databases and Resources

A systems biology consortium can create integrated databases and resources that serve as centralized repositories of multi-omics data, computational models, and other relevant information. These databases provide a valuable resource for researchers to access and share data, tools, and methodologies. By consolidating data and resources, the consortium can facilitate data integration, meta-analyses, and the development of novel approaches for understanding infectious diseases.

6.5.4 Education and Training in Systems Biology

A key objective of a systems biology consortium is to promote education and training in the field of systems biology. The consortium can organize workshops, training programs, and educational resources to equip researchers, clinicians, and students with the necessary knowledge and skills in experimental and computational techniques. This educational effort ensures the wider adoption of systems biology approaches and fosters a new generation of researchers who are proficient in applying systems biology principles to infectious disease research (Cvijovic et al. 2016).

6.5.5 Collaboration with Industry and Policymakers

A systems biology consortium can collaborate with industry partners and policymakers to translate research findings into practical applications and policy recommendations. By fostering partnerships with pharmaceutical companies, diagnostic manufacturers, and public health agencies, the consortium can accelerate the development of diagnostics, therapeutics, and preventive strategies for infectious diseases. Furthermore, collaboration with policymakers ensures that research outcomes inform evidence-based policies and interventions for disease prevention, surveillance, and control (Cvijovic et al. 2016).

6.5.6 Addressing Ethical, Legal, and Social Implications

The scope of a systems biology consortium extends beyond scientific research to address the ethical, legal, and social implications of infectious disease studies. The consortium can actively engage in discussions and initiatives related to data privacy, informed consent, and equitable access to healthcare interventions. By integrating ethical considerations into its activities, the consortium ensures responsible and inclusive research practices that prioritize the welfare of individuals and communities affected by infectious diseases (O'Malley et al. 2007).

By encompassing collaboration, standardization, resource integration, education, industry collaboration, and ethical considerations, a systems biology consortium for infectious diseases can significantly advance the field. Through these objectives, the consortium promotes knowledge exchange, enhances research quality, facilitates translational applications, and contributes to the global efforts to combat infectious diseases.

6.6 Structure and Governance of a Systems Biology Consortium

6.6.1 Leadership and Management

A systems biology consortium requires effective leadership and management to ensure its successful operation. The consortium should have a designated leadership team comprising experienced scientists and professionals who can provide strategic guidance, oversee consortium activities, and foster collaboration among members. The leadership team is responsible for setting goals, defining priorities, and coordinating the overall functioning of the consortium (Mesarovic et al. 2004).

6.6.2 Membership and Collaborations

Membership in a systems biology consortium should be open to researchers, institutions, and organizations involved in infectious disease research and systems biology. The consortium should actively encourage participation from diverse disciplines, including biology, medicine, computational sciences, and public health. Collaboration with other consortia, research networks, and industry partners is essential to leverage complementary expertise, share resources, and facilitate multidisciplinary approaches to address complex research questions.

6.6.3 Funding and Sustainability

A systems biology consortium requires adequate funding to support its activities and ensure its sustainability. Funding can be obtained through various sources, including government grants, philanthropic organizations, industry partnerships, and collaborative research projects. The consortium should develop a clear and transparent funding strategy, which may involve seeking long-term funding commitments, exploring public–private partnerships, and pursuing innovative funding models to support its research, infrastructure, and training initiatives.

To ensure sustainability, the consortium should prioritize the diversification of funding sources and engage in ongoing advocacy efforts to highlight the value and impact of systems biology approaches in infectious disease research. Moreover, the consortium should establish mechanisms for financial accountability and transparent reporting to its funders and stakeholders.

6.6.4 Governance and Decision-Making

A systems biology consortium should have a well-defined governance structure to facilitate decision-making and ensure equitable representation of its members. The governance structure may include a governing board or steering committee composed of representatives from member institutions, elected or appointed by the consortium members. This board or committee should establish policies, guidelines, and procedures for consortium activities, including data sharing, intellectual property, authorship, and conflict resolution (Kuzma 2015).

The decision-making processes within the consortium should be inclusive, participatory, and transparent, allowing input from all stakeholders. Regular meetings, workshops, and communication channels should be established to facilitate dialogue, knowledge sharing, and collaborative decision-making among consortium members (O'Malley and Dupré 2005).

6.6.5 Data Sharing and Intellectual Property

Data sharing is a fundamental principle of a systems biology consortium. The consortium should promote open access and encourage its members to share data, models, and resources in a timely and responsible manner. Data sharing policies should be established to ensure proper data management, protection of intellectual property rights, and adherence to ethical and legal frameworks (Wolstencroft et al. 2011).

Intellectual property issues arising from consortium activities, such as discoveries, inventions, or novel methodologies, should be addressed through clear guidelines and agreements. The consortium should foster an environment that balances the need for scientific openness with the recognition and protection of intellectual contributions from its members.

In summary, a systems biology consortium requires effective leadership, collaboration among diverse stakeholders, sustainable funding strategies, transparent governance, and robust mechanisms for data sharing and intellectual property management. By establishing a well-structured and governed consortium, the field of systems biology in infectious diseases can thrive, leading to advancements in understanding, prevention, and treatment of these complex health challenges.

6.7 Case Studies

6.7.1 Systems Biology Approaches for Viral Infections

Viral infections pose significant threats to global health, and traditional approaches to understanding and treating these infections have often fallen short. However, the advent of systems biology has provided new avenues for studying viral infections at a holistic level, enabling a deeper understanding of viral–host interactions, the dynamics of viral replication, and the host immune response. In this section, we present case studies that highlight the application of systems biology approaches in studying viral infections and their potential for improving diagnosis, treatment, and prevention strategies.

Case Study 1: Systems Biology of Influenza A Virus Infection
Influenza A virus is a major respiratory pathogen responsible for seasonal outbreaks and occasional pandemics. Systems biology approaches have been instrumental in unraveling the complex dynamics of influenza A virus infection and guiding the development of targeted interventions (Kollmus et al. 2014).

- **Multi-omics integration:** Researchers have employed integrative analysis of multi-omics data, including genomics, transcriptomics, proteomics, and metabolomics, to comprehensively characterize the host response to influenza A virus infection. By analyzing these datasets collectively, they have identified key molecular pathways and regulatory networks involved in viral replication, host immune response, and tissue damage.
- **Dynamic modeling of viral replication:** Mathematical models have been developed to simulate the kinetics of influenza A virus replication within infected cells. These models incorporate virus–host interactions, viral protein synthesis, and the dynamics of viral shedding. Such modeling approaches have provided insights into the factors influencing viral replication and the effects of antiviral interventions.
- **Immune response profiling:** Systems biology has facilitated the characterization of the host immune response to influenza A virus infection. Integration of transcriptomic and proteomic data has revealed dysregulated immune pathways, identified key immune effectors, and highlighted potential targets for immune-modulating therapies.
- **Personalized medicine:** Systems biology approaches have contributed to the development of personalized medicine strategies for influenza treatment. By integrating patient-specific clinical data with molecular profiles, researchers have identified biomarkers that can predict patient response to antiviral drugs, enabling tailored treatment regimens.

Case Study 2: Systems Biology of Hepatitis C Virus (HCV) Infection
Hepatitis C virus (HCV) infection affects millions of individuals worldwide and can lead to chronic liver disease. Systems biology approaches have provided valuable insights into the intricate interactions between HCV and the host, paving the way for improved diagnostics, antiviral therapies, and vaccine development (Wilson and Stamataki 2012).

- **Network analysis of host–pathogen interactions:** Systems biology has facilitated the construction of comprehensive interaction networks representing the complex interactions between HCV and host proteins. These networks provide a global view of viral–host interactions, highlighting key pathways involved in viral replication, immune evasion, and host cell responses.
- **Identification of drug targets:** Through the integration of molecular data and computational modeling, systems biology has identified potential drug targets within the HCV replication cycle. By analyzing the interactions between viral

proteins and host factors, researchers have identified critical nodes that could be targeted for therapeutic intervention.

- **Biomarker discovery:** Systems biology approaches have contributed to the identification of biomarkers for HCV infection. By integrating multi-omics data from patient samples, researchers have identified genetic variants, gene expression profiles, and metabolomic signatures associated with disease progression, treatment response, and the development of hepatocellular carcinoma.
- **Vaccine development:** Systems biology has played a crucial role in vaccine development against HCV. By analyzing the host immune response and viral genetic diversity, researchers have identified conserved regions of the virus that can be targeted by vaccines. Systems biology approaches have also aided in understanding the mechanisms of vaccine-induced immune responses, informing strategies for vaccine design and optimization.

Case Study 3: Systems Biology of Human Immunodeficiency Virus (HIV) Infection

Human immunodeficiency virus (HIV) remains a major global health concern, and systems biology approaches have contributed significantly to our understanding of HIV pathogenesis, viral persistence, and host immune response (Kuri-Cervantes et al. 2016; Levy 1989).

- **Dynamics of viral replication:** Systems biology has enabled the characterization of the complex dynamics of HIV replication within the host. Mathematical models have been developed to capture the interactions between viral proteins, host immune cells, and viral reservoirs. These models have shed light on viral latency, the formation of viral reservoirs, and the effects of antiretroviral therapy.
- **Immune response dynamics:** By integrating transcriptomic, proteomic, and immunological data, systems biology has provided insights into the dynamics of the host immune response to HIV infection. This includes the identification of host factors associated with natural resistance or susceptibility to HIV, immune dysregulation during chronic infection, and potential targets for immune-based therapies.
- **Treatment optimization:** Systems biology approaches have contributed to the optimization of antiretroviral therapy for HIV. Mathematical models have been used to simulate the effects of different drug regimens, predict drug resistance development, and optimize treatment schedules. These models have guided the development of combination therapies and strategies to minimize treatment failure and drug resistance.
- **Viral latency and cure strategies:** Systems biology has been instrumental in unraveling the mechanisms of HIV latency and the persistence of the viral reservoir. By combining experimental data and computational modeling, researchers have identified potential targets for latency-reversing agents and strategies for eliminating the viral reservoirs, aiming toward an HIV cure.

These case studies highlight the power of systems biology in unraveling the complexity of viral infections. By integrating multi-omics data, computational modeling, and network analysis, systems biology approaches offer new insights into viral–host interactions, provide a foundation for personalized medicine strategies, and contribute to the development of effective interventions against viral infections.

6.7.2 Systems Biology Approaches for Bacterial Infections

Bacterial infections are a major global health concern, with the emergence of antibiotic resistance posing significant challenges to treatment strategies. Systems biology approaches have revolutionized our understanding of bacterial pathogens, their interactions with the host, and the complex mechanisms underlying antibiotic resistance. In this section, we present case studies that highlight the application of systems biology approaches in studying bacterial infections and their potential for improving diagnosis, treatment, and prevention strategies.

Case Study 1: Systems Biology of Methicillin-Resistant Staphylococcus aureus (MRSA) Infection

Methicillin-resistant *Staphylococcus aureus* (MRSA) is a notorious bacterial pathogen responsible for a wide range of infections, with limited treatment options due to antibiotic resistance. Systems biology approaches have provided valuable insights into the pathogenesis of MRSA and have guided the development of new therapeutic approaches (Gordon and Lowy 2008; Hiramatsu et al. 2001).

- **Network analysis of host–pathogen interactions:** Systems biology has facilitated the construction of comprehensive interaction networks representing the complex interactions between MRSA and host proteins. These networks have revealed key virulence factors, regulatory mechanisms, and pathways involved in MRSA infection. By understanding these interactions, researchers can identify potential targets for therapeutic intervention.
- **Omics profiling:** Integration of genomics, transcriptomics, proteomics, and metabolomics data has allowed researchers to comprehensively characterize MRSA infections. By analyzing the global changes in gene expression, protein abundance, and metabolic pathways, systems biology has provided insights into the adaptation strategies of MRSA to different host environments and antibiotic stress.
- **Antibiotic resistance mechanisms:** Systems biology approaches have been instrumental in understanding the mechanisms of antibiotic resistance in MRSA. By studying the genomic variations, gene expression patterns, and metabolic adaptations associated with antibiotic resistance, researchers have identified key genes and pathways involved in resistance mechanisms. This knowledge aids in the development of novel therapeutic strategies to combat MRSA infections.

Case Study 2: Systems Biology of Mycobacterium Tuberculosis Infection
Mycobacterium tuberculosis, the causative agent of tuberculosis (TB), is a global health threat. Systems biology approaches have played a crucial role in elucidating the complex interactions between *M. tuberculosis* and the host immune system, as well as the mechanisms underlying drug resistance (Akinola 2013).

- **Host–pathogen interactions:** Systems biology has facilitated the study of host–pathogen interactions during *M. tuberculosis* infection. By integrating transcriptomic, proteomic, and immunological data, researchers have identified key molecular pathways involved in immune responses, granuloma formation, and bacterial persistence. This knowledge aids in understanding the dynamics of infection and identifying potential targets for intervention.
- **Latency and reactivation:** *M. tuberculosis* can persist in a dormant state, leading to latent TB infection and subsequent reactivation. Systems biology approaches have provided insights into the mechanisms underlying latency and reactivation. By analyzing gene expression profiles and regulatory networks, researchers have identified molecular markers and pathways associated with latency, offering new avenues for therapeutic interventions.
- **Drug resistance and treatment optimization:** Systems biology has contributed to the understanding of drug resistance mechanisms in *M. tuberculosis*. By analyzing genomic data, researchers have identified genetic mutations associated with drug resistance. Systems biology approaches have also been used to optimize treatment regimens, considering the dynamics of bacterial growth, drug pharmacokinetics, and host immune response, leading to improved treatment outcomes.

Case Study 3: Systems Biology of Escherichia coli Infection
Escherichia coli (*E. coli*) is a versatile bacterial pathogen that can cause a range of infections, including urinary tract infections, bloodstream infections, and gastrointestinal infections. Systems biology approaches have provided insights into the molecular mechanisms of *E. coli* pathogenesis and antibiotic resistance (Stincone et al. 2011).

- **Regulatory networks and virulence factors:** Systems biology has helped unravel the regulatory networks that control virulence gene expression in *E. coli*. By integrating genomic, transcriptomic, and proteomic data, researchers have identified key regulators, signal transduction pathways, and virulence factors. This knowledge has informed the development of therapeutics targeting *E. coli* infections.
- **Antibiotic resistance and resistance mechanisms:** Systems biology approaches have contributed to understanding the mechanisms of antibiotic resistance in *E. coli*. By analyzing genomic data, researchers have identified resistance genes and mutations associated with antibiotic resistance. Systems biology also enables the study of resistance mechanisms, such as efflux pumps, enzymatic modifica-

tions, and target site alterations, providing insights for the development of strategies to combat antibiotic resistance.

- **Metabolic adaptation:** Systems biology has shed light on the metabolic adaptations of *E. coli* during infection. By studying the metabolic pathways and fluxes in infected host tissues, researchers have identified key metabolic adaptations that promote bacterial survival and pathogenesis. Understanding these adaptations can guide the development of therapeutic interventions targeting bacterial metabolism.

These case studies demonstrate the power of systems biology in unraveling the complexity of bacterial infections. By integrating multi-omics data, network analysis, and computational modeling, systems biology approaches offer new insights into bacterial–host interactions, mechanisms of antibiotic resistance, and the development of effective strategies for diagnosis, treatment, and prevention of bacterial infections.

6.7.3 Systems Biology Approaches for Parasitic Infections

Parasitic infections, caused by various parasites such as *Plasmodium*, *Trypanosoma*, and *Leishmania*, continue to be a major global health concern, particularly in resource-limited regions. Systems biology approaches have provided valuable insights into the complex host–parasite interactions, parasite biology, and host immune responses, opening new avenues for understanding and combating parasitic infections. In this section, we present case studies that highlight the application of systems biology approaches in studying parasitic infections and their potential for improving diagnosis, treatment, and prevention strategies.

Case Study 1: Systems Biology of Malaria

Malaria, caused by *Plasmodium* parasites, is a significant cause of morbidity and mortality worldwide. Systems biology approaches have significantly contributed to our understanding of the complex interactions between *Plasmodium* parasites and the human host, as well as the immune responses elicited during infection (Cai et al. n.d.; Fraunholz 2005).

- **Genomic and transcriptomic analysis:** Systems biology has enabled the analysis of genomic and transcriptomic data from both the human host and *Plasmodium* parasites. By integrating these datasets, researchers have identified key genetic variants, gene expression patterns, and regulatory networks associated with malaria susceptibility, disease severity, and drug resistance. These findings have implications for the development of new antimalarial drugs and vaccine candidates.
- **Immune response profiling:** Systems biology approaches have aided in the characterization of host immune responses to *Plasmodium* infection. Through the integration of transcriptomic, proteomic, and immunological data, research-

ers have identified immune pathways, cell types, and molecules involved in the immune response against malaria. This knowledge has guided the development of novel immunotherapeutic strategies and vaccine candidates.

- **Systems pharmacology:** Systems biology has been used to study the interactions between antimalarial drugs and the parasite. By integrating pharmacokinetic data, genomic information, and mathematical modeling, researchers have developed models to predict drug efficacy, identify mechanisms of drug resistance, and optimize treatment regimens for malaria.

Case Study 2: Systems Biology of Trypanosoma cruzi Infection

Trypanosoma cruzi, the causative agent of Chagas disease, is a parasitic infection prevalent in Latin America. Systems biology approaches have provided insights into the host–parasite interactions, parasite biology, and the mechanisms of pathogenesis associated with *T. cruzi* infection (Cardoso et al. 2016).

- **Host–parasite interactions:** Systems biology has facilitated the study of host–parasite interactions during *T. cruzi* infection. By integrating multi-omics data, researchers have identified key molecular pathways, regulatory networks, and virulence factors involved in parasite invasion, intracellular survival, and immune evasion. This knowledge contributes to the understanding of disease progression and identifies potential targets for therapeutic intervention.
- **Immune response dynamics:** Systems biology approaches have aided in the characterization of the host immune response to *T. cruzi* infection. By analyzing transcriptomic and proteomic data from infected individuals, researchers have identified dysregulated immune pathways, immune cell populations, and cytokine profiles associated with disease outcome and progression. This information has implications for the development of immunomodulatory therapies and vaccine strategies.
- **Drug development and resistance:** Systems biology has played a role in the discovery and development of new drugs for Chagas disease. By integrating genomic and proteomic data from *T. cruzi*, researchers have identified potential drug targets and evaluated drug efficacy. Additionally, systems biology approaches have aided in understanding the mechanisms of drug resistance, informing strategies for combating resistance, and optimizing treatment protocols.

Case Study 3: Systems Biology of Leishmania Infection

Leishmaniasis, caused by *Leishmania* parasites, affects millions of people worldwide. Systems biology approaches have contributed to our understanding of the complex interactions between *Leishmania* parasites and the host immune system, as well as the mechanisms underlying parasite survival and pathogenesis (Van Assche et al. 2011).

- **Transcriptomic and proteomic analysis:** Systems biology has enabled the analysis of transcriptomic and proteomic data from *Leishmania* parasites and infected host tissues. By integrating these datasets, researchers have identified key genes, proteins, and pathways involved in parasite survival, virulence, and drug resistance. This knowledge aids in the development of novel therapeutic strategies and the identification of potential vaccine candidates.
- **Host immune response profiling:** Systems biology approaches have provided insights into the host immune response during *Leishmania* infection. By analyzing immune cell populations, cytokine profiles, and signaling pathways, researchers have identified immune factors associated with disease severity, immune evasion mechanisms employed by the parasite, and potential targets for immunomodulatory interventions (Costa-da-Silva et al. 2022).
- **Metabolic adaptation:** Systems biology has shed light on the metabolic adaptations of *Leishmania* parasites during infection. By studying the metabolic pathways and metabolic fluxes within the parasite and the host, researchers have identified essential metabolic processes for parasite survival, potential drug targets, and metabolic vulnerabilities that could be exploited for therapeutic purposes.

These case studies demonstrate the value of systems biology in unraveling the complex interactions between parasites and their hosts. By integrating multi-omics data, computational modeling, and network analysis, systems biology approaches provide valuable insights into parasite biology, host immune responses, and the development of new diagnostic, treatment, and prevention strategies for parasitic infections.

6.8 Challenges and Future Perspectives

6.8.1 Data Integration and Harmonization

One of the key challenges in systems biology research for infectious diseases is the integration and harmonization of diverse data types from multiple sources. Different omics technologies generate large volumes of data with varying formats and standards, making it challenging to combine and analyze them effectively. Data integration requires the development of standardized protocols, ontologies, and data sharing platforms to enable seamless integration, interoperability, and harmonization of data across different studies and research groups (Hwang et al. 2005).

Furthermore, the integration of clinical data with omics data presents additional challenges due to variations in data collection methods, patient cohorts, and ethical considerations. Efforts should be made to establish data sharing frameworks and promote collaboration between researchers, clinicians, and data scientists to overcome these challenges and facilitate comprehensive and meaningful data integration in systems biology research for infectious diseases.

6.8.2 Ethics and Privacy Concerns

As systems biology approaches rely on the analysis of personal health data, including genomic information, ethical and privacy concerns are of utmost importance. Safeguarding patient privacy, ensuring informed consent, and addressing data ownership and sharing policies are critical considerations in systems biology research. Clear guidelines and regulations should be in place to protect patient privacy, maintain data security, and ensure responsible data handling practices.

Moreover, the potential for unintended consequences arising from the use of personal health data should be carefully evaluated. Efforts should be made to raise awareness among researchers, clinicians, and the public about the ethical implications of systems biology research and to engage in transparent and inclusive discussions about the use of personal health data for the advancement of infectious disease research.

6.8.3 Regulatory and Policy Implications

The application of systems biology approaches in infectious disease research may require re-evaluation and adaptation of existing regulatory frameworks and policies. Regulatory agencies and policymakers should be engaged in discussions to ensure that systems biology research meets the necessary standards for safety, efficacy, and ethical considerations (Trump 2017).

The development of new therapeutic interventions, such as personalized treatments and repurposing of existing drugs, may require the establishment of innovative regulatory pathways to expedite their translation from bench to bedside. Collaboration between researchers, regulatory bodies, and industry partners is crucial to ensure that regulatory frameworks keep pace with scientific advancements and enable the timely development and approval of innovative therapies.

6.8.4 Technological Advances and Emerging Opportunities

Technological advancements continue to shape the field of systems biology and offer new opportunities for infectious disease research. Improvements in high-throughput omics technologies, such as next-generation sequencing and single-cell analysis, microfluidics provide a more comprehensive and detailed understanding of host–pathogen interactions. Additionally, advances in computational modeling, artificial intelligence, and machine learning offer powerful tools for data analysis, predictive modeling, and simulation (Kitano 2002; Sun and Lu 2019).

Emerging opportunities also arise from the integration of systems biology with other fields, such as network biology, immunology, and bioinformatics. Interdisciplinary collaborations can lead to new insights and innovative approaches for studying infectious diseases, identifying therapeutic targets, and designing personalized interventions.

In summary, while systems biology holds great promise for advancing our understanding and management of infectious diseases, several challenges need to be addressed. Data integration, ethical considerations, regulatory frameworks, and technological advancements are crucial aspects that require attention. By proactively addressing these challenges and fostering collaborations among researchers, clinicians, policymakers, and industry partners, the field of systems biology can continue to evolve and contribute to the prevention, diagnosis, and treatment of infectious diseases.

6.9 Conclusion

6.9.1 Summary of the Systems Biology Consortium Concept

The concept of a systems biology consortium for infectious diseases represents a collaborative and integrative framework aimed at addressing the complex challenges posed by infectious diseases. The consortium brings together researchers, clinicians, industry partners, and policymakers with the goal of applying systems biology approaches to gain a comprehensive understanding of infectious diseases and accelerate the development of effective interventions. At the core of the consortium is the integration of multi-omics data, computational modeling, and advanced analytical techniques. By combining genomics, transcriptomics, proteomics, and metabolomics data, researchers can unravel the intricate molecular mechanisms underlying infections, identify key biomarkers and therapeutic targets, and decipher the complex interactions between pathogens and the host immune system (Kitano 2002).

The consortium emphasizes collaboration, data sharing, and standardization. It encourages researchers and institutions to work together, pooling their expertise, resources, and data, thereby breaking down silos and facilitating a more holistic approach to infectious disease research. By sharing data, insights, and methodologies, researchers can accelerate discoveries, optimize translational outcomes, and overcome the limitations of individual studies.

The potential impact and benefits of a systems biology consortium are far-reaching. By comprehensively understanding the molecular mechanisms of infectious diseases, researchers can develop more targeted interventions, personalized treatment strategies, and optimized preventive measures. The consortium's collaborative nature ensures that research findings can be rapidly translated into clinical practice, leading to improved diagnostics, therapeutics, and public health policies.

To establish a systems biology consortium, collaborative efforts from multiple stakeholders are required. Researchers, clinicians, industry partners, policymakers, and funding agencies must come together to foster interdisciplinary collaborations, secure sustained funding, promote data sharing and standardization, address ethical and privacy concerns, engage with regulatory bodies, and invest in education and training programs.

By collectively embracing the concept of a systems biology consortium for infectious diseases, the global health community can revolutionize our approach to combating these complex challenges. Through collaborative efforts, data integration, and the application of advanced analytical techniques, we can gain a deeper understanding of infectious diseases, identify novel strategies for prevention and treatment, and ultimately improve the health and well-being of individuals and communities worldwide.

6.9.2 Potential Impact and Benefits

The establishment of a systems biology consortium for infectious diseases holds immense potential for transforming our approach to combating these global health challenges. By harnessing the power of systems biology and fostering collaborative efforts, the consortium can have a significant impact on multiple fronts and bring about a range of benefits (Kumar et al. 2015).

1. **Comprehensive understanding of infectious diseases:** Systems biology approaches enable a holistic understanding of infectious diseases by integrating multi-omics data and computational modeling. This comprehensive view of the molecular mechanisms underlying infections allows researchers to identify key factors influencing pathogenesis, host responses, and disease outcomes. Such knowledge is invaluable for developing targeted interventions, personalized treatment strategies, and optimized preventive measures.
2. **Accelerated discovery and translational outcomes:** The collaborative nature of the consortium facilitates the sharing of data, resources, and expertise, enabling researchers to accelerate discoveries and enhance translational outcomes. By breaking down research silos and promoting collaboration, the consortium can overcome barriers to progress and expedite the translation of research findings into clinical practice. This can lead to the timely development and implementation of improved diagnostics, therapeutics, and preventive strategies.
3. **Optimization of public health policies:** The consortium's efforts can have a direct impact on public health policies and interventions. By engaging policymakers and regulatory bodies, researchers can contribute valuable insights to evidence-based decision-making processes. This collaboration ensures that public health strategies align with the latest scientific advancements and are tailored to the specific challenges posed by infectious diseases. The optimized policies can enhance disease surveillance, outbreak control, and resource allocation, ultimately improving population health outcomes (Chen and Snyder 2012).
4. **Identification of biomarkers and therapeutic targets:** Systems biology approaches aid in the identification of key biomarkers and therapeutic targets for infectious diseases. By analyzing multi-omics data and integrating clinical outcomes, researchers can identify molecular signatures associated with disease progression, treatment response, and patient outcomes. This knowledge can

guide the development of novel diagnostics, targeted therapies, and precision medicine approaches that improve patient outcomes and minimize adverse effects (Weston and Hood 2004).

5. **Innovation and advancements in research and technology:** The consortium stimulates innovation and advancements in research and technology. By promoting collaboration between researchers, clinicians, and industry partners, it fosters the development and application of cutting-edge technologies, computational models, and analytical tools. This enables the refinement and expansion of systems biology approaches, leading to more sophisticated data analysis, predictive modeling, and simulation techniques. These advancements open up new opportunities for studying infectious diseases and designing innovative interventions.

Overall, the potential impact and benefits of a systems biology consortium for infectious diseases are vast. From a deeper understanding of disease mechanisms to accelerated discovery and translation, optimized public health policies, and the identification of biomarkers and therapeutic targets, the consortium has the potential to significantly improve global health outcomes. By embracing collaborative efforts, data sharing, and interdisciplinary research, we can harness the full potential of systems biology to combat infectious diseases and improve the health and well-being of individuals and communities worldwide.

6.9.3 Call to Action: Collaborative Efforts for Global Health

The establishment of a systems biology consortium for infectious diseases requires collaborative efforts from researchers, clinicians, industry partners, policymakers, and funding agencies (Eckhardt et al. 2020; Meyer et al. 2011). To realize the full potential of systems biology in combating infectious diseases, the following actions are crucial:

1. **Foster collaboration:** Encourage interdisciplinary collaboration and establish partnerships among researchers, institutions, and industry stakeholders to share expertise, resources, and data.
2. **Support funding initiatives:** Secure sustained funding from government agencies, philanthropic organizations, and industry partners to support research, infrastructure, and training programs in systems biology for infectious diseases.
3. **Promote data sharing and standardization:** Develop data sharing policies, establish common data formats and standards, and create centralized databases to facilitate seamless integration and harmonization of multi-omics data.
4. **Address ethical and privacy concerns:** Establish guidelines and protocols to protect patient privacy, ensure informed consent, and address ethical considerations related to data sharing and use of personal health information.
5. **Engage policymakers and regulatory bodies:** Collaborate with policymakers and regulatory agencies to adapt existing frameworks, establish innovative regu-

latory pathways, and facilitate the translation of systems biology research into clinical practice.
6. **Invest in education and training:** Develop educational programs, workshops, and training initiatives to equip researchers, clinicians, and students with the necessary skills and knowledge in systems biology approaches for infectious diseases.

By collectively taking these actions, the global health community can establish a robust and sustainable systems biology consortium that drives transformative advancements in understanding, preventing, and treating infectious diseases.

In conclusion, a systems biology consortium for infectious diseases holds immense potential for revolutionizing our approach to combating these global health challenges. By embracing collaboration, data sharing, and interdisciplinary research, we can harness the power of systems biology to gain deeper insights, develop targeted interventions, and ultimately improve the health and well-being of individuals and communities worldwide.

References

Aderem A, Adkins JN, Ansong C, Galagan J, Kaiser S, Korth MJ, Lynn Law G, McDermott JG, Proll SC, Rosenberger C, Schoolnik G, Katze MG (2011) A systems biology approach to infectious disease research: innovating the pathogen-host research paradigm. MBio 2(1):e00325–e00310. https://doi.org/10.1128/mBio.00325-10

Akinola RO (2013) A systems level comparison of mycobacterium tuberculosis, mycobacterium leprae and mycobacterium smegmatis based on functional interaction network analysis. J Bacteriol Parasitol 04:04. https://doi.org/10.4172/2155-9597.1000173

Albert R (2007) Network inference, analysis, and Modeling in systems biology. Plant Cell 19(11):3327–3338. https://doi.org/10.1105/tpc.107.054700

Anne Crock E (2017) HIV and AIDS: an overview of the current issues, treatment and prevention. Nurs Stand 32(15):51–63. https://doi.org/10.7748/ns.2017.e11045

Anosike JC, Keke IR, Uwaezuoke JC, Anozie JC, Obiukwu CE, Nwoke BEB, Amajuoyi OU (2005) Prevalence and distribution of ringworm infections in primary school children in parts of eastern, Nigeria. J Appl Sci Environ Manag 9(3):21–26. https://doi.org/10.4314/jasem.v9i3.17347

Burza S, Croft SL, Boelaert M (2018) Leishmaniasis. Lancet (London, England) 392(10151):951–970. https://doi.org/10.1016/S0140-6736(18)31204-2

Cai H, Zhou Z, Jianying G, Wang Y (n.d.) Comparative genomics and systems biology of malaria parasites plasmodium. Curr Bioinforma 7(4):478–489

Cardoso MS, Reis-Cunha JL, Bartholomeu DC (2016) Evasion of the immune response by Trypanosoma Cruzi during acute infection. Front Immunol 6:659

Chen R, Snyder M (2012) Systems biology: personalized medicine for the future? Curr Opin Pharmacol 12(5):623–628. https://doi.org/10.1016/j.coph.2012.07.011

Costa-da-Silva AC, de Oliveira D, Nascimento JRM, Ferreira KG-P, Freire-de-Lima L, Morrot A, Decote-Ricardo D, Filardy AA, Freire-de-Lima CG (2022) Immune responses in leishmaniasis: an overview. Trop Med Infect Dis 7(4):54. https://doi.org/10.3390/tropicalmed7040054

Cvijovic M, Höfer T, Aćimović J, Alberghina L, Almaas E, Besozzi D, Blomberg A, Bretschneider T, Cascante M, Collin O, de Atauri P, Depner C, Dickinson R, Dobrzynski M, Fleck C, Garcia-Ojalvo J, Gonze D, Hahn J, Hess HM, Hollmann S, Krantz M, Kummer U, Lundh T, Martial G, Martins V, dos Santos A, Mauer-Oberthür BR, Skene B, Stalidzans E, Stelling J, Teusink B, Workman CT, Hohmann S (2016) Strategies for structuring interdisciplinary education in sys-

tems biology: an European perspective. Npj Syst Biol Appl 2(1):1–7. https://doi.org/10.1038/npjsba.2016.11
Dangi T, Jain A (2012) Influenza virus: a brief overview. Proc Natl Acad Sci India Section B 82(1):111–121. https://doi.org/10.1007/s40011-011-0009-6
Davidson SB, Overton C, Buneman P (1995) Challenges in integrating biological data sources. J Comput Biol 2(4):557–572. https://doi.org/10.1089/cmb.1995.2.557
Dix A, Vlaic S, Guthke R, Linde J (2016) Use of systems biology to decipher host-pathogen interaction networks and predict biomarkers. Clin Microbiol Infect 22(7):600–606. https://doi.org/10.1016/j.cmi.2016.04.014
Dräger A, Palsson BØ (2014) Improving collaboration by standardization efforts in systems biology. Front Bioeng Biotechnol 2:61. https://doi.org/10.3389/fbioe.2014.00061
Durek P, Walther D (2008) The integrated analysis of metabolic and protein interaction networks reveals novel molecular organizing principles. BMC Syst Biol 2:100. https://doi.org/10.1186/1752-0509-2-100
Eckhardt M, Hultquist JF, Kaake RM, Hüttenhain R, Krogan NJ (2020) A systems approach to infectious disease. Nat Rev Genet 21(6):339–354. https://doi.org/10.1038/s41576-020-0212-5
Ellebedy AH, Ahmed R (2016) Chapter 15 - antiviral vaccines: challenges and advances. In: Bloom BR, Lambert P-H (eds) The vaccine book (second edition). Academic Press, pp 283–310
Escobedo AA, Cimerman S (2007) Giardiasis: a pharmacotherapy review. Expert Opin Pharmacother 8(12):1885–1902. https://doi.org/10.1517/14656566.8.12.1885
Fraunholz MJ (2005) Systems biology in malaria research. Trends Parasitol 21(9):393–395. https://doi.org/10.1016/j.pt.2005.07.007
Germain RN, Meier-Schellersheim M, Nita-Lazar A, Fraser IDC (2011) Systems biology in immunology: a computational modeling perspective. Annu Rev Immunol 29(1):527–585. https://doi.org/10.1146/annurev-immunol-030409-101317
Gold HS, Moellering RC (1996) Antimicrobial-drug resistance. N Engl J Med 335(19):1445–1453. https://doi.org/10.1056/NEJM199611073351907
Gonty AA, Page LR (1977) Aspergillosis of the maxillary sinus. Review of the literature and report of a case. Oral Surg Oral Med Oral Pathol 43(3):350–356. https://doi.org/10.1016/0030-4220(77)90320-6
Gordon RJ, Lowy FD (2008) Pathogenesis of methicillin-resistant staphylococcus aureus infection. Clin Infect Dis 46(Supplement_5):S350–59. https://doi.org/10.1086/533591
Gryseels B, Polman K, Clerinx J, Kestens L (2006) Human schistosomiasis. Lancet 368(9541):1106–1118. https://doi.org/10.1016/S0140-6736(06)69440-3
Hani U, Shivakumar HG, Vaghela R, Riyaz A, Osmani M, Shrivastava A (2015) Candidiasis: a fungal infection—current challenges and progress in prevention and treatment. Infect Disord Drug Targets 15(1):42–52. https://doi.org/10.2174/1871526515666150320162036
Hiramatsu K, Cui L, Kuroda M, Ito T (2001) The emergence and evolution of methicillin-resistant staphylococcus aureus. Trends Microbiol 9(10):486–493. https://doi.org/10.1016/S0966-842X(01)02175-8
Hotez PJ, Brooker S, Bethony JM, Bottazzi ME, Loukas A, Xiao S (2004) Hookworm infection. N Engl J Med 351(8):799–807. https://doi.org/10.1056/NEJMra032492
Hwang D, Rust AG, Ramsey S, Smith JJ, Leslie DM, Weston AD, de Atauri P, Aitchison JD, Hood L, Siegel AF, Bolouri H (2005) A data integration methodology for systems biology. Proc Natl Acad Sci 102(48):17296–17301. https://doi.org/10.1073/pnas.0508647102
Jarmusch AK, Pirro V, Kerian KS, Graham Cooks R (2014) Detection of strep throat causing bacterium directly from medical swabs by touch spray-mass spectrometry. Analyst 139(19):4785–4789. https://doi.org/10.1039/c4an00959b
Jung SY, Lee SH, Lee SY, Yang S, Noh H, Chung EK, Lee JI (2017) Antimicrobials for the treatment of drug-resistant Acinetobacter baumannii pneumonia in critically ill patients: a systemic review and Bayesian network meta-analysis. Crit Care (London, England) 21(1):319. https://doi.org/10.1186/s13054-017-1916-6
Karahalil B (2016) Overview of systems biology and omics technologies. Curr Med Chem 23(37):4221–4230. https://doi.org/10.2174/0929867323666160926150617

Kelley R (2007) Systems biology in practice: concepts, implementation, and application (Klipp, E., et al; 2005) [book review]. IEEE Eng Med Biol Mag 26(6):9–54. https://doi.org/10.1109/EMB.2007.907086

Kitano H (2002) Systems biology: a brief overview. Science 295(5560):1662–1664. https://doi.org/10.1126/science.1069492

Kollmus H, Wilk E, Schughart K (2014) Systems biology and systems genetics—novel innovative approaches to study host–pathogen interactions during influenza infection. Curr Opin Virol 6:47–54. https://doi.org/10.1016/j.coviro.2014.03.008

Kumar A, Pathak RK, Gupta SM, Gaur VS, Pandey D (2015) Systems biology for smart crops and agricultural innovation: filling the gaps between genotype and phenotype for complex traits linked with robust agricultural productivity and sustainability. OMICS 19(10):581–601. https://doi.org/10.1089/omi.2015.0106

Kuri-Cervantes L, Fourati S, Canderan G, Sekaly R-P (2016) Systems biology and the quest for correlates of protection to guide the development of an HIV vaccine. Curr Opin Immunol 41:91–97. https://doi.org/10.1016/j.coi.2016.06.010

Kuzma J (2015) Translational governance research for synthetic biology. J Responsible Innov 2(1):109–112. https://doi.org/10.1080/23299460.2014.1002055

Levy JA (1989) Pathogenesis of human immunodeficiency virus infection. Ann N Y Acad Sci 567(1):58–68. https://doi.org/10.1111/j.1749-6632.1989.tb16459.x

Medyukhina A, Timme S, Mokhtari Z, Figge MT (2015) Image-based systems biology of infection. Cytometry A 87(6):462–470. https://doi.org/10.1002/cyto.a.22638

Megason SG, Fraser SE (2007) Imaging in systems biology. Cell 130(5):784–795. https://doi.org/10.1016/j.cell.2007.08.031

Meisner M, Reif DM (2015) Chapter 5—computational methods used in systems biology. In: Fry RC (ed) Systems biology in toxicology and environmental health. Academic Press, Boston, pp 85–115

Mesarovic MD, Sreenath SN, Keene JD (2004) Search for organising principles: understanding in systems biology. Syst Biol 1(1):19–27. https://doi.org/10.1049/sb:20045010

Meyer P, Alexopoulos LG, Bonk T, Califano A, Cho CR, de la Fuente A, de Graaf D, Hartemink AJ, Hoeng J, Ivanov NV, Koeppl H, Linding R, Marbach D, Norel R, Peitsch MC, Jeremy Rice J, Royyuru A, Schacherer F, Sprengel J, Stolle K, Vitkup D, Stolovitzky G (2011) Verification of systems biology research in the age of collaborative competition. Nat Biotechnol 29(9):811–815. https://doi.org/10.1038/nbt.1968

Morens DM, Folkers GK, Fauci AS (2004) The challenge of emerging and re-emerging infectious diseases. Nature 430(6996):242–249. https://doi.org/10.1038/nature02759

Ng A, Bursteinas B, Gao Q, Mollison E, Zvelebil M (2006) Resources for integrative systems biology: from data through databases to networks and dynamic system models. Brief Bioinform 7(4):318–330. https://doi.org/10.1093/bib/bbl036

O'Malley MA, Calvert J, Dupré J (2007) Response to open peer commentaries on 'the study of socioethical issues in systems biology. Am J Bioeth 7(4):W7–9. https://doi.org/10.1080/15265160701307720

O'Malley MA, Dupré J (2005) Fundamental issues in systems biology. BioEssays 27(12):1270–1276. https://doi.org/10.1002/bies.20323

Oberg AL, Kennedy RB, Li P, Ovsyannikova IG, Poland GA (2011) Systems biology approaches to new vaccine development. Curr Opin Immunol 23(3):436–443. https://doi.org/10.1016/j.coi.2011.04.005

Oh SJ, Choi YK, Shin OS (2018) Systems biology-based platforms to accelerate research of emerging infectious diseases. Yonsei Med J 59(2):176–186. https://doi.org/10.3349/ymj.2018.59.2.176

Olin SJ, Bartges JW (2015) Urinary tract infections: treatment/comparative therapeutics. Vet Clin North Am Small Anim Pract 45(4):721–746. https://doi.org/10.1016/j.cvsm.2015.02.005

Pallavi K, Sravani D, Durga S et al (2018) Hepatitis-A: review on current and future scenario. J In Silico In Vitro Pharmacol 3(1) Scitechnol Biosoft Pvt. Ltd. https://doi.org/10.21767/2469-6692.100015

Peyvandipour A, Saberian N, Shafi A, Donato M, Draghici S (2018) A novel computational approach for drug repurposing using systems biology. Bioinformatics 34(16):2817–2825. https://doi.org/10.1093/bioinformatics/bty133

Pound MW, Townsend ML, Dimondi V, Wilson D, Drew RH (2011) Overview of treatment options for invasive fungal infections. Med Mycol 49(6):561–580. https://doi.org/10.3109/13693786.2011.560197

Russell DG (2011) Mycobacterium tuberculosis and the intimate discourse of a chronic infection. Immunol Rev 240(1):252–268. https://doi.org/10.1111/j.1600-065X.2010.00984.x

Scarpino SV, Petri G (2019) On the predictability of infectious disease outbreaks. Nat Commun 10(1):898. https://doi.org/10.1038/s41467-019-08616-0

Schröder I (2020) COVID-19: a risk assessment perspective. ACS Chem Health Saf 27(3):160–169. https://doi.org/10.1021/acs.chas.0c00035

Simpson A, Slaymaker M, Gavaghan D (2010) On the secure sharing and aggregation of data to support systems biology research. In: Lambrix P, Kemp G (eds) Data integration in the life sciences. Vol. 6254, lecture notes in computer science. Springer Berlin Heidelberg, Berlin, Heidelberg, pp 58–73

Stincone A, Daudi N, Rahman AS, Antczak P, Henderson I, Cole J, Johnson MD, Lund P, Falciani F (2011) A systems biology approach sheds new light on Escherichia Coli acid resistance. Nucleic Acids Res 39(17):7512–7528. https://doi.org/10.1093/nar/gkr338

Stumpf MP, Robertson BD, Duncan K, Young DB (2007) Systems biology and its impact on anti-infective drug development. Prog Drug Res. Fortschritte Der Arzneimittelforschung Progres Des Recherches Pharmaceutiques 64(1):3–20. https://doi.org/10.1007/978-3-7643-7567-6_1

Sun G, Lu H (2019) Recent advances in microfluidic techniques for systems biology. Anal Chem 91(1):315–329. https://doi.org/10.1021/acs.analchem.8b04757

Talapko J, Škrlec I, Alebić T, Jukić M, Včev A (2019) Malaria: the past and the present. Microorganisms 7(6):179. https://doi.org/10.3390/microorganisms7060179

Trump BD (2017) Synthetic biology regulation and governance: lessons from TAPIC for the United States, European Union, and Singapore. Health Policy 121(11):1139–1146. https://doi.org/10.1016/j.healthpol.2017.07.010

Van Assche T, Deschacht M, da Luz RAI, Maes L, Cos P (2011) Leishmania–macrophage interactions: insights into the redox biology. Free Radic Biol Med 51(2):337–351. https://doi.org/10.1016/j.freeradbiomed.2011.05.011

Vivas R, Barbosa AAT, Dolabela SS, Jain S (2019) Multidrug-resistant bacteria and alternative methods to control them: an overview. Microb Drug Resist (Larchmont, NY) 25(6):890–908. https://doi.org/10.1089/mdr.2018.0319

Wang K, Lee I, Carlson G, Hood L, Galas D (2010) Systems biology and the discovery of diagnostic biomarkers. Dis Markers 28(4):199–207. https://doi.org/10.3233/DMA-2010-0697

Weston AD, Hood L (2004) Systems biology, proteomics, and the future of health care: toward predictive, preventative, and personalized medicine. J Proteome Res 3(2):179–196. https://doi.org/10.1021/pr0499693

Wilson GK, Stamataki Z (2012) In vitro systems for the study of hepatitis C virus infection. Int J Hepatol 2012:e292591. https://doi.org/10.1155/2012/292591

Wolstencroft K, Owen S, du Preez F, Krebs O, Mueller W, Goble C, Snoep JL (2011) Chapter twenty-nine - the SEEK: a platform for sharing data and models in systems biology. In: Jameson D, Verma M, Westerhoff HV (eds) Methods in enzymology. Vol. 500, methods in systems biology. Academic Press, pp 629–655

Wright J, Wagner A (2008) The systems biology research tool: evolvable open-source software. BMC Syst Biol 2(1):55. https://doi.org/10.1186/1752-0509-2-55

Systems Biology and Hospital-Associated Infections

7

Sreejita Ghosh, Dibyajit Lahiri, Moupriya Nag, Rina Rani Ray, and Debasmita Bhattacharya

Abstract

Hospital-associated infections (HAIs) have become a major threat and lead to almost over 90,000 deaths globally per year. Advanced methods of diagnosis and detection would help healthcare workers rapidly determine the most efficient therapeutic process, thereby reducing the use of broad-spectrum antibiotics. Most of the time, HAIs are not detected early, which leads to delays in administering effective treatments. From both clinical and data science points of view, HAI management is a highly complex area. The main aim of integrating systems biology-based approaches such as artificial intelligence (AI) and artificial neural network (ANN), machine learning (ML), big data analysis, and other bioinformatic-based modeling is to regulate the structural and functional system of the biological arrangements. The introduction of systems biology-associated diagnostic, treatment, management, and monitoring methods has led to the unraveling of the underlying genetic associations among the HAI- causative agents and the host. This chapter presents an overview of various types of systems biology-based approaches in controlling the advent of HAIs and how they can be used to monitor and predict the risk of HAI development in hospitalized patients with other comorbidities.

S. Ghosh · R. R. Ray (✉)
Microbiology Research Laboratory, Department of Biotechnology, Maulana Abul Kalam Azad University of Technology, Haringhata, West Bengal, India

D. Lahiri · M. Nag
University of Engineering and Management, Kolkata, Kolkata, West Bengal, India

D. Bhattacharya (✉)
Department of Basic Science and Humanities, Institute of Engineering and Management, University of Engineering and Management, Kolkata, West Bengal, India
e-mail: debasmita.bhattacharya@iem.edu.in

S. Joshi et al. (eds.), *Systems Biology Approaches: Prevention, Diagnosis, and Understanding Mechanisms of Complex Diseases*,
https://doi.org/10.1007/978-981-99-9462-5_7

Keywords

AMR · Systems Biology · Computational technique · AI

7.1 Introduction

Nosocomial pathogenic microbes cause hospital-associated infections (HAI). HAIs not only affect developing countries but also have profound impacts on highly developed countries in the world. They are mainly device-mediated diseases associated with both sessile and planktonic microbial cells (Dadi et al. 2021). Systems biology assures to decode the multiple-scale physiological networks between the microbial cells causing HAIs. The dynamic and structural characteristics of such networks govern and determine the phenotype of a cell. Numerous cells and other tissues work together in a coordinated manner for generating a response at the organ level, thereby regulating the final physiology. The entire network consists of a regulatory hierarchical structure, which, once disturbed, can result in unwanted physiological disturbances known as infections. In this regard, a problem in diagnosing a disease is similar to a faulty diagnosis in the engineering aspect (Somvanshi and Venkatesh 2014).

Studies of HAIs in many of the past decades have produced a qualitative overview of the molecular and biochemical communications contributing to the development of HAIs. Conventionally, reductionist strategies were applied for deducing these communications existing among the biological network of microbial cells. Biological networks are composed of communications between proteins, genes, and metabolites, which coordinate the control of the different cellular-level processes. While communication among numerous cells results in a response at the tissue level, various organs composed of various kinds of tissues further coordinate for eliciting a different biological response. These communications develop into a network of structures coding for the inherent principle of design governing the biological response dynamics. The structural network contains multilevel regulatory motifs aiding the system to evoke a preferred symptom in a coordinated manner. Feedback controls predominantly in such networks are the primary mechanisms that form the basis of the functional features of the networks (Brandman et al. 2005). On the other hand, network quantification is necessary for understanding the fabrication and deducing the functional characteristics of the system. Systems and computational biology approaches can be applied for studying the development of networks at the organ and cellular level, thereby controlling the system of biological dynamics. The main aim of systems biology is to decipher the integrative state of the biological networks, which have been already engineered by nature via evolution, and crack the underlying principles, which are regulating them (Kollef et al. 2021). Until now, it has deciphered with successful, architectures of specific circuits and replicatively indicated their roles according to their fabrications. Particular subnetworks existing in eukaryotic and prokaryotic systems, like cell cycle, chemotaxis among bacteria, the osmotic adaptation among yeasts, and signaling pathways of growth factors among mammals, have been characterized successfully. Expanding this analytical concept of the physiological networks for HAIs and combining it with natural states will aid in extracting useful information about the behavior of a normal phenotypic response and its perturbed state during infections. In this chapter, the application of various engineering approaches for addressing HAIs from a systems biological point of view has been discussed. This chapter throws light on the potential networks and modeling strategies applied to analyze diseases among humans. The application of this analysis is illustrated in the instance of HAIs. The chapter focuses on conceptualizing the cell–human framework, including artificial intelligence (AI), machine learning (ML), the Internet of things (IoT), artificial neural network (ANN), and other bioinformatic tools for addressing the physiology of humans and HAIs on the basis of a paradigm level of system analysis. The chapter explicitly highlights the significance of multiple-scale physiological networks and subsequent analysis and modeling for the fabrication of effective treatments and identification of targets of drugs.

7.2 Hospital-Associated Infections (HAIs)

HAIs are nosocomial-acquired infections or hospital-acquired infections that were not present within patients prior to hospitalization. HAIs may occur in different wards at the time of treatment; they are frequently related to the hospitalization in intensive care units (ICUs). In ICUs, patients are at five to ten times the high risk of developing HAIs because of extrinsic (medical device administration) and intrinsic factors (immunodeficiency). Apart from this, an ICU is frequently a reservoir of multidrug -resistant (MDR) microbes.

HAIs are most often caused by the colonization of microbes arranged inside a structure of biofilm within medical devices. Most of the predominant biofilm-forming microbes include *Staphylococcus epidermidis*, *Staphylococcus aureus*, and *Enterococcus* spp. among Gram-positive bacteria and *Klebsiella pneumoniae*, *Escherichia coli*, and *Pseudomonas aeruginosa* among Gram-negative bacteria. Among the microscopic fungi, *Candida* sp. is one of the major groups to cause HAIs (Vandecandelaere et al. 2012). Biofilms formed from a single species of microbes are mostly found, while multispecies biofilms are not commonly found. However, there is not much information available about them due to the limitations of the techniques of detection. Certain communications among microbes lead to various modifications in the physiology and/or morphology of cells. Various viable but noncultivable microbes (VBNC) and small colony variants can be triggered following competitive communications among a few bacteria, such as between *P. aeruginosa* and *S. aureus* (Nair et al. 2014). Furthermore, there are instances of an elevated quantity of persistent organisms and extracellular polysaccharide (EPS) production in multispecies biofilms. In contrast to this, establishing a direct connection between the microbes residing within a biofilm, medical device colonization by these biofilms, and the onset of diseases in patients is not much easy due to the absence of a guide. The increased resistance of biofilm-residing microbes to

Table 7.1 Microbes and the HAIs caused by them

Microbes	Type of infections	References
Escherichia coli	Surgical site infections (SSIs) and wound infections	Weiner-Lastinger et al. (2020)
Staphylococcus aureus	Sepsis or bacteremia, infections in bone replacements and joint prostheses and wound infections	Edmiston et al. (2015)
Enterococcus spp.	Endocarditis, bloodstream and urinary tract infections (UTIs)	Percival et al. (2015)
Klebsiella pneumoniae	Catheter-associated urinary tract infections (CAUTI), central-line associated bloodstream infections (CLABSI), SSI, and ventilator-associated pneumonia (VAP)	Stewart and Bjarnsholt (2020)
Acinetobacter baumannii	Biofilm-associated infections of the respiratory therapy equipments and indwelling catheters	Howard et al. (2012)
Pseudomonas aeruginosa	VAP, UTIs, burnt site infections, cystic fibrosis (CF), and dental biofilm formation (periodontitis)	Fazeli et al. (2012)

antimicrobials is a contributing factor to making biofilms more pathogenic and frequently leading to the formation of chronic and acute HAIs (Percival et al. 2015). The potential microbes and the HAIs they are causing are listed in Table 7.1 .

7.3 Systems Biology and Its Application in the Health Sector

The identification/detection, reporting/declaration, and registration of HAIs by any healthcare sector are compulsory, and the surveillance actions, limitations, and prevention of HAIs are fractions of the professional compulsions of the personnel and remain included within the job descriptions of the individual employee.

Advancements in diagnostics would help clinicians to more rapidly diagnose the most successful therapy, hence reducing the nontargeted application of broad-spectrum antibiotics, thereby facilitating the application of novel antimicrobial therapies. A newly developed integrated system, known as polarization anisotropy diagnostics (PAD), helps in the quick detection of pathogens causing HAIs. PAD makes use of the modifications in fluorescence anisotropy when the detection probes identify nucleic acids of the target bacteria.

A data-based approach is applied in the selection of the sensors. There exist many limitations regarding the extraction of original data about the spreading of HAIs, like increased cost, sparsity of data, and the capacity of safeguarding the personal information of patients. Because of this, simulated data of the HAI contagion are more relied on. A highly described agent-associated simulation is used, which uses a mobility log extracted from the neighboring hospitals for the production of realistic contagion data. In this type of simulation, human agents (people) move from one place to another (static agents) as explained by the mobility log, and spread HAIs in a stochastic fashion. The simulation was fabricated by using three steps:

1. In silico population or mobility log: Designing of an in silico or computer-associated population and its actions.
2. Disease model: Conceptualizing a disease model for a target pathogen.
3. Simulation: Employing an immensely detailed simulation.

7.3.1 Detection and Rapid Identification of HAIs

The PAD system of detection and identification is a cost-effective and rapid HAI diagnostic system. The fluorescence detection does not depend on the intensity of the fluorescence; rather it is ratiometric, making the assay powerful against environmental conditions. PAD can detect on-site HAIs on multiple levels (Park et al. 2016) (Fig. 7.1).

The PAD assay system can be applied for clinical detection of the most common HAI-associated pathogens such as *K. pneumoniae*, *E. coli*, *P. aeruginosa*, *A. baumannii*, and *S. aureus*. PAD exhibited a sensitivity of detection down to the level of an individual bacterium and also determined the status of pathogenicity and resistance against drugs. In the cases of clinical samples, PAD displayed an accuracy similar to that of the culture of bacteria, but PAD showed a concise turnaround time of about 2 h and can be used in on-site or point-of-care (PoC) detection.

7.3.2 Detection of Contamination Hotspots in the Healthcare Setting

In order to determine the hotspots of microbial contamination inside the hospital, the entire size of the library (read counts) for bacteria leading to HAIs, Gram-negative bacteria (GNB) isolation, and culture-dependent (CD) approach were combined (Sereia et al. 2021).

CD analysis exceptionally focuses on GNB. Amies agar gel comprising the isolated swabs was inoculated within the Brain Heart Infusion broth (BHI broth) and incubated for a period of 12–18 h at a temperature of 36 °C prior to plating on selective MacConkey agar plates, which were further incubated for a period of 18–24 h at a temperature of 36 °C. Various morphologies of colony-forming units (CFUs) were separated and plated on new agar plates of MacConkey agar. Recognition and antimicrobial susceptibility tests (ASTs) of every isolate were carried out by using AST-N239 and GN ID cards. On the basis of the results of AST, the isolates were further classified as multidrug-resistant (MDR) and not multidrug-resistant (not MDR) with respect to the classification of acquired resistivity (Magiorakos et al. 2011). DNA of all the isolates of GNB was extracted through this process and quantified by the Qubit dsDNA BR assay kit.

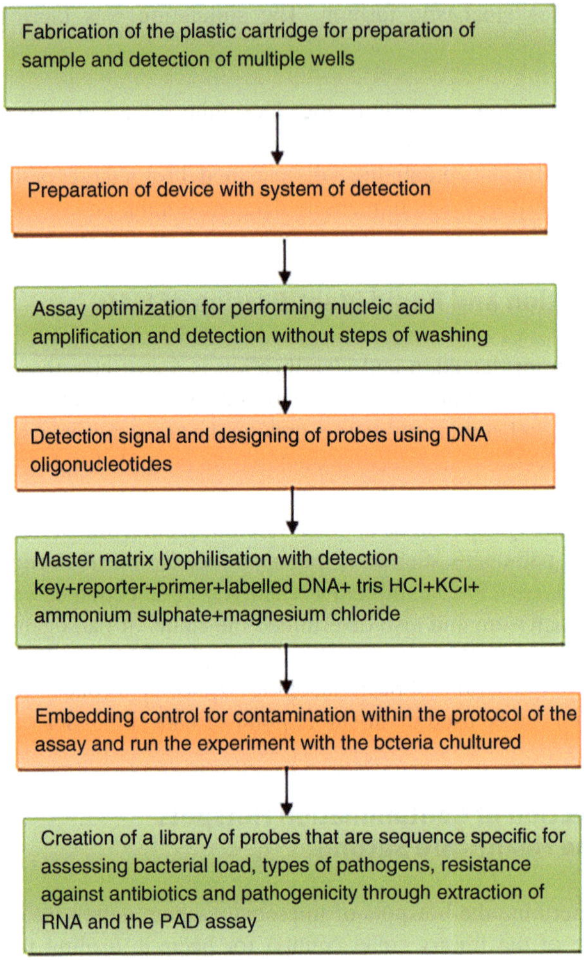

Fig. 7.1 Steps by which the PAD assay is designed and diagnose HAIs

It was observed that patients had the most median read count, GNB isolation rate, and abundance of hospital-associated infection-related bacteria (HAIrB), most of which are associated with rectum samples in comparison to nasal and hand samples. Among healthcare workers, in spite of the higher presence of total bacterial count in their protective clothes, the hospital-associated infection-related bacteria (HAIrB) median distribution was identical for clothing, hands, and mobiles.

Regarding the hospital surroundings, the resting premises of healthcare workers were found to be abundantly contaminated, having a higher abundance of HAIrB and total bacterial count. It is also remarkable that the unit bathrooms and the medical and bed equipment possessed the highest value of the HAIrB Shannon diversity index, which is followed by the patient rectum and the resting areas of healthcare workers. Equipment or areas, which are thought to be cleaner, like the workstations of the healthcare workers utilized during nursing and medical procedures or surgical equipment restricted within general surgery units (GSU), indicated a considerable amount of contamination with HAIrB in spite of possessing very low quantities of the total count of bacteria.

The CD method also demonstrated an identical pattern of distribution of the isolates of GNB, showing a higher abundance of isolates from patients, resting areas of healthcare workers, and bedding in comparison to the mobiles of healthcare workers and the surgical equipment, which exhibited a decreased number of isolates of GNB. Profiles of antimicrobial resistance (AMR) validated these hotspot contaminations, focusing on patients and their surrounding environment as well as the resting areas and workstations of healthcare workers and medical equipments used commonly. All of these areas contained MDR, carbapenem-resistant (CARB-R), and extended-spectrum cephalosporin-resistant (ESC-R) GNB.

For understanding the HAIrB dispersion inside the hospitals, the average presence of the oligotypes with respect to HAIrB isolated by the culture independent (CI) method was used for plotting a heat map. There was a distinct and clear grouping of the patient samples indicating a much high frequency of oligotypes common to various sites of sampling such as the rectum, nasal cavity, and hands, for example, the oligotype_69 from Enterobacteriaceae and the oligotype_47 from *Enterococcus faecalis*. Another grouping of samples collected from the hands and protective clothes of healthcare workers, their resting areas, and bedding was also recognized. The oligotype_239 from Enterobacteriaceae was found in the bedding, resting areas, and hands of healthcare workers. Oligotypes_100, 169, and 48, with respect to Enterobacteriaceae, were detected in the protective clothes and hands of healthcare workers, whereas the oligotype_101 was detected in the hands and resting areas of healthcare workers. It could be even possible to isolate more broadly dispersed oligotypes, like the oligotype_88 found in *Acinetobacter baumannii*, adequately dispersed in the bedding, workstations, and hands of healthcare workers. The oligotype_43, isolated from *Klebsiella pneumoniae*, was screened in the rectum samples of patients and the hands of healthcare workers. The oligotype_6 from *A. baumannii* was detected in the rectum samples of patients and the hands of healthcare workers with identical presence as in the protective clothes of healthcare workers. Dispersion of HAIrB was also examined and compared between the CD and CI approaches, which indicated the presence of various overlapping oligotypes in several CD and CI correspondent samples.

According to the Centers for Disease Control and Prevention (CDC), one out of 25 hospitalized patients gets infected with at least one HAI per day. Prior detection of outbreaks of possible HAIs facilitates medical practitioners in implementing countermeasures before the extensive spreading of the infections. Adhikari et al. (2019), devised a practical model and data-driven approach for outbreak detection with extremely high accuracy. They leveraged mechanistic modeling of infections caused by *Clostridium difficile*, one of the predominant HAI- causative agents, for simulating its spread within the hospital wing and fabricated an effective near-optimum algorithm for selecting locations and people in order to monitor by the use of a formulation for optimization. The results indicated that this approach could detect almost about 95% of *Clostridium difficile* infection outbreaks in the near future. This approach was designed by the introduction of particular hospital practices (such as swab collection from infections). Hence, this approach can outsmart the conventional algorithms used to detect outbreaks. Ultimately, a qualitative analysis of the results obtained from this study showed that the locations and the people selected for monitoring could act as sensors very meaningfully and intuitively.

7.3.2.1 Use of the ANN for HAI Risk Prediction Among Lung Cancer Patients

The ANN can be applied as a tool for forecasting nosocomial infections. It can present us with supplementary data to control and diagnose nosocomial infections (Xu et al. 2006). Statistical methods for predicting and analyzing the associated risk factors of HAIs among patients suffering from lung cancer are numerous. However, the results obtained through them are not much consistent. In a study conducted by Chen et al. (2014), a total of 609 patients having lung cancer were enrolled to help in the comparison of factors by the use of the Mann-Whitney test or Student's *t* test, or the Chi-square test. Variables, which were remarkably associated with the presence of HAIs, were screened as candidates for input within the ultimate model of the ANN. The area under the receiver operating characteristic (ROC) curve (AUC) was utilized for evaluating the efficacy of the logistic regression (LR) and ANN model. The abundance of HAIs acquired through lung cancer in this investigation, in the total population of the study, was 165/609 (20.1%), HAIs from samples of sputum accounted for 85.5% subsequently followed by blood specimens (6.73%), urine samples (6%), and plural fluid effusions (1.82%). It was demonstrated that hospitalization for a long period (more than 21 days), poor clinical condition (stages IIIb and IV), and elderly persons older than 60 years and also the utilization of hormones were associated with HAIs and the ANN model took into consideration all these 4 parameters. The ANN model having variables, such as clinical conditions, age, hormone utilization, and duration of hospitalization, can be applied for the prediction of HAIs among patients suffering from lung cancer.

7.3.2.2 ANN for Predicting the Risk of HAIs in ICUs Using a Didactic Preliminary Model

A didactic preliminary model consists of a neural network for predicting the risks of HAIs in ICUs in accordance with the synthetic set of data, and excellent efficacy can be found having a few patterns, which demonstrates that utilizing a high number of variables and layers of networks for analyzing larger data volumes can lead to the creation of powerful ANNs, which can precisely anticipate HAIs at ICUs. By use of an actual database at the time of simulations can realize the prediction efficacy of this model.

Nistal-Nuño (2020), proposed a preliminary AI model on the basis of ANNs to predict the risk of HAIs in ICUs. In this study, an ANN was fabricated for employing supervised learning. The dataset generation was based on the data obtained from the system of Japanese Nosocomial Infection Surveillance. It is observed the way in which the Java Neural Network Simulator can achieve the potential of categorizing patients in ICUs for prediction of their risks of HAIs. Simulations were performed with the help of numerous learning backpropagation algorithms and with many parameter groups and finally comparing their results via the sum of the square of the errors and mean errors in one pattern. The momentum, along with the backpropagation algorithm, displayed better efficacy than only the backpropagation algorithm. The efficacy got better with the xor: README values of the file parameter in comparison to the preset parameters. There were no shortcomings in patient categorization into their risks of HAIs. While this model is yet based on a set of synthetic data, the excellent efficacy observed with a fewer number of patterns indicates the utilization of a higher number of variables and layers of network, which can precisely anticipate HAIs at ICUs. The use of a real database at the time of simulation has the capacity of realizing the predictive capability of this model.

7.3.2.3 HAI Detection by Use of Gradient Tree Boosting and Support Vector Machines

HAIs pose a severe threat to the health of patients, and their monitoring is an extra workload for healthcare workers. The main objective is to design a reliable system of surveillance that can accurately detect all the risks associated with HAIs. This is mainly done for reducing the workload of healthcare workers. Ehrentraut et al. (2018), conducted a study focusing on the use of the classification of text with the help of gradient tree boosting and support vector machines. Gradient tree boosting and support vector machines have not yet been used for the detection of HAIs from the records of patients. Gradient tree boosting yielded the best results with 79.7% precision, 93.7% recall, and an 85.7% F1 score by the use of stemming. Such tuning of parameters and techniques of preprocessing can produce increased recall with accurate precision for this prediction.

7.4 HAI Transmission

Dynamic models for transmission are overtly in use for improving our knowledge about the epidemiology of HAIs. EMBASE, MEDLINE, CINAHL Plus, Scopus, and databases from Global Health can be searched systematically for dynamic mathematical modeling of HAI transmission (van Kleef et al. 2013). The research was based on the efficacy of infection regulation, various routes of transmission, the effect of the patterns of movement among the healthcare institutions, antimicrobial resistance (AMR) development, and co-colonization of various strains or competitiveness of strains. Methicillin-resistant *S. aureus* (MRSA) was the most frequently modeled HAI-causing bacteria, subsequently followed by vancomycin-resistant enterococci (Braga et al. 2013). Other bacteria usually causing HAIs include *C. difficile*, which is not often modeled. The first-ever model of HAIs included AMR among the bacterial cloud in the hospital surroundings, by the use of a deterministic compartmental approach. Stochastic models, which include the function of probability in the process of transmission, are most often used. Sensitivity analyses and calibration of models (inference of parameters, which were not known by the fitting of models to the data) are relatively not commonly used; however, their application nowadays is escalating. Just 5% of the models can compare their predictions with the exterior data. Models of transmission have been used for understanding complicated systems and for predicting the effect of regulating policies. Approaches generally have been modified with an elevated application of stochastic models and more advanced approaches for the preliminary fitting of the model and sensitivity analysis. Information derived from these models can be widened to a broader range of surroundings and pathogens. Betterment of data availability and statistical techniques can further enhance the prediction capacity of these models.

7.4.1 HAI Modeling of Transmission Dynamics of COVID-19

The COVID-19 pandemic has led to the emergence of an unpredictable threat to human health all over the world. COVID-19 is a respiratory disease, which spread to almost every corner of the world within a very short span of time after its emergence from China in late December of the year 2019 (Aba Oud et al. 2021). It is a very recent type of virus associated with the family of severe acute respiratory syndrome coronavirus-2 (SARS-COV-2) (Ivorra et al. 2020). COVID-19 affected either the upper respiratory passage (sinuses, throat, and nose) or the lower respiratory passage (windpipes and lungs). It broke out as an unseen pneumonia of unidentified etiology in the city of Wuhan in China. Few of the studies indicated that the causative source of the virus may be associated with pangolins (Mahmoudi et al. 2021). COVID-19 was communicated from one individual to the other via respiratory droplets flowing through the mucous membranes (mouth, eyes, and nose) at the time of coughing, sneezing, exhaling, or talking (Peter et al. 2021). Nowadays, attention is more diverted toward researching pandemic diseases including malaria, H1N1, HIV, Ebola, hepatitis B virus (HBV), and others (Naik et al. 2020). The capacity of viruses, bacteria, and parasites, which are responsible for causing infections, keeps modifying and brings about new challenges in strategies of their therapies.

This disease has warned the healthcare setting about the concern of HAIs. Hospital-associated spreading of COVID-19 has been reported and discovered worldwide in various hospitals and healthcare institutes. Superspreaders and asymptomatic patients were considered to be the significant origins of these infections. Hence, Masandawa et al. (2022), conducted a study by formulation of a *SEIHR* mathematical model for gaining a better understanding of HAIs for the transmission dynamics of COVID-19. The role played by personal protective clothing (θ) was analyzed in this model. For the benefit of the next-generation matrix approach, R_0 (basic reproduction number) was calculated. R_0 is the threshold factor that can be applied for regulating the transmission of a certain disease. It aids in providing information about how many on an average a single infected person can infect. With respect to COVID-19, a study indicated that R_0 can be 6.33, which is the largest value during the initial outbreak days of COVID-19 (Masandawa et al. 2021). Another study indicated the value of R_0 to be around 2.37 (Yu et al. 2021). Stable

Metzler's theory of matrix and the Routh-Hurwitz criterion indicated that the point of equilibrium free from COVID-19 was globally and locally stable asymptomatically when $R_0 < 1$. The Lyapunov function indicated that the point of endemic equilibrium was globally stable asymptomatically when $R_0 > 1$. Moreover, the dynamic nature of R_0 was examined by a variation of θ. In θ's absence, the R_0 value was 8.4584, which exhibits that the disease has expanded. When θ is implemented in this model, R_0 becomes 0.4229, demonstrating the reduction of the disease within the community. Numerical computations were simulated by the use of the Runge-Kutta fourth-order approach. Global sensitivity analyses were carried out for presenting the most remarkable parameter. The numerical solutions mathematically illustrated that personal protective clothing equipment could diminish HAIs induced by COVID-19 (Park et al. 2021). Research in one of the German hospitals demonstrated a total of 48 confirmed COVID-19 cases associated with HAIs, out of which, 28 of them were healthcare workers, 13 were hospitalized patients, and the rest of them accompanied their family members in the hospital (Du et al. 2021). Out of these mentioned cases, only four infected individuals had a close interaction of around 15 min with the medical personnel, who did not wear personal protective equipment (PPE). Additionally, seven of them got infected after providing medication to COVID-19-affected patients without wearing PPE (Du et al. 2021). Oke et al. (2021), designed a new model, SEIHRD mathematical model, to evaluate the effect of saturated therapies in Nigeria. In this study, it was concluded that a reduction of HAI infectivity rate can be achieved by stringent follow-up of the disease protocols. Martos et al. (2020), modeled the COVID-19 transmission dynamics within hospital units in the UK focusing mainly on HAIs caused within hospitalized patients.

7.4.2 Mathematical Modeling Improvement to Check the Transmission of HAIs

Mathematical models are being applied for gauging the effects of checkpoints for HAIs. Just like all the analytical procedures, these procedures also need some assumptions. Two general assumptions are that asymptomatic colonization within individuals needs to be hospitalized immediately and that they must spend a longer duration in hospital according to their status of colonization. These assumptions do not have any biological basis and could affect the estimated impacts of checkpoints in an unprecedented manner. Hence, Gowler et al. (2022), devised a model for the transmission of methicillin-resistant *S. aureus* (MRSA) for the detailed evaluation of the effects of these assumptions. It was observed that assumptions about the asymptomatic colonization of the individuals were hospitalized immediately or spent a long time in hospitals, in comparison to the individuals who were uncolonized. Such biased outcomes, in comparison with a more realistic model, did not demand any more assumptions. The results were intensely biased while estimating the effects of an intervention, which decolonized hospitalized patients.

Estimating the efficacies of the preventive measures adopted for HAI transmission, however, is very challenging (van Kleef et al. 2013). The major barrier is that there is usually a low infection rate in a certain hospital. Hence, each of the facilities has little chance of quantifying the effects of an intervention in a short survey. Multicenter investigations have been performed to aid in addressing this issue, but they may have exorbitant prices, and the duration of follow-up is really very short (Huang et al. 2013). Furthermore, this problem gets complex when patients are transferred between nursing homes or hospitals, or other healthcare units. Therefore, transmission and the onset of the very first syndrome in a nonhospital atmosphere can complicate the attempts for evaluating the effects of hospital-associated interventions (Rubin et al. 2018).

In order to address such limitations, researchers nowadays make use of simulations for aiding in the evaluation of whether an initial intervention was successful or if a new intervention is needed (Cremin et al. 2018). Models of transmission are likely to be cost-effective and can be rapidly created; however, they require extra assumptions in comparison to conventional epidemiologic strategies such as cohort studies and randomized trials. Interestingly, even the assumptions, which are innocuous, also appear to be introduced in the simplification of the analysis and can modify the outcomes in ways that are not expected by designers. For example, designers frequently simulate those individuals who were colonized asymptomatically because they have higher chances of admission to hospitals and a longer duration of stay at hospitals in comparison to those who were uncolonized asymptomatically (Lee et al. 2013). On the other hand, it is proved that individuals who were colonized spent more time in hospitals on average per admission and are more frequently hospitalized. There are many underlying factors, which are governing this relationship. For example, asymptomatically colonized persons who have been colonized with HAI-associated pathogens are likely to be sicker and older with respect to the noncarriers (De Angelis et al. 2011). Subsequently, a more plausible cause is individuals with poor health conditions tend to be hospitalized for a longer period frequently and hence appear to be more colonized with HAI-associated pathogens. A converse explanation can further describe the crude relationship between periods of stay and the status of the carriage. As a result, the models of transmission could give wrong estimations of the impacts of intervention when the asymptomatic carriage is simulated using these methods. Hence, there is a need to examine whether the assumptions about admission to hospitals and the period of hospital stay in the case of asymptomatic carriers can impact the simulation results after evaluating the intervention effects, which function by inhibiting the intervention of HAI transmission.

7.5 HAI Management by Systems Biology

Systems biology offers varied ways in which HAIs can be regulated effectively. Surveillance forms the basis for successful HAI control and prevention (Scardoni et al. 2020). However, surveillance through manual approaches requires skilled labor, is expensive, and is not properly standardized. Therefore, systems biology aid in facilitating the fabrication of algorithms for the surveillance of HAIs for a better understanding of potential risk parameters, enhancement of risk stratification of patients suffering from HAIs, transmission process identification, and real-time or timely diagnosis and detection. Some of the approaches through which systems biology can effectively manage HAIs are listed below.

7.5.1 HAI Management by the System of the ANN and Artificial Intelligence (AI)

The factors that characterize the impacts on the bodies are described by imprecision (substantial decrease and significant increase), which was found to be worthy of analyzing such variations via ANN principles. Weights of the organs under study can be considered as the input parameter, and bacterial species can be considered as an output parameter (Saddek et al. 2013). This established system can be used to make a match among the two spaces of the outputs and inputs. Neural networks were fabricated for mimicking the function of the brains of humans. There exist input levels, output levels, and a varied number of hidden or internal layers. The input levels remain connected with the internal layer, and all of these are connected with the output levels. Since the neural network takes input from a set of data, weights in connection get adjusted. Data are fed into the input levels, followed by data processing in the internal layer(s), and nodes of connection with the output levels get adjusted. Neural nets are classified on the basis of the paradigm of learning. Neural networks have the capacity of detecting unexpected and otherwise undiagnosable patterns in case of a large set of data. The significant limitation in the solutions provided by neural networks is the fact that the approaches via which an association is detected remain hidden and hence not easily explained or understood. In a very simple way, a cooperative model can be fabricated in the form of a preprocessor, in which the ANN mechanism of learning can be used to determine the training of the data.

AI is widely used in our daily lives and ranges from the recognition of voices on smart speakers to identifying new kinds of music from streaming applications, which could predict new singers for new listeners, and computerized cancer detection from mammograms (McKinney et al. 2020). AI makes use of mathematical models for iterating learning patterns, which are present in new data, followed by a translation into a decision by the AI tool. The application of AI in medicine has been increased by the presence of huge sets of health-related data because of health record digitization in combination with sharing of anonymized health-related data. In the case of HAI management, AI has been widely introduced by the World Health Organization (WHO) (Storr et al. 2017). Because healthcare information technology (IT) systems generate large quantities of data from discrete sources and get immensely implemented, AI systems can detect data patterns triggering HAI outbreak detection and generate enriched sets of data for follow-up analysis. From a data science and clinical perspective, HAI management is a highly complicated premise, wherein information from many different clinical sources and specialties needs to be implemented within a single patient. Simultaneously, there is an increased requirement for optimization of HAI management in terms of adequate and timely therapeutic administration, survival of the patient, and curbing antimicrobial resistance development. Data-generated, evidence-based, and automated AI facilitates clinical practitioners and teams of antibiotic stewardship to tackle these problems of HAIs (De Corte et al. 2022). AI can support the change in the system by pointing out cost-effectiveness, solution modeling by simulation of the behavior of the various kinds of agents inside a complicated system, and supporting the modification by data gathering and production of analytics. In HAI management, AI-associated simulations provide a link to bridge the gap between professional experience, which does not compromise the safety of patients, and recurring cycles of objective feedback and evaluation, which are the key parameters regulating the process of AI-based learning.

Radiography of the chest is the basic component for the diagnosis and screening of tuberculosis (TB) caused in patients suffering from HAI-associated cystic fibrosis (CF). Enhancements in the detection of TB facilitate the timely administration of anti-TB therapeutics and proper precautions to prevent HAIs. AI provides the chance to improve and standardize this process, particularly among CF patients, who are susceptible to TB with suboptimal access to radiologists. Deep learning along with convolutional neural networks has been applied for the classification of TB using radiography of the chest, which provides discrepant results, and those results get reviewed by radiologists (Lakhani and Sundaram 2017). This kind of approach presents opportunities to develop those hospitals in the world, which lack the expertise of radiologists and use AI to interpret a large number of investigations with the reviews of radiologists in only the equivocal cases.

In laboratories of medical microbiology, algorithms of AI are designed from population genomics, and they can be applied in the prediction of risks of infection from the genomic characteristics of *Staphylococcus epidermidis* and effectively recognize genotypes with high risks preoperatively on the target and post- and preoperatively on prevention strategies of HAIs. AI-induced microscopy in laboratories has the capability of streamlining the quick detection of HAIs among patients, thereby aiding in preventing AMR development and helping in targeted antimicrobial therapies. In a study, a convolutional neural network (a kind of AI applied for analyzing visual data) was trained to classify bacteria in specimens collected from blood samples at the stage of Gram staining with almost 90% correctness (Smith et al. 2018). Interpretation via Gram staining is time-taking and is strongly dependent on skilled professionals for accuracy in interpreting the obtained results. AI-based microscopy throws a plethora of options where there is no need for skilled professionals with microbiological expertise. This approach only requires the image to be sent to the central facility for reviewing and appropriation of medical liaison regarding HAI management among patients. AI has also been used in laboratories of medical microbiology for the diagnosis of bacterial vaginosis at the molecular level and can be carried out with efficacy against conventional Gram stain differentiation (Drew et al. 2020). The basic principle of AI-associated microscopy can hence be expanded to different specimens requiring interpretation via Gram stain (such as cerebrospinal fluid [CSF] from patients presumed to be suffering from meningitis), other different pathogenic microbes, including TB, in which diagnosis using a microscope is a key factor in preventing HAI pathways to AMR pathogen diagnosis at the molecular level for management of HAIs caused by them.

7.5.2 HAI Management by the System of Mathematical Modeling

Diverse types of microbes are being transferred within hospital settings, posing severe threats to healthcare workers as well as to patients. Few pathogens like Ebola and SARS-related coronavirus can widely disseminate among the populations within a community. Other kinds of emerging threats are represented by AMR bacteria, such as *Enterococcus,* resistant against vancomycin and MRSA, and they have become common HAI-associated pathogens in hospitals. Even though a considerable amount of knowledge has been acquired regarding the epidemiology of HAIs, the transmission modes of HAIs are yet to be known. Hospitals routinely maintain and collect intensive records of the discharge and admission times from and to units and wards. These records are readily available for sampling, and the data collected from the tests being conducted in the laboratories for the detection of colonizing pathogens, which infect patients. Colonization is the process by which bacteria live on the skin or other sites in the body without causing any diseases. Bacteria, which are responsible for causing HAIs, colonization is the first step in HAIs. The ultimate aim of mathematical modeling is to decrease the rates of HAIs (Thomas et al. 2018). Since clinical data cannot be efficiently analyzed, so it is hard to differentiate between HAI and asymptomatic colonization, records from hospitals have been utilized in a number of ways for estimating colonization rates and therefore evaluating the impacts of interventions fabricated for decreasing rates of acquisition. Thomas et al. (2018), focused on the models of patient levels in comparison to the compartmental models.

Modeling of HAIs at the patient level is a conventional hidden, statistically variable problem; although data regarding admission and swab tests are informative, the exact colonization time cannot be precisely observed. Cooper et al. (2008), concisely reviewed initial attempts in addressing the limitation prior to the implementation of an approach using augmented data. In this study, the data observed were augmented by the numerous times of colonization, which remained unobserved, and the values of which were constantly fed on the basis of the current values of the parameters and the data observed. The updates on the parameters and the imputations were made by the use of Markov chain Monte Carlo stochastic integration (MCMC), which, with the utmost care, assured that the values of the parameters that were generated constituted a correlated sample from the posterior densities of those parameters in the Bayesian analysis. Such methods can be typically also applied for stochastic optimization in order to find the highest likelihood or highest posterior estimates.

The advantage of the hidden variable method is that given the augmented values of data, the problem of estimation has been immensely simplified. The cost included the huge samplings of the augmented data, which would have otherwise recurred computational expenses and often involved intricate program designing. Even though the model proposed by Cooper et al. (2008), was specific for continued intervals, the approach integrated a disparate approximation of time. On the other hand, Thomas et al. (2015), presented a somewhat more effective disparate implementation of time and designed a continuous and true time-based solution. It was shown that the continued implementation was not demanding computationally, in comparison to the disparate time approach, thereby eliminating some of the biases in the estimation of the parameters.

Thomas et al. (2018), analyzed the data being implemented in a continuous mode of time. They expanded the susceptible-infected (SI) model, which supposes that once an individual is colonized, he/she rapidly becomes infectious and remains in that state of infectivity for the remaining period of study, to allow the models for infectiousness loss or decolonization, i.e., susceptible-infected-susceptible (SIS) model, and also for a transient duration of exposure preceding infectiousness, i.e., susceptible-exposed-infected-susceptible (SEIS) model. The model presumption that the state of infectiousness never subsides is in contradiction with the available pool of information regarding the behavior of colonization. Consequent empirical studies have demonstrated a diverse range of microbes that colonize human hosts, such as MRSA can get lost and get again acquired during the period of follow-up (Shenoy et al. 2014). On contrary, these studies did not presume that all the surveillance negative tests following a positive test are every time known to produce false negatives and hence it is more realistic to facilitate the probability of decolonization. In most of the studies, knowledge for estimation of the rate of decolonization comes from the various tests conducted on patients having prior positive tests, and therefore, this will be better predicted within the scope of continuous testing and surveillance. The incorporation of a transient period enriches the realism of these models because in nature, always, there lie some drawbacks among onset and infectiousness transmission in the host, who has been colonized newly. For SEIS and SIS models, the transition numbers in various imputations in the augmented results can get modified and hence also alter the problem dimensionality. Green (1995), addressed this problem through the use of reversible jump MCMC (RJMCMC) updates for histories of patients that make up the augmented data.

Nowadays, data from multiple units or wards and facilities within a single hospital are taken into consideration. The transfer of patients between units and the time spent outside any of the units can be tracked, and the status of colonization on readmission is dependent on the status during the initial discharge and the time between these two incidences. A three or two-state continuous time-based Markov process is used for modeling the colonization out of the facilities depending on whether the model facilitates a transient state or not. Models, which presume that no patients were hospitalized more than once, cannot provide a realistic approach that occurs with readmission having quite a high number of frequencies. The logic is this simplifying presumption poses disadvantages because readmission produces a direct relationship between forthcoming importation and previous transmission. The fact that this is an epidemiologically significant association has been supported in previous investigations of the colonization mediated by MRSA; acquisition of MRSA lagging at the facility level was observed to be correlated temporally with the prevalence of admission (Jones et al. 2015). The rate of decolonization out of the units is estimated by the status of colonization upon readmission in the case of patients who have been seemingly colonized after their discharge and will get reduced if the interval between frequent hospitalization gets increased.

A new parameter has also been introduced for the possibility that a patient inside the unit at the beginning of the study is colonized. In the previous models, such patients were typically considered to be time zero admissions, and the real impact of this seems to be a minimum, indicating that it is totally a wrong presumption since the patient might get colonized as a result of the previous exposure to the unit. It can be noted that this parameter can be considered insignificant and has been implemented only to avoid complications in parameter-of-interest estimation in comparison to something fascinating on its own.

Ultimately, three kinds of patient-to-patient models of transmission can be considered: models wherein the risk of a patient being colonized is fixed, is dependent on the frequency of patients being colonized within the facility/unit, and this model is known as the density-dependent model. The model that is dependent on the proportion of patients being colonized is usually known as the frequency-dependent model (McCallum et al. 2001).

For informing about the choice of model, the deviance information criterion (DIC) and cross-validation can both be used as a consolidated framework requiring extra MCMC above and over those required for the estimation of parameters (Gelman et al. 2014). A little expansion was made by leaving out one of the parameters through cross-validation. A different version was adopted to allow arbitrary test sets to be eliminated in the phase of estimation preceding prediction. Celeux et al. (2006), took into consideration a number of probable versions of the DIC for unavailable problems of data that were used to treat the augmented variables differently as parameters or data. Since the DIC needs deviance calculation in the presence of constant estimates of parameters, it was found incorrect in the pretext, where the augmented variables are complete histories of patients as compared to the simplified numerical problems, so applying posterior mean estimates, which are not well-defined. Celeux et al. (2006), implemented the DIC_6 version of augmented data variables, which is called the Watanabe-Akaike information criteria (WAIC). This model possesses the computational interest of being calculable from the statistical data produced in the simulation course run for the estimation of parameters; hence unlike the DIC and cross-validation, this requires no extra effort.

7.5.3 HAI Management by the System of Machine Learning (ML)

Nowadays, healthcare workers need to interpret and process vast quantities of complicated data. ML-associated methods have widely started to invade the medical literature in such a way that the proper utilization of tools and algorithms can aid in therapies and diagnosis. The availability of vast amounts of data, which are of high quality, could result in an enriched understanding of risk factors within a community or hospital. For example, in the field of antibiotic stewardship, scientists used data from the Massachusetts statewide antibiogram for predicting the 3 upcoming years of antibiotic vulnerabilities by the use of ML regression-associated approaches (Mancini et al. 2020). Till now, international guidelines prescribed to utilize institutional antibiograms in developing empirical antibiotic treatments, and the methods of ML could aid medical practitioners in empirical therapies for urinary tract infections (UTIs). These are generally considered the most usual HAIs with a remarkable financial load on society (Naber et al. 2001). In hospitals, almost 25% of HAIs are bacteriuria, and 40% are UTIs, which are mainly caused by biofilm-associated infections in the urinary catheters, which have been used for more than a week. The choice of enough therapies for managing HAIs is very painstaking because of the resistivity profiles of the bacteria, their etiology, and the evolution of various adaptation processes within the bacterial genome. Furthermore, antibiotic resistance in bacteria has increased significantly with rising lesser therapeutic approaches. One of the main reasons for recurrent HAIs is the emergence of multidrug-resistant (MDR) bacteria. Various risk factors are related to HAIs, such as age and sex (Foxman 2010).

Mancini et al. (2020), designed an ML model to predict the risk of patients being affected with MDR UTIs followed by long-term hospitalization. This study implemented a readily usable cloud platform known as Data Science as a Service (DSaaS), which is well based on hospital infrastructure, in which healthcare workers might not have particular expertise in applying programming languages. Yet, they are required to analyze data continuously. Moreover, this DSaaS model allowed the validation of the models of data analysis on the basis of supervised ML classification and regression algorithms. Zoabi et al. (2021), devised an electronic clinical record-associated ML model for predicting the impacts of bloodstream infections (BSIs) among hospitalized patients. The area within the operator-receiver characteristics curves was found to be 0.82 for a fully characterized inclusive model and 0.81 for a compact model utilizing only 25 characteristics. These models were made to learn the electronic medical records, including blood tests, demographics and diagnosis, and clinical history of 7889 patients who were hospitalized and suffering from BSIs. This model can work on the principle of rapid, selective identification of the microbes and help in the timely administration of proper antibiotic treatment. In addition to this, this particular model may also reduce BSI development and its related adverse health complications and impacts. Barchitta et al. (2021), identified patients who are at increased risk of developing HAIs in intensive care units (ICUs). ML has the capacity to enhance the stratification of patient risk leading to targeted prevention of HAIs and other control strategy development. This study evaluated the efficacy of the Simplified Acute Physiology Score (SAPS) II for the prediction of HAI risks in ICUs by the use of ML and conventional statistical approaches. The algorithm known as Support Vector Machine (SVM) was used in classifying patients according to their origin, sex, surgical history, nonsurgical therapies for acute coronary diseases, presence of indwelling medical devices, SAPS II utilization during admission, compromised immunity, trauma, and antibiotic therapies within 48 h before admission in ICUs. The performance of SAPS II in predicting the risk of HAI development provided a receiver-mediated characteristic curve having an area within the curve of 0.612 ($p < 0.001$) and 56% accuracy. Taking SAPS II into account with other features of the admission in ICUs, the SVM classifier was observed to have 88% accuracy and an AUC value of 0.90 ($p < 0.001$) for this particular test. The prediction capacity was less than when the SVM model was taken into consideration but the SAPS II variable was eliminated (78% accuracy and an AUC value of 0.66). This study indicated that the SVM model is a helpful tool for the early forecast of patients who are at a higher risk of being affected by HAIs during their stay at ICUs.

7.5.4 HAI Management by the System of Data Analysis

The analysis of HAIs caused by AMR pathogens is very complex. First of all, typical AMR pathogens generally colonize more asymptomatically than through overt infections, so the transmission could not be observed perfectly via a sequential surveillance swab, which themselves are not perfectly sensitive. Any given series of results obtained from the swabs can therefore remain consistent with numerous different kinds of patterns of transmission. Second, data are frequently increasingly dependent on the status of colonization in one patient and the chance of infecting other patients by the colonized patient, and in a few wards, recurring admissions are usual. Cooper et al. (2008), devised a model to analyze typical HAI data comprising results from arbitrary timely screening of swabs, which overcomes these limitations and helps in simultaneously estimating importation and transmission parameters, colonization period, the sensitivity of swabs, and patient- and ward-level covariates. The model also considers dependencies by the use of a stochastic mechanistic model of transmission, and it facilitates uncertainty within the data by feeding the imperfectly calculated status of colonization in the patient in the case of recurrent admissions. This model works on the basis of the Markov chain Monte Carlo algorithm facilitating inference from the Bayesian framework. The model was used to illustrate data from the interrupted series of time-based studies of the transmission of enterococci resistant against vancomycin in a hematology unit.

Analysis of risk factors and surveillance of infections are the critical parameters for treating and managing HAIs, in both noninfected and infected patients. Cheng et al. (2020), explored the risk factors of HAIs among people affected by COVID-19 and provided a theoretical principle for scientific management and control of HAIs. Out of 212 confirmed cases of COVID-19, 31 of them developed HAIs with an occurrence rate of 14.62%. The most commonly developed HAIs included UTIs (32.26%), lower respiratory passage infections (38.71%), upper respiratory passage infections (3.23%), BSIs (22.8%), and infections of the gastrointestinal tract (3.23%). The prevalence of HAIs was remarkably dependent on arteriovenous catheterization, age, white blood cell count (WBC), urinary catheterization, and procalcitonin. Furthermore, multivariate analyses demonstrated that procalcitonin, WBC, and urinary catheterization were independent risk parameters of HAIs among patients suffering from COVID-19. Gaining a better insight into HAI development among COVID-19 patients and strong monitoring of different risk factors could aid in managing the incidence of HAIs in the isolation wards of COVID-19.

7.5.5 HAI Management by the System of Internet of Things (IoT)

During the COVID-19 outbreak, a huge number of patients have been hospitalized posing logistical limitations. Wen et al. (2022), used 5G technology for diagnosing HAIs under the supervision of clinicians. The wireless feature and high bandwidth enabled high-definition real-time virtual consultations, psychological therapies, and other virtual services. Electronic prescriptions were given to intelligently screen post-COVID-19 HAIs. These systems consisted of automated algorithms fabricated with a decision tree for classifying patients with respect to their symptoms. Effective ward management can also reduce the risk of HAIs. A semi-active radio frequency identification (RFI) technology can make personnel and objects readily identifiable in the form of virtual representations in an Internet-like structure (IoT). This enabled a virtual electronic boundary to be formed outside a previously set security zone. When patients are out of the security area or unauthorized people gain entry into the zone, the system alarm gets triggered. Simultaneously, the whereabouts of hospital workers can also be tracked by the use of software for face recognition. A virtual reality (VR) approach facilitated the families of patients to visit in a remote manner. When the families of patients entered the visiting room, the clinicians informed the nurses in the wards, who would set up an intelligent trolley at the side of the beds of patients. The intelligent trolley comprises a computer, camera, speaker, and microphone. The families needed to wear VR glasses for watching patients within the wards, and the clinicians could explain the conditions of patients to their families. Healthcare workers also used mobile phones for undertaking small rounds of the wards anywhere and at any time by the use of VR technology.

Robots were used to disinfect and for delivering drugs in high-risk zones such as isolation wards. A disinfection robot can be described as a disinfection machine, which can implement ultra-dry fog hydrogen peroxide sterilization, ultraviolet disinfection, and filtration of air. By the use of laser navigation and 5G network technology, the robot could move automatically in its set route and could autonomously carry out thorough and regular disinfection within complicated environments assuring all the zones are appropriately disinfected. The robot for delivery of drugs can automatically reach patients' bedsides according to the pre-set bed number. An intelligent health watch can be worn by the patient for ensuring that they get access to all the medicines.

Nonventilator hospital-acquired pneumonia (NV-HAP) is a fatal HAI and cannot be easily documented, tracked, and prevented in a majority of hospitals. The successful management of NV-HAP depends on updated data surveillance. Surveillance can analyze, target, and quantify the management system by detecting patients who are at a high risk of developing NV-HAP. On the other hand, monitoring of pneumonia is complicated because of subjective criteria, nonprecise and inconsistent recordings, and labor-extensive nature. NV-HAP needs to be standardized and monitored by the use of cutting-edge approaches and the fabrication of advanced sensors. Abubeker and Baskar (2022), developed a wireless body area network (WBAN) structure that is equipped with wireless biosensors, and real-time biosensor data are fed into it through a cloud platform. The study also fabricated a wireless sensor network (WSN) for tracking volatile organic compounds (VOC) and other real-time atmospheric features for curbing NV-HAP spreading. The ESP-32 IoT and Raspberry Pi 4B graphical processing unit platforms become the residents of the final WSN and WBAN. In order to decrease the NV-HAP-associated rate of mortality, this biosensor-mediated IoT-based framework can help in monitoring the hospitals' and clinics' indoor and outdoor quality of air. It also maintains a tab for the condition of each patient, treatment zones, rooms, and ICUs in real time. This method proved to be cost-effective by reducing the cost of storage, communication, and computational cost and less utilization of energy. The WSN monitored the quality of air inside ICUs, indoor units, and other hospital areas by tracking toxic VOC. The WBAN was fitted in a wearable watch for continuous tracking of health conditions of the ICU and indoor patients and health care workers, and this information was finally passed to the IoT cloud platform. A mobile application could monitor the IoT-based WBAN and WSN-associated biosensor factors for avoiding NV-HAP by early diagnosis of symptoms of pneumonia and by reducing the chances of the droplet and air-borne diseases. Wearable biosensors enabled the diagnostic kit to continuously track a patient's biochemical and physiological parameters in the presence of natural physiological conditions in any location. WSNs and WBANs form the major parts of healthcare IoT. The WBAN can monitor three key respiratory parameters, including the rate of heartbeat, the temperature of the body, and the level of oxygen in the blood (SpO_2). Wireless sensor networks report and measure parameters of the quality of air, such as humidity, temperature, and VOC (carbon dioxide, ozone, benzene, carbon monoxide, and acetone). Wireless sensor network-based smart wearable devices presented innovative prospects for the upcoming tailored respiratory therapies, prevention, and early detection of NV-HAP.

7.5.6 HAI Management by the System of Bioinformatics

Whole genome sequencing (WGS) is a cutting-edge infection surveillance technique that enables a high-resolution grouping of associated microbes and detection of outbreaks. However, efficient communication and visualization of genomic data back to healthcare workers are critical in implementing targeted responses for managing HAIs. CATHAI (Cluster Analysis Tool for Healthcare-associated Infections) is a communicative web-mediated visualizing platform that combines WGS-based clustering and related metadata, hence, transforming sequence data into accessible and informative clinical knowledge for managing HAIs (Cuddihy et al. 2022). Mustapha et al. (2022), detected HAI outbreaks using WGS-based surveillance of the pathogenic bacteria among patients in hospitals. The bacterial biodiversity of patients in the hospitals were described by systematically analyzing their genomes. Genomes of 3000 isolated bacteria from hospitalized patients were collected for 25 months and the isolates were differentiated by the average nucleotide identity (ANI) and principal component analysis of the accessory genes (PCA-A). Genetic distances among the isolates and evolution rates were not constant for each species implicating the selection of distance cut-offs for analyzing outbreaks. AMR genes and mobile genetic element (MGE) sharing among species were extensively observed. The study described the structure of the pathogenic population circulating in a single hospital setting and indicated that studying the population dynamics of the microbes could provide insight into genomic epidemiology.

The causative microbes and the involved processes of sepsis are multifaceted. Sepsis involves a range of genetic and pathophysiological features, which make managing sepsis quite challenging. Hence, there remains a gap in the understanding of the critical gene interactions and/or pathway relations behind the formation of these diseases. Enhanced understanding is required for early diagnosis of sepsis in order to improve its medical impacts and the health of patients. Wise gene biomarker selection designed for developing diagnostic tests could ease sepsis detection. Tong et al. (2020), described a strategic framework to validate and identify sepsis biomarkers among patients, who are at high risk of developing septic shock by the use of a two-tier gene screening, ANN technique of data mining from initially published genetic expression sets of data. Eight important hub markers have been detected, which can separate distinct and core processes of the disease showing a capacity to inform about the underlying pathological and immunological processes, thereby stratifying the therapies of patients. This method did not present much change in fold differences between the various kinds of diseases, which can be used as initial detection biomarkers but can also be used to identify candidate processes and other related biomarkers for more exploration. The four major objectives of this study were to detect a series of gene expressive profile biomarkers that can identify those patients who are at high risk of developing sepsis, and investigate the immune pathogenesis by comparison of the various profiles in order to uncover the related cellular types related with critically identified biomarkers.

An alternative methodology for plasma cell-free DNA (CfDNA) sequencing is clinical metagenomic (CMg) whole-blood sequencing. This method can directly identify blood pathogens by overcoming the disadvantages of CfDNA tests, i.e., a nonspecified diagnosis from bacteria from the mouth, gut and the inability of detecting AMR. In contrast, these methods are limited by the increased levels of the host DNA in relation to pathogenic DNA, which decreases the sensitivity unless depletion of the host is not done. Nucleic acid (RNA and DNA) is extracted from whole blood samples or plasma (after treatment with DNase for removing CfDNA) followed by cDNA synthesis (only for RNA) and multiple displacement amplification (MDA) and subsequent preparation of library and sequencing through an Ion Proton instrument. Human reads are deleted by mapping an indwelling medically relevant genomic database of bacteria that also allot a score to the various reads observed on the basis of the coverage of the genome, depth of sequencing, and the quality of alignment; only those reads with high scores are thought to be positive. The total time of this test was 2–3 days and is only applicable for samples collected from immunosuppressed (neutropenic) patients (Parize et al. 2017). The WBC count in the immunocompromised patient blood samples is less in healthy people, thereby proving the sensitivity of the test. Moragues-Solanas et al. (2021), conducted a study on immunosuppressed patients at a high risk of developing sepsis, and the results indicated that this approach could clinically identify relevant pathogens in a remarkably higher number of samples in comparison to the conventional culture methods. Out of all the positive culture samples, nine were confirmed by sequencing with a sensitivity of 81.8%. The specificity was comparatively less than 70% due to the lack of confirmation polymerase chain reactions (PCR). The only limitation of this method is that it cannot identify viruses or parasites.

Nanopore sequencing is another method for real-time analysis with a reduced turnaround time. This method can identify viruses and parasites from the infected patients by using CMg. This approach included an initial extraction of CfDNA from plasma and subsequent preparation of library and Illumina or nanopore sequencing. Following sequencing, the data were analyzed by means of the sequence-based ultra-rapid pathogen identification (SURPI+) platform of bioinformatics. With nanopore sequencing, pathogen detection takes place within 6 h. Apart from this, there exist other various tools for HAI management with the help of metagenomic data, such as Epi2Me, NanoOK, BugSeq, and IDSeq analytical tools. IDSeq is an open-source, easily accessible bioinformatic tool used for the analysis of CMg data. This tool can readily eliminate the host data and provides a rapid taxonomical classification of the metagenomic reads. It is devised to make analysis using CMg more easy-to-use and open access. However, the accuracy of such tools at the level of species is not fixed, and most of the tools reported low levels of pathogenic reads, which are not present inside the samples.

CRISPR-based approaches can help in point-of-care nucleic acid detection and can be used as diagnostic platforms to detect and manage the HAI outbreak. Cas proteins such as Cas9, Cas3, Cas13, Cas12, and Cas14 can be used to devise highly sensitive and accurate detection tools in synergy with fluorescence and amplification of signals, colorimetric, potentiometric, and lateral flow assay detection tools. Particularly, the most highly developed platforms such as DETECTR, SHERLOCK/v2, CRISPR-Chip, and CARMEN enabled the detection of attomolar quantities of nucleic acids of the pathogens with more specificity than PCR (Kostyusheva et al. 2022). Further advancement of CRISPR-associated tools could transform molecular diagnostics by making them more open access and user friendly. Nowadays, CRSIPR-based tools can help in remodeling epidemiological surveillance and molecular diagnostics.

7.6 Conclusion and Future Prospects

The application of systems biology approaches in the field of HAI management, detection, diagnosis, and treatment is huge. Sustainable enhancements of systems biology in the field of HAIs require a change of behavior and culture and governance infrastructure support. The saying that "correlation does not imply causation" is specifically meaningful while using systems biology. Systems biology is mainly driven by big data in order to identify the associations, which may imply conditions that are clinically relevant and recognize potential risk parameters of HAI development among hospitalized patients. On the other hand, sometimes, systems biology can overlook minute clusters, which may be relevant clinically, and currently, systems biology cannot make use of deep knowledge of the controlling pathways for reasoning about small sets of data. Systems biology-based approaches need to be more developed and goal-oriented. Few organizations, which have successfully implemented digital transformations, can now use these approaches for understanding the management of HAIs. Additionally, including insiders, which are accustomed to the hospital environment, considering that the same approach may not be feasible for all HAIs, and using a flat stratification for supporting readily iterative alterations are critical parameters to be considered (Tabrizi et al. 2019). Clinical practitioners have to be careful about the disadvantages of systems biology and its biases to offload the tasks at hand and be overconfident in its capabilities. Systems biology-based approaches are required to be tested and then introduced into real-life medical laboratories and hospitals, and careful investment in building the infrastructure of the data is also needed to realize the capacities of systems biology-based approaches in real life.

References

Aba Oud MA, Ali A, Alrabaiah H, Ullah S, Khan MA, Islam S (2021) A fractional order mathematical model for COVID-19 dynamics with quarantine, isolation, and environmental viral load. Adv Differ Equ 2021(1):106. https://doi.org/10.1186/s13662-021-03265-4

Abubeker KM, Baskar S (2022) Wireless sensor and wireless body area network assisted biosensor network for effective monitoring and prevention of non-ventilator hospital-acquired pneumonia. Frontiers in Sustainable Cities 4:1063067

Adhikari B, Lewis B, Vullikanti A, Jiménez JM, Prakash BA (2019) Fast and near-optimal monitoring for healthcare acquired infection outbreaks. PLoS Comput Biol 15(9):e1007284. https://doi.org/10.1371/journal.pcbi.1007284

Barchitta M, Maugeri A, Favara G, Riela PM, Gallo G, Mura I, Agodi A, SPIN-UTI Network (2021) A machine learning approach to predict healthcare-associated infections at intensive care unit admission: findings from the SPIN-UTI project. J Hosp Infect 112:77–86. https://doi.org/10.1016/j.jhin.2021.02.025

Braga TM, Pomba C, Lopes MF (2013) High-level vancomycin resistant enterococcus faecium related to humans and pigs found in dust from pig breeding facilities. Vet Microbiol 161(3–4):344–349. https://doi.org/10.1016/j.vetmic.2012.07.034

Brandman O, Ferrell JE Jr, Li R, Meyer T (2005) Interlinked fast and slow positive feedback loops drive reliable cell decisions. Science 310(5747):496–498. https://doi.org/10.1126/science.1113834

Celeux G, Forbes F, Robert CP, Titterington DM (2006) Deviance information criteria for missing data models. Bayesian Anal 1(4)

Chen J, Pan QS, Hong WD, Pan J, Zhang WH, Xu G, Wang YM (2014) Use of an artificial neural network to predict risk factors of nosocomial infection in lung cancer patients. Asian Pac J Cancer Prev 15(13):5349–5353. https://doi.org/10.7314/apjcp.2014.15.13.5349

Cheng K, He M, Shu Q, Wu M, Chen C, Xue Y (2020) Analysis of the risk factors for nosocomial bacterial infection in patients with COVID-19 in a tertiary hospital. Risk Manag Healthc Policy 13:2593–2599. https://doi.org/10.2147/RMHP.S277963

Cooper BS, Medley GF, Bradley SJ, Scott GM (2008) An augmented data method for the analysis of nosocomial infection data. Am J Epidemiol 168(5):548–557. https://doi.org/10.1093/aje/kwn176

Cremin Í, Watson O, Heffernan A, Imai N, Ahmed N, Bivegete S, Kimani T, Kyriacou D, Mahadevan P, Mustafa R, Pagoni P, Sophiea M, Whittaker C, Beacroft L, Riley S, Fisher MC (2018) An infectious way to teach students about outbreaks. Epidemics 23:42–48. https://doi.org/10.1016/j.epidem.2017.12.002

Cuddihy T, Harris PNA, Permana B, Beatson SA, Forde BM (2022) CATHAI: cluster analysis tool for healthcare-associated infections. Bioinform Adv 2(1):vbac040. https://doi.org/10.1093/bioadv/vbac040

Dadi NCT, Radochová B, Vargová J, Bujdáková H (2021) Impact of healthcare-associated infections connected to medical devices-an update. Microorganisms 9(11):2332. https://doi.org/10.3390/microorganisms9112332

De Angelis G, Allignol A, Murthy A, Wolkewitz M, Beyersmann J, Safran E, Schrenzel J, Pittet D, Harbarth S (2011) Multistate modelling to estimate the excess length of stay associated with meticillin-resistant *Staphylococcus aureus* colonisation and infection in surgical patients. J Hosp Infect 78(2):86–91. https://doi.org/10.1016/j.jhin.2011.02.003

De Corte T, Van Hoecke S, De Waele J (2022) Artificial intelligence in infection management in the ICU. Crit Care 26(1):79. https://doi.org/10.1186/s13054-022-03916-2

Drew RJ, Murphy T, Broderick D, O'Gorman J, Eogan M (2020) An interpretation algorithm for molecular diagnosis of bacterial vaginosis in a maternity hospital using machine learning: proof-of-concept study. Diagn Microbiol Infect Dis 96(2):114950. https://doi.org/10.1016/j.diagmicrobio.2019.114950

Du Q, Zhang D, Hu W, Li X, Xia Q, Wen T, Jia H (2021) Nosocomial infection of COVID19: a new challenge for healthcare professionals (review). Int J Mol Med 47(4):31. https://doi.org/10.3892/ijmm.2021.4864

Edmiston CE, McBain AJ, Roberts C, Leaper D (2015) Clinical and microbiological aspects of biofilm-associated surgical site infections. In: Donelli G (ed) Biofilm-based healthcare-associated infections. Springer International, Cham, pp 47–67

Ehrentraut C, Ekholm M, Tanushi H, Tiedemann J, Dalianis H (2018) Detecting hospital-acquired infections: a document classification approach using support vector machines and gradient tree boosting. Health Informatics J 24(1):24–42. https://doi.org/10.1177/1460458216656471

Fazeli H, Akbari R, Moghim S, Narimani T, Arabestani MR, Ghoddousi AR (2012) Pseudomonas aeruginosa infections in patients, hospital means, and personnels specimens. J Res Med Sci 17(4):332–337

Foxman B (2010) The epidemiology of urinary tract infection. Nat Rev Urol 7(12):653–660. https://doi.org/10.1038/nrurol.2010.190

Gelman A, Hwang J, Vehtari A (2014) Understanding predictive information criteria for Bayesian models. Stat Comput 24(6):997–1016

Gowler CD, Slayton RB, Reddy SC, O'Hagan JJ (2022) Improving mathematical modeling of interventions to prevent healthcare-associated infections by interrupting transmission or pathogens: how common modeling assumptions about colonized individuals impact intervention effectiveness estimates. PloS One 17(2):e0264344. https://doi.org/10.1371/journal.pone.0264344

Green PJ (1995) Reversible jump Markov chain Monte Carlo computation and Bayesian model determination. Biometrika 82(4):711–732

Howard A, O'Donoghue M, Feeney A, Sleator RD (2012) *Acinetobacter baumannii*: an emerging opportunistic pathogen. Virulence 3(3):243–250. https://doi.org/10.4161/viru.19700

Huang SS, Septimus E, Kleinman K, Moody J, Hickok J, Avery TR, Lankiewicz J, Gombosev A, Terpstra L, Hartford F, Hayden MK, Jernigan JA, Weinstein RA, Fraser VJ, Haffenreffer K, Cui E, Kaganov RE, Lolans K, Perlin JB, Platt R, AHRQ DECIDE Network and Healthcare-

Associated Infections Program (2013) Targeted versus universal decolonization to prevent ICU infection. N Engl J Med 368(24):2255–2265. https://doi.org/10.1056/NEJMoa1207290

Ivorra B, Ferrández MR, Vela-Pérez M, Ramos AM (2020) Mathematical modeling of the spread of the coronavirus disease 2019 (COVID-19) taking into account the undetected infections. The case of China. Commun Nonlinear Sci Numer Simul 88:105303. https://doi.org/10.1016/j.cnsns.2020.105303

Jones M, Nielson C, Gupta K, Khader K, Evans M (2015) Collateral benefit of screening patients for methicillin-resistant *Staphylococcus aureus* at hospital admission: isolation of patients with multidrug-resistant gram-negative bacteria. Am J Infect Control 43(1):31. https://doi.org/10.1016/j.ajic.2014.09.016

Kollef MH, Torres A, Shorr AF, Martin-Loeches I, Micek ST (2021) Nosocomial infection. Crit Care Med 49(2):169–187. https://doi.org/10.1097/CCM.0000000000004783

Kostyusheva A, Brezgin S, Babin Y, Vasilyeva I, Glebe D, Kostyushev D, Chulanov V (2022) CRISPR-Cas systems for diagnosing infectious diseases. Methods 203:431–446. https://doi.org/10.1016/j.ymeth.2021.04.007

Lakhani P, Sundaram B (2017) Deep learning at chest radiography: automated classification of pulmonary tuberculosis by using convolutional neural networks. Radiology 284(2):574–582

Lee BY, Yilmaz SL, Wong KF, Bartsch SM, Eubank S, Song Y, Avery TR, Christie R, Brown ST, Epstein JM, Parker JI, Huang SS (2013) Modeling the regional spread and control of vancomycin-resistant enterococci. Am J Infect Control 41(8):668–673. https://doi.org/10.1016/j.ajic.2013.01.013

Magiorakos A, Srinivasan A, Carey RB, Carmeli Y, Falagas ME, Giske CG et al (2011) Bacteria: an international expert proposal for interim standard definitions for acquired resistance. Microbiology 18:268–281. https://doi.org/10.1111/j.1469-0691.2011.03570.x

Mahmoudi MR, Heydari MH, Qasem SN, Mosavi A, Band SS (2021) Principal component analysis to study the relations between the spread rates of COVID-19 in high risks countries. Alex Eng J 60(1):457–464. https://doi.org/10.1016/j.aej.2020.09.013

Mancini A, Vito L, Marcelli E, Piangerelli M, De Leone R, Pucciarelli S, Merelli E (2020) Machine learning models predicting multidrug resistant urinary tract infections using 'DsaaS'. BMC Bioinform 21(Suppl 10):347. https://doi.org/10.1186/s12859-020-03566-7

Martos DM, Parcell BJ, Eftimie R (2020) Modelling the transmission of infectious diseases inside hospital bays: implications for COVID-19. Math Biosci Eng 17(6):8084–8104. https://doi.org/10.3934/mbe.2020410

Masandawa L, Mirau SS, Mbalawata IS (2021) Mathematical modeling of COVID-19 transmission dynamics between healthcare workers and community. Results Phys 29:104731. https://doi.org/10.1016/j.rinp.2021.104731

Masandawa L, Mirau SS, Mbalawata IS, Paul JN, Kreppel K, Msamba OM (2022) Modeling nosocomial infection of COVID-19 transmission dynamics. Results Phys 37:105503. https://doi.org/10.1016/j.rinp.2022.105503

McCallum H, Barlow N, Hone J (2001) How should pathogen transmission be modelled? Trends Ecol Evol 16(6):295–300. https://doi.org/10.1016/s0169-5347(01)02144-9

McKinney SM, Sieniek M, Godbole V, Godwin J, Antropova N, Ashrafian H, Back T, Chesus M, Corrado GS, Darzi A, Etemadi M, Garcia-Vicente F, Gilbert FJ, Halling-Brown M, Hassabis D, Jansen S, Karthikesalingam A, Kelly CJ, King D, Ledsam JR, Shetty S et al (2020) Addendum: international evaluation of an AI system for breast cancer screening. Nature 586(7829):E19. https://doi.org/10.1038/s41586-020-2679-9

Moragues-Solanas L, Scotti R, O'Grady J (2021) Rapid metagenomics for diagnosis of bloodstream and respiratory tract nosocomial infections: current status and future prospects. Expert Rev Mol Diagn 21(4):371–380. https://doi.org/10.1080/14737159.2021.1906652

Mustapha MM, Srinivasa VR, Griffith MP, Cho ST, Evans DR, Waggle K, Ezeonwuka C, Snyder DJ, Marsh JW, Harrison LH, Cooper VS, Van Tyne D (2022) Genomic diversity of hospital-acquired infections revealed through prospective whole-genome sequencing-based surveillance. mSystems 7(3):e0138421. https://doi.org/10.1128/msystems.01384-21

Naber KG, Bergman B, Bishop MC, Bjerklund-Johansen TE, Botto H, Lobel B, Jinenez Cruz F, Selvaggi FP, Urinary Tract Infection (UTI) Working Group of the Health Care Office (HCO) of the European Association of Urology (EAU) (2001) EAU guidelines for the management of urinary and male genital tract infections. Urinary tract infection (UTI) working Group of the Health Care Office (HCO) of the European Association of Urology (EAU). Eur Urol 40(5):576–588. https://doi.org/10.1159/000049840

Naik PA, Yavuz M, Qureshi S, Zu J, Townley S (2020) Modeling and analysis of COVID-19 epidemics with treatment in fractional derivatives using real data from Pakistan. Eur Phys J Plus 135(10):795. https://doi.org/10.1140/epjp/s13360-020-00819-5

Nair N, Biswas R, Götz F, Biswas L (2014) Impact of *Staphylococcus aureus* on pathogenesis in polymicrobial infections. Infect Immun 82(6):2162–2169. https://doi.org/10.1128/IAI.00059-14

Oke II, Oyebo YT, Fakoya OF, Benson VS, Tunde YT (2021) A mathematical model for Covid-19 disease transmission dynamics with impact of saturated treatment: Modeling, analysis and simulation. Open Access Library Journal 8(5):1–20

Nistal-Nuño B (2020) A neural network for prediction of risk of nosocomial infection at intensive care units: a didactic preliminary model. Einstein (Sao Paulo) 18:eAO5480. https://doi.org/10.31744/einstein_journal/2020AO5480

Parize P, Muth E, Richaud C, Gratigny M, Pilmis B, Lamamy A, Mainardi JL, Cheval J, de Visser L, Jagorel F, Ben Yahia L, Bamba G, Dubois M, Join-Lambert O, Leruez-Ville M, Nassif X, Lefort A, Lanternier F, Suarez F, Lortholary O, Eloit M et al (2017) Untargeted next-generation sequencing-based first-line diagnosis of infection in immunocompromised adults: a multicentre, blinded, prospective study. Clin Microbiol Infect 23(8):574.e1–574.e6. https://doi.org/10.1016/j.cmi.2017.02.006

Park KS, Huang CH, Lee K, Yoo YE, Castro CM, Weissleder R, Lee H (2016) Rapid identification of health care-associated infections with an integrated fluorescence anisotropy system. Sci Adv 2(5):e1600300. https://doi.org/10.1126/sciadv.1600300

Park Y, Sylla I, Das AK, Codella J (2021) Agent-based modeling to evaluate nosocomial COVID-19 infections and related policies. Nature 3:4

Percival SL, Suleman L, Vuotto C, Donelli G (2015) Healthcare-associated infections, medical devices and biofilms: risk, tolerance and control. J Med Microbiol 64(Pt 4):323–334. https://doi.org/10.1099/jmm.0.000032

Peter OJ, Qureshi S, Yusuf A, Al-Shomrani M, Idowu AA (2021) A new mathematical model of COVID-19 using real data from Pakistan. Results Phys 24:104098. https://doi.org/10.1016/j.rinp.2021.104098

Rubin MA, Samore MH, Harris AD (2018) The importance of contact precautions for endemic methicillin-resistant Staphylococcus aureus and vancomycin-resistant enterococci. JAMA 319(9):863–864. https://doi.org/10.1001/jama.2017.21122

Saddek B, Hacen B, Fateh A, Chara B, Mustapha B (2013) Artificial neural networks in prevention of nosocomials infections. Int J Sci Technol Res 2(10)

Scardoni A, Balzarini F, Signorelli C, Cabitza F, Odone A (2020) Artificial intelligence-based tools to control healthcare associated infections: a systematic review of the literature. J Infect Public Health 13(8):1061–1077. https://doi.org/10.1016/j.jiph.2020.06.006

Sereia AFR, Christoff AP, Cruz GNF, da Cunha PA, da Cruz GCK, Tartari DC, Zamparette CP, Klein TCR, Masukawa II, Silva CI, Vieira MLV, Scheffer MC, de Oliveira LFV, Sincero TCM, Grisard EC (2021) Healthcare-associated infections-related Bacteriome and antimicrobial resistance profiling: assessing contamination hotspots in a developing country public hospital. Front Microbiol 12:711471. https://doi.org/10.3389/fmicb.2021.711471

Shenoy ES, Paras ML, Noubary F, Walensky RP, Hooper DC (2014) Natural history of colonization with methicillin-resistant Staphylococcus aureus (MRSA) and vancomycin-resistant enterococcus (VRE): a systematic review. BMC Infect Dis 14:177. https://doi.org/10.1186/1471-2334-14-177

Smith KP, Kang AD, Kirby JE (2018) Automated interpretation of blood culture gram stains by use of a deep convolutional neural Network. J Clin Microbiol 56(3):e01521–e01517. https://doi.org/10.1128/JCM.01521-17

Somvanshi PR, Venkatesh KV (2014) A conceptual review on systems biology in health and diseases: from biological networks to modern therapeutics. Syst Synth Biol 8(1):99–116. https://doi.org/10.1007/s11693-013-9125-3

Stewart PS, Bjarnsholt T (2020) Risk factors for chronic biofilm-related infection associated with implanted medical devices. Clin Microbiol Infect 26(8):1034–1038. https://doi.org/10.1016/j.cmi.2020.02.027

Storr J, Twyman A, Zingg W, Damani N, Kilpatrick C, Reilly J, Price L, Egger M, Grayson ML, Kelley E, Allegranzi B, WHO Guidelines Development Group (2017) Core components for effective infection prevention and control programmes: new WHO evidence-based recommendations. Antimicrob Resist Infect Control 6:6. https://doi.org/10.1186/s13756-016-0149-9

Tabrizi B, Lam E, Girard K, Irvin V (2019) Digital transformation is not about technology. Harv Bus Rev 13(March):1–6

Thomas A, Redd A, Khader K, Leecaster M, Greene T, Samore M (2015) Efficient parameter estimation for models of healthcare-associated pathogen transmission in discrete and continuous time. Math Med Biol 32(1):79–98. https://doi.org/10.1093/imammb/dqt021

Thomas A, Khader K, Redd A, Leecaster M, Zhang Y, Jones M, Greene T, Samore M (2018) Extended models for nosocomial infection: parameter estimation and model selection. Math Med Biol 35(suppl_1):29–49. https://doi.org/10.1093/imammb/dqx010

Tong DL, Kempsell KE, Szakmany T, Ball G (2020) Development of a bioinformatics framework for identification and validation of genomic biomarkers and key immunopathology processes and controllers in infectious and non-infectious severe inflammatory response syndrome. Front Immunol 11:380

van Kleef E, Robotham JV, Jit M, Deeny SR, Edmunds WJ (2013) Modelling the transmission of healthcare associated infections: a systematic review. BMC Infect Dis 13:294. https://doi.org/10.1186/1471-2334-13-294

Vandecandelaere I, Matthijs N, Van Nieuwerburgh F, Deforce D, Vosters P, De Bus L, Nelis HJ, Depuydt P, Coenye T (2012) Assessment of microbial diversity in biofilms recovered from endotracheal tubes using culture dependent and independent approaches. PloS One 7(6):e38401. https://doi.org/10.1371/journal.pone.0038401

Weiner-Lastinger LM, Abner S, Edwards JR, Kallen AJ, Karlsson M, Magill SS, Pollock D, See I, Soe MM, Walters MS, Dudeck MA (2020) Antimicrobial-resistant pathogens associated with adult healthcare-associated infections: summary of data reported to the National Healthcare Safety Network, 2015-2017. Infect Control Hosp Epidemiol 41(1):1–18. https://doi.org/10.1017/ice.2019.296

Wen L, Ou Z, Duan W et al (2022) Using a 5G network in hospitals to reduce nosocomial infection during the COVID-19 pandemic. Commun Med 2:51. https://doi.org/10.1038/s43856-022-00118-3

Xu LY, Bai Y, Hu M, Xu YY, Sun ZQ (2006) Artificial neural network in the prediction of nosocomial infection risk. Zhong nan da xue xue bao. Yi xue ban= Journal of Central South University Medical Sciences 31(3):404–407

Yu CJ, Wang ZX, Xu Y, Hu MX, Chen K, Qin G (2021) Assessment of basic reproductive number for COVID-19 at global level: a meta-analysis. Medicine 100(18):e25837. https://doi.org/10.1097/MD.0000000000025837

Zoabi Y, Kehat O, Lahav D et al (2021) Predicting bloodstream infection outcome using machine learning. Sci Rep 11:20101. https://doi.org/10.1038/s41598-021-99105-2

Systems Biology Approach in Understanding Mitochondrial Disease

8

Moupriya Nag, Debarati Chatterjee, Sumana Paul, Arina Guha, Debasmita Bhattacharya, and Dibyajit Lahiri

Abstract

The term "mitochondrial diseases" refers to a broad category of hereditary disorders characterized by failure in the mitochondria, the cell's energy-producing organelles. The molecular mechanisms underlying mitochondrial diseases have benefited from traditional reductionist approaches, but a more comprehensive understanding is required due to the complex interplay of various factors within the mitochondria and their interactions with the larger cellular environment. The dynamic character of mitochondrial functions and their interconnections with other cellular components are now also taken into account by systems biology. In this article, we discuss the major methods and elements of the systems biology approach concerning mitochondrial illnesses. To establish a connection between the genes, proteins, and metabolites involved in mitochondrial function, we must first explain the multi-omics datasets. We also emphasized several mathematical formulae that explain the dynamic behavior of mitochondria and how they react to therapeutic treatments. So now our main goal of this research review is to elaborate some or the other applications of systems biology associated with bio-

M. Nag (✉)
University of Engineering and Management, Kolkata, Kolkata, West Bengal, India
e-mail: moupriya.nag@uem.edu.in

D. Chatterjee · S. Paul · A. Guha · D. Lahiri
Department of Biotechnology, University of Engineering and Management, Kolkata, West Bengal, India

D. Bhattacharya (✉)
Department of Basic Science and Humanities, Institute of Engineering and Management, University of Engineering and Management, Kolkata, West Bengal, India
e-mail: debasmita.bhattacharya@iem.edu.in

S. Joshi et al. (eds.), *Systems Biology Approaches: Prevention, Diagnosis, and Understanding Mechanisms of Complex Diseases*,
https://doi.org/10.1007/978-981-99-9462-5_8

chemistry for studying challenges of the uncharacterized proteins and to open the broader aspects of mitochondrial disorders.

Keywords

Mitochondrial diseases · Systems biology · Multi-omics datasets · Therapeutic treatments

8.1 Introduction

All eukaryotic cells include mitochondria, a membranous cell organelle, whose main job is to produce more energy in the form of adenosine triphosphate (ATP). They have originated from prokaryotic cells. They have a distinct genome, and two distinct membranes, and use binary fission for reproduction. Now through the outer membrane, small molecules can pass, as it has specialized channels that can carry molecules. On the other hand, the inner membrane is much less permeable, permitting only a few tiny molecules to enter the gel-like matrix. The matrix mainly contains DNA of mitochondrial genes and some enzymes required for the TCA cycle and can be used for energy production. The cellular power plants, mitochondria are mostly known for their involvement in oxidative phosphorylation (OXPHOS), which catabolizes foods in producing adenosine triphosphate (ATP) (Rahman and Rahman 2018). Mitochondrial DNAs are more prone to mutations as it does not possess robust DNA repair pathways such as nuclear DNA. Mutations in the mtDNA are the root cause of several illnesses and disorders connected to mitochondrial malfunction (Khan et al. 2017). A large number of mitochondrial disorders are brought on by changes to the nuclear genome's genes, which produce a wide range of proteins. However, several mitochondrial illnesses are significantly influenced by mtDNA mutations. The development of cancers and other illnesses, as well as aging, all are thought to be significantly influenced due to the accumulation of mtDNA mutations throughout an organism's lifetime. Researchers came across that variations in mtDNA by various generations have helped to decipher the geographical origins of different human populations.

Recent advancements in high-throughput omics technologies, like next-generation sequencing and mass spectrometry, have allowed for unusually in-depth molecular physiology studies of whole-tissue and organismal equilibrium. Only around 2000 of the roughly 19,000 genes in the human genome are the focus of most research on human genes, and only 100 genes are responsible for more than one-quarter of the publications as categorized by the National Library of Medicine (Stoeger et al. 2018). On the other hand, one-third of the genes have not yet attained their full potential, unlike model organisms like *E. coli*. So it can be even estimated that out of 5000 entries of enzymes in the Enzyme Commission, 1000 known enzymes still lack associated proteins (Sorokina et al. 2014).

The reason behind these is that they either might have redundant functions having a lack of standard laboratory conditions, or it might affect various other cellular processes during characterization (Hillenmeyer et al. 2008). In reality, a large number of uncharacterized (x) proteins (MXPs) in the mitochondria have lately been identified, and new mitochondria-related activities are continually being identified. So nowadays, scientists blend systems analysis with classical biology and biochemical approaches. This method seems to have been made especially suitable for mitochondria because of their known proteome and their controllable complexity and verifiability. Recent systems biochemistry methods have sped up the discovery of previously "missing" proteins that carry out essential tasks as well as novel mitochondrial proteins linked to illness. When taken as a whole, this research advances our knowledge of mitochondrial functions and provides a molecular foundation for the study of mitochondrial disease.

The search for disease-causing genes is a long-term goal for scientists. Despite having several molecular-level ways to identify a disease, the search for the gene that is causing the disease was not found. But this systems biology approach has made it somewhat possible in disease identification as well as in identifying the genes of interest (e.g., identification the cause of cystic fibrosis) (Rommens et al. 1989). The traditional way medical science approaches disease identification is far more different from that of systems aspects. Systems try to fill the gap between the phenotypic and genotypic approaches to disease identification. The systems approach has a higher potential in biochemical research than in studying the disease at molecular levels and considers the whole as an operating system. Due to the computational blending of many kinds of genome-wide measurements, systems methods are taking on a greater significance in this field.

The next-generation sequencing and mass spectrometry-based high-throughput omics technologies have made significant advancements recently, enabling exceptionally in-depth molecular physiology investigations of whole-tissue and organismal homeostasis. By consolidating supporting evidence from various levels and integrating diverse levels of these data (multi-omics), strong insights into complicated molecular functional processes are made possible. The multi-omics techniques are anticipated to be extremely valuable for revealing the pathogenic pathways in mitochondrial illnesses, a disease group extending to practically all disciplines of medicine since mitochondrial processes are implicated at several levels of regulatory pathways (Hasin et al. 2017). The diagnosis of mitochondrial diseases is difficult since the symptoms vary. Given that up to 40–60% of patients have a known genetic abnormality, the introduction of next-generation sequencing tools has been very beneficial (Maldonado et al. 2019). The genetic diagnosis enables counseling and occasionally guides treatment decisions. It also serves as the starting point for studying the illness processes. To encourage the identification of novel pathways underlying tissue-specific symptoms, extensive molecular and physiological studies using appropriately fabricated disease models and patient materials are necessary.

8.2 Advances in Systems Mitochondrial Biology

The diagnosis of mitochondrial illnesses has grown more difficult as a result of their diverse symptoms. Since up to 40% to 60% of patients have a known genetic defect, the development of next-generation sequencing technologies has proven extremely helpful (Maldonado et al. 2019). The pathophysiology and function of mitochondria must be better understood to improve diagnostic and therapeutic results. Systems biology, which is the modeling of large biological systems using computational and mathematical techniques, has recently become a useful tool for analyzing and characterizing complex cellular and organellar connections in both healthy and pathological states. Biological networks must be constructed using omics data in order to quickly uncover new biomarkers and treatment targets for mitochondrial disease (Suomalainen et al. 2011). So nowadays there are simplified systems-based approaches that are being categorized into "top-down" and "bottom-up" approaches. The top-down workflow entails the collection of samples, their processing using high throughput techniques, and their analysis using bioinformatics tools. On the other hand, the bottom-up can be defined as the identification of molecular data, the structuring of these data into a genome-scale metabolic model (GEM), and the use of constraint-based modeling (CBM) to forecast solutions and improve processes. Hence together by offering more information on mitochondrial disorders, therapy options, and eventually enhancing patient healthcare, these systems techniques can support mitochondrial research as a whole. Again in one of the research works, it is proved that to get a comprehensive knowledge of a complex biological system, the combined use of top-down and bottom-up techniques has shown to be potent instruments for successfully integrating experimental data (Sun et al. 2018). The multi-omics approaches such as integrative genomics, multiomics datasets, interactive models, mitochondrial GEMs and CBMs, and a few more systems biology models for mitochondrial research.

It has now turned out to be extremely fortunate that the emergence of new technologies, most notably mass spectrometry (MS)-based proteomics, which allowed researchers to examine the minute details of mitochondria in much more depth, corresponded with the surge in mitochondrial biology (Aebersold and Mann 2003). Now in our recent studies, the reviews not only cover the mitochondrial proteomes but also extend to revealing the submitochondrial properties. Predictive computational biology has more recently developed as a key component of systems methods. It offers a logical progression within experimental biology to clarify intricate, cooperative actions that support features that originate from a biological system as a whole (Kitano 2002). The network-based approaches enable the observation and manipulation of many interactions between genes, transcripts, proteins, and metabolites, and are effective tools for studying mitochondrial function as well as malfunction. This makes it possible to understand integrative mitochondrial activities and might hasten the identification of new connections that could otherwise go unnoticed when utilizing conventional experimental methods. One of the researchers (Suomalainen et al. 2011) has mentioned that this method not only allows for the discovery of novel pathomechanisms but also the identification of novel biomarkers for the treatment of mitochondrial diseases.

8.3 Comprehensive Analysis of Proteins Found in Mitochondria

A well-established protein library for a biological system overcomes the major challenge in two ways: it discloses the system's unrecognized components. It establishes the foundation for a methodical process to give those parts their functions. For instance, the MitoCarta study used the extended compendium to identify additional genes associated with the Complex I (CI) disorder after revealing that the functional characterization of nearly 300 mammalian proteins found in mitochondria was minimal to nonexistent (Pagliarini et al. 2008). In MitoCarta 2.0, the number of citations from the NCBI or National Centre for Biotechnology Information PubMed that referenced each human protein's GeneID was counted (Calvo and Mootha 2010). These patterns demonstrate an established mitochondrial system's continuous effects on accelerating the study of its understudied proteins. The proteins were then sorted into deciles according to the total number of citations that each protein received. About three times as many publications are cited for the leading 10% of the proteins found in mitochondria, or the "on trend" proteins, as for the entire proteome's bottom half, or the "unpopular proteins. However, an optimistic pattern appears when the analysis's findings are split between references to the MitoCarta study and other sources. Before this study, the mitochondrial proteins in the top tenth accounted for over half of the citations about human mitochondria, whereas the bottom decile accounted for just 16%. Following that decade, this distribution started to change. Between 2009 and 2019, the top decile of citations about human mitochondria declined to about 40%, whereas the lower half climbed to 20%. In contrast to all earlier citations, the share of citations is attributable to each decile over the previous 10 years, this pattern becomes much more obvious. Compared to all citations in that decile before 2009, the number of sources for the bottom decile of proteins has increased by more than double during the past 10 years. Given the growing divide between the top decile and the bottom half of the proteome, the reverse trend in the general distribution of PubMed citations for the entire human proteome is particularly fascinating.

Systems biochemistry approaches, which have been the inspiration for a number of these studies, usually take one of two shapes: a top-down strategy looking for an absence of a routine procedure (known or unknown), or a bottom-up strategy beginning by purposefully disrupting poorly defined genes (known unknowns).

8.3.1 Insights from Structural Genomics into Coenzyme Q Biosynthesis

For the majority of proteins found in nature, a bottom-up approach termed structural genomics tries to simulate various information on structure. Hoping that structural data would offer insights into how proteins work and aid in future protein structure prediction from primary sequence, targets for determining structure are commonly selected before their biochemical roles are clarified (Skolnick et al. 2000). A variety of proteins present in mitochondria that includes COQ4, COQ8A, and COQ9 have benefited from recent initiatives like the protein structure initiative (Montelione 2012; Stefely and Pagliarini 2017). Even though all of these three proteins are essential for the creation of CoQ, nothing is known about how they work on a molecular level (Stefely and Pagliarini 2017). It was first believed that COQ8A (also called ADCK3) was a protein kinase (Poon et al. 2000). However, as shown by further structural and biochemical research, it appears more likely that its function as an ATPase protects the stability of a biosynthetic compound with additional CoQ-related proteins (Stefely et al. 2015; Reidenbach et al. 2018; Rhee et al. 2014; Floyd et al. 2016). Unexpectedly, when the ligand is attached, an analysis of the COQ9 structure revealed that it transforms into the vintage TetR shape, which is typically used by bacteria to control transcription.

Further investigation showed that like CoQ, COQ9 binds prenyl lipids preferentially and interacts strongly physically and functionally with COQ7, an enzyme for the catalysis of the CoQ production final phase (Lohman et al. 2014). In order to help overcome the biophysical challenges of synthesizing a highly hydrophobic molecule, the combination of the COQ9 structure work raised its identification as an innovative prenyl lipid chaperone that links CoQ intermediaries and presents these to other peripheral-membrane-associated biosynthetic enzymes. (Lohman et al. 2019). Even though the functioning of COQ4 is yet to be researched, the bacterial homolog, Alr8543, demonstrated an alike lipid-binding domain and associated transition metals, leading to fresh mechanistic stories on its role in the system. Overall, the systems biochemistry method is best demonstrated by structural genomics, where the "screen" provides quick information that aids in the development of particular novel mechanistic theories (Rea et al. 2010).

8.3.2 Cristae Coordinating System and Mitochondrial Interface Site

The maintenance of cristae junctions, inner membrane architecture, and the formation of contact sites between the inner and outer membranes are all made possible by the mitochondrial contact site and cristae organizing system, which is conserved through yeast to humans. The description of the *Saccharomyces cerevisiae* MICOS component was established by four distinct investigations in 2011 and 2012 (Pfanner et al. 2014). Here, two of the system's biochemistry methods are described—one top-down and another bottom-up—that were used to identify the components of the MICOS group in yeast. In a groundbreaking study, Hoppins et al. (2011) created a bottom-up statistical genetic association map that concentrated exclusively on genes that code for proteins present in the mitochondria as well as a small number of genes whose functions are expected to affect mitochondria. Tracking genetic interactions systematically can be performed to determine the pairwise development losses in cells with two mutations and the related gene function (Schuldiner et al. 2005). The generated MITO-MAP tied MXPs successfully to their functions in building mitochondrial membrane organization, both physically and functionally. Their research located a group of genes that produce six proteins that together make up a sizable hetero-oligomeric inner mitochondrial membrane protein family. At the same time, to get the same conclusion, Harner et al. (2011) applied an orthogonal top-down technique. The researchers set out to find the proteins that make these junctions after observing that both outer and interior membranes remained connected at various points throughout studies on mitochondrial expansion and contraction. At these points of interaction, they first developed an indicator protein that crossed both the exterior and interior membranes of the mitochondria.

8.3.3 Integrating Top-Down and Bottom-Up Approach

By itself, top-down and bottom-up techniques of systems biochemistry reveal significant perception in as-of-yet exceptional proteins and processes. Even though one technique has been highlighted in each of the vignettes above, of course, these two methods are not antagonistic to one another. Instead, they cooperate to develop CI construction and the organization of the cristae and contact sites are two distinct subfields of mitochondrial biology. One of the most typical inborn metabolic errors globally is isolated CI insufficiency (Kirby et al. 1999). Even though the structure and biological activity of mature CI have been thoroughly investigated (Agip et al. 2018). Mostly, CI diseases are directly not connected with structural CI subunit mutations but rather by disruption of genes called CI assembly factors (CIAFs), which play a crucial role in the construction and/or operation of the developed complex (McKenzie et al. 2011). CIAFs have proven difficult to locate because they are not a component of the finished complex (Guerrero-Castillo et al. 2017). Different approaches in systems biochemistry have helped locate these elusive proteins. A top-down system first, using a phylogenetic technique, mitochondrial proteins were discovered in the initial MitoCarta analysis whose patterns of presence or absence resembled the patterns of the bacteria-specific CI subunits. Six among the best 19 candidates found in this investigation that include disease-causing mutations have recently been confirmed to be real CIAFs (Pagliarini et al. 2008; Sugiana et al. 2008). Second, by doing MS-based proteomics on cells that were deficient in one of either the 25 auxiliary CI subunits or the 7 assembly factors, Stroud et al. (2016) discovered two CIAFs. Third, through the use of an interaction proteomics approach, researchers were able to pinpoint a novel CIAF by way of its mutually beneficial interactions with many essential CI subunits and previously recognized CIAFs (Guarani et al. 2014). Fourth, Heide et al. (2012) developed a method dubbed "complexion" profiling that utilizes native gel electrophoresis and MS to find proteins that co-migrated with CI subunits throughout the CI assembly phase, identifying an additional assembly component. Guerrero-Castillo et al. (2017) improved on this method by defining the CI assembly intermediates and deducing the sequential assembly process that results in mature CI. These extensive initiatives in each case utilized the well-established mitochondrial proteome and "omic''technology.

The discovery of new CIAFs and the addition of mechanistic depth to already discovered CIAFs have both benefited greatly from bottom-up approaches. The former is demonstrated by using the research of protein–protein interactions (PPI) to extrapolate the functions of uncharacterized proteins. Using deep affinity enrichment mass spectrometry investigations, Floyd et al. (2016) mapped the PPIs of 50 MXPs. One of the best matches was the orphan protein C17orf89, which eventually proved to be a CIAF and was given the new name NDUFAF8. It interacted with a known assembly factor. The consequences of NDUFAF7 suppression on the methylation state of the putative substrate were compared by the researchers, NDUFS2, using an MS technique. Their research established NDUFAF7's function in NDUFS2's post-translational maturation by demonstrating that it was required for the symmetric demethylation of NDUFS2 (Rhein et al. 2013). Combining the top-down and the bottom-up systems biology has led to the discovery of over 15 CIAFs, expanding the scope and depth of our knowledge of CI assembly (Formosa et al. 2018).

8.4 Various Approaches for Multi-Omics Data Interpretation

Large, complex datasets have proliferated as a result of the introduction of multi-omics approaches in mitochondrial biology (Rahman and Rahman 2018). Many of these datasets are available online, including the Integrated Mitochondrial Protein Index (IMPI) Q2 2018, MitoBreak, The EMPOP database, MitoMiner 4.0MitoProteome, and many more (Table 8.1).

Table 8.1 List of some mitochondrial databases

Name of the database	Main uses	Last update of the database	Website
Integrated Mitochondrial Protein Index (IMPI) Q2 2018	It has a group of genes that are responsible for encoding proteins	2018	http://www.mrc-mbu.cam.ac.uk/impi
MitoMiner 4.0	Confirmation of mitochondrial location and knowledge of yeast, zebrafish, and mammal traits	2018	http://mitominer.mrc-mbu.cam.ac.uk
MitoBreak	Collection of mDNA breakpoints	2017	http://mitobreak.portugene.com/cgi-bin/Mitobreak_home.cgi
The EMPOP database	It has the worldwide collection, evaluation, and searchable display of mtDNA haplotypes	2015	https://empop.online/
HmtDB	It includes population and variation data annotated human mitochondrial genome sequences	2018	www.hmtdb.uniba.it

However, it is extremely difficult to make significant conclusions given the size of this data. Large complicated datasets may now be managed and analyzed, and meaningful biological interpretation is made possible because of the advent of sophisticated bioinformatics pipelines (Luscombe et al. 2001). The capacity of bioinformatics to anticipate the behavior of complex systems is constrained, even though it can offer extra insight by reanalyzing experimental data. Predictive computational biology has more recently evolved into a core component of systems methods. To clarify complex, synergistic, interacting behaviors that support emergent features from a biological system evaluated as a whole provide a natural progression within experimental biology (Pagliarini and Rutter 2013). To include and harmonize the increasingly accessible independent, disparate datasets, systems methods can utilize some complementary techniques, such as multiple dataset analyses, GEMs, omics integration tools, and constraint-based modeling (CBM).

8.4.1 Omics as a Tool in Mitochondrial Research

The pathophysiology of primary mitochondrial disease and the fundamental biology of mitochondrial circuits, as well as in models of neoplasms and aging, are some remaining pathophysiological mitochondrial topics that may be addressed using systems biology. Using global, unbiased information from "omics' 'datasets or high-throughput experiments is one way to improve our comprehension of intricate biological systems. The bulk of this research has made conclusions and focused solely on one form of analysis. The target molecules for AMPK's mitochondrial phosphorylation or the sequential assembly of eukaryotic mitochondrial complex I are only two examples of newly discovered aspects of mitochondrial structure and function (Stroud et al. 2016; Toyama et al. 2016).

Omics-based studies have proven helpful in comprehending illness pathways beyond OXPHOS dysfunction or energy deficit, which is typically inadequate to explain the clinical phenotypes seen in patients (Legault et al. 2015; Esterhuizen et al. 2018).

In mice models of adult and pediatric mitochondrial illness, the alteration of the metabolism of one carbon in aberrant mtDNA preservation is one well-known example (Nikkanen et al. 2016). When *Saccharomyces cerevisiae* genome-scale data became available, yeast researchers began integrating massive functional genomics datasets to identify probable mitochondrial genes associated with disease more than 10 years ago (Aiyar et al. 2008). A Bayesian statistical model was used at the same time by the Harvard Mootha team to incorporate data from various genome-scale sets of data and predict the chances that a specific candidate protein will work in the mitochondria (Calvo et al. 2006).

The resultant compendium, dubbed "Maestro," employed 8 sets of data to calculate the possibility that the Ensemble human genome database found 33,860 proteins would localize in mitochondria: the TargetP program predicted the existence of a mitochondrial import sequence at the N-terminus; the MitoPred program projected that protein domains would exist indicating mitochondrial activity;

cis-regulatory regions within 2 kb upstream of the gene that contain any of the binding sites for the three transcription factors Err, Gapba, and NRF1; similarity to one or more of the 749 proteins in mitochondria that *S. cerevisiae* genome encodes; proteins in *Rickettsia prowazekii*, which is regarded to be the most closely related bacterial source of mitochondria, have ancient bacterial homology; co-expression of genes with mitochondrial protein-encoding genes; an analysis of the mouse mitochondrial proteome by tandem mass spectrometry; transcriptional activation during a cell-based simulation of biogenesis of mitochondria, in which access to the mitochondrial transcriptional coactivator PGC1-induced mitochondrial growth.

When Maestro was initially developed, it accurately predicted 71% of mitochondrial proteins. The mitochondrial genomic diagnostic revolution has been driven by more current iterations of the complication called MitoCarta and MitoCarta 2.0 (Calvo et al. 2016). In the future, it is hoped that new information and methods will make it easier to create a comprehensive inventory of the mitochondrial proteome, including tissue-specific databases that could explain why particular mitochondrial gene abnormalities present differently in different tissues.

8.4.2 Bringing Together Multi-Omic Datasets

Multi-omics research is becoming increasingly important as a tool for understanding how the mitochondrion functions dynamically. An integrative omics strategy is utilized when different compatible omics signatures (transcriptomic, genomic, and proteomic) are used to verify a biological discovery—as opposed to one-omics-based techniques—is advantageous because it raises the assurance of the result. Making endpoint phenotypic assumptions based solely on untargeted transcriptome analysis without previous connections might be erroneous or deceptive since transcriptomic and proteomic markers frequently do not match (Wu et al. 2014).

This is partially accounted for by the reality that high-throughput finding techniques primarily concentrate on endpoint analyses (transcript/protein level) and are unable to precisely recognize transcriptional and translational control activities that may alter the expression of proteins and have downstream functional ramifications (Fisher-Wellman et al. 2018). Specialist assays that can work with omics approaches to track real-time phenotypic changes in response to global transcriptomic/proteomic remodeling can help in meaningful functional interpretations. One such method that has been applied in combination with genomics for finding new genes that are involved in OXPHOS is genome-wide CRISPR-Cas9 screening (Arroyo et al. 2016). Multi-omics research can also benefit from functional phenotyping methods that are consistent with high-throughput screens, including a new assay set created for identifying mitochondrial bioenergetics variables in real time. Recent advances in multi-layered omics have illuminated the molecular mechanisms underlying numerous intricate mitochondrial processes, such as the induction of the response of unfolded proteins in mitochondria, the identification of transcription factor ATF4 as a mitonuclear stress regulator, the importance of mitoribosomes and complex IV subunits in T-cell activation, and the role of mitoprotease Oct1p as a

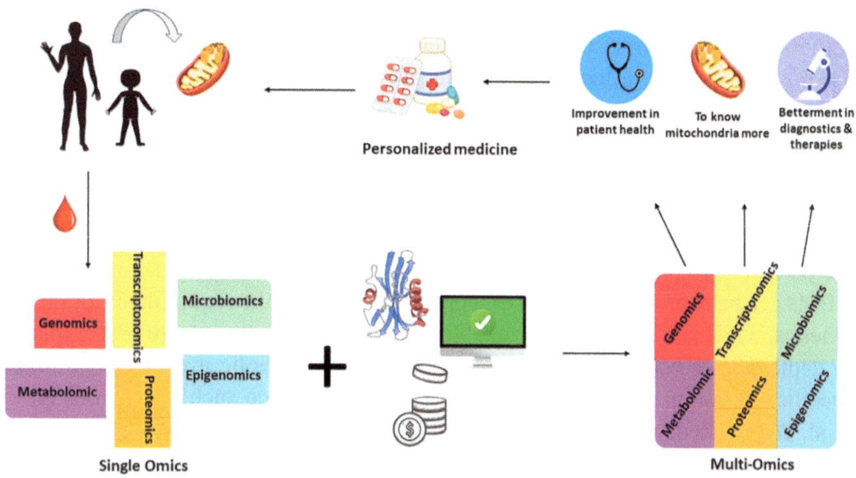

Fig. 8.1 Importance of multi-omics in MD. Patients with MD get skin or muscle biopsies or provide blood samples. From tissue biopsies, fibroblasts or myotubes are produced and cultivated. For identifying the genes causing the disease, several omics analyses are conducted independently. These analyses, however, which are not always decisive. Despite the single-omics approach's validity, the diagnosis' effectiveness is constrained by MD's complexity. Multi-omics can help in this situation. Data from several omics analyses may be combined to provide superior findings. With the use of databases, literature, and multi-omics, mitochondrial illnesses are better understood

novel regulator of Coenzyme Q10 biosynthesis. In mitochondrial physiology, integrated omics can be used to distinguish between main disease-causing processes and coping mechanisms (Wu et al. 2014; Quirós et al. 2017; Tan et al. 2017; Veling et al. 2017) (Fig. 8.1).

This was recently proven by research, in which multi-omic profiling of benign renal tumors rich in mitochondria revealed a selective loss-of-function of complex I, which in turn led to compensatory glutathione production (Gopal et al. 2018). According to many researchers, these integrated investigations are currently only conducted using cell and animal models. However, this body of work underscores the significance of systems methods and integrative omics in uncovering novel pathways when more pertinent patient populations and animal disease models become accessible. It will surely come in handy when studying the pathophysiology of mitochondrial diseases in these patient populations and these animal disease models as they are employed more frequently.

8.4.3 Genome-Scale Metabolic Models

To handle the diversity of data that has been accumulated over a decade, datasets have been transformed into biochemically and genetically coherent forms, mathematically structured "understanding bases," like GEMs (Papin et al. 2003; (O'Brien et al. 2015; Orth et al. 2010; Maldonado et al. 2019). An organism's whole metabolic data is used to create a GEM, which is a categorized list of metabolic activities. To reconstruct GEMs into a mathematically ordered network, a formula, $S = (m \times n)$, may be utilized. Here, m stands for the amount of metabolic products and n for the certain number of processes needed to carry out CBM. The system is frequently thought to be in a state of equilibrium, in which case zero is the net flux, or Sv is equal to 0 and v is the flux vector. Enzymatic capacity, nutritional availability, and thermodynamic constraints—which permit irreversible reactions—are other often employed restrictions. These can also be used to target metabolic events that are particular to phenotypic behavior such as growth or energy generation, so that GEMs can be examined using CBM approaches without the requirement for precise kinetics, which are typically unavailable. These representations were first employed to depict single-celled organisms with just a few metabolic events to optimize a specific phenotype or a function of objective significance such as biomass (Price et al. 2003). Recon3D, the most current global reconstruction of human metabolism, shows how the breadth and scope of metabolic reconstructions have expanded to include multicellular animals with noticeably larger genomes (Brunk et al. 2018). There are certain drawbacks to employing GEMs for CBM approaches, including the fact that metabolic behavior is frequently modeled using arbitrary rates and that traditional CBM methods frequently struggle to capture stochastic, complicated behavior like whole-cell metabolism. This may be because of a variety of CBM limitations, but one significant area of debate in the industry is the application of a single, precisely stated objective function. In the past, since it was simpler to identify a certain desirable trait, this method was applied to biotechnological research of single cellular creatures, such as growth or biomass. This becomes considerably less obvious when simulating complex organisms when it is desirable to explore several phenotypes. Although there is still controversy about this, more recent developments in CBM consist of the creation of more complicated algorithms that permit the modeling of more complex behavior, as will be briefly covered in the next section of this study. Even while this discipline is also growing, the majority of genome-scale modeling is more concerned with metabolism than signaling (Hyduke and Palsson 2010; Münzner et al. 2017).

8.4.4 Models from Other Systems in Biology for Mitochondrial Research

The Vo mitochondrial model was one of the first GEMs that presented the entire cell rather than merely the mitochondrion. Over the last 10 years, whole-cell GEMs have seen a significant evolution, with Recon3D (Brunk et al. 2018) being the most recent global GEM to be released (Vo et al. 2004; Duarte et al. 2007).

The Virtual Metabolic Human (VMH) provides access to Recon3D, which is visually shown in ReconMap. Recon has been linked to several primary mitochondrial disorders, including phosphate carrier deficit, SUCLA2-related mtDNA depletion syndrome, and mitochondrial trifunctional protein deficiency (Noronha et al. 2017, 2019; Sahoo et al. 2012). Multi-tissue systems have been represented using global GEM reconstructions for predictive modeling. This technique may be especially helpful in illustrating the main mitochondrial failure in any number of tissues, interactions between tissues, and the effects of these relationships on the network as a whole. A brain model with several cell types is an illustration of a more in-depth mitochondrion-centric model. The ultimate rebuilding of multi-tissue mitochondrial models may be aided by such tissue-specific models (Bordbar et al. 2011; Lewis et al. 2010).

Another usual approach to computational modeling is the use of mathematical models, such as regular differential equations with kinetic data as variables, to predict physiological occurrences.

Beyond the purview of this study, the employment of mathematical frameworks for research in the field of fundamental mitochondrial science has a long history. However, a few notable cases will be briefly discussed below. Numerous mathematical research have focused on modeling aspects of the metabolism of mitochondrial energy and communication, calcium dynamics and regulation, ROS and redox, apoptosis, and fission and fusion.

Magnus and Keizer created a well-known mitochondrial model based on OXPHOS, and subsequent independent studies using various datasets and alternative biophysical theories, all of which focused on OXPHOS, resulted in additional models (Magnus and Keizer 1997; Korzeniewski and Zoladz 2001). The Magnus and Keizer model has been widely applied and modified, for example, to incorporate metabolic pathways including the TCA cycle, calcium dynamics, and ROS production (Dudycha 2000; Cortassa et al. 2003, 2004). Since then, further model iterations have been created to study these behaviors, such as ROS, pH control, and ion dynamics (Gauthier et al. 2013; Wei et al. 2011). Using this method, several studies on mitochondrial apoptosis have been conducted by researchers like Albeck, Bertaux, Fussenegger, and Rehm. More recently, these models have been developed to capture more mitochondrial behaviors, such as Dash and Zhang's "dynamic regulation of cellular metabolism and energetics." This is due to improvements in statistical and computational technologies as well as the growing availability of kinetic data. Furthermore, models have made an effort to represent variation within a mitochondrial population to examine mitochondrial spatial and temporal patterns (fission, fusion, mass, and motility) and heterogeneity by identifying the sources of within-cell variations in mitochondrial morphology and energetic stress states (Kowald and Kirkwood 2011; Johnston et al. 2015; Dalmasso et al. 2017). Primarily, metabolic and signaling model combinations have been included in kinetic modeling.

It is a potent approach that can represent metabolic concentrations and reaction rates in great detail and can depict dynamic behavior over time. The parameters may be constrained or unsuitable for the conditions under study, making this technique incompatible with modeling at the full cell or tissue size.

Leigh Map is a newly developed sort of predictive mitochondrial model, which was painstakingly curated, containing more than 500 papers up to November 2016 as well as 89 genes and 237 phenotypes from the senior author's clinical library. This model was developed as a Leigh syndrome diagnostic aid. It may be used to search a gene for symptoms that are connected or to query phenotypes to determine the gene(s) most likely to be the cause. In 20 cases of Leigh syndrome, the effectiveness of the Leigh Map was determined to be accurate up to 80% in gene identification (Rahman et al. 2017). Future research is being done to broaden the use of this approach to identify more main mitochondrial disorders (Rahman et al. 2017).

8.5 Systems Biochemistry and Mitochondrial Disease

Basic mitochondrial disorders, which can be brought on via both genomic and mtDNA anomalies, are some of the most common and incapacitating hereditary metabolic ailments and commonly result in premature mortality (Gorman et al. 2015). It is a diverse collection of diseases marked by a malfunction in oxidative phosphorylation (OXPHOS), which is thought to have a hereditary basis. It is a diverse collection of disease marked by a malfunction in oxidative phosphorylation (OXPHOS), which is thought to have a hereditary basis, even though clinical evidence frequently leads one to suspect the diagnosis. To assess if a particular patient is definitively suffering from a mitochondrial ailment, two diagnostic methods have been suggested for evidence of mitochondrial impairment (Bernier et al. 2002). Primary mitochondrial illnesses may exhibit either traditional genetic Mendelian inheritance patterns or maternal inheritance trends, which are mtDNA mutations. Moreover, the majority occurs due to nDNA mutations followed by an autosomal recessive type inheritance. The challenging aspect now is identifying the genes dispersed over two genomes that are in charge of mitochondrial dysfunction in a certain patient. Nowadays on a worldwide basis, various laboratories employ various techniques, and diagnostic facilities do not consistently perform fundamental enzymatic tests that are essential for the objective assessment of mitochondrial activity (Gellerich et al. 2004). The system's biochemistry method provides a paradigm for bridging this gap between genotype and phenotype in the post-genomic era, which enhances the identification and management of mitochondrial illnesses. So according to the research work of (Frazier et al. 2019), the pathogenesis of a disease can be better identified by pinpointing its cellular process that is being disrupted. The ability to characterize MXPs to understand protein function and then place them into biological pathways, akin to bottom-up systems biochemistry, enables us to choose genes for further in-depth diagnostic investigation. With the inclusion of other "omic" metrics like lipidomics and metabolomics, the diagnostic effectiveness of this approach will likely rise. In the long run, we expect that an extensive library of multi-omic profiles for various pathophysiologic asserts can be used to efficiently and precisely detect mitochondrial ailments by analyzing someone's multi-omic "signature," similar to how these signatures frequently facilitated protein categorization.

The great majority of mitochondrial illnesses lack efficient treatments in addition to difficulties in diagnosis. The bulks of existing therapies are calming, and they typically fall short of halting the progression of the illness. Therefore, additional systematic attempts to characterize protein activity and understand the underlying causes of malfunctioning mitochondria are needed to fuel the development of tailored therapeutics.

8.6 Challenges

Despite the developments, there are still some difficulties with the usage of systems biology in the research of mitochondria. Having two genomes presents a unique problem for the research of mitochondria. Variations in the mtDNA copy numbers and varying heteroplasmy levels among various tissues and cells present further challenges (Guantes et al. 2016). When examining the various characteristics of basic mitochondrial abnormalities that are multi-systemic, these difficulties became more evident. Systems biology techniques can be used to address some of these issues, such as whole-cell GEMs, which have been developed to map various important mitochondrial diseases as well as mitochondrial and nuclear genomes necessary for trafficking of membranes (Sahoo et al. 2012). Furthermore, whole-cell models can be used to simulate various tissues, and they may be improved for the study of mitochondria by creating more intricate models focused on the mitochondria (Lewis et al. 2010). The variety within a population of mitochondria as well as between individual cells has also been attempted to be captured by various computational techniques, such as agent-based modeling (Dalmasso et al. 2017). Even within the same family, there is notable phenotypic variability in mitochondrial diseases. A Leigh syndrome predictive diagnostic knowledge base has started to address these issues, which are brought on by heterogeneity, for the identification of mitochondrial illnesses (Rahman et al. 2017). To understand the biology and pathophysiology of mitochondria, systems-level data have been utilized, and analytical tools have constantly been improved. Methods based on networks have been essential for systems biology development as they make complicated molecular interactions easier to understand. But recently, these networks had a drawback: the modalities of the network were based on one datum type study and manual comparisons of diverse data type networks (Barabasi et al. 2011; Stevens et al. 2014; Wang et al. 2014). This strategy unsuitably prompted the creation of integrative network tools. The nature of the data integration network has also changed, favoring patient (population/study cohort)-centric approaches like SNF, which are yet to be implemented with the datasets of mitochondria. In the end, this would be done to create more effective methods for integrating data from many omics to better understand mitochondrial biology and pathology. It has been demonstrated that the creation and analysis of computer-based models contribute to a deeper understanding of illness mechanisms and the search for new therapeutic options. The drawbacks of these network models are that their accuracy depends on the data that is fed into them. As a result, these models frequently only depict existing pathways. However, future development of well-known models like MitoCore may be possible when data and integration technologies become more widely available. Future computer models and related analysis instruments will be crucial in improving diagnostic and treatment results for primary mitochondrial illnesses as our understanding of mitochondrial biology and pathophysiology grows (Maldonado et al. 2019).

8.7 Future Perspectives and Conclusion

Although there are still some issues to be resolved, multi-omics techniques are quite promising for enhancing diagnostic capabilities. They are often created for cancer research, which has access to vast quantities of samples, whereas MD does not. Therefore, multi-omics techniques that are relevant to small cohorts must be developed. Additionally, similar approaches in oncology search for common signatures, but in MD, the majority of signatures are "private" and pertain to a single patient or family, such as a patient's changed genes. For small patient groups, it could be possible to find common signs, such as a lack of a certain respiratory chain component.

The creation of MD-specific databases presents another difficulty. Since there are so few individuals with rare diseases, there are not much data on them, and even fewer patients have the same pathogenic variation. Because data from a single institution are insufficient, it would be feasible to increase the number of available cohorts and advance our understanding of these diseases by creating interoperable national and European clinical-biological databases. The RD-Connect project is one of the several efforts that have been established. Since it was founded in 2012 with funding from the European Union, RD-Connect has been working to create tools and data transmission protocols for omics science and bioinformatics analysis to be integrated with a centralized system that connects patient databases, biobanks, and medical computational biology information into one repository for the study of rare diseases (Johnston et al. 2014).

International and multidisciplinary partnerships will be crucial in the future to create more efficient tools, share data, and overcome the diagnostic deadlock. This research has thus allowed us to create a new categorization by condensing the primary multi-omics integration techniques, which will assist the entire scientific community by making it easier for them to select a technique that is appropriate for each type of data collection.

To better understand human disorders, develop methods for prognosticating outcomes, discover biomarkers, and find molecular markers, the use of multi-omics in computational biology is currently developing and may soon surpass the usage of single omics. However, there are still a few things that need to be developed in the future. Further comparative studies will be required to evaluate the effectiveness of tools in settings other than cancer to select the best tool depending on the available data, even if a certain instrument may not always be preferred. To estimate false positives and false negatives more precisely, reference data should also be created using simulation techniques. It is necessary to create two more ideas. On the one hand, creation of instruments can be utilized for smaller cohorts which can handle a greater number of variables than patients. On the other hand, creation of databases for healthy people will be utilized as an indicator for verifying the technologies.

The fusion of numerous layers of knowledge also requires collaboration between researchers in other domains. The creation of new treatment strategies as well as the clarification of how diseased processes take place benefit greatly from this superimposition of knowledge.

References

Aebersold R, Mann M (2003) Mass spectrometry-based proteomics. Nature 422:198–207

Agip ANA, Blaza JN, Bridges HR, Viscomi C, Rawson S, Muench SP, Hirst J (2018) Cryo-EM structures of complex I from mouse heart mitochondria in two biochemically defined states. Nat Struct Mol Biol 25(7):548–556

Aiyar RS, Gagneur J, Steinmetz LM (2008) Identification of mitochondrial disease genes through integrative analysis of multiple datasets. Methods 46:248–255. https://doi.org/10.1016/j.ymeth.2008.10.002

Arroyo JD, Jourdain AA, Calvo SE, Ballarano CA, Doench JG, Root DE et al (2016) A genome-wide CRISPR death screen identifies genes essential for oxidative phosphorylation. Cell Metab 24:875–885. https://doi.org/10.1016/j.cmet.2016.08.017

Barabasi AL, Gulbahce N, Loscalzo J (2011) Network medicine: a network-based approach to human disease. Nat Rev Genet 12:56–68. https://doi.org/10.1038/nrg2918

Bernier FP et al (2002) Diagnostic criteria for respiratory chain disorders in adults and children. Neurology 59(9):1406–1411

Bordbar A, Feist AM, Usaite-Black R, Woodcock J, Palsson BO, Famili I (2011) A multi-tissue type genome-scale metabolic network for analysis of whole-body systems physiology. BMC Syst Biol 5:180. https://doi.org/10.1186/1752-0509-5-180

Brunk E, Sahoo S, Zielinski DC, Altunkaya A, Drager A, Mih N et al (2018) Recon3D enables a three-dimensional view of gene variation in human metabolism. Nat Biotechnol 36:272–281. https://doi.org/10.1038/nbt.4072

Calvo SE, Mootha VK (2010) The mitochondrial proteome and human disease. Annu Rev Genomics Hum Genet 11:25–44

Calvo S, Jain M, Xie X, Sheth SA, Chang B, Goldberger OA et al (2006) Systematic identification of human mitochondrial disease genes through integrative genomics. Nat Genet 38:576–582. https://doi.org/10.1038/ng1776

Calvo SE, Clauser KR, Mootha VK (2016) MitoCarta2.0: an updated inventory of mammalian mitochondrial proteins. Nucl Acids Res 44:D1251–D1257. https://doi.org/10.1093/nar/gkv1003

Cortassa S, Aon MA, Marban E, Winslow RL, O'Rourke B (2003) An integrated model of cardiac mitochondrial energy metabolism and calcium dynamics. Biophys J 84:2734–2755. https://doi.org/10.1016/S0006-3495(03)75079-6

Cortassa S, Aon MA, Winslow RL, O'Rourke B (2004) A mitochondrial oscillator dependent on reactive oxygen species. Biophys J 87:2060–2073. https://doi.org/10.1529/biophysj.104.041749

Dalmasso G, Marin Zapata PA, Brady NR, Hamacher-Brady A (2017) Agent-based modeling of mitochondria links sub-cellular dynamics to cellular homeostasis and heterogeneity. PLoS One 12:e0168198. https://doi.org/10.1371/journal.pone.0168198

Duarte NC, Becker SA, Jamshidi N, Thiele I, Mo ML, Vo TD et al (2007) Global reconstruction of the human metabolic network based on genomic and bibliomic data. Proc Natl Acad Sci U S A 104:1777–1782. https://doi.org/10.1073/pnas.0610772104

Dudycha S (2000) A detailed model of the tricarboxylic acid cycle in heart cells. Johns Hopkins University, Baltimore, MD

Esterhuizen K, Lindeque JZ, Mason S, van der Westhuizen FH, Suomalainen A, Hakonen AH et al (2018) A urinary biosignature for mitochondrial myopathy, encephalopathy, lactic acidosis and stroke like episodes (MELAS). Mitochondrion 45:38. https://doi.org/10.1016/j.mito.2018.02.003

Fisher-Wellman KH, Davidson MT, Narowski TM, Lin C-T, Koves TR, Muoio DM (2018) Mitochondrial diagnostics: a multiplexed assay platform for comprehensive assessment of mitochondrial energy fluxes. Cell Rep 24:3593.e10–3606.e10. https://doi.org/10.1016/j.celrep.2018.08.091

Floyd BJ, Wilkerson EM, Veling MT, Minogue CE, Xia C, Beebe ET, Pagliarini DJ et al (2016) Mitochondrial protein interaction mapping identifies regulators of respiratory chain function. Mol Cell 63(4):621–632

Formosa LE, Dibley MG, Stroud DA, Ryan MT (2018) Building a complex complex: assembly of mitochondrial respiratory chain complex I. Semin Cell Dev Biol 76:154–162

Frazier AE, Thorburn DR, Compton AG (2019) Mitochondrial energy generation disorders: genes, mechanisms, and clues to pathology. J Biol Chem 294:5386–5395

Gauthier LD, Greenstein JL, O'Rourke B, Winslow RL (2013) An integrated mitochondrial ROS production and scavenging model: implications for heart failure. Biophys J 105:2832–2842. https://doi.org/10.1016/j.bpj.2013.11.007

Gellerich FN et al (2004) The problem of inter-lab variation in methods for mitochondrial disease diagnosis: enzymatic measurement of respiratory chain complexes. Mitochondrion 4(5–6):427–439

Gopal RK, Calvo SE, Shih AR, Chaves FL, McGuone D, Mick E et al (2018) Early loss of mitochondrial complex I and rewiring of glutathione metabolism in renal oncocytoma. Proc Natl Acad Sci U S A 115:E6283–E6290. https://doi.org/10.1073/pnas.1711888115

Gorman GS, Schaefer AM, Ng Y, Gomez N, Blakely EL, Alston CL et al (2015) Prevalence of nuclear and mitochondrial DNA mutations related to adult mitochondrial disease. Ann Neurol 77:753–759. https://doi.org/10.1002/ana.24362

Guantes R, Diaz-Colunga J, Iborra FJ (2016) Mitochondria and the non-genetic origins of cell-to-cell variability: more is different. Bioessays 38:64–76. https://doi.org/10.1002/bies.201500082

Guarani V, Paulo J, Zhai B, Huttlin EL, Gygi SP, Harper JW (2014) TIMMDC1/C3orf1 functions as a membrane-embedded mitochondrial complex I assembly factor through association with the MCIA complex. Mol Cell Biol 34(5):847–861

Guerrero-Castillo S, Baertling F, Kownatzki D, Wessels HJ, Arnold S, Brandt U, Nijtmans L (2017) The assembly pathway of mitochondrial respiratory chain complex I. Cell Metab 25(1):128–139

Harner M, Korner C, Walther D, Mokranjac D, Kaesmacher J, Welsch U, Griffith J, Mann M, Reggiori F, Neupert W (2011) The mitochondrial contact site complex, a determinant of mitochondrial architecture. EMBO J 30:4356–4370

Hasin Y, Seldin M, Lusis A (2017) Multi-omics approaches to disease. Genome Biol 18(1):83

Heide H, Bleier L, Steger M, Ackermann J, Dröse S, Schwamb B, Brandt U et al (2012) Complexome profiling identifies TMEM126B as a component of the mitochondrial complex I assembly complex. Cell Metab 16(4):538–549

Hillenmeyer ME, Fung E, Wildenhain J, Pierce SE, Hoon S, Lee W, Proctor M, St Onge RP, Tyers M, Koller D et al (2008) The chemical genomic portrait of yeast: uncovering a phenotype for all genes. Science 320:362–365

Hoppins S, Collins SR, Cassidy-Stone A, Hummel E, Devay RM, Lackner LL, Westermann B, Schuldiner M, Weissman JS, Nunnari J (2011) A mitochondrial-focused genetic interaction map reveals a scaffold like complex required for inner membrane organization in mitochondria. J Cell Biol 195:323–340

Hyduke DR, Palsson BO (2010) Towards genome-scale signalling network reconstructions. Nat Rev Genet 11:297–307. https://doi.org/10.1038/nrg2750

Johnston L, Thompson R, Turner C, Bushby K, Lochmüller H, Straub V (2014) The impact of integrated omics technologies for patients with rare diseases. Expert Opin Orphan Drugs 2:554

Johnston IG, Burgstaller JP, Havlicek V, Kolbe T, Rulicke T, Brem G et al (2015) Stochastic modelling, bayesian inference, and new in vivo measurements elucidate the debated mtdna bottleneck mechanism. Elife 4:e07464. https://doi.org/10.7554/eLife.07464

Khan NA, Nikkanen J, Yatsuga S, Jackson C, Wang L, Pradhan S et al (2017) mTORC1 regulates mitochondrial integrated stress response and mitochondrial myopathy progression. Cell Metab 26:419.e5–428.e5. https://doi.org/10.1016/j.cmet.2017.07.007

Kirby D, Crawford M, Cleary MA, Dahl HH, Dennett X, Thorburn DR (1999) Respiratory chain complex I deficiency: an underdiagnosed energy generation disorder. Neurology 52(6):1255–1255

Kitano H (2002) Computational systems biology. Nature 420:206–210. https://doi.org/10.1038/nature01254

Korzeniewski B, Zoladz JA (2001) A model of oxidative phosphorylation in mammalian skeletal muscle. Biophys Chem 92:17–34. https://doi.org/10.1016/S0301-4622(01)00184-3

Kowald A, Kirkwood TB (2011) Evolution of the mitochondrial fusionfission cycle and its role in aging. Proc Natl Acad Sci U S A 108:10237–10242. https://doi.org/10.1073/pnas.1101604108

Lewis NE, Schramm G, Bordbar A, Schellenberger J, Andersen MP, Cheng JK et al (2010) Large-scale in silico modeling of metabolic interactions between cell types in the human brain. Nat Biotechnol 28:1279–1285. https://doi.org/10.1038/nbt.1711)

Lohman DC, Forouhar F, Beebe ET, Stefely MS, Minogue CE, Ulbrich A, Stefely JA, Sukumar S, Luna-Sánchez M, Jochem A et al (2014) Mitochondrial COQ9 is a lipid-binding protein that associates with COQ7 to enable coenzyme Q biosynthesis. Proc Natl Acad Sci U S A 111:E4697

Lohman DC, Aydin D, Von Bank HC, Smith RW, Linke V, Weisenhorn E, Mcdevitt MT, Hutchins P, Wilkerson EM, Wancewicz B, Russell J, Stefely MS, Beebe ET, Jochem A, Coon JJ, Bingman CA, Dal Peraro M, Pagliarini DJ (2019) An isoprene lipid-binding protein promotes eukaryotic coenzyme Q biosynthesis. Mol Cell 73:763–774

Luscombe NM, Greenbaum D, Gerstein M (2001) What is bioinformatics? A proposed definition and overview of the field. Methods Inf Med 40:346–358. https://doi.org/10.1055/s-0038-1634431

Magnus G, Keizer J (1997) Minimal model of beta-cell mitochondrial Ca2+ handling. Am J Physiol 273(2 Pt 1):C717–C733. https://doi.org/10.1152/ajpcell.1997.273.2.C717

Maldonado EM, Taha F, Rahman J, Rahman S (2019) Systems biology approaches toward understanding primary mitochondrial diseases. Front Genet 10:19

McKenzie M, Tucker EJ, Compton AG, Lazarou M, George C, Thorburn DR, Ryan MT (2011) Mutations in the gene encoding C8orf38 block complex I assembly by inhibiting the production of the mitochondria-encoded subunit ND1. J Mol Biol 414(3):413–426

Montelione GT (2012) The protein structure initiative: achievements and visions for the future. F1000 Biol Rep 4:7

Münzner U, Lubitz T, Klipp E, Krantz M (2017) Toward genome-scale models of signal transduction networks. In: Nielsen J, Hohmann S (eds) Systems biology. Wiley-VCH Verlag, Weinheim

Nikkanen J, Forsstrom S, Euro L, Paetau I, Kohnz RA, Wang L et al (2016) Mitochondrial DNA replication defects disturb cellular dNTP pools and remodel one-carbon metabolism. Cell Metab 23:635–648. https://doi.org/10.1016/j.cmet.2016.01.019

Noronha A, Danielsdottir AD, Gawron P, Jannsson F, Jonsdottir S, Jarlsson S et al (2017) ReconMap: an interactive visualization of human metabolism. Bioinformatics 33:605–607. https://doi.org/10.1093/bioinformatics/btw667

O'Brien EJ, Monk JM, Palsson BO (2015) Using genome-scale models to predict biological capabilities. Cell 161:971–987. https://doi.org/10.1016/j.cell.2015.05.019

Pagliarini DJ, Rutter J (2013) Hallmarks of a new era in mitochondrial biochemistry. Genes Dev 27:2615–2627. https://doi.org/10.1101/gad.229724.113

Pagliarini DJ, Calvo SE, Chang B, Sheth SA, Vafai SB, Ong SE, Walford GA, Sugiana C, Boneh A, Chen WK et al (2008) A mitochondrial protein compendium elucidates complex I disease biology. Cell 134:112–123

Papin JA, Price ND, Wiback SJ, Fell DA, Palsson BO (2003) Metabolic pathways in the post-genome era. Trends Biochem Sci 28:250–258. https://doi.org/10.1016/S0968-0004(03)00064-1

Pfanner N, van der Laan M, Amati P, Capaldi RA, Caudy AA, Chacin-ska A, Darshi M, Deckers M, Hoppins S, Icho T et al (2014) Uniform nomenclature for the mitochondrial contact site and cristae organizing system. J Cell Biol 204:1083–1086

Poon WW, Davis DE, Ha HT, Jonassen T, Rather PN, Clarke CF (2000) Identification of *Escherichia coli* ubiB, a gene required for the firstmonooxygenase step in ubiquinone biosynthesis. J Bacteriol 182:5139–5146

Price ND, Papin JA, Schilling CH, Palsson BO (2003) Genome-scale microbial in silico models: the constraints-based approach. Trends Biotechnol 21:162–169. https://doi.org/10.1016/S0167-7799(03)00030-1

Quirós PM, Prado MA, Zamboni N, D'Amico D, Williams RW, Finley D et al (2017) Multi-omics analysis identifies ATF4 as a key regulator of the mitochondrial stress response in mammals. J Cell Biol 216:2027–2045. https://doi.org/10.1083/jcb.201702058

Rahman J, Rahman S (2018) Mitochondrial medicine in the omics era. Lancet 391:2560–2574. https://doi.org/10.1016/S0140-6736(18)30727-X

Rahman J, Noronha A, Thiele I, Rahman S (2017) Leigh map: a novel computational diagnostic resource for mitochondrial disease. Ann Neurol 81:9–16. https://doi.org/10.1002/ana.24835

Rea SL, Graham BH, Nakamaru-Ogiso E, Kar A, Falk MJ (2010) Bacteria, yeast, worms, and flies: exploiting simple model organisms to investigate human mitochondrial diseases. Dev Disabil Res Rev 16:200–218

Rhein VF, Carroll J, Ding S, Fearnley IM, Walker JE (2013) NDUFAF7 methylates arginine 85 in the NDUFS2 subunit of human complex I. J Biol Chem 288(46):33016–33026

Rommens J, Iannuzzi M, Kerem B, Drumm M, Melmer G et al (1989) Identification of the cystic fibrosis gene: chromosome walking and jumping. Science 245:1059–1065

Sahoo S, Franzson L, Jonsson JJ, Thiele I (2012) A compendium of inborn errors of metabolism mapped onto the human metabolic network. Mol Biosyst 8:2545–2558. https://doi.org/10.1039/c2mb25075f

Schuldiner M, Collins SR, Thompson NJ, Denic V, Bhamidipati A, Punna T, Ihmels J, Andrews B, Boone C, Greenblatt JF et al (2005) Exploration of the function and organization of the yeast early secretory pathway through an epistatic miniarray profile. Cell 123:507–519

Skolnick J, Fetrow JS, Kolinski A (2000) Structural genomics and its importance for gene function analysis. Nat Biotechnol 18:283–287

Sorokina M, Stam M, Médigue C, Lespinet O, Vallenet D (2014) Profiling the orphan enzymes. Biol Direct 9:10

Stefely JA, Pagliarini DJ (2017) Biochemistry of mitochondrial coenzyme Q biosynthesis. Trends Biochem Sci 42:824–843

Stoeger T, Gerlach M, Morimoto RI, Nunes Amaral LA (2018) Large-scale investigation of the reasons why potentially important genes are ignored. PLoS Biol 16:e2006643

Stroud DA, Surgenor EE, Formosa LE, Reljic B, Frazier AE, Dibley MG, Ryan MT et al (2016) Accessory subunits are integral for assembly and function of human mitochondrial complex I. Nature 538(7623):123–126

Sugiana C, Pagliarini DJ, McKenzie M, Kirby DM, Salemi R, Abu-Amero KK, Thorburn DR et al (2008) Mutation of C20orf7 disrupts complex I assembly and causes lethal neonatal mitochondrial disease. Am J Hum Genet 83(4):468–478

Sun BB, Maranville JC, Peters JE, Stacey D, Staley JR, Blackshaw J et al (2018) Genomic atlas of the human plasma proteome. Nature 558:73–79. https://doi.org/10.1038/s41586-018-0175-2

Suomalainen A, Elo JM, Pietilainen KH, Hakonen AH, Sevastianova K, Korpela M et al (2011) FGF-21 as a biomarker for muscle-manifesting mitochondrial respiratory chain deficiencies: a diagnostic study. Lancet Neurol 10:806–818. https://doi.org/10.1016/S1474-4422(11)70155-2

Tan H, Yang K, Li Y, Shaw TI, Wang Y, Blanco DB et al (2017) Integrative proteomics and phosphoproteomics profiling reveals dynamic signaling networks and bioenergetics pathways underlying T cell activation. Immunity 46:488–503. https://doi.org/10.1016/j.immuni.2017.02.010

Thompson Legault J, Strittmatter L, Tardif J, Sharma R, TremblayVaillancourt V, Aubut C et al (2015) A metabolic signature of mitochondrial dysfunction revealed through a monogenic form of leigh syndrome. Cell Rep 13:981–989. https://doi.org/10.1016/j.celrep.2015.09.054

Toyama EQ, Herzig S, Courchet J, Lewis TL, Losón OC, Hellberg K et al (2016) AMP-activated protein kinase mediates mitochondrial fission in response to energy stress. Science 351:275–281. https://doi.org/10.1126/science.aab4138

Veling MT, Reidenbach AG, Freiberger EC, Kwiecien NW, Hutchins PD, Drahnak MJ et al (2017) Multi-omic mitoprotease profiling defines a role for oct1p in coenzyme Q production. Mol Cell 68:970.e11–977.e11. https://doi.org/10.1016/j.molcel.2017.11.023

Vo TD, Greenberg HJ, Palsson BO (2004) Reconstruction and functional characterization of the human mitochondrial metabolic network based on proteomic and biochemical data. J Biol Chem 279:39532–39540. https://doi.org/10.1074/jbc.M403782200

Wang B, Mezlini AM, Demir F, Fiume M, Tu Z, Brudno M et al (2014) Similarity network fusion for aggregating data types on a genomic scale. Nat Methods 11:333–337. https://doi.org/10.1038/nmeth.2810

Wei AC, Aon MA, O'Rourke B, Winslow RL, Cortassa S (2011) Mitochondrial energetics, pH regulation, and ion dynamics: a computational-experimental approach. Biophys J 100:2894–2903. https://doi.org/10.1016/j.bpj.2011.05.027

Wu Y, Williams EG, Dubuis S, Mottis A, Jovaisaite V, Houten SM et al (2014) Multilayered genetic and omics dissection of mitochondrial activity in a mouse reference population. Cell 158:1415–1430. https://doi.org/10.1016/j.cell.2014.07.039

Systems Approaches in Identifying Disease-Related Genes and Drug Targets

9

Sohini Chakraborty and Satarupa Banerjee

Abstract

Target discovery is the first crucial stage in the target drug identification and discovery pipeline. For a better understanding of the underlying causes of diseases and the therapeutic efficacies of medicines, systems-based techniques are gaining significance. The conventional method of compiling data from single isolated layers does not always yield useful conclusions. A developing subject that aids in the identification of novel therapeutic targets and biomarkers is systems biology. It uses new, cutting-edge strategies to predict probable molecular interactions by combining omics data from the genomic, proteomic, transcriptional, and metabolic layers in order to overcome the aforementioned challenges, thereby, mitigating the spread of disease via enhancing treatment and drug delivery techniques. Combining predictive and computational models with drug delivery planned by control regulations can enhance therapeutic performance. In this chapter, different systemic approaches identify disease biomarkers and therapeutic targets.

Keywords

Protein · Genes · Interaction · Network · Modeling · Molecular · Regulatory · Tools · Databases

S. Chakraborty · S. Banerjee (✉)
Department of Biotechnology, School of Biosciences and Technology, Vellore Institute of Technology, Vellore, Tamil Nadu, India
e-mail: satarupa.banerjee@vit.ac.in

S. Joshi et al. (eds.), *Systems Biology Approaches: Prevention, Diagnosis, and Understanding Mechanisms of Complex Diseases*,
https://doi.org/10.1007/978-981-99-9462-5_9

Abbreviations

BN	Bayesian network
BP	Biological process
CA	Clustering analysis
CAD	Computer Aided Design
CC	Cellular components
CCo	Clustering coefficient
ChEA3	ChIP-X Enrichment Analysis 3 Database
CPD	Conditional probability distribution
CRISPR	Clustered regularly interspaced short palindromic repeats
CTD	Comparative Toxicogenomics Databases
DAG	Directed acyclic graph
DGI	Disease-gene interaction
DIP	Database of interacting proteins
GCN	Gene co-expression network
GIN	Gene interaction network
GO	Gene ontology
GRN	Gene regulatory network
GUI	Graphical user interface
GWAS	Genome Wide Association Studies
lncRNA	Long non-coding RNA
mbSUS	Mating-based ubiquitin system
MCC	Maximal clique centrality
MF	Molecular function
MINT	The molecular interaction database
miRNA	microRNA
ML	Machine learning
MNC	Maximum neighborhood component
mRNA	Messenger RNA
NLP	Natural language processing
ODE	Ordinary differential equation
OS	Operating system
PCR	Polymerase chain reaction
PDE	Partial differential equation
PIPs	Human protein-protein interaction prediction database
PPI	Protein-protein interactions
RCC	Residue cluster classes
SBML	Systems Biology Markup Language
SNP	Single nucleotide polymorphism
STN	Signal transduction network
SVM	Support vector machines
TCSBN	Database for tissue/cancer-specific biological networks
TF	Transcription factor
TPA	Topological parameter analysis

WEGO	Web Gene Ontology Annotation Plot Database
WGS	Whole genome sequencing
Y2H	Yeast two-hybrid screenings

9.1 Introduction

Patient survival is limited by the mortality rate of diseases, which brings us to the fact that there is an utmost need to develop remedies to recover from the aberrant stages of the diseases (Lindsay 2003). It is well known that a diseased state is not only restricted to a single gene or protein dysregulation but to various signaling pathway reactions collectively leading to the diseased consequence (Egner et al. 2005). Incorporating enriched gene ontology (GO) terms into biological networks helps clinicians to focus on strategizing therapeutic strategies (Tomczak et al. 2018).

Numerous target recognition strategies, including affinity-based approaches, genetic-based strategies, computational approaches, and chemical proteomics, have been established as a result of breakthroughs and developments in novel methodologies and tools for identification and biological analysis. These approaches are categorized as target detection methods, different molecular target verification techniques, and their operational models (Yang et al. 2008; Tomczak et al. 2018).

For instance, the collaborative effect of single nucleotide polymorphisms (SNPs), that may cause stroke, cannot be revealed using only the traditional marker or the haplotype connection. Instead, the systemic perspective on dynamic gene regulation demonstrates that genes function as a component of complex networks rather than operating independently to carry out cellular functions (Wang et al. 2023). When analysis and prediction based on complex cellular networks need to be done transcriptional factors (TFs), genes and their expression profiles offer a thorough route of metabolism and regulation as depicted in Fig. 9.1 (Chen et al. 2015).

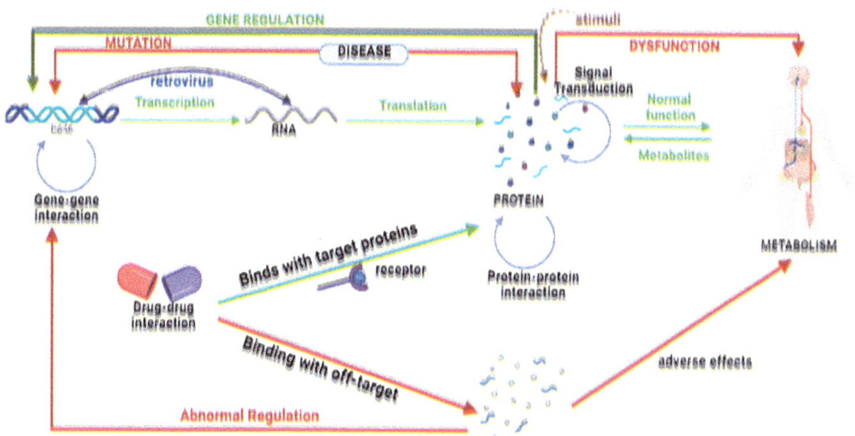

Fig. 9.1 Interactions of bioentities on different levels demonstrate the inter-regulation of organisms. Drug molecules (ligands) bind with specific target molecules

9.2 Systems Biology: The Concept

Instead of focusing on the traits of individual cells or organisms, the structural and dynamical activity at a cellular and organ level needs to be analyzed in order to comprehend biology at the system level. System-level properties such as robustness come in handy in identifying targets. However, prior to the success of systems biology to anticipated levels, numerous advancements in experimental apparatus, sophisticated software, and analytical techniques are needed (Qin et al. 2014).

An understanding of four fundamental characteristics of a biological system can be obtained by analyzing (Kitano 2002b):

- *System architecture*: Biochemical routes and network of gene connections as well as altered physical intra and multicellular traits.
- *Dynamics of the system*: It is possible to comprehend how a system acts over time and under different situations by using tools like phase picture and bifurcation analysis.
- *The monitoring technique*: To reduce malfunctions and provide possible therapeutic targets for disease treatment, mechanisms that methodically regulate the state of the cell can be regulated.
- *The design approaches*: Transparent design principles and simulations might benefit from advancements in our comprehension of computational sciences, genetics, and measurement technologies, as well as from integrating these discoveries with current knowledge.

A number of initiatives are required to build a sizable and thorough database on regulatory and biochemical networks. Although these databases are good sources of information, a great share of network patterns are still unknown. Significant research has been done on clustering analysis-based expression profiling to explore genes that co-express with known fungal genes.

9.3 Systems Biology Research Cycle

A systems biology cycle of research (Fig. 9.2) is generated after the selection of contrasting significant biological concerns and the development of a model for the phenomena. Models can be generated manually or mechanically. The effectiveness of assumptions and hypotheses incorporated into each model is revealed by computational "dry" experiments on models, such as simulation. Lacking models would reveal discrepancies with proven experimental findings and thus have to be improved or eliminated (Kitano 2002a). A complete system analysis is performed on models that pass these criteria, based on a number of predictions. A suitable model is chosen among the competing ones using a set of predicts, which are then validated in a wet lab setup.

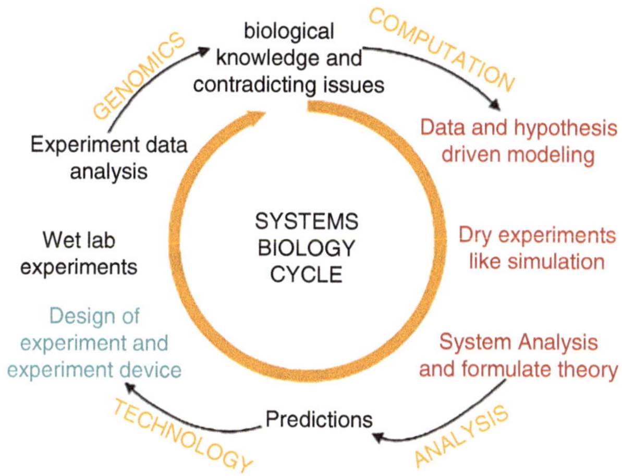

Fig. 9.2 Hypothesis-based research approach in systems biology

9.4 The Conventional Strategies to Identify Targets

Numerous target recognition strategies, including affinity-based approaches, genetic-based strategies, computational approaches, and proteomics have been established as a result of breakthroughs and developments in novel methodologies and tools for identification and biological analysis. These approaches are categorized as target detection methods, different molecular target verification techniques, and their operational models.

(a) Affinity purification and mass spectrometry combination: Mass spectrometry in conjunction with affinity purification mostly detects stable contacts, which better identifies functional in vivo protein–protein interactions (PPIs) (Wodak et al. 2013). The tagged protein usually at a cellular level circulates at vivo concentrations, and its related proteins are first purified based on affinity property (Fig. 9.3). Hence, PPIs can quantify and qualify based on mass spectrometry using various methods such as chemical constitution, physiological or enzymatic integration (SILAC) (Jones and Thornton 1996), and label-free approaches.
(b) Genetic-based strategies: In a number of human diseases, including cancer and autoimmune disorders, genomic and proteomic approaches have started to elucidate novel functional pathways and therapeutic targets. Phenotype-oriented target identification, as opposed to gene-profiling methods, enables a direct connection to be made between genetic changes and a disease phenotype. There are numerous methods for controlling gene expression in both setups (in vivo and in vitro).

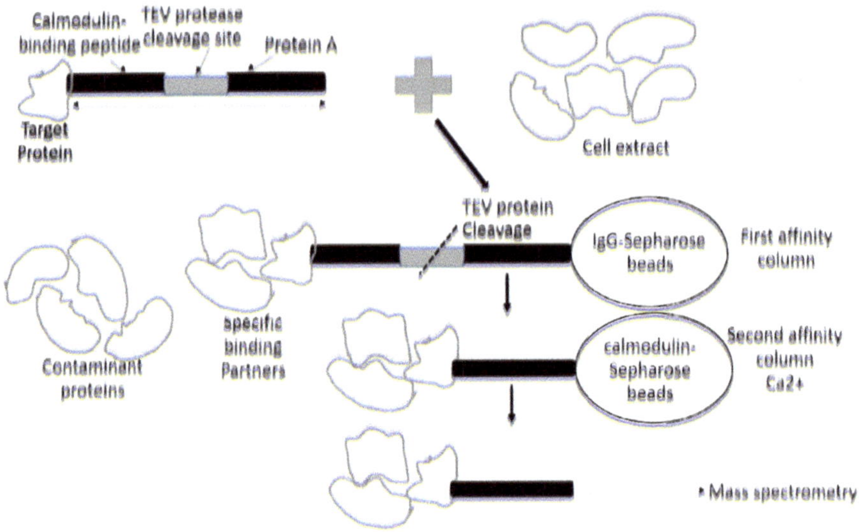

Fig. 9.3 Tandem affinity purification principle

Gene expression in both healthy and pathological cells and tissues can be examined using potent methods made possible by molecular profiling, which also encompasses genomics and proteomics (Segal et al. 2005). An apparent example is mRNA expression profiling, which compares mRNA expression levels to conduct a large-scale examination of biological transcripts. Breast cancer research has made use of *microarray technology* to identify tumor-specific genes as diagnostic and prognostic indicators (Sun et al. 2013). It is a "closed" technique (Perou et al. 2000) to discover targets because it can only detect what is seen on a slide and has been effectively used to identify possible targets. Based on the fact that the number of sequence tag samples (STSs) from a cDNA library is directly proportional to the expression levels of the relevant gene, *expressed sequence tag* (*EST*)-derived expression analysis and SAGE were developed (Lash et al. 2000). *Analyzing complex pathways*, as opposed to single genes, is a more effective method for identifying therapeutic targets (Brazhnik et al. 2002). Once a disease-related route has been discovered, it is next possible to evaluate each of the pathway's signaling proteins to see if they might be druggable. Finding and validating pharmacological target genes should rely heavily on the analysis of gene networks following gene disruption or drug therapy. This method also makes it possible to pinpoint the cellular pathways that a drug affects either directly or indirectly. The synthetic mutagen *N*-ethylnitrosourea (ENU) has been used widely in *forward genetics* to create random mutations in the genome. Through the genome, ENU causes single base-pair DNA alterations (De Angelis et al. 2000; Balling 2003). Alongside chemical variants, random integration of retroviruses or transposons may be the best screening technique because the vector sequence functions as a molecular fingerprint for the transformed gene, simplifying identification. Characterizing genes implicated in important biological pathways including apoptosis and cell signaling has been accomplished with the use of *reverse genetics*, which is based on functional mutation studies in animals (biochemistry and 1998 n.d.; Zambrowicz et al. n.d.).

(c) Proteomics. A promising method of logical identification of new pharmaceutical targets is proteomics, the study of proteins and protein processes. Additionally, proteomics allows for the independent examination of several subcellular protein fractions, which lessens the complexity of the proteome (Resing and Ahn 2005). Recent research on target and biomarker identification emphasizes the value of combining various techniques to identify insightful pharmacological targets. An objective elucidation of new cancer biomarkers, for instance, was made possible by the integration of DNA microarray, SAGE, and proteomics. Proteomics methods can also detect post-translational alterations that are crucial for the detection of tumor antigens (Sahin et al. 1995). Yeast two-hybrid (Y2H) system, protein chips, and phage display can be used in conjunction with proteomics efforts to profile patient immune responses and identify tumor antigens (Abu-Elheiga et al. 2001).

The Y2H system (Fig. 9.4) is an effective experimental technique for PPI identification that may be used in a high-throughput manner to find interactions throughout an organism's complete proteome. Smaller size, more focused interaction maps have been created considering the high potential of human PPIs for comprehending disease processes and signaling cascades (Rives and Galitski 2003).

It has been used to examine interactions among proteins and different molecules such as those between proteins and antigens and/or nucleic acids, lipids, small molecules, and enzyme substrates (Zhu and Snyder 2003).

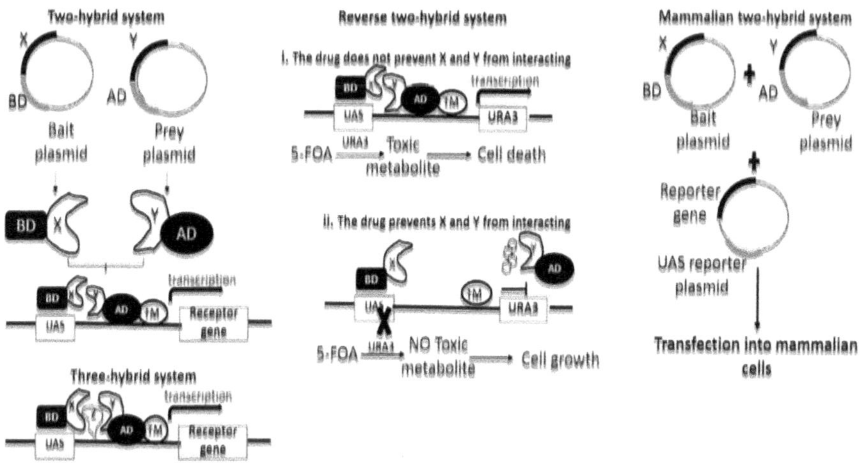

Fig. 9.4 Principle behind yeast and mammalian two-hybrid system

Differential gene expression in disease is an analysis of the literature and high-throughput investigations. High-throughput microarray expression research and the scientific literature are the two main sources of DEG (differentially expressed gene) data. In order to do that, we examined microarray datasets on a variety of diseases and explored the scientific (Loging et al. 2007) literature to compare and contrast these two types of data in particular biological situations. Information regarding DEGs is primarily provided in unorganized form and dispersed throughout papers in the research literature. Studies on DEGs frequently lack precision or depth, making it difficult for computers as well as humans to automatically gather these DEGs (Campbell et al. 2012).

Recent advances in deep sequencing have revolutionized biology and medicine by enabling high throughput nucleic acid sequence analysis with base-level accuracy. An increasingly popular technique for examining gene expression and discovering new RNA species is *RNA sequencing* or *RNA-Seq*. Specialized techniques for creating cDNA libraries can be used to investigate aspects of RNA biosynthesis and metabolism.

However, numerous networks can also be generated *computationally*, frequently beginning with experimental evidence. Nevertheless, techniques are additionally developed to predict interaction networks to be "de novo," regardless of any prior proof of such connections. Such computational techniques include:

- *Text-mining*: Text mining is substantially more affordable and time-efficient than other high-throughput methods. Machine learning techniques and rule/pattern-based data extraction obtain binary connections between interacting proteins from individual phrases. A wide range of text mining apps can be used for PPI extraction and/or prediction and repositories that usually contain PPIs manually vetted and/or forecasted cognitively (Roberts 2006; Fluck and Hofmann-Apitius 2014). Text mining can be accomplished in two stages (Fig. 9.5): retrieving and extraction of data.

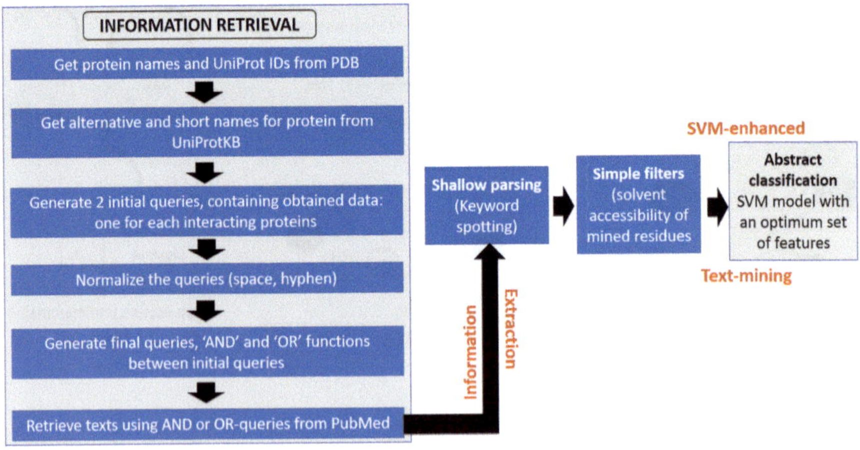

Fig. 9.5 Text-mining methodology/protocol

- Methodologies involved in text-mining include phylogenetic profiling, Natural language processing (NLP), and kernel methods to predict PPI (Kong and Gerstein 2018).
  - *Machine learning (ML) methods*: Prediction methods can be categorized into the following groups according to the data utilized to support them: protein pattern, comparative genome research, protein regions, tertiary arrangement, and interacting network topography (Zhang et al. 2019). Based on labeling the input parameters corresponding to the anticipated outcome, predicted models using ML approaches can be roughly categorized into two primary categories (Fig. 9.6): "supervised" and "unsupervised" (Sidak et al. 2022).
  - The mating-based ubiquitin system (mbSUS) was used to examine the integral proteins of the membrane from *Saccharomyces cerevisiae* in 2005 (Sidak et al. 2022). To organize and categorize interactions, a support vector machine (SVM) was used to refer to high, intermediate, and low-confidence interactions. Transcriptional reporters are used by the split-ubiquitin membrane Y2H to recognize yeast transformants that encode combinations of interdependent proteins (Obrdiik et al. 2004). Random forest algorithm, a monitored strategy, has been used to identify protein interactions in the human membrane proteins and the interactome of proteins. The system looks for connections between extracellular signaling proteins and membrane proteins. Instances of protein–protein interactions were categorized accurately in 96–99% of cases using an algorithm built in 2020 utilizing residue cluster classes (RCCs) (Grefen et al. 2007).
  - *Genomic methods*: The Rosetta Stone or Domain Fusion approach is based on the idea of interacting proteins occasionally combining into a single protein in a different genome. The conserved neighborhood approach relies on genes encoding two proteins that share adjacent chromosomal locations across multiple genomes, they are presumably functionally associated. The phylogenetic profile technique involves two or more proteins that are probably substantially linked if they are simultaneously present or absent across many genomes (Ferrer et al. 2010).

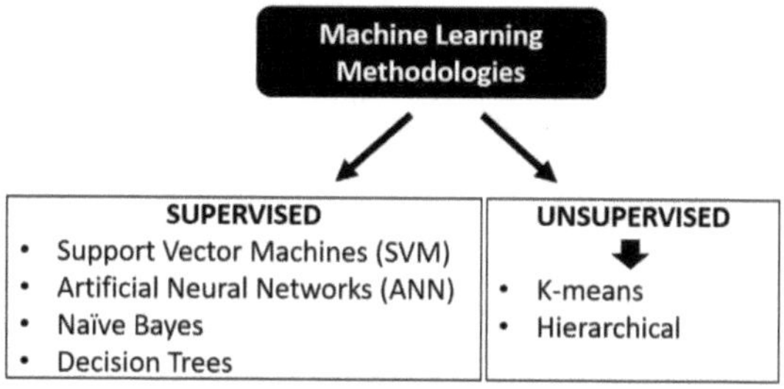

Fig. 9.6 ML methodologies classification hierarchy

9.5 Systems Biology Approaches to Discover Disease Genes

9.5.1 Biological Network

The traditional approach for representing molecular interactions is to view biochemical reactions as a substrate acted upon by an enzyme to produce a product, or through mass action kinetics of two or more molecular entities forming a complex. The post-genomic view of biological molecular systems is to look at molecular entities as nodes and their relationships as edges/links (Fig. 9.7). Molecular entities can be proteins, mRNAs, or small molecules not encoded by genes; interactions can be direct physical, enzymatic, or indirect, such as transcriptional activation or indirect effects of downstream cell signaling components (Zaghloul Salem 2018).

Systems biology aspects are either data or model driven. The data model uses data mining techniques to understand designs in massive datasets, whereas the latter is a model approach. In order to better understand convoluted human diseases like cancer, the models incorporate a network of biological components. Integrating the network approach and gene ontology data will boost the comprehension of the built-in network models. The impact of these alterations within the biological system can be analyzed using systems biology approaches at both protein level and nucleic acids (DNA/RNA) levels, alongside post-translational modifications. By identifying enhanced phenomena, functions, and cellular elements during interaction network assessment, functionally saturated GO keywords enable researchers to focus on a particular biological process. By emphasizing the topological attributes, biological networks determine pharmacological candidates. Unique traits reveal the relevance and function of the nodes involved in the network, highlighting the significance of entities in the biological system. According to the biological entities being studied, either of the distinct biological networks such as metabolic networks or cell signaling networks can be built. In metabolic pathways, reactions can be directional or bidirectional while interactions define the flow of information (activation or inhibition) in cell signaling pathways.

Metabolic networks are commonly represented by two types of nodes such as enzyme-substrate leading to a reaction. These networks are commonly represented as bipartite graphs where reactions are connected to substrates. Most of the interactions in such networks are always direct and physical. The goal of interaction network analysis is to develop new systems-based therapeutics and identify novel pharmacological targets (Jeong et al. 2000; Wagner and Fell 2001).

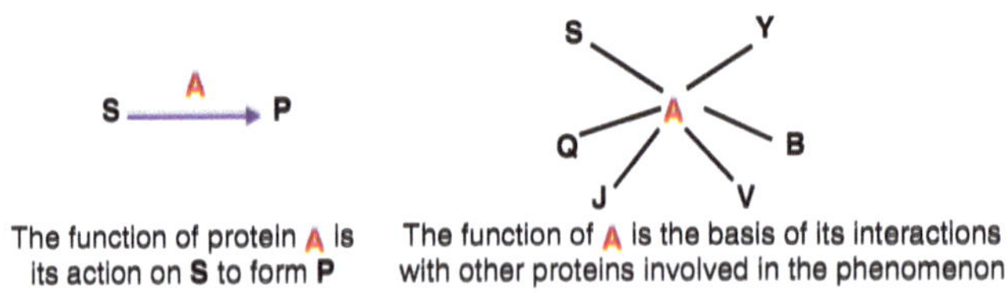

Fig. 9.7 The evolution of protein modeling. *Left*: traditional view; *right*: post-genomic view

Cell signaling pathways are networks combining of nodes attributing to different proteins and metabolites. They can be merged into large-scale networks connecting ligand–receptor interactions to their downstream effectors terminating at components from different cellular types of machinery (Rives and Galitski 2003). The interactions are typically enzymatic or binding.

The following portion of this chapter deals with the different interaction networks, commonly known as network or cluster approaches in the field of systems biology.

9.5.1.1 Protein–Protein Interaction (PPI)

PPIs serve as regulatory nodes in numerous cell-signaling networks that are connected to the "hallmarks of cancer." Numerous PPIs that are closely related to cell signaling and cell survival have been found to be therapeutic targets for cancer and are now the focus of drug discovery initiatives in both academia and business (Zhu and Snyder 2003). By using high-throughput screening, rational design, virtual screening, or fragment screens, small-molecule PPI inhibitors have been found.

Networks of protein–protein interactions offer a useful foundation for understanding how the proteome is functionally organized. All biological activities depend on protein–protein interactions (PPIs). Thus, assembling PPI networks offers a wealth of fresh perspectives on how proteins function (Stelzl et al. 2005). Additionally, interaction networks are important from a systems biology perspective because they may reveal the general organizing principles of functioning cellular networks when the interaction's time and space dimensions are considered. It is a continuous process to create precise biological protein interaction networks utilizing various experimental and computational techniques such as mass spectroscopy and high-throughput yeast two-hybrid screenings (Y2H), text-mining, and ML strategies (Jansen et al. 2002).

PPI databases, along with genomic and proteomic databases, have grown into important tools for molecular and systems biology. PPI data, however, are dispersed throughout a very large number of studies and are explored by a broad diversity of experimental and computational methodologies, probably more so than other data types in systems biology. As a result, PPI databases have made extensive integration efforts, such as curating hundreds of journal articles, developing a regulated vocabulary for PPI experiment descriptions, and establishing standard formats for PPI data. In addition, they have created techniques for assessing interactions, methodologies for linking interactions to context, and quality control criteria for curation (Kotlyar et al. 2019) (Fig. 9.8).

9.5.1.2 Gene–Interaction Network (GIN)

ijtMacArthur et al. showed, in this study, that both embryonic stem cells and osteoblasts are regulated by circuits made of nested positive feedback loops, while components from each circuit inhibit components from the other positive feedback circuit (Bebek 2012). Using dynamical simulations, MacArthur et al. showed how cells typically go from the undifferentiated state to the differentiated state smoothly. However, if noise is added to the two connected circuits, differentiated cells can "jump" back to the embryonic state, suggesting a role for noise in iPS reprogramming.

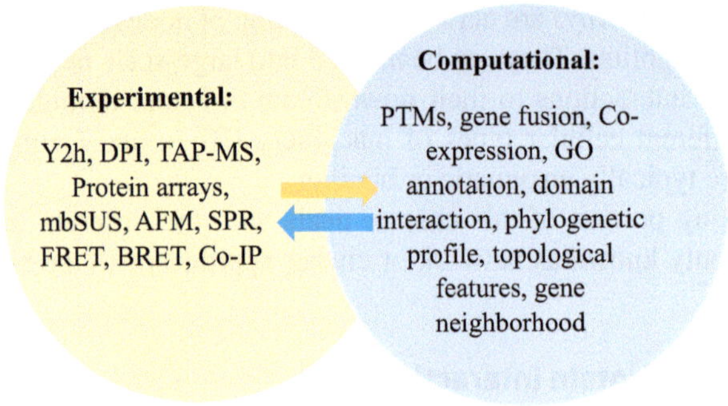

Fig. 9.8 Prediction of PPIs by combining various methods

In contrast to PPIs, GINs are genetic connections involving two genes. The significant findings from GIN suggest that the interacting genes have a consistent functional connection: biological processes or pathways. The genes are nodes and the interactions are edges (Cui et al. 2022). It is crucial to assess the prime processes employing interactions that regulate the significant biological processes (Paci et al. 2021). Understanding the intricate biological mechanisms by which proteins function and identifying new treatment targets for various diseases will be made possible through the analysis of these protein interactions. The sequential steps for the detection of drug targets and using GIN are represented in Fig. 9.9.

9.5.1.2.1 Gene Co-expression Network (GCN)

In GCNs, nodes represent the genes and are only linked if the genes exhibit a considerable degree of tissue-specific co-expression. The underlying idea of GCN is to use a network to define pairwise interactions between nodes. Additionally, it is utilized to link genes with unidentified biological activities to those of known genes. The development of GCN has enabled identifying the genes that control specific traits (Montenegro 2022). Microarray or RNA Seq evidence can create GCNs for both labeled and unlabeled networks. GCNs are created using a variety of techniques (Paci et al. 2021). The two steps they all take are computing the co-expression value and choosing a relevant threshold (Fig. 9.10). A co-expression value is chosen in the first stage, and a similarity score is computed employing this criterion for every pair of genes. A threshold is then established, and gene pairings with similarity scores greater than the chosen threshold are considered to have a substantial link with respect to co-expression and are connected by an edge in the network (van Dam et al. 2018). The four prevalent co-expression techniques for building GCNs are Pearson's correlation coefficient, mutual information, Spearman's rank correlation coefficient, and Euclidean distance.

9.5.1.2.2 Gene Regulatory Network (GRN)

The stages of development like cell cycles and signal transduction networks are represented as GRNs. Regulatory components play a critical role in creating cancer networks since they monitor how genes are expressed in various systems (Liu et al.

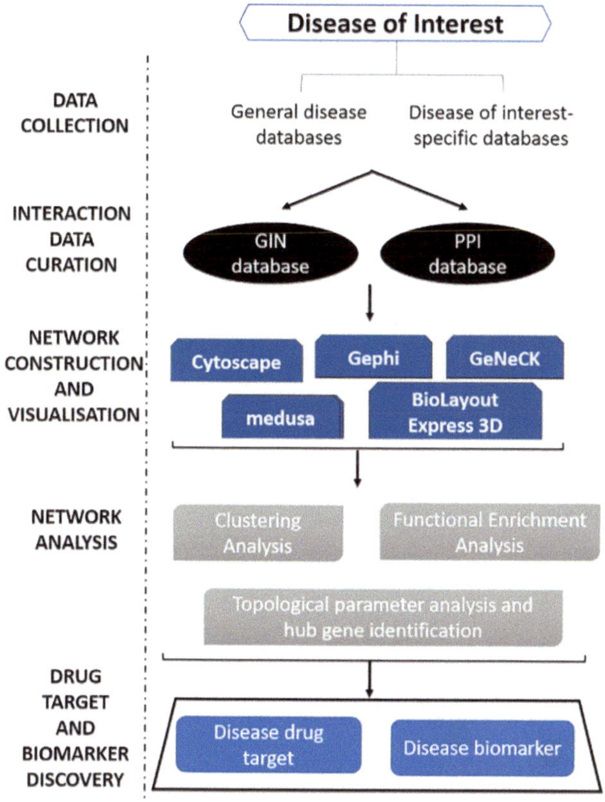

Fig. 9.9 Shows a schematic representation of the creation and analysis of GIN

2019). TFs are the principal regulatory proteins. Networks displaying how the hub genes interact with the TFs in cancer aid in assessing the function of the relevant hub genes (Peter 2020). Understanding the method by which TFs control the expression profile of the targeted gene is made easier with the aid of GRN. When many TFs influence the target gene, things get complicated. By contrasting GRNs that control related processes in distinct evolutionary contexts, it is possible to dissociate from particular factors and signals, which are often tailored to specific contexts, and instead disclose the function of network circuitry that connects these molecules (Fig. 9.11). The functional analysis of GRNs, which are made out of different combinations of TFs and molecules, allows us to differentiate between the network-level function of regulatory circuits in development and the function of regulatory molecules (Davidson and Peter 2015).

9.5.1.3 Signal Transduction Network (STN)

Signal transduction networks control the basic biological functions that enable cells to respond to both internal and external inputs as surroundings change. Small groups of networks that are frequently impacted by the diseased state as a result of genetic

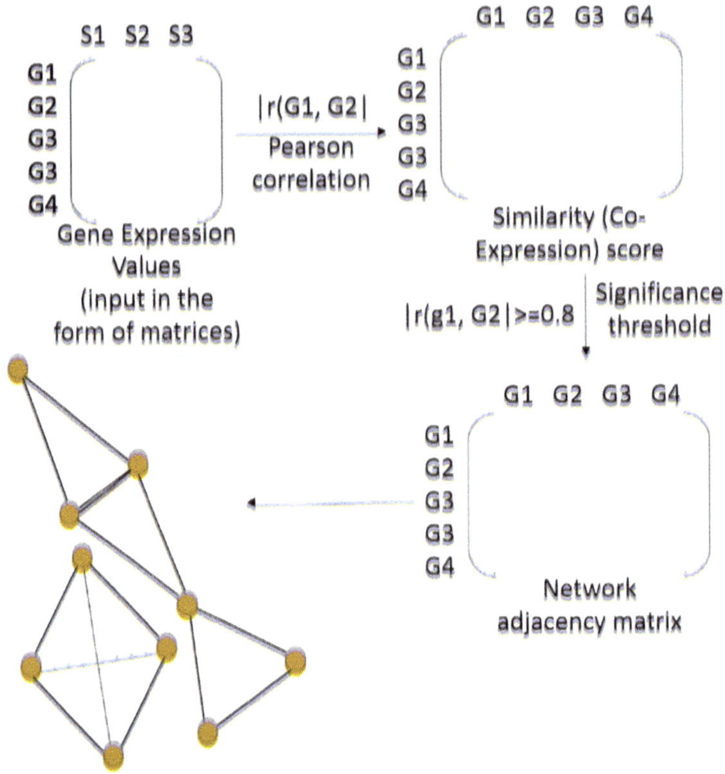

Fig. 9.10 The two-step strategy to construct a GCN

abnormalities carry out this signaling (Jang et al. 2022). These variants have a significant impact on innate epigenetic alterations, which have an impact on pharmacological therapies. A powerful technique for studying the consequences of STNs and their modifications in disease therapies is computational modeling (Kolch et al. 2015).

9.5.1.4 Evolutionary Network

Employing different topological analysis strategies combined with an understanding of the network structure gives an insight into the evolutionary details responsible for the topologies displayed by the complex regulatory networks (Jang et al. 2022). Owing to the advancement in techniques, various algorithms are developed that generate such realistic topologies utilizing a few measures that constantly monitor the expansion of the network. The thought behind the network design can be innately studied by utilizing the network evolution models (Soyer 2012).

9.5.1.5 Disease–Gene Interaction Network (DGI)

Nodes in biological networks represent diseases. Each node is color coordinated based on the disorder class. The size of each node is proportional to the number of

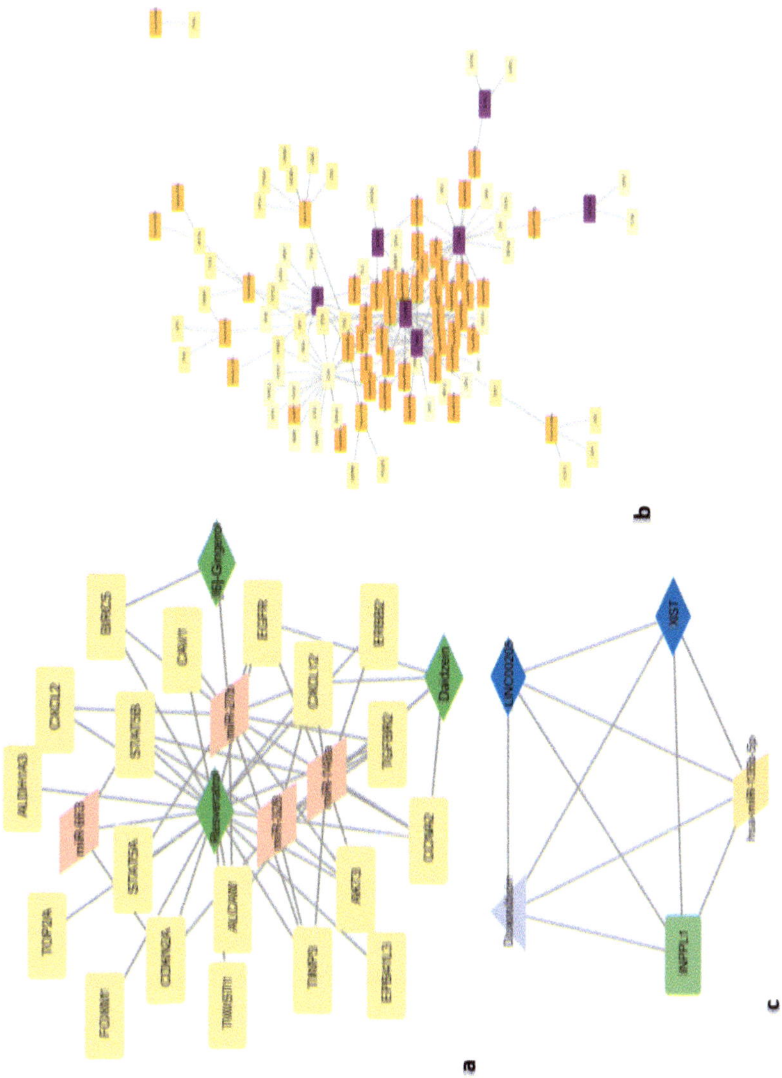

Fig. 9.11 GRNs: **(a)** miRNA-mRNA-drug **(b)** mRNA-TF-miRNA **(c)** mRNA-miRNA-lncRNA-drug

genes with mutations and the link (connecting lines) thickness correlates to the number of genes shared by two disorders (Goh et al. 2007). Systems biology employs the same PPI strategy to map DGI networks just like it enabled the prediction of drug–drug interactions. Drugs, target genes, or side effects represent the nodes and relationships are the edges. DGI demonstrates the relationship between a medicine and a genetic variation and measures how well a patient responds to drug therapy (Halu et al. 2019). Drug–drug–gene interactions (DDGI) are designed to pinpoint the interactions that cause phenoconversion, induction, and inhibition. The pharmacokinetic properties of drugs can either boost or decrease drug dose concentrations as a result of the inhibitory and inductive interactions. DGIs make it easier to develop cancer therapies by demonstrating the effectiveness of repurposed drugs against the target gene.

9.5.1.6 Functional Analysis of Disease Genes

Systems biology does not apply to genome-scale studies that are purely exploratory. It serves as a platform for conducting genome-scale studies that are hypothesis-driven and predictive in nature (Cooper-Knock et al. 2012). It is vital that comprehensive single-molecule experiments and studies match systems-level models that act as supporting data. Many diseases with quantifiably significant hereditary features still lack thorough genomic underpinnings notwithstanding a few instances of success (Khatri et al. 2012; Huntley et al. 2014).

Gene ontology (*GO*): An ontology is a systematic illustration of an ensemble of data within an area of research. In most cases, ontologies comprise a number of categories and the associations that exist among them (Ashburner et al. 2000; Carbon et al. 2019). Our understanding of the biological domain is described in three semantics of GO annotation:

- *Molecular function* (*MF*): Concepts like "transport" or "catalysis" in molecular function terminology refer to activities that take place at the molecular level. In contrast to the entities that regulate the actions, GO molecular function (MF) phrases refer to functions rather than the specific times, places, or circumstances in which the action occurs. Most molecular functions can be carried out through single gene products (such as a protein or RNA). Still, some functions are carried out by molecular complexes comprising several gene products.
- *Cellular component* (*CC*): Refers to the cyto structural regions where a gene product functions. In contrast to other GO components, cellular component categories refer to cellular anatomy rather than processes.
- *Biological Processes* (*BPs*): The BPs are basically broader programs executed by many molecular mechanisms. A pathway is not the same as a biological process. Currently, the GO does not try to depict the interconnections or dynamics necessary to define a pathway completely.

Each GO term is a node that represents the structure of GO, and the connections among the terms are the edges that connect the nodes. There are numerous methods to refer to and depict logical relations (Huntley et al. 2016). The following standards are applied in the GO relations documentation (Fig. 9.12a):

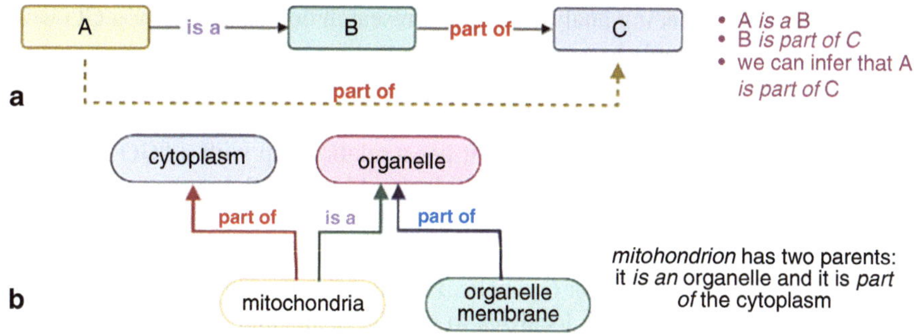

Fig. 9.12 GO conventions describing annotations among entities. (**a**) "*is a*" means "A *is a* B" means that "node A" is a subtype of "node B," (**b**) "*part of*" means *part of* A: B exists as *part of* A, hence the presence of B implies A is present

- A node denotes a GO trait.
- A "parent" denotes the closest node to the graph's root(s), and a "child" denotes the node closest to the leaf nodes;
- For, e.g., for the relation a and part of, the parent would be a more general term.
- The direction of the connection direction is shown by the arrowhead.
- Dashed lines signify implied relationships, or those the ontology does not explicitly claim exist.

Nodes can have any number and type of relationship with other nodes. A node might interact with one or more child nodes, but unlike them, nodes can have multiple parent nodes and different relationships with its different parents (Fig. 9.12b). GO annotation defines the functionalities of a specific gene. In order to understand the underlying mechanisms and to forecast the effects of perturbations, such as pharmaceutical therapies, on these biological systems, we can foresee the behaviors of biological systems by building computational models of them (Bard and Rhee 2004). A GO annotation can only be identified by four pieces of data: gene product, GO term, reference, and evidence. The different annotation qualifiers are:

- The ***NOT*** qualifier: The "NOT" function is employed to make a clear statement that a gene product has lost function over the course of evolution, as evidenced by experimental evidence that it is unable to perform a specific activity or by rapid divergence following a duplication event. This is especially crucial when linking a GO term to a gene product should be avoided. For instance, a protein annotated as GO:ABCXYZ has experimental evidence indicating that it lacks the enzymatic activity of an enzyme despite sharing sequence similarity with the enzyme (GO: ABCXYZ). When a linked reference specifically stipulates that a certain protein "is not present in the cytoplasm" NOT is used. Annotators can say that a specific gene product is not connected to the specific GO annotation by qualifying an annotation with the qualifier "NOT." It is employed when an experiment, sequencing analysis, or other research demonstrate that a GO word does not, in fact, apply to a gene product.

- The ***contributes_to*** qualifier: If a certain gene product "is part" of an irreducible complex—where a specific MF cannot be ascribed to an individual or small set of subunits, then it can be said that it has a relationship with a "GO MF" term owing to the qualifier "contributes_to." It should be noted that the "contributes_to" attribute only applies to molecular activities.
- The c***olocalizes_with*** qualifier: Refers to the protein's brief or marginal connection with an entire system or organelle. For instance, human KIF2A microtubule depolymerase dynamically localizes to spindle points and controls microtubule breakdown throughout mitotic advancement. Consequently, PMID:18411309 represents "KIF2A" that is annotated to GO: 0000922: spindle pole using the colocalizes_with qualification.

9.6 Modeling

One of the most crucial aspects of systems biology connecting it to mathematical biology is biological systems modeling. In order to simulate biological processes on computers, computer-aided biology seeks to create and employ effective algorithms, data structures, visualization techniques, and tools for communication (Koch et al. 2011). It includes the analysis and visualization of the intricate relationships between these cellular processes through computational simulations of biological mechanisms including cellular intrasystems.

9.6.1 Ordinary Differential Equation (ODE)

The traditional method for creating mathematical models is based on translating existing biological interactions into rate laws, such as the dynamic form of the law of mass (LOM) action. ODE models are generated as a result and they are referred to as mechanical as each changing variable and factor has an analogue in the process. Dynamic variables often indicate biological compound concentrations, whereas parameters are utilized for setting up unknown state and rate constants. Proper model structures must be derived from experimental evidence, and a large number of parameters must be calculated (Nieto 2022). To this purpose, a sizable and complete amount of experimental data are necessary to develop an ODE model. The diplomacies of signaling in live cells are typically described using ODE models. ODE models are frequently defined in systems biology as the conversion of pertinent biochemical interactions into rate equations. These mechanistic models have the benefit that every dynamic variable and model parameter has a biological counterpart, which enhances interpretations and permits physiologically pertinent conclusions (Bonotto et al. 2021).

Fig. 9.13 The four conditions when B → 0 and A + B → A

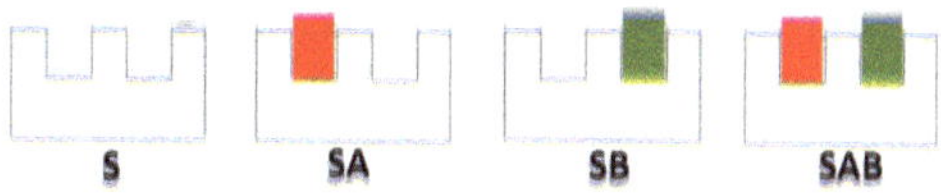

Often, ODE are used to determine the dynamics and make predictions about the outcomes as well as to pinpoint the function of particular pathway components. PDE is employed to investigate cell structural characteristics such cell deformation, polarization of cells, cell metastasis, as well as pattern generation, such as the development of stripes in different organisms (Koch et al. 2011).

$$\text{ODE} = da / dt = f,$$

a = a quantifiable variable (no. of cells, rate of cell damage)
f = maybe a complex formula
ODE, hence, can be interpreted a,

$$da = f \times dt$$

It indicates that the variation in "a" over a brief time frame "dt" equals to "f" times of "dt." Little phases of a dimension of "dt" are used to represent the evolution of 'a' throughout time. According to the equation,

$$A(t + dt) = a(t) + f \times dt$$

Some of the common syntax to prepare model:

- A → 0 (A implying to zero, means 'A dies').
- A → B ("A changes to B"—e.g., if A is macrophage, B is the activated version of the macrophage).
- A + B → C (When A and B interact to give the product C).

When B → 0 (When B implies to zero, B degrades) and A + B → A (B degrades in the presence of A), then (Fig. 9.13)

$$\text{S} + \text{A} \leftrightarrow \text{SA}; \quad \text{SA} + \text{B} \leftrightarrow \text{SAB}; \quad \text{S} + \text{B} \leftrightarrow \text{SB} \quad \text{and} \quad \text{SB} + \text{A} \leftrightarrow \text{SAB}$$

ODE modeling reactions The LOM action in ODE modeling implies that the rate of change of products is proportional to the "product" of the reactant concentration (Verdière et al. 2020; Tu et al. 2022).

- A → 0, the only reactant "a": rate of change of "a." Hence, ODE is–

$$da / dt = -k \times a\left(\text{"}-\text{" is because } a \text{ is lost}\right)$$

- A → B, here B (product) is created with each A disappearing. ODE is–

$$da / dt = -k \times a\left(\text{just like before}\right)$$

- But with the presence of B,

$$da/dt = k \times a\left(\text{"+" because B product is formed}\right)$$

- A + B → C, (reactants: A and B; Product: C)

$$dc/dt = k \times a \times b$$

- The rate of production of C is directly proportional to reactants' concentration

$$da/dt = k \times a \times b$$

- The rate of change of A is corresponds to the rate of change of C (each C is created for each loss of A).

$$db/dt = k \times a \times b, \text{similar to A}$$

- A + B → A, (Loss of in the presence of A). Here, the reactants are A and B. A is constant in both the reactant and product sides.

$$da/dt = 0$$

- The product of A and B employed by the LOM action is proportionately changes B,

$$db/dt = -k \times a \times b$$

- When B → 0, (loss of B).

$$db/dt = -k \times b$$

9.6.2 Partial Differential Equation (PDE)

When some physical or physiological knowledge of the medium it is available, they may be employed instead of ODE models for the physiological modeling of the evolution of a biological substance (for example, a drug in an organism) subjected to modifications. Derivatives are thus partial, which means that they are computed with regard to both time and a structure variable important for the representation of the medium (Raviprakash et al. 2022). Space, which denotes the use of spatial coordinates in one, two, or three dimensions, as well as physiological factors like age in the cell division cycle (q.v.) or molecular content, such as DNA or cyclins, may be included in this list of structure variables. By including a TF in the equations, PDEs are frequently created from ODEs (Abdulla et al. 2023).

Different mathematical and physical behaviors involved in PDE are (Anco and Cheviakov 2020; Carbinatto and Rybakowski 2023):

- *Elliptic Type* (Diffusion, equilibrium problems):
  - Laplace/Poisson (second-order linear).

$$\blacktriangledown^2 u = \delta^2 u/\delta x^2 + \delta^2 u/\delta y^2 = f(x,y) \quad [f = 0: \text{Laplace}, f \neq 0: \text{Poisson}]$$

  - Helmholtz equation (second-order linear).

$$\nabla^2 u = c + c^2 u = 0 (\text{time} - \text{dependent harmonic waves})$$

- *Parabolic Type*: time- or space marching.
  - Burger's equation (second-order nonlinear).

$$\delta u / \delta t + u \times (\delta u / \delta t) = v \delta^2 u / \delta x^2 \, (\text{Diffusion / dispersion})$$

  - Fourier equation (second-order linear).

$$\delta T / \delta t = \alpha \delta^2 T / \delta x^2$$

- *Hyperbolic Type* (Propagation).
  - Advection equation (first-order linear).

$$\delta u / \delta t + c \times (\delta u / \delta x) = 0$$

  - Wave equation (second-order linear).

$$\delta^2 u / \delta t^2 + c^2 (\delta^2 u / \delta x^2) = 0.$$

In a nutshell, the general form of second-order PDE involves two variables—

$$A(\delta^2 u / \delta x^2) + B(\delta^2 u / \delta x \delta y) + C(\delta^2 u / \delta x^2) + D\delta u / \delta x + E\delta u / \delta y + Fu + G = 0.$$

Hence, PDEs can be classified as—

- B^2 – 4.4C <0 (*elliptic*).
- B^2 – 4.4.C = 0 (*parabolic*).
- B^2 – 4.4C >0 (*hyperbolic*).

9.7 Databases Involved in Systems Biology

Large amounts of complex systems data and disparate bits of information are the fundamental obstacles in systems biology and as a result, systematic, computational methods are essential in systems biology. In order to understand the underlying molecular mechanisms and to forecast the effects of perturbations, such as pharmaceutical therapies, on these biological systems, we can foresee the behaviors of biological systems by building computational models of them.

9.7.1 Databases for Ontology Analysis

- *The Gene Ontology Resource* (http://geneontology.org/): The aim of the GO Consortium is to generate a complete computer model of biological systems that span the diversity of species in the chronology of life, from the molecular to the organism level (Mi et al. 2016) (Fig. 9.14).
- *AmiGo 2* (http://amigo.geneontology.org/amigo): It is the compilation of a set of tools that help users to search and browse the GO database. The components of the database are divided into two repositories consisting of three groups: GOlr

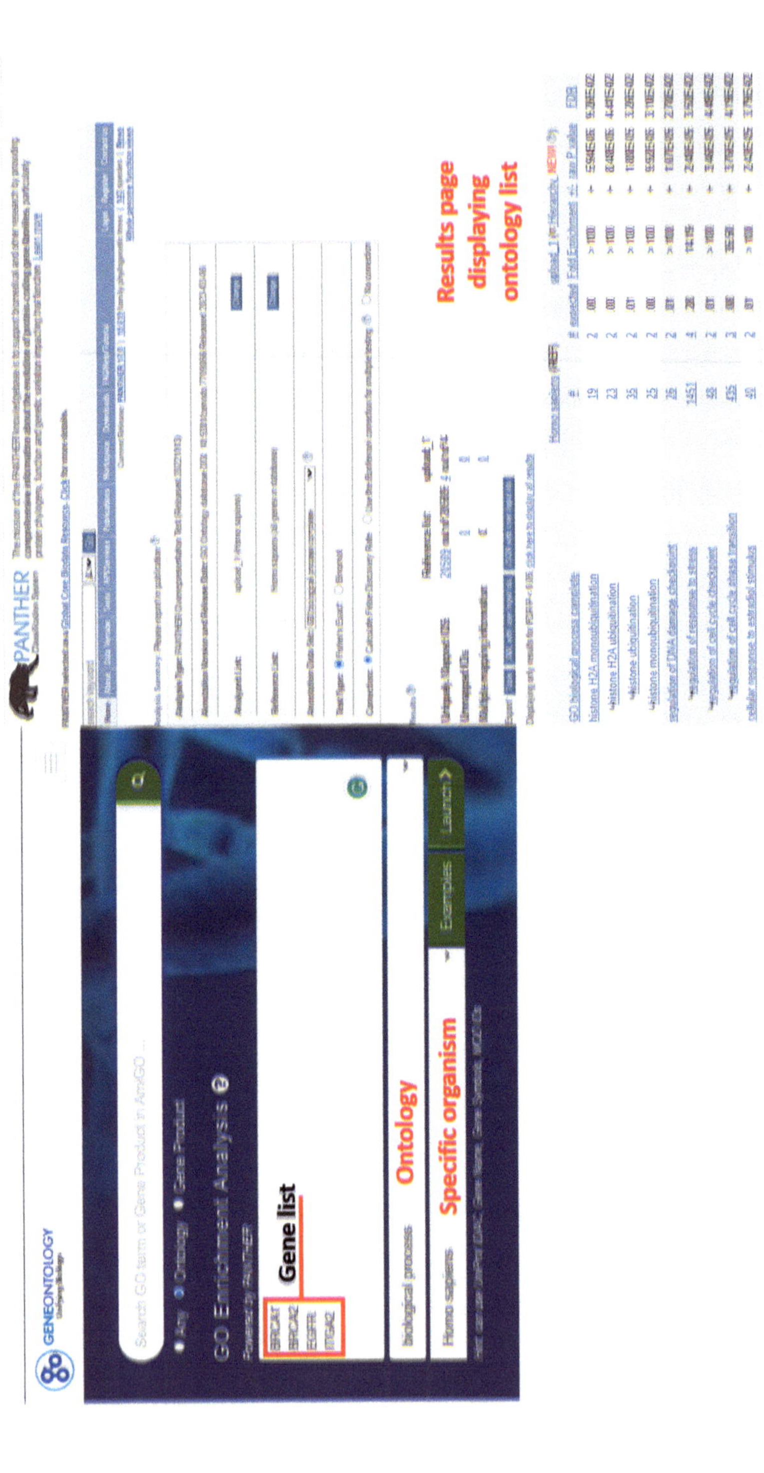

Fig. 9.14 Steps involved in Gene Ontology database

(backend and loading) and the Perl and JavaScript employed to make the frontend (Goto et al. 2022) (Fig. 9.15).

- *EnrichR* (https://maayanlab.cloud/Enrichr): Enrichr is a comprehensive database of gene sets that have been carefully selected, as well as a biological knowledge-accumulating search engine (Kuleshov et al. 2016) (Fig. 9.16).
- *NaviGO* (https://kiharalab.org/web/navigo/views/help.php): An interactive tool for visualizing, retrieving, and computing the relationships and functional similarity between GO keywords and genes is called NaviGO. Six separate scores are used to measure how closely GO keywords and gene activities resemble one another (Wei et al. 2017) (Fig. 9.17).
- *WEGO* (https://wego.genomics.cn/): A helpful tool for visualizing the results of GO annotation is WEGO (Web Gene Ontology Annotation Plot). It has been extensively utilized in numerous significant biological research initiatives, including the rice genome project (Ye et al. 2006) (Fig. 9.18).
- *iPathwayGuide* (https://advaitabio.com/ipathwayguide/): It enables analysis of gene set using least common denominator pruning techniques and high specificity techniques to get right to the few key activities (Ahsan and Drăghici 2017) (Fig. 9.19).
- *DAVID* (https://david.ncifcrf.gov/): A well-known web server and web service for functional annotation and enrichment studies of gene lists are included in the DAVID resource system for bioinformatics. It has a large knowledge base and several tools for functional analyses (Huang et al. 2009) (Fig. 9.20).
- *ChEA3* (https://maayanlab.cloud/chea3/): A similar approach was developed for TFs/target–gene interactions from genome-wide ChIP studies. Data tables from publications that profiled mammalian transcription factors using ChIP-chip or ChIP-seq reporting putative lists of target genes for transcription factors were integrated into a database called ChEA. Consists of 118 unique transcription factors, 107 publications, 35,286 genes, more than 150 ChIP-X assays (ChIP-chip, ChIP-seq, ChIP-PET), and approximately 1300 average targets per TF resulting in a total of 254,854 interactions. ChEA takes as input a list of genes and uses the database to rank transcription factors by their likelihood to regulate gene expression (Keenan et al. 2019) (Fig. 9.21).

9.7.2 PPI Databases

- *DIP* (https://dip.doe-mbi.ucla.edu/dip/Main.cgi): The DIP database can be searched in several ways to identify the initial protein of interest before exploring the protein–protein interaction network. It is possible to look for whole protein groups that meet certain requirements, such as sequence similarity to a particular protein, specific function or cellular location, or the presence of a particular domain (Xenarios et al. 2000).
- *STRING database* (https://string-db.org/): The STRING database collects, evaluates, and integrates all publicly available sources of data regarding PPIs with the aim of augmenting this understanding with computer calculations. Its goal is to

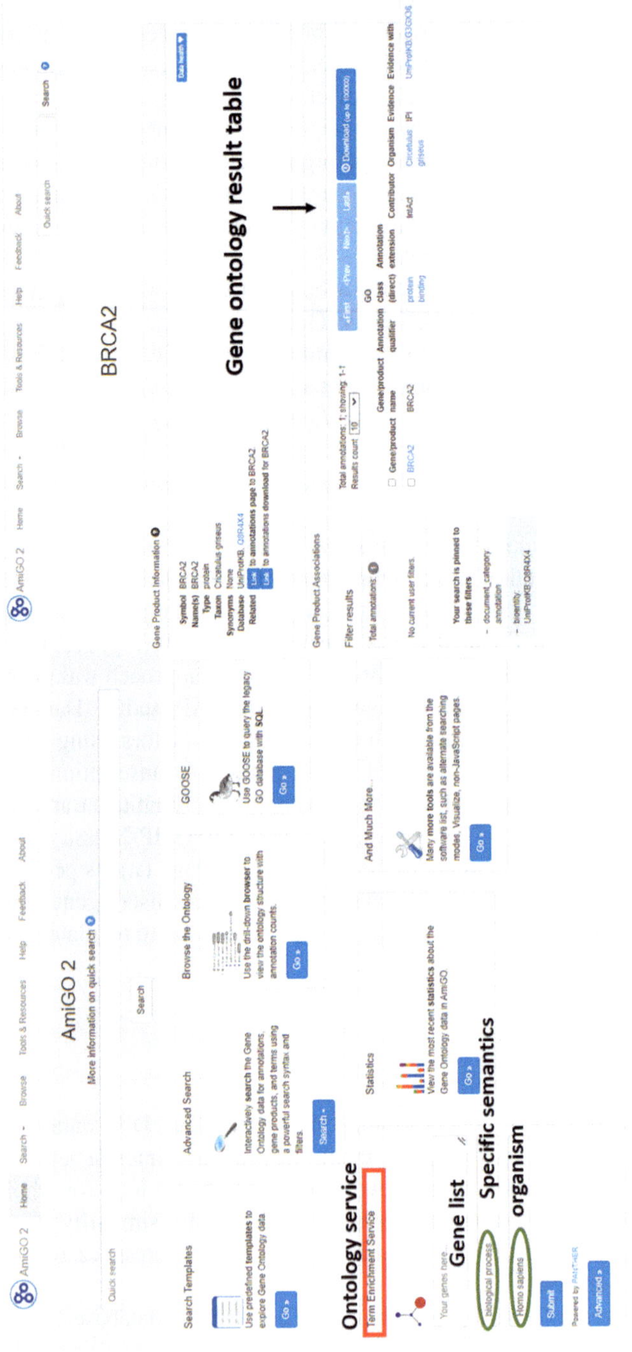

Fig. 9.15 Steps involved in AmiGO2 database

Fig. 9.16 Steps involved in Enrichr database

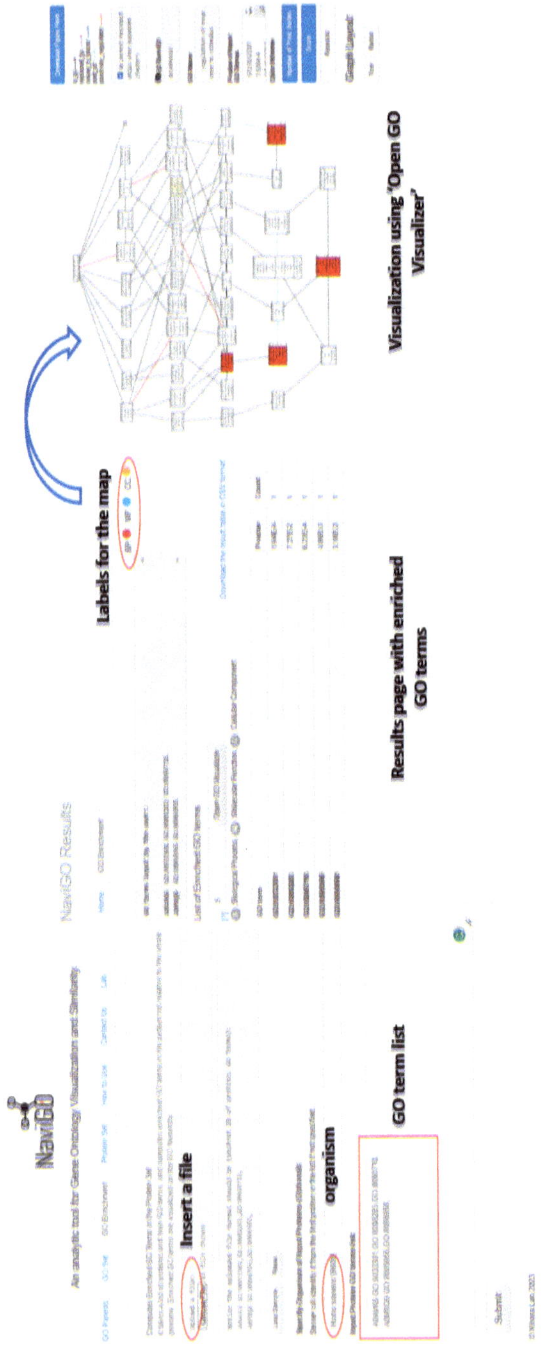

Fig. 9.17 Steps involved in NaviGO database

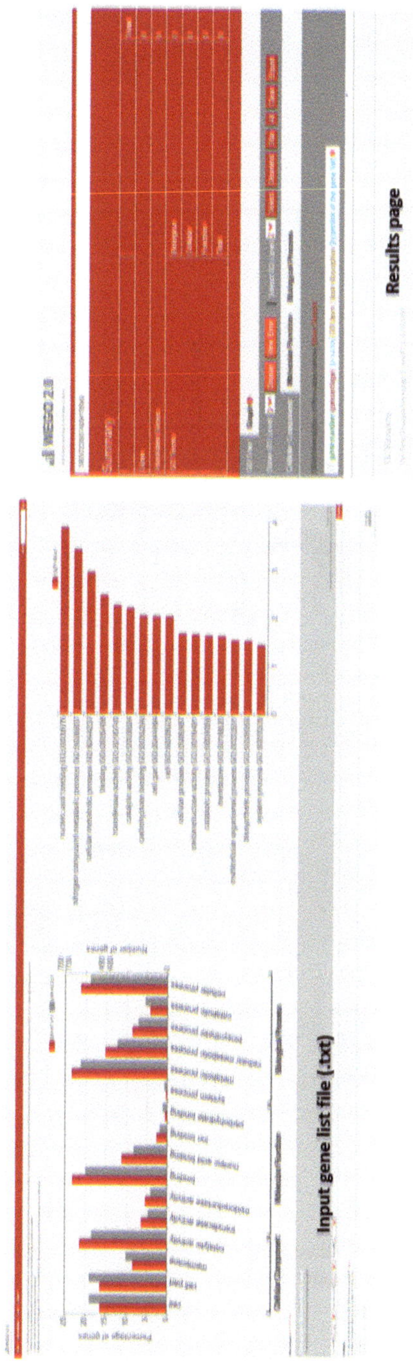

Fig. 9.18 Steps involved in WEGO database

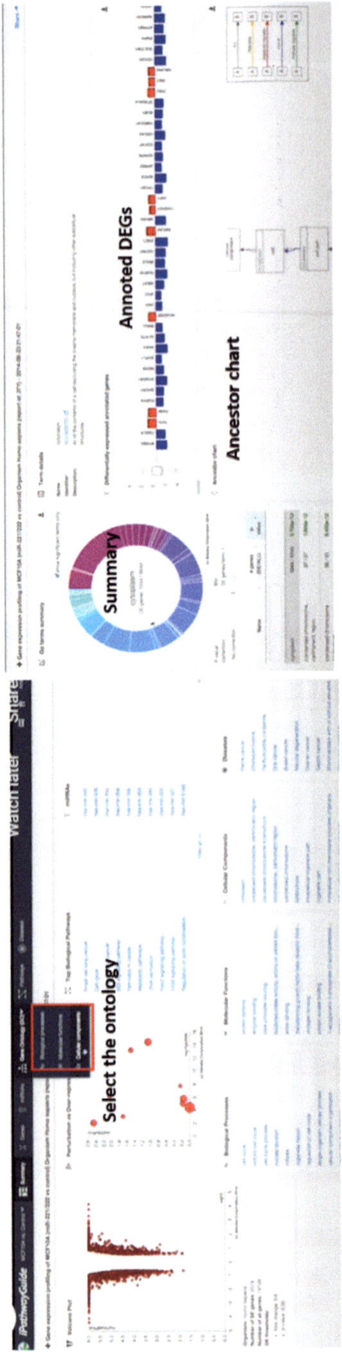

Fig. 9.19 Steps involved in iPathwayGuide database

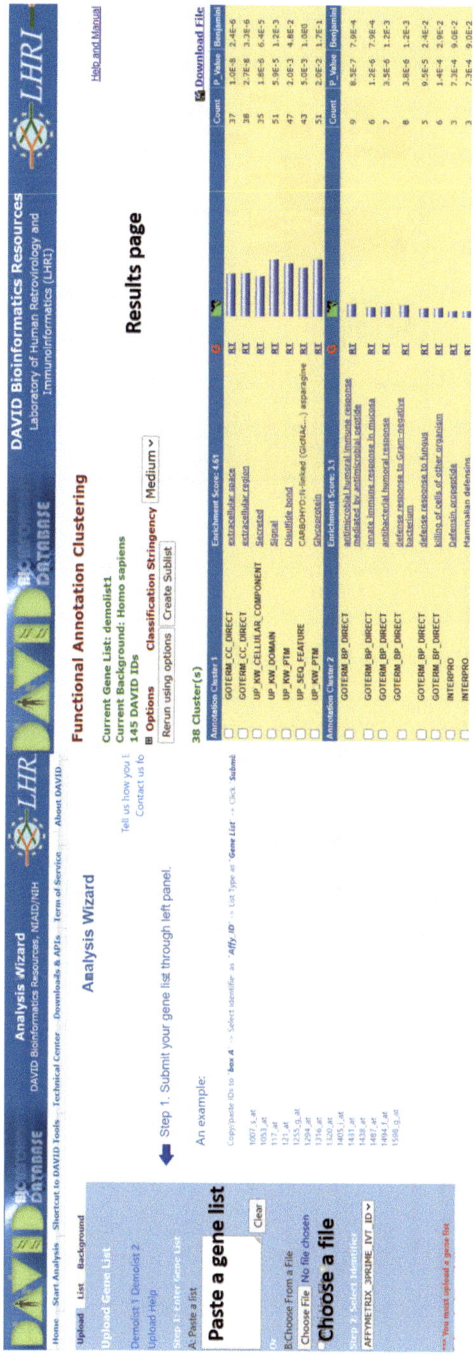

Fig. 9.20 Steps involved in DAVID database

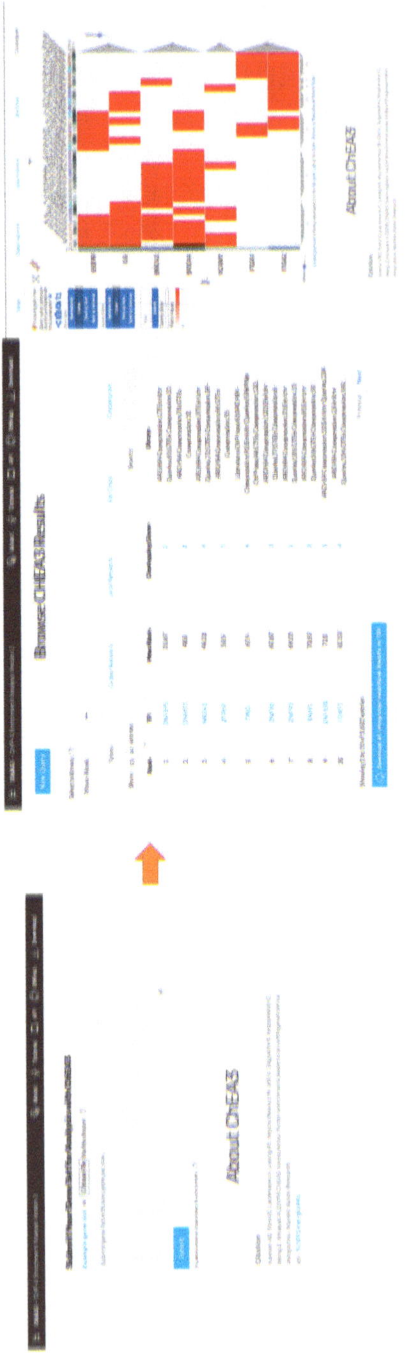

Fig. 9.21 Steps involved in ChEA3 database

establish a comprehensive, unbiased global network that connects both functional and physiological links (Szklarczyk et al. 2021) (Fig. 9.22).

- *GPS-Prot* (http://gpsprot.org/): By combining genetic and protein-protein connections between HIV and its human host into a unified representation, GPS-Prot enables even inexperienced users to swiftly construct networks. The platform can eventually be expanded to include additional host-pathogen systems (Fahey et al. 2011) (Fig. 9.23).
- *PIPS* (http://www.compbio.dundee.ac.uk/www-pips/): The PIPs approach employs a naive Bayesian approach to determine the probability of PPI by combining the predictive power of multiple characteristics. Co-expression, orthology, co-occurring domains, post-translational alteration, and a new feature that examines the semantic similarity of GO terms are among the features considered by the predictor (McDowall et al. 2009) (Fig. 9.24).
- *MINT database* (https://mint.bio.uniroma2.it/): MINT focuses on PPIs that have been experimentally validated and are culled by knowledgeable curators from existing research. The information on effective protein interactions is maintained in this database. MINT was designed to retain several kinds of biological interactions, such as enzymatic changes of one of the partners, in addition to binary complexes (Calderone et al. 2020) (Fig. 9.25).

9.7.3 GIN Databases

- *BioGRID* (https://thebiogrid.org/): A free and accessible database of CRISPR screens called the BioGRID Open Repository of CRISPR Screens (ORCS) was created after rigorous curation of every genome-wide CRISPR screening of information disclosed in the scientific literature (Oughtred et al. 2021) (Fig. 9.26).
- *GeneMANIA* (https://genemania.org/): It develops theories about how genes work and gives the gene top priority for function tests. By adding related functional genes that are made available through genomics or proteomics data, it expands the query list. Using GeneMANIA, a genomic screen with phenotypic output is carried out. It facilitates the discovery of interconnected with respect to the input genes. GeneMANIA is an acronym for a Multiple link Network Integration technique that consists of two components: (1) a linear regression technique that determines the functional link between various pieces of data. The propagation algorithm (2) uses the functional association network to forecast the gene function. Currently, 166,691 genes from 9 model organisms are mapped to 660,554,667 interactions in 2830 linked networks in GeneMANIA (Franz et al. 2018) (Fig. 9.27).
- *WeiBI* (http://weislab.com/WeiDOCK/?page=PKPD): WeiBI version 1.0 gathers data by transmitting interactions between more than 115,570 entries, using an upgraded method, and enables statistical analysis with an automated background for the provided inputs for functional enrichment. This method also enables the visualization of subsets as interaction networks, the successful completion of the

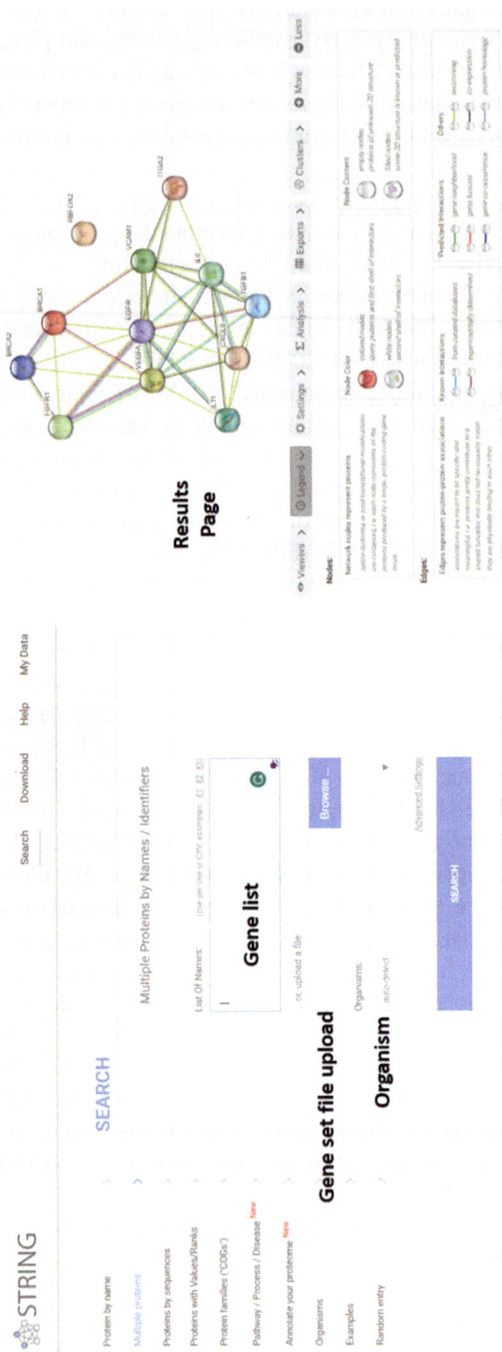

Fig. 9.22 Interface of the STRING database

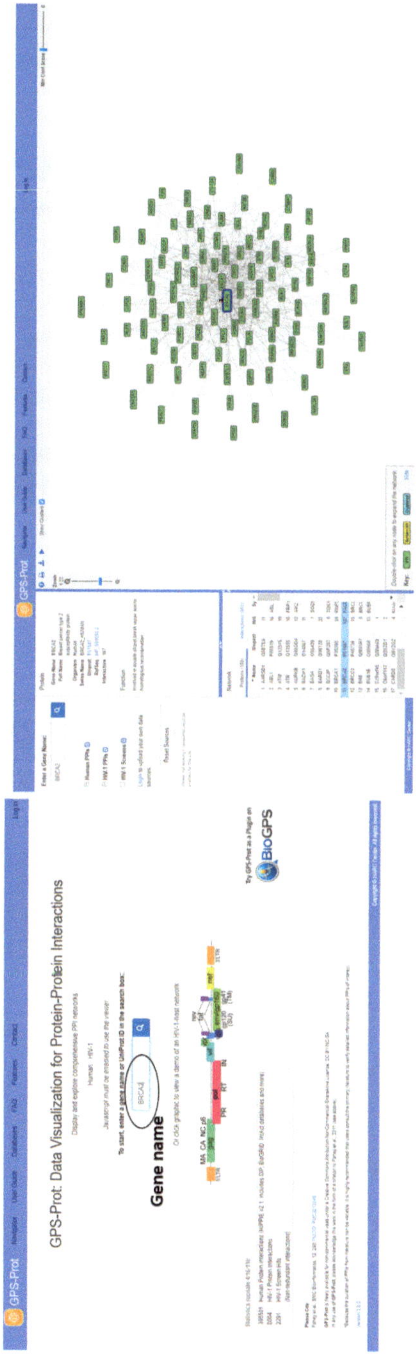

Fig. 9.23 Interface of the GPSProt database

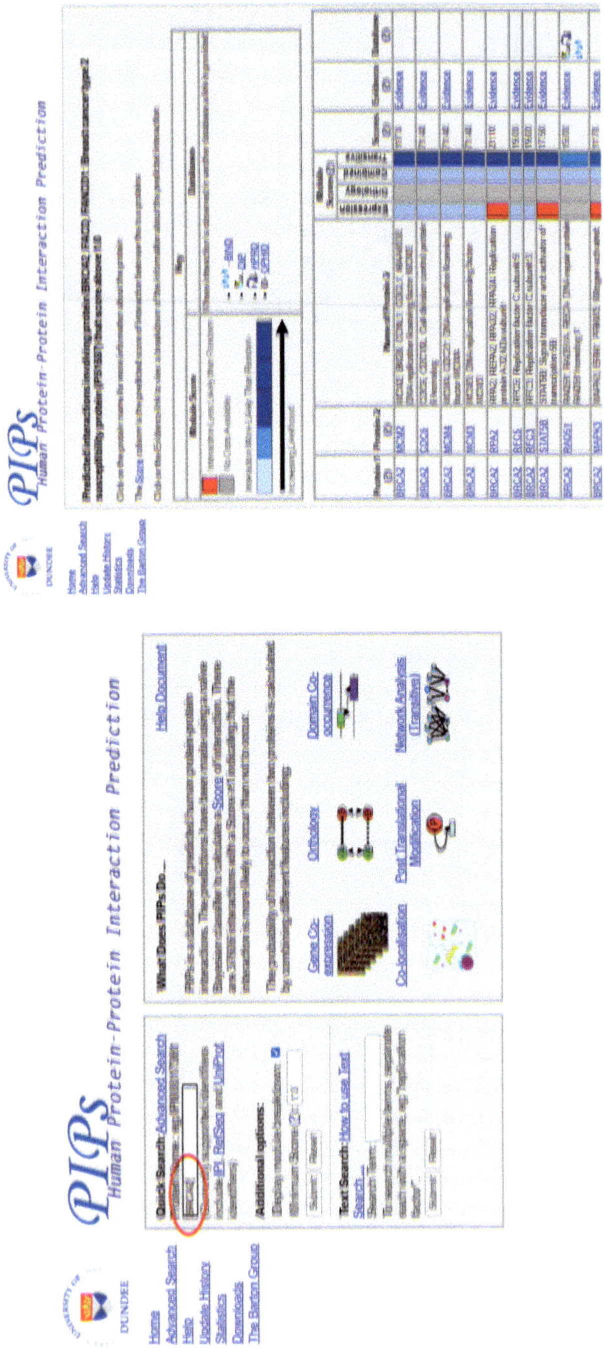

Fig. 9.24 Interface of the PIPs database

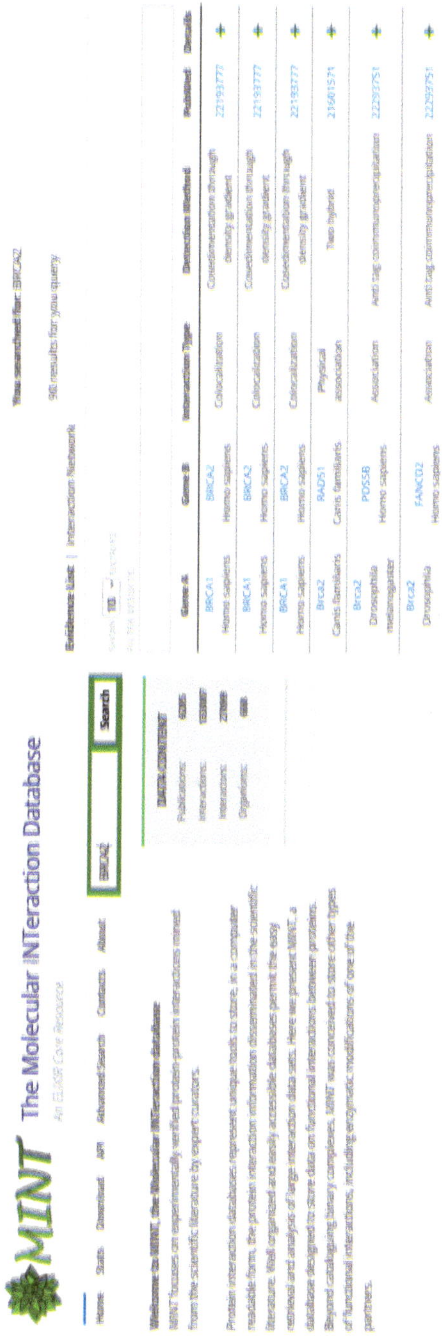

Fig. 9.25 Interface of the MINT database

Fig. 9.26 Interface of the BioGRID database

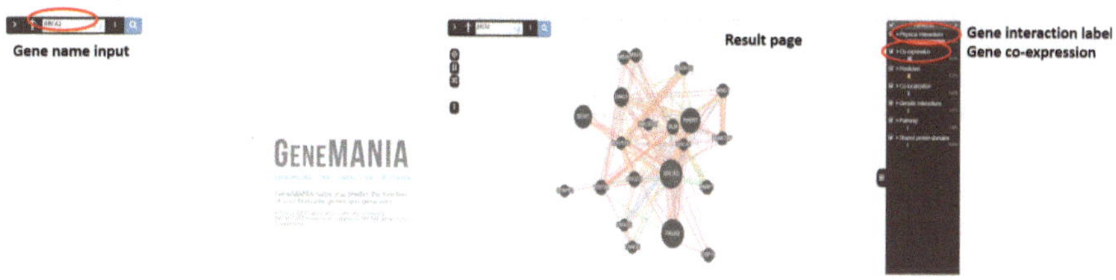

Fig. 9.27 Interface of the GeneMANIA database

enrichment analysis for a gene set, and the input of an entity list from a database (Kaushik et al. 2020).

- *BENviewer* (http://benviewer.bmeonline.cn/): BENviewer is a recent graph embedding model-based server for visualizing gene interaction networks online. More than 2000 biological pathways are visualized in a user-friendly 2D format using sophisticated graph embedding methods, and the results intuitively show not only the genes involved but also the tightness of relationships (Liu et al. 2021).
- *TFLink* (https://tflink.net/): With cross-links to numerous additional databases, TFLink offers straightforward access to information on transcription factor-target gene interactions, interactive network visualizations, and transcription factor binding locations. It also provides data for gene set enrichment analysis and makes systems biology and comparative gene regulation investigations easier. "Browse and search" can be accessed using the TFLink gateway, and then Entry page of the selected transcription factor or target gene opens (Liska et al. 2022).

9.7.4 GCN Databases

- *GWENA*(https://bioconductor.org/packages/release/bioc/html/GWENA.html): With the help of gene set enrichment, phenotypic association, hub gene detection, topological metric computation, and differential co-expression, the new R package GWENA integrates the building of gene co-expression networks and complete characterization of the found modules (Lemoine et al. 2021).
- *GeNET* (http://bengi.cs.mun.ca/genet): GeNET is a web tool for disseminating, viewing, and examining GCN data. GeNET allows researchers to upload their own GCN data by simply filling out a web form and uploading as little as three text files in a straightforward tabular format is one of its key advantages (Desai et al. 2017). GeNET automatically does the following upon data submission:
  - Step 1: data validation.
  - Step 2: computation of the correlation and adjacency matrices.
  - Step 3: obtains gene functional annotations from web sources like KEGG and Pfam.
  - Step 4: carries out the modules' Over-Representation Analysis (ORA).

  - Step 5: uses this data to build and populate a relational database.
  - Step 6: connects this database to an online user interface that enables searching, visualizing, and downloading of GCNA data from the database.
- *TCSBN* (https://inetmodels.com/): TCSBN integrates metabolic, regulatory, and protein–protein interaction networks to provide tissue-specific integrated networks (INs) for adipose, liver, and muscle tissues. In order to explore the functional relationships between genes as well as their relationships with biological functions, one needs to look into the synergies between operational and physical interactions provided by CNs and INs, respectively for human gene co-expression networks (CNs) respective to 46 normal tissues and 17 cancers are embedded (Lee et al. 2018).

9.8 Tools Involved in Systems Biology Approaches

9.8.1 Tools for Construction and Visualization of Networks

- *Cytoscape*: Cytoscape is a popular tool for the comprehensive visualization and analysis of systems networks. It is an easy-to-use piece of software that is publicly downloadable. It allows for the creation and visualization of biological components such as genes, proteins, and minute metabolites, which are portrayed as nodes, and the relationships among them, which are depicted as edges. Gene expression or methylation datasets, for example, can be imported into Cytoscape in order to improve the network and reveal more information alongside the network data already there. The program offers a superior network layout choice, making it a better candidate for a huge input dataset.
- The program offers a superior network design choice, making it a user friendly tool for a huge set of input data. This software's various plugins, which help incorporate additional analysis including gene set enrichment analysis, pathway analysis, and gene ontology analysis, are another benefit. The combination of such networks provides a deeper perspective for additional investigation. The program gives users the ability to obtain high-quality numbers for use in publications (Smoot et al. 2011) (Fig. 9.28).
- *Gephi*: Gephi is a free and convenient network analysis tool for all types of networks. It is an open graph visualization platform. Gephi has extra layout plugins like GeoLayout, NoverlapLayout, and Multimode Networks Transformation. With 100,000 nodes, the program has the benefit of processing huge datasets. Although the application calculates topological characteristics like degree and other centrality parameters, it focuses primarily on the visualization aspect of networks (Bastian et al. 2009) (Fig. 9.29).
- *Medusa*: The Fruchterman-Reingold technique that generates a 2D representation of network nodes is the main attribute of the open-source Java program known as Medusa. The instrument works well with tiny datasets that have medium-sized nodes and edges. The tool supports weighted graphs and denotes the significance of the interaction on the basis of different thickness of the edges

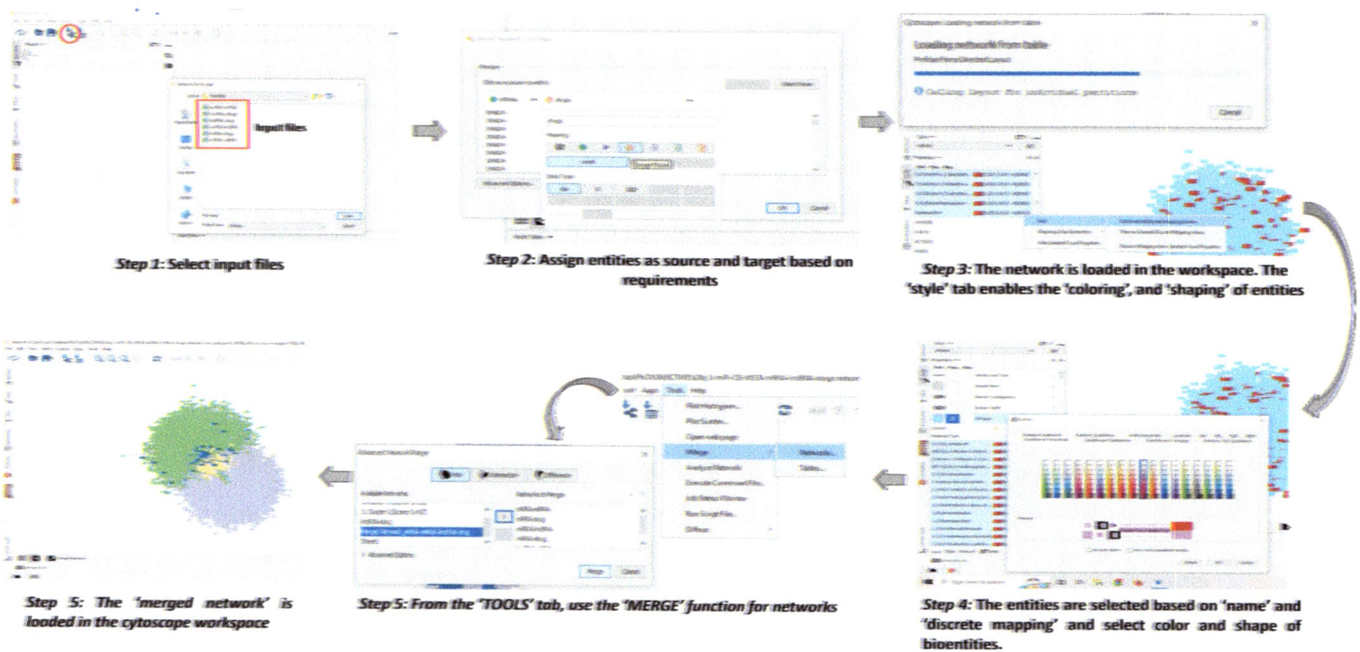

Fig. 9.28 The workflow of Cytoscape to construct interaction networks

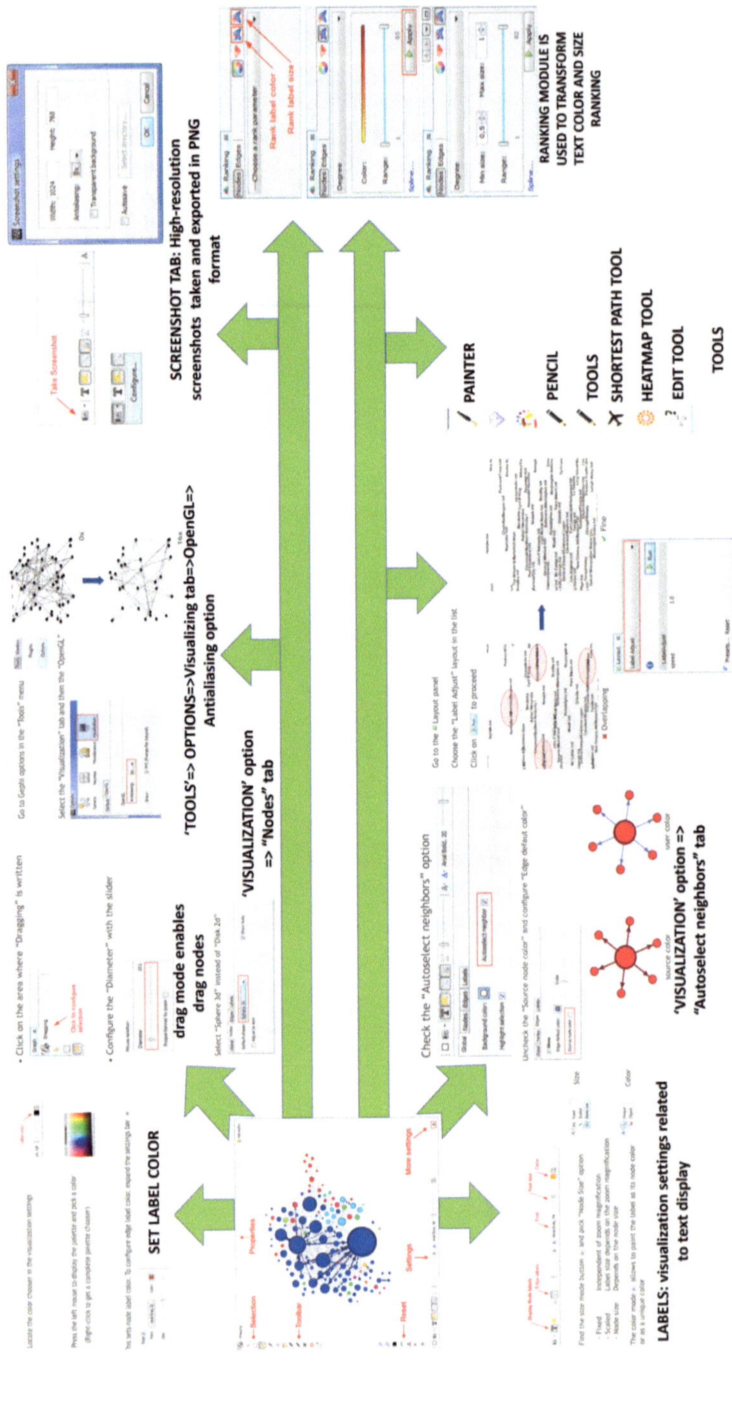

Fig. 9.29 The framework underlying GEPHI to construct interaction networks

connecting the nodes. The tool works better for interactions between multidimensional edges that reflect various types of data. Additionally, the program offers expression-based research, and the network can be saved for later reloading (Pavlopoulos et al. 2011).

- *BioLayout3D*: Visualization and analysis of biological data, particularly based on gene expression or pathway analysis, the network analysis tool BioLayout3D was developed. With the most recent revisions, the application enables the fusion, import, and display of pathways and protein networks in BioPAX Level 3 standard exchange format. Both weighted and unweighted nodes are supported by the tool. The shortcoming of BioLayout Express3D is that it uses fairly tiny node datasets. The tool's main benefits are its user-friendly and straightforward input of files that contain a list of interactions. The utility offers several methods for microarray analysis and the opportunity to integrate with Cytoscape (Theocharidis et al. 2009).
- *Genes2Networks.* This software tool was developed and can be used to build subnetworks that connect seed proteins with intermediates extracted from databases of protein interactions. Involves hash function for fast loading of the datasets. It implements AJAX for the changing of the page without reloading (Berger et al. 2007) (Fig. 9.30).
- *GeNeCK* (https://lce.biohpc.swmed.edu/geneck/): Gene Network Construction Kit is a publicly available tool that incorporates methods for building networks. GeNeCK uses gene expression data to form networks. Ten alternative network creation techniques, including partial correlation, likelihood, Bayesian, mutual information-based methods, are supported by the program. The resulting network will be made by combining several techniques. The outcomes of the job

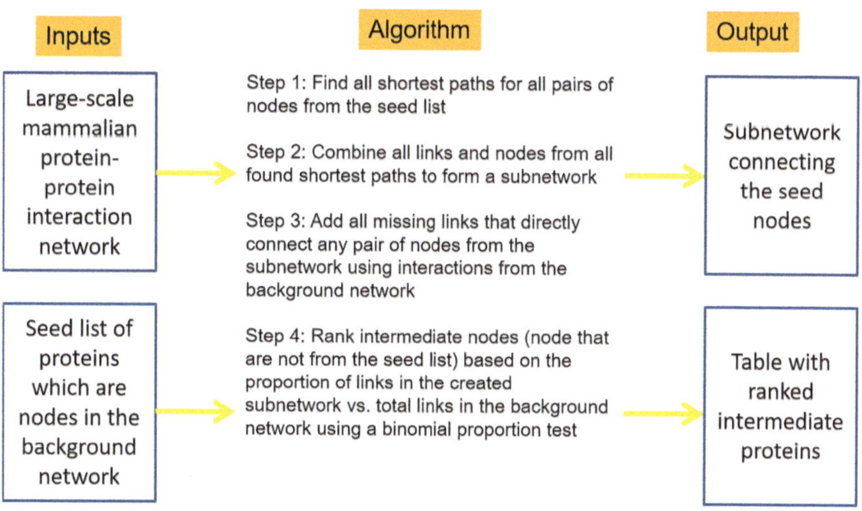

Fig. 9.30 The algorithm underlying Genes2Network

will be presented on the Internet after it is complete, and the online tool provides additional choices for choosing the defined approach (Zhang et al. 2019).

- *CytoTalk Algorithm*: The purpose of CytoTalk is to create an STN between two cell types from scratch, which is defined as the combination of many STNs. It starts by building a weighted integrated gene network made up of functional gene–gene interactions that occur both within and outside of cells. Using the combined data shared by two genes, intracellular functional gene interactions are calculated and weighted. By means of cross-talk edges, two intracellular networks are linked. The merged network nodes are weighted according to a combination of their proximity to the ligand/receptor genes and the function of genes relevant to their kind of cell. To assess a gene's proximity to ligand/receptor genes, we employ a network propagation approach. Identification of the signaling network is formulated as a PCSF issue using the integrated network as the input (Tuncbag et al. 2013).

9.8.2 Tools Used for Network Analysis

The strategy is less prejudicial and provides more coverage for recognized genes and proteins. Both clustering and topological analysis can be used to analyze networks, with topological analysis creating a graph based on the various network properties.

Clustering Analysis (*CA*): A group of genes or proteins are grouped into several clusters using the CA technique. Understanding biological systems with complicated natures is made easier with the use of CA in biological networks. The CA on the PPI network is significant as it gives functional annotations of subclusters generally following certain biological processes, aiding in the identification of stable complexes (Tan et al. 2005).

- *ClusterMaker*: It is a Cytoscape plugin that is publicly accessible and frequently used for CA utilizing almost all clustering methods, including MCODE, community clustering, hierarchical clustering, and others. The technique divides itself into two categories: attribute and network clustering (Utriainen and Morris 2023). In the case of attribute, clustering combines nodes with similar node qualities whereas network clustering is primarily utilized to find PPI complexes (Fig. 9.31).
- *MCODE*: A free Cytoscape plugin called molecular complex detection uses the molecular complex detection method to perform CA. Large PPI networks that represent chemical complexes are used by the technique to identify highly connected complexes (Fig. 9.32). The fact that MCODE is unaffected by false positives is a significant advantage. It makes it simple to visualize vast networks and highlights the key areas (Tan et al. 2005).
- *ClusterViz*: Cytoscape plugin for CA is called ClusterViz, and it focuses on the trio modules: interface, algorithms, and visualization. The tool enables compari-

Fig. 9.31 Steps to use clusterMaker plugin on merged network

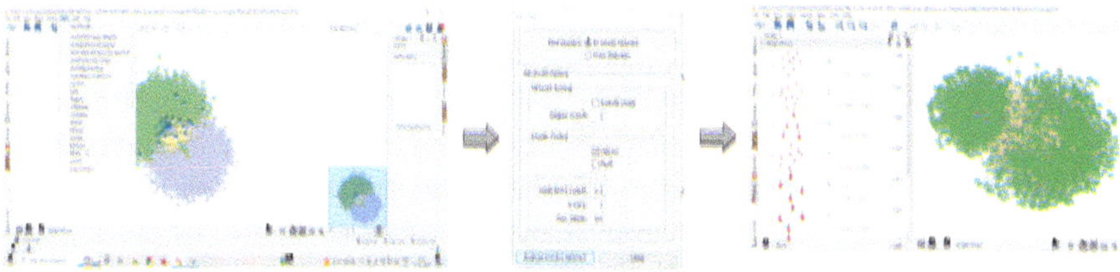

Fig. 9.32 Steps to use MCODE plugin on merged network

Fig. 9.33 Steps to use ClusterViz plugin on merged network

sons between the outcomes with those from various algorithms. Due to its easy-to-use interface, the plugin offers users a very simple and straightforward method for CA (Wang et al. 2015) (Fig. 9.33).

- *VisANT* (http://www.visantnet.org/visantnet.html): The small-footprint Java code or browser-based applet, VisANT is a tool for network visualization. By mapping each node to separate gene designations, its name resolution function aims to ensure that two proteins encoded by a single gene are always mapped as a single entity. It has a user-friendly and streamlined name-mapping feature. On a system having 1 Gb of random-access memory (RAM), VisANT has been demonstrated to display more than 200,000 nodes for huge data sets. This represents meta-graphs in which a single node may contain a subgraph. Additionally, VisANT is linked to an online record that has more than 450,00 interactions between various organisms (Hu et al. 2004).

9.8.3 Tools for Functional Enrichment Analysis

Huge amounts of biological data are derived from high-throughput data, which need a suitable pipeline for decoding and obtaining useful information. It takes a lot of work to functionally annotate each gene, associate the genes to the specific pathways of which they are a part of, eliminate redundant information, and see the linkages that a specific group of genes shares. A number of tools that aid in the facilitation of the aforementioned points have been developed for this aim.

- *FunRich* (http://www.funrich.org/): For network and functional enrichment investigations, FunRich is a straightforward, open-access independent tool. The FunRich tool gives users the choice of selecting a customized dataset or using an already-built FunRich dataset that is human specific. Content from databases like UniProt and others can also be downloaded. Users are able to supply the gene set and carry out the necessary analysis. Plots and spreadsheets can be downloaded for functional enrichment analysis visualization. The application is simple to use, and the output may be downloaded as a variety of charts, including bar plots, Venn diagrams, column graphs, pie charts, and heat maps (Pathan et al. 2015).
- *BiNGO*: A Cytoscape plugin called Biological Networks GO is available for free download via the app manager. The tool thoroughly evaluates whether GO phrases are overrepresented in a biological network. The integration of interactions with dynamically enriched routes gives BiNGO a significant advantage over competing tools. The program accepts a variety of identities and offers data for several organisms with pre-built annotations from NCBI. The network-based viable enrichment can later be preserved in a variety of ways. The outcome will be shown as a series of colored gradient nodes, with the node with the strongest color representing the most important information (Maere et al. 2005) (Fig. 9.34).
- *GeneWeaver* (https://geneweaver.org/): The tool is a curated database of experimental genomic data that enable users to determine the types of biological processes in which a particular gene is involved. Data on 80,000 genes from seven distinct species are included in the tool. The program also incorporates information from several databases including the GeneNetwork database, Comparative Toxicogenomics Databases (CTD), and other drug-associated databases, in addition to the gene's curated biological links and functional characteristics.

Fig. 9.34 Steps to use BINGO plugin on merged network

GeneWeaver employs gene set homology clustering to organize big datasets into tiered clusters. This enables the identification of groups of genes or proteins from the input set that are not related to one another (Baker et al. 2012).

- *GeneCodis3* (https://genecodis.genyo.es/): In order to find the gene set with a comparable biological nature in MF, BP, or CC, GeneCodis3 compiles data from several sources. The program has tried to lessen background noise and redundancy from the output data with the most recent release. Additionally, the program has the ability to analyze exosomes by including exosome IDs. The most recent release includes comprehensive route annotations, functional annotations, pharmacological annotations, comparative analysis, interactive visualization, and single-node enrichment analyses (Tabas-Madrid et al. 2012).

9.8.4 Topological Parameter Analysis (TPA) and Hub Gene Identification

By figuring out the centrality characteristics that help configure the key nodes in a network, one may determine the topology of the network. The network's nodes are ranked according to their centrality, which establishes their relative importance. Node degree, average shortest path length, clustering coefficient, and betweenness centrality are the typical centrality parameters utilized in GIN. Radiality, proximity, centrality, and stress are further factors (Zhao and Liu 2019). A node's degree is equal to the total number of edges that connect it. The degree is straightforward and often used to measure and assess a node's importance in a network. The clustering coefficient (CCo) is a measure of how cohesive or aggregative a network's nodes are (Masuda et al. 2018). It is used to determine how much a gene or protein, for example, has control over other genes in a biological network. Resources to identify TPA and hub genes:

- *NeAT*: To analyze networks or clusters, it offers an accessible and open web server. The online server offers two collections of analytical tools (Fig. 9.35). The first category supports the basics of the graph, such as path discovery, graph comparison. A second set aids in linking the nodes in clusters. The program is

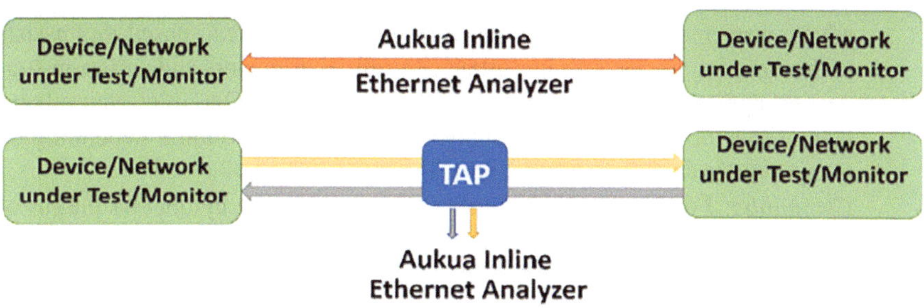

Fig. 9.35 Protocol for a binary set of device/data association analysis

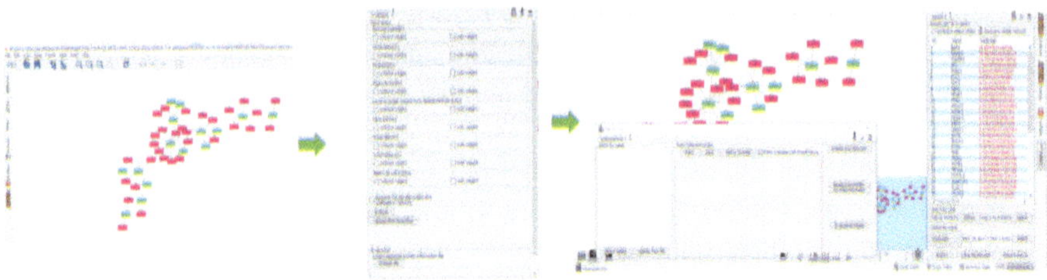

Fig. 9.36 Steps to use CytoNCA plugin on merged network

adaptable in that it can analyze data collected from different databases and handle huge datasets (Brohée et al. 2008).

- *CytoNCA*: Eight distinct centrality measures can be calculated, assessed, and visualized using the CytoNCA plugin of cytoscape. Both weighted and unweighted networks can use the plugin. In order to combine biological details with structural evidence, the program enables users to import supporting information for both nodes and edges. The program also provides numerous methods for illustrating the data as graphs, charts, and tables (Tang et al. 2015) (Fig. 9.36).
- *NetworkAnalyzer*: NetworkAnalyzer is a Cytoscape plugin that offers an effortless way to conduct network analysis. The tool decides if the edges are directed or undirected edges. Additionally, it calculates topological factors including the distribution of shared neighbors, median clustering coefficient dispersion, betweenness centralities, closeness centralities, and topological variables. A limited number of local attributes can also be examined by users. This plugin works in the 3.8 version of Cytoscape.
  - *Network analysis Profiler* (https://bib.fleming.gr:8084/app/nap): This easily accessible web application uses a systems biology approach to analyze multifaceted biological situations like tumors. The tool evaluates the intra- and internetwork designs and executes network profiling. The instrument bridges the knowledge gap between network analysis, statistics, and graph theory. Based on the igraph library, NAP is built in R to analyze small to medium-sized networks and produces plots of the highest quality (Theodosiou et al. 2017).
  - *CentiScape*: It is another Cytoscape plugin that facilitates calculating network centrality characteristics and enables users to examine the nodes and their edges, was created to enhance network data. The Boolean-based tool assists in determining nodes with important centrality parameters and assists in suggesting appropriate drug targets for illness networks. One important advantage of this tool is its ability to analyze topological importance depending on many metrics (Scardoni et al. 2009) (Fig. 9.37).
  - *Cytohubba*: For locating hub genes in a network, the cytohubba plugin of Cytoscape is frequently used. The plugin calculates 11 topological characteristics, including degree, the density of the maximum neighborhood compo-

nent (MNC), the maximal clique centrality (MCC), and 6 other centralities. The graphical tool then presents the hub genes in accordance with the parameter once the tool has computed the parameters. The nodes are typically displayed using a color gradient format, which facilitates the effective visualization of computed hub genes VisANT. The tool's compatibility with additional plugins, which can produce original analysis and offer fresh perspectives on the fundamental networks, is a significant asset (Chin et al. 2014) (Fig. 9.38).

- *CyNetSVM*: The NetSVM approach, which was created to find the probable biomarkers by analyzing expression profile data and PPI data, was implemented into CyNetSVM plugin of Cytoscape plugin. CyNetSVM forecasts therapeutic outcomes for patients and provides valuable biological network insights. The Java-written utility can manage large datasets. The plugin displays the sensitivity and specificity of the prediction together with a graphical representation of the predicted network. The method is especially helpful for cancer studies since it can forecast how the identified biomarker would perform clinically. Integrating the findings with functionally enriched signaling pathways. This plugin is compatible with the 3.1 version of Cytoscape (Shi et al. 2017).

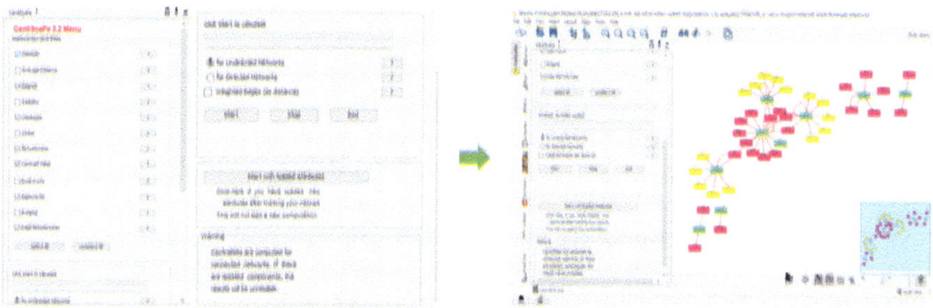

Fig. 9.37 Steps to use CentiScape plugin on merged network

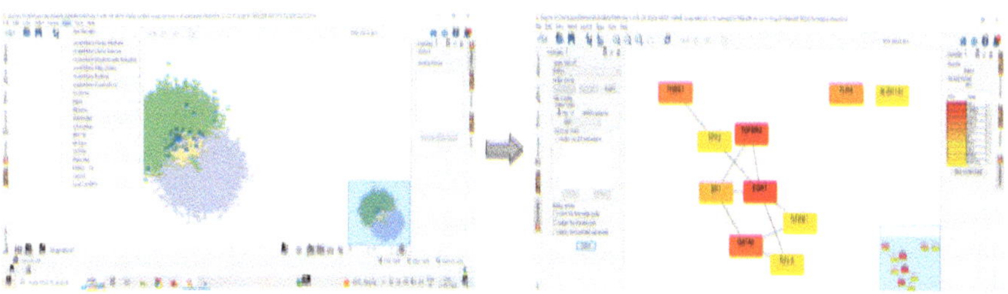

Fig. 9.38 Steps to use Cytohubba plugin on merged network

9.8.5 Tools Involved in Modeling

Building mathematical models is a key component of systems biology, which is used to anticipate and better understand biological processes. Since the invention of the first digital computers, specialized software has been created to aid in the construction of models. The supported software programs that are currently accessible to researchers are listed below (Table 9.1). SBML, the de facto standard for exchanging models of biological cellular processes, is supported by the great majority of contemporary systems biology modeling tools. The standard CellML, which is used to model physiological processes, is also supported by several tools. PDEs are used in biomedical modeling to depict spatiotemporal systems; resolving PDEs requires additional work than resolving ODEs. The finite element method (FEM), a numerical approach that offers approximative solutions for PDEs, is typically used to solve PDEs. Tools like ANSYS, FreeFEM++, and OpenFEM support FEM. PDE Toolbox in MATLAB is another option for analyzing PDEs.

Table 9.1 A resource matrix of software tools and data resources

Software	Description	Support modeling paradigms	References
iBioSim	A CAD tool for modeling, analyzing, and designing genetic circuits	ODE and stochastic	Myers et al. (2009)
CompuCell3D	Building and simulating multicellular models using a GUI/scripting tool	ODE	Cickovski et al. (2005)
COPASI	SBML model analysis and simulation	ODE and stochastic	Hoops et al. (2006)
Cytosim	Flexible cytoskeletal filaments and motor proteins spatial simulation	Stochastic and spatial	Lugo et al. (2022)
libroadrunner	High-performance software library for SBML model modeling and evaluation	ODE and stochastic	Welsh et al. (2023)
massPy	A simulation program compatible with COBRApy	Employs libroadrunner	Haiman et al. (2021)
MCell	Singular molecule-based particle-based spatially randomized simulation using a GUI tool	Spatial (particle)	Tapia et al. (2019)
OpenCOR	A cross-platform modeling environment for Windows, Linux, and macOS that may be used to organize, edit, simulate, and analyze CellML files	ODE	Garny and Hunter (2015)
PhysiBoSS	A unique variation of the PhysiCell agent-based modeling system that incorporates cell agents' Boolean signaling networks immediately		Letort et al. (2019)

(continued)

Table 9.1 (continued)

Software	Description	Support modeling paradigms	References
PhysiCell	Framework for multicellular systems biology modeling using agents	Employs libroadrunner	Gianlupi et al. (2023)
PySCeS	A Python program for creating and examining SBML models	ODE	PySCeS: The Python Simulator for Cellular Systems (n.d.)
pySB	A Python-based framework that focuses on models fueled by rules	ODE	Lopez et al. (2013)
ReaDDy	A statistical simulator based on particles that incorporate cross-molecular potential		Hoffmann et al. (2019)
SBSCL	Java library with thorough and effective SBML support	ODE	Sauro et al. (2003)
SBW (software)	A distributed workbench with a variety of modeling instruments	ODE	Panchiwala et al. (2022)
Smoldyn	Spatial stochastic simulations with individual molecules using a particle-based simulator	Spatial (continuous)	Andrews (2017)
Spatiocyte	A fine lattice with up to one molecule per site is used in spatial modeling software		Arjunan et al. (2020)
SpringSaLaD	A particle-based coordinate simulator where molecules are connected by spring-loaded spheres		Michalski and Loew (2016)
STEPS	Heterogeneous meshes with a stochastic reaction-diffusion and membrane potential solver		
Tellurium (software)	A system for simulation that combines several libraries into a single platform	Employs libroadrunner	Choi et al. (2018)
URDME	Simulation of probabilistic response propagation on unorganized meshes		Drawert et al. (2012)
VCell	A robust modeling platform for nonspatial, spatial, predictable, and dynamic simulations, including interaction networks and reaction rules	ODE	Resasco et al. (2012)

9.9 Additional Concepts in Systems Biology Approaches

9.9.1 Graph Theory-Basic Concepts

(a) *Graph Theory-Basic Concepts*: Networks are well-defined mathematical objects as part of the sub-field of mathematics called graph theory. Typically, networks are annotated with the letter G which is a mathematical object that

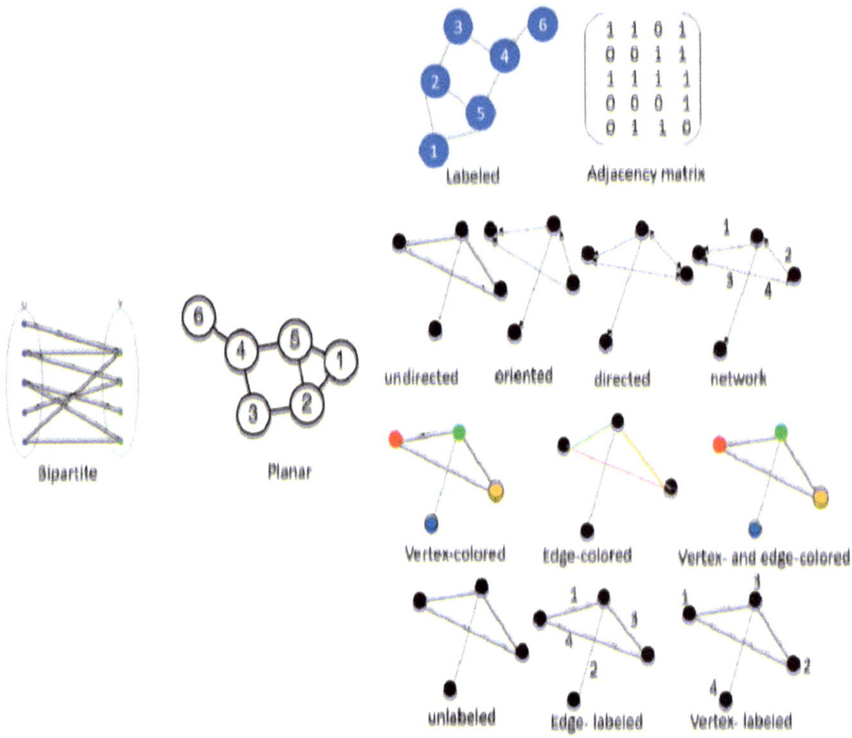

Fig. 9.39 Concepts of graph theory

contains V vertices/nodes, E edges/links and/or A arcs/directed edges/arrows. Such a graph can be also represented as an adjacency symmetric matrix made of zeros and ones where 1 corresponds to the presence of an edge/link between two nodes, which are the rows and columns. Graphs can be undirected, directed, or mixed (Jungck and Viswanathan 2015). Nodes and edges can have different "colors" representing common properties shared by subsets of nodes or edges. Below are the different types of graphs in systems biology (Fig. 9.39):

- Planar Graphs: When there is no edge crossing.
- Bipartite graphs: Two sets of nodes; links only between entities of the same set.

(b) *Clustering coefficient*: Clustering coefficient is a measure of how neighbors of a node are connected to each other. For example, the red node has 4 neighbors, if all the neighbors are connected with the 6 possible links, the clustering coefficient for that node is 1 (6/6). If only 3 out of the 6 links are implemented then the clustering coefficient is 0.5 (3/6). Clustering coefficient is computed for all nodes in the network and the average is the coefficient (Fig. 9.40). It can be computed using the Dijkstra algorithm or the Floyd algorithm (Heer et al. 2020).

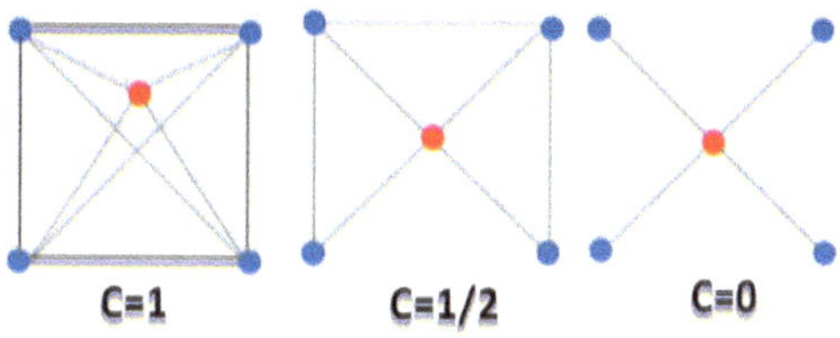

Fig. 9.40 Concepts of clustering coefficient

(c) *Importance of the hubs*: One of the advantages of scale-free networks is that they are robust to random attacks since most such attacks would delete lowly connected nodes. However, when hubs are attacked, the network topology would rapidly collapse. It was shown that protein connectivity correlates with network robustness based on genetic deletion screens of yeast proteins and the yeast protein interactome (Palva 2018).

(d) *Steiner Trees*: Steiner Trees can be applied to construct subnetworks with the minimum number of intermediate protein nodes used to connect seed lists of nodes. The Steiner Tree algorithm is NP-hard, which means it is exponentially more time consuming to compute as the seed list increases in size. Hence, heuristics to approximate a solution are developed. Steiner Trees idea was used to build a cell signaling network of the yeast pheromone response by Huang and Fraenkel (Gupta et al. n.d.).

(e) *Perturbations and Bayesian Networks*: Mutations and other molecular alterations can be thought of as perturbations of a biological network. The network's connectedness is utilized to forecast how the perturbation will affect the system. To calculate the system-level impact of certain perturbation patterns, such as mutational profiles or differentially expressed genes in tumor cells, we use graph diffusion algorithms. The core network topology can also be understood via network perturbations. Computational techniques are developed for learning networks from perturbation data. For instance, using data from RNA interference and CRISPR-Cas9 screening, multiple stacked effects models are designed for the reconstruction of intracellular signaling networks (Caldera et al. 2019).

A Bayesian network (BN) compactly represents a combined probability distribution. BN has two parts: A graph that is directed and acyclic and a set of conditional distributions. In a directed acyclic graph (DAG), the nodes denote random variables and the edges establish parent-child relationship on the basis of statistical dependencies between such variables. Each node (e.g., X_i) has a conditional probability distribution (CPD) representing $P(X_i$ |Parents|). The CPD can be represented in either of the many forms: tabular and a tree representation. BN provides an efficient way to work with large joint distributions (Fig. 9.41).

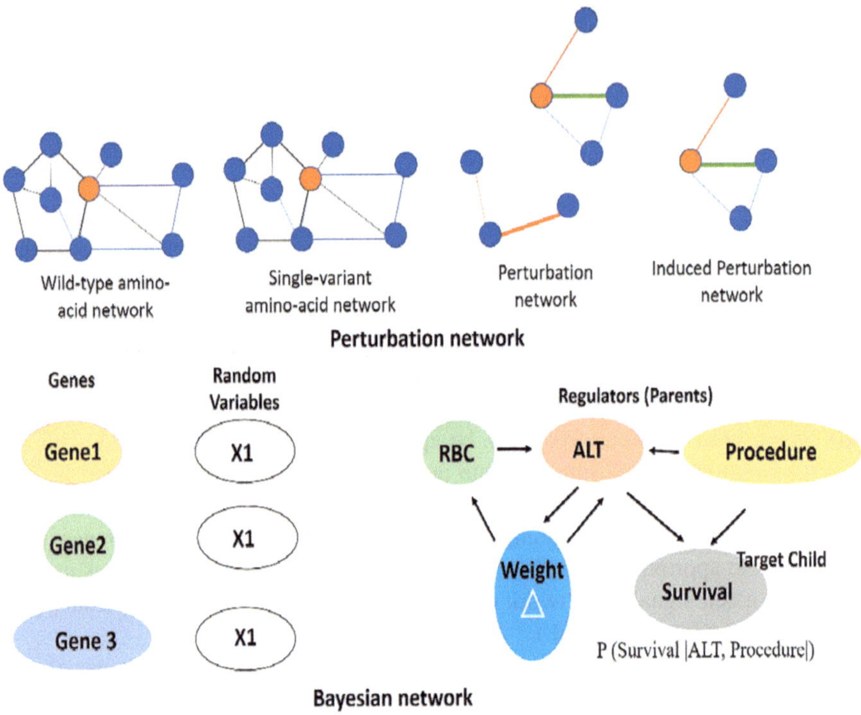

Fig. 9.41 Schematic representation of perturbation and Bayesian networks

9.10 Current Problems in Conventional Strategies

(a) *GWAS*: One of the major limitations of GWAS is to take into consideration the high level of significance of multiple tests. This limitation is likely to be mitigated with increasing genomic coverage of GWAS with a concurrent rise in the adoption of WGS data and an increased number of independent tests. Increasing the sample size and reducing the number of test runs are either strategy to get around the GWAS multiple testing restriction. The possibility of too many hits contributing to the causation in GWAS is another difficulty. GWAS is unlikely to ever fully account for the heritability of complex traits. This restriction does not only apply to GWAS because there is currently no technology or method that can fully determine the genetic basis of complex traits. Population stratification is a major issue in genetic association research because, if not properly accounted for, it can lead to erroneous results. In GWAS and post-GWAS investigations in humans, identifying significant gene–gene interactions has proven difficult primarily due to a lack of statistical power and methodological difficulties. SNP array-based GWAS relies on the accuracy of the sequencing studies and reference panels to create genotyping arrays and impute untyped variants in GWAS. SNP array-based GWAS lacks to find extremely uncommon variants linked to illness (Manolio 2013).

(b) *ODEs*: Inferred by model training methods from measurement data such as immunoblotting, proteomics, quantitative PCR, or cell viability, the majority of ODE models contain unknown parameters, such as reaction rate constants. More data are needed for larger models in order to guarantee the accuracy of parameter estimations and model predictions. Public databases can be used to find models of cancer signaling (Stapor et al. 2022). On the other hand, data from various perturbation experiments correlate to various initial value problems, which we refer to as "experimental conditions."

The ODE system's input parameters for each experimental condition are different vectors, necessitating independent simulation for each iteration of model training. In light of this, the computing time increases linearly with the number of experimental conditions. Even with the most advanced techniques, such as adjoint sensitivity analysis and hierarchical optimization, large-scale ODE models with thousands of testing criteria and several hundred types of chemicals might require tens of thousands of computation hours to complete. The most effective techniques currently in use for the optimization of ODE models are gradient-based strategies like multi-start local optimization or hybrid dispersion screening.

(c) *Network-based approach*: Biological networks mediate interactions between specific DNA sequences and specific phenotypic characteristics. However, this does not imply that specific DNA sequences and specific phenotypic characteristics in complex biological systems have any direct causal relationships. The "causality" of regulatory links is not revealed by clustering analysis, despite the fact that it sheds light on the "correlation" between genes and biological processes (Fig. 9.42). On the basis of microarray data alone, several techniques have been proposed to automatically identify regulatory connections. There is

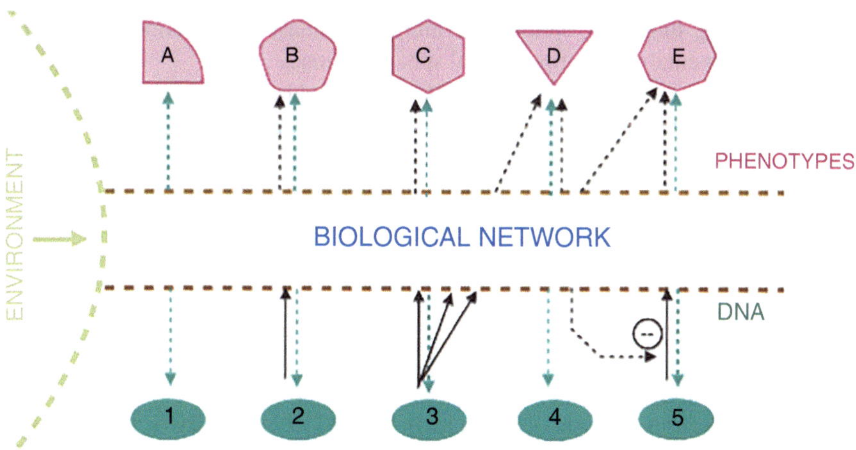

Fig. 9.42 Overview of correlation between genes and disease phenotypes. The dashed arrows are not mere extensions of the solid-line arrows. Environmental factors (whether "external" or "internal" with respect to biological networks) have an impact on DNA sequences, their expression, and the development of phenotypic features, as indicated by the green arrows

currently limited scope to infer causation based on transcriptional regulation because these methods rely on data collected from mRNA abundance. As large-scale data become accessible, post-transcriptional, and post-translational processes of regulation must be considered, yet many features have not yet been evaluated accurately or in a high-throughput manner. Each causal link between a DNA sequence and a trait has been depicted as varying (from solid to dashed) as it passes through and is altered by biological interaction networks. In a conventional manner, the causal arrows interact with one another inside the network in addition to transforming (Barabási et al. 2011).

9.11 Conclusion

The comprehensive characterization and modeling of a living system as a biological network is known as systems biology. Systems techniques aim to give a comprehensive model of a process by quantification of all observable components and their connections, in contrast to reductionist approaches, which strive to simplify or isolate the impact of a single component on a broader biological process. The resulting models are hence incredibly effective tools for discovering new linkages between components, understanding the role of previously unrecognized components and identifying multivariate emergent features that would otherwise go unnoticed.

It is crucial to understand that no solo biology technique currently works best from biological research for best results. The specialized nature of these developments requires close interdisciplinary collaboration from the initial phases of experimental design through data analysis, model construction, and application even though they have produced unique prospects for understanding human health and illness. Rethinking of the long-standing conventions and incentives is necessary in order to close these gaps and enable systems biology to attain its full potential along with providing financing options for multidisciplinary cooperation, particularly across experimental and computational specializations.

Finally, we must broaden experimental and computational training strategies to equip the upcoming generation of biologists with the skills necessary to use these resources to their fullest potential in order to address the most urgent open challenges.

References

Abdulla SO, Abdulazeez ST, Modanli M (2023) Comparison of third-order fractional partial differential equation based on the fractional operators using the explicit finite difference method. Alex Eng J 70:37–44. https://doi.org/10.1016/J.AEJ.2023.02.032

Abu-Elheiga L, Matzuk MM, Abo-Hashema KAH, Wakil SJ (2001) Continuous fatty acid oxidation and reduced fat storage in mice lacking acetyl-CoA carboxylase 2. Science 291:2613–2616. https://doi.org/10.1126/SCIENCE.1056843

Ahsan S, Drăghici S (2017) Identifying significantly impacted pathways and putative mechanisms with iPathwayGuide. Curr Protoc Bioinformatics 57:7.15.1–7.15.30. https://doi.org/10.1002/CPBI.24

Anco SC, Cheviakov AF (2020) On the different types of global and local conservation laws for partial differential equations in three spatial dimensions: review and recent developments. Int J Non Linear Mech 126:103569. https://doi.org/10.1016/J.IJNONLINMEC.2020.103569

Andrews SS (2017) Smoldyn: particle-based simulation with rule-based modeling, improved molecular interaction and a library interface. Bioinformatics 33:710–717. https://doi.org/10.1093/BIOINFORMATICS/BTW700

Arjunan SNV, Miyauchi A, Iwamoto K, Takahashi K (2020) pSpatiocyte: a high-performance simulator for intracellular reaction-diffusion systems. BMC Bioinformatics 21:33. https://doi.org/10.1186/S12859-019-3338-8

Ashburner M, Ball CA, Blake JA, Botstein D, Butler H, Cherry JM et al (2000) Gene ontology: tool for the unification of biology. The Gene Ontology Consortium. Nat Genet 25:25–29. https://doi.org/10.1038/75556

Baker EJ, Jay JJ, Bubier JA, Langston MA, Chesler EJ (2012) GeneWeaver: a web-based system for integrative functional genomics. Nucleic Acids Res 40:D1067. https://doi.org/10.1093/NAR/GKR968

Balling R (2003) ENU mutagenesis: analyzing gene function in mice. Annu Rev Genomics Hum Genet 2:463–492. https://doi.org/10.1146/ANNUREV.GENOM.2.1.463

Barabási AL, Gulbahce N, Loscalzo J (2011) Network medicine: a network-based approach to human disease. Nat Rev Genet 12:56–68. https://doi.org/10.1038/nrg2918

Bard JBL, Rhee SY (2004) Ontologies in biology: design, applications and future challenges. Nat Rev Genet 5:213–222. https://doi.org/10.1038/nrg1295

Bastian M, Heymann S, Jacomy M (2009) Gephi: an open source software for exploring and manipulating networks. Proc Int AAAI Conf Web Soc Media 3:361–362. https://doi.org/10.1609/ICWSM.V3I1.13937

Bebek G (2012) Identifying gene interaction networks. Methods Mol Biol 850:483–494. https://doi.org/10.1007/978-1-61779-555-8_26

Berger SI, Posner JM, Ma'ayan A (2007) Genes2Networks: connecting lists of gene symbols using mammalian protein interactions databases. BMC Bioinformatics 8:1–6. https://doi.org/10.1186/1471-2105-8-372/FIGURES/2

biochemistry, T. M.-S., and 1998, undefined (n.d.) Lessons from keratin transgenic and knockout mice. europepmc.org. https://europepmc.org/article/med/9932492. Accessed 26 May 2023

Bonotto EM, Federson M, Gadotti MC (2021) Recursive properties of generalized ordinary differential equations and applications. J Differ Equ 303:123–155. https://doi.org/10.1016/J.JDE.2021.09.013

Brazhnik P, De La Fuente A, Mendes P (2002) Gene networks: how to put the function in genomics. Trends Biotechnol 20:467–472. https://doi.org/10.1016/S0167-7799(02)02053-X

Brohée S, Faust K, Lima-Mendez G, Sand O, Janky R, Vanderstocken G et al (2008) NeAT: a toolbox for the analysis of biological networks, clusters, classes and pathways. Nucleic Acids Res 36:W444. https://doi.org/10.1093/nar/gkn336

Caldera M, Müller F, Kaltenbrunner I, Licciardello MP, Lardeau CH, Kubicek S et al (2019) Mapping the perturbome network of cellular perturbations. Nat Commun 10(1):5140. https://doi.org/10.1038/s41467-019-13058-9

Calderone A, Iannuccelli M, Peluso D, Licata L (2020) Using the MINT database to search protein interactions. Curr Protoc Bioinforma 69:e93. https://doi.org/10.1002/CPBI.93

Campbell SJ, Gaulton A, Marshall J, Bichko D, Martin S, Brouwer C et al (2012) Visualizing the drug target landscape. Drug Discov Today 17 Suppl:S3. https://doi.org/10.1016/J.DRUDIS.2011.12.005

Carbinatto MC, Rybakowski KP (2023) Partial functional differential equations and Conley index. J Differ Equ 366:20–41. https://doi.org/10.1016/J.JDE.2023.04.015

Carbon S, Douglass E, Dunn N, Good B, Harris NL, Lewis SE et al (2019) The Gene Ontology Resource: 20 years and still GOing strong. Nucleic Acids Res 47:D330. https://doi.org/10.1093/NAR/GKY1055

Chen G, Zhao J, Cohen T, Tao C, Sun J, Xu H et al (2015) Using Ontology Fingerprints to disambiguate gene name entities in the biomedical literature. Database 2015:bav034. https://doi.org/10.1093/DATABASE/BAV034

Chin CH, Chen SH, Wu HH, Ho CW, Ko MT, Lin CY (2014) cytoHubba: identifying hub objects and sub-networks from complex interactome. BMC Syst Biol 8:S11. https://doi.org/10.1186/1752-0509-8-S4-S11

Choi K, Medley JK, König M, Stocking K, Smith L, Gu S et al (2018) Tellurium: an extensible python-based modeling environment for systems and synthetic biology. Biosystems 171:74–79. https://doi.org/10.1016/J.BIOSYSTEMS.2018.07.006

Cickovski TM, Huang C, Chaturvedi R, Glimm T, Hentschel HGE, Alber MS et al (2005) A framework for three-dimensional simulation of morphogenesis. IEEE/ACM Trans Comput Biol Bioinform 2:273–287. https://doi.org/10.1109/TCBB.2005.46

Cooper-Knock J, Kirby J, Ferraiuolo L, Heath PR, Rattray M, Shaw PJ (2012) Gene expression profiling in human neurodegenerative disease. Nat Rev Neurol 8:518–530. https://doi.org/10.1038/NRNEUROL.2012.156

Cui T, El Mekkaoui K, Reinvall J, Havulinna AS, Marttinen P, Kaski S (2022) Gene–gene interaction detection with deep learning. Commun Biol 5:1238. https://doi.org/10.1038/s42003-022-04186-y

Davidson EH, Peter IS (2015) Gene regulatory networks. In: Genomic control process. Elsevier, Amsterdam, pp 41–77. https://doi.org/10.1016/B978-0-12-404729-7.00002-2

De Angelis MH, Flaswinkel H, Fuchs H, Rathkolb B, Soewarto D, Marschall S et al (2000) Genome-wide, large-scale production of mutant mice by ENU mutagenesis. Nat Genet 25(4):444–447. https://doi.org/10.1038/78146

Desai AP, Razeghin M, Meruvia-Pastor O, Peña-Castillo L (2017) GeNET: a web application to explore and share Gene Co-expression Network Analysis data. PeerJ 2017:3678. https://doi.org/10.7717/PEERJ.3678/SUPP-1

Drawert B, Engblom S, Hellander A (2012) URDME: a modular framework for stochastic simulation of reaction-transport processes in complex geometries. BMC Syst Biol 6:1–17. https://doi.org/10.1186/1752-0509-6-76/FIGURES/7

Egner U, Krätzschmar J, Kreft B, Pohlenz HD, Schneider M (2005) The target discovery process. Chembiochem 6:468–479. https://doi.org/10.1002/CBIC.200400158

Fahey ME, Bennett MJ, Mahon C, Jäger S, Pache L, Kumar D et al (2011) GPS-Prot: a web-based visualization platform for integrating host-pathogen interaction data. BMC Bioinformatics 12:1–13. https://doi.org/10.1186/1471-2105-12-298/FIGURES/5

Ferrer L, Dale JM, Karp PD (2010) A systematic study of genome context methods: calibration, normalization and combination. BMC Bioinformatics 11:493. https://doi.org/10.1186/1471-2105-11-493

Fluck J, Hofmann-Apitius M (2014) Text mining for systems biology. Drug Discov Today 19:140. https://doi.org/10.1016/j.drudis.2013.09.012

Franz M, Rodriguez H, Lopes C, Zuberi K, Montojo J, Bader GD et al (2018) GeneMANIA update 2018. Nucleic Acids Res 46:W60–W64. https://doi.org/10.1093/NAR/GKY311

Garny A, Hunter PJ (2015) OpenCOR: a modular and interoperable approach to computational biology. Front Physiol 6:26. https://doi.org/10.3389/FPHYS.2015.00026/BIBTEX

Gianlupi JF, Sego TJ, Sluka JP, Glazier JA (2023). PhenoCellPy: a Python package for biological cell behavior modeling. bioRxiv 2023.04.12.535625. https://doi.org/10.1101/2023.04.12.535625

Goh KI, Cusick ME, Valle D, Childs B, Vidal M, Barabási AL (2007) The human disease network. PNAS 104:8685–8690. https://doi.org/10.1073/pnas.0701361104

Goto K, Morimoto M, Osaki M, Tanio A, Izutsu R, Fujiwara Y et al (2022) The impact of AMIGO2 on prognosis and hepatic metastasis in gastric cancer patients. BMC Cancer 22:280. https://doi.org/10.1186/S12885-022-09339-0

Grefen C, Lalonde S, Obrdlik P (2007) Split-ubiquitin system for identifying protein-protein interactions in membrane and full-length proteins. Curr Protoc Neurosci Chapter 5. https://doi.org/10.1002/0471142301.ns0527s41

Gupta A, Könemann J, Leonardi S, Ravi R, Schäfer G, Könemann J, et al (n.d.) An efficient cost-sharing mechanism for the prize-collecting Steiner forest problem. iris.uniroma1.it. https://iris.uniroma1.it/retrieve/handle/11573/844718/630014/Gupta_preprint_Efficient_2015.pdf. Accessed 20 May 2023

Haiman ZB, Zielinski DC, Koike Y, Yurkovich JT, Palsson BO (2021) MASSpy: building, simulating, and visualizing dynamic biological models in Python using mass action kinetics. PLoS Comput Biol 17:e1008208. https://doi.org/10.1371/JOURNAL.PCBI.1008208

Halu A, De Domenico M, Arenas A, Sharma A (2019) The multiplex network of human diseases. NPJ Syst Biol Appl 51(5):1–12. https://doi.org/10.1038/s41540-019-0092-5

Heer H, Streib L, Schafer RB, Ruzika S (2020) Maximising the clustering coefficient of networks and the effects on habitat network robustness. PLoS One 15:e0240940. https://doi.org/10.1371/JOURNAL.PONE.0240940

Hoffmann M, Fröhner C, Noé F (2019) ReaDDy 2: fast and flexible software framework for interacting-particle reaction dynamics. PLoS Comput Biol 15:e1006830. https://doi.org/10.1371/JOURNAL.PCBI.1006830

Hoops S, Gauges R, Lee C, Pahle J, Simus N, Singhal M et al (2006) COPASI—a COmplex PAthway SImulator. Bioinformatics 22:3067–3074. https://doi.org/10.1093/BIOINFORMATICS/BTL485

Hu Z, Mellor J, Wu J, DeLisi C (2004) VisANT: an online visualization and analysis tool for biological interaction data. BMC Bioinformatics 5:17. https://doi.org/10.1186/1471-2105-5-17

Huang DW, Sherman BT, Lempicki RA (2009) Systematic and integrative analysis of large gene lists using DAVID bioinformatics resources. Nat Protoc 4:44–57. https://doi.org/10.1038/NPROT.2008.211

Huntley RP, Harris MA, Alam-Faruque Y, Blake JA, Carbon S, Dietze H et al (2014) A method for increasing expressivity of Gene Ontology annotations using a compositional approach. BMC Bioinformatics 15:155. https://doi.org/10.1186/1471-2105-15-155

Huntley RP, Sitnikov D, Orlic-Milacic M, Balakrishnan R, D'Eustachio P, Gillespie ME et al (2016) Guidelines for the functional annotation of microRNAs using the Gene Ontology. RNA 22:667–676. https://doi.org/10.1261/RNA.055301.115

Jang KH, Heras CR, Lee G (2022) m6A in the Signal Transduction Network. Mol Cells 45:435–443. https://doi.org/10.14348/MOLCELLS.2022.0017

Jansen R, Greenbaum D, Gerstein M (2002) Relating whole-genome expression data with protein-protein interactions. Genome Res 12:37. https://doi.org/10.1101/gr.205602

Jeong H, Tombor B, Albert R, Oltval ZN, Barabásl AL (2000) The large-scale organization of metabolic networks. Nature 407:651. https://doi.org/10.1038/35036627

Jones S, Thornton JM (1996) Principles of protein-protein interactions. Proc Natl Acad Sci U S A 93:13. https://doi.org/10.1073/pnas.93.1.13

Jungck JR, Viswanathan R (2015) Graph theory for systems biology: interval graphs, motifs, and pattern recognition. In: Algebraic and discrete mathematical methods for modern biology. Elsevier, Amsterdam, pp 1–27. https://doi.org/10.1016/B978-0-12-801213-0.00001-0

Kaushik AC, Mehmood A, Dai X, Wei DQ (2020) WeiBI (web-based platform): enriching integrated interaction network with increased coverage and functional proteins from genome-wide experimental OMICS data. Sci Rep 10:5618. https://doi.org/10.1038/S41598-020-62508-8

Keenan AB, Torre D, Lachmann A, Leong AK, Wojciechowicz ML, Utti V et al (2019) ChEA3: transcription factor enrichment analysis by orthogonal omics integration. Nucleic Acids Res 47:W212–W224. https://doi.org/10.1093/NAR/GKZ446

Khatri P, Sirota M, Butte AJ (2012) Ten years of pathway analysis: current approaches and outstanding challenges. PLoS Comput Biol 8:e1002375. https://doi.org/10.1371/JOURNAL.PCBI.1002375

Kitano H (2002a) Computational systems biology. Nature 420:206–210. https://doi.org/10.1038/nature01254

Kitano H (2002b) Systems biology: a brief overview. Science 295:1662–1664. https://doi.org/10.1126/SCIENCE.1069492

Koch I, Reisig W, Schreiber F (eds) (2011) Modeling in systems biology. Springer, London, p 16. https://doi.org/10.1007/978-1-84996-474-6

Kolch W, Halasz M, Granovskaya M, Kholodenko BN (2015) The dynamic control of signal transduction networks in cancer cells. Nat Rev Cancer 159(15):515–527. https://doi.org/10.1038/nrc3983

Kong X, Gerstein MB (2018) Text mining systems biology: turning the microscope back on the observer. Curr Opin Syst Biol 11:117. https://doi.org/10.1016/j.coisb.2018.09.004

Kotlyar M, Pastrello C, Rossos AEM, Jurisica I (2019) Protein–protein interaction databases. Encycl Bioinforma Comput Biol ABC Bioinforma 1–3:988–996. https://doi.org/10.1016/B978-0-12-809633-8.20495-0

Kuleshov MV, Jones MR, Rouillard AD, Fernandez NF, Duan Q, Wang Z et al (2016) Enrichr: a comprehensive gene set enrichment analysis web server 2016 update. Nucleic Acids Res 44:W90–W97. https://doi.org/10.1093/NAR/GKW377

Lash AE, Tolstoshev CM, Wagner L, Schuler GD, Strausberg RL, Riggins GJ et al (2000) SAGEmap: a public gene expression resource. Genome Res 10:1051–1060. https://doi.org/10.1101/GR.10.7.1051

Lee S, Zhang C, Arif M, Liu Z, Benfeitas R, Bidkhori G et al (2018) TCSBN: a database of tissue and cancer specific biological networks. Nucleic Acids Res 46:D595–D600. https://doi.org/10.1093/NAR/GKX994

Lemoine GG, Scott-Boyer MP, Ambroise B, Périn O, Droit A (2021) GWENA: gene co-expression networks analysis and extended modules characterization in a single bioconductor package. BMC Bioinformatics 22:267. https://doi.org/10.1186/S12859-021-04179-4

Letort G, Montagud A, Stoll G, Heiland R, Barillot E, MacKlin P et al (2019) PhysiBoSS: a multi-scale agent-based modelling framework integrating physical dimension and cell signalling. Bioinformatics 35:1188–1196. https://doi.org/10.1093/bioinformatics/bty766

Lindsay MA (2003) Target discovery. Nat Rev Drug Discov 210(2):831–838. https://doi.org/10.1038/nrd1202

Liska O, Bohár B, Hidas A, Korcsmáros T, Papp B, Fazekas D et al (2022) TFLink: an integrated gateway to access transcription factor–target gene interactions for multiple species. Database 2022:baac083. https://doi.org/10.1093/DATABASE/BAAC083

Liu E, Li L, Cheng L (2019) Gene regulatory network review. In: Encyclopedia of bioinformatics and computational biology: ABC of bioinformatics, vol 1–3. Elsevier, Amsterdam, pp 155–164. https://doi.org/10.1016/B978-0-12-809633-8.20218-5

Liu Y, Zhu Y, He C, Lu Z (2021) BENviewer: a gene interaction network visualization server based on graph embedding model. Database 2021:baab033. https://doi.org/10.1093/DATABASE/BAAB033

Loging W, Harland L, Williams-Jones B (2007) High-throughput electronic biology: mining information for drug discovery. Nat Rev Drug Discov 6:220–230. https://doi.org/10.1038/NRD2265

Lopez CF, Muhlich JL, Bachman JA, Sorger PK (2013) Programming biological models in Python using PySB. Mol Syst Biol 9:646. https://doi.org/10.1038/MSB.2013.1

Lugo CA, Saikia E, Nedelec F (2022) A typical workflow to simulate cytoskeletal systems with Cytosim. https://arxiv.org/abs/2205.13852v1. Accessed 20 May 2023

Maere S, Heymans K, Kuiper M (2005) BiNGO: a cytoscape plugin to assess overrepresentation of gene ontology categories in biological networks. Bioinformatics 21:3448–3449. https://doi.org/10.1093/BIOINFORMATICS/BTI551

Manolio TA (2013) Bringing genome-wide association findings into clinical use. Nat Rev Genet 148(14):549–558. https://doi.org/10.1038/nrg3523

Masuda N, Sakaki M, Ezaki T, Watanabe T (2018) Clustering coefficients for correlation networks. Front Neuroinform 12:7. https://doi.org/10.3389/fninf.2018.00007

McDowall MD, Scott MS, Barton GJ (2009) PIPs: human protein-protein interaction prediction database. Nucleic Acids Res 37:D651. https://doi.org/10.1093/nar/gkn870

Mi H, Poudel S, Muruganujan A, Casagrande JT, Thomas PD (2016) PANTHER version 10: expanded protein families and functions, and analysis tools. Nucleic Acids Res 44:D336–D342. https://doi.org/10.1093/NAR/GKV1194

Michalski PJ, Loew LM (2016) SpringSaLaD: a spatial, particle-based biochemical simulation platform with excluded volume. Biophys J 110:523. https://doi.org/10.1016/J.BPJ.2015.12.026
Montenegro JD (2022) Gene co-expression network analysis. Methods Mol Biol 2443:387–404. https://doi.org/10.1007/978-1-0716-2067-0_19
Myers CJ, Barker N, Jones K, Kuwahara H, Madsen C, Nguyen NPD (2009) iBioSim: a tool for the analysis and design of genetic circuits. Bioinformatics 25:2848. https://doi.org/10.1093/bioinformatics/btp457
Nieto JJ (2022) Solution of a fractional logistic ordinary differential equation. Appl Math Lett 123:107568. https://doi.org/10.1016/J.AML.2021.107568
Obrdiik P, El-Bakkoury M, Hamacher T, Cappellaro C, Vilarino C, Fleischer C et al (2004) K+ channel interactions detected by a genetic system optimized for systematic studies of membrane protein interactions. Proc Natl Acad Sci U S A 101:12242. https://doi.org/10.1073/pnas.0404467101
Oughtred R, Rust J, Chang C, Breitkreutz BJ, Stark C, Willems A et al (2021) The BioGRID database: a comprehensive biomedical resource of curated protein, genetic, and chemical interactions. Protein Sci 30:187–200. https://doi.org/10.1002/PRO.3978
Paci P, Fiscon G, Conte F, Wang RS, Farina L, Loscalzo J (2021) Gene co-expression in the interactome: moving from correlation toward causation via an integrated approach to disease module discovery. NPJ Syst Biol Appl 71(7):1–11. https://doi.org/10.1038/s41540-020-00168-0
Palva S (2018) The importance of hubs in large-scale networks. Nat Hum Behav 2(10):724–725. https://doi.org/10.1038/s41562-018-0438-9
Panchiwala H, Shah S, Planatscher H, Zakharchuk M, Konig M, Drager A (2022) The systems biology simulation core library. Bioinformatics 38:864–865. https://doi.org/10.1093/BIOINFORMATICS/BTAB669
Pathan M, Keerthikumar S, Ang CS, Gangoda L, Quek CYJ, Williamson NA et al (2015) FunRich: an open access standalone functional enrichment and interaction network analysis tool. Proteomics 15:2597–2601. https://doi.org/10.1002/PMIC.201400515
Pavlopoulos GA, Hooper SD, Sifrim A, Schneider R, Aerts J (2011) Medusa: a tool for exploring and clustering biological networks. BMC Res Notes 4:384. https://doi.org/10.1186/1756-0500-4-384
Perou CM, Sørile T, Eisen MB, Van De Rijn M, Jeffrey SS, Ress CA et al (2000) Molecular portraits of human breast tumours. Nature 406(6797):747–752. https://doi.org/10.1038/35021093
Peter IS (2020) The function of architecture and logic in developmental gene regulatory networks. Curr Top Dev Biol 139:267–295. https://doi.org/10.1016/BS.CTDB.2020.04.001
PySCeS: The Python Simulator for Cellular Systems (n.d.). https://pysces.sourceforge.net/. Accessed 20 May 2023
Qin T, Matmati N, Tsoi LC, Mohanty BK, Gao N, Tang J et al (2014) Finding pathway-modulating genes from a novel ontology fingerprint-derived gene network. Nucleic Acids Res 42:e138. https://doi.org/10.1093/NAR/GKU678
Raviprakash K, Huang B, Prasad V (2022) A hybrid modelling approach to model process dynamics by the discovery of a system of partial differential equations. Comput Chem Eng 164:107862. https://doi.org/10.1016/J.COMPCHEMENG.2022.107862
Resasco DC, Gao F, Morgan F, Novak IL, Schaff JC, Slepchenko BM (2012) Virtual Cell: computational tools for modeling in cell biology. Wiley Interdiscip Rev Syst Biol Med 4:129–140. https://doi.org/10.1002/WSBM.165
Resing KA, Ahn NG (2005) Proteomics strategies for protein identification. FEBS Lett 579:885–889. https://doi.org/10.1016/J.FEBSLET.2004.12.001
Rives AW, Galitski T (2003) Modular organization of cellular networks. Proc Natl Acad Sci U S A 100:1128–1133. https://doi.org/10.1073/PNAS.0237338100
Roberts PM (2006) Mining literature for systems biology. Brief Bioinform 7:399. https://doi.org/10.1093/bib/bbl037
Sahin U, Türeci Ö, Schmitt H, Cochlovius B, Johannes T, Schmits R et al (1995) Human neoplasms elicit multiple specific immune responses in the autologous host. Proc Natl Acad Sci 92:11810–11813. https://doi.org/10.1073/PNAS.92.25.11810

Sauro HM, Hucka M, Finney A, Wellock C, Bolouri H, Doyle J et al (2003) Next generation simulation tools: the Systems Biology Workbench and BioSPICE integration. OMICS 7:355–372. https://doi.org/10.1089/153623103322637670

Scardoni G, Petterlini M, Laudanna C (2009) Analyzing biological network parameters with CentiScaPe. Bioinformatics 25:2857. https://doi.org/10.1093/bioinformatics/btp517

Segal E, Friedman N, Kaminski N, Regev A, Koller D (2005) From signatures to models: understanding cancer using microarrays. Nat Genet 376(37):S38–S45. https://doi.org/10.1038/ng1561

Shi X, Banerjee S, Chen L, Hilakivi-Clarke L, Clarke R, Xuan J (2017) CyNetSVM: A cytoscape App for cancer biomarker identification using network constrained support vector machines. PLoS One 12:e0170482. https://doi.org/10.1371/journal.pone.0170482

Sidak D, Schwarzerová J, Weckwerth W, Waldherr S (2022) Interpretable machine learning methods for predictions in systems biology from omics data. Front Mol Biosci 9:926623. https://doi.org/10.3389/fmolb.2022.926623

Smoot ME, Ono K, Ruscheinski J, Wang PL, Ideker T (2011) Cytoscape 2.8: new features for data integration and network visualization. Bioinformatics 27:431–432. https://doi.org/10.1093/BIOINFORMATICS/BTQ675

Soyer OS (2012) Evolutionary systems biology. Springer, Cham, p 751. https://doi.org/10.1007/978-1-4614-3567-9

Stapor P, Schmiester L, Wierling C, Merkt S, Pathirana D, Lange BMH et al (2022) Mini-batch optimization enables training of ODE models on large-scale datasets. Nat Commun 13(1):1–17. https://doi.org/10.1038/s41467-021-27374-6

Stelzl U, Worm U, Lalowski M, Haenig C, Brembeck FH, Goehler H et al (2005) A human protein-protein interaction network: a resource for annotating the proteome. Cell 122:957–968. https://doi.org/10.1016/J.CELL.2005.08.029

Sun H, Chen GYJ, Yao SQ (2013) Recent advances in microarray technologies for proteomics. Chem Biol 20:685–699. https://doi.org/10.1016/j.chembiol.2013.04.009

Szklarczyk D, Gable AL, Nastou KC, Lyon D, Kirsch R, Pyysalo S et al (2021) The STRING database in 2021: customizable protein-protein networks, and functional characterization of user-uploaded gene/measurement sets. Nucleic Acids Res 49:D605–D612. https://doi.org/10.1093/NAR/GKAA1074

Tabas-Madrid D, Nogales-Cadenas R, Pascual-Montano A (2012) GeneCodis3: a non-redundant and modular enrichment analysis tool for functional genomics. Nucleic Acids Res 40:W478. https://doi.org/10.1093/nar/gks402

Tan P-N, Steinbach M, Kumar V (2005) Cluster analysis: basic concepts and algorithms. In: Introduction to data mining. Pearson Addison Wesley, Boston. https://doi.org/10.1016/0022-4405(81)90007-8

Tang Y, Li M, Wang J, Pan Y, Wu FX (2015) CytoNCA: a cytoscape plugin for centrality analysis and evaluation of protein interaction networks. Biosystems 127:67. https://doi.org/10.1016/j.biosystems.2014.11.005

Tapia JJ, Saglam AS, Czech J, Kuczewski R, Bartol TM, Sejnowski TJ et al (2019) MCell-R: a particle-resolution network-free spatial modeling framework. Methods Mol Biol 1945:203. https://doi.org/10.1007/978-1-4939-9102-0_9

Theocharidis A, van Dongen S, Enright AJ, Freeman TC (2009) Network visualization and analysis of gene expression data using BioLayout Express(3D). Nat Protoc 4:1535–1550. https://doi.org/10.1038/NPROT.2009.177

Theodosiou T, Efstathiou G, Papanikolaou N, Kyrpides NC, Bagos PG, Iliopoulos I et al (2017) NAP: the Network Analysis Profiler, a web tool for easier topological analysis and comparison of medium-scale biological networks. BMC Res Notes 10:278. https://doi.org/10.1186/s13104-017-2607-8

Tomczak A, Mortensen JM, Winnenburg R, Liu C, Alessi DT, Swamy V et al (2018) Interpretation of biological experiments changes with evolution of the Gene Ontology and its annotations. Sci Rep 81:1–10. https://doi.org/10.1038/s41598-018-23395-2

Tu HT, Phan TT, Nguyen KP (2022) Modeling information diffusion in social networks with ordinary linear differential equations. Inf Sci (N Y) 593:614–636. https://doi.org/10.1016/J.INS.2022.01.063

Tuncbag N, Braunstein A, Pagnani A, Huang SSC, Chayes J, Borgs C et al (2013) Simultaneous reconstruction of multiple signaling pathways via the prize-collecting steiner forest problem. J Comput Biol 20:124–136. https://doi.org/10.1089/CMB.2012.0092. https://home.liebertpub.com/cmb

Utriainen M, Morris JH (2023) clusterMaker2: a major update to clusterMaker, a multi-algorithm clustering app for Cytoscape. BMC Bioinformatics 24:134. https://doi.org/10.1186/S12859-023-05225-Z/FIGURES/10

van Dam S, Võsa U, van der Graaf A, Franke L, de Magalhães JP (2018) Gene co-expression analysis for functional classification and gene–disease predictions. Brief Bioinform 19:575. https://doi.org/10.1093/BIB/BBW139

Verdière N, Manceau D, Zhu S, Denis-Vidal L (2020) Inverse problem for a coupling model of reaction-diffusion and ordinary differential equations systems. Application to an epidemiological model. Appl Math Comput 375:125067. https://doi.org/10.1016/J.AMC.2020.125067

Wagner A, Fell DA (2001) The small world inside large metabolic networks. Proc R Soc Lond Ser B Biol Sci 268:1803–1810. https://doi.org/10.1098/RSPB.2001.1711

Wang J, Zhong J, Chen G, Li M, Wu FX, Pan Y (2015) ClusterViz: a cytoscape APP for Cluster Analysis of Biological Network. IEEE/ACM Trans Comput Biol Bioinform 12:815–822. https://doi.org/10.1109/TCBB.2014.2361348

Wang Y, Zong H, Yang F, Tong Y, Xie Y, Zhang Z et al (2023) A knowledge empowered explainable gene ontology fingerprint approach to improve gene functional explication and prediction. iScience 26:106356. https://doi.org/10.1016/J.ISCI.2023.106356

Wei Q, Khan IK, Ding Z, Yerneni S, Kihara D (2017) NaviGO: interactive tool for visualization and functional similarity and coherence analysis with gene ontology. BMC Bioinformatics 18:1–13. https://doi.org/10.1186/S12859-017-1600-5/FIGURES/9

Welsh C, Xu J, Smith L, König M, Choi K, Sauro HM (2023) libRoadRunner 2.0: a high performance SBML simulation and analysis library. Bioinformatics 39:btac770. https://doi.org/10.1093/BIOINFORMATICS/BTAC770

Wodak SJ, Vlasblom J, Turinsky AL, Pu S (2013) Protein-protein interaction networks: the puzzling riches. Curr Opin Struct Biol 23:941. https://doi.org/10.1016/j.sbi.2013.08.002

Xenarios I, Rice DW, Salwinski L, Baron MK, Marcotte EM, Eisenberg D (2000) DIP: the database of interacting proteins. Nucleic Acids Res 28:289–291. https://doi.org/10.1093/NAR/28.1.289

Yang K, Bai H, Ouyang Q, Lai L, Tang C (2008) Finding multiple target optimal intervention in disease-related molecular network. Mol Syst Biol 4:228. https://doi.org/10.1038/MSB.2008.60

Ye J, Fang L, Zheng H, Zhang Y, Chen J, Zhang Z et al (2006) WEGO: a web tool for plotting GO annotations. Nucleic Acids Res 34:W293. https://doi.org/10.1093/nar/gkl031

Zaghloul Salem MS (2018) Biological networks: an introductory review. J Proteomics Genomics Res 2:41. https://doi.org/10.14302/issn.2326-0793.jpgr-18-2312

Zambrowicz B, Discovery, A. S.-N. Reviews D., and 2003, Undefined (n.d.). Knockouts model the 100 best-selling drugs—will they model the next 100? nature.com. https://www.nature.com/articles/nrd987. Accessed 26 May 2023

Zhang M, Li Q, Yu D, Yao B, Guo W, Xie Y et al (2019) GeNeCK: a web server for gene network construction and visualization. BMC Bioinformatics 20:12. https://doi.org/10.1186/s12859-018-2560-0

Zhao X, Liu ZP (2019) Analysis of topological parameters of complex disease genes reveals the importance of location in a biomolecular network. Genes (Basel) 10:143. https://doi.org/10.3390/genes10020143

Zhu H, Snyder M (2003) Protein chip technology. Curr Opin Chem Biol 7:55–63. https://doi.org/10.1016/S1367-5931(02)00005-4

Systems Approach for Identifying Drug Targets by Computational Approaches

10

Sonali S. Shinde, Komal B. Padule, Swapna L. Sawant, and Aniket P. Sarkate

Abstract

Large datasets are frequently gathered, stored and analysed in the big data age with the goal of guiding biological discoveries and validating hypotheses. There is no question that the introduction of new technologies and open data initiatives has significantly enhanced the volume and diversity of data. The whole drug development process uses big data, from identifying targets and mechanisms of action to finding new leads and therapeutic candidates. With the intention of giving readers a broad overview of the computing resources and databases accessible, these approaches are shown and explored. We believe that big data leveraging should prioritize personalized care and be cost-effective. On the basis of their synergy, we suggest combining information technologies with (chemo) informatics tools to accomplish this.

Keywords

Drug · Computational · Target identification · Validation · Databases

S. S. Shinde (✉) · A. P. Sarkate
Department of Chemical Technology, Dr. Babasaheb Ambedkar Marathwada University, Aurangabad, Maharashtra, India

K. B. Padule
Department of Quality Assurance Technique, Dr. D. Y. Patil Institute of Pharmaceutical Sciences and Research, Pune, Maharashtra, India

S. L. Sawant
Department of Pharmaceutics, Dr. D. Y. Patil Institute of Pharmaceutical Sciences and Research, Pune, Maharashtra, India

S. Joshi et al. (eds.), *Systems Biology Approaches: Prevention, Diagnosis, and Understanding Mechanisms of Complex Diseases*,
https://doi.org/10.1007/978-981-99-9462-5_10

10.1 Introduction

Drug design and development are made easier by recent advancements in computer technology, software and algorithms. The creation of efficient and economically viable treatments requires an interdisciplinary approach due to the complexity of the drug development process. Using computational tools, the discovery of drugs may be speeded up and made cheaper. Bioinformatics can assist in identifying the pathophysiology functions of therapeutic-able genes from a vast amount of genomic data, offering possible targets proteins for drug discovery and development (Stumm et al. 2002). Protein structure prediction techniques can help to accurately elucidate protein structures. The structural and thermodynamic properties of target proteins can be obtained using multidisciplinary biomolecular simulations, which may be useful for identifying drug-binding sites and clarifying the mechanisms behind therapeutic action (Bakheet and Doig 2009; Moult et al. 2018).

One of the conventional approaches to medication design is a random screening of natural or lab-made molecules. This approach's drawbacks are its high cost and protracted design cycle. In this regard, a few crucial issues need to be addressed, like which chemical environment regions should be investigated to identify potential therapeutic options. There haven't been many new chemical entities discovered through searches in the conventional, drug-like chemical space.

Although various computational techniques are used in the design of medicines, many challenging conditions have also been observed. A crucial phase in the development of drugs is the identification and confirmation of molecular targets with therapeutic applications. Identifying the chemical space that is physiologically relevant is one of the most important concerns. Allosteric sites, transient binding sites and drug-able binding pockets can all be found using computational methods and employed in drug development (Keiser et al. 2009).

10.2 Target Identification

The process of identifying a drug's efficacy target that is hypothesized and connected to the illness is known as target identification. The first phase in drug discovery is target identification, next comes hit identification. Generally speaking, there are two approaches to computational drug design: structure- and ligand-based. Target identification uses computational and experimental techniques. Drug target identification uses a computational approach that takes into account structure, pharmacology and genomic space. Structure-based and ligand-based approaches are used in computational drug design to identify pharmacological targets. Several computational chemical methods are accessible in silico drug design. Using panel docking, algorithms based on bioactivity spectra, and searches for chemical structural similarities these tools, data mining and machine learning are continuously used (Nidhi et al. 2006; Cheng et al. 2011; Li et al. 2006; Clementi et al. 2012).

There are various computational methods for the various target compounds. To find potential targets like p38 MAP kinase, the ligand-based interaction fingerprint

(LIFt) methodology utilizes docking and sampling techniques. GPR17 can be identified utilizing the protein–ligand interaction fingerprinting (PLIF) between ligands and proteins. Network-based drug discovery focuses on identifying targets by integrating all data from drug–protein and protein–disease networks. It links databases with pharmacogenomics, proteomics, transcriptomics, metabolomics and genomics. Drug target identification uses computational frameworks that are developed by correlating pharmacological and genomic spaces. Recently, significant structural genetics and health association research were merged to construct a three-dimensional human interactome. As a result, potential genes for previously unknown disease–gene correlations with proposed molecular pathways were discovered. Several approach have different target such as shape-based screening (CatShape, Catalyst)-Transforming growth factor-b 1 receptor kinase (TGFβ), artificial neural network (ANN) quantitative structure–activity relationship (QSAR)-Metabotropic glutamate receptor 5 (mGlu5), Proteochemometrics modeling (PCM)-prostaglandin D2 receptor 2 (CRTH2), Molecular mechanics energies combined with the Poisson-Boltzmann surface area (MM-PBS)-HUMAN immunodeficiency virus 1 reverse transcriptase (HIV-1 RT), Molecular dynamics/free energy perturbation (FEP)-Biotin, Linear interaction energy (LIE)-β-Secretase (BACE), Bidimensional pharmacophoric fingerprints (ChemAxon and CCG's GpiDAPH3 fingerprints)-T-type calcium channel (CaV), Docking virtual screening (PyPx and AutoDock Vina)-Chemo-attractant receptor (OXE-R), Ligand based pharmacophore modelling (Catalyst)-Angiotensin II receptor type 1 (AT1), Docking virtual screening (Glide)-Pim-1 kinase and Epidermal growth factor receptor (EGFR)/Bromodomain-containing protein 4 (BRD4), Structure based pharmacophore virtual screening (Discovery Studio)-Calcineurin.

Ligand-based computational strategy is founded on the idea of chemical structural similarity, which holds that like-minded ligands or molecules will attach to like-minded targets with almost the same binding affinities and have like-minded biological effects. Due in large part to the fact that it takes into account the fact that medications have a variety of pharmacological effects, the concept of complementarity has become utilized frequently in lead identification. It is also required for rapid analysis on both main and secondary objectives, as well as for targeting family selection (Jenkins et al. 2006). In order to anticipate physiological targets, the method described here uses the interaction of known targets for proteins and known ligands exhibiting comparable chemical traits, traits and pharmacophoric elements. This is achieved by tracing the chemical compositions of chemicals that have been shown to influence cellular morphologies. (mainly orphan or natural products) onto chemogenomics databases comprising substances with known targets and biological profiles (Agamah et al. 2020).

Different aspects of the miRNA–target interaction are considered in the first scenario. Analysis and evaluation can be done on pairings between the miRNA and mRNA seed sequences. By calculating the pair formation's free energy and thermodynamic stability, the thermodynamic analysis may be carried out. It is possible to evaluate the target sequence's evolutionary conservation among similar species. It is also possible to assess the structural accessibility of the 3′-UTR for miRNAs and determine the several miRNA target. Instead of creating 'de novo' predictions based on sequence properties, the goal of machine learning is to find miRNA targets that refer to miRNA–mRNA duplexes with known biological importance (Riolo et al. 2020). In general, machine learning is a use of artificial intelligence (AI) that gives systems the capacity to develop through practice; these systems 'learn' from example datasets and make use of the acquired knowledge to anticipate outcomes for unknowable data (Bishop and Nasrabadi 2006).

Developing reliable models is still a challenging endeavour despite the fact that there are some excellent computational drug repositioning models. As a result of the complexity involved in adapting theoretical computer methods to imitate the behaviour of actual organisms and additional hurdles such as incomplete, skewed and erroneous data, putting theoretical computational methods into practice is one of the key challenges. For example, due to a number of variables, including deviations in experimental circumstances (such as environmental factors and patient demographics across different experiments), which can result in an information variation in the expression patterns and lead to prejudiced information, it may be difficult to define reliable gene expression signature profiles. We think that incorporating multiple sources of information about medications (like molecular structure), disorders (like clinical data), as well as how these illnesses and medicines influence people's bodies (such as gene characteristic patterns and adverse effects) is crucial in computerized drug repositioning studies, improving their performance and advancing them to the next level. Additionally, there are many illnesses for which there are currently no effective therapies that stop, slow down or change their development. This encourages and motivates researchers and multidisciplinary investigators to perform research, especially in the fields of cancer and the numerous orphan and uncommon diseases (Jarada et al. 2020).

Through computational methods like as in silico screening and molecular docking, peptides can be discovered. The target site, often referred to as a docking programme, predicts a protrusion that contains hydrogen bond donors and acceptors with hydrophobic properties and diverse molecular structures. The strongest bonding protein is produced by a collection of proteins docking with such pockets (Nevola and Giralt 2015). For instance, peptides are being created through computation through artificial docking with the Zika virus (ZIKV) membrane protein for detection. Tetra, penta, hexa and heptapeptide library have been docked against the ZIKV envelope protein using Open Eye Scientific Software; eight peptides were then selected. They performed additional testing using direct enzyme-linked immunosorbent assay (ELISA), and three of them were identified as having the greatest performance for Zika detection (Mascini et al. 2019). A different approach relies on peptidomimetics, which copies the intended targets and blocks protein interaction through competitive binding. Peptidomimetics was used to create four potential peptide inhibitors against Nipah virus (NiV) proteins, and MD modelling was used to assess the peptide–protein interactions' equilibrium. While the peptide sequence can be predicted using in silico techniques (Docking and peptidomimetics), further confirmation using in vitro/in vivo procedures is required to demonstrate its physiological function (Agarwal and Gabrani 2021).

The Biological General Repository for Interaction Datasets (BioGRID) (Stark et al. 2006) is an interaction repository that links software tools for visualizing intricate interaction networks, such as Osprey (Breitkreutz et al. 2003) and Cytoscape (Shannon et al. 2003) with assembled biological data that is freely available in standardized formats. BioMart is a public-driven project that solicits data contributions from researchers and provides free data and software tools for researchers in an effort to encourage scientific interactions and the process of research discoveries (Guberman et al. 2011). Oncomine, a cancer-related array dataset and online data analysis tool, provides queries and for either one or more genes throughout every analysis with the aim of assisting discovery through genome-wide activity research (Rhodes et al. 2004). The online Cancer-Related Analysis of Variants Toolkit (CRAVAT) can assist with the high-throughput evaluation and selection of genes and missense alterations crucial for cancer tumour development through offering forecasting scores for genetic variations, somatic alterations and compared gene significance (Douville et al. 2013). Through a web server, the Sorting Intolerant from Tolerant (SIFT) method forecasts how coding changes will affect how proteins operate. It is popular for describing missense variations and gives users predictions about their variants (Sim et al. 2012). Software called PROVEAN (Protein Variation Effect Analyzer) makes predictions about how the physiological function of a protein will be impacted by a modification to an amino acid (Choi and Chan 2015). Additionally, GenePattern offers tools for network analysis, sequence variation analysis and gene expression analysis. To create interconnected metabolic networks, an freely available programme called MetaMapR combines enzymatic processes with quantitative linkages, mass spectrum analogy and metabolites structure homology (Grapov et al. 2015). Software for visualizing proteins and DNA, like VMD (Humphrey et al. 1996) and Chimaera (Pettersen et al. 2004), is frequently used to analyse biomolecules and drug interactions in three dimensions. The exchange of data for target identification has been started by a number of networks (caBIG, http://cabig.cancer.gov; BIRN, http://www.nbirn.net) and projects (Genotype-Tissue Expression Project (Lonsdale et al. 2013); RD-Connect (Thompson et al. 2014)).

Semantics is a crucial area of study in information technology that has the potential to provide light on correlations between disparate data on diseases and potential medication targets. Such network-based computational methods have become more and more popular recently, offering fresh treatment targets and figuring out how diseases work. However, there hasn't been much work put into comprehensively examining correlations between medications, illnesses and genes. In like an investigation, Zhang et al. constructed a correlation networking by spotting in pairs connections between diseases, treatments and genes in Semantic MEDLINE and then employed a networking-based approach to mine the nearby network framework (Zhang et al. 2014). This can lead to the formulation of original research hypotheses, which is essential for the study of personalized and translational medicine (Tables 10.1 and 10.2).

Table 10.1 Approaches in computational drug discovery

Fragment-based (Langrea 2022)	Ligand-based (Acharya et al. 2011)	Structure-based (Meng et al. 2011)
• Fragment-based docking • Fragment-based QSAR • Ligand growing	• ADMET • QSAR • Pharmacophore • Statistical molecular design	• Molecular modelling • Docking reverse docking • Protein structure prediction • Normal mode analysis • Binding cavity analysis • Pharmacophore • Molecular dynamics

Table 10.2 Databases (Muresan et al. 2012)

Target	Databases
Small molecules	DrugBank, ChEMBL, Pubchem, BindingOB, Zinc
Proteins	PDB, UniProt, SCOP
Protein–protein	MINT, STITCH, STRING
Pathway	KEGG, Reactome

10.3 Properties Preferred in Drug Target

The biological basis for a pharmacological target and its use is the primary property. A macromolecule, most frequently a protein that can be altered to eliminate the causes of or alleviate the symptoms brought on by the underlying pathophysiology could be described as a therapeutic target. Small molecules are typically used for manipulation, while biologicals like peptides and antibodies are quickly catching up. The act of manipulation itself could take the shape of the protein's natural function being increased or decreased. However, even if biological relevance is a crucial factor, it is insufficient on its own to determine a good pharmacological target. Criteria that should have in an ideal target are (1) essentiality—It is necessary for the pathophysiology-related system; (2) process/condition specificity—It should be particular to the illness; (3) species/family specificity—If appropriate, it should be unique to the pathogen species or family, as shown by the protein's conservation in related organisms; (4) druggability or chemical tractability—It should consider if a suitable small molecule can change how it functions; (5) biological tractability and assay capacity that determine if the target is present in adequate amounts in vivo and whether appropriate techniques are available to evaluate the protein's function and then determine the degree of inhibition or activation by potential lead compounds (Lindsay 2003). In addition, in recent years, other criteria such as (6) virulence variables as innovative therapeutic intervention techniques; (7) low mutability to lower chances of drug resistance (Tan et al. 2000); (8) modification in the quantitative characteristics of the target/reaction/pathway; (9) resemblance of a side effect to a recognized drug/relevant target (Levy 2000); and (10) All of these factors are being investigated as additional target selection criteria. Another crucial component is the (11) combination of two or a small number of targets from the same disease system, so that the effects are synergistic (Chandra 2009).

10.4 Target Validation

It is the procedure used to confirm a target that has been identified for a biological route, chemical process or illness. We agree and believe rigorous data filtering and statistics can greatly boost the effectiveness of target validation because relevant instances of disease using high throughput screening responses from cells. Network validation can be done by contrasting the network significance to 100 networks that are randomly produced through randomized graph shuffling with degrees kept as provided in the randomized network module in Cytoscape 2.6.3. Using functional screens like RNAi and CRISPR-Cas9 as well as genome-wide techniques and gene regulatory networks, networks regulating genes and/or functioning of genes can be confirmed. As e-health records and clinical information trials have just been available, it is now possible to track and analyse inter-individual variability following drug delivery or intervention. Data published in the literature they include to ensure full validity, potential miRNA targets should always be confirmed experimentally, regardless of the method employed to predict miRNA–mRNA interactions. Depending on the information provided, experimental methodologies can be split into two categories: direct approaches and indirect approaches. Methods for examining the miRNA–mRNA pair directly or inserting a particular site that is bound by miRNAs, which is known as a microRNA responsive element (mRE), into a reporter gene to look for potential miRNA-induced changes at the protein level belong in the preliminary type (Barrangou et al. 2015; Wessel et al. 2015; Chen and Butte 2016). These methods, in particular, are typically restricted to mREs obtained via computational forecasts. Alternatively, another category consists of methods that use high-throughput technology (such as mass spectrometry and sequencing) to examine the effects of changing miRNA expression. Some of these techniques, nevertheless, have negative knock-on effects. This happens as a result of the fact that in addition to altering the protein that is encoded by its target once a miRNA links with it, many additional genes that activity is controlled by the miRNA's targeted sequence are also indirectly affected (Ritchie and Rasko 2014). Because of this, the modification of some products acquired using these methods should not always be attributed to miRNA–mRNA interaction. The combination of projected target mRNA and miRNA should first be proven. Second, it must be demonstrated that the target mRNA and the miRNA of interest directly interact. Third, tests demonstrating gain and loss of function are required to show how miRNAs control target protein expression. Fourth, it must confirm if the anticipated modifications in protein expression correspond to altered biological functions (Elton and Yalowich 2015).

In contrast with the typical de novo pharmaceutical development and discovery strategy, computational drug repositioning studies aim to improve the pre-clinical phase of drug discovery and development of generating new pharmaceuticals by identifying new applications for currently available drugs and saving time and money. Researchers accomplish their models and validate/evaluate their findings by suggesting a list of potential medication repositioning candidates. However, specific validation models may not be accurate and reliable, or they may differ in some instances from the proposed computational models. Identifying and choosing the right validation models have a significant impact on the effectiveness of any suggested computational approaches. Due to a number of factors, including a high price tag, high degree of harmful effects, lowered bioavailability and the reality that certain treatments have already been discontinued or aren't favoured by doctors or scientists, selecting the right group of medication repositioning candidates for validation. Despite certain well-known drawbacks, in vitro and in vivo validation approaches have been widely used to verify pharmaceutical repositioning candidates. Both 'in vitro' and 'in vivo' validation systems refer to the conduct of studies in a laboratory that is set apart from a living entity, such as molecular biology examines apart from living things or tissues, and in a living organism as a whole, such as experiments involving animals and human. For instance, Albendazole was identified by Lim et al. as a potential drug repositioning option for cancer-fighting effects, and they gave in vitro and in vivo elements of data in favour of utilizing it to treat ovarian and hepatic cancers.

10.5 Scope of Computational Approaches

Given that biology is studied at all of these different levels of the biological organization hierarchy, it is evident that the biological information obtained in literature and databases crosses these boundaries. An easy way to address specificity is to study genomic sequences. Specificity can be addressed at a finer level when three-dimensional structures or high-confidence predictions of structural models are available, by contrasting the three-dimensional structures of the target protein(s) with those of proteins in both the target cell and any additional cells that may be relevant. Substructure or functional site comparison is frequently more illuminating for this purpose (Blundell et al. 2006). Structural models may be used to analyse druggability, another significant parameter, in addition to offering insights into specificity (Al-Lazikani et al. 2007). All elements of drug action are based on molecular recognition events that enable precise binding of the medication to the target. Knowing the target macromolecule's structure aids in determining if it can precisely bind to a small molecule and whether its function can be changed by using the right small molecule ligands. In order to do this, the binding site(s) are located and their characteristics are described. The targets would be druggable if they had deep grooves with adequate cavity convexity, interaction potential and strategic placement to affect the function of the protein (Pérot et al. 2010). Ligand docking is one structural bioinformatics technique that can be useful for calculating ligand binding potential (Chandra et al. 2010; Gao et al. 2008). A related idea that was later described and known as a chemical systematic biology technique uses ligand-binding sites to identify off-targets (Xie et al. 2011). Since essentiality is a systemic trait, it cannot be understood by examining proteins separately. Such challenges require an understanding of systems. It is possible to do this via a variety of methods. One of the most effective methods to date for gaining a quantitative understanding is kinetic modelling, which can typically be used at the level of specific metabolic or signalling processes (Voit 2000). Numerous constraint-based techniques have been used since the parameters needed for kinetic modelling are typically unavailable, especially at the genome scale. One such technique is flux balance analysis, which analyses genome-scale metabolic models through methodical perturbations to find deadly gene deletions (Oberhardt et al. 2009). The set of proteins that are crucial for upholding network structure may be found by analysing networks that capture structural or functional protein–protein interactions (Zhao and Li 2010). Topological analysis using graph theoretical approaches may be used to find hubs and chokepoints and identify targets (Hasan et al. 2006). Network topologies' resilience may also be examined by eliminating nodes or edges (Strong and Eisenberg 2007). By adding genomic data from gene-expression profiling or related studies, conditional specificities, such as the importance of a particular protein for a particular state of the disease or disease-causing organism, may be investigated using networks (Forst 2002). Drug resistance is a significant issue we currently face for many illnesses, especially with antibacterial and antiviral drugs. The use of proteins that are less changeable and hence less likely to develop resistance is one of the suggested countermeasures (Tan et al. 2000). Bacterial virulence factors are thought to fall into this group and might be targeted.

It has been proposed using network studies that some proteins may act as hub nodes in causing the establishment of resistance and might be co-targeted with the primary target to hinder the resistance mechanism itself (Raman et al. 2008). Instead of the traditional strategy of attacking one target at a time, it has been suggested that a more helpful method for target identification would be to uncover protein combinations that perturb the resilience frequently found in disease-causing phenotypes (Hopkins 2008). The widespread clinical practice of prescribing a mix of medications for numerous disorders is a fact that supports this strategy. The best combinations to consider as potential therapeutic targets can be found via systems-level analysis (Raman et al. 2009). For researchers in the discipline, databases that include a number of these elements are proving to be valuable tools. One such example is the TDR database, which provides information on chemical desirability as well as druggability predictions for possible drug targets in viruses that cause tropical diseases (Agüero et al. 2008).

10.6 Future Aspects

In computational biology, network-based methods are frequently used to identify precise targets and interactions of novel medicinal compounds. In these models, the interactions or links between the nodes in the networks—which stand in for drugs, diseases, or gene products—and the edges are represented by the latter. These networks have a variety of representations that are computationally constructed from a variety of data sources in the pharmaceutical industry. These include protein-protein interactions, transcriptional networks, disease–disease, drug–target, drug–disease and interactions. From the side effect, transcriptional, drug-disease and other phenotypes data, analogous homogeneous or models of bipartite networks have been provided (Oberhardt et al. 2009).

A bioactive substance that has been demonstrated according to accumulating scientific evidence, to be safe and effective in people, it is expected to serve a number of therapeutic applications. Given the growing popularity of this 'simple' method of repositioning, there is an urgent need for more effective and organized computational methods to deal with and first organize the existing genetic and pharmacological databases for representation and information disclosure, and then use these datasets and examples disclosure instruments to predict the potential new uses for existing medications. It is clear that a shift in perspective is needed in the techniques of genomic, bio-pharmacological and computational-based approaches for a more informed and effective rediscovery of medications ('orderly good fortune') while taking into account all available information resources (Zhao et al. 2010).

Despite the fact that a number of recently developed network-based approaches can be used to find prospective drug–target interactions for both recognized medications and novel chemical entities, they are unable to forecast potential drug–target interactions (DTIs) for those targets in the absence of known ligands. Secondly, network-based techniques currently lack quantitative precision. For each prospective DTI, they merely provide a predictive score, with a larger number indicating a higher likelihood of occurrence. The expected DTIs' binding affinities are unknown. The sort of engagement is also not taken into account yet.

10.7 Conclusion

The advancement of computational techniques facilitates the investigation of medications. The computer-aided drug design, novel therapies based on nucleoside acid and artificial intelligence have produced great benefits for predicting future compounds and medicinal products against life threatening diseases. A powerful set of tools for target identification, discovery and optimization of drug candidate molecules has been made available. Insight into the fundamental causes of disease and its phenotypic manifestations is provided by information technology in conjunction with statistics and cheminformatic approaches, providing potential treatment targets that can subsequently be further verified by high throughput screening techniques Combinatorial techniques are often used in the drug research industry, but new ones are also emerging, such as proteochemometrics. The dynamic nature of therapeutic targets may be examined, binding sites can be anticipated and characterized, novel active compounds can be discovered and existing active molecules can be enhanced. Additionally, the development of network-based multi-target pharmaceuticals or the identification of novel therapeutic targets may be facilitated by the combining of '-omics' technologies. In the field of drug development, methods based on structure and ligands are most frequently employed; nevertheless, new combinatorial techniques, including proteochemometrics, are emerging. Recently, interest in miRNA research has risen to a high level and the regulatory role of miRNAs has been exhaustively investigated. It is vital to find the detailed principles of interactions among microfinites and their targets, given that research on target genes has been more challenging than expected. This has encouraged the creation of numerous computational algorithms for the prediction of miRNA targets using various methodologies. The accuracy of the forecast and the number of prospective targets may both be greatly increased by integrating prediction algorithms based on various rules. To confirm their function, all potential miRNA targets must be empirically confirmed. Four requirements must be met for a miRNA–mRNA combination to be functional. First, the target mRNA gene and miRNA must both be expressed. Second, it is important to confirm the miRNA–mRNA interaction. Third, miRNA mimics and miRNA inhibitors must be used to show how miRNA affects target protein expression. Fourth, altered biological functions should be connected to how miRNAs regulate the expression of their target genes. The validation can be carried out using a number of other techniques. High-throughput techniques could be a valuable resource for this kind of study. The invaluable tools for the management of all these data could be available through transcriptome and proteomic data this situation as well.

References

Acharya C, Coop A, Polli JE, Mackerell AD Jr (2011) Recent advances in ligand-based drug design: relevance and utility of the conformationally sampled pharmacophore approach. Curr Comput Aided Drug Des 7(1):10–22. https://doi.org/10.2174/157340911793743547; PMID: 20807187; PMCID: PMC2975775

Agamah FE, Mazandu GK, Hassan R, Bope CD, Thomford NE, Ghansah A, Chimusa ER (2020) Computational/in silico methods in drug target and lead prediction. Brief Bioinform 21(5):1663–1675

Agarwal G, Gabrani R (2021) Antiviral peptides: identification and validation. Int J Pept Res Ther 27:149–168

Agüero F, Al-Lazikani B, Aslett M, Berriman M, Buckner FS, Campbell RK, Verlinde CL (2008) Genomic-scale prioritization of drug targets: the TDR targets database. Nat Rev Drug Discov 7(11):900–907

Al-Lazikani B, Gaulton A, Paolini G, Lanfear J, Overington J, Hopkins A (2007) The molecular basis of predicting druggability. Bioinform Genom Ther 1:1315–1334

Bakheet TM, Doig AJ (2009) Properties and identification of human protein drug targets. Bioinformatics 25(4):451–457

Barrangou R, Birmingham A, Wiemann S, Beijersbergen RL, Hornung V, Smith AVB (2015) Advances in CRISPR-Cas9 genome engineering: lessons learned from RNA interference. Nucleic Acids Res 43(7):3407–3419

Bishop CM, Nasrabadi NM (2006) Pattern recognition and machine learning (Vol 4, No 4). Springer, New York, NY, p 738

Blundell TL, Sibanda BL, Montalvão RW, Brewerton S, Chelliah V, Worth CL, Burke D (2006) Structural biology and bioinformatics in drug design: opportunities and challenges for target identification and lead discovery. Philos Trans R Soc B Biol Sci 361(1467):413–423

Breitkreutz BJ, Stark C, Tyers M (2003) Osprey: a network visualization system. Genome Biol 4:1–4
Chandra N (2009) Computational systems approach for drug target discovery. Expert Opin Drug Discovery 4(12):1221–1236
Chandra N, Anand P, Yeturu K (2010) Structural bioinformatics: deriving biological insights from protein structures. Interdiscip Sci 2:347–366
Chen B, Butte A (2016) Leveraging big data to transform target selection and drug discovery. Clin Pharmacol Ther 99(3):285–297
Cheng T, Li Q, Wang Y, Bryant SH (2011) Identifying compound-target associations by combining bioactivity profile similarity search and public databases mining. J Chem Inf Model 51(9):2440–2448
Choi Y, Chan AP (2015) PROVEAN web server: a tool to predict the functional effect of amino acid substitutions and indels. Bioinformatics 31(16):2745–2747
Clementi E, André JM, McCammon JA (eds) (2012) Theory and applications in computational chemistry: the first decade of the second millennium. American Institute of Physics, College Park, ML
Douville C, Carter H, Kim R, Niknafs N, Diekhans M, Stenson PD, Karchin R (2013) CRAVAT: cancer-related analysis of variants toolkit. Bioinformatics 29(5):647–648
Elton TS, Yalowich JC (2015) Experimental procedures to identify and validate specific mRNA targets of miRNAs. EXCLI J 14:758
Forst CV (2002) Network genomics–a novel approach for the analysis of biological systems in the post-genomic era. Mol Biol Rep 29:265–280
Gao Z, Li H, Zhang H, Liu X, Kang L, Luo X, Jiang H (2008) PDTD: a web-accessible protein database for drug target identification. BMC Bioinformatics 9:1–7
Grapov D, Wanichthanarak K, Fiehn O (2015) MetaMapR: pathway independent metabolomic network analysis incorporating unknowns. Bioinformatics 31(16):2757–2760
Guberman JM, Ai J, Arnaiz O, Baran J, Blake A, Baldock R, Kasprzyk A (2011) BioMart central portal: an open database network for the biological community. Database 2011:bar041
Hasan S, Daugelat S, Rao PS, Schreiber M (2006) Prioritizing genomic drug targets in pathogens: application to mycobacterium tuberculosis. PLoS Comput Biol 2(6):e61
Hopkins AL (2008) Network pharmacology: the next paradigm in drug discovery. Nat Chem Biol 4(11):682–690
Humphrey W, Dalke A, Schulten K (1996) VMD: visual molecular dynamics. J Mol Graph 14(1):33–38
Jarada TN, Rokne JG, Alhajj R (2020) A review of computational drug repositioning: strategies, approaches, opportunities, challenges, and directions. J Chem 12(1):1–23
Jenkins JL, Bender A, Davies JW (2006) In silico target fishing: predicting biological targets from chemical structure. Drug Discov Today Technol 3(4):413–421
Keiser MJ, Setola V, Irwin JJ, Laggner C, Abbas AI, Hufeisen SJ, Roth BL (2009) Predicting new molecular targets for known drugs. Nature 462(7270):175–181
Langrea R (2022) Computational approaches in fragment based drug design. Drug Des 11:221
Levy J (2000) The effects of antibiotic use on gastrointestinal function. Am J Gastroenterol 95(1):S8–S10
Li H, Gao Z, Kang L, Zhang H, Yang K, Yu K, Jiang H (2006) TarFisDock: a web server for identifying drug targets with docking approach. Nucleic Acids Res 34(suppl_2):W219–W224
Lindsay MA (2003) Target discovery. Nat Rev Drug Discov 2(10):831–838
Lonsdale J, Thomas J, Salvatore M, Phillips R, Lo E, Shad S, Moore HF (2013) The genotype-tissue expression (GTEx) project. Nat Genet 45(6):580–585
Mascini M, Dikici E, Robles Mañueco M, Perez-Erviti JA, Deo SK, Compagnone D, Daunert S (2019) Computationally designed peptides for Zika virus detection: an incremental construction approach. Biomol Ther 9(9):498
Meng XY, Zhang HX, Mezei M, Cui M (2011) Molecular docking: a powerful approach for structure-based drug discovery. Curr Comput Aided Drug Des 7(2):146–157. https://doi.org/10.2174/157340911795677602; PMID: 21534921; PMCID: PMC3151162

Moult J, Fidelis K, Kryshtafovych A, Schwede T, Tramontano A (2018) Critical assessment of methods of protein structure prediction (CASP)—round XII. Proteins 86:7–15
Muresan S, Sitzmann M, Southan C (2012) Mapping between databases of compounds and protein targets. Methods Mol Biol 910:145–164. https://doi.org/10.1007/978-1-61779-965-5_8; PMID: 22821596; PMCID: PMC7449375
Nevola L, Giralt E (2015) Modulating protein–protein interactions: the potential of peptides. Chem Commun 51(16):3302–3315
Nidhi, Glick M, Davies JW, Jenkins JL (2006) Prediction of biological targets for compounds using multiple-category Bayesian models trained on chemogenomics databases. J Chem Inf Model 46(3):1124–1133
Oberhardt MA, Palsson BØ, Papin JA (2009) Applications of genome-scale metabolic reconstructions. Mol Syst Biol 5(1):320
Pérot S, Sperandio O, Miteva MA, Camproux AC, Villoutreix BO (2010) Druggable pockets and binding site centric chemical space: a paradigm shift in drug discovery. Drug Discov Today 15(15–16):656–667
Pettersen EF, Goddard TD, Huang CC, Couch GS, Greenblatt DM, Meng EC, Ferrin TE (2004) UCSF chimera—a visualization system for exploratory research and analysis. J Comput Chem 25(13):1605–1612
Raman K, Chandra N (2008) Mycobacterium tuberculosis interactome analysis unravels potential pathways to drug resistance. BMC Microbiol 8:1–13
Raman K, Vashisht R, Chandra N (2009) Strategies for efficient disruption of metabolism in mycobacterium tuberculosis from network analysis. Mol BioSyst 5(12):1740–1751
Rhodes DR, Yu J, Shanker K, Deshpande N, Varambally R, Ghosh D, Chinnaiyan AM (2004) ONCOMINE: a cancer microarray database and integrated data-mining platform. Neoplasia 6(1):1–6
Riolo G, Cantara S, Marzocchi C, Ricci C (2020) miRNA targets: from prediction tools to experimental validation. Methods Protoc 4(1):1
Ritchie W, Rasko JE (2014) Refining microRNA target predictions: sorting the wheat from the chaff. Biochem Biophys Res Commun 445(4):780–784
Shannon P, Markiel A, Ozier O, Baliga NS, Wang JT, Ramage D, Ideker T (2003) Cytoscape: a software environment for integrated models of biomolecular interaction networks. Genome Res 13(11):2498–2504
Sim NL, Kumar P, Hu J, Henikoff S, Schneider G, Ng PC (2012) SIFT web server: predicting effects of amino acid substitutions on proteins. Nucleic Acids Res 40(W1):W452–W457
Stark C, Breitkreutz BJ, Reguly T, Boucher L, Breitkreutz A, Tyers M (2006) BioGRID: a general repository for interaction datasets. Nucleic Acids Res 34(suppl_1):D535–D539
Strong M, Eisenberg D (2007) The protein network as a tool for finding novel drug targets. Prog Drug Res 64:191–215
Stumm G, Russ A, Nehls M (2002) Deductive genomics: a functional approach to identify innovative drug targets in the post-genome era. Am J Pharmacogenomics 2(4):263–271
Tan YT, Tillett DJ, McKay IA (2000) Molecular strategies for overcoming antibiotic resistance in bacteria. Mol Med Today 6(8):309–314
Thompson R, Johnston L, Taruscio D, Monaco L, Béroud C, Gut IG, Lochmüller H (2014) RD-connect: an integrated platform connecting databases, registries, biobanks and clinical bioinformatics for rare disease research. J Gen Intern Med 29:780–787
Voit EO (2000) Computational analysis of biochemical systems: a practical guide for biochemists and molecular biologists. Cambridge University Press
Wessel J, Chu AY, Willems SM, Wang S, Yaghootkar H, Brody JA, Bottinger EP (2015) Low-frequency and rare exome chip variants associate with fasting glucose and type 2 diabetes susceptibility. Nat Commun 6(1):5897
Xie L, Xie L, Bourne PE (2011) Structure-based systems biology for analyzing off-target binding. Curr Opin Struct Biol 21(2):189–199

Zhang Y, Tao C, Jiang G, Nair AA, Su J, Chute CG, Liu H (2014) Network-based analysis reveals distinct association patterns in a semantic MEDLINE-based drug-disease-gene network. J Biomed Semantics 5(1):1–13

Zhao S, Li S (2010) Network-based relating pharmacological and genomic spaces for drug target identification. PLoS One 5(7):e11764

Systems Biology Approaches for Precision Medicine

11

Subrahmanya Hegde, S. M. Andrian, Harshith Sajal,
Anil K. K. Mallikarjunappa, Angel Benny,
and Sumitha Elayaperumal

Abstract

System biology studies the large-scale measurement of biomolecules in intricate biological systems. High-throughput omics, such as proteomics, transcriptomics, and metabolomics, are included in the datasets. These may be utilized to discover alterations in molecular pathways through the use of techniques such as pathway enrichment and network analysis, including machine learning. The ability to reveal novel biological insights is crucially dependent on the integration of various omic datasets. There remains a lot of hype concerning the personalized medicine concept. Each person differs by distinct characteristics including at the (1) physiological, (2) behavioral, and (3) environmental levels. Hence, there arises a need to treat diseases with different strategies to cater large population. The interindividual variations were confirmed by sequencing the whole genome and monitoring health using wireless devices, and imaging protocols which have shown inter variation in disease progression to drug response. Interactions between triad factors including environmental, neurological, and immunological play a vital role in complex disorders. Hence, there arises a need to understand the complexity of these disorders from a novel perspective. Epilepsies have been better understood as a result of recent developments in genetics and neuroscience. These discoveries have opened the door for a cutting-edge method of personalized treatment that tries to reverse or prevent the negative consequences of epilepsy-causing gene abnormalities. The efficacy and safety of epilepsy medi-

S. Hegde · S. M. Andrian · H. Sajal · A. Benny · S. Elayaperumal (✉)
Department of Biotechnology and Bioinformatics JSS Academy of Higher Education and Research, Mysore, Karnataka, India

A. K. K. Mallikarjunappa
Department of Environmental Science JSS Academy of Higher Education and Research, Mysore, Karnataka, India

S. Joshi et al. (eds.), *Systems Biology Approaches: Prevention, Diagnosis, and Understanding Mechanisms of Complex Diseases*,
https://doi.org/10.1007/978-981-99-9462-5_11

cations might be considerably improved by selectively addressing the molecular mechanisms that cause seizures in each individual. These systems methods are used to answer a range of biological questions, such as how newborns develop during their first week of life, how hosts and pathogens interact, how to predict disease, how to find biomarkers, and how to discover and repurpose drugs. Overall, systems biology-based omics data analysis has been used to treat several complex diseases.

Keywords
System biology · Precision medicine · Personalized medicine · Treatment response

11.1 Introduction

The way cancer is identified and treated is influenced by molecular findings from genomic and systems biology. We do a thorough analysis of the big data's challenges for precision medicine. Two distinct kinds of melanoma have been identified: the mitogen-activated protein kinase driver pathway and persistent UV damage. Remarkably, membrane receptors contain non-genomic drivers of mitogen-activated protein kinase that are associated with melanoma, even if the frequency of mutations in these drivers is quite low. Similar mitogenic effectors can be triggered by using other pathways, such as metabolism or epigenetic signaling, to get around the main mitogen-activated protein kinase molecules. Numerous aspects are taken into account while looking for therapeutic targets and resistance mutations, including genomic and non-genomic disease development and progression pathways, mutation hotspots, structural modeling, UV signature, and disease-progression pathways. By taking into account these variables, scientists hope to pinpoint particular mutations that are more likely to result in treatment resistance and investigate potential new targets for therapeutic approaches. The study of mutation patterns, knowledge of protein structures, evaluation of UV-induced DNA damage, and a more thorough investigation of the genetic and non-genetic processes involved in the beginning and course of the disease are all included in this comprehensive approach. Logic-based approaches for pharmaceutical treatment with tailored medicines are revealed by a full precision medicine profile of a patient with malignant melanoma. Independent molecular pathways that are suited for drug targeting can be found with precise insights. To combat immune evasion, adverse effects, and drug resistance, synergistic treatment combinations of orthogonal modalities, such as immunotherapy, mitogen-triggered protein kinase enzyme inhibitors, epigenetic inhibitors, and metabolic inhibitors, may all be employed. We now go on to the current therapeutic strategies, which are often ineffectual in individuals with established diabetic kidney disease (DKD), an epidemic that affects one in three people with diabetes. Diabetes-related kidney problems may be lessened by early detection of patients who are at high risk of progression and by individualized therapy. To understand the molecular mechanisms behind neurodegenerative diseases (ND), a systems biology approach that incorporates large-scale genomic data is ideally suited, and it may offer fresh perspectives on tailoring therapy. Studies have demonstrated that the integration of genome-scale datasets produced from prospectively designed clinical cohort studies with model systems using cutting-edge bioinformatic analysis revealed important molecular pathways in ND and resulted in the development of prognostic candidate molecular biomarkers. A significant advance in the effectiveness and safety of epileptic treatments could come from focusing on the molecular pathways that underlie particular epilepsy problems. The use of a systems biology approach for individualized treatment in refractory epilepsy will be highlighted in this chapter, with a particular focus on understanding the mechanisms of medication resistance in epilepsy. The concept of systems medicine has gained significant attention in medical research, although its precise definition and interpretation remain ambiguous. It is often referred to as P4 medicine, encompassing the principles of predictive, preventative, personalized, and participatory medicine. However, the lack of a clear and universally accepted definition leads to complexity and unfounded expectations in both scientific and public discussions regarding the prospects, risks, and limitations of systems medicine.

In light of this, our objective is to establish a comprehensive and widely acceptable definition of systems medicine. The term has sparked numerous points of debate, but three key aspects currently dominate the discourse:

(a) Systems medications are the heirs to personalized drugs.
(b) Systems medications are substitutes for precision drugs.
(c) Systems medicines are pharmaceuticals that put Systems Biology into operation.

An "assemblage of scientific concepts and endeavors incorporating bioinformatics approaches to human biology" is systems medicine (Goetz and Schork 2018).

This misunderstanding, however, is problematic because it hinders rational arguments about the opportunities, hazards, and limits of systems medicine. Additionally, it is challenging to recognize and address moral and legal difficulties that can arise in the context of systems medicine. It is, therefore, not feasible to discuss its issue and (future) treatment. Finally, the terms under the specification may give patients false optimism, such as on the likelihood of systems medicines methods. Against this backdrop, it's easy to see why it's still viewed as one of the primary obstacles to designing systems medicine.

11.2 Next-Generation Sequencing (NGS): Precision Medicine

NGS technology, which is widely considered the foundation of precision medicine, has been successfully used in the diagnosis and treatment of cancer (Morash et al. 2018). With improvements in gene diagnostics and immunotherapy, it is feasible to limit the progression of cancer and ease the suffering of chemotherapy patients. Promote the adoption of genetic testing for personalized medicine and facilitate the seamless transfer of precision medicine insights from a centralized bank to individual clinical sites. New approaches for analyzing NGS and genomic data are required. The bioinformatics workflows for analyzing NGS data, encompassing genomics and transcriptomics analysis, are depicted in Fig. 11.1. The NGS panel, for example, differs from whole genome sequencing (WGS) in that it focuses on fewer genes or areas yet demands more precision and efficiency. Drive genes are often clusters of genes within a regulatory network in complicated diseases such as cancer (Qin 2019). Shortest path analysis and random walk algorithms, for example, will aid in the research of genome-wide connections in important modules or signaling pathways whose failure is related to disease development.

Systems biology uses a comprehensive scientific approach and will offer answers to difficult issues involving intricate biological systems, such as the brain (Rai et al.

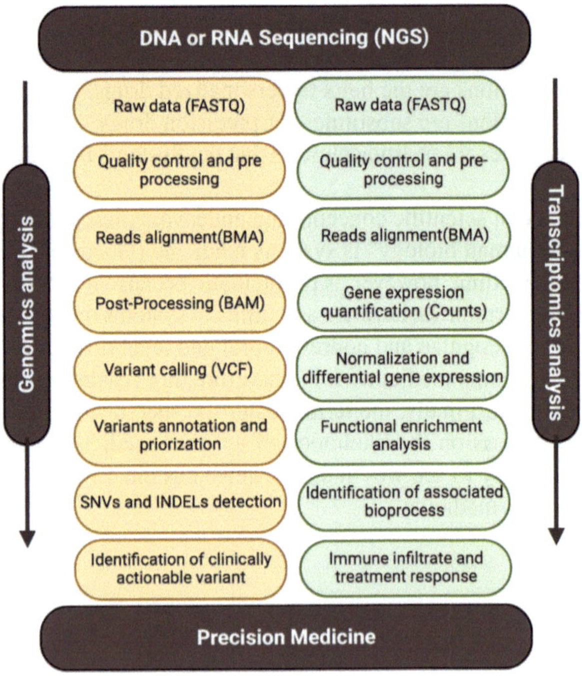

Fig. 11.1 Techniques used in the field of bioinformatics to analyze data from next-generation sequencing. *FASTQ* TXT file with the sequence and quality of the read, *BAM* binary alignment map, *SNV* single nucleotide variants, *INDEL* insertion or deletion variants

2018). Desegregation data from genetic, genomic, neurobiological, or proteomic sources enables the identification of networks and pathways involved in the etiopathogenesis of medical specialty disorders.

In this chapter, we'll go through how to use a systems biology approach to comprehend the pathophysiology of a brain condition and develop customized diagnoses and treatments. System biology studies complex biological systems, including interactions between genes, proteins, and other molecules. This approach emphasizes the need to understand the underlying mechanisms of disease at a molecular level, rather than just treating symptoms (Dhillon et al. 2020). Healthcare delivery through precision medicine is patient-centered that takes into consideration the unique variations in genes, environment, and lifestyle among individuals. Rather than using a standardized approach, precision medicine aims to customize medical treatment according to the specific needs of each patient. The integration of system biology and precision medicine holds great potential for improving the effectiveness and personalization of treatments for various diseases. For instance, utilizing system biology techniques to analyze the molecular characteristics of a patient's cancer cells can enable the identification of specific genetic mutations responsible for driving the disease. This knowledge can then be leveraged to develop targeted therapies that are more likely to be successful for that particular patient. Similar to this, based on a patient's genetic profile and other pertinent criteria, precision medicine approaches can assist in identifying which individuals are most likely to benefit from a particular medication. This can help reduce the risk of adverse side effects and improve outcomes. As seen in Table 11.1, precision medicine is commonly used in Western nations (Rai et al. 2018). With the growth and connectivity of our knowledge base, precision medicine will further change how we approach patient decisions. Table 11.2 shows some examples of precision medicine strategies, featuring increasing levels of complexity (Rai et al. 2018).

11.3 System Biology: Diabetic Kidney Disease

A significant proportion of individuals suffering from diabetic kidney disease (DKD), a prevalent condition affecting one out of every three diabetes patients, do not exhibit positive responses to current therapeutic approaches (Thomas et al. 2015). To mitigate kidney complications associated with diabetes, early identification of individuals at high risk of disease progression and the implementation of personalized treatment plans are crucial. To achieve this, a comprehensive understanding of the complex pathophysiology of diabetic nephropathy (DN) is necessary. Employing a systems biology approach that incorporates comprehensive omics data can help unravel the molecular mechanisms underlying DN and provide novel insights into tailoring treatment strategies (Pinu et al. 2019).

Recent research has shown that it is possible to assemble genome-scale datasets. Obtained from well-structured clinical cohort studies with advanced bioinformatics analysis using model systems. This approach has successfully unveiled important molecular pathways involved in DN and led to the identification of potential

Table 11.1 Examples of applications for precision medicine (Jameson and Longo 2015)

Medical field	Disease	Biomarker	Intervention
Hematology	Thrombosis	Protein C deficiency	Avoid prothrombotic medications
Cancer	Acute promyelocytic leukemia	T(15:17)	Daunorubicin and idarubicin
Neurology	Spinocerebellar ataxia 17	TBP	Immunotherapy
Hematology	Hepatitis B	Hepatitis B viral load	Direct-acting antiviral agents
Renal disease	Graft rejection	CXCL9, BAFF, and sC5b-9	Antirejection drugs
Cardiovascular disease	Acute decompensated heart failure	Blood galectin-3 level	ACE inhibitors and ARBs
Endocrine disease	Hashimoto's disease	Thyroid peroxidase antibody	Levothyroxine
Pharmacogenomics	Smoking cessation	CYP2A6	Varenicline
Metabolic disease	Diabetes type 2	Blood glucose level	Insulin
Infectious disease	Zika virus infection	L18 and CXCL10	Chloroquine
Pulmonary disease	Chronic obstructive pulmonary disease (COPD)	CRP (C-reactive protein), tumor necrosis factor-α, IL-6, IL-8, fibrinogen, and neutrophil	Fluticasone, budesonide, prednisolone
Psychiatry	Bipolar disorder	Brain-derived neurotrophic factor (BDNF)	Fluoxetine/ olanzapine

prognostic molecular biomarkers. Significant progress has been made in the field of molecular biomarkers for DKD using integrative biology techniques. By observing the systems and biological methods of precision medicine, valuable insights can be gained in understanding and managing DKD effectively.

11.4 System Biology: Treatment Response

Research can be conducted by utilizing multi-omics profiling to examine treatment response in vitro and in vivo model systems. Specifically, relevant models for studying kidney-related conditions include organoids, primary kidney cell types (preferred over established cell lines for their closer resemblance to in vivo conditions), and animal models of diabetic kidney disease (DKD) (J. Kim et al. 2020). These molecular profiles offer valuable insights into the mechanisms of action (MOA) of tested drugs by linking them to observable morphometric and phenotypic changes.

By integrating intrarenal transcriptome data with non-invasive molecular profiles specific to individual patients, it becomes possible to generate signatures that accurately represent the progression of kidney pathology (J. Kim et al. 2020). Employing

Table 11.2 Examples of precision medicine delivery methods in practice (Faulkner et al. 2020)

Approach	Example	Essential premise	Focus of approach
Single genetic test	EGFR-TK detection	Response to therapies is impacted by genetic differences in solid tumors	To determine which lung cancer patients may benefit from gefitinib
Prescribing algorithm	'Biologic calculator'	Response to a biological substance is influenced by genetic variation as well as other elements including immune system function, alcohol and tobacco use, family history, sex, and age	Facilitate the selection of appropriate biologics and optimal dosages for individuals with autoimmune diseases, exclusively within a research setting. Currently, solely in a research environment
Single genetic test	Identification of TMPT	Genetic variations in the drug's metabolizing enzymes affect the possibility of adverse effects	To identify those who might potentially suffer from severe neutropenia from azathioprine
Gene expression profiling test	Oncotype DX has about 30 genes	The likelihood of a solid tumor recurring is influenced by genetic differences	To evaluate whether systemic anticancer treatment is necessary and the likelihood of tumor recurrence in breast cancer patients
Next-generation sequencing	Foundation a single CDx to look for a genetic change in one of 324 genes microsatellite instability or tumor mutational burden (TMB) estimation in multimarker testing	Treatment response is influenced by genetic differences in solid tumors	To choose appropriate targeted therapy for a variety of malignancies
Gene panel test	Diagnostic procedures for numerous uncommon disorders, including retinal dystrophies, cancer, and other genetic conditions	Genetic diversity in a particular germline	To determine if inherited retinal dystrophies are genetic. There are currently no cures for some genetic mutations, although gene therapies are being explored

pattern-matching techniques allows for the identification of similarities between gene expression signatures, which has significantly contributed to advancing the understanding of diseases and the development of innovative treatments, particularly in the field of oncology. The best potential is considered to be the chemical profile of a drug that has the strongest relationship to the molecular profile of the patient. The "appropriate medicine for the right patients" guiding principle is the type of integrated systems biology approach that physicians will require to progress DKD treatment. As far as we know, no report has yet been applied (Komorowsky et al. 2012). However, the Biomarker Enterprise to Combat Diabetic Kidney Disease (BEAt-DKD) Collaboration is now working on several DKD-related projects. This initiative's results are anxiously expected. The warning is that many aspects of DKD that exist in people cannot always be correctly modeled using animal or cell culture models; therefore, model system limits must be taken into consideration (Komorowsky et al. 2012).

Consequently, it is necessary to use human biospecimens from clinical trial studies that have a solid design and enough power. By comparing the molecular profiles of responders and non-responders after the fact, it is possible to identify potential markers that are connected to patients' pharmaceutical responses. These potential indicators will be an effective tool for more accurate therapy prediction when combined with the pharmaceutical chemical fingerprints stated above.

11.5 Personalized Medicine for Treatment-Resistant Epilepsy Using a Structural Biology Approach

Completion of the Human Genome Project reveals the genetic diversity of humans, highlighting that it is crucial to bring pharmacogenomics into clinical practice to achieve successful therapy. Pharmacogenomics aims to identify groups of patients with refractory diabetes with similar genetic makeup, allowing for the development of targeted treatment approaches specifically tailored to those subgroups (Komorowsky et al. 2012).

In the future, using a systems biology approach to managing medicine may be beneficial owing to the complexity of drug resistance in epilepsy. It is possible to identify and target critical elements within complex networks that contribute to multidrug resistance by integrating genomic, proteomic, and metabolomic data to overcome drug resistance. So, pharmacogenomics is a crucial tool for determining a person's vulnerability to disease or prospective drug toxicity. Pharmacogenomics chronological progression is illustrated in Fig. 11.2 from monogenic to complicated genetic research analysis (Antonarakis 2001). The incorporation of pharmacogenomics provides clinicians with valuable assistance in predicting how a patient will respond to medications based on identified genetic variants. Recognizing the significance of pharmacogenomics, the Clinical Pharmacogenetics Implementation Consortium (CPIC) developed guidelines in 2014 specifically addressing the utilization of CYP2C9 pharmacogenetic testing for phenytoin dosing (Caudle et al. 2014).

The AED package leaflets also contain information on the side effects and treatment effectiveness when genetic mutations (SNPs) take place. The complex issues surrounding the mechanism of drug resistance in living systems like the brain can be resolved by systems biology, which takes a broad scientific approach: Integration of data from genetic, genomic, neurophysiology, and proteomic sources enables the identification of pathways and networks involved in the etiopathogenesis of neurological diseases (Diaz-Beltran et al. 2013). Moreover, by emphasizing the network interactions of molecular elements, systems biology makes it easier to find new

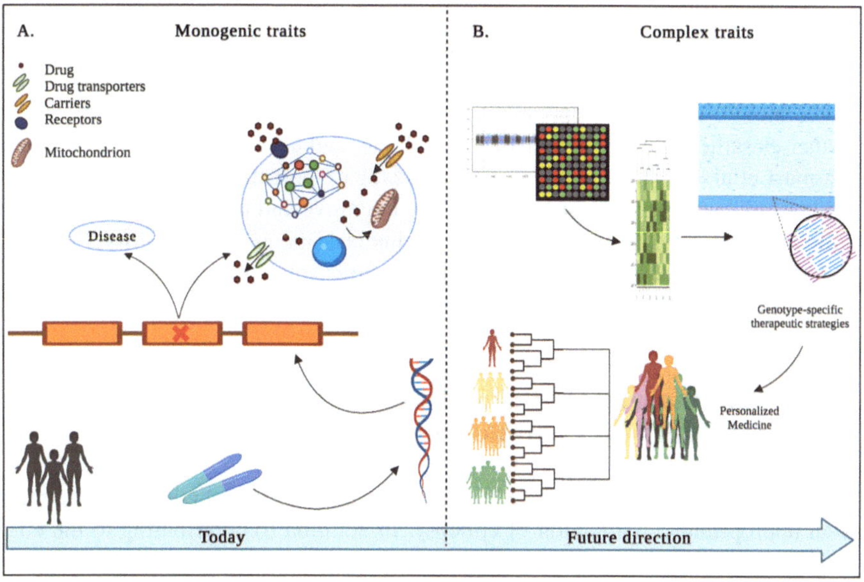

Fig. 11.2 A diagram showing the development of pharmacogenomics over time, from monogenic to complex genetic analyses. (**a**) Genetic testing revealed variations connected to the medication resistance mechanism in epilepsy. (**b**) The establishment of disease networks and patient categorization for personalized medicine and therapy are made possible by the combination of computational approaches and genetic technologies (NGS and CGH) (Naimo et al. 2019)

target drugs to treat resistance. Several aspects of patient care are impacted by the variety of epilepsy conditions, most frequently in the form of a delayed or incorrect diagnosis. As a result, the main cause of refractory epilepsy is represented by inaccurate epilepsy diagnosis, which prevents effective treatment. According to the International League Against Epilepsy's criteria, clinicians must perform a differential examination of the 36 various types of epilepsy based on the etiology, seizure type, epilepsy type, and epilepsy syndrome (Engel 2014). The variety of epilepsy conditions has an impact on many facets of patient care, most frequently in the form of a delayed or inaccurate diagnosis. As a result, incorrect epilepsy diagnosis, which hinders efficient treatment, is the primary cause of refractory epilepsy (Fattorusso et al. 2021). Clinicians are required to conduct a differential examination of the 36 different types of epilepsy based on the etiology, seizure type, epilepsy type, and epilepsy syndrome following the criteria set forth by the International League Against Epilepsy. Focused seizures are categorized as either "focal aware seizures" or "focal impaired awareness seizures" depending on whether a person retains consciousness or loses it during an epileptic seizure. Tonic, atonic, and clonic epileptic spasms are classified as "motor" seizures, while cognitive, autonomic, and affective elements are classified as "non-motor" seizures. The term "focal bilateral tonic-clonic convulsions" refers to a condition in which focal tonic-clonic seizures spread to both hemispheres (Naimo et al. 2019).

Generalized seizures can be categorized into two types: "motor" seizures and "non-motor" seizures, specifically absences. Motor generalized seizures encompass various subtypes such as clonic, tonic, tonic-clonic, myoclonic tonic-clonic, myoclonic atonic, and atonic epileptic spasms. Absences, on the other hand, can be further classified as typical, atypical, myoclonic, and absent with eyelid myoclonia (Sarmast et al. 2020). It is important to investigate potential morbidity issues that may resemble epileptic disorders, including cognitive and behavioral delays, metabolic changes, or head injuries. However, this classification system overlooks the underlying hereditary basis of epilepsy and focuses solely on the specific subpopulation of brain cells from which the seizures originate.

11.6 Personalized Medicine Using a Systems Biology Approach for Pharmacoresistant Epilepsy

The first step toward a more accurate diagnosis and successful course of treatment is an appropriate classification of epilepsy. In addition to contributing to the etiopathogenesis of epilepsy, genetic differences may have an impact on pharmacoresistance (Sarmast et al. 2020). Understanding the causes of the emergence of resistance to antiepileptic drug treatment is crucial for ensuring better therapeutic methods. Research on the role of eux transporters and how these pathways affect antiepileptic drug resistance has drawn increasing attention in recent years (Löscher et al. 2020). The most well-known eux transporters are members of the adenosine triphosphate-binding cassette (ABC) proteins superfamily, whose genes are highly conserved. These proteins carry a range of substrates across extracellular and intracellular membranes despite concentration gradients while acting as ATP-driven membrane pumps that hydrolyze ATP (Löscher et al. 2020). These transporters come in a wide variety that includes amino acids, peptides, minerals, carbohydrates, pigments, and other compounds.

First, the function and involvement of eux transporters in treatment resistance were investigated in a range of tumor types. It was assumed that the overexpression of multidrug transporter proteins, which are known to be constitutively expressed at the blood-brain barrier, would be associated with antiepileptic drug resistance, similar to chemotherapy resistance (Lazarowski et al. 2007). Several lines of evidence suggest that the blood-brain barrier's increased expression of these proteins appears to promote drug extrusion from the target site, preventing the treatment from reaching the epileptic center at the lowest therapeutic dose.

The blood-brain barrier's enhanced expression of these proteins may increase medication extrusion from the target site, preventing the treatment's minimum effective dose from reaching the epileptic focus, according to several lines of evidence. Currently, ABCD (ALD), ABCE (OABP), ABCF (GCN20), and ABCG make up the ABCA (ABC1), ABCB (MDR/TAP), ABCC (MRP/CFTR), and ABCG subfamily genes for eux proteins (Pardridge 2012). Pharmacoresistance and the upregulation of MDR1 mRNA levels in the brain tissue of individuals with refractory epilepsy have been linked, according to a study by Tishler and colleagues

(Pardridge 2012). Many antiepileptic drugs, including phenytoin, topiramate, carbamazepine, phenobarbital, lamotrigine, felbamate, and gabapentin, can be delivered by P-glycoprotein (P-gp) transporters, which are encoded by MDR1 or ABCB1 (Pardridge 2012).

In the early connection studies of pharmacological resistance in epilepsy and genetic alterations, the MDR1 CC genotype was discovered to be common, and the C3435T SNP was connected to an increase in P-glycoprotein 170 expression. This polymorphism then affected the pharmacological response to antiepileptic medications, as various evidence proved. The most common homozygous genotypes in the ABCB1 gene were T1236C and G2677A when compared to the control group. Moreover, among epileptic patients using phenytoin, the homozygous genotype G2677T/A has been reported to be common (Sun et al. 2016). Instead of the individual polymorphisms alone having a major impact on treatment results, as currently established by the evidence, the interaction between ABCB1 polymorphisms may decide intractability in epilepsy. C3435T, T1236C, and G2677T showed considerable linkage disequilibrium in ABCB1, suggesting that the three loci collectively may signal medication responsiveness in epilepsy. These results raise the possibility that the haplotypes of patients could be utilized to predict medication resistance in epilepsy.

Even though the association research between pharmacoresistant and multidrug transporters with altered function revealed ABCB1 as a potential possibility, several investigations looked at a variety of additional polymorphisms in different ABC genes (Kim et al. 2009). It's interesting to note that correlation research showed that patients with the ABCC2-24 T variation could not be treated with first-line antiepileptic drugs. The variation in medication digestion is responsible for a significant portion of the therapy of adverse effects or toxic effects of various drug mixes (Kim et al. 2009).

11.7 Cytochrome P450-Pharmacoresistant Epilepsy

Around 90% of all controlled drugs undergo biochemical pathway reactions catalyzed by various cytochrome P450 chemicals (CYPs). In the human genome, 57 genes encode CYP proteins, organized into families (CYP1, CYP2, CYP3) and subfamilies (CYP1A, CYP2C), with some existing in multiple isoforms (CYP2C19, CYP2D6, CYP3A4) (Nebert et al. 2013). According to a new theory on pharmacoresistance, changes in drug targets, voltage-gated ion channel characteristics, or genes encoding target molecules and receptors can reduce sensitivity to drug ligands, thereby impacting the response to pharmacological treatment (Tang et al. 2017). The pathophysiological mechanisms underlying seizures involve alterations in the balance between inhibitory GABAergic and excitatory glutamatergic signaling pathways (Shao et al. 2019). Anticonvulsant drugs currently used to manage seizures act by increasing the gamma-aminobutyric acid (GABA) signal, enhancing chloride ion flow, inhibiting neurotransmitter metabolism and reuptake, or modifying voltage-gated Na^+ sensitivity. These medications may also affect other voltage-gated channels, such as Ca^{2+} channels, by interfering with pathways or reducing the glutamatergic signal (Greenfield 2013). One of the primary targets of antiepileptic drugs, used both in monotherapy and polytherapy, is the voltage-gated Na^+ channel. Experimental models, both in vitro and in vivo, have demonstrated an association between changes in Na^+ channel shape and sensitivity and a reduced response to antiepileptic medications (Abdelsayed and Sokolov 2013). The voltage-gated Na^+ channel consists of two modulating subunits joined by a single subunit, forming a transmembrane pore. These channels can exist in three states: active, inactive, and disabled. While at resting membrane potential, the channels are inactive, and they rapidly switch to the active state upon depolarization, mediating the inward Na^+ current. With prolonged depolarization, the channels briefly become inactive before returning to the resting state after the hyperpolarization phase (Din et al. 2017). The shift between these functional states is of clinical significance because antiepileptic drugs bind differently to voltage-gated Na^+ channels depending on their conformation (Wei et al. 2022). Several first-line drugs for epilepsy treatment, such as carbamazepine, valproic acid, phenytoin, lamotrigine, topiramate, and oxcarbazepine, accelerate the Na^+ channel's exit from the inactivation phase (Wei et al. 2022). Consequently, these medications have a lower affinity for the resting channel state when it is inactive and exhibit a more pronounced blockage of sodium inward current during heightened neuronal excitability.

Changes in the genes that code for Na^+ channels may have led to the development of subunits that are resistant to medication or a protein that did not respond to anticonvulsant therapy, claim several studies. SCN1A, SCN2A, and SCN3A are the three primary genes that encode subunit Na^+ channels. Genetic screening of these three genes revealed a favorable connection between SCN1A IVS5–91 G > A, SCN2A IVS7-32A > G, and c.56 G!A and C (Lakhan et al. 2009). Pharmacoresistance and R19K polymorphisms indicate a higher prevalence of specific genotypes in individuals who cannot be treated with Na^+ channel-blocking antiepileptic drugs. Alternative splicing, which creates different domains that are either non-functional or drug-sensitive, is also a crucial mechanism in the development of Na^+ channels. Three intronic SCN1A polymorphisms, rs6730344, rs6732655, and rs10167228, can change the final shape of Na^+-voltage channels by either causing the loss of splice sites or establishing new ones, according to an intriguing inquiry into the multidrug resistance mechanism (Lakhan et al. 2009). Even though the subunit is the sodium channel's most important functional component, experimental data suggested that mutations in the genes that code for the accessory subunit may have an impact on how well antiepileptic medications work (Lakhan et al. 2009).

Epilepsy is a neurological illness that is challenging to categorize due to the many etiologies, types of convulsions, epileptic syndromes, locations, and amplitudes of the CNS involved. Given the complex origins of epilepsy, the diagnostic criteria necessitate specific studies involving clinical examination, electroencephalography (EEG) to characterize the type of seizure, and neuroimaging to detect structural abnormalities (Stafstrom and Carmant 2015).

11.8 Heterogeneity in Pharmacoresistant Epilepsy

Due to the heterogeneity of epilepsy, choosing a therapeutic approach to manage seizures while reducing side effects is an ongoing research topic. Pharmacogenomics research, which makes it possible to identify genes that may affect drug response and toxicity, is essential for directing treatment selection (Oates and Lopez 2018). One partner in this process is genetic counseling. Nowadays, the drug used to treat epileptogenic episodes is chosen based on empirical assessments of the patient's state. A first-generation antiepileptic drug is typically used first, with a dose based on the severity of the epileptic state: large doses to control severe seizures immediately; starting at low doses and progressively increasing them until the dose that controls minor epileptic episodes is established (Oates and Lopez 2018). Unfortunately, only a tiny proportion of epileptic participants had no further seizures after receiving the first antiepileptic medicine as monotherapy. Patients who are believed to be pharmacologically resistant to treatment are therefore given a second or third antiepileptic medication until they transition to polytherapy (Park et al. 2019).

High-resolution genomic technologies like array-comparative genomic hybridization (aCGH) and next-generation sequencing have enabled a breakthrough in the identification of copy number variations (CNVs) and genomic alterations linked to the emergence of a wide range of disorders and their therapeutic implications (NGS). These technologies allow for an unparalleled deep understanding of the complete human genome as well as the creation of personalized profiles for each patient (Moreno-De-Luca and Cubells 2011). Even if the use of genomic high-throughput technology for diagnosis is already common in clinical practice, using the same genetic data to address pharmacogenomics is still a compelling method. Subjects with epilepsy caused by the SLC2A1 mutation and coding for glucose transporter type 1 provide a tangible example of the immense potential of pharmacogenomics in the treatment of epilepsy (GLUT-1). This mutation interferes with the correct transport of glucose to the brain (CNS). The discovery of this mutation can reduce epileptic seizures by supplying the brain with alternate energy sources through a ketogenic diet (Dello et al. 2014).

Pharmocoresistance and CNVs in epileptic patients have recently been connected. When characterization of CNVs across the entire genome was done, recurrent microdeletions and microduplications were discovered (Dello et al. 2014). A variety of CNVs at loci 1q21.1, 15q11.2, 15q13.3, 16p11.2, and 22q11.2 were specifically associated with a range of neurodevelopmental diseases, including autism spectrum disorder, schizophrenia, dysmorphism, speech delay, and epilepsy. The highest correlation between neurobehavioral and mental health problems, such as childhood absence epilepsy (CAE) and idiopathic generalized epilepsies (IGEs), was found in the CNV at chr15q11.1–11.2 (Moreno-De-Luca and Cubells 2011).

Evaluation of genetic changes (SNPs or CNVs) without a definite correlation to the phenotype is, without a doubt, meaningless. However, due to compelling scientific evidence that certain genetic changes play a critical part in the drug resistance mechanism, taking these connections into account is now required

(Moreno-De-Luca and Cubells 2011). Synergistic integration, annotation, and analysis of genetic data led to the emergence of Systems Biology, a new field of functional biology. Each chemical signature is highlighted and considered a part of a system's network in this branch (Bebek et al. 2012). The systems biology method is based on incorporating a variety of data, from straightforward analyses of biological processes to intricate network analyses. True, the saying "one" examines genes, the function of coded proteins, and the interaction of numerous interrelated pathways can provide an overview of the intricacy of biological occurrences (Ideker et al. 2001). We were able to corroborate our earlier discovery that a pharmacoresistant phenotype may be brought on by several seemingly unrelated chromosomal abnormalities using a systems biology method. More specifically, a change in molecules that were working along different routes was what led to the development of self-sustaining convulsions that did not respond to treatment (Ideker et al. 2001). One of the contributing factors was the imbalance between excitatory and inhibitory states. Based on this evidence, we propose that the integration of genomic data with systems biology analysis enables us to anticipate individualized treatments based on health rather than disease by developing a better understanding of the roles that critical factors play in the interaction between complex pathways (Esplin et al. 2014).

Given the significant multifactorial of medication resistance in epilepsy, the systems biology approach may prove to be a beneficial tool for pharmacological management in the future. By combining genomic, proteomic, and metabolomic data, it is possible to pinpoint crucial elements that operate within complex networks and aid in the formation of multidrug resistance (Murgia et al. 2017). Then, to overcome refractoriness, these components could be focused on. This technique is therefore a helpful tool for determining a person's risk of illness or drug toxicity.

11.9 Cancer System Biology Drives Precision Medicine

The discipline of oncology has been obliged to address the disease from a comprehensive systems-based perspective due to the availability of high-throughput omics data, such as deep sequencing of enormous patient cohorts. Data-driven science has been utilized in clinical research. In a perfect world, judgments about precision medicine would be made following the needs of the patient, much like a key would be made to fit into a lock (Murgia et al. 2017). Each cancer tissue has a unique molecular profile of genomic and epigenomic alterations. Hence, incorporating comprehensive genomic and systemic insights for targeted cancer treatment represents the logical and inevitable progression toward data-driven disease management, ultimately contributing to improved survival outcomes (Chakravarthi et al. 2016).

The availability of high-throughput omics data, such as deep sequencing of huge patient cohorts, has forced the field of oncology to approach the disease from a holistic systems-based perspective. The field of systems biology over the past few years has grown and established rules for building multiscale, predictive models. These systems biology models allow us to comprehend biochemical networks, identify biomarkers, and categorize patients based on unique genetic and non-genomic (e.g., transcriptomic, epigenomic, proteomic, metabolic) profiles. Yet, without a mechanism, it is impossible to pinpoint the underlying genetic etiology of a group of symptoms (Mathé et al. 2018).

It might not be possible to foresee which course of treatment will be most effective. Several network vulnerabilities are highlighted by master regulators. Their identification is necessary for both systems-based rational medication design and illness management through precision medicine (Litman 2019). System biologists extract synergy from complicated, multidimensional datasets to recognize that biological networks are more potent than the sum of their nodes by hypothesizing, testing, verifying, improving, and hypothesizing again in a cyclical activity. Systems biology has been able to benefit from hitherto unheard-of opportunities because of data from the cancer genome, which has helped with the mystery of this complex disease (Esplin et al. 2014).

Yet, it will be important to solve large data difficulties and quickly acquisition of cancer genome data to quickly transfer genomic insights into innovative cancer therapeutics. Malignant melanoma, like other cancers, is brought on by unchecked cell proliferation brought on by accumulating mutations to our genomes and epigenomes (Berger and Mardis 2018). By the disruption of signals that regulate cell division and migration, these alterations may eventually lead to invasive malignancies that can metastasize and colonize distant areas (Van Zijl et al. 2011). The molecular remnants of alterations in the genome and epigenome will reveal the elements that distinguish melanoma as such a severe and aggressive cancer. This knowledge will also affect how we view the causes, symptoms, and methods of melanoma treatment, diagnosis, and prevention (Van Zijl et al. 2011). We will show how understanding malignant melanoma and responding to it are accelerated by systems biology and genomic biology-driven approaches to cancer. In addition, high-throughput single-nucleotide polymorphism (SNP) arrays have sped up developments in the field of somatic copy number changes (SCNAs). SCNAs can be recorded using low-coverage WGS or SNP arrays. The reported mutation rates for different tumor types and within each of them are separated by more than five orders of magnitude (Mao et al. 2007). One of the first tumor forms to be sequenced was melanoma, which aided in the development of cancer genomics. WGS used the COLO-829 metastatic melanoma patient-derived cell line to compile the first thorough record of somatic events (Craig et al. 2016).

Cancer genomics studies from The Cancer Genome Atlas (TCGA), which were supported by the National Cancer Institute and the general public, developed and peaked in 2012. As part of a collaborative effort, thousands of tumor specimens were sequenced, processed at the individual data level, combined, and evaluated (Weinstein et al. 2013). Similarly to this, the genomes of hundreds of melanomas were sequenced in the same year. Unfortunately, melanoma genomics' ability to match the pace and trends of other cancer genomics studies was briefly limited by the high data requirements and frequent somatic mutation calls (Diaz-Beltran et al. 2013). The 2015 TCGA skin cutaneous melanoma (SKCM) investigations (average rate of 18/Mbp; a greater number of individual specimens with rates surpassing 1000/Mbp) required more somatic mutation calls than all other TCGA genomics studies combined (Guan et al. 2015a).

The bioinformatics tool Introns Vs Exons (InVEx), which examines the particular background mutation rate per patient and per gene to identify driver genes, is now the gold standard in melanoma genomics. To record important somatic events across the genome in melanoma, a lot of computing is needed. Now, there are more than 700 WES-equipped melanoma specimens available, or around 4% of all cancer genomics initiatives (Guan et al. 2015a). Given the present 1.6-fold increase in data production and exome sequencing, it is projected that cancer research on all tissues will soon be able to gather 10,000 specimens annually, and melanoma research will soon be able to collect 1000 specimens annually. It will be simpler to perform further meta-studies of both public and private melanoma genomics research thanks to Oncotator and other sharable data formats. According to the saturation study of cancer drivers, a comprehensive database of cancer drivers with intermediate mutation frequencies of 2–5% in melanoma would require more than 5000 specimens to obtain more than 99% power (Martincorena et al. 2017). Although the quest for melanoma driver genes is not done, the cancer research community has already established a consistent list of high-frequency driver genes. There are vast, complicated data structures in the big biomedical data of today. For thorough coverage, additional data levels, formats, and sources are also needed. Volume, velocity, variety, and veracity are examples of V-type barriers that must be surmounted to link, match, clean, and convert data across systems (Raghupathi and Raghupathi 2014).

The difficulties presented by large data are readily visible at the molecular level for each patient when looking at a precision medicine profile. The evaluation of huge patient cohorts of precision medicine data also calls for connection and correlation, hierarchies, and multilevel linkages. Systems biology can link data from nucleotide- or gene-based next-generation sequencing with other functional levels like protein structures or metabolite fragments using high-throughput spectroscopy or spectrometry (Pereira et al. 2020). Future big data in precision medicine requires creative ways to be able to collect, curate, store, share, maintain, update, query, and visualize data in compliance with established privacy rules.

Many projects have been initiated globally in recent years to convert innovative scientific findings into prospective clinical trials. One important strategy is the stratification of PDAC patients based on the tumor transcriptome to forecast therapy response (Fig. 11.3) (Regel et al. 2020). Other approaches focus on genetic changes and the discovery of customized targeted therapies. More experimental research is being done to find new biomarkers for cancer diagnosis, subtyping, predicting

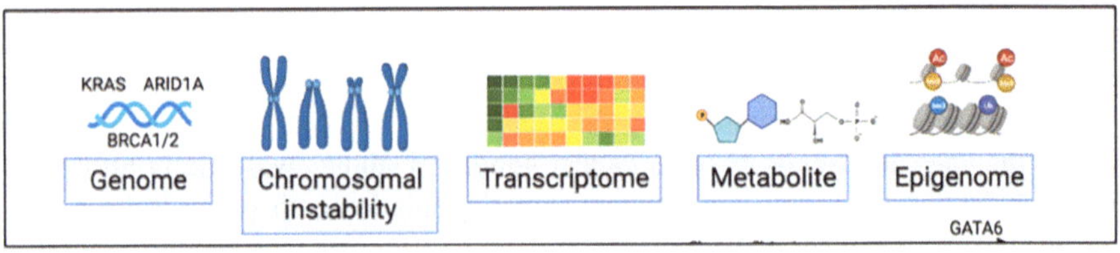

Fig. 11.3 Pancreatic ductal adenocarcinoma subtyping according to numerous molecular tumor characteristics (Regel et al. 2020)

therapy response, and clinical outcomes. Transferring the knowledge into clinical practice, nevertheless, still presents a barrier. Only through seamless collaboration between clinicians, researchers, and providers will melanoma precision medicine be successful. Although insights from genetics are increasingly incorporated into the clinical decision-making process, the infrastructure for genomics and precision medicine still has to be created (Valenti et al. 2021). Deep sequencing efforts of melanoma specimens are not well integrated into clinical practice, and cohort-based genomics tools like GISTIC, MutSigCV, and InVEx are designed more for discovery research than for everyday diagnoses (Vanni et al. 2020). The patient's permission is necessary for clinical carers, biobanks, omics data acquisition systems, cloud-based processing, and genomics researchers to securely communicate data (Antoniades et al. 2021).

Unsurprisingly, long before precision medicine existed, direct-to-consumer genetic tests and population genome research programs questioned traditional privacy norms. With the ability to link longitudinal health records, future repositories will shift away from generalizable knowledge and toward completely consented models. Contrary to the early hyperbole around personalized medicine, a first step—rebranding it as precision medicine—has been accomplished and has won the acceptance of a large portion of the general public. Increased therapy success rates can be achieved by doing genome and epigenome tests. Although the probability usefulness of genetic markers is well-defined, clinical application is essential to resolving ethical issues, such as the stigma associated with genetic determinism. It is necessary to review and update ethical principles to stay up with the quickly evolving innovations in population genomics, forensics, and precision medicine.

11.10 System Biology: Neuroblastoma

Early identification of activating point mutations in the B-Raf proto-oncogene (BRAF, serine/threonine kinase, Gene ID: 673) and the RAS viral oncogene homolog for neuroblastoma (NRAS, Gene ID: 4893) has made it possible to use precision medicine methods. According to Guan et al. (2015b), these particular kinase mutations have made it possible to utilize kinase inhibitors that are matched to the genotyped mutations. As a highly sensitive subtype of melanoma, BRAF-activated melanomas initially responded favorably to targeted treatment with kinase inhibitors. These tumors do, however, frequently become resistant to such treatments over time. Alternative therapy methods are therefore developing quickly, especially those that use the immune system to fight cancer cells, as mentioned by Kong et al. (2016). Theoretically, melanoma might be treated using a variety of methods, including immunological checkpoint inhibitors (IMMUNOi), which aid in triggering the immune system's defenses against cancer cells. Additionally, melanoma can be treated with targeted treatment using inhibitors that target certain signaling pathways, such as mitogen-activated protein kinase inhibitors (MAPKi), as well as epigenetic inhibitors (EPIi) and metabolic inhibitors (METABi) (Kong et al. 2016).

These multifaceted strategies seek to overcome resistance and improve the efficacy of available melanoma therapy options.

The survival rates of patients with advanced metastatic melanoma have been markedly improved by second-generation clinical trials of melanoma medicines, which include regimens with immune checkpoint antibodies or targeted therapy (Achkar and Tarhini 2017). Combinations performed better for both groups than single-agent regimens: In terms of survival, the combined inhibition of mitogen-activated protein kinase kinase 7 (MAP2K7, MEK, Gene ID:) and BRAF (Gene ID: 5609) demonstrated superior efficacy compared to a single MAPKi intervention. Additionally, the survival outcomes of simultaneously targeting immunological checkpoints with humanized antibodies were notably higher than those achieved with a singular IMMUNOi against cytotoxic T-lymphocyte-related protein 4 (CTLA4, Gene ID: 1493). This effect was observed in genes associated with programmed cell death 1 (PDCD1, PD1, CD279, Gene ID: 5133) and in molecules like CD274 (PDL1, PDCD1L1, Gene ID: 29126) (Cheng and Tian 2017).

A small-molecule drug targeting BRAF, known as MAPKi, has undergone specific trials, examining its effects on the combination of MAP2K7 (MEK) and BRAF, particularly in conjunction with the MAPKi vemurafenib (PLX4032), dabrafenib (GSK2118436), encorafenib (LGX818), cobimetinib (GDC-0973), and trametinib (GSK-1120212). There are also ImMUNOi that combine CTLA4 and PDCD1 (Proietti et al. 2020).

The treatments with the worst results were single-agent CTLA4 blocking or conventional chemotherapy. Yet, early clinical data that assess the effectiveness of MAPKi and IMMUNOi combos are optimistic. Despite frequent relapses and severe side effects, targeted medicines show a remarkable ability to limit tumor development. This may increase the longevity of the immunotherapy response in upcoming MAPKi and IMMUNOi combo trials (C. Yu et al. 2019). Third-generation studies that incorporate knowledge from precision medicine, new schedules or combinations, and the less studied classes of EPIi and METABi are likely to improve previously successful regimens. In particular, cancer systems biology sheds light on the immunosuppressive consequences of metabolic, epigenomic, and genomic factors.

While relapse, disease progression, or high rates of autoimmune toxicity still need to be defeated, precision medicine in cancer offers hope for enhanced long-term benefits and decreased side effects (Maeda and Khatami 2018). Even though genome sequencing captures the somatic landscape and the clinical importance of immune cells in the control of human tumors is well established, tumor-specific mutations that lead to the emergence of neoantigens play a significant role in the activity of immunotherapies. Neoantigens can also be used as targets in tailored vaccinations and cell treatments that target cancer mutations unique to a patient. Neoantigens can be identified with the help of modeling or prediction techniques based on structure (Peng et al. 2019). Predictions of treatment responses in melanoma will eventually become more precise by looking at gene expression characteristics linked to immunological sensitivity or resistance.

The abundance of recently identified epigenetic players in the genetics of melanoma suggests that environmental variables may have effects beyond UV-induced photoproducts that cause cancer. In contrast to genomics, the young discipline of epigenomics only makes up approximately 2% of genome-scale investigations but is growing far more quickly (Santourlidis et al. 2022). The co-occurrence of BRAF and EZH2 hyperactivation in the melanoma precision medicine scenario outlined above opens up new pathways for targeted therapy. It is well established that using MAPKi alone to treat BRAF or MAP2K7 can lead to medication resistance and disease progression (Bernocchi et al. 2021). Although they are effective, combining several MAPKi compounds in a single therapy has many drawbacks. A targeted therapy that benefits from the independence of signaling and epigenetic pathways could combine MAPKi and EPIi (Bernocchi et al. 2021). Small-molecule medications can target increased EZH2 activity in cancer, which is connected to susceptibility to apoptosis. One such medication is GSK126, PubChem CID 68210102.

Downregulation of well-known tumor suppressors, such as RASSF5, BOK, NDRG1, and N-myc downstream regulated 1, occurs in cells deficient in EZH2 (Nepali and Liou 2021). However, the downregulation of these genes can be reversed through inhibitor therapy. In an in vitro experiment targeting EZH2 with drugs, the validation of genes associated with immune recognition, cytokines, chemokines, and major histocompatibility complex class II antigen presentation was demonstrated. Through a combination of transcriptome and ChIP-Seq analyses, target genes related to antigen processing and presentation pathways, tumor suppression, apoptosis, cell differentiation, cell cycle inhibition, and metastasis repression were identified (Zhou et al. 2020).

The regulatory role of EZH2 in modulating the interaction between natural killer cells and tumor cells suggests that it can inhibit both innate and adaptive immunity. Animal models and cell cultures treated with EZH2 inhibitors demonstrate significant reductions in spheroid growth, cell cycle arrest, apoptosis, and tumor regression. By inhibiting specific gene expression pathways related to immune receptors and apoptotic signaling, EZH2 promotes immune evasion and tumor growth. This raises the question of whether the effectiveness of IMMUNOi treatment can be enhanced in patients with rewired epigenetics and altered tumor metabolism through the use of EPIi or METABi. Notably, lactate dehydrogenase, a metabolic enzyme central to the Warburg effect in cancer, serves as a predictive biomarker for overall survival and a negative predictor of the immune response in patients with malignant melanoma.

When used as a whole, therapy combinations involving EZH2 inhibitors may be beneficial for patients with somatic activation of EZH2 (Zhou et al. 2020). Because genomic and non-genomic somatic activation of EZH2 occurs more frequently than 60% of the time in melanoma patients, this instance of precision medicine may be more generalizable, and the synergy between epigenetic and immune checkpoint inhibition has a strong potential for a broader treatment regimen. The most effective individualized approach is detailed but comprehensive. Precision medicine methods raise serious concerns about the generalizability of isolated case studies and recognized somatic occurrences. The heterogeneity of tumors and the high prevalence of background mutations in malignancies like melanomas further complicate one-dimensional disease management that can be assumed to be effective for all patients.

Comparisons made with genome datasets, such as The Cancer Genome Atlas (TCGA), provide evidence supporting the significant co-occurrence pattern of BRAF with epigenetic regulators. Specifically, EZH2 (also known as CREB3L2 or cAMP-responsive element binding protein 3 like 264,764) and KDM7A, located near the genome, exhibit strong co-amplification with significant statistical significance (p values below 0.001). Furthermore, their mutational profiles suggest a concurrent occurrence (Ishak et al. 2016). In melanoma, essential epigenetic regulators that function as tumor suppressors are frequently lost due to somatic copy number deletions or inactivating somatic mutations. When considering all genomic, transcriptomic, and proteomic data, it is anticipated that the observed precision medicine profile, characterized by rewired MAPK signaling and altered epigenetic regulators, will apply to a significant portion of the melanoma cohort.

To generate proteomic data, two primary methods have been developed: top-down proteomics and bottom-up proteomics. These techniques, along with additional subcategories, are summarized in Table 11.3 (Krzyszczyk et al. 2018).

11.11 Use of Bioinformatics and Systems Biology in Immuno-Oncology and Precision Medicine

Oncology diagnostics and immunotherapy have greatly benefited from next-generation sequencing (NGS) technology, which is regarded as a crucial element of precision medicine. It shows potential for limiting tumor development and lessening the misery endured by chemotherapy patients by improving gene diagnostics and immunotherapy (Krzyszczyk et al. 2018). However, new NGS methodologies and genetic data processing techniques must be developed to make it easier to translate precision medicine from the laboratory to clinical practice and enable the use of genetic testing in personalized therapies. While whole genome sequencing (WGS) gives more precision and efficiency, NGS panels, for instance, concentrate on fewer genes or areas (Krzyszczyk et al. 2018). Driver genes frequently form a regulatory network in complicated illnesses like cancer.

The deconstruction of genome-wide interactions into crucial modules or pathways that are important in the emergence of illness is made possible by graph theories, such as random walk algorithms and shortest route analysis (Choobdar et al. 2019). The first to find gene network modules linked to survival were H. Choi and K. J. Na. By choosing representative genes from these survival-related modules, they built a deep learning-based lung cancer risk classification model. In separate datasets, the model showed strong prognostic prediction, and its predictive value was unaffected by the clinical and pathological characteristics of lung cancer (Wang et al. 2022). A brand-new computational technique called SRMDAP for predicting miRNA-disease connections was developed by X. Li and colleagues. The density-based clustering recommender model, SimRank, and the SRMDAP showed precise

Table 11.3 Proteomics workflows and strategy overview—a list of methods for gathering proteome data or details on the kinds of proteins found in a patient's malignant tissues (Krzyszczyk et al. 2018)

Strategy	Purpose	Data collection	Data analysis
Top-down proteomics	Whole-protein analysis with a focus on post-translational modifications	Proteins are ionized, brought into the gaseous state, separated using LC, and then subjected to MS analysis using methods like MALDI and ESI	Protein databases, such as ProSight PTM, which provides a free windows program for sequence identification, are matched to mass spectra
Middle-down proteomics	Fewer complicated solutions are produced, making it simpler to identify proteins and analyze high-level protein characteristics	Only enough protein is digested to create big peptide fragments for MS analysis	Protein composition and the presence of high-level characteristics like protein isoforms and alterations are both identified by MS analysis
Bottom-up proteomics with labelling	Allows for the investigation of proteome alterations in many samples simultaneously, such as changes brought on by biological disturbances	Using techniques like SILAC, ICAT, and iTRAQ, C, H, N, and O isotopes are added to peptide samples, and peptides are then examined using tandem mass spectrometry (MS/MS)	Comparing the intensities of the various isotope species in the MS/MS data allows for the measurement of relative peptide abundances
Bottom-up (shotgun) proteomics	Big mixed protein sample analysis and composition determination, for example, in biomarker identification investigations	Trypsin proteolysis converts proteins into peptides, which are then sorted in a mass spectrometer according to their size or charge	For protein identification, mass spectra are compared to databases like Andromeda or PeptideAtlas

predictions of the associations between miRNAs and disease relationships in leave-one-out cross-validation and case studies (Wang et al. 2022).

Researchers under the direction of S. Zhu and colleagues have created a method for creating an early warning index (EWI) that can detect the pre-disease stage in a sample using individual-specific dynamical network biomarkers (DNB). In this method, DNB genes are chosen using gene information from microarrays for influenza A illness. The inquiry is divided into various parts, including GO and KEGG pathway exploration, PPI network analysis, survival analysis of hub genes, and prediction of miRNA targets. According to Niemira et al. (2019), the initial stage focuses on finding genes that demonstrate variable expression between patients who respond favorably to chemotherapy and those who do not. This research provides valuable insights into gene expression patterns in drug-resistant samples, which can contribute to the treatment of osteosarcoma.

In a separate work, Y. Liang and colleagues created the improved binary differential evolution technique (BDEP) for employing the fluorescent in situ hybridization (FISH) technology to infer the phylogenetic tree of malignancies. The created phylogenetic trees perform better in explaining the development of malignancies than other comparable techniques. Additionally, a method known as PEDD, developed by H. Cui and associates, makes use of gene expression patterns from disease-related tissues and cell lines treated with pharmaceuticals to foretell the effects of treatments on illnesses (Hughes and Friedman 2009). These developments offer useful tools for studying the development of cancer and predicting the impact of medications on diverse disorders.

MGB1's possible contribution to the onset and spread of lung cancer, as well as its prognostic and clinicopathologic relevance, was examined by X. Wu and colleagues. According to their research, HMGB1 expression is strongly linked to lung cancer progression and may serve as a diagnostic and prognostic marker for the disease (Feng et al. 2016). H.-B. Xing and associates looked into whether IL-6 deletion contributes to the elevated gemcitabine sensitivity of PANC-cells. They found that IL-6 suppression increased gemcitabine's anticancer efficacy considerably, triggered apoptosis, and decreased cell proliferation and tumorigenicity. According to the research, gemcitabine and IL-6-targeting shRNA may be combined in a clinical context to treat pancreatic cancer.

Among Chinese patients with metastatic colorectal cancer (mCRC), R. Liu and colleagues discovered the RAS mutation, Immunocore, and expression of PD-L. They hypothesized that these findings would serve as predictors of outcomes for mCRC patients having palliative surgery (Xu et al. 2021).

Research on hypoxia-induced ovarian cancer was carried out by X. Yu and associates to look into the AEG-controlled metastasis regulatory mechanism. They hypothesized that AEG controls the HIF-alpha/NF-kappa-B/VEGF pathway, which is connected to hypoxia in ovarian cancer, by controlling the protein levels of AEG, HIF, and VEGF in ovarian cancer tissue samples (X. Yu et al. 2018).

11.12 Conclusion

Personalized medicine strategy is being developed in the field of systems biology for diabetic kidney disease (DKD), which is revolutionizing healthcare (X. Yu et al. 2018). Patients can be divided into several subgroups according to their illness features through the examination of molecular profiles. Early detection of those who are very susceptible to illness development is made possible by the discovery of precise and trustworthy prognostic biomarkers. Additionally, accurate predictive biomarkers make it easier to link a patient's molecular profile with the molecular profile of a particular medicine, enabling customized treatment options (Nalejska et al. 2014).

Utilizing systems biology can enhance clinical trials by selecting appropriate patients, resulting in smaller, more efficient, and ethically conducted trials with improved success rates. Collaborative initiatives such as the BEAt-DKD and

DYNAMO studies aim to provide knowledge and technologies that facilitate the development of novel personalized treatments for DKD. These international consortia possess the potential to create additional cohorts for Diabetic Kidney Disease (DKD), providing valuable resources for the identification of molecular biomarkers and therapeutic targets. Moreover, these collaborations aim to unveil the molecular mechanisms driving the early progression of DKD through comprehensive multi-omics data analysis conducted on biospecimens obtained from the same patients (Nalejska et al. 2014).

Furthermore, a systems biology approach can enhance the utilization of pharmacogenomic data by establishing causal relationships between genotype and phenotype profiles. This approach aids in improving the understanding of how genetic variations influence individual responses to medications.

The proliferation of this new paradigm will be aided by the growing accessibility of extensive clinical data, large-scale genetic data, and bioinformatic tools. Biomarkers that forecast therapy response in oncology serve as the foundation for customized care. Despite neurology lagging behind in some aspects, improved patient categorization holds the promise of advancing personalized treatment. The integration of pharmacogenomics, particularly the widespread adoption of this innovative paradigm in refractory epilepsy, represents a significant step forward in this direction. Biomarkers that forecast therapy response in oncology serve as the foundation for customized care (Nalejska et al. 2014). Systems biology and pharmacogenomics are now being used to treat refractory epilepsy, which has the potential to drastically reduce morbidity and death. Large-scale sequencing efforts have also greatly aided precision medicine for melanoma. Exome- and genome-wide sequencing resulted in the identification of genetic biomarkers, targeted medicines, and positive clinical outcomes. Clinical practice will soon use collaborative, multiplatform genome sequencing techniques for cancer. Several new findings will have an impact on melanoma prevention, diagnosis, and therapy. The development of a precision medicine framework entails spending money on better diagnostic methods and therapeutic regimens.

Innovative genotype-phenotype connections in human melanoma are made possible by cutting-edge molecular, high-throughput platforms that aid in the systems biology interpretation. Before cancer precision medicine can effectively integrate genomic technology into clinical oncology practice, there are still challenges that need to be addressed. One such challenge is the genomic heterogeneity and frequent genomic alterations observed in melanoma. To address this, it is crucial to combine genomic precision data with cancer databases and clinical statistics. This integration allows the precision medicine team to prioritize therapeutic decisions and deliver personalized cancer therapies based on each patient's genome. It is essential to distinguish between phenotypically neutral events and cancer drivers with clinically actionable targets to determine the most appropriate treatment approach. The existing licensed mono and combination therapies show promise and molecular information from precision medicine can recommend orthogonal therapies that focus on metabolism and epigenetics to combat drug resistance and immune evasion.

References

Abdelsayed M, Sokolov S (2013) Voltage-gated sodium channels: pharmaceutical targets via anticonvulsants to treat epileptic syndromes. Channels (Austin) 7:146–152

Achkar T, Tarhini AA (2017) The use of immunotherapy in the treatment of melanoma. J Hematol Oncol 10:1–9

Antonarakis SE (2001) The search for allelic variants that cause monogenic disorders or predispose to common, complex polygenic phenotypes. Dialogues Clin Neurosci 3:7

Antoniades A, Papaioannou M, Malatras A, Papagregoriou G, Müller H, Holub P, Deltas C, Schizas CN (2021) Integration of biobanks in national eHealth ecosystems facilitating long-term longitudinal clinical-omics studies and citizens' engagement in research through eHealthBioR. Front Digit Health 3:628646

Bebek G, Koyutürk M, Price ND, Chance MR (2012) Network biology methods integrating biological data for translational science. Brief Bioinform 13:446

Berger MF, Mardis ER (2018) The emerging clinical relevance of genomics in cancermedicine. Nat Rev Clin Oncol 15:353

Bernocchi O, Sirico M, Corona SP et al (2021) Tumor type agnostic therapy carrying BRAF mutation: case reports and review of literature. Pharmaceuticals (Basel) 14(2):159

Caudle KE, Rettie AE, Whirl-Carrillo M, Smith LH, Mintzer S, Lee MT, Klein TE, Callaghan JT (2014) Clinical pharmacogenetics implementation consortium guidelines for CYP2C9 and HLA-B genotypes and phenytoin dosing. Clin Pharmacol Ther 96:542

Chakravarthi BVSK, Nepal S, Varambally S (2016) Genomic and epigenomic alterations in cancer. Am J Pathol 186:1724

Cheng Y, Tian H (2017) Current development status of MEK inhibitors. Molecules 22:1551. https://doi.org/10.3390/MOLECULES22101551

Choobdar S, Ahsen ME, Crawford J et al (2019) Assessment of network module identification across complex diseases. Nat Methods 169(16):843–852

Craig DW, Nasser S, Corbett R et al (2016) A somatic reference standard for cancer genome sequencing. Sci Rep 6:24607. https://doi.org/10.1038/SREP24607

Dello RC, Di GG, Cignini P, Padula F, Mangiafico L, Mesoraca A, D'Emidio L, McCluskey MR, Paganelli A, Giorlandino C (2014) Comparative study of aCGH and next generation sequencing (NGS) for chromosomal microdeletion and microduplication screening. J Prenat Med 8:57

Dhillon BK, Smith M, Baghela A, Lee AHY, Hancock REW (2020) Systems biology approaches to understanding the human immune system. Front Immunol 11:1683

Diaz-Beltran L, Cano C, Wall DP, Esteban FJ (2013) Systems biology as a comparative approach to understand complex gene expression in neurological diseases. Behav Sci (Basel) 3:253

Din FU, Aman W, Ullah I, Qureshi OS, Mustapha O, Shafique S, Zeb A (2017) Effective use of nanocarriers as drug delivery systems for the treatment of selected tumors. Int J Nanomedicine 12:7291

Engel J (2014) Approaches to refractory epilepsy. Ann Indian Acad Neurol 17:S12

Esplin ED, Oei L, Snyder MP (2014) Personalized sequencing and the future of medicine: discovery, diagnosis and defeat of disease. Pharmacogenomics 15:1771

Fattorusso A, Matricardi S, Mencaroni E, Dell'Isola GB, Di Cara G, Striano P, Verrotti A (2021) The pharmacoresistant epilepsy: an overview on Existant and new emerging therapies. Front Neurol 12:1030

Faulkner E, Holtorf AP, Walton S et al (2020) Being precise about precision medicine: what should value frameworks incorporate to address precision medicine? A report of the personalized precision medicine special interest group. Value Health 23:529–539

Feng A, Tu Z, Yin B (2016) The effect of HMGB1 on the clinicopathological and prognostic features of non-small cell lung cancer. Oncotarget 7:20507

Goetz LH, Schork NJ (2018) Personalized medicine: motivation, challenges, and progress. Fertil Steril 109:952

Greenfield LJ (2013) Molecular mechanisms of antiseizure drug activity at GABAA receptors. Seizure 22:589
Guan J, Gupta R, Filipp FV (2015a) Cancer systems biology of TCGA SKCM: efficient detection of genomic drivers in melanoma. Sci Rep 5:7857. https://doi.org/10.1038/SREP07857
Guan J, Gupta R, Filipp FV (2015b) Cancer systems biology of TCGA SKCM: efficient detection of genomic drivers in melanoma. Sci Rep 51(5):1–10
Hughes AL, Friedman R (2009) A phylogenetic approach to gene expression data: evidence for the evolutionary origin of mammalian leukocyte phenotypes. Evol Dev 11:382
Ideker T, Galitski T, Hood L (2001) A new approach to decoding life: systems biology. Annu Rev Genom Hum Genet 2(1):343
Ishak CA, Marshall AE, Passos DT et al (2016) An RB-EZH2 complex mediates silencing of repetitive DNA sequences. Mol Cell 64:1074
Jameson JL, Longo DL (2015) Precision medicine—personalized, problematic, and promising. N Engl J Med 372:2229–2234
Kim DW, Lee SK, Chu K, Jang IJ, Yu KS, Cho JY, Kim SJ (2009) Lack of association between ABCB1, ABCG2, and ABCC2 genetic polymorphisms and multidrug resistance in partial epilepsy. Epilepsy Res 84:86–90
Kim J, Koo BK, Knoblich JA (2020) Human organoids: model systems for human biology and medicine. Nat Rev Mol Cell Biol 21:571
Komorowsky CV, Brosius FC, Pennathur S, Kretzler M (2012) Perspectives on systems biology applications in diabetic kidney disease. J Cardiovasc Transl Res 5:491
Kong BY, Carlino MS, Menzies AM (2016) Biology and treatment of BRAF mutant metastatic melanoma. Melanoma Manag 3:33
Krzyszczyk P, Acevedo A, Davidoff EJ et al (2018) The growing role of precision and personalized medicine for cancer treatment. Technology 6:79
Lakhan R, Kumari R, Misra UK, Kalita J, Pradhan S, Mittal B (2009) Differential role of sodium channels SCN1A and SCN2A gene polymorphisms with epilepsy and multiple drug resistance in the North Indian population. Br J Clin Pharmacol 68:214
Lazarowski A, Czornyj L, Lubienieki F, Girardi E, Vazquez S, D'Giano C (2007) ABC transporters during epilepsy and mechanisms underlying multidrug resistance in refractory epilepsy. Epilepsia 48(Suppl 5):140–149
Litman T (2019) Personalized medicine—concepts, technologies, and applications in inflammatory skin diseases. APMIS 127:386
Löscher W, Potschka H, Sisodiya SM, Vezzani A (2020) Drug resistance in epilepsy: clinical impact, potential mechanisms, and new innovative treatment options. Pharmacol Rev 72:606
Maeda H, Khatami M (2018) Analyses of repeated failures in cancer therapy for solid tumors: poor tumor-selective drug delivery, low therapeutic efficacy and unsustainable costs. Clin Transl Med 7:11. https://doi.org/10.1186/S40169-018-0185-6
Mao X, Young BD, Lu Y-J (2007) The application of single nucleotide polymorphism microarrays in cancer research. Curr Genomics 8:219
Martincorena I, Raine KM, Gerstung M, Dawson KJ, Haase K, Van Loo P, Davies H, Stratton MR, Campbell PJ (2017) Universal patterns of selection in cancer and somatic tissues. Cell 171:1029–1041.e21
Mathé E, Hays JL, Stover DG, Chen JL (2018) The omics revolution continues: the maturation of high-throughput biological data sources. Yearb Med Inform 27:211
Morash M, Mitchell H, Beltran H, Elemento O, Pathak J (2018) The role of next-generation sequencing in precision medicine: a review of outcomes in oncology. J Pers Med 8:30. https://doi.org/10.3390/JPM8030030
Moreno-De-Luca D, Cubells JF (2011) Copy number variants: a new molecular frontier in clinical psychiatry. Curr Psychiatry Rep 13:129
Murgia F, Muroni A, Puligheddu M et al (2017) Metabolomics as a tool for the characterization of drug-resistant epilepsy. Front Neurol 8:459

Naimo GD, Guarnaccia M, Sprovieri T, Ungaro C, Conforti FL, Andò S, Cavallaro S (2019) A systems biology approach for personalized medicine in refractory epilepsy. Int J Mol Sci 20:3717. https://doi.org/10.3390/IJMS20153717

Nalejska E, Mączyńska E, Lewandowska MA (2014) Prognostic and predictive biomarkers: tools in personalized oncology. Mol Diagn Ther 18:273

Nebert DW, Wikvall K, Miller WL (2013) Human cytochromes P450 in health and disease. Philos Trans R Soc B Biol Sci 368:20120431. https://doi.org/10.1098/RSTB.2012.0431

Nepali K, Liou JP (2021) Recent developments in epigenetic cancer therapeutics: clinical advancement and emerging trends. J Biomed Sci 281(28):1–58

Niemira M, Collin F, Szalkowska A, Bielska A, Chwialkowska K, Reszec J, Niklinski J, Kwasniewski M, Kretowski A (2019) Molecular signature of subtypes of non-small-cell lung cancer by large-scale transcriptional profiling: identification of key modules and genes by weighted gene co-expression network analysis (WGCNA). Cancers (Basel) 12(1):37

Oates J, Lopez D (2018) Pharmacogenetics: an important part of drug development with a focus on its application. Int J Biomed Investig 1:1–16

Pardridge WM (2012) Drug transport across the blood–brain barrier. J Cereb Blood Flow Metab 32:1959

Park KM, Kim SE, Lee BI (2019) Antiepileptic drug therapy in patients with drug-resistant epilepsy. J Epilepsy Res 9:14

Peng M, Mo Y, Wang Y et al (2019) Neoantigen vaccine: an emerging tumor immunotherapy. Mol Cancer 18:128. https://doi.org/10.1186/S12943-019-1055-6

Pereira R, Oliveira J, Sousa M (2020) Bioinformatics and computational tools for next-generation sequencing analysis in clinical genetics. J Clin Med 9:132. https://doi.org/10.3390/JCM9010132

Pinu FR, Beale DJ, Paten AM, Kouremenos K, Swarup S, Schirra HJ, Wishart D (2019) Systems biology and multi-omics integration: viewpoints from the metabolomics research community. Metabolites 9:76. https://doi.org/10.3390/METABO9040076

Proietti I, Skroza N, Michelini S et al (2020) BRAF inhibitors: molecular targeting and immunomodulatory actions. Cancers (Basel) 12:1–13

Qin D (2019) Next-generation sequencing and its clinical application. Cancer Biol Med 16:4

Raghupathi W, Raghupathi V (2014) Big data analytics in healthcare: promise and potential. Health Inf Sci Syst 2:3. https://doi.org/10.1186/2047-2501-2-3

Rai S, Raj U, Varadwaj PK (2018) Systems biology: a powerful tool for drug development. Curr Top Med Chem 18:1745–1754

Regel I, Mayerle J, Mukund Mahajan U (2020) Current strategies and future perspectives for precision medicine in pancreatic cancer. Cancers (Basel) 12:1024. https://doi.org/10.3390/CANCERS12041024

Santourlidis S, Schulz WA, Araúzo-Bravo MJ, Gerovska D, Ott P, Bendhack ML, Hassan M, Erichsen L (2022) Epigenetics in the diagnosis and therapy of malignant melanoma. Int J Mol Sci 23:23

Sarmast ST, Abdullahi AM, Jahan N (2020) Current classification of seizures and epilepsies: scope, limitations and recommendations for future action. Cureus 12:e10549. https://doi.org/10.7759/CUREUS.10549

Shao LR, Habela CW, Stafstrom CE (2019) Pediatric epilepsy mechanisms: expanding the paradigm of excitation/inhibition imbalance. Children (Basel) 6(2):23

Stafstrom CE, Carmant L (2015) Seizures and epilepsy: an overview for neuroscientists. Cold Spring Harb Perspect Med 5:1–19

Sun F, Cao BQ, Wang B, Wu SQ, Jiang DH (2016) Association between ABCB1 genetic polymorphism and the effect on epilepsy following phenytoin treatment. Exp Ther Med 12:1780–1784

Tang F, Hartz AMS, Bauer B (2017) Drug-resistant epilepsy: multiple hypotheses, few answers. Front Neurol 8:301

Thomas MC, Brownlee M, Susztak K, Sharma K, Jandeleit-Dahm KAM, Zoungas S, Rossing P, Groop PH, Cooper ME (2015) Diabetic kidney disease. Nat Rev Dis Prim 11(1):1–20

Valenti F, Falcone I, Ungania S et al (2021) Precision medicine and melanoma: multi-omics approaches to monitoring the immunotherapy response. Int J Mol Sci 22:22

Van Zijl F, Krupitza G, Mikulits W (2011) Initial steps of metastasis: cell invasion and endothelial transmigration. Mutat Res 728:23

Vanni I, Casula M, Pastorino L et al (2020) Quality assessment of a clinical next-generation sequencing melanoma panel within the Italian melanoma intergroup (IMI). Diagn Pathol 15:1–18

Wang S, Zhang H, Liu Z, Liu Y (2022) A novel deep learning method to predict lung cancer long-term survival with biological knowledge incorporated gene expression images and clinical data. Front Genet 13:800853. https://doi.org/10.3389/FGENE.2022.800853/FULL

Wei X, Yohannan S, Richards JR (2022) Physiology, cardiac repolarization dispersion and reserve. StatPearls

Weinstein JN, Collisson EA, Mills GB et al (2013) The cancer genome atlas pan-cancer analysis project. Nat Genet 4510(45):1113–1120

Xu J, Zhao W, Liao K, Tu L, Jiang X, Dai H, Yu Y, Xiong Q, Xiong Z (2021) Clinical retrospective study on the expression of the PD-L1 molecule in sporadic colorectal cancer and its correlation with K-ras gene mutations in Chinese patients. Am J Transl Res 13:6142

Yu X, Wang Y, Qiu H, Song H, Feng D, Jiang Y, Deng S, Meng H, Geng J (2018) AEG-1 contributes to metastasis in hypoxia-related ovarian cancer by modulating the HIF-1alpha/NF-kappaB/VEGF pathway. Biomed Res Int 2018:3145689. https://doi.org/10.1155/2018/3145689

Yu C, Liu X, Yang J, Zhang M, Jin H, Ma X, Shi H (2019) Combination of immunotherapy with targeted therapy: theory and practice in metastatic melanoma. Front Immunol 10:990. https://doi.org/10.3389/FIMMU.2019.00990

Zhou L, Mudianto T, Ma X, Riley R, Uppaluri R (2020) Targeting EZH2 enhances antigen presentation, antitumor immunity and circumvents anti-PD-1 resistance in head and neck cancer. Clin Cancer Res 26:290

Systems Biology Approaches for Autoimmune Diseases

12

Sonali S. Shinde, Gresi D. Mate, Sneha S. Mali, Neha D. Naikwadi, Mohammad Abuzar, and Aniket P. Sarkate

Abstract

Autoimmune diseases are complicated conditions defined by immune system dysregulation, which leads to abnormal immunological reactions to self-antigens. A systems biology approach is needed to fully understand the underlying processes and interactions between the many immune system components. The importance of each of the four genomics disciplines, namely, genomics, transcriptomics, proteomics, and metabolomics provided a new path to the study of autoimmune diseases. The integration of data from different omics disciplines is a necessary step toward developing a thorough understanding of autoimmune diseases. New developments in autoimmune disease systems biology methods such as the use of single-cell technologies, the function of systems biology in individualized treatment for autoimmune disorders, the use of machine learning techniques, the blending of multi-omics data, and the use of computational/in silico modeling made it possible to modify the treatment possibilities. This review describes systems biology and provides the function and significance of systems biology in the understanding of autoimmune diseases.

Keywords

Autoimmune disease · Systems biology · Multi-omics · Biomarkers · Network biology

S. S. Shinde · A. P. Sarkate (✉)
Department of Chemical Technology, Dr. Babasaheb Ambedkar Marathwada University, Aurangabad, Maharashtra, India

G. D. Mate · S. S. Mali · N. D. Naikwadi · M. Abuzar
Department of Pharmaceutical Quality Assurance, Dr. D. Y. Patil Institute of Pharmaceutical Sciences and Research, Pimpri, Pune, Maharashtra, India

S. Joshi et al. (eds.), *Systems Biology Approaches: Prevention, Diagnosis, and Understanding Mechanisms of Complex Diseases*,
https://doi.org/10.1007/978-981-99-9462-5_12

12.1 Introduction

Autoimmune diseases (AIDs) are the third most prevalent disease group after cancer and heart disease, affecting around 58% of the entire population (Whitacre 2001). Autoimmune illnesses encompass a wide spectrum of diseases from organ-specific (e.g., Hashimoto's thyroiditis) to systemic (multi-organ) involvement (Theofilopoulos et al. 2017).

As a result, the worldwide burden of chronic diseases is enormous. Autoimmune disorders (AIDs) are caused by an intricate interaction of genetic and epigenetic variables, as well as an abnormal immune response to endogenous or foreign antigens (Goldblatt and O'Neill 2013). In a genetically vulnerable host, the beginning of autoimmune responses (the trigger stage) is brought on by a confluence of environmental triggers and epigenetic changes. An immune-mediated onslaught against either exogenous or endogenous self-antigens initiates the autoimmune response. The period of exposure to antigen, the effects of epigenetic alterations or sensitive genes, and the function of effectors may all have an impact on the clinical stage of AIDs. The US National Institutes for Allergy and Infectious Diseases estimates that this group of illnesses causes an annual cost to the country's healthcare system of more than $100 billion, even after accounting for the cost of those without health insurance, which is estimated to be at least an additional $25 billion. A significant number of autoimmune conditions are linked to eight diseases: rheumatoid arthritis (RA), type I diabetes, multiple sclerosis, psoriasis, systemic lupus erythematosus (SLE), ulcerative colitis, Crohn's disease, and scleroderma (Blumberg et al. 2012).

Systems biology is a method for analyzing complicated biological systems by quantifying multiple biomolecules on a broad scale (Dhillon et al. 2020). Omics technologies enable such measurements to be made in an impartial, high-throughput, and cost-effective manner (Pulendran et al. 2010). Using biological networks to combine multi-omics data to identify unique biological insights is a valuable strategy (Tavassoly et al. 2018). It is suggested that immunological research employ a multi-omics approach in order to comprehend the intricacy of the immune system. This review offers a comprehensive summary of the methodological approaches that are currently available for the various omics data layers—genomics, transcriptomics, proteomics, and metabolomics—that are pertinent to immunological research (Mardinoglu et al. 2018) (Fig. 12.1).

Systems biology studies have also revealed mechanisms driving prognosis for diseases and the estimation of diagnostic biomarkers (Schreiber et al. 2011). In each stage of the disease, systems biology has involved expression profiles and functional evaluation of DNA, RNA, protein, and metabolite-specific cell subsets (Kitteringham et al. 2009). Inherited polymorphisms (or DNA sequence variants) have a substantial influence on the development of autoimmunity. The enormous variety in genetic variations, on the other hand, leads to susceptibility to AIDs (Trowsdale 2011). From a clinical perspective, biomarkers can be divided into many categories depending on how they are used, including predictive, diagnostic,

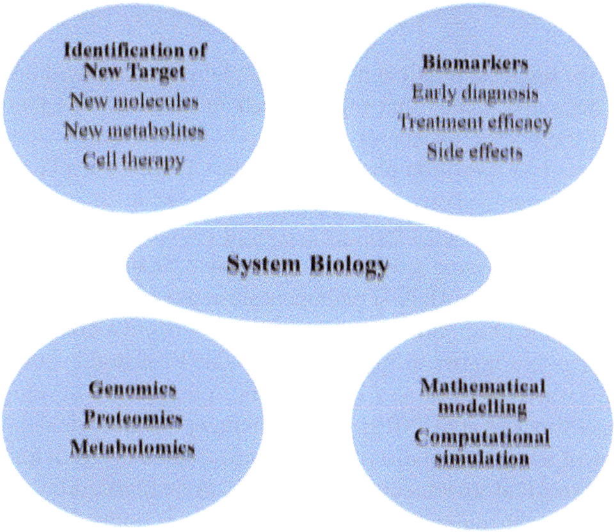

Fig. 12.1 Components of systems biology

prognostic, mechanistic, pharmacodynamic, safety, and surrogate endpoint biomarkers (Tektonidou and Ward 2010). Further assisting in the identification of various valuable biomarkers for systemic AIDs, the utilization of multi-omics platforms and multiplexed techniques will also highlight possibilities for future development in translational research.

The development of new biomarkers and recent discoveries into the etiology of complicated diseases have been made possible by using microarray technology for blood transcriptome research (Ghosh and Poisson 2009). This technology allows for the simultaneous analysis of thousands of genes and their products, making it possible to identify potential genetic markers for the development of diseases. Successful biomarker translation into the clinic will lead to the introduction of personalized medications that benefit AIDs patients. More standardized multi-omics methodologies are required, as well as network analysis for massive data mining, to uncover the most predictive biomarkers across DNA, RNA, protein, phenotype, and metabolites (Steensberg and Simons 2015).

12.2 Systems Biology Approaches for Autoimmune

Systems biology strategies are frequently employed as autoimmune diseases are caused by an intricate combination of epigenetic and genetic variables, along with a distorted immune reaction to endogenous or foreign antigens. The utilization of multi-omics methods, including genomics, transcriptomics, proteomics, and metabolomics, not only allows for the discovery of a variety of biomarkers but also opens up new avenues for future application in translational AIDs (Kim et al. 2014).

12.2.1 Genomics

Genomics, in the broadest possible meaning, explains the codification of an organism through the set of genes that represent chemical structures that function as actual parts of the body (Johnson et al. 2005). With the Human Genome Project, biology has come to a transition phase over the last 15 years, explaining to us how biological data is organized on a global scale (Auffray et al. 2003). The great development in the study of systems biology has been caused by the concurrent rise of genomics, which is the area oriented to a massive analysis of genetic features (Conesa and Mortazavi 2014). AID pathogenesis is thought to be complex. Several studies, including the observation that the levels of concord among numerous autoimmune disorders are higher in monozygotic twins than in dizygotic twins, have proven genetic variables to be key contributors to the risk of developing autoimmune diseases (Simakou et al. 2019). Furthermore, epidemiological data reveal that the prevalence of autoimmune diseases varies geographically, providing support to this concept. Recent advancements in genomics have expanded our understanding of disease's biological foundations. Microarrays, for example, have enabled unbiased exploratory investigations, such as genome-wide association studies to be conducted to better understand the relationship of genetic variables to disease. These developments have increased our understanding of the genetic risk factors for AIDs. However, the inheritance of risk to AIDs is complicated due to genetic polymorphisms, differing penetrance of genes in males and females, and environmental variables (Amur et al. 2012).

12.2.2 Transcriptomics

Transcriptomics is the study of RNA, a single-stranded nucleic acid that was not distinguished from the DNA world until Francis Crick formulated the fundamental dogma in 1958. The transcriptome is the set of RNAs that the genome produces from a certain type of tissue or cell at a particular developmental stage and/or under specific physiological conditions Dong & Chen (2013). After the genome has been organized, transcriptome analysis allows us to understand how the genome is expressed at the transcription stage, providing details on the structure of genes, the control of gene expression, the function of gene products, and the dynamics of the genome. Transcriptome analysis will eventually provide some assistance in disease diagnosis, clinical therapy, and enhancement of crops. It will further disclose the regulatory network of biological processes (Jacquier 2009).

The identification of pathogenic markers in human autoimmunity and autoinflammation has been made possible by transcriptomics, the high-throughput analysis of RNAs (Banchereau et al. 2017). In the clinical management of autoimmune patients, quantifying autoantibodies enables disease diagnosis, monitoring, and even prognosis. Transcriptomics datasets were examined to reveal the autoantigens messenger RNA expression in 32 tissues and organs. According to this analysis, autoantigens belong to one of three categories depending on where they are expressed: in the tissue that is most severely impacted by the disease (Group I), generally throughout the body with predominance in immunological tissues (Group II), as well as other types of tissue (Group III), which are not often associated with clinical findings (Burbelo et al. 2016).

12.2.3 Proteomics

Proteomics studies give information that may be challenging or impossible to collect with alternative approaches about alterations to gene expression, protein localization, protein-protein interactions, or post-translational modifications, any of which may influence the function of proteins and consequently lead to disease (Song and Lin 2017). Proteomics is necessary for early disease detection, prognosis, and disease progression monitoring. It also plays an important role in medication development as a target compound (Aslam et al. 2017). Proteins are biologically important effectors, and the levels are affected by host translational control and regulation in addition to the related mRNA levels. Proteomics would thereafter be viewed as the most relevant dataset for describing a biological system (Cox and Mann 2007).

Proteomics frequently makes use of two-dimensional gel electrophoresis to separate complex protein mixtures, such as cell lysates derived from primary cell cultures, cell lines, or tissue. Individual proteins are visualized via the analysis of Western blots or silver staining on an antibody or autoimmune serum sample. This strategy lends itself well to the discovery of novel reactivities, including the identification of novel autoantigens by autoimmune disease patient serum (Robinson et al. 2002). Transcriptomics is only one aspect of proteomics; structural and post-translational alterations are also included. The pathogenesis and specificity of autoimmune reactions can be studied with the help of protein arrays and other multiplex screening technologies. Potentially, each patient's antigenic fingerprint may be located. This could be a key step in identifying the main pathogenetic antigen, revealing that several proteins or peptides that have undergone intrinsic or environmental modification are capable of triggering specific autoimmune illnesses in addition to a person's genetic makeup (Koepenik et al. 2004).

The etiology of the majority of autoimmune disorders is significantly affected by autoantigens or autoantibodies. Crucial disease markers for diagnosis, prognosis, and treatment include some autoantigens or autoantibodies. The development of proteomics technology has given researchers the perfect tools for studying the autoantigens or autoantibodies system in autoimmune disorders (Xiang and Kato 2006). Proteomic technologies have the potential to revolutionize clinical treatment by offering tools for the discovery of biomarkers for assessment, prognosis of disease progression, guiding therapeutic choice, and monitoring therapeutic response. Numerous proteomics techniques are used to study autoimmune diseases, such as (1) two-dimensional gel electrophoresis (2-DE) and mass spectrometry for the discovery of autoantigens and biomarkers; (2) autoantigen microarrays to characterize autoantibody responses; (3) antibody array techniques to profile cytokines and other biomolecules; (4) reverse phase protein array studies to analyze phosphoproteins; and (5) flow cytometric analysis (Hueber and Robinson 2006).

12.2.4 Metabolomics

In order to gain insights into various facets of cellular physiology, metabolomics creates a small molecule profile resulting from cellular metabolism and may represent the results of intricate networks of biochemical events (X. Liu and Locasale 2017). To define the endpoints of physiology and pathology, metabolomics examines the activity and status of cellular as well as organismal metabolism on a global or network scale (Cajka and Fiehn 2016). A metabolomics study provides critical information on the fundamental biological status of the system under study and directly represents the functioning of the metabolic network that generates these metabolites. Within systems biology, functional genomics, and biomarker discovery, metabolomics has a unique place (Fell and Cornish-Bowden 1997). Though it had been predicted that alterations in the levels of individual enzymes would have minimal effect on metabolic fluxes, the theory of metabolic control analysis and empirical data demonstrate that these changes can and do have a major impact on the concentrations of a range of individual metabolites (Kell 2004). The amount of transcripts, proteins, and metabolites found in humans shows that metabolomics is "the way forward" (Goodacre 2018).

Metabolomics can play an important role in pharmacogenetics, the study of individual differences in drug metabolism and toxicity. Direct genotyping for identified variant alleles of genes encoding drug-metabolizing transporters and enzymes has historically been used in clinical pharmacology to study genetic differences in drug metabolism. Metabolomics can be used to conduct a more comprehensive examination of variation between individuals in metabolism and toxicity susceptibility. A proof-of-concept study in an animal model demonstrated the utility of metabolomics analysis in predicting medication reactions (Idle and Gonzalez 2007).

Metabolomics is crucial in agriculture, environmental science, illness, and health systems biology. Small molecule biochemistry has been forced into the shadows by genomes, transcriptomics, and proteomics over the last two decades (Clayton et al. 2006).

The two types of general analytical methodologies for accomplishing a metabolomic study are targeted and untargeted. The detection and precise quantification (in nM or mg/mL) of a narrow set of known molecules is referred to as targeted metabolomics. It has been driven by a specific biochemical query or hypothesis, and the set of metabolites associated with one or more pathways is previously identified. One disadvantage of the focused strategy is that the compounds of interest must be known in advance and be available in pure form. Instead, the untargeted approach, also known as metabolite fingerprinting, is utilized for total metabolome comparison, in which as many metabolites as possible are measured and compared between samples (Cambiaghi et al. 2017).

12.3 Network Biology/Network Analysis

Network science, often known as systems biology, is an interdisciplinary study that studies biological systems by investigating and simulating the relationships and interactions between their constituent parts. Networks can be used to represent biological connections at many different levels of detail, including interactions between atoms in the structure of a folded protein and the relationships between animals in a population or ecosystem (X. Liu and Locasale 2017). Every node in the molecular interaction network represents a metabolite, gene, or gene product, and every edge connecting the nodes reflects a functional association or a direct physical connection (Alm and Arkin 2003).

Unexpected non-random global organizational patterns have been discovered in studies of large networks as different as the World Wide Web and the scientific co-authorship network (Arita and Costa 2011). Similar topological characteristics have lately been revealed for biological networks, particularly metabolism. The study of large-scale characteristics is best suited to metabolic networks since they provide the most complete picture of any molecular network, at least for a few model organisms (Alm and Arkin 2003).

Metabolites, enzymes, proteins, and genes are the vertices of biological networks, which are connected based on their ability to interact. Power law connection distributions, small-world characteristics, the central concept and community structures, and other fundamental characteristics of biological networks have all been identified using complex network theory. The key ideas for comprehending biological systems, such as emergence, robustness, and modularity, are also components of the complex networks theory (Barabási and Oltvai 2004) (Fig. 12.2).

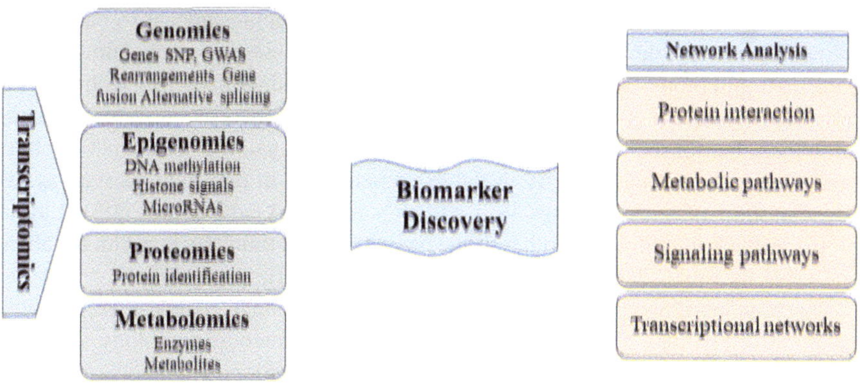

Fig. 12.2 Systems biology approaches for AIDs in combination with multi-OMICs and network analysis

12.4 Recent Advancements in Systems Biology Approaches for Autoimmune Diseases

Systems biology is a wide branch that links molecular components across various biological scales to physiological functions and organismal traits by using models based on computation, high-throughput methods of experimentation, and quantitative analysis. To interpret the data flow between genes, proteins, and other subcellular components of signaling, regulatory, and functional pathways to govern cell, tissue, organ, and organismal level activities, systems biology employs an extended range of quantitative experimental and computational approaches (Tavassoly et al. 2018).

12.4.1 Single-Cell Technologies

The fundamentals for advancing treatment to a greater step of individualized and precise care are systems-level understanding. At the moment, systems biology methods are used in both drug development and clinical practice. Pharmacology has evolved into a systems science because of systems biology. Pharmacogenomics is a component of systems pharmacology, which is demonstrated to be helpful for drug development and for expecting potential responses and pharmacological side effects. Quantitative systems pharmacology (QSP) uses network and dynamical models linked with pharmacodynamics and pharmacokinetics to determine the most effective therapies for particular patients with a certain ailment, in contrast to pharmacogenomics, which analyzes genomic data of drug-metabolizing enzymes for estimation and determination of drug action and effects (Hansen et al. 2011).

These developments made it possible to modify medication regimens and dosages for specific individuals depending on their molecular markers (Zhao et al. 2013).

Systems biology-based precision medicine methodologies appear to be a possibility, thanks to the development of biosensors that can assemble patient time-course data, do liquid biopsies, and identify biomarkers. These data-gathering technologies provide the core components for prediction models that use systems biology methodologies. Biosensors are devices that can monitor physiological signals such as heart rate and electrical activity in the brain and heart in real time, as well as the concentration of various drugs in a patient's blood (Asada et al. 2003). Liquid biopsy gathers tumor cells or other tissue constituents in fluids such as blood, saliva, and urine for omics data and biomarker identification (Turner 2000).

Machine learning can help develop high-resolution image analysis, which needs quantitative estimation of a large number of pictures, such as pathology slides, to give an effective diagnostic tool for illness status and development models (Crowley et al. 2013).

12.4.2 Role of Systems Biology in Personalized Medicine for Autoimmune Diseases

Personalized medicine is a large and expanding discipline that is notified by a patient's unique data and biomarker profile (Ginsburg and Willard 2009). Although autoimmune disorders are complicated, common genetic variables and pathways tend to demonstrate disease burden, a concept known as an autoimmune tautology (Anaya et al. 2012). There are both optimistic and pessimistic outlooks on disease approaches directed toward the development of personalized medicine models. The first proposes that the greater the amount of data available, the closer research will be to the creation of highly efficient treatments, therapeutic targets, and the understanding of the molecular causes of disease. Additionally, proper medications along with lifestyle modifications, are sufficient to minimize disease burden with no requirement for expensive genetic investigations and pharmaceuticals. Even though environmental factors are a major prevention approach for expensive autoimmune disorders, they cannot be fully controlled or healed. The following is a concise overview of the recent state of personalized medications in the treatment of two major autoimmune diseases (Anaya et al. 2016; Coote and Joyner 2015; Zhang et al. 2015).

12.4.3 Machine Learning Algorithms

One of the biggest challenges associated with systems biology is converting large amounts of data at various scales into useful information. This information is critical for improving approaches for studying biological systems, precisely understanding and diagnosing diseases at various stages, and designing innovative therapy modalities centered on the person. Predictive models can be created by combining biomedical data from hypothesis-based experiments with computational modelings, such as dynamic models that rely on the causality of links between components. Such simulations frequently shed light on mechanisms. The breadth of biology and medical knowledge is increasingly outstripping our existing ability to develop large-scale mechanistic models. Statistical models based on correlation are another method for generating predictive models from vast amounts of information. This approach can benefit from artificial intelligence (AI), which employs statistical reasoning to uncover previously unknown correlations, co-occurrences, and dependencies in broad-scale datasets (Baker et al. 2018).

Building expert systems based on machines that can evaluate data and predict future developments in computer science is also called artificial intelligence (AI). Machine learning, deep learning, and artificial neural networks are a few examples of AI techniques. Artificial neural networks can learn certain task-oriented classifications modeled after real brain neural networks when tested with a training set (Baxt 1995). Machine learning applies to a class of approaches that analyze large datasets and make predictions depending on them. Three approaches to using machine learning are supervised, unsupervised, and semi-supervised. Training datasets in the form of labeled input or output relations are provided in supervised machine learning, and a function is deduced that may be used to examine recent examples and anticipate the result based on input data and categorization. When there are no labels on the data and the objective is to identify underlying patterns without utilizing a manual, unsupervised machine learning is utilized. When there is a lack of labeled data, semi-supervised learning fills the gap between supervised and unsupervised learning (Deo 2015; Leung et al. 2016; Sommer and Gerlich 2013). Deep learning is a type of machine learning that uses multiple layers of computational processing to characterize data and identify inherent patterns in massive amounts of data (Janowczyk and Madabhushi 2016; LeCun et al. 2015).

AI is an effective tool for developing and improving models. AI employs a dataset and an algorithm. The dataset frequently consists of two components: a significant number of measurements (such as molecular markers) and the resulting prediction (for instance, the phenotype). A test set is used to evaluate the performance of a machine learning model after it has been trained on a dataset. The test set is separate from the training set and is used to assess the model's ability to generalize to new data. The algorithm used in machine learning is typically a statistical model that modifies itself to produce better predictions (Leung et al. 2016; Sommer and Gerlich 2013; Camacho et al. 2018). Metabolomics amplifies changes in the proteome and represents more closely the phenotype of an organism. Recent advances enable the production (and computer-readable encoding as SBML) of metabolic network models reconstructed from genome sequences, as well as experimental measurements of much of the metabolome. There is increasing convergence between the number of human metabolites estimated via genomics (~3000) and the number measured experimentally (Kell 2006).

Due to the size of the Numerical datasets, they work with machine learning methodologies that are also used in looking into the parameter spaces of dynamical models and performing sensitivity analysis (Moles et al. 2003). As they can identify genetic markers connected to therapeutic interventions and treatments, prediction models developed through AI and machine learning have the potential to be extremely effective tools in precision medicine. Large-scale clinical and biomolecular datasets are employed as training sets with predetermined responses. These models' predictions are put to the test, and those that pass are used to add additional data to the training set. For instance, molecular biomarkers and pathology-based cancer outcomes have been accurately predicted using AI (Catto et al. 2003).

12.4.4 Integration of Multi-Omics Data

The enhanced characterization of biological processes across molecular levels is promised by multi-omics investigations. Combining the data from transcriptomics and genome investigations can provide a wealth of data and new theories (Musunuru et al. 2017). Therefore, it is anticipated that the integration of data produced by several platforms, such as transcriptomics, genomes, and proteomics, will increase our understanding of the mechanisms behind multiple sclerosis. Concluded that the genes differently expressed by lymphocytes in multiple sclerosis, insulin-dependent diabetes, and rheumatoid arthritis are grouped within chromosomal areas in the genome. Surprisingly, they discovered that genetic linkage studies could map the chromosomal regions containing the genes differently expressed in autoimmune disorders to disease vulnerability linked to those diseases (Nazliel et al. 2002).

12.4.5 Computational/In Silico Modeling

Systems biology is a technique that combines experimental and computational study to better understand complex biological systems. Through logical modeling and theoretical investigation, computational biology offers a strong foundation from which to approach important scientific topics. Computational or in silico modeling approaches have acquired greater significance in the realm of systems biology for understanding the complex dynamics of autoimmune diseases. These approaches utilize mathematical and computational techniques to simulate and analyze the behavior of biological systems, enabling researchers to gain insights into disease mechanisms, identify key molecular interactions, and explore potential therapeutic interventions (Kitano 2002). Systems biology modeling techniques may be roughly categorized as qualitative and mechanistic. Models of chosen biological processes that match available data are referred to be models that are qualitative or guided by hypotheses. They are often logical or statistical in nature. The assignment of Boolean parameters to specific links has been used mostly to examine gene networks (Lee et al. 2003). In the biomedical field, computer-aided or in silico design is utilized to improve absorption, distribution, metabolism, excretion, and toxicity profiles, speed and simplify hit detection and hit-to-lead selection, and lower safety hazards. Ligand-based drug design (pharmacophore, a 3D spatial arrangement of chemical characteristics crucial for biological action), structure-based drug design, and quantitative structure-activity and quantitative structure-property connections are all common computational methodologies. Computational technologies are becoming an essential part of drug discovery and development, and they are thought to offer ways for the industry to increase efficiency (Kapetanovic 2008). Genome-scale metabolic models have proved helpful since they rely on accurate reconstructions of metabolic networks. These reconstructions are primarily based on a sequenced genome (Papp et al. 2011). To compile and analyze quantitative data, computational models formally codify a complicated biological or scientific process. Therefore, modeling is essential for disciplines like proteomics, genomics, and systems biology. The goal of proteomics research is to elucidate additional biological information inside networks, including novel proteins, interactions, and complexes. However, the level of detail for computer models of such networks must be limited by the accessibility of supporting quantitative data. The fact that these decisions are programmed into mathematical formulas of computer model and it is a benefit of computationally studying proteome networks. Different types of models, like network definition models, network function models, and network measurement models, are utilized in computer modeling for proteome analysis. By training models on measurements of the upstream signaling network, cellular function predictions can also be made more quantitatively (Janes and Lauffenburger 2006). A dynamic link between information from several fields is created by an *Escherichia coli* single-cell model to simulate the growth of a single bacterial cell, such as biochemistry, microbiology, physiology, and genetics, and has been used as a base for further development to solve additional issues. So, if systems biology is to be able to forecast and create dynamic behavior in biologically complex systems, it must apply mathematical structures and methodologies. These build virtual systems by combining actual biological information collected from the biochemical characteristics of gene products and known biological interactions into a set of equations that describe the system to be simulated (Lee et al. 2003). Making a system or design as efficient or useful as feasible is the goal of optimization. Systems biology optimization is categorized by the kind of optimization challenge. Programming techniques include linear programming (LP), nonlinear programming (NLP), semidefinite programming (SDP), bilevel optimization (BLO), mixed integer linear programming (MILP), and mixed integer nonlinear programming (MINLP). The ideal flux distributions are established through linear optimization and used to illustrate the metabolic phenotype under certain circumstances. Optimization has also been the driving force behind metabolic flux balance studies. This flux balancing technique offers a tool for optimizing bioprocesses as well as a road map for metabolic engineering. Success stories include the in silico forecasts of *E. coli*'s metabolic capacity and the genome-scale reconstruction of the *Saccharomyces cerevisiae* metabolic network (Banga 2008).

12.5 Application of Systems Biology Approaches for Autoimmune Diseases

12.5.1 Rheumatoid Arthritis

Millions of people worldwide suffer from the autoimmune disease rheumatoid arthritis (RA). The disease can be understood on a molecular level as an enhanced immune response, involving interactions between different cell types, intracellular messengers like cytokines, and systemic implications like cell recruitment to the inflamed location. Systems biology can make a difference in this situation by creating models that make it possible to place data in the pertinent human, multicellular context (Rullmann et al. 2005). It is a persistent, systemic autoimmune illness that impacts 0.5–1% of people worldwide and is characterized by inflammatory polyarthritis with no known cause. The clinical characteristics and molecular markers used to subtype RA patients have been discovered over time (Najafzadeh et al. 2021). It is discovered that the presence or absence of anti-citrullinated protein antibodies is associated with specific RA risk profiles. For managing RA, a systems biology approach has been recommended. To identify specific patient groups with rheumatoid arthritis, a systems biology approach is investigated, paving the way for the advancement of more individualized therapy. Systems biology examines the numerous interactions involving a large number of genes, proteins, and metabolites at once to better understand biological systems (Wang et al. 2011). RA genomics in the inflammatory processes associated with RA, CD+T cells are crucial. According to some theories, RA patients' immunodeficiency as well as their autoimmune response may be caused by a change in CD+T cell homeostasis (Smigiel et al. 2014). A significant number of metabolites are typically measured as a result of metabolomics investigations. The functional gene annotation tool of DAVID was used to identify functionally linked genes to better understand regarding biology behind RA and the subgroups RA Heat and RA Cold. According to CM theory, patients with rheumatoid arthritis might be split into two groups. It was discovered that there are molecular variations between the RA Cold and RA Heat groups, which supports the subtyping. Molecular profiling of genes and metabolites has shed light on the connections between many indicators, offered insight into the mechanism underlying RA, and served as the first building blocks for a biological explanation of the CM groups RA Cold and RA Heat (Van Wietmarschen et al. 2009). Using bioinformatics techniques, we analyzed and compared the transcriptional patterns of RA, OA (osteoarthritis), and ND (normal donors) patients in order to identify probable molecular networks and crucial genes that are involved in the development of RA and OA. Initial classification of the total transcriptional patterns was done using hierarchical clustering. We discovered differentially expressed genes (DEGs) in patients with ND, RA, and OA. Additionally, functional modules were retrieved, PPI networks were built, and functional annotation was used (Liu et al. 2015). The study of autoimmunity using systems biology approaches is made possible by a thorough research plan that is provided. The utilization of cutting-edge functional and comparative genomics approaches is presented as part of an integrative scientific framework for identifying the molecular causes of complicated disorders. To understand the characteristics of autoimmune rheumatic illnesses, a continual exchange of hypothesis-driven and data-driven research is added. Today, many research studies get RNA and protein microarray profiles instead of focusing on individual genes and proteins, yielding readouts known as gene signatures instead of DNA and/or protein markers. Many tools have been created recently in the field of transcriptomics to analyze the RNA expression patterns of whole genomes (Wolf 2013). The most widely used system is the microarray-chip technology created and offered by Affymetrix. However, quantitative RT-PCR methods (TaqMan and light cycler procedures) cannot be used with this platform to accurately identify low abundantly expressed genes. Analysis of protein abundances, recognition of proteins and post-translational modifications, the study of protein-protein interactions, and amino acid polymorphisms are the main topics of proteomics (Julka and Regnier 2004). Using mass spectrometric techniques, quantitative proteomics can be carried out, for instance, using stable isotope labeling Stable isotope labeling with amino acids in cell culture (SILAC) or isotope-coded affinity tagging (ICAT). Since metabolites are by-products of cellular regulatory mechanisms, their concentrations can be thought of as the biological systems' final response to alterations in genetic makeup and/or environment. Similar to the terms "transcriptome" and "proteome," the term "metabolome" refers to the collection of metabolites produced by a biological system (Ryan and Robards 2006). To manage and preserve data obtained from international screening and monitoring initiatives, an integrated bioinformatic IT platform is required. Functional genomics data analysis includes the following steps: pre-processing of the data, detection of differentially expressed genes, proteins, and metabolites, supervised and unsupervised learning of patterns associated with predefined states, and reverse engineering of interaction networks (Glocker and Guthke 2005). A cell's RNA molecules are dynamic, functional components whose by-products can recognize, engage with, and alter other transcripts. In addition to acting as passive templates from which genes are transcribed, RNA molecules are active, functional components of the cell whose products can detect, interact with, and modify other transcripts (Jarvis and Frank 2010).

12.5.2 Multiple Sclerosis

Multiple sclerosis (MS) is known as a neuroinflammatory disease characterized by autoimmune-mediated inflammation that damages myelin and results in an axonal loss in the central nervous system. Genomics uses instruments to collect genetic data and bioinformatics approaches to evaluate these often massive datasets. In the case of MS, genetics has revealed the key to a better understanding of illness susceptibility, immunological and neurodegenerative systems that cause the illness, treatment goals, and possible biomarkers. Furthermore, unique bioinformatic methodologies and computer analytic pipelines are required for each methodology to overcome inherent technical and biological constraints, which have been explored in length elsewhere (for both genomic and proteomic methods) (Hollenbach and Oksenberg 2015; Huizar et al. 2020). Regarding both disease development and the treatment response to DMTs (disease-modifying therapies), biomarkers may be useful in the therapeutic management of patients with MS. The choice-making process of doctors caring for MS patients would be greatly impacted by the discovery of a disease progression biomarker that is predictive. A patient with CIS (clinically isolated syndrome), for instance, may not consider current DMTs to be a top priority if they are anticipated to make slow progress over the coming years. However, if a biomarker correctly estimates that a patient will probably acquire clinically confirmed MS within months, it may be best to begin a first-line DMT regimen as soon as possible. Recent research, in which soon after diagnosis $CD4^+$ T lymphocytes from CIS patients were subjected to transcriptional profiling and statistical analysis (Corvol et al. 2008), lends credence to this hypothetical situation. Transcriptional profiling has been used by several groups to try and find biomarkers of therapy response. Our teams discovered in 2005 that specific blood transcript combinations were very good indicators of how patients would respond to interferon (IFN)-beta therapy. Comabella and colleagues then showed that since changes to the type I IFN signaling pathway in monocytes are associated with failure of response to IFN-beta treatment, type I IFN-regulated genes may be used as response markers. The same researchers also discovered a natural killer cell subset (CD56bright) that was associated with an IFN-beta response (Martínez-rodríguez et al. 2011). The dynamics of the underlying pathogenetic events must be combined with static data to produce an accurate understanding of the physiological importance of MS biomarkers (Villoslada and Oksenberg 2006). MS is known for its relapsing-remitting dynamics. Although there has been substantial advancement in our knowledge of the cellular and molecular mechanisms behind the pathophysiology of autoimmune disorders, it is still unclear how these mechanisms' activation results in these fundamental dynamics. Mendizábal and colleagues used this biological knowledge to build a computational model that sheds light on the mechanisms underlying these relapsing-remitting dynamics. They proposed that the relapsing dynamics in autoimmune disease can result from a breakdown in the mechanisms regulating the cross-regulation between regulatory and effector T cells with the interaction of stochastic events (such as a breakdown in central tolerance or immune system activation by pathogens). They altered the model specifically to simulate the way immunotherapies affect populations of effector and regulatory cells (Vélez de Mendizábal et al. 2011).

12.6 Future Directions and Challenges

Data Integration from Multi-Omics: The integration of several layers of omics data, such as genomes, transcriptomics, proteomics, and metabolomics, will help systems biology techniques to continue to improve. A more thorough knowledge of the molecular pathways behind autoimmune disorders will be possible because of the integration of these databases (Pinu et al. 2019).

Single-Cell Analysis: In order to understand the variability among the cell populations implicated in autoimmune disorders, single-cell technologies will be essential. Researchers can discover unusual cell subgroups, examine cell-to-cell variability, and identify dynamic cellular interactions that contribute to disease pathogenesis by profiling individual cells. Network analysis and modeling are key components of systems biology techniques that help to understand complicated interactions between genes, proteins, and biological processes (Corvol et al. 2008).

Integration of Clinical Data: Our knowledge of the interactions between genetic, environmental, and clinical variables in autoimmune disorders will be improved by combining clinical data with molecular data, such as patient demographics, disease phenotypes, and treatment responses (Chawla and Davis 2013). This integration will enable personalized medicine approaches and the identification of biomarkers for disease diagnosis, prognosis, and treatment response.

Functional Validation: Functional validation is essential as systems biology techniques provide vast volumes of data and hypotheses. In vivo, models and CRISPR-Cas gene editing organoid cultures will be used as experimental tools to test potential therapeutic interventions and confirm the functional significance of discovered molecular targets and pathways.

12.7 Challenges

1. Data Integration and Standardization: Integrating diverse omics datasets from multiple sources remains a significant challenge due to differences in experimental protocols, platforms, and data formats (Pinu et al. 2019). Developing standardized data repositories and harmonizing data analysis pipelines will be crucial for maximizing the utility of systems biology approaches in autoimmune disease research.
2. Computational Complexity: Systems biology approaches generate complex datasets that require sophisticated computational tools and algorithms for data processing, integration, and analysis. Developing scalable and efficient computational methods to handle big data and extract meaningful insights poses a challenge (Kohl et al. 2000).
3. Biological Complexity and Variability: Autoimmune diseases are highly complex and exhibit significant inter-individual and intra-individual variability. Capturing this complexity in computational models and experimental systems is challenging, requiring advanced techniques that can capture the dynamic nature of disease progression and response to therapies (Berg et al. 2006).
4. Validation of Predictive Models: A significant problem is creating precise and predicting models that can direct clinical decision-making and find efficient treatment interventions. To verify these models' clinical value and dependability, independent datasets and clinical trials must be used for validation.

12.8 Conclusion

This review study emphasizes the important contribution of systems biology methods to the development of our understanding of autoimmune disorders. Using high-throughput data integration techniques, computational modeling, and network analysis, these approaches have provided valuable insights into the complex nature of autoimmune diseases, shedding light on their underlying mechanisms and potential therapeutic targets. By analyzing large-scale omics data, including genomics, transcriptomics, proteomics, and metabolomics, researchers have been able to unravel intricate molecular networks that contribute to disease development and progression. Moreover, systems biology approaches have facilitated the identification of potential biomarkers for autoimmune diseases. By integrating various types of data and employing sophisticated computational algorithms, researchers have discovered molecular signatures that can distinguish between different disease subtypes or predict disease outcomes. These biomarkers hold great promise for early diagnosis, personalized treatment strategies, and monitoring of disease progression, thereby enhancing clinical management and patient care.

References

Alm E, Arkin AP (2003) Biological networks. Curr Opin Struct Biol 13(2):193–202. https://doi.org/10.1016/S0959-440X(03)00031-9

Amur S, Parekh A, Mummaneni P (2012) Sex differences and genomics in autoimmune diseases. J Autoimmun 38(2–3):J254–J265. https://doi.org/10.1016/j.jaut.2011.12.001

Anaya JM, Rojas-Villarraga A, García-Carrasco M (2012) The autoimmune tautology: from poly autoimmunity and familial autoimmunity to the autoimmune genes. Autoimmune Dis 2012:1. https://doi.org/10.1155/2012/297193

Anaya JM, Duarte-Rey C, Sarmiento-Monroy JC, Bardey D, Castiblanco J, Rojas-Villarraga A (2016) Personalized medicine. Closing the gap between knowledge and clinical practice. Autoimmun Rev 15(8):833–842. https://doi.org/10.1016/j.autrev.2016.06.005

Arita A, Costa M (2011) Environmental agents and epigenetics. Handb Epigenetics:459–476

Asada HH, Shaltis P, Reisner A, Rhee S, Hutchinson RC (2003) Mobile monitoring with wearable photoplethysmographic biosensors. IEEE Eng Med Biol Mag 22(3):28–40. https://doi.org/10.1109/MEMB.2003.1213624

Aslam B, Basit M, Nisar MA, Khurshid M, Rasool MH (2017) Proteomics: technologies and their applications. J Chromatogr Sci 55(2):182–196. https://doi.org/10.1093/chromsci/bmw167

Auffray C, Imbeaud S, Roux-Rouquié M, Hood L (2003) From functional genomics to systems biology: concepts and practices. C R Biol 326(10–11):879–892. https://doi.org/10.1016/j.crvi.2003.09.033

Baker RE, Pena JM, Jayamohan J, Jerusalem A (2018) Mechanistic models versus machine learning, a fight worth fighting for the biological community? Biol Lett 14:20170660. https://doi.org/10.1098/rsbl.2017.0660

Banchereau R, Cepika AM, Banchereau J, Pascual V (2017) Understanding human autoimmunity and autoinflammation through transcriptomics. Annu Rev Immunol 35:337–370. https://doi.org/10.1146/annurev-immunol-051116-052225

Banga JR (2008) Optimization in computational systems biology. BMC Syst Biol 2:1–7. https://doi.org/10.1186/1752-0509-2-47

Barabási A, Oltvai ZN (2004) Network biology: understanding the cell's functional organization. Nat Rev Genet 5(2):101. https://doi.org/10.1038/nrg1272

Baxt WG (1995) Application of artificial neural networks to clinical medicine. Lancet 346:1135–1138. https://doi.org/10.1016/S0140-6736(95)91804-3

Berg EL, Kunkel EJ, Hytopoulos E (2006) Biological complexity and drug discovery: a practical systems biology approach. Syst Biol (Stevenage) 152:201. https://doi.org/10.1049/ip-syb:20050036

Blumberg RS, Dittel B, Hafler D, Von Herrath M, Nestle FO (2012) Unraveling the autoimmune translational research process layer by layer. Nat Med 18(1):35–41. https://doi.org/10.1038/nm.2632

Burbelo PD, Iadarola MJ, Alevizos I, Sapio MR (2016) Transcriptomic segregation of human autoantigens useful for the diagnosis of autoimmune diseases. Mol Diagn Ther 20(5):415–427. https://doi.org/10.1007/s40291-016-0211-6

Cajka T, Fiehn O (2016) Toward merging untargeted and targeted methods in mass spectrometry-based metabolomics and lipidomics. Anal Chem 88(1):524–545. https://doi.org/10.1021/acs.analchem.5b04491

Camacho DM, Collins KM, Powers RK, Costello JC, Collins JJ (2018) Next-generation machine learning for biological networks. Cell 173:1581–1592. https://doi.org/10.1016/j.cell.2018.05.015

Cambiaghi A, Ferrario M, Masseroli M (2017) Analysis of metabolomic data: tools, current strategies and future challenges for omics data integration. Brief Bioinform 18(3):498–510. https://doi.org/10.1093/bib/bbw031

Catto JW, Linkens DA, Abbod MF, Chen M, Burton JL, Feeley KM et al (2003) Artificial intelligence in predicting bladder cancer outcome: a comparison of neuro-fuzzy modeling and artificial neural networks. Clin Cancer Res 9:4172–4177

Chawla NV, Davis DA (2013) Bringing big data to personalized healthcare: a patient-centered framework. J Gen Intern Med 28(SUPPL.3):660–665. https://doi.org/10.1007/s11606-013-2455-8

Clayton TA, Lindon JC, Cloarec O, Antti H, Charuel C, Hanton G, Net L, Baker D, Walley RJ, Everett JR, Provost J, Nicholson JK (2006) Pharmaco-metabonomic phenotyping and personalized drug treatment. Nature 440(7087):1073–1077. https://doi.org/10.1038/nature04648

Conesa A, Mortazavi A (2014) The common ground of genomics and systems biology. BMC Syst Biol 8(Suppl 2):S1. https://doi.org/10.1186/1752-0509-8-S2-S1

Coote JH, Joyner MJ (2015) Is precision medicine the route to a healthy world? Lancet 385(9978):1617. https://doi.org/10.1016/S0140-6736(15)60786-3

Corvol J, Pelletier D, Henry RG, Caillier SJ, Wang J, Pappas D, Casazza S, Okuda DT, Hauser SL, Oksenberg JR, Baranzini SE (2008) Abrogation of T cell quiescence characterizes patients at high risk for multiple sclerosis after the initial neurological event. Proc Natl Acad Sci U S A 105(33):11839. https://doi.org/10.1073/pnas.0805065105

Cox J, Mann M (2007) Is proteomics the new genomics? Cell 130(3):395–398. https://doi.org/10.1016/j.cell.2007.07.032

Crowley E, Di Nicolantonio F, Loupakis F, Bardelli A (2013) Liquid biopsy: monitoring cancer-genetics in the blood. Nat Rev Clin Oncol 10:472. https://doi.org/10.1038/nrclinonc.2013.110

Deo RC (2015) Machine learning in medicine. Circulation 132:1920–1930. https://doi.org/10.1161/CIRCULATIONAHA.115.001593

Dhillon BK, Smith M, Baghela A, Lee AH, Hancock RE (2020) Systems biology approaches to understanding the human immune system. Front Immunol 11:1683. https://doi.org/10.3389/fimmu.2020.01683

Dong Z, Chen Y (2013) Transcriptomics: advances and approaches. Sci China Life Sci 56:960–967. https://doi.org/10.1007/s11427-013-4557-2

Fell D, Cornish-Bowden A (1997) Understanding the control of metabolism, vol 2. Portland Press, London

Ghosh D, Poisson LM (2009) "Omics" data and levels of evidence for biomarker discovery. Genomics 93(1):13–16. https://doi.org/10.1016/j.ygeno.2008.07.006

Ginsburg GS, Willard HF (2009) Genomic and personalized medicine: foundations and applications. Transl Res 154(6):277–287. https://doi.org/10.1016/j.trsl.2009.09.005

Glocker MO, Guthke R (2005) Rheumatoid arthritis, a complex multifactorial disease: on the way toward individualized medicine. Med Res Rev 26(1):63–87. https://doi.org/10.1002/med.20045

Goldblatt F, O'Neill SG (2013) Clinical aspects of autoimmune rheumatic diseases. Lancet 382(9894):797–808. https://doi.org/10.1016/S0140-6736(13)61499-3

Goodacre R (2018) Metabolomics of a superorganism. J Nutr 137:259S–266S. https://doi.org/10.1093/jn/137.1.259S

Hansen J, Zhao S, Iyengar R (2011) Systems pharmacology of complex diseases. Ann N Y Acad Sci 1245:E1. https://doi.org/10.1111/j.1749-6632.2011.06382.x

Hollenbach JA, Oksenberg JR (2015) The immunogenetics of multiple sclerosis: a comprehensive review. J Autoimmun 64:13. https://doi.org/10.1016/j.jaut.2015.06.010

Hueber W, Robinson WH (2006) Proteomic biomarkers for autoimmune disease. Proteomics 6:4100–4105. https://doi.org/10.1002/pmic.200600017

Huizar CC, Raphael I, Forsthuber TG (2020) Genomic, proteomic, and systems biology approaches in biomarker discovery for multiple sclerosis. Cell Immunol 104219:104219. https://doi.org/10.1016/j.cellimm.2020.104219

Idle JR, Gonzalez FJ (2007) Metabolomics. Cell Metab 6:348–351. https://doi.org/10.1016/j.cmet.2007.10.005

Jacquier A (2009) The complex eukaryotic transcriptome: unexpected pervasive transcription and novel small RNAs. Nat Rev Genet 10(12):833–844. https://doi.org/10.1038/nrg2683

Janes KA, Lauffenburger DA (2006) A biological approach to computational models of proteomic networks. Curr Opin Chem Biol 10(1):73–80. https://doi.org/10.1016/j.cbpa.2005.12.016

Janowczyk A, Madabhushi A (2016) Deep learning for digital pathology image analysis: a comprehensive tutorial with selected use cases. J Pathol Inform 7(1):29. https://doi.org/10.4103/2153-3539.186902

Jarvis JN, Frank MB (2010) Functional genomics and rheumatoid arthritis: where have we been and where should we go? Genome Med 2:1–5. https://doi.org/10.1186/gm165

Johnson JD, Case DO, Andrews JE, Allard SL (2005) Genomics—the perfect information-seeking research problem. J Health Commun 10(4):323–329. https://doi.org/10.1080/10810730590950048

Julka S, Regnier F (2004) Quantification in proteomics through stable isotope coding: a review. J Proteome Res 3:350–363. https://doi.org/10.1021/pr0340734

Kapetanovic IM (2008) Computer-aided drug discovery and development (CADDD): in silico-chemical-biological approach. Chem Biol Interact 171(2):165–176. https://doi.org/10.1016/j.cbi.2006.12.006

Kell DB (2004) Metabolomics and systems biology: making sense of the soup. Curr Opin Microbiol 7:296–307. https://doi.org/10.1016/j.mib.2004.04.012

Kell DB (2006) Metabolomics, modeling and machine learning in systems biology–towards an understanding of the languages of cells. FEBS J 273:873–894. https://doi.org/10.1111/j.1742-4658.2006.05136.x

Kim H-Y, Kim H-R, Lee S-H (2014) Advances in systems biology approaches for autoimmune diseases. Immune Netw 14(2):73. https://doi.org/10.4110/in.2014.14.2.73

Kitano H (2002) Systems biology: a brief overview. Science 295(5560):1662. https://doi.org/10.1126/science.1069492

Kitteringham NR, Jenkins RE, Lane CS, Elliott VL, Park BK (2009) Multiple reaction monitoring for quantitative biomarker analysis in proteomics and metabolomics. J Chromatogr B 877(13):1229–1239. https://doi.org/10.1016/j.jchromb.2008.11.013

Koepenik A, Blind C, Krenn V, Petersen I, Thomas H, Dietel M, Konthur Z, Skriner K (2004) Array technology and proteomics in autoimmune diseases. Pathol Res Pract 200:95–103. https://doi.org/10.1016/j.prp.2004.02.005

Kohl P, Noble D, Winslow RL, Hunter PJ (2000) Computational modeling of biological systems: tools and visions. Philos Trans R Soc A Math Phys Eng Sci 358(1766):579–610. https://doi.org/10.1098/rsta.2000.0547

LeCun Y, Bengio Y, Hinton G (2015) Deep learning. Nature 521(7553):436–444. https://doi.org/10.1038/nature14539

Lee KH, Aggarwal K, Lee KH (2003) Functional genomics and proteomics as a foundation for systems biology. Brief Funct Genomic Proteomic 2(3):175. https://doi.org/10.1093/bfgp/2.3.175

Leung MK, Delong A, Alipanahi B, Frey BJ (2016) Machine learning in genomic medicine: a review of computational problems and data sets. Proc IEEE 104:176–197. https://doi.org/10.1109/JPROC.2015.2494198

Liu X, Locasale JW (2017) Metabolomics: a primer. Trends Biochem Sci 42(4):274–284. https://doi.org/10.1016/j.tibs.2017.01.004

Liu T, Lin X, Yu H (2015) Identifying genes related to rheumatoid arthritis via system biology analysis. Gene 571:97. https://doi.org/10.1016/j.gene.2015.06.058

Mardinoglu A, Boren J, Smith U, Uhlen M, Nielsen J (2018) Systems biology in hepatology: approaches and applications. Nat Rev Gastroenterol Hepatol 15(6):365–377. https://doi.org/10.1038/s41575-018-0007-8

Martínez-rodríguez JE, López-botet M, Munteis E, Rio J (2011) Natural killer cell phenotype and clinical response to interferon-beta therapy in multiple sclerosis. Clin Immunol 141(3):348–356. https://doi.org/10.1016/j.clim.2011.09.006

Moles CG, Mendes P, Banga JR (2003) Parameter estimation in biochemical pathways: a comparison of global optimization methods. Genome Res 13(11):2467–2474. https://doi.org/10.1101/gr.1262503

Musunuru K, Ingelsson E, Fornage M, Liu P, Murphy AM, Newby LK, Newton-Cheh C, Perez MV, Voora D, Woo D (2017) The expressed genome in cardiovascular diseases and stroke: refinement, diagnosis, and prediction: a scientific statement from the American Heart Association. Circ Cardiovasc Genet 10(4):e000037. https://doi.org/10.1161/HCG.0000000000000037

Najafzadeh L, Mahmoudi M, Ebadi M, Shasaltaneh MD, Masoudinejad A (2021) Identification of key genes and pathways for differential diagnosis between rheumatoid arthritis and osteoarthritis based on systems biology approaches introduction. Rheumatol Res 6(1):23–34. https://doi.org/10.22631/rr.2021.69997.1112

Nazliel BİJEN, Taşkiran DİLEK, Irkec CEYLA, Kutay FZ, Pöğün Ş (2002) Serum nitric oxide metabolites in patients with multiple sclerosis. J Clin Neurosci 9(5):530–532

Papp B, Notebaart RA, Pál C (2011) Systems-biology approaches for predicting genomic evolution. Nat Rev Genet 12(9):591–602. https://doi.org/10.1038/nrg3033

Pinu FR, Beale DJ, Paten AM, Kouremenos K, Swarup S, Schirra HJ, Wishart D (2019) Systems biology and multi-omics integration: viewpoints from the metabolomics research community. Meta 9(4):76. https://doi.org/10.3390/metabo9040076

Pulendran B, Li S, Nakaya HI (2010) Systems vaccinology. Immunity 33(4):516–529. https://doi.org/10.1016/j.immuni.2010.10.006

Robinson WH, Steinman L, Utz PJ (2002) Proteomics technologies for the study of autoimmune disease. Arthritis Rheum 46(4):885–893. https://doi.org/10.1002/art.10129

Rullmann JAC, Struemper H, Defranoux NA, Ramanujan S, Meeuwisse CML, Van Elsas A (2005) Systems biology for battling rheumatoid arthritis: application of the Entelos PhysioLab platform. Syst Biol (Stevenage) 152(4):256. https://doi.org/10.1049/ip-syb:20050053

Ryan D, Robards K (2006) Metabolomics: the greatest omics of them all? Anal Chem 78(23):7954–7958. https://doi.org/10.1021/ac0614341

Schreiber F, Lynn DJ, Houston A, Peters J, Mwafulirwa G, Finlay BB, Gordon MA (2011) The human transcriptome during nontyphoid salmonella and HIV coinfection reveals attenuated NFκB-mediated inflammation and persistent cell cycle disruption. J Infect Dis 204(8):1237–1245. https://doi.org/10.1093/infdis/jir512

Simakou T, Butcher JP, Reid S, Henriquez FL (2019) Alopecia areata: a multifactorial autoimmune condition. J Autoimmun 98:74–85. https://doi.org/10.1016/j.jaut.2018.12.001

Smigiel KS, Srivastava S, Stolley JM, Campbell DJ (2014) Regulatory T-cell homeostasis: steady-state maintenance and modulation during inflammation. Immunol Rev 259(1):40–59. https://doi.org/10.1111/imr.12170

Sommer C, Gerlich DW (2013) Machine learning in cell biology–teaching computers to recognize phenotypes. J Cell Sci 126(24):5529–5539. https://doi.org/10.1242/jcs.123604

Song X, Lin Q (2017) Genomics, transcriptomics, and proteomics to elucidate the pathogenesis of rheumatoid arthritis. Rheumatol Int 37(8):1257–1265. https://doi.org/10.1007/s00296-017-3732-3

Steensberg A, Simons TD (2015) Beyond biomarkers in drug discovery and development. Drug Discov Today 20(3):289–291. https://doi.org/10.1016/j.drudis.2014.12.009

Tavassoly I, Goldfarb J, Iyengar R (2018) Systems biology primer: the basic methods and approaches. Essays Biochem 62(4):487–500. https://doi.org/10.1042/EBC20180003

Tektonidou MG, Ward MM (2010) Validity of clinical associations of biomarkers in translational research studies: the case of systemic autoimmune diseases. Arthritis Res Ther 12(5):1–10. https://doi.org/10.1186/ar3143

Theofilopoulos AN, Kono DH, Baccala R (2017) The multiple pathways to autoimmunity. Nat Immunol 18(7):716. https://doi.org/10.1038/ni.3731

Trowsdale J (2011) The MHC, disease, and selection. Immunol Lett 137(1–2):1–8. https://doi.org/10.1016/j.imlet.2011.01.002

Turner AP (2000) Biosensor–sense and sensitivity. Science 290:1315–1317. https://doi.org/10.1126/science.290.5495.1315

Van Wietmarschen H, Yuan K, Lu C, Gao P, Wang J (2009) Systems biology guided by Chinese medicine reveals new markers for sub-typing rheumatoid arthritis patients. J Clin Rheumatol 15(7):330–337. https://doi.org/10.1097/RHU.0b013e3181ba3926

Vélez de Mendizábal N, Carneiro J, Solé RV, Goñi J, Bragard J, Martinez-Forero I, Villoslada P (2011) Modeling the effector-regulatory T cell cross-regulation reveals the intrinsic character of relapses in multiple sclerosis. BMC Syst Biol 5(1):1–15. https://doi.org/10.1186/1752-0509-5-114
Villoslada P, Oksenberg JR (2006) Neuroinformatics in clinical practice: are computers going to help neurological patients and their physicians? vol 1, pp 159–170. https://doi.org/10.2217/14796708.1.2.159
Wang N, Shen N, Vyse TJ, Anand V, Gunnarson I, Sturfelt G, Rantapää-Dahlqvist S, Elvin K, Truedsson L, Andersson BA, Dahle C, Örtqvist E, Gregersen PK, Behrens TW, Hammarström L (2011) Selective IgA deficiency in autoimmune diseases. Mol Med 17:1383–1396
Whitacre CC (2001) Sex differences in autoimmune disease. Nat Immunol 2(9):777–780. https://doi.org/10.1038/ni0901-777
Wolf JBW (2013) Principles of transcriptome analysis and gene expression quantification: an RNA-seq tutorial. Mol Ecol Resour 13:559–572. https://doi.org/10.1111/1755-0998.12109
Xiang Y, Kato T (2006) Use of proteomics in the analysis of autoimmune diseases. Lupus 15:431–435. https://doi.org/10.1191/0961203306lu2329oa
Zhang X, Kuivenhoven JA, Groen AK (2015) Forward individualized medicine from personal genomes to interactomes. Front Physiol 6:1–11. https://doi.org/10.3389/fphys.2015.00364
Zhao S, Nishimura T, Chen Y, Azeloglu EU, Gottesman O, Giannarelli C, Iyengar R (2013) Systems pharmacology of adverse event mitigation by drug combinations. Sci Transl Med 5(206):206ra140. https://doi.org/10.1126/scitranslmed.3006548

Systems Biology: A New Era for Vaccine Development; from Deductive Toward Inductive

13

Usnik Debnath and Sahana Ghosh

Abstract

Routine vaccination has always been a center of controversy due to its lack of acceptability among the masses. Adverse event (AE) study relating to vaccine making is termed 'adversomics,' focusing on studying the adverse events associated with vaccination. As the technology improves, with the advancement of recombinant DNA technology, we get to see biology from the molecular perspective and manipulate it according to our wishes for betterment. The complete knowledge of the human genomic sequence now presents new unexploited fields. The advent of high-dimension assay technology and system biology presents a plethora of potential in developing vaccines with a newer approach that incorporates the recent discoveries about iRNA, transcriptome, immunogenetics, and many more. The major problem associated with complex diseases is low immunogenicity and difficulty to isolate antigens. For example, in the case of meningococcus B, an endless four decades of efforts using conventional methods of microbiology and biochemistry could not provide us with an effective and universal vaccine. It was later decided to obtain genomic sequencing of the serogroup B *Neisseria meningitidis* (MenB) and use this information for vaccine manufacturing. This was a novel approach denoted as "reverse vaccinology."

Keywords

Empirical vaccination · Adversomics · Genomics · Reverse vaccinology · iRNA · RDT · Transcriptome · Immunogenetics

U. Debnath · S. Ghosh (✉)
Department of Microbiology, Vijaygarh Jyotish Ray College, Jadavpur, Kolkata, West Bengal, India

S. Joshi et al. (eds.), *Systems Biology Approaches: Prevention, Diagnosis, and Understanding Mechanisms of Complex Diseases*,
https://doi.org/10.1007/978-981-99-9462-5_13

13.1 Introduction

The biological assessment of any living being is an arduous and time-consuming task to perform, mentioning every detail and intricacy of a biological system. Long ago without the technological leverage that we have today, scientist Berzelius in 1809 predicted that a vital spark drives all living entities. He commented organic compounds must have their origin in a living organism and could never be synthesized from inorganic materials in the laboratory. Later on, further investigations by Friedrich Wohler in 1823 disapproved of the vital force theory by synthesizing the first organic compound urea from ammonium cyanate.

Drawing conclusions from observations in system biology inherently differs from other sciences. Living entities belong to a very complex system where a single bacterial cell comprises a colossal amount of molecules at the very basic level. These biomolecules participate in numerous biochemical reactions. Highly regulated enzymes drive these biological reactions and external signal molecules interfere with the cellular mechanism in the form of hormones and drugs. The human mind cannot keep up with so many processes in parallel. With the help of modern computation technology, we can create highly accurate and detailed computer models and statistical trends that can aid our visualization. From the mathematical models of biological systems, we can evaluate the interaction of cells and predict system behavior. These models are constructed by considering the physiological constraints of the biological systems; hence, they are reliable and diagnostically conclusive. The overwhelming complexity of a biological system is brought down to a computer simulation that can predict the entire biological process by gaining a better understanding of the system. This contributes to the vaccine development procedure where we can elucidate the intricacies of pathogenic incursion inside the cell and activation of the human immune system by gene-to-gene interaction and also determine the adverse effects of a potential vaccine on an individual basis.

13.2 From Traditional Vaccination to Reverse Vaccinology

The term "vaccine" is derived from the Latin word Vacca (cows) as Dr. Edward Jenner in 1796 used infected materials isolated from cows to immunize against smallpox (Jenner 1801). Hundreds of years passed by, and then the father of microbiology, Louis Pasteur, discovered that infections are caused by microbes (Pasteur 1880), establishing the basic rules of vaccinology. To make a vaccine, the disease-causing microorganism is to be isolated, inactivated, and injected (Rappuoli 2004). The vaccine developers followed these rules for a century. Jonas Salk developed a vaccine containing a poliovirus that had been killed by formaldehyde treatment. Albert Sabin used a poliovirus that had been attenuated by serial passage in vitro (Levine et al. 2009). Hillman developed vaccines against mumps, measles, and rubella by attenuating the viruses causing the diseases (Offit 2007). Ramon and Glenny isolated essential components from bacterial or viral cultures such as *Neisseria meningitidis*, *Streptococcus pneumoniae*, and *Haemophilus influenzae*, inactivated them, and paved the way for the development of vaccines against diphtheria and tetanus (Glenny and Hopkins 1923). However, when it comes to "hepatitis B," it was found that it could not be cultured in vitro. So, the vaccine was prepared by inactivating the viral antigen present in the plasma of chronically infected people (Fleischmann et al. 1995). Later on, the vaccines were developed using Pasteur's rules, and thus the history book of medicine notes the event as the end of some of the most devastating diseases globally.

At this point in time when the world was heading to wireless technology, scientific discoveries like radio, television, transistor, and laser were skyrocketing, and most of the vaccines that could be developed by these traditional technologies of 'isolate, inactivate, and inject' were already developed, and new technologies became necessary to conquer the remaining pathogens. Incredible progress, however, was put in order during this period through the introduction of new technologies such as recombinant DNA technology and the chemical conjugation of proteins to polysaccharides. The advances in the use of novel adjuvants added another feather in the cap (Fleischmann et al. 1995).

At the end of the twentieth century, a powerful tool was introduced to access the genomes of microorganisms, a new technology that became available in 1995 when Craig Venter published the genome of the first free-living organism *H. influenzae* (Fleischmann et al. 1995). This technological revelation in vaccine preparation allowed for the first time the capacity to move beyond the rules of Pasteur and use computers to design vaccines starting with information present in the genome, which doesn't require cultivation of specific microorganisms. This new approach was depicted as "reverse vaccinology" (Rappuoli 2000).

13.3 Reverse Vaccinology

Revere vaccinology encourages the idea of screening pathogenic genome sequences and identifying the protein-coding gene sequence that contributes to a good vaccine target. The whole-cell genomic sequence analysis of the pathogenic organisms involves either a single pathogenic strain isolate or a pan-genome sequencing meaning the genomic information from several strains of a pathogenic species. All antigens encoded by the genome are available, even if not highly immunogenic during disease. The most conserved protective antigens can also be identified. The novel antigens are screened against the human genome, and antigens with homology to self-antigens are removed upfront to prevent any incident regarding auto-immunity (Rappuoli 2001).

However, the entire genomic sequence of the pathogen provides an exhaustive amount of data that includes essentially every protein antigen that the pathogen expresses at the surface at any time. Reverse vaccinology thus deals with bioinformatics and statistics to manage such a vast amount of data to evaluate the antigens in silico that are then tested experimentally to induce an immune response in the host without any adverse effect. This is a reverse workflow of a standard traditional routine vaccine development procedure requiring initial pathogenic organism growth culture and subsequently bioinformatics analysis (Rappuoli 2001). Virtually every single T cell epitope is available. Screening of the total T cell immunity can be done by overlapping peptides.

For example, in the case of meningococcus B, an endless four decades of efforts using conventional methods of microbiology and biochemistry could not provide us with an effective and universal vaccine. It was later decided to obtain genomic sequencing of the serogroup B *N. meningitidis* (MenB) and use this information for vaccine manufacturing. Two independent laboratories found that the GNA1870 antigen binds to the human complement regulator factor H, which leads to the downregulation of the complementation activation (Schneider et al. 2009). This was a fast approach and relatively easy to identify the potential antigenic component (GNA1870), which provides a new horizon to the difficulties faced in vaccine production. Several curated databases are now being developed which provide comprehensive information about the experimental validity of antigens, for example, Protegen (Yang et al. 2011), IEDB-AR (Immune Epitope Database Analysis Resource) (Hang et al. 2008), and AntigenDB (Ansari et al. 2010).

In the meantime, the study of B streptococcus using the reverse vaccinology method led to the identification of eight genomes from which 312 surface proteins were expressed and the development of a vaccine containing four vaccine proteins capable of providing protection against all types of serotypes (Centers for Disease Control and Prevention 1999).

13.4 Adversomics

Vaccines are historically proven to be a cost-effective, easy-to-use, and most efficient solution to any disease such as smallpox (Centers for Disease Control and Prevention 1999). However, vaccines have also raised public health issues with various side effects. Immune response to vaccines varies greatly depending on the quantity of vaccine antigen(s), route of immunization, the number of doses administered, host genetics, and host individual characteristics such as age, gender, race, medical condition, etc. Although, major studies and research focus on understanding the gene-mediated immune response to the invading infectious pathogen (Kaslow et al. 2008), little we know about vaccine-mediated adverse events. However, in the case of live attenuated vaccines, studying the administration of vaccines as an infectious agent and an associated immune response is regarded as the same. For example, rotavirus (RV) is an enteric pathogen which causes severe, watery diarrhea and vomiting in infants and young children. Comparing the transcriptome of the children affected by rotavirus infection to those immunized with live attenuated RV vaccine shows that RV vaccination causes change in transcriptome in the infected children causing similar symptoms like diarrhea, nausea, vomiting, intussusception, and abnormal morphology of midgut (Gómez-Carballa et al. 2021).

"Adverse events" of vaccination accounts for low public morale which is very dangerous for countering pandemics or epidemics such as COVID-19 and Ebola. In the past, the controversies around the use of the measles-mumps-rubella (MMR) vaccine and thimerosal have poised great public health concerns and questions about the vaccine's safety. Vaccine adverse reactions (AEs), thus so far, have emerged as a future field of research for vaccine development (Pratt and Jachna 2014).

Scientists consider AEs to be genetically predetermined incident and may vary from person to person and from one ethnicity to another. For example, studies have shown that certain populations of 11 different Amerindian groups (Native Americans) are more susceptible to the measles vaccine reactions than Caucasian populations and exhibit 0.4 °C higher post-vaccine fever (Black et al. 1971). It is speculated that Amerinds are genetically different, and this causes intensified reactions to the measles vaccine. Later studies showed that Native American children have a higher risk of *H. influenzae* type b infection than Caucasian children. Decreased IgG2 and IgG3 response against the *H. influenzae* type b polysaccharide vaccine was observed, which explains the higher risk of the pathogen (Siber et al. 1990). Subsequent studies pointed out that a lesser Km genetic allotype is also one of the reasons for poor immune response (Goldblatt et al. 1994).

Recent discoveries led scientists to believe an important role of cytokines in AEs of live attenuated viral vaccines such as the smallpox vaccine. These side effects are generally manifested as lymphadenopathy, fever, and localized or generalized rash. After the smallpox immunization, increased levels of IFN-γ, TNF-α, IL-2, IL-5, and IL-10 were observed; however, individuals who were not affected demonstrated high levels of IFN-γ after immunization (Rock et al. 2004). Another concern regarding the side effects of AEs is myopericarditis following smallpox vaccination, which is the inflammation of the pericardium and heart muscle, considered to have a possible genetic association (Halsell et al. 2003; Wilson et al. 2001).

Additional research hypothesized that individuals experiencing AEs exhibited unique genetic polymorphism, which is associated with the AEs reactions that affect the immune and physiological responses following smallpox vaccination (Stanley et al. 2007). To prove this hypothesis, 346 smallpox (Dryvax)-immunized individuals were genotyped for single-nucleotide polymorphisms (SNPs) in 19 candidate genes that could assess the development of febrile fever associated with the administration of the vaccine. This study proposed a view that the causes for the post-vaccine fever that later develops febrile seizures are specific haplotypes in the IL1 gene complex and in the IL18 and IL4 genes. This insight into immunogenetics enlightened the possibility that these genes can also be involved in other live attenuated vaccine-generated AEs including MMR (Stanley et al. 2007; Usonis et al. 1999). The knowledge of a genetic association could allow the development of predictive tests or preventive therapies that could be administered with vaccines to prevent such AEs.

In another study, 131 healthy volunteers in two independent groups were genotyped across 386 genes following smallpox vaccination. The authors mentioned genetic polymorphism in genes, enzymes, various pharmacological agents (i.e., the methylenetetrahydrofolate reductase, MTHFR, gene), and immunologic transcription factors (i.e., the interferon regulatory factor-1, IRF1) that are associated with generalized skin eruptions, or enlarged or tender regional lymph nodes after vaccination (Reif et al. 2008).

Scientists have done extensive hard work to identify genetic associations between HLA, cytokine, cytokine receptor, innate receptors, innate immune response genes, and signaling molecules and both humoral and cell-mediated immune responses (Poland et al. 2007, 2008a). Nowadays, genetic polymorphism and variations in the immune response are very well understood and are used to identify the association of adverse events with vaccination. Recently, another study put forward that children between 12 and 13 months vaccinated with MMR have a high risk of immune thrombocytopenic purpura (ITP) and anaphylaxis, with an estimated occurrence rate of one case per forty thousand vaccinated subjects (Stratton et al. 1994). France et al. showed that 76% of the IIT cases were children aged 1 year, indicating that this age group is more prone to get the disease (France et al. 2008).

Advanced technologies currently available in the field of biotechnology and molecular biology such as whole-genome sequencing, proteomics, epigenetics, transcriptomics, and new biostatistical approaches contribute to huge databases of information. We can now better understand the genetic regulations and interactions that promote vaccine-mediated response and elucidate how reactions occur on a cellular level. Mathematical modeling and in silico presentation of such huge data are organized and rigorously analyzed and are utilized in vaccine production so that the AEs can be predicted beforehand and minimize or eliminate serious vaccine-induced side effects to provide a more personalized or individualistic approach to vaccine production. This is what we call adversomics. Adversomics utilizes immunogenetics and immunogenomics of vaccine-induced AEs from an individual level to the mass population.

13.5 Vaccinology: Basic Principles

Vaccines have transformed the public health sector drastically. In countries with high coverage of vaccination, many vicious diseases that were responsible for childhood deaths have essentially been eradicated (Fig. 13.1). WHO predicts that each year vaccination saves 2–3 million lives! It is estimated that the mortality of children less than 5 years of age reduced globally from 93 deaths per 1000 live births in 1990 to 39 deaths per 1000 live births in 2018 (World Health Organization 2020).

A vaccine is a biological product that contains antigens derived from the pathogen or synthetically produced components of the pathogen that immunize the body system and protect against the infection or disease upon subsequent exposure to the pathogen.

The introduction of a vaccine antigen initiates and stimulates the body's immune system that identifies the foreign object and induces specific immune responses (immunogenicity) against that object. Vaccines on the basis of effectiveness are divided into two parts: (1) prophylactic (i.e., to prevent or reduce the effect of a future invasion by a pathogen) and (2) therapeutic (vaccines against cancer) (Levine and Sztein 2004).

Vaccines on the basis of biological characteristics are classified as living, which may contain an attenuated replicating strain of a pathogenic organism and nonliving

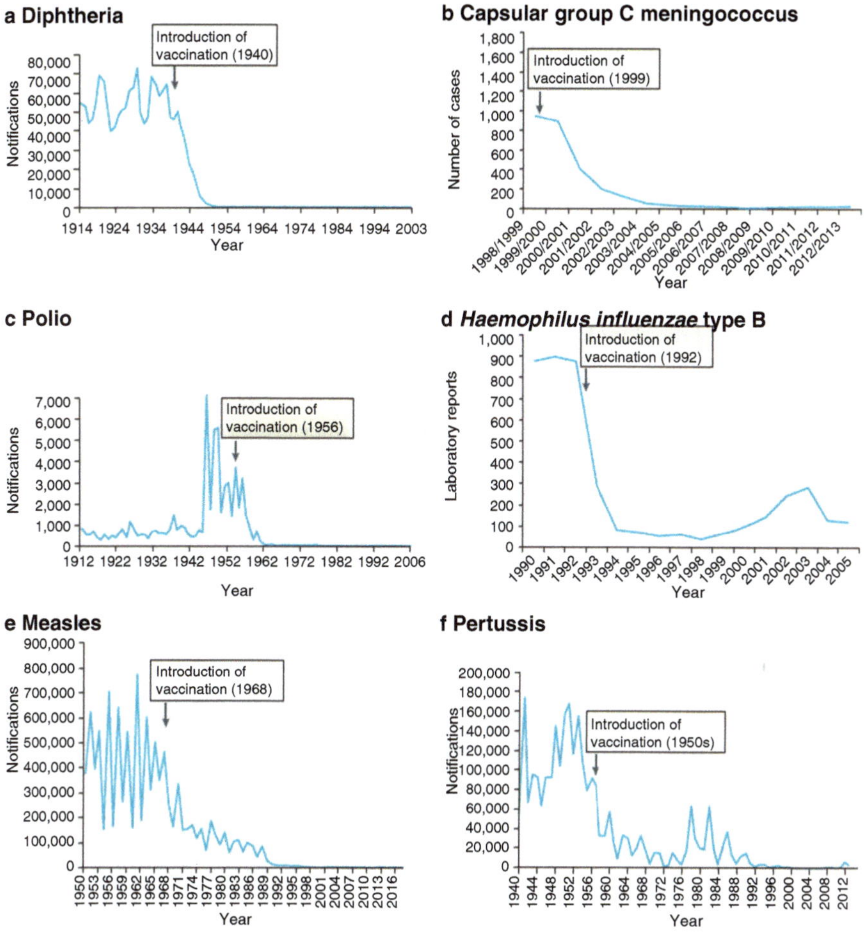

Fig. 13.1 The impact vaccination brought in the UK (Pollard and Bijker 2021)

(also referred to as “inactivated”), which contain only a component of a pathogenic organism or killed whole organism.

The antigenic materials of non-live vaccines generally include whole organisms (e.g., whole-cell pertussis vaccine and inactivated polio vaccine), purified proteins from the organism *Bordetella pertussis* (acellular pertussis vaccine) that generate immunogenicity, recombinant proteins (e.g., hepatitis B virus (HBV) vaccine), or polysaccharides (e.g., the pneumococcal vaccine against *S. pneumoniae*). Toxoid vaccines (e.g., for tetanus and diphtheria), which are toxins of the pathogenic organism, are formaldehyde-treated inactivated protein toxins that can also be introduced into the body to efficiently induce immunogenicity.

To induce a better immune response (immunogenicity) non-live vaccines are often treated with adjuvants. There are pieces of evidence that support the theory that adjuvants provide a danger signal to the innate immune system and further enhance the effectiveness of the vaccine in inducing specific immune responses. Generally, adjuvants are of many types as follow:

(a) liposome-based adjuvants
(b) oil-in-water emulsions
(c) aluminum salts (alum)

Although alum is extensively used as an adjuvant for more than 80 years, its mechanism of action is incompletely understood (Marrack et al. 2009). However, MF59, which is used in some influenza vaccines (Wilkins et al. 2017); AS01, which is used in one of the shingles vaccines and the licensed malaria vaccine (Kaslow and Biernaux 2015); and AS04, which is used in a vaccine against human papillomavirus (HPV) (Pedersen et al. 2007), are all examples of novel adjuvants classified as an oil-in-water emulsion. Vaccines also contain other components that function as preservatives, emulsifiers (such as polysorbate 80), or stabilizers (e.g., gelatine or sorbitol). Other than that various chemical components are also used in the manufacture of vaccines including antibiotics, egg or yeast proteins, latex, formaldehyde and/or glutaraldehyde, and acidity regulators (such as potassium or sodium salts). Except in the case of allergy to any of these components, there is no potential evidence of risk to human health, and are completely safe to use (Mitkus et al. 2013; Eldred et al. 2006).

13.6 Vaccine-Induced Immunogenic Protection

The antigenic vaccine establishes sufficient immune memory as an adaptive immune response in the target individual, which helps to identify and eliminate pathogenic organisms. The adaptive immune response is mediated by T cell (cellular immunity) and B cell (humoral immunity) compartments, which play an important part in tackling infectious challenges (prophylactic vaccines) or a tumor (therapeutic vaccines). The design of the vaccine is done according to the immune competence of the host and provides protection against the pathogen.

After designing the vaccine, system biology helps to evaluate the immune response of the individual at multiple levels. The evaluation of the host's innate or adaptive immune system provides us with predictions and models, which is helpful to validate and refine the vaccine (Fig. 13.2).

The vaccination process synergizes both the innate and the adaptive immune system and confers an effective immune response. Indeed, after the inoculation of the antigenic vaccine, the antigen-presenting pathogenic cells are taken up by the dendritic cells, which are activated through the binding of adjuvants to the pattern recognition receptors (PRR). Then it is transported to the draining lymph nodes where the presentation of the processed vaccine protein antigens by the MHC molecules attached to the dendritic cells activates the naïve CD4+ and CD8+ T lymphocytes (CTL) through their T cell receptor (TCR). Naïve T cells are stimulated to proliferate and differentiate into effector and memory T cells. Activated, effector,

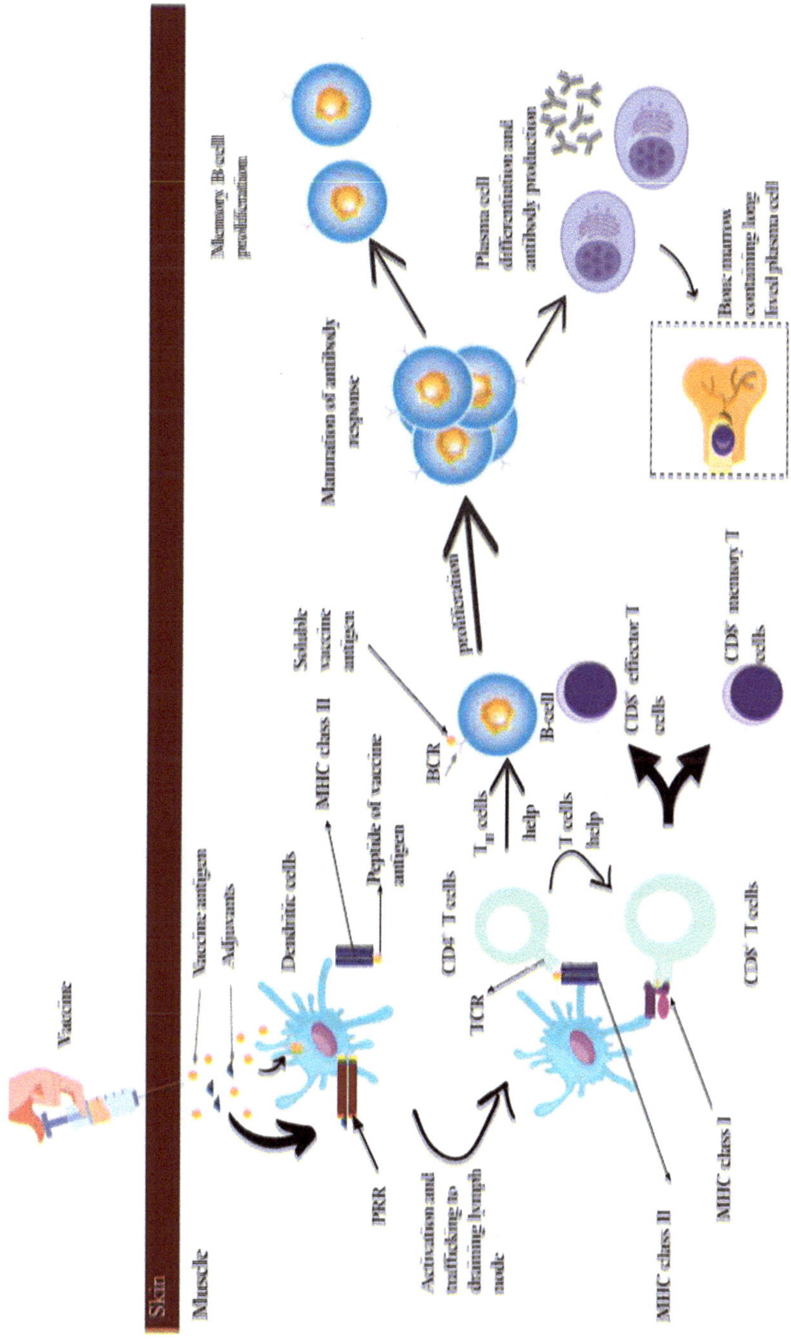

Fig. 13.2 The generation of immune response to the vaccine (Pollard and Bijker 2021)

and memory CD4+ T cells provide help to B cells to mount antibody responses, and to naïve CD8+ T cells to enhance their antigen-driven clonal expansion and differentiation into cytotoxic CTL. Simultaneously, as the soluble antigens signal through the B cell receptor (BCR), the T cells drive B cell development in the lymph node. The T cell-dependent B cell activation results in increased antibody affinity and induces different antibody isotypes. The short-lived plasma cell drives the upsurge of antibody production for the vaccine protein for approximately 2 weeks, whereas long-lived plasma cells reside in the bone marrow and continue antibody production lifelong. Proliferated B cells also produce an immune memory cell line (Pollard and Bijker 2021).

Most antigens and vaccines trigger B- and T-cell responses to achieve high-level antibody production that can comprehend humoral immunity, as well as T-cell mediated cellular immunity. Nevertheless, the nature of the vaccine components influences the type of immune effectors of the host system that is predominantly elicited to mediate protective efficacy. The quality of the vaccine-induced immune response depends on several factors, e.g., nature of the antigen, route of administration, quality of antigen presentation, vaccine preparation adjuvants, and timing between challenges. Moreover, different individuals have different lymphocyte repertoires or antigen-presenting cell potentials that affect the state of immunocompetence. The subjects receiving the vaccine can be selected according to the response efficacy (Goriely and Goldman 2007). Recently, scientists have discovered the importance of the innate immune system that determines the orientation of quality and quantity of the adaptive immune response. Apparently, the discovery of the pattern recognition receptors such as toll-like receptors (TLRs) elicited on the surface of most of the immune cells upon ligation with various pathogenic compartments places the initial recognition event as key to the overall immunogenicity of the individual (Schenten and Medzhitov 2011).

The antigen carried in a vaccine preparation is directly taken up by the innate immune system cells, namely dendritic cells. Then the vaccine antigen is presented to T lymphocytes through the close association between MHC and TCR, which initiates the adaptive immune response (Pollard and Bijker 2021). Therefore, the choice of antigen is very carefully drawn out to avoid any toxicity and produce maximum antigen production.

System biology applied in the field of vaccinology helps prioritize the personalization of the vaccination process where each individual can be cured of the diseases by understanding their immunogenetics. With the help of methods in system biology, we can elucidate the involvement of associated immune components. For example, in some individuals experiencing immunological defects in the complement system, susceptibility to the pathogen *Neisseria meningitides* is observed (Fijen et al. 1999), whereas the complement is directed to the bacterial surface by IgGantibodies. Pneumococcal disease is particularly common in patients with hyposplenism (Wara 1981). Antibody-deficient individuals are susceptible to varicella zoster virus, which results in chickenpox and other viral infections; however, once they have been infected with the virus, they can become resistant and can control the disease the same way as a nonimmune deficient individual does as long as they have a normal T cell response (Gershon et al. 2015).

13.7 Immunological Data

Immunological data provides insights into the immune system and helps us to understand how vaccines work. For example, polysaccharide vaccines, which are made up of the surface polysaccharide of a pathogenic organism such as meningococci (*N. meningitidis*) (Patel and Lee 2005) and pneumococci (*S. pneumoniae*) (Moberley et al. 2013), provide sufficient immunity against the invasive pathogen. These vaccines do not induce T cell responses and rely solely on antibody-dependent mechanisms. However, protein-polysaccharide conjugate vaccines contain immunogenic low molecular weight protein carrier (mostly tetanus toxoid, or diphtheria toxoid or a mutant protein derived from it, known as CRM197) and polysaccharide chemically bonded with each other (Andrews et al. 2014; Ramsay et al. 2003). The protein carrier induces T cells that help B cells recognize the polysaccharide, but the polysaccharide part does not induce T cells.

13.8 Vaccine Development and Immunoinformatics

The immune system is a dynamic and response-mediated tissue system that is made up of huge, diversified, circulating, and interconnected cells with loop/circuit types of interactions. System biology aims to not only study the individual components involved but also understand the localized and temporal effectiveness of each individual component and model the complex interactions among the individual immune component. System biology is the development of tools for studying such complexity based on large data input analysis (Benoist et al. 2006).

Historically, vaccine development has been evolving with technological evolutions. Recent developments in vaccinology translate it as a combinatorial science that studies the diversity of pathogens and the complexity of the immune system, through screening or analyzing immunoinformatic tools. Future development will be based on a system biology approach and the creation of a virtual simulation capable of exhibiting complex interactions (Flower and Timmis 2007). The major aim of immunoinformatics is to develop computational vaccinology and increase the success rate of vaccine preparations. However, in present situations, it is applied to study vaccine immunogenicity and its suitability to clinical trials.

13.9 Bioinformatics

Bioinformatics is a broad subject considering information about the biological system and its interaction with others. In vaccinology, however, bioinformatics is divided into three groups: pathogen biology, host biology, and the interaction between the two. The focus of bioinformatics is not just a sequence of a gene and its regulations but whole interactions with other genes and the eventual phenotype that these interactions of genes bring about. Bioinformatics is the key that changed the prospect of vaccine development from classic (isolate, inactivate, inject) to a targeted approach with the preparation of mathematical modeling and extensive in silico screening.

The analysis of pathogenic organisms such as Brucella (He and Xiang 2010) and flavivirus (Koraka et al. 2010) are based on in silico docking (Khan and Ranganathan 2010), consensus epitope prediction (Sanchez-Burgos et al. 2010), and multiple epitope prediction (Larsen et al. 2010) by identifying a set of highly immunogenic genes based on prior vaccine data, structural and localization predictions, and comparative cross-species/strain virulence. These novel approaches rely on curated databases of growing genomic and proteomic datasets (Heinz et al. 2009).

Recent development in pathogen biology presents genome sequencing data and bioinformatics analysis such as dynamic plots, which can be used to map a population drift of rapidly evolving pathogenic species (Steinbruck and McHardy 2010; Chong et al. 2010). This type of development is very useful in order to systematically improve vaccine efficacy in response to specific challenges.

Large-scale protein-protein interactions and RNAi knockdown screening techniques are used to identify the virulence factors, candidate genes essential for the infections, and critical pathways in host pathogenesis (Wang et al. 2010; Guo et al. 2010). These methods are important to predict the successful outcome of vaccines.

Innate immune response has been the center of studies for the last few years, and scientists have been able to extract many valuable data regarding the involvement of pattern recognition receptors (PRR) and their subfamilies, e.g., toll-like receptors (TLRs). These recent discoveries along with the detailed database of gene expression and regulation have contributed to creating a model simulation that can predict the innate immune system response (Teuwen et al. 2009). These methods can also be applied to construct individual immune profiling by customizing the adjuvants and also the development of vaccines to counter autoimmune diseases with adaptive responses (Jan et al. 2010).

In silico modeling of the adaptive immune system has also been employed to improve our understanding of the host immune system by predicting the epitope binding site of the T cells. Apart from that, recent studies have proclaimed the application of deep sequencing of escape-prone epitopes of SIV for modeling their effect on $CD8^+$ T cell response (Love et al. 2010). Host population modeling has provided us with insights into the transmission of pathogens. Understanding of the genetic difference between individuals has gained significant importance for pharmacogenomics and vaccine development.

Pathogen strains are studied and monitored for the selection of appropriate vaccine targets. This is achieved by the application of bioinformatics and microarray technologies (Kaewpongsri et al. 2010).

The consequent research in the field of bioinformatics is the reason for the availability of high-dimensional data convoluted by biotechnology and statistical analysis, without these subdisciplines the expansion of bioinformatics would not be possible from simple gene targeting to host-pathogen modeling.

13.10 Immunomics

Immunomics or computational vaccinology consists of modeling of antigen binding and processing for T cell epitope mapping using web-accessible computational methods. The antigen processing steps generally include proteasome cleavage, transport by the transporter associated with antigen processing, binding of peptides to MHC molecules, and presentation on the cell surface (Brusic et al. 2004; De Groot 2006; Flower 2007; Flower et al. 2010). For example, PEPVAC (promiscuous epitope-based VACine) is a tool that effectively works for the formulation of multi-epitope-based vaccines that generally cover the mass population. This method uses probabilistic modeling of HLA binding profile matrices to evade immunoproteasome cleavage (Reche and Reinherz 2005). In contrast, OptiTope employs a step-by-step approach in designing epitope-based vaccines that rely on the maximization of the overall immunogenicity of an epitope set. Although using such a prediction tool is not affordable for many research institutions and independent endeavors, the personalization of the vaccination is especially successful for cancer vaccination and discovering novel T-cell epitopes across various targets, including pathogen antigens, cancer antigens, autoantigens, and allergens (Toussaint and Kohlbacher 2009).

Immunosignaturing is an immunoproteomics technology that has been developed to measure serum antibody reactivity spectrum against random sequence peptide arrays. In the immune system, serum antibodies are an essential part of innate and adaptive immunity that take part as immunological memory. This statistical analysis of the immune signature of serum antibodies explains the effectiveness of vaccines and the immune potency of an individual (Legutki et al. 2010). Computer-based mathematical analysis of such technologies is being developed; however, they face great interpretation challenges and labor to process such a piece of vast information (Brown et al. 2011).

Immunomics is now the leading branch of bioinformatics that combines both immunoinformatics algorithms and in silico protein immunogenicity prediction to predict T- and B-cell epitopes, data and literature mining, systematic transcriptomics and proteomics gene expression analyses, and Vaccine Ontology formalism in order to offer computer-based strategies for automated vaccine development (He et al. 2010) (Fig. 13.3).

13.11 Vaccinomics

Vaccinomics is a branch of omics that combines immunogenomics and immunogenetics which uses computer-based prediction tools to identify the variability of pathogen genomic sequence and/or HLA complex (Poland et al. 2008b). Adverse

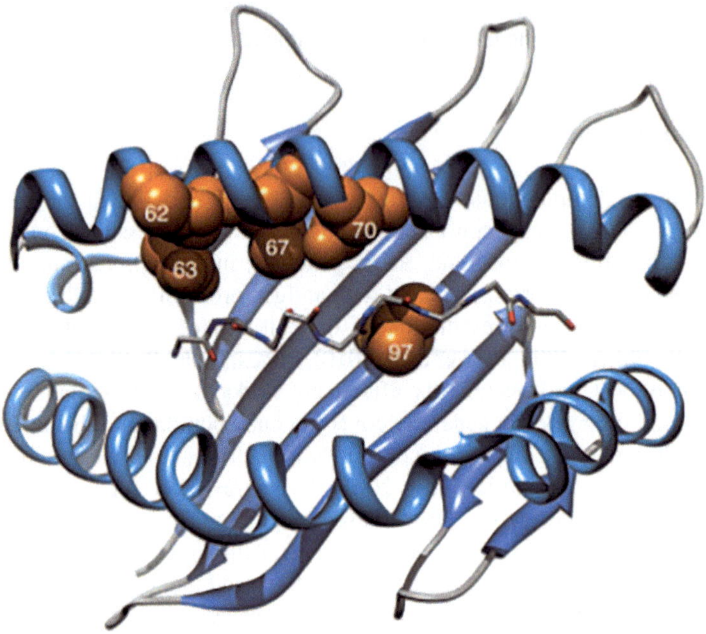

Fig. 13.3 Three-dimensional ribbon representation of the HLA-B protein based on Protein Data Bank entry 2bvp, highlighting amino acid positions 62, 63, 67, 70, and 97 lining the peptide binding pocket (The International HIV Controllers Study 2010)

effects of a vaccine and minimizing the possible failure of a vaccine are the major targets of vaccinomics, achieved by investigating heterogeneity in the host genetic marker that results in a variation of the vaccine-induced immune response. A study in HIV vaccine preparation analyzed a genome-based association in a multi-ethnic cohort of HIV-1 controllers and progressors. The report states that 313 genome-wide significant SNPs (single-nucleotide polymorphisms) have been located in the MHC locus. Statistical analysis (Fig. 13.4) shows some specific changes in the HLA-B peptide binding groove, which are responsible for the loss of control of HIV-1 infection (The International HIV Controllers Study 2010).

13.12 Systems Vaccinology

Genomics and molecular biology techniques are employed to discover novel vaccine antigens, while the analysis and evaluation of host-mediated vaccine-induced immune responses within the framework of systems biology establish it as a promising tool. This framework facilitates the accurate assessment of vaccine efficacy.

Pulendran and his colleagues were among the first to apply a system biology framework as a potential method to successfully analyze the immune response induced by the yellow fever vaccine. It was one of the most successful vaccines ever

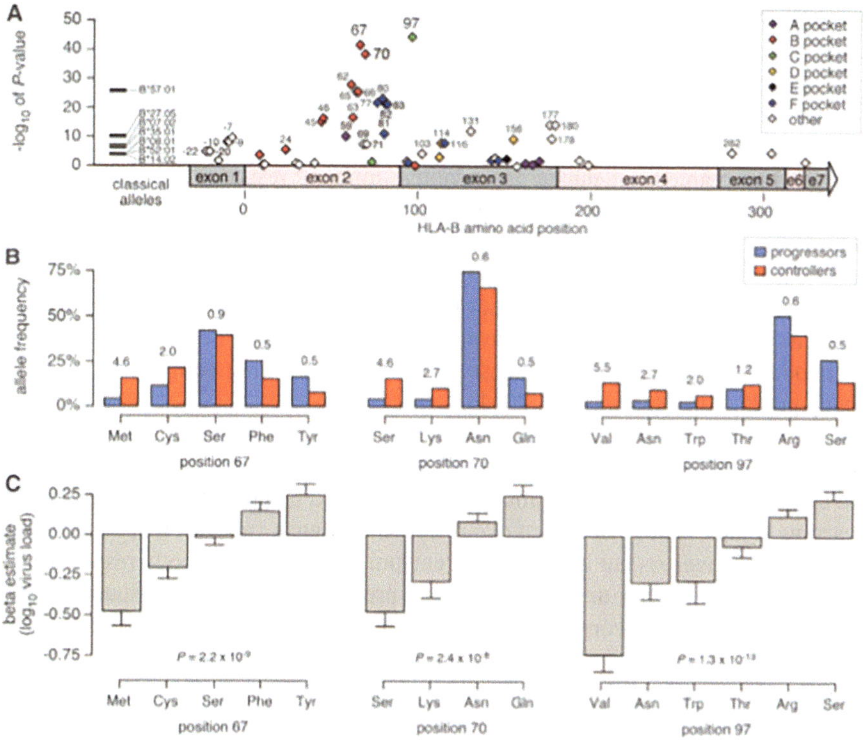

Fig. 13.4 Associations of amino acids in HLA-B in the European sample. (**a**) Association results for all variable amino acid positions, as calculated by the omnibus test. Colors denote conventional pocket positions. P values for significant classical HLA-B alleles are shown for comparison. (**b**) Marked allele frequency differences between controllers and progressors for amino acids at positions 67, 70, and 97. Numbers above the bars indicate odds ratios (values >1 indicate a protective effect). (**c**) Associations between allelic variants at amino acid positions 67, 70, and 97 and quantitative virus load set point in the independent Swiss HIV cohort study. Effect estimates (beta coefficients from a linear-regression model) are given in log10 units of virus load set point. P values refer to the omnibus test for association at each position. Error bars indicate the standard error of the beta coefficient (The International HIV Controllers Study 2010)

made (Teuwen et al. 2009). It was their strategy to involve bioinformatics, genomics, and immunology in order to gain a broader picture where the function of approximately 30,000 genes involved in the expression of proteins and cells participating in immune responses to vaccination was statistically analyzed. This approach could predict up to 90% of the immunogenic signature in the bloodstream of yellow fever.

On the other hand, Sékaly and colleagues made similar observations; however, they employed a different approach using functional genomics coupled with polychromatic flow cytometry, which showed a strong and coordinated initial plasma cell production in response to the antigen which is determinant of an imminent polyfunctional T cell and long-lasting adaptive response (Gaucher et al. 2008).

Both studies prioritize early innate immunity-related events and protective vaccine response. The occurrence of early CD8+ T cell and antibody responses to the yellow fever vaccine establishes broader applicability for different types of immunogens designed to protect the body against the invading pathogen.

Pulendran and his group recently published another systems biology-based study where they pointed out innate and adaptive immune responses in healthy adults after receiving either trivalent inactivated (which contains three inactivated influenza virus strain; two influenza type A and one influenza type B strain) or live attenuated influenza vaccines, and they predicted early innate response-related molecular signatures with an accuracy of 90%. Their study expanding parameters of vaccine response mechanism unravels the previously unknown role of calcium/calmodulin-dependent protein kinase type IV in controlling antigen-specific antibody production (Nakaya et al. 2011).

There exist significant commonality in the early innate immune response between yellow fever and influenza vaccine studies; however, there remains little to distinguish whether such immune signature could be predictive of other various vaccine setups. A comparison of two different influenza vaccines that have similar clinical effectiveness but induced different antigen-specific antibody responses led to further studies that unraveled an early predictive molecular signature of the immune system that correlates with the same antibody titer. These early gene expressions of the innate immune system activation could very well be used to predict the ensuing adaptive immune response; therefore, the system biology framework permits the observation of the entire picture of vaccine-induced early immune responses.

Apparently, not all predictive early gene expressions are related to the inflammatory immune response. The analysis and prioritization of potential tool for the assessment of the vaccine efficacy system biology can serve as an initial step in developing new vaccines aimed at discovering parameters that control the memory T cell and antibody production. Similar studies are being carried out for another successful attempt at constructing vaccines and creating an immunome database of vaccine-induced responses. This can very well be used as the baseline measurement before vaccination (Germain 2010). This concerns, for example, HIV (Fonseca et al. 2011) or adjuvant research (Pulendran 2004). Another field in which system biology can be useful is restoring failed vaccine candidates, such as the Merck MRKAd5 HIV vaccine (Sekaly 2008), or looking for new algorithms for hypo- or non-responsiveness of some vaccines to the recipients, e.g., Pneumococcal and Herpes zoster vaccine response in the elderly, and nonresponders to Hepatitis B virus vaccine.

13.13 Mathematical and Computer Modeling

Mathematical or in silico modeling of host-pathogen interactions has been one of the key methods in vaccine design and development (Cohn and Mata 2007). For example, ImmSim is an automata-based simulator of immunogenic response at a cellular level capable of comparing 64 virtual viral invasions as well as evaluate speed of growth, infectivity level, and lethal load. Different vaccination strategies can be tested to identify how different viruses are susceptible to T cell-mediated responses (Kohler et al. 2000).

C-ImmSim is a modern-day extension of Seiden and Celada's agent-based model of T- and B-cell epitope prediction, making use of bioinformatics in predicting the antigenic side of pathogen and lymphocyte receptors. The authors of C-ImmSim carried out simulations of several classical immunological experiments that pointed out the role of MHC haplotype heterozygosity during influenza. Heterozygosity at the MHC loci enhances the resistance against infectious diseases that increase the number of antigens presented to the T cell. It increases the diversity of the T cell repertoire and offers a high-affinity clonal expansion of the antibody during chronic infection. This increases various possibilities in silico to elucidate the immune response (Rapin et al. 2010).

Most recently, technological achievements in mathematical modeling and computational algorithms have allowed us to view immune components as the language of statecharts (Vainas et al. 2011), which provides us with visual simulations of their behavior with multi-agent technologies (MAS). MAS is an advanced AI and a core area of research that consists of multiple decision-making units or agents that interact with each other in a common environment to achieve common or conflicting goals (Chavali et al. 2008). Nowadays, a graphical communication medium, designed by a biologist, can automatically generate mathematical modeling of a biological system. This development opens up new potential for increasing vaccine efficacy and success (McEwan et al. 2011).

13.14 The Rationale of Vaccine Evaluation

A central goal of the vaccination process is to provide immunity against future pathogenic invasion. Researchers attempt to identify early vaccine-induced immune response to predict the later protection efficiency from diseases or infections. After vaccination various factors such as pathogen prevalence levels, host factor deficiency, or HLA polymorphism are considered for the protective system. Correlates of immunity are depicted as a standard for vaccine development and its further refinement by calculating vaccine efficacy.

Although traditional vaccinology methods were effective at a certain point in time, they were not efficient enough to predict the early immune response in the cell. For example, the assays for antibodies are totally based on antigen binding. If the antibody loses the function to bind with the antigen, the efficacy of the vaccine fails to induce protective immunity. Apart from that the production of antibodies after vaccination completely depends on the activation of cell-mediated immunity. In almost all diseases, the activation of CD4+ cells is necessary for the development of B cell and antibody production. For example, the efficacy of Bacillus Calmette-Guérin (BCG) vaccine against tuberculosis is totally dependent on the T cell activation, which requires the identification of the T cell epitope; however, at this moment no true correlates of immunity against the disease are known, which significantly affects BCG vaccine production. IFNγ production by CD4+ T cells is needed to prevent tuberculosis but is not an adequate correlate of BCG vaccine-induced protection (Black et al. 2002).

The challenge, therefore, lies in identifying the early correlates of vaccination, which would greatly enhance the capability to produce an effective vaccine. The molecular mechanism of the immune system of our body, the availability of the high throughput genomic and proteomic technologies, and the evaluation of the integration of the components of the immune system can be measured by modeling the vaccine-induced immune response, which offers a new approach for the establishment of predictive signature of an effective immune response.

In this line, a plethora of complementary immuno-monitoring technologies have emerged that successfully predict the immunological parameters related to and measure vaccine-induced immune responses. These include the evaluation of

1. Innate immune responses, e.g., dendritic cell activation (Pulendran 2004), inflammatory response, and complement activation.
2. Adaptive antibody responses, e.g., antibody and B-cell immune responses (Bondada and Robertson 2003), neutralizing antibodies (Ochsenbauer and Kappes 2009), non-neutralizing antibodies, and antibody-dependent cell cytotoxicity.
3. Adaptive T-cell immune responses, e.g., CTL activity, T-cell specificity using tetramer/peptides, and cytokine production (Seder et al. 2008).
4. Lymphocyte repertoire diversity, e.g., immunoscope/CDR3 spectratyping, TCR/Ig rearrangement quantification, and TCR/Ig deep sequencing (Boudinot et al. 2008).

The current advancement in the field of flow cytometry implies that an increasing number of parameters can be looked at, which allows a more comprehensive assessment of lymphocyte characteristics and their multifunctional profile. The challenge is to analyze these multiparametric high-throughput data in order to fully extract and analyze the characteristic signatures of the immune response.

13.15 Conclusion

Modern-day technologies do solve problems to some extent; however, there are a plethora of reasons why the mechanisms describing how effective vaccines stimulate protective immune responses are still very poorly understood. Probably, the most genuine problem is to determine the efficacy of a vaccine. System biology and its high throughput technologies, for example, multiplex analysis of cytokines gene expression profiling, and multiparametric flow cytometry, combined with computational modeling pave the way for a new horizon. Recent developments in the field of genomics by the discovery and elucidation of clusters of codon signature enable scientists to link immunogenic response to codon usage bias among individuals from different ethnicities, which increases the vaccine efficacy many fold as well as helps in rapid screening of vaccines that can mediate long-term protective immune response. The key to the success of vaccine immunogenicity lies in understanding the interaction between the correlates of the immune system, and get such results we need a better understanding of immunology, vaccinology, computer science, bioinformatics, biostatistics. System biology thus offers a great promise for future development of vaccines stemming from rigorous immunology research. The modern system biology approach towards vaccine development is more expeditious, effecient and safer compared to the traditional classical approach. It also opens up the possibility of using the knowledge to make vaccines on an individual approach that can possibly revolutionize the human understanding of disease treatment.

References

Andrews NJ et al (2014) Serotype-specific effectiveness and correlates of protection for the 13-valent pneumococcal conjugate vaccine: a postlicensure indirect cohort study. Lancet Infect Dis 14:839–846

Ansari HR, Flower DR, Raghava GP (2010) AntigenDB: an immunoinformatics database of pathogen antigens. Nucleic Acids Res 38:D847–D853

Benoist C, Germain RN, Mathis D (2006) A plaidoyer for 'systems immunology'. Immunol Rev 210:229–234

Black FL et al (1971) Intensified reactions to measles vaccine. J Infect Dis 124:306–317

Black GF, Weir RE, Floyd S, Bliss L, Warndorff DK, Crampin AC (2002) BCG-induced increase in interferon-gamma response to mycobacterial antigens and efficacy of BCG vaccination in Malawi and the UK: two randomised controlled studies. Lancet 359:1393–1401

Bondada S, Robertson DA (2003) Assays for B lymphocyte function. Wiley, Hoboken, pp 333–334

Boudinot P, Marriotti-Ferrandiz ME, Pasquier LD, Benmansour A, Cazenave PA, Six A (2008) New perspectives for large-scale repertoire analysis of immune receptors. Mol Immunol 45:2437–2445

Brown J, Stafford P, Johnston S, Dinu V (2011) Statistical methods for analyzing immunosignatures. BMC Bioinformatics 12:349

Brusic V, Bajic VB, Petrovsky N (2004) Computational methods for prediction of T-cell epitopes – a framework for modelling, testing, and applications. Methods 34:436–443

Centers for Disease Control and Prevention (1999) Ten great public health achievements—United States, 1900–1999. MMWR 48:241–243

Chavali AK, Gianchandani EP, Tung KS, Lawrence MB, Peirce SM, Papin JA (2008) Characterizing emergent properties of immunological systems with multi-cellular rule-based computational modeling. Trends Immunol 29:589–599

Chong YL, Padhi A, Hudson PJ, Poss M (2010) The effect of vaccination on the evolution and population dynamics of avian paramyxovirus-1. PLoS Pathog 6:e1000872

Cohn M, Mata J (2007) Quantitative modeling of immune responses. Immunol Rev 216:5–8

De Groot AS (2006) Immunomics: discovering new targets for vaccines and therapeutics. Drug Discov Today 11:203–209

Eldred BE, Dean AJ, McGuire TM, Nash AL (2006) Vaccine components and constituents: responding to consumer concerns. Med J Aust 184:170–175

Fijen CA, Kuijper EJ, teBulte MT, Daha MR, Dankert J (1999) Assessment of complement deficiency in patients with meningococcal disease in The Netherlands. Clin Infect Dis 28:98–105

Fleischmann RD, Adams MD, White O, Clayton RA, Kirkness EF, Kerlavage AR, Bult CJ, Tomb JF, Dougherty BA, Merrick JM et al (1995) Whole-genome random sequencing and assembly of Haemophilus influenzae Rd. Science 269:496–512

Flower DR (2007) Immunoinformatics and the in silico prediction of immunogenicity. An introduction. Methods Mol Biol 409:1–15
Flower DR, Timmis J (2007) In silico immunology. Springer, New York, pp 1149–1152
Flower D, Macdonald I, Ramakrishnan K, Davies M, Doytchinova I (2010) Computer aided selection of candidate vaccine antigens. Immunome Res 6:S1
Fonseca SG, Procopio FA, Goulet JP, Yassine-Diab B, Ancuta P, Sékaly RP (2011) Unique features of memory T cells in HIV elite controllers: a systems biology perspective. Curr Opin HIV AIDS 6:188–196
France EK et al (2008) Risk of immune thrombocytopenic purpura. Pediatrics 121:e687–e692
Gaucher D, Therrien R, Kettaf N, Angermann BR, Boucher G, Filali-Mouhim A (2008) Yellow fever vaccine induces integrated multilineage and polyfunctional immune responses. J Exp Med 205:3119–3131
Germain RN (2010) Ron Germain: towards a grand unified theory – interview by Amy Maxmen. J Exp Med 207:266–267
Gershon AA et al (2015) Varicella zoster virus infection. Nat Rev Dis Primers 1:15016
Glenny AT, Hopkins BE (1923) Diphtheria toxoid as an immunizing agent. Br J Exp Pathol 4:283–288
Goldblatt D et al (1994) Association of GM allotypes with the antibody response. J Immunol 153:5316–5320
Gómez-Carballa A, Barral-Arca R, Cebey-López M, Currás-Tuala MJ, Pischedda S, Gómez-Rial J, Habgood-Coote D, Herberg JA, Kaforou M, Martinón-Torres F, Salas A (2021) Host transcriptomic response following administration of rotavirus vaccine in infants' mimics wild type infection. Front Immunol 11:580219
Goriely S, Goldman M (2007) From tolerance to autoimmunity: is there a risk in early life vaccination? J Comp Pathol 137:S57–S61
Guo X, Xu Y, Bian G, Pike AD, Xie Y, Xi Z (2010) Response of the mosquito protein interaction network to dengue infection. BMC Genomics 11:380
Halsell JS et al (2003) Myopericarditis following smallpox vaccination. JAMA 289:3283–3289
Hang Q, Wang P, Kim Y, Haste-Andersen P, Beaver J, Bourne PE (2008) Immune epitope database analysis resource (IEDB-AR). Nucleic Acids Res 36:W513–W518
He Y, Xiang Z (2010) Bioinformatics analysis of Brucella vaccines and vaccine targets using VIOLIN. Immunome Res 61:S5
He Y, Rappuoli R, De Groot AS, Chen RT (2010) Emerging vaccine informatics. J Biomed Biotechnol 2010:218590
Heinz E, Tischler P, Rattei T, Myers G, Wagner M, Horn M (2009) Comprehensive in silico prediction and analysis of chlamydial outer membrane proteins reflects evolution and life style of the Chlamydiae. BMC Genomics 10:634
Jan M, Meng S, Chen NC, Mai J, Wang H, Yang XF (2010) Inflammatory and autoimmune reactions in atherosclerosis and vaccine design informatics. J Biomed Biotechnol 2010:459798
Jenner E (1801) The origin of the vaccines inoculation. Shury, London
Kaewpongsri S, Sukasem C, Srichunrusami C, Pasomsub E, Zwang J, Pairoj W, Chantratita W (2010) An integrated bioinformatics approach to the characterization of influenza A/H5N1 viral sequences by microarray data: implication for monitoring H5N1 emerging strains and designing appropriate influenza vaccines. Mol Cell Probes 24:387–395
Kaslow DC, Biernaux S (2015) RTS,S: toward a first landmark on the Malaria Vaccine Technology Roadmap. Vaccine 33:7425–7432
Kaslow R et al (2008) Genetic susceptibility to infectious diseases. Oxford University Press, New York, pp 1–447
Khan JM, Ranganathan S (2010) pDOCK: a new technique for rapid and accurate docking of peptide ligands to Major Histocompatibility Complexes. Immunome Res 61:S2
Kohler B, Puzone R, Seiden PE, Celada F (2000) A systematic approach to vaccine complexity using an automaton model of the cellular and humoral immune system. I. Viral characteristics and polarized responses. Vaccine 19:862–876

Koraka P, Martina BE, Osterhaus AD (2010) Bioinformatics in new generation flavivirus vaccines. J Biomed Biotechnol 2010:864029
Larsen MV, Lelic A, Parsons R, Nielsen M, Hoof I, Lamberth K, Loeb MB, Buus S, Bramson J, Lund O (2010) Identification of CD8+ T cell epitopes in the West Nile virus polyprotein by reverse-immunology using NetCTL. PLoS ONE 5:e12697
Legutki JB, Magee DM, Stafford P, Johnston SA (2010) A general method for characterization of humoral immunity induced by a vaccine or infection. Vaccine 28:4529–4537
Levine MM, Sztein MB (2004) Vaccine development strategies for improving immunization: the role of modern immunology. Nat Immunol 5:460–464
Levine MM, Lagos R, Esparza J (2009) Vaccines and vaccination in historical perspective. In: Levine MM, Dougan G, Good MF, Liu MA, Nabel GJ, Nataro JP, Rappuoli R (eds) New generation vaccines, 4th edn. Informa Healthcare, New York, pp 1–11
Love TM, Thurston SW, Keefer MC, Dewhurst S, Lee HY (2010) Mathematical modeling of ultradeep sequencing data reveals that acute CD8+ T-lymphocyte responses exert strong selective pressure in simian immunodeficiency virus-infected macaques but still fail to clear founder epitope sequences. J Virol 84:5802–5814
Marrack P, McKee AS, Munks MW (2009) Towards an understanding of the adjuvant action of aluminium. Nat Rev Immunol 9:287–293
McEwan CH, Bersini H, Klatzmann D, Thomas-Vaslin V, Six A (2011) A computational technique to scale mathematical models towards complex heterogeneous systems. In: Stepney S, Welch PH, Andrews PS, Ritson CG (eds) CoSMoS 2011 proceedings on the 2011 workshop on complex systems modelling and simulation. Luniver Press, Bristol
Mitkus RJ, Hess MA, Schwartz SL (2013) Pharmacokinetic modeling as an approach to assessing the safety of residual formaldehyde in infant vaccines. Vaccine 31:2738–2743
Moberley S, Holden J, Tatham DP, Andrews RM (2013) Vaccines for preventing pneumococcal infection in adults. Cochrane Database Syst Rev 1:CD000422
Nakaya HI, Wrammert J, Lee EK, Racioppi L, Marie-Kunze S, Haining WN (2011) Systems biology of vaccination for seasonal influenza in humans. Nat Immunol 12:786–795
Ochsenbauer C, Kappes JC (2009) New virologic reagents for neutralizing antibody assays. Curr Opin HIV AIDS 4:418–425
Offit PA (2007) Vaccinated: one man's quest to defeat the world's deadliest diseases. Harper Collins, New York
Pasteur L (1880) De l'attenuation du virus du Choléra des poules. CR Acad Sci Paris 91:673–680
Patel M, Lee CK (2005) Polysaccharide vaccines for preventing serogroupA meningococcal meningitis. Cochrane Database Syst Rev 3:CD001093
Pedersen C et al (2007) Immunization of early adolescent females with human papillomavirus type 16 and 18 L1 virus-like particle vaccine containing AS04 adjuvant. J Adolesc Health 40:564–571
Poland GA et al (2007) Heterogeneity in vaccine immune response. Clin Pharmacol Ther 82:653–664
Poland GA et al (2008a) Vaccine immunogenetics. Vaccine 26:6183–6188
Poland GA, Ovsyannikova IG, Jacobson RM (2008b) Personalized vaccines: the emerging field of vaccinomics. Expert Opin Biol Ther 8:1659–1667
Pollard AJ, Bijker EM (2021) A guide to vaccinology: from basic principles to new developments. Nat Rev Immunol 21:83–100
Pratt SD, Jachna BR (2014) Care of the clinician after an adverse event. Beth Israel Deaconess Medical Center, Boston
Pulendran B (2004) Modulating vaccine responses with dendritic cells and toll-like receptors. Immunol Rev 199:227–250
Ramsay ME, McVernon J, Andrews NJ, Heath PT, Slack MP (2003) Estimating Haemophilus influenzae type B vaccine effectiveness in England and Wales by use of the screening method. J Infect Dis 188:481–485

Rapin N, Lund O, Bernaschi M, Castiglione F (2010) Computational immunology meets bioinformatics: the use of prediction tools for molecular binding in the simulation of the immune system. PLoS ONE 5:e9862
Rappuoli R (2000) Reverse vaccinology. Curr Opin Microbiol 3:445–450
Rappuoli R (2001) Reverse vaccinology, a genome-based approach to vaccine development. Vaccine 19:2688–2691
Rappuoli R (2004) From Pasteur to genomics: progress and challenges in infectious diseases. Nat Med 10:1177–1185
Reche PA, Reinherz EL (2005) PEPVAC: a web server for multi-epitope vaccine development based on the prediction of supertypic MHC ligands. Nucleic Acids Res 33:W138–W142
Reif DM et al (2008) Genetic basis for adverse events after smallpox vaccination. J Infect Dis 198:16–22
Rock MT et al (2004) Adverse events after smallpox immunizations. J Infect Dis 189:1401–1410
Sanchez-Burgos G, Ramos-Castaneda J, Cedillo-Rivera R, Dumonteil E (2010) Immunogenicity of novel Dengue virus epitopes identified by bioinformatic analysis. Virus Res 153:113–120
Schenten D, Medzhitov R (2011) The control of adaptive immune responses by the innate immune system. Adv Immunol 109:87–124
Schneider MC, Prosser BE, Caesar JJ, Kugelberg E, Li S, Zhang Q, Quoraishi S, Lovett JE, Deane JE, Sim RB et al (2009) Neisseria meningitidis recruits factor H using protein mimicry of host carbohydrates. Nature 458:890–893
Seder RA, Darrah PA, Roederer M (2008) T-cell quality in memory and protection: implications for vaccine design. Nat Rev Immunol 8:247–258
Sekaly RP (2008) The failed HIV Merck vaccine study: a step back or a launching point for future vaccine development? J Exp Med 205:7–12
Siber GR et al (1990) Impaired antibody response to Haemophilus influenzae type B polysaccharide. N Engl J Med 323:1387–1392
Stanley SL et al (2007) The immunogenetics of smallpox vaccination. J Infect Dis 196:212–219
Steinbruck L, McHardy AC (2010) Allele dynamics plots for the study of evolutionary dynamics in viral populations. Nucleic Acids Res 39(1):e4
Stratton KR et al (1994) Adverse events associated with childhood vaccines other than pertussis and rubella. JAMA 271:1602–1605
Teuwen D, Pirani A, Gernert K, Deng J, Marzolf B, Kennedy K, Wu H, Bennouna S, Oluoch H, Miller J, Vencio RZ, Mulligan M, Aderem A, Ahmed R, Pulendran B (2009) Systems biology approach predicts immunogenicity of the yellow fever vaccine in humans. Nat Immunol 10:116–125
The International HIV Controllers Study (2010) The major genetic determinants of HIV-1 control affect HLA class I peptide presentation. Science 330:1551–1557
Toussaint NC, Kohlbacher O (2009) OptiTope – a web server for the selection of an optimal set of peptides for epitope-based vaccines. Nucleic Acids Res 37:W617–W622
Usonis V et al (1999) Reactogenicity and immunogenicity of a new live. Pediatr Infect Dis J 18:42–48
Vainas O, Harel D, Cohen IR, Efron S (2011) Reactive animation: from piecemeal experimentation to reactive biological systems. Autoimmunity 44:271–281
Wang A, Johnston SC, Chou J, Dean D (2010) A systemic network for Chlamydia pneumoniae entry into human cells. J Bacteriol 192:2809–2815
Wara DW (1981) Host defense against Streptococcus pneumoniae: the role of the spleen. Rev Infect Dis 3:299–309
Wilkins AL et al (2017) AS03- and MF59-adjuvanted influenza vaccines in children. Front Immunol 8:1760
Wilson CB et al (2001) Vaccine safety–vaccine benefits. Nat Rev Immunol 1:160–165
World Health Organization (2020) Child mortality and causes of death. WHO, Geneva. https://www.who.int/gho/child_health/mortality/mortality_under_five_text/en/
Yang B, Sayers S, Xiang Z, He Y (2011) Protegen: a web-based protective antigen database and analysis system. Nucleic Acids Res 39:D1073–D1078

The Important Role of Systems Biology in Neuroscience and Neurology and Its Associated Diseases

14

M. Haripriyaa and K. Suthindhiran

Abstract

Neurodegenerative diseases are a huge global health concern in the Western world due to our ageing population and other factors. Research into these illnesses is therefore essential. The scientific status on neurodegenerative illnesses is summarised in this contemporary book chapter. Understanding the biological complexity of neuronal networks by annealing several physiological processes initiated the role of systems biology in neuroscience and its associated diseases. The overall neurological mechanism encompasses the influence of the memory system over behavioural control, switching mechanisms initiated by several neurotransmitters, and access to motivational and effector systems. This access has fostered the use of systems biology in internal medicine using integrative tools. Lately, the input of systems biology in neuroscience has been more fruitful in marking the topographical mappings localised to the brain region, thereby targeting disease-specific regions. They have enabled several alternatives for treating neurological disorders by identifying new pharmacological targets leading to novel interventional strategies. Therefore, the potential of computational neuroscience might pave the way for comprehensive methods for evaluating and optimising the efficacy of drugs. Also, neuroimaging studies reveal the synaptic impulses underlying the mechanism of depression, addiction, alcoholism, schizophrenia, bipolar disorder, and several other neurological complications. Systems biology in neurosciences should be encouraged in order to further attempts to bridge the gap and strengthen a comprehensive understanding of the NeuroPhysiome.

M. Haripriyaa (✉) · K. Suthindhiran
Marine Biotechnology and Bioproducts Lab, Department of Biomedical Sciences, School of Biosciences and Technology, Vellore Institute of Technology, Vellore, India
e-mail: haripriyaa.2020@vitstudent.ac.in

S. Joshi et al. (eds.), *Systems Biology Approaches: Prevention, Diagnosis, and Understanding Mechanisms of Complex Diseases*,
https://doi.org/10.1007/978-981-99-9462-5_14

Keywords

Neurotransmitters · Topographical · Internal medicine · Computational · Schizophrenia

14.1 Introduction

Despite the complexity of each topic, modern science has begun to specialise by gathering a vast amount of data. Grounding a particular area at the molecular level and supplying quantitative statistics that represent the overall system are the first steps in this approach. From this perspective, "neuroscience in system biology" has been depicted as structural frameworks, combining physiological and technically separate subdisciplines (Grant 2003). Integrating system biology and neuroscience has been one of the most intriguing and challenging frontiers, as represented in Fig. 14.1. Here, data integration and mining tools have been involved in merging the phenotypic and genotypic datasets and have been relating biological pathways to nervous system function at many levels on the basis of systems biology (Geschwind and Konopka 2009).

An information-rich environment is required for systems biology to function, whereas neuroscience involves the genetic and molecular network of which data retrieval can be attempted at the final stages. The primary agenda of neuroscience involves the fundamental principles of perception and functional brain organisation using a simple paradigm with a basic stimulus. By focusing on membrane simulations of diverse transmitter–receptor interactions, modulators' effects on synaptic transmission, and their influence on network behaviour, computational neuroscience further refines the field (Bassett et al. 2018).

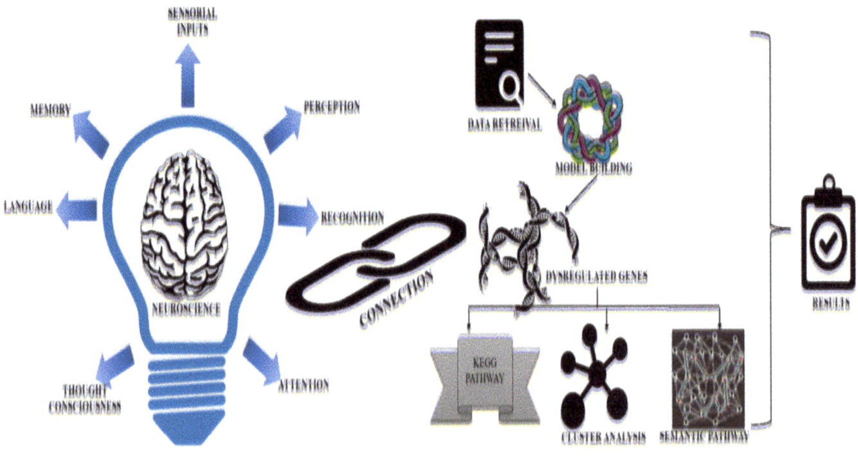

Fig. 14.1 Integration of neuroscience and systems biology involving the combination of a genetic-molecular network and a data-rich environment

More attention is paid to signalling and genetic mechanisms in systems biology (Morris and Schwaber 2012). It is a field of study that focuses on the complex interactions between biological components including molecules and cells in varied contexts. Systems biology has made it considerably simpler to create new medications and reposition those that already exist while comprehending complex illnesses. Figure 14.2 illustrates how systems biology techniques employ a variety of high-throughput/omics technologies to examine the behaviour of thousands of biological components under various environmental conditions, stressors, and medical interventions (Mei et al. 2012).

The scientific cultures of these two disciplines differ significantly, with computational neuroscience clearly having a more "conventional" beginning. Data-driven modelling and standard building based on the community should be used to take a few parameters into account. As previously said, systems biology mostly operates in an information-rich environment where it is harder to distinguish between significant areas and unknown aspects by regulating the dynamics. Conversely, the situation in neuroscience frequently requires a lot of supposition and minimal evidence).

The different neural network topologies can be used to demonstrate this concept. The use of graph theory in the analysis of the topologies of intricate genetic and molecular networks is one of the most active areas in systems biology. Although neurology also uses similar techniques, most work on neural networks mimics randomly connected networks to study their dynamics. This approach might facilitate analysis in some circumstances, but not every time because comprehensive data is also not available.

Keeping all the conditions aside, the systems biology of neural function has taken a very long time to develop, as quoted by Le Novère (2007). They have made

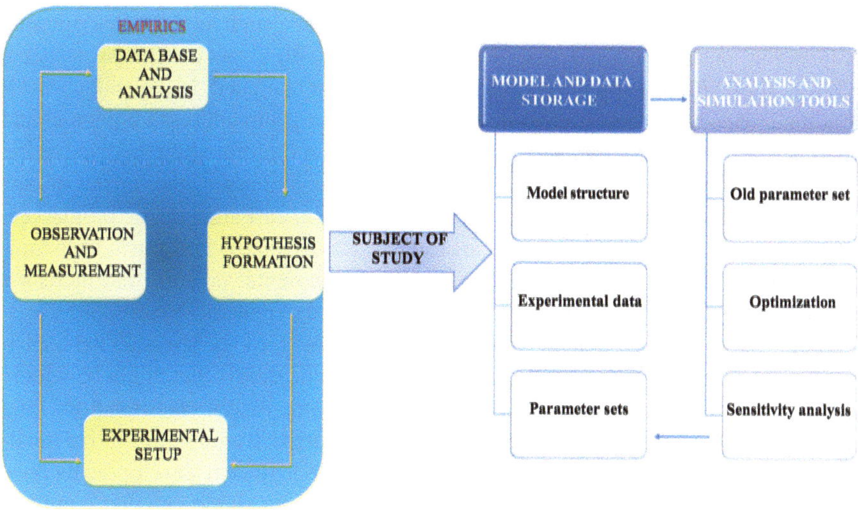

Fig. 14.2 A cycle of empirical research and a scheme of a workflow developed by systems biology-related automated tools

great strides in elucidating the neurobiological mechanisms underlying the decision-making process. This combination has kickstarted their achievement by assessing the composition of synapses and other neuronal processes and molecular mechanisms of learning through neural network models (Santos et al. 2021).

Here, standard statistical model-selection procedures are often used to evaluate the validity of a specific network model. These neural networks have resulted in several advancements in cognitive science and information technology. These techniques strike a compromise between complexity penalties and goodness of fit to the data to prevent overfitting and produce economical and generalisable findings. However, given the variety of objectives and applications of network models, achieving this balance has been difficult (Bassett et al. 2018). The difficulties encountered are relatable as animals have been trained to perform stereotyped behaviours in a lab setting and so do not describe the decisions made in a natural environment (Dennis et al. 2020).

The Hodgkin–Huxley model, which quantitatively depicts cellular behaviour resulting from the interaction of two molecular components, a potassium channel and a sodium channel, can be seen as the precursor to computational systems biology (Le Novère 2007). Imaging has been one of the driving forces behind neuroinformatics activities over the last 10 years, from magnetic resonance imaging (MRI) to the localisation of individual macromolecules by microscopy, by quantitatively defining the Hodgkin–Huxley model through a system biology approach. There are currently a wide variety of web-accessible services available, from simple data collection to sophisticated databases. Neuroinformatics advocates' vision of the new electronic environment has numerous barriers that must be overcome before neuroscience can fully integrate it, but the actions launched have successfully codified the circumstances for effective data interchange and integration from various sources (Martone et al. 2004).

By integrating and combining various large datasets, such as the transcriptome, epigenome, proteome, and metabolome associated with a specific illness, systems biology applications in the biomedical area have been launched. In the end, unravelling the intricacy of neurological diseases paved the path to address unmet medical needs for neurological disorders. Systems biology can therefore help in discovering pathophysiology to develop novel therapeutic targets and assess the impact of current medicines on the progression of the disease. Also, early diagnostic biomarkers have been utilised to recognise neurological diseases that have a protracted subclinical phase and the exposome, a collection of environmental poisons that increases the risk of acquiring particular neurological disorders (Yuan and Nixon 2021).

When regarded from a medical perspective, systems biology offers a vital contribution to performing functional evaluations of genomic events that might be widely employed in gene discovery, biomarker identification, illness classification, drug development, and therapy techniques, as well as in predictive and preventive medicine (Diaz-Beltran et al. 2013; Fernández et al. 2009). Recent developments have made it possible to investigate human cancer even at the single cell and complete genome levels. They offer statistical methods for creating networks from massive amounts of data, as well as numerical mathematical network models to examine the dynamic behaviour of biological systems (Mei et al. 2012).

Although there is a strong case for studying nervous system traits that vary over time, the existing funding and review procedures make it challenging to gather longitudinal data (Geschwind and Konopka 2009). It is time to increase the scope of the project and develop a precise systems biology of nervous system illnesses. To accurately represent the neuronal function, many challenges must be addressed, straining computational biology's techniques and resources (Le Novère 2007). Efforts on the multiscale and model integration fronts can be made as the newest technology meets the computational power. In order to examine the underlying biological processes and eventually comprehend the aetiopathogenesis of neurological illnesses, this chapter aims to highlight the importance of system biology in neuroscience, the role of systems biology in discovering new biomarkers, or the identification of therapeutic targets in complex disorders.

14.2 Overview of Systems Biology in Neuroscience

Systems neuroscience must apply sophisticated, selective, and recurrent analyses to high-dimensional data on brain activity to understand the activity of the brain. Therefore, it is crucial to investigate genetic variants and neural substrates using a systems approach to comprehend the neurobiological mechanisms underlying cognition. The central nervous system (CNS) has the most intricately specialised functions, according to a 2017 study by a team led by Joseph Dougherty of the Department of Genetics. It is made up of numerous morphologically and molecularly distinct subtypes of glia and neurons, all arranged into intricate but stereotyped networks of circuits. Decades of post-mortem pathological research have linked distinct behavioural abnormalities to the deterioration of particular circuits in late-onset illnesses (Dougherty et al. 2017). Therefore, understanding the molecular mechanism of the different components of the nervous tissues will aid in untangling the complexity involved in the NeuroPhysiome (Green et al. 2008).

14.3 Insights of Systems Biology About Cellular Components of Nervous Tissue

Networks of interacting parts form a scaffold for cellular systems that adapt throughout time in response to both internal and external stimuli. Therefore, understanding cellular processes and disease pathways begins with an investigation of the dynamic functioning of networks (Assmus et al. 2006). It is known that neurons, glial cells, blood arteries, and extracellular matrix make up the nervous system, as mentioned in Fig. 14.3. The initiation, processing, and transmission of information are predominantly carried out by neurons, which are electrically excitable cells. However, their interactions with glial cells, which support synaptic formation and survival, influence the activity of neuronal ensembles and systems that are important for

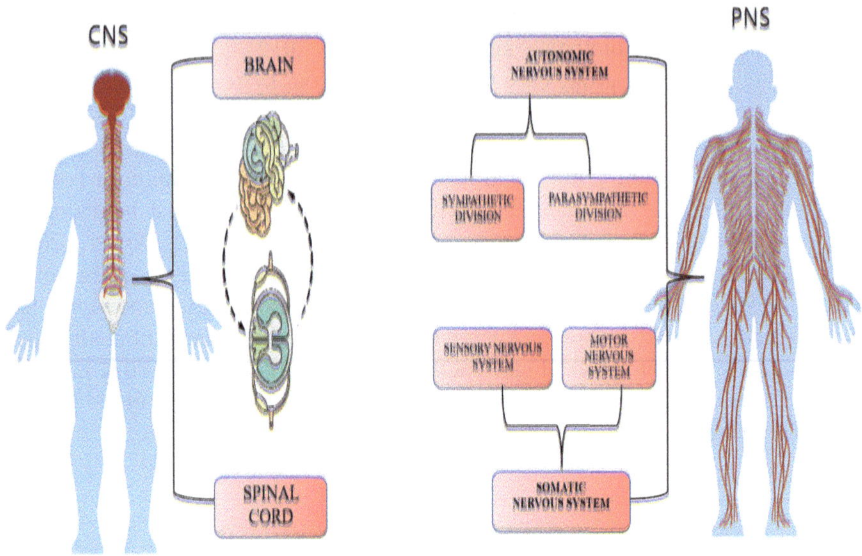

Fig. 14.3 Functions of the central nervous system (CNS) and peripheral nervous system (PNS)

cognition and behaviour (Benarroch et al. 2020). The complex molecular networks control these cell types to maintain their physiological roles.

This technology is revolutionising neuroscience because it allows for the generation of high-throughput data (HTD) that can account for changes at the transcriptome, proteome, or epigenome level. Because HTD must be understood in the context of the spatiotemporal interactions among the numerous cell types and molecular components that form the CNS, one is required to carefully study the data. HTD data are analysed and combined in systems biology utilising a combination of bioinformatics and mathematical modelling methods. These results are used to clarify how molecular networks that control cellular function and phenotype during development, normal cellular function, and pathological circumstances are organised and managed (Braun 2014)

Understanding common variant linkages and rare variant processes, which yield multiple insights, is basically how systems biology works. The novel discoveries of common locus and rare genic changes causing the disease were brought about by this investigation, leading to multiple distinct and overlapping concerns. Finding the relevant gene in a shared, non-coding genetic link is the first challenge. For instance, recent studies have connected CD33 and Trem2 common and low-frequency coding variants to Alzheimer's disease (AD), respectively (Jonsson et al. 2013). In the brain's microglia, a sizable phagocytic cell type that is currently believed to facilitate synaptic pruning, both of these genes are highly expressed. The expression of these risk genes has now highlighted this hitherto hidden aspect of AD, and a new cell type focusing on therapy has also been proposed. The systems biology technique has been extensively used to investigate the possibility that certain types of brain cells may express risk genes more frequently than others (Skene and Grant 2016).

14.4 Glial Cells

A century of research has disproved the old belief that glial cells serve only as the "glue" that holds neurons together. Though many aspects of these cells have already been extensively defined, the activities of the diverse glial populations in the brain in both healthy and diseased conditions are still largely unclear (Jäkel and Dimou 2017). These glial cells support the operation of synapses and maintain homeostasis in close proximity to the neuronal cells of the nervous system by assisting in the formation of the myelin sheaths around the axons, which are important for cell-to-cell transmission. As shown in Fig. 14.4, the four primary subgroups of the total population of glial cells are microglia, astrocytes, oligodendrocytes, and their progenitors, NG2-glia (Jäkel and Dimou 2017).

Studies on purely neuronal or glial populations show how gathering substantial amounts of microarray data from a single study paves the way for eventual work on establishing neural classes. In a similar vein, the recently created "bacTRAP" approach allows for the purification of translating ribosomes by using BAC-transgenic mice that have a ribosomal protein coupled to a green fluorescent protein that is under the control of a particular cell type promoter. The mRNAs of a certain cell type can be extracted from the translation-active state by utilising an antibody against the green fluorescent protein and ribosomes. This method protects against the possibility of stress- or injury-induced alterations in gene expression, making it appropriate for high-throughput research after the mice are fully developed. This approach prevents the possibility of stress- or injury-induced alterations in gene expression, making it suitable for high-throughput research after the mice are fully developed. Thousands of new markers and cell-specific targets for further study were provided by an unusual generation of bacTRAP mice that developed deep expression profiles. Future research will be necessary to verify this tactic and assess

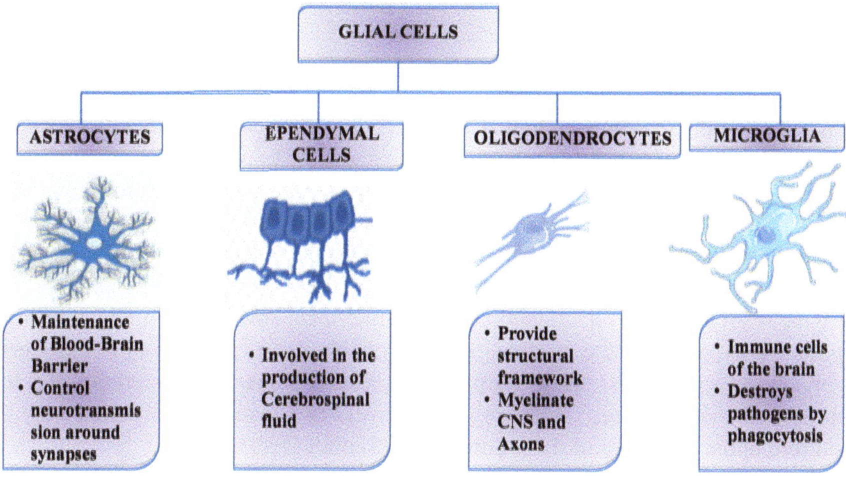

Fig. 14.4 Groups of glial cells population

its superiority to the previously fully investigated methods (Geschwind and Konopka 2009).

14.5 Neuroglia and Microglia

A variety of non-neuronal cells involved in the peripheral, central, enteric, or autonomic nervous systems make up neuroglia. Nearly half of the entire volume of the brain may be made up of these cells. The nervous system's homeostatic and protective cells, which are represented by a wide range of cellular populations, define them. In particular, in neurodegenerative illnesses, these cells play a progressive role in neuropathologies (Verkhratsky et al. 2019).

The healthy functioning of the brain depends on the specialised macrophages, or microglia, in the brain. Growing data suggests that these cells control the inflammatory response in connection with a number of neurodegenerative illnesses and mental disorders. Over the past 10 years, investigating these cells' functions and mechanisms for control has thus emerged as a crucial area of study in the field of neuroscience. In this regard, recent studies describing the epigenomic mechanisms regulating microglial gene transcription have shed light on the processes determining the ontogeny and cellular capacity of microglia. Systems biology has revealed that, in contrast to other tissue-resident macrophages, a significant fraction of the microglial repertoire of promoter-distal genomic regulatory elements, or enhancers, is primarily specific to these cells. Notably, the mechanisms that preserve this specialisation in the brain—transcription factors in microglia—are those that regulate the activity of target axes of signalling pathways. Because the microglial enhancer repertoire is highly responsive to changes in the brain tissue environment, it has a considerable impact on how these cells behave in neurological diseases (Xavier et al. 2014).

14.6 Synapses

Synapses in a brain circuit quickly transport information between neurons while also transforming that information. The connections between pre- and postsynaptic neurons influence the many computational characteristics of synapses. It is mainly unknown how synapses come together to create a brain circuit and how synaptic connections are produced. The interaction of the presynaptic and postsynaptic terminals is mentioned in Fig. 14.5. Synaptic adhesion molecules (SAMs) are proposed to direct the development of synapses. The great specificity and adaptability of synapses result from the collaboration of numerous SAMs, each of which brings a unique perspective to the process. SAMs probably function as signal transducers in organising synapse assembly. Surprisingly, several SAMs have hereditary ties to neuropsychiatric illnesses, indicating that defects in synapse construction play a vital role in the aetiology of these conditions (Südhof 2021).

An innovative and fascinating model of the structural and functional changes related to synaptic transmission and brain processing has been developed as a result

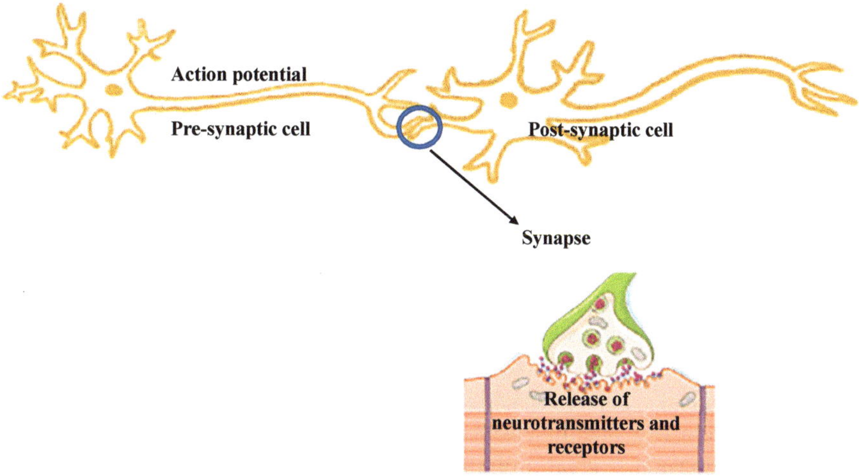

Fig. 14.5 Initiation of a synaptic response

of the extensive study of the synaptic cleft over the past several decades. The traditional neurocentric model, which takes into account the pre- and postsynaptic terminals of neurons, only partially describes the carefully controlled plastic changes that occur in both diseased and healthy conditions. The role of oligodendrocytes, astrocytes, and microglia in the development and remodelling of synapses in both the developing and adult brain is now beyond dispute. Through the study of synaptic plasticity and its pathological counterpart, maladaptive plasticity, the extracellular matrix (ECM), which was previously only thought to serve as an extracellular structural scaffold along with cellular glue, has also revealed a close relationship with other molecular components. The ECM makes the contemporary concept of the synapse much more complex, especially in terms of long-term plasticity and circuit preservation (Glasgow et al. 2019).

Recent studies have confirmed these elements as structural and significant actors in synaptic plasticity, despite the fact that they were previously thought to be significantly distinct from the plasticity of the central nervous system (CNS), at least in physiologic situations. According to theories and studies, this situation serves as a consistent demonstration of both adaptive and maladaptive plasticity. Here, the systems biology approach can provide a better understanding of the complexity of the CNS through the development of prognostic molecular models that allow illumination of the regulatory logic of the intricate molecular networks supporting brain function in both health and disease, paving the way for more successful treatments.

The systems biology method uses behaviour and perception that are based on synapses that link neurons. By clustering together, synapses commonly create stereotypical layers in space. Here, the initial step in the polarisation process is symmetry breaking, which leads to the creation of a single axon and many dendrites that are crucial for brain function. A growing corpus of research has revealed numerous substances that regulate neuronal polarity and act asymmetrically in neurons.

Systems biology has therefore concentrated on comprehending the asymmetric signals generated throughout the symmetry-breaking process. The developed models may contain a positive feedback loop to implement local activation, but they have different implementations of global inhibition. A robust approach in systems biology is quantitative testing coupled with computational modelling, and investigations in this area have produced a more in-depth knowledge of how neurons lose their symmetry during polarity development (Inagaki et al. 2011).

Behaviour and perception are built on synapses that connect neurons in the systems biology approach. Synapses frequently form stereotyped layers in space by grouping together. In neuromorphic systems, artificial synapses imitate biological synaptic impulses to produce computing and autonomous learning capabilities similar to those of the brain. In this area, a range of materials have been used to induce different biological synaptic capacities. Due to their molecular-level thickness and nanoscale feature size, one-dimensional (1D) and two-dimensional (2D) nanomaterials have demonstrated great promise through artificial synapses. In 2018, Chen and colleagues reported that advanced artificial synapses based on nanomaterials and research based on the history of artificial synapses can replicate biological synaptic signals in neuromorphic systems to produce brain-like computing and autonomous learning behaviours in non-von-Neumann systems (Chen et al. 2019).

14.7 Axons and Dendrites

Axons serve as the presynaptic compartment comprising clusters of synaptic vesicles at release locations. Research has shown that a considerable number of proteins have contributed to the axons by axoplasmic trafficking. Here, axonal local translation plays a crucial role in the neuron's cell biology. Numerous studies demonstrate how local protein production mediates communication between the homologous axon and the nucleus of neurons. The combination of omics analysis at several layers of gene expression was first done in this study. The local synthesis of one-third of the axonal proteome occurs at baseline circumstances. Previous research has shown that neurofilaments, made from heavy neurofilaments by transfecting them with green fluorescent protein (GFP) under the control of an inducible promoter, have a half-life of 3 months in cultured neurons (Farias et al. 2019).

Clusters of synaptic vesicles at release sites are housed in presynaptic compartments called axons. Axoplasmic trafficking has been found to contribute a significant amount of proteins to the axons. Here, the cell biology of the neuron is greatly influenced by axonal local translation. Numerous studies show that communication between the homologous axon and the nucleus of neurons is mediated by local protein synthesis. This work was the first to combine omics analysis across many layers of gene expression. At baseline conditions, one-third of the axonal proteome is locally synthesised. Previous studies have demonstrated that the Green fluorescent Protein (GFP)-transfected heavy neurofilaments used to create neurofilaments have a 3-month half-life in cultivated cells (Ramirez-Suarez et al. 2019) (Table 14.1).

Table 14.1 Major differences between axons and dendrites

Axons	Dendrites
One axon arises from each neuron at the time of early neuronal development	From their cell bodies, numerous dendrites emerge, but they do not separate until the axon has developed
In vertebrates, large axons are myelinated, and the microtubules there have consistent polarity	Dendrites lack myelination, and the proximal dendrites' microtubules have mixed polarity
Axonal branches often create obtuse angles with diameters that are identical to the parent and are distal to the body	Dendrites produce acute angles and have numerous branches close to the perikaryon, which are smaller than the parent stem
Stem	Action potentials may be produced by dendrites, although they mostly modify the electrical state in the perikaryon's early segments
Normally, action potentials are produced at the axon hillock and carried away from the cell body	Dendritic architecture is ideal for integrating synaptic responses from many neural inputs
Axons have distinct conduction and synaptic transmission functions	Although dendrites carry MAP2, unlike axons, they are also high in tau protein

Numerous studies have combined other imaging modalities, anatomical imaging, and stereological analysis of neuronal structure, including dendritic spines, with electrophysiology, which involves injecting fluorescent dye into cells through the recording pipette, to depict the complex changes in disease models. The main obstacle in these investigations is the precise quantification and classification of anatomical characteristics. Although it is now possible to obtain enormous volumes of 3D anatomical data or data series required to feed systems-level models utilising high-resolution imaging techniques. The potential incorporation of imaging data into models for the prediction of neuronal function and structure is significantly hampered by the need for new computational tools to interpret, analyse, and quantify different types of imaging data. The development of automated or partially automated software systems that recognise and classify dendritic spines in labelled neurons has advanced significantly. To fully automate the process of precisely finding spines in high-density or poor signal-to-noise images, however, additional work remains to be done (Mancuso et al. 2014).

14.8 Astrocytes

Astrocytes are a collection of cells that have various morphologies and functional properties depending on where they are positioned in the brain. After conception, steroid progenitors go to the tissues surrounding the brain. They play a regulating role in the formation of neurons and synapses in the brain, the management of the permeability of the blood-brain barrier, and the maintenance of extracellular homeostasis (Rosalba et al. 2019). Despite a wealth of evidence suggesting astrocytes control local brain circuits, networks, and complex behaviours, systems neuroscience is still primarily a neuronal field. Researchers therefore tried to identify calculations utilising mathematical techniques that support coding and higher-brain operations in addition to their well-documented homeostatic and metabolic tasks.

In system-like investigations, astrocytes were employed to examine putative computational processes in which these cells might use Ca^{2+} transients as their encoding language. The analysis demonstrates that astrocytes are capable of performing conventional computations in sensory processing, neuromodulation, brain state, memory formation, fear, and sophisticated homeostatic reflexes in the time range of sub-seconds to seconds. Here, fundamental objectives require employing machine learning-based statistical analyses, such as dimensionality reduction and decoding in the context of complex behaviours, in conjunction with connecting omics of astrocyte-neuronal networks. Astrocytes have also been incorporated into intricate models of brain circuits while technical and analytical methods for analysing neuronal and astrocyte populations have been used concurrently (Ung et al. 2020).

Despite the fact that astrocytes have been extensively discussed in works on systems neuroscience, very few investigations have been done to examine how they affect higher-order brain functions. When memories are being formed, cholinergic neuromodulation plays a role in the encoding of new information. By expressing opsins (light-sensitive proteins) and stimulating them with precise timing of light, a technology called optogenetics enables control of the membrane potential or intracellular signalling in genetically defined cell populations. The optogenetic techniques that directly influence membrane potentials, such as ion pumps or ion channels, are those that are utilised the most frequently in neuroscience. For neurons, this type of therapy makes sense, but not for astrocytes since the changes in ionic concentration and membrane potential that are created do not correspond to any known physiological activity (Ferenczi et al. 2019).

Studies have revealed that respiration can be controlled in response to changes in the pH of the respiratory system by optogenetic stimulation of astrocytes employing ChR1/2, Arch, and OptoGq, of which less than 1% of the 3000 communications involved non-neuronal cells. Even little is known about how changing the light stimulation frequency affects astrocytes. Recent research has used light-sensitive G-coupled receptors to modify astrocytic activity, which may better imitate physiology. For instance, the CA1 astrocytes, in combination with photoreceptor Gq and melanopsin, enhance cognitive function by promoting de novo plasticity in nearby neurons. Machine learning has been used to decode strategies for reward-associated decision-making in a variety of contexts as a way to tackle the problem (Kastanenka et al. 2019).

It is intriguing to consider how astrocytes might be crucial to the development of artificial intelligence. Artificial neural networks that are interconnected are the foundation of connectionist artificial intelligence systems. Recent studies have demonstrated that artificial astrocytes enhance the functionality of constructed networks. This is in line with recent experimental results that indicate astrocytes have a direct part in the brain's information processing. Neuroscientists can now better understand how astrocytes perceive, assess, and control synaptic transmission thanks to this new information. It was also described how synthetic astrocytes performed on artificial neural networks. In order to understand how astrocytes affect intricate brain networks involving inhibitory neurons and neuronal feedback transmission, more research is required (Perea et al. 2014).

14.9 Mechanism of Action of Neurotransmitters and Neuromodulators: A System Biology Approach

A key model in neurosciences for understanding how arthropods perceive sensory stimuli and exhibit particular behaviours is the crayfish nervous system. For more than 200 years, crayfish embryos have also been important research topics. The widespread application of immunohistochemistry against neurotransmitters, neuromodulators, and neurohormones to investigate the ontogeny of neurons in the developing brain of various species of crustaceans represents one of the most efficient approaches to date for investigating various facets of brain development in this group of organisms (Harzsch and Viertel 2019).

Neurotransmitters are substances that move electrical messages along nerves or from nerves to effector organs. Animals have a wide variety of neurotransmitters, including acetylcholine, amino acids, amines, peptides, and gases. Presynaptic vesicles, specialised neurotransmission system receptors, synthetic and degradative enzymes, and other neurotransmission system elements may all be affected by toxicants. Strong poisons that disrupt the cholinergic system include anticholinesterases (organophosphates and carbamates), substances that act on receptors like nicotine, and neonicotinoid insecticides like imidacloprid. The toxin that causes amnesic shellfish poisoning, domic acid, has a major impact on the glutamatergic system. While strychnine and fipronil are excitatory antagonists, glycine and 4-aminobutyric acid (GABA) are inhibitory neurotransmitters. Abnormalities characterised in Parkinson's disease are due to dopamine neurotransmission and several drugs that affect this system can cause Parkinson's symptoms and clinical signs, most notably 1-methyl-4-phenyl-1,2,3,6-tetrahydropyridine, which is the precursor to 1-methyl-4-phenylpyridinium. There are some instances of less common chemicals that can disrupt adrenergic, histaminergic, or serotoninergic neurotransmission. The only known hazardous substances in peptide neurotransmission systems are opioid agonists (Marrs and Maynard 2013).

The importance of neurotransmitter signalling and its early function before synaptogenesis is supported by numerous studies. Numerous synapses serve as electrical synapses as well as neurotransmitter receptors to join together neurons. The brainstem contains the majority of neurons, and it is here that studies of the norepinephrinergic development of the CNS are being conducted on the cell bodies. Two neuromodulators produced by other cells, adenosine triphosphate (ATP) and prostaglandin, can affect neuronal activity. Small, permeable lipophilic or gaseous molecules have demonstrated to be effective neuroactive agents throughout the previous 10 years (Ojeda and Ávila 2019).

The body's two main interfaces for identifying, analysing, and responding to internal or external disruptions are the immune and neurological systems. Neurotransmitters in the local tissue microenvironment have an autocrine or paracrine effect on these types of cellular communication and the following signal transduction events during cross talk between the nervous system and the immune system. Thus, neurotransmitters have an impact on both innate and adaptive immune responses. Additionally, immune cells in the brain influence neurological reactions and use cytokines to send messages to the brain. Alterations in the brain and immune systems' ability to communicate are a typical feature of immunopathological and neurodegenerative diseases. Recent systems biology research has looked at how neurotransmitter-driven neuroimmune networks may play a role in the development of cancer, autoimmune diseases, and neurodegenerative diseases. This analysis can eventually narrow down the potential clinical ramifications of regulating neuroimmune cross talk through therapeutic regulation (Hodo et al. 2020).

One cell releases neurotransmitters (NTs) and neuromodulators (NMs) into the synaptic cleft in the chemical synapse, typically from synaptic vesicles localised in the presynaptic cell. These chemicals subsequently bind to the receptors of the targeted postsynaptic cells. Even though it has been shown that knocking out all synaptic trafficking in the mouse does not prevent normal brain development, including the formation of layered structures, fibre pathways, and morphologically defined synapses, the intricate wiring of the neuronal circuits can be attributed to neurotransmitters. Transmitters and synaptic activity must both be present for synaptic connections to persist; after this, neurons that have created synapses undergo apoptosis. This is because neurotransmitter secretion is necessary for maintenance (Herlenius and Lagercrantz 2001).

The release of neurotransmitters (NTs) and neuromodulators (NMs) occurs in the G-protein coupled receptors and ligand-gated ionotropic receptors modulate the effects of conventional NTs on postsynaptic cells, including glutamate, GABA, and acetylcholine (ACh) (GPCRs). Contrary to NMs, which predominantly establish chemical signalling cascades through GPCRs, such as monoamines, nucleotides, neurolipids, and neuropeptides, NTs interact with ionotropic receptors and often start quick, spatially confined, point-to-point synaptic transmission. Numerous physiological processes, including arousal, attention, perception, learning, and other complex behaviours, are regulated by chemical neurotransmission. Numerous brain conditions, including Alzheimer's disease, Parkinson's disease, schizophrenia, depression, and drug addiction, have been linked to impaired neurotransmission. Because of this, it will be possible to gain vital insights into the intricate mechanisms governing intercellular communication in both health and illness by carefully monitoring the concentration and dynamics of certain NTs and NMs in the extracellular space. It has been reported that many techniques and indicators, both genetically and non-genetically encoded, have been created to monitor neurotransmission. These complementary technologies are essential for understanding the dynamic regulation of different NTs and NMs in the extraordinarily complex nervous system. In addition, research utilising these technologies is enabling the investigation of novel strategies for the avoidance, recognition, and treatment of a variety of diseases and conditions (Wu et al. 2022).

14.10 Acetylcholine

Acetylcholine (ACh) has an immediate and moment in time action at the neuromuscular junction and in the autonomic ganglia. It is hydrolysed by acetylcholinesterase into choline and acetate (AChE). It primarily dwells at animal synapses and neuromuscular junctions, where it inhibits earlier ACh-released signalling. Despite being the primary excitatory neurotransmitter in the peripheral nervous system, ACh functions as a neuromodulator in the brain. Acetylcholine transmission is seen in Fig. 14.6. The two main sources of ACh in the brain are local interneurons, which are scattered among their cellular targets, and projection neurons, which innervate distant regions of the brain. These regions represent the medial habenula (MHb) and the basal forebrain (BF) complex, which includes the medial septum, pedunculopontine, and laterodorsal tegmental, as well as other brain regions (Colangelo et al. 2019).

These cholinergic neurons diffusely and broadly project to innervate neurons throughout the CNS. Despite the fact that the underlying neurological underpinnings and cellular mechanisms responsible for these effects are still being studied, it has been discovered that elevations in ACh signalling appear to contribute to stress-related disorders. Despite substantial progress, we still have a very poor grasp of the dynamics of ACh release from diverse neuronal populations, their sensitivity to environmental factors like stress or metabolic need, and how far ACh can diffuse in different brain locations from the point of release. New approaches will allow for more precise ACh release stimulation, but the physiological mechanism must still be considered when creating the release patterns. For determining the physiological consequences of cholinergic neuromodulation and separating the real from potential effects of ACh in the brain, the capacity to imitate patterns of ACh release in vivo will be an essential tool (Picciotto et al. 2012).

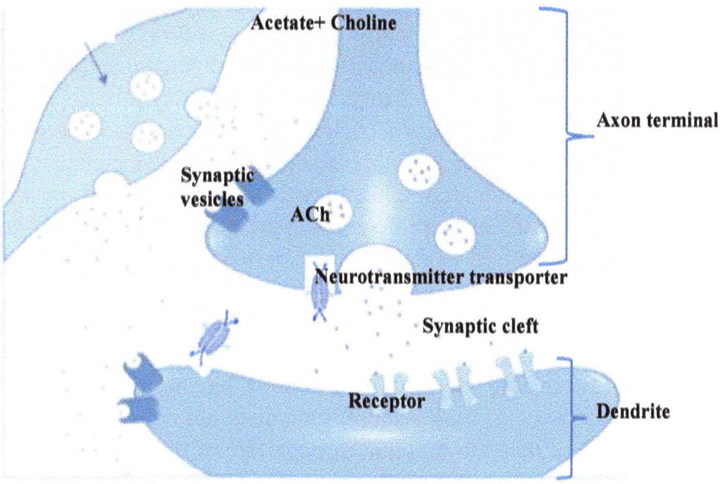

Fig. 14.6 Neurotransmission of acetylcholine

Existing methods either only allow indirect detection of ACh or have a low signal-to-noise ratio and interfere with cellular signalling. From a practical perspective, it is also essential to directly quantify anticholinesterase compounds (ACs), or molecules that target AChE, such as drugs for disorders associated with AChE, pesticides, and chemical weapons. The competition between a tethered ligand and a free analyte alters the fluorescence resonance energy transfer (FRET) efficiency within the probe, according to the SNIFIT database. The semisynthetic FRET-based probe ACh-SNIFIT operates in this manner. AChE, CLIP, and SNAP-tag fusion protein are the components of the SNIFIT sensor. Functional ACh-SNIFIT is produced following site-specific labelling using synthetic compounds. Specifically, CLIP is labelled with a FRET donor, and SNAP is labelled with a compound that contains a FRET acceptor and an AChE ligand. In conclusion, a real-time fluorescence probe has been designed that is reliable and adaptable for cholinergic and anticholinesterase chemicals. Both in homogeneous experiments and when tethered to the outer cell surface of live cells, ACh-SNIFIT demonstrated exemplary performance. This probe's ratiometric response to ACs makes it a desirable instrument for imaging the biodistribution of ACs as well as for quantifying medicines, insecticides, nerve agents, or their antidotes. From this systems biology approach, it is anticipated that ACh-SNIFIT in the future can be implemented either through direct expression by the host cell or the implantation of AChSNIFIT cell lines, following the recent demonstrations of SNAP-tag expression and labelling in animals and imaging of implanted fluorescent cell lines (Schena and Johnsson 2013).

14.11 Glutamate

Using mass spectrometric analysis of multiprotein complexes isolated under natural conditions, the known and novel interactors of 118 proteins, including glutamate receptors, K+ channels, scaffolding, and signalling proteins, were discovered. Figure 14.7 mentions the glutamate neurotransmission pathway. Network clustering of protein interactions resulted in the formation of five interconnected clusters, two of which had voltage-dependent K^+ channels and all of the key ionotropic glutamate receptors. Through the annotation of clusters with human disease correlations, including schizophrenia, which is highly associated with the glutamate receptor clusters, various diseases, including schizophrenia, are mapped to the network. The systems biology and mammalian proteomics approaches to disease are generally appropriate to this customised tandem affinity purity tagging method (Fernández et al. 2009).

The excitatory neurotransmitter glutamate has an important role in the mammalian brain and regulates memory and cognition, among other aspects of neuronal activity. However, excessive glutamate stimulation has been connected to brain deterioration and cell death. Furthermore, glutamate activates G-protein-coupled metabotropic receptors that can function through the cyclic adenosine monophosphate (cAMP) pathway, which is linked to neurodevelopment and melatonin regulation. The excitatory neurotransmitter glutamate is the main focus of contemporary

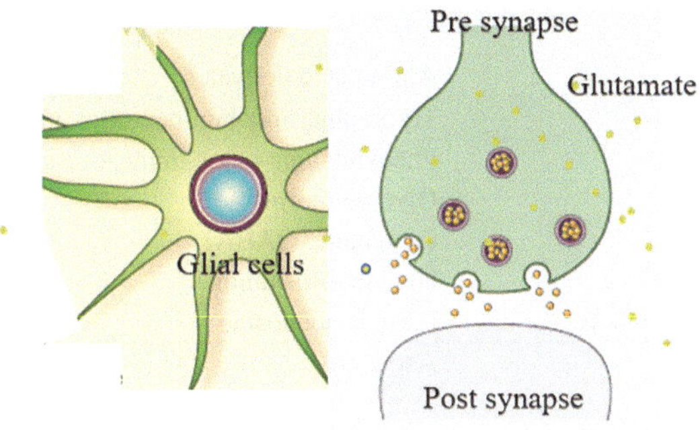

Fig. 14.7 Mechanism of glutamate neurotransmission

molecular mode. Postsynaptic receptors that transmit electrical depolarisation (AMPA receptors) and those that trigger signalling and plasticity pathways are both activated by this neurotransmitter (NMDA and mGluR). NMDA and mGluR are organised into massive complexes of 186 proteins totalling 2–3 MDa, and AMPA receptors are assembled into much smaller complexes of ten proteins, according to proteomic profiling of glutamate receptors isolated from the brain (Pocklington et al. 2006).

It is well-known that foetal alcohol syndrome (FAS) can harm neurodevelopment and disrupt vitamin metabolism. According to a study, ethanol exposure in the brain may have a deleterious impact on the cAMP pathway, leading to abnormalities in G-protein-coupled receptor activation in a mouse model of FAS when glutamate levels are abnormally high. The systems biology analysis revealed the transcriptome of the genes in this scenario to contain 19 genes that were under expressed in the prenatal, late gestation-exposed postnatal, and postnatal-exposed networks (Feltes et al. 2014).

By integrating proteomic data and looking for answers about glutaminergic signalling, systems biology research supports the idea that animal models can contribute to studying the biology and treatments of depression. In this instance, the HRPD and IntAct mice are employed as model protein-protein interaction databases. The networks' functional enrichment analysis found a connection to biochemical processes, like those involved in the immunological response, that are connected to human depression. Clinically effective antidepressants impact several pathways, including glutamatergic signalling and neurotrophic responses. Furthermore, proteins that control energy metabolism and circadian rhythms are implicated in dysregulations. Significant overlaps were found when compared to altered protein pathways in depressive patients. This research concludes the molecular foundations of depressive behaviour with an unbiased approach (Carboni et al. 2016).

Table 14.2 ■

Neuropeptides	Amino acid derivates
Enkalphins, ß-endorphins, dynorphins	Tryptophan-indolamines-serotonin
Vasopressin, oxytocin, neurophysins	Phenylalanine-octopamine, dopamine
Tachykinins-neurokinin A, kassinin	Glycine-taurine
Gastrins	Glutamate: GABA
Somatostatin 14 and somatostatin 28	Histidine-histamine
Glucagon-related vasoactive intestinal polypeptides	Purines-adenosine
Calcitonin gene-related peptide	Aspartate

14.12 Neuropeptides

For cell-cell communication in neurotransmission and the control of endocrine function, neuropeptides, a class of peptide neurotransmitters and peptide hormones, are essential. A list of neuropeptides and the amino acids they are produced from are given in Table 14.2. From protein precursors that must go through proteolytic processing, secretory vesicles produce neuropeptides that eventually target cellular and organ systems. As peptide neurotransmitters, neuropeptides mediate neurotransmission. To regulate endocrine activity, they act as a conduit for cell-cell communication. The difficulty of their controlled coordination arises because the production of neuropeptides uses the unique cathepsin L and proprotein convertase routes. Numerous physiological processes, including analgesia, appetite regulation, blood pressure control, cognition, stress, water balance, and glucose metabolism, are mediated by neuropeptides (Hook et al. 2018).

Proteolytic processing is a crucial step in producing many active neuropeptides from inactive precursors. Current and upcoming studies will examine the characteristics of protease systems that might offer particular pharmacological targets for later therapeutic control of neuropeptide synthesis and activity. Aspects of the findings support earlier research in this area: Additionally, enkephalin brain levels in PC2 defective mice are decreased by around 50%. These findings suggest that cathepsin L and PC2 play complementary roles in enkephalin synthesis. Neuropeptides govern a wide range of physiological and pathological circumstances. Thus the understanding mechanisms of small molecule approaches for creating therapeutic medicines can influence the production of particular peptide neurotransmitters (Hook et al. 2008).

In mice fed high-calorie, high-fat, and high-sugar diets, stress potentiates the development of obesity and other symptoms of metabolic syndrome by working through the neuropeptide Y (NPY) and glucocorticoid systems. In both animals and people who have experienced acute, life-threatening, or catastrophic stress, changes to the NPY and glucocorticoid systems have been demonstrated to have an effect on behavioural response to stress. In order to characterise variations in neuropeptide expression and comprehend their roles in a number of biological issues, such as metabolic function and neurodegenerative diseases, this work uses mass spectrometry data to enable systems biology (Sauer et al. 2021). According to system biology, the study demonstrates a link between stress and post-traumatic stress disorder, metabolic syndrome, and obesity (post-traumatic stress disorder). The pathogenesis of the comorbidities of metabolic syndrome and post-traumatic distress in this instance appears to be most likely influenced by the NPY and stress steroid systems. It may be possible to improve interdisciplinary approaches to the prevention and treatment of metabolic syndrome by paying particular attention to these brain systems and reducing stress in general (Rasmusson et al. 2010).

14.13 Dopamine and Serotonin

Dopamine (DA) and serotonin are two of the most important neurotransmitters in the central nervous system (5-hydroxytryptamine or 5-HT). Serotonin is an inhibitory neurotransmitter, whereas dopamine is an excitable member of the catecholamine family of neurotransmitters motor control, reward-based learning, excitation, addiction, activeness, motivation, and upper cognitive function are a few of the pathways it has a significant impact on. Along with its many other roles, serotonin affects mood and feelings as well as activity, aggressiveness, addiction, analgesia, anorexia, cognition, and stress tolerance. It is known that the cortical and subcortical regions are innervated by two serotonergic pathways that come from the dorsal (DRN) and medial reticular formation nuclei. A comparative computational analysis has revealed the potential for the simultaneous detection of dopamine and serotonin, which is necessary in order to properly identify and measure the amounts of dopamine (DA) and serotonin (5-HT) in mixtures. Numerous studies have confirmed both the label-free detection of DA and 5-H as well as their 1:1 compound interaction. Any primate's behaviour and psychomotor abilities revolve around them. According to their research, there is a possibility that they could interact in intriguing ways due to their close biological proximity and physiological cohabitation within the striatum. This study validated the interaction, which was more obvious in their combinations at larger doses (Manciu et al. 2018).

Given the complexity of the neurological basis for all mental processes, it is possible for multiple genes to have an impact on a single cognitive process. For instance, it is anticipated that changes in SLC6A4 and other serotonin system and neurotransmitter-related genes may affect many of the same brain functions. Variations in acetylcholine transporter and tryptophan hydroxylase 2 (TPH2), the gene that generates the rate-limiting enzyme for serotonin production, influence amygdala reactivity (CHT1) (Green et al. 2008).

A computer model was proposed to study major depressive disorder and explore the connections between dopaminergic and serotonergic-kynurenine metabolism under cortisolemia and their effects on the onset of depression. Biomedical databases, additional general knowledge, and experimental data from the literature were used to choose parameter values, which were then optimised. According to model simulations, high levels of cortisol can alter the kynurenine pathway, which raises the risk of neurotoxicity and results in higher concentrations of

3,4-dihydroxyphenylaceltahyde (DOPAL) and 5-hydroxy indole acetaldehyde (5-HIAL). A new understanding of the roles played by DOPAL and 5-HIAL, two serotonin and dopamine metabolites, in the development of depression, which is aided by high cortisol levels, was supplied by the mathematical model. Last but not least, this model showed that the use of kynurenine-3-monooxygenase and serotonin transport inhibitors may be a useful pharmaceutical strategy to restore the slowing of monoamine neurotransmission usually brought on by inflammation. These systems biology insights suggest distinct potential drug targets and fresh hypotheses for model-driven research (Dalvi-Garcia et al. 2021).

14.14 Gama-Aminobutyric Acid (GABA)

GABA is a dominant inhibitory neurotransmitter in the central nervous system. The rhythm-producing networks in the brain must include GABA interneurons and neuronal oscillation. They must be in sync for consciousness, perception, and memory to function. The detrimental effects of GABA, which interfere with glutaminergic and dopaminergic transmission in schizophrenia, were highlighted using systems biology. The GABA neurotransmitter system has been found to be aberrant in schizophrenia research, both in human clinical and basic neuroscience as well as in animal models. These changes could have an impact on schizophrenia's pathogenesis. Schizophrenia and this disorder may be connected, because schizophrenia has notable impacts on parvalbumin-positive chandelier subtype GABA neurons. The possible therapeutic role of GABA in the management of schizophrenia must be made clear (Yang and Tsai 2017).

GABA neurotransmitter dysfunction also contributes to major depressive illness. The pathophysiology and growth of this illness may be attributed to GABAergic neurons in the frontal brain. Additional proof that GABAergic malfunction may cause mood and cognitive symptoms comes from the discovery of the GABAergic synapse pathway through computational research (Fan et al. 2019).

The most frequent inherited form of intellectual disability and the genetic aetiology of autism is fragile X syndrome (FXS), which involves an Fmr-1 deficit. Given the crucial function that GABA plays in the transition of GABA from an excitatory to an inhibitory state, reduced GABA levels may hinder the greatest possible development of the cortex and cerebellum, especially in children. Several mRNA targets and proteins interacting with FMRP are thought to be responsible for the FXS metabolic response, according to a metabolomic and systems biology investigation of FMRP and its metabolic markers (a neuronal messenger). This interaction network was developed using data from databases and books. With the use of the innovative "integrated metabolome and interactome mapping" (iMIM) technique, it has been made clear how important specific cellular pathways are to the pathophysiology of FXS by combining novel metabolic findings with previously unrelated knowledge. These metabolomic and integrative systems biology approaches can help identify prospective drug targets and create fresh treatment approaches that will ultimately help FXS patients (Davidovic et al. 2011).

14.15 Memory and Learning: A System Biology Approach

The processes of learning and memory entail the interaction of broad, dispersed brain networks and are critical to an individual's interaction with their environment. Our understanding of human learning and memory has advanced due to recent technological developments investigating the neurological correlations of neuropsychological paradigms. Cognitive processes like memory and learning have many different components. There are various ways to arrange these elements. For instance, we can concentrate on their temporal aspect or distinguish between distinct types of memory based on their content or mechanisms of acquisition. These features make it evident that learning and memory have a complex cognitive structure and that neither neuropsychological nor neurobiological models can provide a fully satisfying taxonomy due to the numerous interactions and overlaps between important subcomponents (Brem et al. 2013).

14.16 Neurology of Memory

We can connect events, learn new things, and make sense of our existence because of our capacity for memory. The neuroanatomical, neurobiological, neurophysiological, and psychological mechanisms underpinning this complex capacity are complex, and psychologists and neuroscientists struggle to fully comprehend them. Various categories have been developed utilising the most recent scientific studies on the subject in order to better understand memory. They include sensory memory, short-term memory, and long-term memory. In Fig. 14.8, the classification of memory is highlighted. Information from the environment is initially stored in sensory

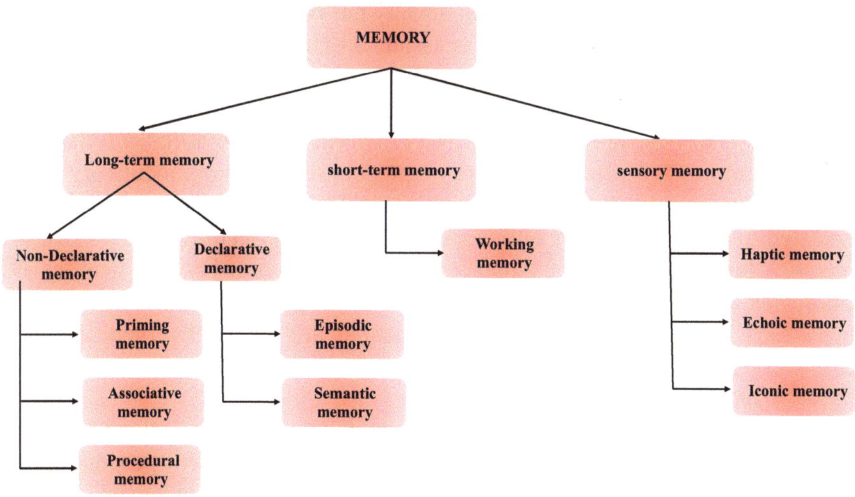

Fig. 14.8 Classification of memory

memory, allowing for later storage and use of that information. Unlike short-term memory, which refers to information processed in a little period of time, long-term memory allows us to store knowledge for longer periods of time, including knowledge that can be recovered consciously or unconsciously (Camina and Güell 2017).

Both genetic and epigenetic elements affect how memories are formed. The transcription of genetic information from DNA to mRNA involves the methylation of DNA, histones, and histone acetylation. These intracellular changes are believed to have contributed to memory formation, and maintenance cannot account for memory without consideration of synapse function. The electrical activity and information flow in the brain that are necessary for the creation of experience-dependent memory cannot be changed by such alterations. Rather, these changes regulate the activity of a number of genes, the results of which re-weight synapses and allow a long-term rerouting of neuronal activity (Kim and Kaang 2017).

Finding the fundamental mechanisms governing memory creation and storage has been one of the main topics of study for neuroscientists the past 20 years. The molecular and cellular findings in vertebrates, particularly in mammalian systems, were made possible by research in invertebrate systems like Aplysia californica and Drosophila melanogaster. The majority of research has focused on temporal lobe-dependent memories. The development of long-term memory is significantly influenced by post-translational, translational, and transcriptional molecular alterations, according to research in genetics, molecular biology, electrophysiology, and anatomy. These include both general long-term plasticity mechanisms—found throughout the brain and implicated in all paradigms requiring long-lasting changes—and selective long-term plasticity mechanisms, which are found only in particular brain regions (Bisaz et al. 2014).

Dopamine has a critical role in stabilising the integration of new information into long-term hippocampal memory, according to numerous research studies. Even when there is no related reward, novelty has been demonstrated to stimulate dopamine neurons. This stimulation is assumed to be mediated by the polysynaptic pathway that links the hippocampus to the dopamine cells in the VTA. However, there are still a plethora of unanswered concerns regarding this procedure (Tryon et al. 2017). Its role in memory has been connected to the long-held "Hebbian learning" notion. It is discovered that synaptogenesis and neuromodulatory activities promote certain mechanisms involved in Hebbian plasticity. Additionally, research has found that synaptogenesis is crucial for learning and associative memory during development. Contacts between neurons or between neurons and astrocytes can start synaptogenesis. Cell-adhesion proteins, including neurexin/neuroligin, Eph receptors, and cadherins, mediate this process by triggering intracellular signalling pathways including cofilin, GTPases, and other proteins (Nelson and Alkon 2015).

Ecphory, a remarkable cognitive process that involves the interaction of engrams from previously stored memory with outside sensory input or self-generated stimuli, allows us to retrieve memories. Ecphoria has been studied in human cognitive neuroscience, but little is known about the underlying brain mechanisms. Given that ecphory includes the "reawakening" of engrams, current technological advancements that can identify and alter engrams in rats present a rich opportunity for studying retrieval at the level of neuronal ensembles. It will be easier to organise and comprehend early results if we review the growing body of this type of neuroscientific research using cognitive theory as a guiding principle. Engrams and retrieval cues need to work together for memory access and retrieval to be successful. These findings also demonstrate the close connection between the mechanisms crucial for engram formation and recovery, as reflected in the cognitive concept of "encoding specificity" (Frankland et al. 2019).

14.17 Systems Biology of Memory

Numerous signalling paths exist for a neuron to use to connect with other neurons, and signalling networks are formed when elements of these pathways interact. Instead of focusing only on the peculiarities of particular brain regions, Understanding the structure and dynamics of cellular and biochemical function is necessary to comprehend learning and memory in brain systems at the systems level. A system-level comprehension of a neurological system can be represented as the network of gene connections and metabolic pathways, as well as the ways in which these interactions change the electrophysiological properties of neurons. Computational systems biology techniques are used to analyse biochemical signalling networks and predict how they will operate in complex biological processes (Kandel et al. 2014). Sadly, there has not been much collaboration between the two fields to investigate the neural underpinnings of cognition. Computational systems biology and computational computer neuroscience employ computational methods to foresee and investigate the properties of the brain (Mendis et al. 2016).

Thanks to network approaches, our understanding of the dispersed processes enabling effective memory performance has begun to advance. In one fMRI study employing a yes-no smell identification test, it was discovered that the hippocampus, caudate nucleus, and anterior cingulate gyrus all belong to the same functional module more frequently after certain hits than under other circumstances. The relationships between network-wide modularity values and bias scores in the hit and false alarm conditions as well as between network-wide modularity values and memory in the hit condition show that accuracy declined as network segregation increased. Therefore, this memory subsystem may need to incorporate certain discriminatory decision-making procedures (Zhang et al. 2020).

The working memory-related cognitive processes have also been elucidated through network analysis. They have demonstrated that the effective organisation of functional networks throughout the brain, in particular frequency domains, supports working memory function. Compared to delta and theta band networks, which were more concentrated and small-world like, alpha and beta band networks had a lower global visual memory maintenance efficiency. Variations in working memory performance and capacity are related to specific functional network configuration components. Working memory capacity was predicted using network heterogeneity in intrinsic functional connections after a test. Individuals who performed better on a

2-back working memory test had more efficient functional network architecture, according to MEG in the beta frequency range (Medaglia et al. 2015).

The systems biology approach to memory consolidation, which speaks of a slower post-encoding reorganisation of long-term memory over distributed brain circuits into remote memory lasting months to years, is frequently studied in the context of the cortico-hippocampal system, which supports explicit memory. The main sources of inspiration for the current models of systems consolidation in humans are behavioural and anatomical investigations of amnesic patients and fMRI research, which tracks time-dependent changes in recollection-correlated brain activity in healthy human volunteers. Models that support the "standard consolidation theory" and those that do not, such as the "many trace theory" and the "trace transformation theory," fall into two categories. Animal models that can now make use of contemporary technologies allow for a clearer comprehension of some aspects of system consolidation. Numerous studies have demonstrated that while hippocampal lesions significantly impair recall after learning, they have no impact on recall after contextual fear training in mice. Similar to the inconsistent reports regarding amnesic gradients in human declarative memory, there have been some (Shah et al. 2019).

By swiftly optogenetically silencing the hippocampus, halorhodopsin, a light-gated chloride pump that functions on a millisecond timescale to hyperpolarise neurons, prevents the generation of action potentials. The hippocampus is still engaged in long-term memory, as shown by the effect of hippocampal silencing using halorhodopsin on memory recall after contextual fear training. It is interesting to note that, in order to replicate earlier pharmacological and lesion-based research, memory was unaffected by extending the silence for an additional 30 min before the recall session. This illustrates that, despite the fact that the hippocampus is frequently activated and necessary for retrieval at long after learning times, there are compensatory mechanisms that enable retrieval without the hippocampus in the case of a persistent loss of the hippocampal pathway (Liu et al. 2012).

The development of novel, creative algorithms represented the breakthrough in memory and learning in the field of systems biology. These techniques may play a significant role in the development of more advanced cognitive development robotics (CDR) systems. Large brain networks, synaptic dynamics, and the chemical underpinnings of memory and learning make up the majority of our knowledge. The development of creative methods involving the balance of excitatory and inhibitory neurons in neural systems as well as the simplicity of knowledge about these fundamental brain activities is required for the construction of novel CDRs using spiking neural networks. In particular, computational systems with biology-based learning and memory algorithms enhance the development of cognitive processes in CDR. Additionally, the design and testing of trials in experimental studies may use these algorithms in CDR systems (Wheeler et al. 2015; Faghihi and Moustafa 2017).

14.18 Neurology of Learning

In the active learning process, meaning is produced based on experience, and an individual image is built by mixing the information a body receives from the sense organs. The relationship between memory and learning makes it common for people to misunderstand it. However, the researchers who study them believe that they are two distinct occurrences. These scholars contend that learning is a process that modifies future behaviour. The ability to recall the past is the opposite of memory. Since memory enables the storing and retrieval of learned information, all learning is dependent on it. Actually, memory is just a record of what has been taught. Learning and memory are hence interconnected (Maraver et al. 2022).

It is well-established that learning affects the activity of neurons in several brain regions across a wide range of species. Research on mice has revealed some information regarding the encoding of fear learning and reveals that fear memories are divided between the thalamus and amygdala in rodents (Keck and Josselyn 2021). Another part of learning that appears to go beyond the long-term storage of individual facts is the creation of links between facts (concepts). A semantic framework is a term used to describe this type of information storage. Some authors contend that the cornerstone of thought and learning may be associative processing and associations and that a computer modelling study into these processes could benefit from understanding neuroscience studies. The idea that analogical thinking can be utilised to build connections between facts may be supported by an understanding of these mechanisms from the standpoint of education. The control of semantic memory selection and retrieval has been associated with the ventral lateral prefrontal cortex, which displays higher activity when a learning task comprises concepts or facts that are only marginally related. The posterior associated cortex, in particular the temporal pole, is involved in the consolidation of semantic memory (Binder and Desai 2011).

14.19 Systems Biology of Learning

A significant and a relatively new method in computational neuroscience is to directly and quantitatively analyse the link between structural connectivity (SC) and functional connectivity (FC). Here, the blood oxygenation level-dependent (BOLD) signal in fMRI studies shows substantial correlation values between structural brain connectivity and functional connectivity, with values in the range of 0.6–0.8 in simulations and around 0.3 in experimental data. When the network structure acts as a "template" for the self-organisation of dynamical processes, resulting in a set of patterns known as functional correlation, these correlations are only compared in one direction. A broader perspective considers learning, or the modification of SC caused by FC's continual action (Turnbull et al. 2018).

Many in silico methods have been proposed to predict drug hepatotoxicity. Drug structural properties were used to train machine learning models, and these models have shown good accuracy. Additionally, deep neural networks have shown to have excellent predictive accuracy when combined with gene- and pathway-level indicators using transcriptomics data. The International Conference on Critical Assessment of Massive Data Analysis (CAMDA), in response to the current increase in interest in machine learning approaches to forecast drug-induced toxicity, has been hosting the Connectivity Map (CMap) Drug Safety Challenge since 2018. The challenge's goal was to assess the existing drug-induced liver damage (DILI) prediction techniques using a range of data sources, such as transcriptomics information, chemical structures, and cellular pictures (Sumsion et al. 2020).

According to systems biology analysis of learning neural pathways, teaching will be more effective if the techniques are in accordance with how the brain chooses to comprehend and retain knowledge. Current studies and the thinking patterns that are assessed across highly precise regions of the brain support a "whole brain" approach to learning. Through learning, the nervous system encodes the data that the sense organs gather about the "environment." Systems biology uses data retrieved from the CNS to encode the process of collecting signals from the sense organs (Maraver et al. 2022).

Many atypical substrates, including cells and even subcellular components like chemical processes and gene regulatory networks, have been postulated to have played a role in the learning process. The ability to learn is described as specific adjustments made to future acts in the light of prior experiences. Additionally, it is becoming easier to find micro-designed or emergent active living structures. The space of prospective subjects for learning experiments is huge and is constantly growing thanks to developments in fields like smart materials, synthetic bioengineering, brain-computer interfaces, and other areas. The ability to programme artificial living machines for positive reasons will be a vital new tool in the bioengineer's toolbox, in addition to creating new bodies with hardwired functionality. Since cutting-edge learning architectures found in neural and aneural systems may improve performance when applied to silicon-based or other media, it might also have an effect on robotics and machine learning. Research on unusual substrates has the ability to clarify a variety of disagreements about the location and nature of memory. By pursuing this approach in the fields of bioengineering and regenerative medicine, it is possible to produce effects that are too complex for physical (genetic, route) rewiring to be in control of. In order to implement biomedical strategies that could shift a lot of the complexity onto the system itself by utilising the learning process, it will be useful to use stimuli retrieved by systems biology rather than hardware rewiring to achieve desired end points like particular morphogenetic outcomes (Abramson and Levin 2021).

14.20 A Systems Biology Approach to Neurological Complications and Molecular Basis of Neural Pathologies

The term "developmental neuropathology" refers to a broad range of abnormalities and functional limitations brought on by disruptions in brain growth that can appear from the embryonic stage through adolescence and early adulthood. Depending on the compromised brain systems, the neurologic and psychiatric symptoms of neurodevelopmental disorders can take many different forms. Current classification schemes for neurodevelopmental abnormalities place a strong emphasis on the disturbed developmental processes. This pragmatic approach has been mainly supported by recent developments in human genetics and developmental neurobiology, but they have also shown that even single gene mutations and chromosomal changes can lead to a variety of malformations, and multiple genes or other aetiologies can contribute to the development of the same malformation (Flanagan et al. 2018). The hallmarks of neurodegenerative diseases are listed in Fig. 14.9.

Systems biology can be utilised by allowing the integration of readily available genetic, physiological, and clinical information in the context of a quantitative framework after a complete understanding of the current pathophysiology (Schleidgen et al. 2017). Systems biology is frequently used by engineers to supplement the conventional reductionist methods in the biomedical sciences. The main goal of a systems approach to biology is to deal with problems caused by the complexity of biological systems, such as the brain, which cannot be adequately explained by the molecular biology and genomics techniques currently in use. As of right now, RNA splicing-related pathways have been identified as a unique

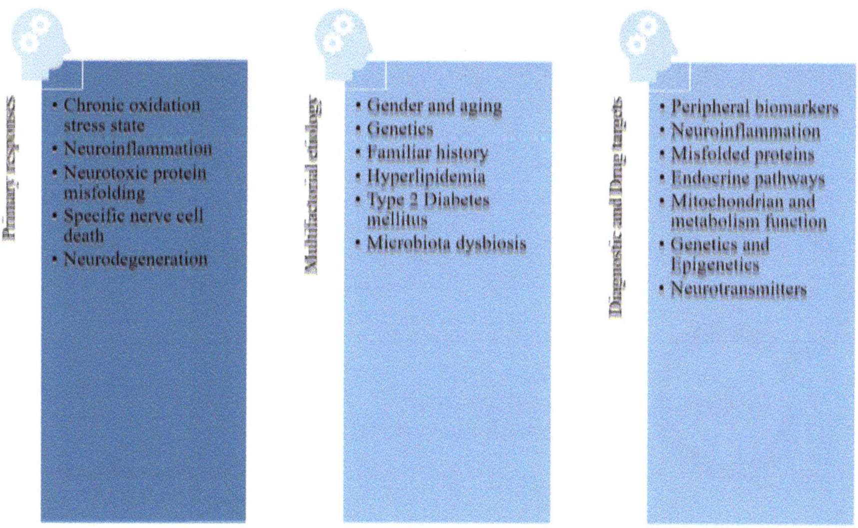

Fig. 14.9 Trademarks of neurodegenerative diseases

pathogenic mechanism for hereditary ataxias, according to the network-based analysis of the genes involved in these conditions (Villoslada et al. 2009).

14.21 Systems Biology of Alcoholism and Addiction

The characteristics of addiction point to adaptive brain dynamics with shared connections in intricate neural networks. Repeated drug use induces adaptation via process reduction or process amplification, which accelerates the dynamics of the addictive spiral towards increasingly intense and unpleasant experiences. Initial research on astroglia and addiction identified drug-mediated disturbances in glutamate homeostasis as a key risk factor for recurrence (Kruyer and Scofield 2021). It is recognised that dopamine or glutamate initiates well-established signal transduction pathways in which mitogen-activated protein kinases are involved. Also, abstinence comes with the possibility of relapse. Positive drug-induced experiences that are repeated indicate the beginning of addiction. Addiction is stabilised and adaptation takes place throughout that time. When those processes are interrupted, polysymptomatic withdrawal symptoms result (MacNicol 2016).

Due to their ability to mimic human behaviour and their emphasis on the role of genes and proteins in addictive behaviour, animal models have been employed in addiction research (Sibili et al. 2001). Figure 14.10 discusses the brain's response in regard to addiction. Recent years have seen a massive accumulation of data across many different disciplines because of high-throughput technology. Systems It is

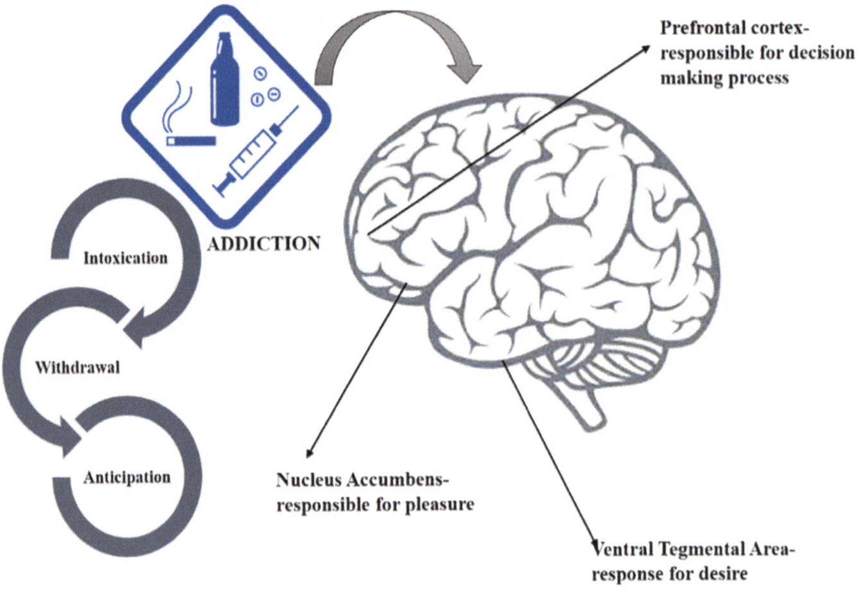

Fig. 14.10 The interconnective process of an addictive brain

possible to think of theoretical biology as a branch of biology that reconstructs cells and cellular networks as intricate, dynamic, self-organising systems and looks for patterns in enormous, complex datasets. Systems research uses this method, which is drawn from theoretical neuroscience, as part of a multidisciplinary endeavour to comprehend complicated dynamical systems. By increasing knowledge of the molecular biology of addiction at the cellular and network levels, "computational molecular systems biology of the neuron" could help in the study of addiction (Bernardi 2015).

Alcoholism, one of the major addictive disorders, is considered as a heterogeneous psychiatric illness. The way in which the addiction develops with comorbid conditions and the phenomenon of abstinence difficulties are examples of how different phenotypic traits are exhibited. When it comes to the aetiology of alcoholism, it is believed that both environmental and genetic variables interact to cause the condition (Matošić 2016).

There are enormous unmet requirements that affect our healthcare system and society, including treatment for alcoholism as well as prevention and intervention for excessive alcohol drinking by adolescents. By using a systems medicine approach, it is possible to identify novel pharmacological targets and molecules for the treatment of alcoholism as well as distinct neurobehavioural risk profiles in adolescents that are indicative of future alcohol use issues. In this example, the data came from omics-based sources such as transcriptomics, genetics, neurodynamics, and global neurochemical connectomes. This method has uncovered novel molecules for relapse prevention that will be explored in experimental human research and new early intervention studies (Spanagel et al. 2013; Noble and Weimer 2014).

From the perspective of behavioural neuroscience, neuronal ensembles are groups of neurons that are precisely and consistently activated by a stimulus, though the precise mechanism of co-activation of the neurons in an ensemble can take many different forms, such as synchronous action potential firing, synfire chains, or activation at some point while performing a behavioural task. With regard to drinking, the infralimbic (IL) cortex and amygdala have shown seeking-related ensembles. Another alcohol reward-related ensemble, using the lacZ transgene as a driver and the nuclear factor kappa-light chain of activated B cells (NF-B) as an activity marker, was discovered at the brain-body interface. Activated B cells produce nuclear factor-B, also known as NF-B. Although the activity increased seeking behaviour, the region was not entirely deactivated. Ablation was induced by a certain cue-related ensemble's activity in the IL. The same group, nevertheless, was not linked to alcohol consumption brought on by stress (Russo et al. 2018; Körber and Sommer 2022).

Systems biology started to be applied in medical contexts in 2009. Researchers identified brain regions and functional networks that can be specifically targeted by treatment methods using functional magnetic resonance imaging data from humans and animals that exhibit behaviours similar to alcohol addiction. They also developed mathematical models of the relapse-prone network states. The research serving as proof of concept shows that network dynamics can be regulated to improve animal resilience. However, an initial attempt to include this in a clinical study that focused on the insula was unsuccessful. The study contributed to the mapping of numerous biomarker network properties and treatment entry points for human alcoholism. This inspired the use of systems biology to fully understand circuit-specific brain modelling (Sommer 2017).

Researchers identified brain regions and functional networks that can be specifically targeted by treatment methods using functional magnetic resonance imaging data from humans and animals that exhibit behaviours similar to alcohol addiction. They also developed mathematical models of the relapse-prone network states. The research serving as proof of concept shows that network dynamics can be regulated to improve animal resilience. However, an initial attempt to include this in a clinical study that focused on the insula was unsuccessful. The study contributed to the mapping of numerous biomarker network properties and treatment entry points for human alcoholism. This inspired the use of systems biology to fully understand circuit-specific brain modelling (Guo and Zakhari 2008).

On the basis of information gathered from magnetic resonance imaging, electrophysiological, and neurochemical testing carried out on humans and laboratory animals, predictive models of the "relapse-prone" state of brain networks were created. To completely evaluate the mathematical models and strengthen the network's relapse resistance, experimental techniques were used. The major objectives of the study were to identify nodes that enhance "relapse-proneness" and to forecast how anomalous network conditions can be remedied. Animal proof of concept research demonstrating how the theoretical models particularly alter functional brain networks will need to be used to substantiate this claim. The development of a theoretical and experimental framework for making predictions based on fMRI and mathematical modelling, tested in animals, and then transferable to people, can thus attain this translational purpose (Sommer 2017; Furuya et al. 2020).

Metabolomics is a comprehensive quantitative examination of low-molecular-weight metabolites in a biological system under various environmental conditions (such as alcohol administration) (i.e., metabolome). Clear evidence of an organism's reaction to genetic or environmental changes can be seen in the levels and regulation of metabolites. The intermediate and final outcomes of cellular processes include metabolite levels and regulation. Numerous bioinformatic, statistical, and computational methods, such as mass spectrometry, nuclear magnetic resonance spectroscopy, capillary electrophoresis, and high-performance liquid chromatography, can be used to identify these metabolites. There are available virtual representations of the numerous tiny molecules and their distinctive time-dependent modulations that make up the biological system (Guo and Zakhari 2008). Metabolomic studies in alcohol research are complementary to studies of the genome, transcriptome, or proteome because they can be used to extract latent biochemical data with diagnostic or prognostic significance, representing actual biological activities, and because they can act as a sentinel for disorders (Sommer 2017).

In metabolomics, low-molecular-weight metabolites must be thoroughly quantitatively examined in a variety of environmental settings. Therefore, an effective strategy for studying illnesses connected to alcohol will be comprehensive systems biology. Systems biology can provide the integrative approach required for the prognosis, diagnosis, and treatment of alcohol-induced illnesses in the future with the help of ongoing technological advancements and multidisciplinary teamwork.

14.22 Systems Biology of Stress and Depression

Stressful situations have the capacity to either start or worsen a depression illness. The development of depression and stress-related sorrow can begin in adolescence or adulthood, despite the fact that data suggests that early adverse events, even during the prenatal stage, may predispose the kid to later depression. The hypothalamus-pituitary-adrenal axis, which is crucial for regulating the stress response in both humans and animal models, has been linked to depression (Abdul Aziz et al. 2012; Sickmann et al. 2014).

Skeletal muscle, bone, and adipose tissue in humans are markers of chronic stress exposure that influence how the body feels. Increased adiposity, particularly its visceral component, sarcopenia, and osteopenia/osteoporosis are responsible for the extraordinarily high prevalence of the so-called medically unexplained symptoms (MUS), which include fatigue, discomfort, gastrointestinal issues, dizziness, and sleeplessness. Additionally, these situations exhibit an ongoing cycle of chronic stress system activation, cortisol, catecholamine, and inflammatory cytokine increases (Chrousos et al. 2018).

How the brain and body respond to stress together is a topic of growing research. In these once separate bodies of study, it has been crucial to show the reciprocal and compounding links between cognitive and physiological responses to stress and depressive symptoms. In past studies on the relationship between cognitive and physical stress reactivity, the brain regions that support cognition and emotion have garnered the majority of attention (LeMoult et al. 2016). The neuroanatomy of a depressed brain is shown in Fig. 14.11.

Numerous studies in the field of stress research have concentrated on a single or a small number of measurements taken at a limited number of time points of the response to a stressor or stressors, whether this be a single behaviour, a single

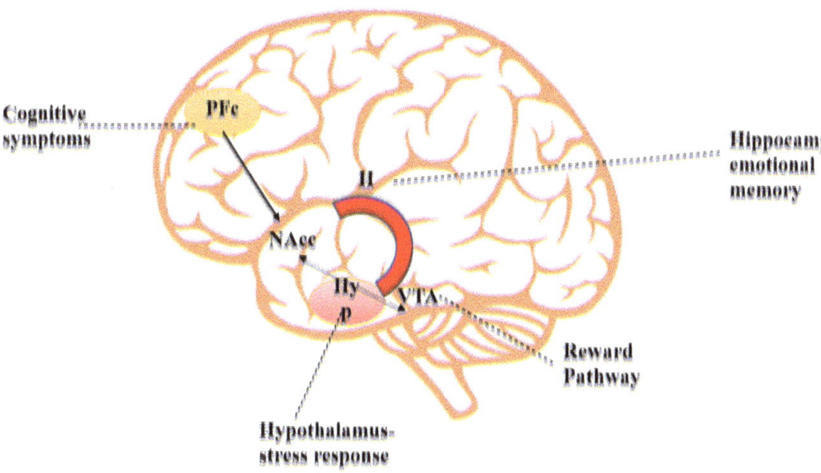

Fig. 14.11 Neuroanatomical region associated with symptoms of stress and depression

physiological measure like metabolic rate, heart rate, the abundance of a particular protein (e.g., HSP70), a steroid hormone like cortisol/corticosterone, or the expression of a small number of genes. A multimodal viewpoint is necessary for characterising a specific stress response and determining the precise short- and long-term impacts on organismal fitness. It is becoming increasingly clear that a single measure cannot effectively represent the complexity and context of the stress response (Schwartz 2020).

It is believed that a complicated interaction between many molecular systems and the environment is what causes stress-related diseases to have their underlying aetiology. The next natural step in defining this may be to take a more all-encompassing approach that combines several omics layers with environmental exposures. A systems biology technique can be used to research complicated phenotypes since a panel of inbred strains has a variety of related characteristics. Even if the data collection needs terminal procedures, comparisons and correlations between proteomic, transcriptional, metabolic, physiological, genotypic, and behavioural data can be done with ease because the same genetic background can be retested several times.

Scientists identified a new, more precise model of the human metabolic syndrome from a collection of 48 inbred strains by gathering ten molecular and physiological characteristics relevant to metabolic function and homeostasis. The characterisation of the common inbred strains has expanded quickly because of the clear benefits of this form of multilevel phenotypic collection and comparison. Inbred strain genomics has a lot of promise for improving our understanding of intricate biological systems. The examination of depression and the effectiveness of antidepressants should be similarly susceptible to this type of systems biology approach, as the ability to establish links and correlations across various datasets has already led to a wealth of new insights for different phenotypes. Using forced swim test (FST) and tail suspension test (TST) models of behavioural despair, some small-scale inbred stain screens have been carried out, with up to 11 strains in one instance. A large phenotypic variety was seen within these few inbred strain groups, as would be expected (Akil et al. 2018).

Theoretical categorisation of SNPs discovered by genome-wide association studies of depressed symptoms into biologically significant polygenic profile scores that can be evaluated in genetic epidemiology designs is one notable application of a systems biology method. The incorporation of these replicating discoveries into a more substantial polygenic process has received relatively little attention thus far. An increase in the genetic influence on mood and behaviour would be expected if environmental circumstances can alter these links and life experiences, as is being supported by growing research (Uffelmann et al. 2021).

The molecular basis of depressive-like behaviours has also recently been the subject of a plethora of proteomics investigations in mouse models of depression. The project intends to integrate the results of these proteome investigations into depression models in order to pinpoint the most pertinent molecular pathways involved in the disorder. Network analysis is performed while using previously collected proteome data from rodent depression models. The IntAct mouse and the

HRPD are used as common protein-protein interaction databases. We then seek overrepresented GO biological process keywords and pathways to assess the functionality of the networks (Carboni et al. 2016).

By offering a new method for selecting which genes or combinations of genes to examine inside genetic association designs, neurobiological comprehension of the stress response systems can work in parallel with genome-wide techniques. It is well-known that 5HTT plays a crucial role in controlling 5HT availability for post-synaptic reception in the mesocorticolimbic system (MCLS), and monoamine oxidase-A (MAO-A) inhibitors are developed to enhance 5HT's capacity for neurotransmission. It is still unclear whether 5HTT and monoamine oxidase-A (MAO-A) can independently function as biomarkers for depression risk, despite the fact that both have received extensive research as possible genes for gene-environment interactions with early life stress in depression. The 5HT1A receptor, a well-known anxiolytic medication target site, is another MCLS gene product that mood-regulating medicines target. By using NE-selective tricyclic antidepressants, which act as a blocker of NE-transporters activity in the reuptake of NE to delay NE clearance from the synaptic cleft and decrease the availability of functional NE for neurotransmission, it is possible to prioritise potential genes through the Hindbrain Autonomic Regulatory System (HARS) system. Functional studies that concentrate on the effects and mechanisms of action of particular gene variants show that genes like catechol-oxymethyltransferase COMT, 5HT transporter, ACE, mineralocorticoid receptor (MR), and the glucocorticoid receptor all play essential roles in anxiety and depression. These genes may continue to receive top priority in the next candidate gene investigations in the light of the findings (Eekelen et al. 2011).

A multi-gene profile scoring method incorporating three specific polymorphisms in NR3C1 and SLC6A4 provided the strongest effect sizes, with an increasing number of unfavourable alleles being associated with unfavourable changes in the TTS sub-score of up to three standard deviations, according to research on the polygenic risk for stress-related traits of depression and anxiety driven by systems biology. Another instance of synergistic hormone action at the intersection of the HPA axis and HARS might be included in the aforementioned integrated neuronetwork using a similar polygenic risk analysis programme. Stress affects both the corticolimbic brain region excitability and amygdala-related CORT binding to GR. When taken as a whole, these incidents have an impact on how memories are processed under stress that seems to be more intense than usual and can lead to the malignant memory formations that constitute the basis of post-traumatic stress disorder. A heightened emotional response to stimuli that, while not immediately hazardous, nevertheless vividly remember a traumatic incident in the past, is the defining feature of this type of stress-related mental disease (Eekelen et al. 2012).

Recent theories corroborate the original theory that maladaptive cognitive reactions to stress may be caused by malfunction in brain areas that promote emotional processing and control. However, there are good reasons to concentrate on how cognitive and biological stress reactivity interact with one another as well as to look into a larger range of biological systems that put people in danger when exposed to stressors from the outside world. Experimental research that examines whether changes in cognition result in biological changes provide some of the greatest evidence for how cognition influences the biological stress response. Then, scientists saw alterations in the individuals 'bodies' reactions to stress. In order to do this, researchers have trained individuals to have more flexible emotion regulation systems or less prejudice (Monroe et al. 2019).

Rumination was linked to a prolonged neuroendocrine response, according to experimental data on cortisol increases that persisted for around 60 min after a laboratory stressor had passed. Other groups also showed the advantages of cognitive training, and these advantages included reduced inflammation (lower proinflammatory cytokines) and autonomic stress responses. Evidence demonstrating persistent autonomic, neuroendocrine, and inflammatory dysregulation over time also affects cognition and lends support to the idea that aberrant physiological stress responses and cognitive stress responses have reciprocal effects on one another. Memory, learning, and emotion regulation can all be hampered particularly by chronic physiologic dysregulation(Beevers et al. 2015; Jopling et al. 2020).

According to studies that support this theory, physiological reactions to stress, particularly prolonged cortisol production, might negatively affect the parts of the brain that control emotions and deal with everyday stress. Chronic stress-related increases in inflammation disturb the same brain regions as well, which over time can have a number of physiologic and cognitive impacts in addition to an increase in depressive symptoms. Together, the data show that cognitive and physiological reactions to stress continue to interact, raising the likelihood of depression and unhelpful stress reactivity (Beevers et al. 2015).

Thus, systems biology employs genetic data at multiple levels that also contain information on environmental factors in order to examine pathophysiology in greater detail than reductionist techniques (via the use of methylation data and its relationship with gene expression). Systems biology studies occasionally need a lot of data points, so they need to be powered properly to prevent generating false positive and false negative results (Dalvie et al. 2021).

14.23 Systems Biology of Schizophrenia

Schizophrenia is a debilitating condition that causes both desirable and unpleasant cognitive and emotional symptoms. Numerous studies have looked closely at the various factors that affect the condition. Numerous genes, including neuregulin 1 and dystrobrevin binding protein 1 (DTNBP1), are implicated in a potent genetic aetiology (NRG1). Although there is some debate over the relationship between neurotransmitter changes and the pathophysiology of schizophrenia, research has shown that subcortical dopamine insufficiency is the main contributing factor. Specific sections of the grey and white matter are altered in schizophrenia patients, with the grey matter abnormalities being more pronounced after the start of psychosis. These disease-related changes could impact executive function, working memory, and attention (Luvsannyam et al. 2022).

A dynamical system that may be responsible for these symptoms was found to have schizophrenia-specific effects by using a systems biology method. An integrate-and-fire network model was employed in the computer model, which concentrated on changes to synaptic channels activated by NMDA and GABA receptors. The study showed that by lowering neuronal firing rates and the stability of attractor states, a decrease in NMDA receptor conductance could cause both cognitive and unpleasant symptoms. The attractor states linked to pleasure experiences become unstable when both NMDA and GABA conductance are reduced (Loh et al. 2007).

Despite the fact that many genes have been shown to be associated with schizophrenia, little is known about how these genes physically interact with the proteins they express. System-level analysis of these genes was carried out to see whether the risk genes for schizophrenia are grouped on relevant biological pathways and encode a strongly connected disease module. By combining the protein–protein interactions (PPI) information from the meticulously characterised databases, a list of high-confidence physical interactions was produced. Afterwards, a thorough analysis of the physical connections between each protein that the schizophrenia risk genes encode was conducted. The final gene ontology (GO) study will look for improved common disease pathways. The schizophrenia risk genes in the interactome perform noticeably better than the background genes in terms of degree, proximity, centrality, clustering, and average shortest path length. It is interesting to note that the highly connected PPI network identified 48 hub genes and 4 modules of schizophrenia-related genes. This demonstrated that the hub genes for schizophrenia encode proteins that interact physically considerably more frequently (Liu et al. 2018).

Proteomics and metabolomics are, in this regard, widely acknowledged as essential methods for identifying illnesses. Even though all metabolites are produced and degraded by proteins, understanding which metabolic processes are typically active or inactive in a given healthy and/or pathological environment might aid in the identification of crucial pathways for understanding the pathophysiology of disease. It was interestingly shown that metabolomic analysis, utilising a range of techniques, such as proton NMR (1H-NMR) or mass spectrometry, could differentiate between persons with schizophrenia and healthy control (HC) individuals (MS). It also showed how different biological samples, such as blood, urine, or cerebrospinal fluid, may be utilised to find irregularities in metabolism (CSF). Following serum metabolomic analyses, second-generation antipsychotics were given to schizophrenic patients, illuminating the specificity of the pharmacometabolome as a clinical diagnostic tool. The goal of the current work was to gain fresh insight into the pathophysiology of these mental diseases by using a systems-level approach to analyse the altered metabolites in schizophrenia. An approach at the systems level will most likely make up for some missing data and eliminate noisy data from the process. It has been established that genes and proteins sharing a close correlation with the biomarker metabolites and coupled by trustworthy PPI are useful in this procedure. Using a network that was already constructed, the graph-clustering algorithm DPClusO sought clusters in the network. This made it easier to find the statistically significant clusters that represent substantial and widespread pathways underlying schizophrenia.

To progress diagnosis and develop more effective treatments, it is imperative to completely comprehend the likely origin and pathophysiology of schizophrenia. Systems biology methods can be used to understand networks of gene-gene interactions and epigenetic alterations in order to better understand the mechanism behind schizophrenia (Huang et al. 2014). Recent advancements in analytical techniques, including next-generation sequencing, the IPSC neuron model, the SZ gene database, expression pattern analysis, and protein-protein interaction analysis, have the potential to provide fresh perspectives on pathophysiology and, ultimately, new schizophrenia treatment options. The fundamental genetic and epigenetic causes of the horrible disease schizophrenia may be discovered by identifying related complexes or pathways and identifying novel therapeutic targets (Rudan 2010).

By studying differential expression and genetic linkages in schizophrenia, transcriptome analysis of post-mortem brain samples offers additional insights into how biological dysfunctions are evaluated. This is because research has shown that environmental factors including stress, pollutants, and oxidative processes can affect epigenetic changes like DNA methylation and histone modification. In order to identify potential protein complexes connected to the aetiology of schizophrenia, the gene groups from the schizophrenic differential methylation genes (SDMG) database and schizophrenia candidate marker genes were analysed and searched against the vast database of mammalian protein complexes, including CORUM, a comprehensive library of mammalian protein complexes (Haroutunian 2014).

In order to aid in phenotypic prediction and comprehension of underlying mechanisms in psychiatric diseases, as well as their link to other comorbid conditions, systems biology must be able to provide suitable quantitative frameworks.

14.24 Systems Biology of Epilepsy

Epileptogenesis is the process by which epilepsy develops into a condition that can support seizures. Then, a series of things happen that change the healthy brain into one that can. Clusters of neurons are intended to get overexcited and prepare for erroneous release. Another important element of neurobiology that can suggest a potential seizure cause is the ionic base of the action potential. The complicated coordination of sodium and potassium channels that rely on chemical and ionic gradients across the cell membrane is how neurons are built to discharge. Anomalies in the sodium channel may lower the threshold for an action potential if the mechanism controlling sodium channel activation changes. It has been established that mutations can result in epilepsy in the voltage-dependent sodium channel subunits (Scharfman 2007).

Several medications can halt seizures, but they have little effect on either preventing or treating epilepsy. Due to the fact that acute seizures are caused by different molecular pathways than epileptogenesis, new, truly "antiepileptic" treatments will need to be created. Thus, identifying and evaluating potential antiepileptic drug targets, first in animal models and then in humans, is a critical next step in the fight against epilepsy. The vast range of brain problems that can cause a condition of chronic, repeated seizures, however, makes the fight harder in focused forms of human epilepsy (Loeb 2011).

Researchers have created a database that combines several data sources, such as clinical, electrophysiological, tissue, and gene data, into a relational database as part of the systems biology approach. Among the patient-specific information gathered are age, sex, and seizure medication (Huang et al. 2014). Additional electrode-specific data include spike frequency, location, if tissue was collected, histology, RNA and protein extract quantities, and gene expression levels. The positioning of each electrode into a three-dimensional model of the patient's brain is essential to this database since each patient's neocortical anatomy is unique. Each electrode must be incorporated into a three-dimensional model of each patient's neocortical architecture since each patient has a unique neocortical anatomy (Loeb 2011).

It has been discovered that a network of 320 genes controls epilepsy susceptibility in both uncommon and common forms of the condition. By merging information from numerous sources, Delahaye-Duriez and colleagues were able to pinpoint the gene networks that control epilepsy in their study. They discovered 18 consensuses and 80 divergent modules that could be repeated in various brain expression datasets (Ramasamy et al. 2014). These important gene networks were thought to be essential for healthy brain function. A total of 34 co-expression modules were shared by all brain regions. A total of 13 "differential" modules exhibited different degrees of co-expression in diverse parts of the brain. These modules were subsequently examined in two more publicly accessible datasets on human gene expression, including GTEx, which contains genotype and tissue expression data across a number of brain areas, and Brainspan, which spans prenatal to postnatal human brain development. The final step in comparing between species in healthy mouse hippocampal RNA-seq datasets was the search for conserved modules. This study discovered that across mice and humans, more than 58% of modules were preserved. No modules were enriched for the other neurodevelopmental diseases despite their genes. This study found that more than 58% of modules were conserved in both mice and humans. A recent GWAS meta-analysis found that genes associated with more prevalent kinds of epilepsy are enriched in module M30. Both human temporal lobe epilepsy resections and gene expression data from the hippocampus of mice with epilepsy showed the enrichment of Module M30 genes. Despite the fact that this was one of the features used to design the module, other M30 members have not yet been directly linked to epilepsy. It possesses several genes linked to monogenic epilepsy. The discovery of co-expression might be explained by the fact that NRSF/REST controls several M30 genes (Miller et al. 2014). By focusing on rebalancing the brain networks towards the non-disease state, it may be possible to produce anti-epileptogenic drugs in the future rather than anti-seizure ones (Zhao et al. 2017).

Figure 14.12 highlights the systems biology approach to treating epilepsy. Using a multivariate interactome and quantitative electrical, brain imaging, and genetics for refractory epilepsy, researchers have developed a discovery machine for human

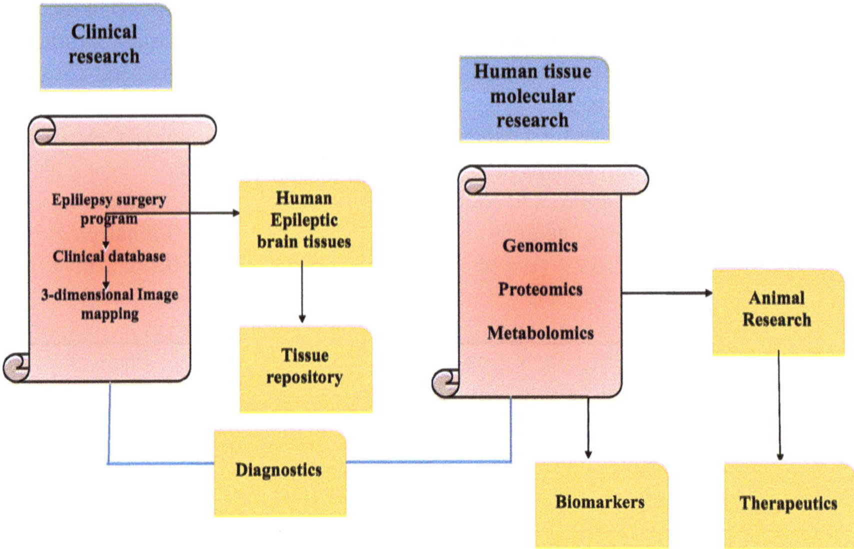

Fig. 14.12 Systems biology workflow for epilepsy treatment

epilepsy. This was made using a systems biology construction strategy to fully utilise these resections. The brain tissues have also changed in order to uncover new biomarkers and therapeutic targets as well as comprehend the basic mechanisms underlying neocortical epilepsy. The neurorepository, which holds electrically mapped epileptic tissues and related data, offers quantitative assessments of ictal and interictal activity corresponding to a specific intracranial electrode for use in clinical, histological, imaging, genomic, proteomic, and metabolomic studies. Thanks to this completely documented data and tissue collection, it is now able to explore the underlying electrical, cellular, and molecular mechanisms of the human epileptic brain (Kirchner et al. 2020).

It is crucial for improving patient care to quantify various types of data from patients who have undergone epilepsy surgery. Furthermore, it offers a quantitative framework for investigating fundamental issues like why some brain regions are epileptic while others are not, or why some persons experience seizures. Despite the fact that many lesions are known to be connected to epileptic disorders, it is frequently noted that the histopathology in the areas where seizures occur and/or where there is a lot of interictal spiking is essentially normal. The revolution in genomics and other high-throughput measurements of biological molecules has made it possible to examine hundreds of chemical variables in resected human brain tissues at once (Beaumont et al. 2012).

The hypothesis that all epileptic patients with a common electrical pattern have similar changes in gene transcription has been put to the test by researchers (the quantities of messenger RNA [mRNA] molecules for each gene). Both experimental approaches resulted in the identification of a group of genes with differential expression. Unsurprisingly, a large number of these genes are shared by the brain areas associated with seizures and spiking. Given the regular occurrence of interictal spikes in seizure-onset regions, this might be anticipated (Englot et al. 2012).

It is crucial for improving patient care to quantify various types of data from patients who have undergone epilepsy surgery. Furthermore, it offers a quantitative framework for investigating fundamental issues like why some brain regions are epileptic while others are not. Researchers have studied a large number of genes that do not code for proteins but instead produce long noncoding RNAs that have been shown to have crucial regulatory functions in order to map the expression of genes that code for proteins as a means of understanding the cellular building blocks that produce epileptic circuits. Additionally, genes unique to spiking and seizures have been reported, indicating that these two electrical events have distinct transcriptional pathways and may hence necessitate distinct therapeutic modalities (Mittal et al. 2013).

14.25 Systems Biology of Alzheimer's Disease

It is crucial for improving patient care to quantify various types of data from patients who have undergone epilepsy surgery. Furthermore, it offers a quantitative framework for investigating fundamental issues like why some brain regions are epileptic while others are not. Alzheimer's disease is the most prevalent and harmful neurological illness that causes dementia (AD). Extracellular-amyloid [amyloid precursor protein (APP) deposits, intracellular neurofibrillary tangles (NFT)], dystrophic neuritis, and amyloid angiopathy are some of the neuropathological features of AD. Frequently, the illness presents as the mild cognitive impairment (MCI) syndrome, which is characterised by mild cognitive impairments on formal tests, memory complaints, and maintenance of everyday functioning in the absence of overt dementia. MCI is hypothesised to occur between early Alzheimer's disease and healthy ageing. Functional magnetic resonance imaging (fMRI) frequently reveals progressive brain shrinkage in AD individuals, with the entorhinal cortex and hippocampal regions being notably affected. Atrophy rates may be predictive and related to changes in clinical status for an effective treatment (Mohandas et al. 2009).

The central homeostatic networks of nerve cells and damaged pathways interact constantly and dynamically in this condition. Inherent chromosomal vulnerability is another feature. New comprehensive systems-level approaches at both the experimental and computational levels are required for a thorough analysis of AD's complexity, and they depict the brain condition of an Alzheimer's patient. Techniques in systems biology can uncover novel fundamental concepts, fundamental mechanisms, networks, and network interactions (Castrillo and Oliver 2016). These could provide a better understanding of mechanism-based molecular signatures, early molecular and cellular hallmarks of AD, as well as later stages and subtypes of AD, opening the door to customised treatments in line with the emerging precision medicine paradigm. This includes the most recent advances in single-cell and genome editing technologies, as well as in relevant disease models, neuroimaging, multi-omics, experimental, and computational techniques. Their gradual incorporation

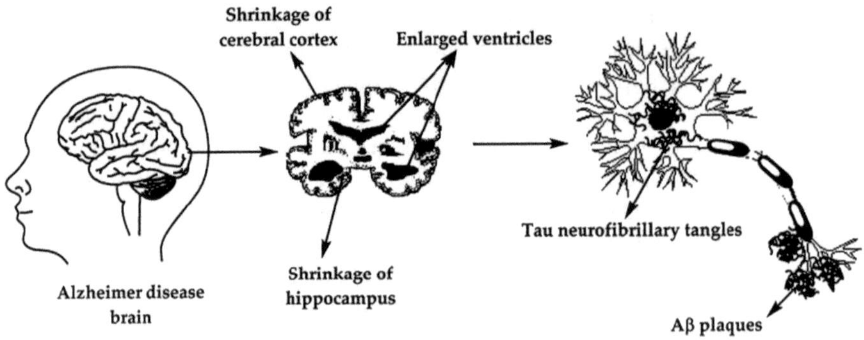

Fig. 14.13 Neurobiology of an Alzheimer's disease brain

into basic research, cohort studies, and clinical trials is beginning to provide new insights into the underlying mechanisms, molecular markers, and signs of AD that will eventually result in mechanism-based staging and categorisation as well as specialised treatments (Castrillo et al. 2018; Hampel et al. 2021) (Fig. 14.13).

By exposing the protein network of axonal transport and the changes caused by Aß, more proteomic research should be done to address the discrepancies in this area. The final analysis of a proteomic investigation on animal models of AD revealed that each model had altered proteins that affected the cytoskeletal structural integrity, antioxidant defence, and energy metabolism. Proteomic research on human AD brain tissues and animal models suggests that intracellular Aß most likely causes a mitochondrial metabolic modification as the first change in cell function in AD. The mitochondrial consequences of AD may be linked to the process of memory impairment through their modulation of synaptic function. It suggests that, on the one hand, significant alterations in mitochondrial metabolism and the production of free radicals trigger cell death mechanisms in neurons, causing them to self-destruct, and, on the other hand, even a small alteration in mitochondrial processes close to synapses can affect synaptic function and impair memory (Juhász et al. 2011).

Systems biology is a field of study that tries to understand how intricate molecular assemblages function in biological systems. Recent systems biology studies have demonstrated that proteins can carry out specific tasks when assembled into protein complexes. Similar to real protein networks, these artificial protein machines are less complex than synapse or mitochondria. A number of studies have found that the multimolecular targets of Aß and their effects on the cellular machinery of brain cells and on the synapse pose the biggest obstacles to advancement in the development of early diagnosis, prevention, and successful treatment of AD (Bashor et al. 2010; Sleegers et al. 2010).

A clinical test that can identify AD in its early stages could be a crucial tool for enabling interventions to limit the disease's progression. Diagnostics of cerebrospinal fluid and easily accessible neuroimaging, however, are relatively expensive. As a result, using blood transcripts that demonstrate how brain transcripts are expressed throughout the disease to diagnose AD may be more accurate. To accomplish this, researchers used cis-eQTL data in conjunction with transcriptional analysis of the affected organs. Researchers examined blood and brain cell microarray gene expression data from AD patients and healthy controls in this investigation. Differentially expressed genes have been found in both blood cells and brain tissue (DEGs). The pathway and gene ontology enrichment analysis was used to identify the most likely common genes and biological pathways.

These potentially disease-associated genes were corroborated by the findings from the most well-known benchmarking databases (gene SNP-disease linkage). Important biological pathways and gene ontologies illustrate how AD progresses. Additionally, regulatory elements like transcription factors (TFs), microRNAs, and potential medications were discovered by this research. New potential connections between failing brain tissue and blood cells in AD have therefore been discovered. These connections may one day allow for the assessment of the AD condition utilising blood samples. Thus, the established techniques show the efficacy of data from transcriptomics, cis-eQTL, and epigenetics for diseases associated with the brain (Kuleshov et al. 2016; Chowdhury et al. 2020).

A total of 30 gene co-expression modules related to Late-onset Alzheimer's disease (LOAD) have been found by a recent multi-cohort analysis of post-mortem human brain transcriptomes, creating a molecular library of important endophenotypes. This tool provides precise gene-based alignment of novel animal models and molecular markers for human disease. This article described an innovative technique for assessing the LOAD relevance of mouse models. Important human disease processes and pathways were linked with mRNA from mouse brains using a brand-new NanoString nCounter® Mouse AD panel. When functionally enriched in significant disease-related pathways, three LOAD-genetic animal models with the APOE4 and TREM2*R47H alleles displayed similarities to distinct human AD modules. These models comprised the TREM2*R47H mouse and the well-known amyloid pathology model, 5×FAD mice. With the same-sample RNA-Seq data, the entire transcriptome was thoroughly examined to reveal a substantial link between changes in gene expression that was independent of the experimental platform. Therefore, it is evident that this panel provides a straightforward, low-cost, and remarkably consistent method for determining the disease relevance of potential LOAD animal models (Preuss et al. 2020).

14.26 Systems Biology of Parkinson's Disease

Due to abnormal protein-synuclein deposition in intraneuronal Lewy inclusions among vulnerable populations of brain neurons, Parkinson's disease (PD) is a neurobiological illness that kills nigrostriatal dopamine neurons. The pathologic changes in PD occur gradually and over a long period of time, with a large interval of clinically silent cellular failure and, for some populations of neurons, cell death. Due to the long lag between the initial damage to dopaminergic cells and the emergence of clinical symptoms, PD is rarely diagnosed in its early stages. The motor and non-motor PD deficiencies are listed in Table 14.3. Finding reliable genetic biomarkers that can track the development of PD, separate it from other illnesses, or

Table 14.3 Crucial motor and non-motor impairments of a Parkinson's disease individual

Motor impairments	Non-motor impairments
Bradikinesia	Cognitive impairment
Hypersalivation	Sleeping disorders
Postural imbalance	Mood disturbances
Trembor	Gastrointestinal issues
Muscular rigidity	Sweat and olfactory disorders
Walking difficulties	Anxiety and depression
Freezing of movements	Pain

demonstrate whether a patient is benefiting from a therapy intervention is challenging (Halliday et al. 2014).

Comparing the dysregulated cellular or gene pathways in the affected tissues is another method to look at the connections between PD and co-morbidities. In order to find significant relationships between Parkinson's disease (PD) and other critical health markers that could be altered to achieve this, researchers created a quantitative methodology. This work used PD, type II diabetes, age, sedentary lifestyle, high dietary fat, hypercholesterolemia, high body fat, high red meat diet, high alcohol use, smoking, and control datasets in the analysis of gene expression microarray data. They have discovered genetic connections between these variables and PD using neighbourhood-based benchmarking and multilayer network structure. In the PD patients compared to healthy controls, this network analysis found 544 upregulated genes and 779 downregulated genes, totalling 1323 differentially expressed genes (DEGs). Using gold standard datasets from Mendelian Inheritance in Man (OMIM), dbGaP, and a validation gene expression dataset, the identified PD-related genes and biochemical pathways were further validated (Sakib et al. 2020).

Systems biology analyses of PD that take into account genes and their regulators may assist in clarifying the intricate molecular relationships that underlie this disease. Of the eleven hub genes discovered by the co-expression network, eight are new to PD. The significance of these genes in a number of neurodegenerative illnesses was discovered after further research into the eight unique hub genes' roles. These additional hub genes' histone modification patterns and tissue-specific expression patterns were also discovered. The bulk of these genes have several histone modification sites that have previously been linked to neurodegenerative disorders (Jouan et al. 2013).

Systems biology analyses of PD that take into account genes and their regulators may assist in clarifying the intricate molecular relationships that underlie this disease. Among the eleven hub genes, eight are the lncRNA-mediated regulatory network showed the two-way regulation pattern of PD-specific miRNAs, in which both TFs and lncRNAs can control miRNAs. This network also found unique lncRNAs of PD. Twenty SNPs were found after SNP analysis of the co-expression network's most important genes. The miRNAs that are also involved in PD regulate these SNPs, which are located in the 3′ UTR of discovered PD genes. Eight novel hub genes were discovered in this study, and they could be considered potential candidates for PD biomarker studies in the future (Chatterjee et al. 2017).

Understanding changes in molecular activity in the context of well-established cellular pathways and molecular processes is typically the first step in understanding systems-level modifications in omics datasets for complicated diseases. Several open-access websites with numerous hand-verified route and process descriptions are the Kyoto Encyclopedia of Genes and Genomes, the Gene Ontology database, BioCarta, and WikiPathways. Disease-specific resources with specialised pathway maps for neurodegenerative illnesses have recently been established in addition to these general route libraries. Over-representation analysis (ORA), Geneset enrichment analysis (GSEA), Network module-based pathway analysis (NMPA), and Network topology-based pathway analysis are the four core categories into which these omics-based pathways and geneset enrichment analysis methodologies can be separated. A general suggestion for omics-based pathway analysis is to compare at least a couple of the various sorts of techniques outlined above in order to find diverse biologically significant modifications. As a result, systems biology is now being used extensively in Parkinson's disease treatment.

14.27 Conclusion

Understanding how biological systems process information is a challenge for both systems biology and neurology. Even if the idea of the genome as a map has developed in genetics from a simple one to one that emphasises more complicated gene networks, the kinds of analysis that are now used do not put a lot of emphasis on the information content. Dynamical systems are the approach that is most frequently used to analyse genetic and molecular networks. Conversely, theoretical and computational neuroscientists investigate the operation of the brain's information processing systems using state-of-the-art techniques. In addition, a wide range of neural coding levels, including synapses, individual neuronal spikes, and information processing at the network and system levels, have been extensively investigated. In conclusion, a thorough understanding of brain coding principles may inspire the development of innovative solutions to the high-dimensional information processing issues that are present in virtually every biological system. Two elements that contributed to the total amount of knowledge learned were the diversity of interests and the predisposition for work that has to be improved. As a result, this new industry will shed light on how to treat a number of neurodegenerative disorders and pave the way for innovative therapeutic strategies.

References

Abdul Aziz NH, Kendall DA, Pardon MC (2012) Prenatal exposure to chronic mild stress increases corticosterone levels in the amniotic fluid and induces cognitive deficits in female offspring, improved by treatment with the antidepressant drug amitriptyline. Behav Brain Res 231(1):29–39. https://doi.org/10.1016/j.bbr.2012.02.040

Abramson CI, Levin M (2021) Behaviorist approaches to investigating memory and learning: a primer for synthetic biology and bioengineering. Commun Integr Biol 14(1):230–247. https://doi.org/10.1080/19420889.2021.2005863

Akil H, Gordon J, Hen R, Javitch J, Mayberg H, McEwen B, Meaney MJ, Nestler EJ (2018) Treatment resistant depression: a multi-scale, systems biology approach. Neurosci Biobehav Rev 84:272–288. https://doi.org/10.1016/j.neubiorev.2017.08.019

Assmus HE, Herwig R, Cho KH, Wolkenhauer O (2006) Dynamics of biological systems: role of systems biology in medical research. Expert Rev Mol Diagn 6(6):891–902. https://doi.org/10.1586/14737159.6.6.891

Bashor CJ, Horwitz AA, Peisajovich SG, Lim WA (2010) Rewiring cells: synthetic biology as a tool to interrogate the organizational principles of living systems. Annu Rev Biophys 39:515–537. https://doi.org/10.1146/annurev.biophys.050708.133652

Bassett DS, Zurn P, Gold JI (2018) On the nature and use of models in network neuroscience. Nat Rev Neurosci 19(9):566–578. https://doi.org/10.1038/s41583-018-0038-8

Beaumont TL, Yao B, Shah A, Kapatos G, Loeb JA (2012) Layer-specific CREB target gene induction in human neocortical epilepsy. J Neurosci 32(41):14389–14401. https://doi.org/10.1523/JNEUROSCI.3408-12.2012

Beevers CG, Clasen PC, Enock PM, Schnyer DM (2015) Attention bias modification for major depressive disorder: effects on attention bias, resting state connectivity, and symptom change. J Abnorm Psychol 124(3):463–475. https://doi.org/10.1037/abn0000049

Benarroch L, Bonne G, Rivier F, Hamroun D (2020) The 2021 version of the gene table of neuromuscular disorders (nuclear genome). Neuromuscul Disord 30(12):1008–1048. https://doi.org/10.1016/j.nmd.2020.11.009

Bernardi RE (2015) Advances in nicotine research in addiction biology. Addict Biol 20(5):877–889. https://doi.org/10.1111/adb.12269

Binder JR, Desai RH (2011) The neurobiology of semantic memory. Trends Cogn Sci 15(11):527–536. https://doi.org/10.1016/j.tics.2011.10.001

Bisaz R, Travaglia A, Alberini CM (2014) The neurobiological bases of memory formation: from physiological conditions to psychopathology. Psychopathology 47(6):347–356. https://doi.org/10.1159/000363702

Braun R (2014) Systems analysis of high-throughput data. Adv Exp Med Biol 844:153–187. https://doi.org/10.1007/978-1-4939-2095-2_8

Brem AK, Ran K, Pascual-leone A (2013) Learning and memory. Handb Clin Neurol 116:693–737. https://doi.org/10.1016/b978-0-444-53497-2.00055-3

Camina E, Güell F (2017) The neuroanatomical, neurophysiological and psychological basis of memory: current models and their origins. Front Pharmacol 8:438. https://doi.org/10.3389/fphar.2017.00438

Carboni L, Nguyen TP, Caberlotto L (2016) Systems biology integration of proteomic data in rodent models of depression reveals involvement of the immune response and glutamatergic signaling. Proteomics Clin Appl 10(12):1254–1263. https://doi.org/10.1002/prca.201500149

Castrillo JI, Oliver SG (2016) Alzheimer's as a systems-level disease involving the interplay of multiple cellular networks. Methods Mol Biol 1303:3–48. https://doi.org/10.1007/978-1-4939-2627-5_1

Castrillo JI, Lista S, Hampel H, Ritchie CW (2018) Systems biology methods for Alzheimer's disease research toward molecular signatures, subtypes, and stages and precision medicine: application in cohort studies and trials. Methods Mol Biol 1750:31–66. https://doi.org/10.1007/978-1-4939-7704-8_3

Chatterjee P, Roy D, Bhattacharyya M et al (2017) Biological networks in Parkinson's disease: an insight into the epigenetic mechanisms associated with this disease. BMC Genomics 18:721. https://doi.org/10.1186/s12864-017-4098-3

Chen Y, Yu H, Gong J, Ma M, Han H, Wei H, Xu W (2019) Artificial synapses based on nanomaterials. Nanotechnology 30(1):012001. https://doi.org/10.1088/1361-6528/aae470

Chowdhury UN, Babul Islam M, Ahmad S, Moni MA (2020) Systems biology and bioinformatics approach to identify gene signatures, pathways and therapeutic targets of Alzheimer's disease. Inf Med 21:100439. https://doi.org/10.1016/j.imu.2020.100439
Chrousos GP, Pervanidou P, Dalla C (2018) Mini-reviews based on the first conference of the institute of stress biology & medicine "systems biology-medicine and stress". Hormones 17(1):3–4. https://doi.org/10.1007/s42000-018-0027-3
Colangelo C, Shichkova P, Keller D, Markram H, Ramaswamy S (2019) Cellular, synaptic and network effects of acetylcholine in the neocortex. Front Neural Circuits 13:24. https://doi.org/10.3389/fncir.2019.00024
Dalvie S, Chatzinakos C, Al Zoubi O, Georgiadis F, Lancashire L, Daskalakis NP (2021) From genetics to systems biology of stress-related mental disorders. Neurobiol Stress 15:100393. https://doi.org/10.1016/j.ynstr.2021.100393
Dalvi-Garcia F, Fonseca LL, Vasconcelos ATR, Hedin-Pereira C, Voit EO (2021) A model of dopamine and serotonin-kynurenine metabolism in cortisolemia: implications for depression. PLoS Comput Biol 17(5):e1008956. https://doi.org/10.1371/journal.pcbi.1008956
Davidovic L, Navratil V, Bonaccorso CM, Catania MV, Bardoni B, Dumas ME (2011) A metabolomic and systems biology perspective on the brain of the fragile X syndrome mouse model. Genome Res 21(12):2190–2202. https://doi.org/10.1101/gr.116764.110
Dennis EJ, El Hady A, Michaiel A, Clemens A, Tervo DRG, Voigts J, Datta SR (2020) Systems neuroscience of natural behaviors in rodents. J Neurosci 41(5):911–919. https://doi.org/10.1523/jneurosci.1877-20.2020
Diaz-Beltran L, Cano C, Wall D, Esteban F (2013) Systems biology as a comparative approach to understand complex gene expression in neurological diseases. Behav Sci 3(2):253–272. https://doi.org/10.3390/bs3020253
Dougherty JD, Yang C, Lake AM (2017) Systems biology in the central nervous system: a brief perspective on essential recent advancements. Curr Opin Syst Biol 3:67–76. https://doi.org/10.1016/j.coisb.2017.04.011
Eekelen JAM, Olsson CA, Ellis JA, Ang W, Hutchinson D, Zubrick SR, Pennell CE (2011) Identification and genetic determination of an early life risk disposition for depressive disorder: atypical stress-related behaviour in early childhood. Aust J Psychol 63(1):6–17. https://doi.org/10.1111/j.1742-9536.2011.00002.x
Eekelen JA, Ellis JA, Pennell CE, Craig J, Saffery R, Mattes E, Olsson CA (2012) Stress-sensitive neurosignalling in depression: an integrated network biology approach to candidate gene selection for genetic association analysis. Ment Illn 4(2):e21. https://doi.org/10.4081/mi.2012.e21
Englot DJ, Berger MS, Barbaro NM, Chang EF (2012) Factors associated with seizure freedom in the surgical resection of glioneuronal tumors. Epilepsia 53(1):51–57. https://doi.org/10.1111/j.1528-1167.2011.03269.x
Faghihi F, Moustafa AA (2017) Combined computational systems biology and computational neuroscience approaches help develop of future "cognitive developmental robotics". Front Neurorobot 11:63. https://doi.org/10.3389/fnbot.2017.00063
Fan T, Hu Y, Xin J, Zhao M, Wang J (2019) Analyzing the genes and pathways related to major depressive disorder via a systems biology approach. Brain Behav 10(2):1502. https://doi.org/10.1002/brb3.1502
Farias J, Sotelo JR, Sotelo-Silveira J (2019) Toward axonal system biology: genome wide views of local mRNA translation. Proteomics 19(16):1900054. https://doi.org/10.1002/pmic.201900054
Feltes BC, de Faria Poloni J, Nunes IJ, Bonatto D (2014) Fetal alcohol syndrome, chemo-biology and OMICS: ethanol effects on vitamin metabolism during neurodevelopment as measured by systems biology analysis. OMICS 18(6):344–363. https://doi.org/10.1089/omi.2013.0144
Ferenczi EA, Tan X, Huang CLH (2019) Principles of optogenetic methods and their application to cardiac experimental systems. Front Physiol 10:1096. https://doi.org/10.3389/fphys.2019.01096
Fernández E, Collins MO, Uren RT, Kopanitsa MV, Komiyama NH, Croning MDR, Zografos L, Armstrong JD, Choudhary JS, Grant SGN (2009) Targeted tandem affinity purification of

PSD-95 recovers core postsynaptic complexes and schizophrenia susceptibility proteins. Mol Syst Biol 5(1):269. https://doi.org/10.1038/msb.2009.27

Flanagan M, Sonnen JA, Keene CD, Hevner RF, Montine TJ (2018) Molecular basis of diseases of the nervous system. Mol Pathol 2018:651–690. https://doi.org/10.1016/b978-0-12-802761-5.00029-8

Frankland PW, Josselyn SA, Köhler S (2019) The neurobiological foundation of memory retrieval. Nat Neurosci 22(10):1576–1585. https://doi.org/10.1038/s41593-019-0493-1

Furuya S, Argemi J, Uehara T, Katou Y, Fouts DE, Schnabl B, Dubuquoy L, Belorkar A, Vadigepalli R, Kono H, Bataller R, Rusyn I (2020) A novel mouse model of acute-on-chronic cholestatic alcoholic liver disease: a systems biology comparison with human alcoholic hepatitis. Alcohol Clin Exp Res 44:87–101. https://doi.org/10.1111/acer.14234

Geschwind DH, Konopka G (2009) Neuroscience in the era of functional genomics and systems biology. Nature 461(7266):908–915. https://doi.org/10.1038/nature08537

Glasgow SD, McPhedrain R, Madranges JF, Kennedy TE, Ruthazer ES (2019) Approaches and limitations in the investigation of synaptic transmission and plasticity. Front Synaptic Neurosci 11:20. https://doi.org/10.3389/fnsyn.2019.00020

Grant S (2003) Systems biology in neuroscience: bridging genes to cognition. Curr Opin Neurobiol 13(5):577–582. https://doi.org/10.1016/j.conb.2003.09.016

Green AE, Munafò MR, DeYoung CG, Fossella JA, Fan J, Gray JR (2008) Using genetic data in cognitive neuroscience: from growing pains to genuine insights. Nat Rev Neurosci 9(9):710–720. https://doi.org/10.1038/nrn2461

Guo QM, Zakhari S (2008) Commentary: systems biology and its relevance to alcohol research. Alcohol Res Health 31(1):5–11

Halliday GM, Leverenz JB, Schneider JS, Adler CH (2014) The neurobiological basis of cognitive impairment in Parkinson's disease. Mov Disord 29(5):634–650. https://doi.org/10.1002/mds.25857

Hampel H, Nisticò R, Seyfried NT, Levey AI, Modeste E, Lemercier P, Baldacci F, Toschi N, Garaci F, Perry G, Emanuele E, Valenzuela PL, Lucia A, Urbani A, Sancesario GM, Mapstone M, Corbo M, Vergallo A, Lista S, Alzheimer Precision Medicine Initiative (APMI) (2021) Omics sciences for systems biology in Alzheimer's disease: state-of-the-art of the evidence. Ageing Res Rev 69:101346. https://doi.org/10.1016/j.arr.2021.101346

Haroutunian V (2014) A systems biology approach toward understanding schizophrenia. Schizophr Res 153:S10. https://doi.org/10.1016/s0920-9964(14)70031-3

Harzsch S, Viertel C (2019) Immunolocalization of neurotransmitters and neuromodulators in the developing crayfish brain. Methods Mol Biol 2047:271–291. https://doi.org/10.1007/978-1-4939-9732-9_15

Herlenius E, Lagercrantz H (2001) Neurotransmitters and neuromodulators during early human development. Early Hum Dev 65(1):21–37. https://doi.org/10.1016/s0378-3782(01)00189-x

Hodo TW, de Aquino MTP, Shimamoto A, Shanker A (2020) Critical neurotransmitters in the neuroimmune network. Front Immunol 11:1869. https://doi.org/10.3389/fimmu.2020.01869

Hook V, Funkelstein L, Lu D, Bark S, Wegrzyn J, Hwang SR (2008) Proteases for processing proneuropeptides into peptide neurotransmitters and hormones. Annu Rev Pharmacol Toxicol 48:393–423. https://doi.org/10.1146/annurev.pharmtox.48.113006.094812

Hook V, Lietz CB, Podvin S, Cajka T, Fiehn O (2018) Diversity of neuropeptide cell-cell signaling molecules generated by proteolytic processing revealed by neuropeptidomics mass spectrometry. J Am Soc Mass Spectrom 29(5):807–816. https://doi.org/10.1007/s13361-018-1914-1

Huang KC, Yang KC, Lin H, Tsao TT, Lee SA (2014) Transcriptome alterations of mitochondrial and coagulation function in schizophrenia by cortical sequencing analysis. BMC Genomics 15(9):S6. https://doi.org/10.1186/1471-2164-15-S9-S6

Inagaki N, Toriyama M, Sakumura Y (2011) Systems biology of symmetry breaking during neuronal polarity formation. Dev Neurobiol 71(6):584–593. https://doi.org/10.1002/dneu.20837

Jäkel S, Dimou L (2017) Glial cells and their function in the adult brain: a journey through the history of their ablation. Front Cell Neurosci 11:24. https://doi.org/10.3389/fncel.2017.00024

Jonsson T, Stefansson H, Steinberg S, Jonsdottir I, Jonsson PV, Snaedal J, Bjornsson S, Huttenlocher J, Levey AI, Lah JJ, Rujescu D, Hampel H, Giegling I, Andreassen OA, Engedal K, Ulstein I, Djurovic S, Ibrahim-Verbaas C, Hofman A, Stefansson K (2013) Variant of TREM2 associated with the risk of Alzheimer's disease. N Engl J Med 368(2):107–116. https://doi.org/10.1056/nejmoa1211103

Jopling E, Tracy A, LeMoult J (2020) Childhood maltreatment, negative self-referential processing, and depressive symptoms during stress. Psychol Res Behav Manag 13:79–87. https://doi.org/10.2147/PRBM.S231505

Jouan L, Girard SL, Dobrzeniecka S, Ambalavanan A, Krebs MO, Joober R, Gauthier J, Dion PA, Rouleau GA (2013) Investigation of rare variants in LRP1, KPNA1, ALS2CL and ZNF480 genes in schizophrenia patients reflects genetic heterogeneity of the disease. Behav Brain Funct 9:9. https://doi.org/10.1186/1744-9081-9-9

Juhász G, Földi I, Penke B (2011) Systems biology of Alzheimer's disease: how diverse molecular changes result in memory impairment in AD. Neurochem Int 58(7):739–750. https://doi.org/10.1016/j.neuint.2011.02.008

Kandel E, Dudai Y, Mayford M (2014) The molecular and systems biology of memory. Cell 157(1):163–186. https://doi.org/10.1016/j.cell.2014.03.001

Kastanenka KV, Moreno-Bote R, De Pittà M, Perea G, Eraso-Pichot A, Masgrau R, Poskanzer KE, Galea E (2019) A roadmap to integrate astrocytes into systems neuroscience. Glia 68(1):5–26. https://doi.org/10.1002/glia.23632

Keck T, Josselyn SA (2021) Editorial overview: neurobiology of learning and plasticity. Curr Opin Neurobiol 67:iii–v. https://doi.org/10.1016/j.conb.2021.04.002

Kim S, Kaang BK (2017) Epigenetic regulation and chromatin remodeling in learning and memory. Exp Mol Med 49(1):e281. https://doi.org/10.1038/emm.2016.140

Kirchner A, Dachet F, Loeb JA (2020) Identifying targets for preventing epilepsy using systems biology of the human brain. Neuropharmacology 168:107757. https://doi.org/10.1016/j.neuropharm.2019.107757

Körber C, Sommer WH (2022) From ensembles to meta-ensembles: specific reward encoding by correlated network activity. Front Behav Neurosci 16:977474. https://doi.org/10.3389/fnbeh.2022.977474

Kruyer A, Scofield MD (2021) Astrocytes in addictive disorders. Adv Neurobiol 26:231–254. https://doi.org/10.1007/978-3-030-77375-5_10

Kuleshov MV, Jones MR, Rouillard AD, Fernandez NF, Duan Q, Wang Z, Koplev S, Jenkins SL, Jagodnik KM, Lachmann A, McDermott MG, Monteiro CD, Gundersen GW, Ma'ayan A (2016) Enrichr: a comprehensive gene set enrichment analysis web server 2016 update. Nucleic Acids Res 44(1):W90–W97. https://doi.org/10.1093/nar/gkw377

Le Novère N (2007) The long journey to a systems biology of neuronal function. BMC Syst Biol 1(1):28. https://doi.org/10.1186/1752-0509-1-28

LeMoult J, Joormann J, Kircanski K, Gotlib IH (2016) Attentional bias training in girls at risk for depression. J Child Psychol Psychiatry 57(11):1326–1333. https://doi.org/10.1111/jcpp.12587

Liu X, Ramirez S, Pang PT, Puryear CB, Govindarajan A, Deisseroth K, Tonegawa S (2012) Optogenetic stimulation of a hippocampal engram activates fear memory recall. Nature 484(7394):381–385. https://doi.org/10.1038/nature11028

Liu J, Li M, Luo XJ, Su B (2018) Systems-level analysis of risk genes reveals the modular nature of schizophrenia. Schizophr Res 201:261–269. https://doi.org/10.1016/j.schres.2018.05.015

Loeb JA (2011) Identifying targets for preventing epilepsy using systems biology. Neurosci Lett 497(3):205–212. https://doi.org/10.1016/j.neulet.2011.02.041

Loh M, Rolls ET, Deco G (2007) A dynamical systems hypothesis of schizophrenia. PLoS Comput Biol 3(11):e228. https://doi.org/10.1371/journal.pcbi.0030228

Luvsannyam E, Jain MS, Pormento M, Siddiqui H, Balagtas A, Emuze BO, Poprawski T (2022) Neurobiology of schizophrenia: a comprehensive review. Cureus 14(4):e23959. https://doi.org/10.7759/cureus.23959

MacNicol B (2016) The biology of addiction. Can J Anesth 64(2):141–148. https://doi.org/10.1007/s12630-016-0771-2

Manciu F, Manciu M, Ciubuc J, Sundin E, Ochoa K, Eastman M, Durrer W, Guerrero J, Lopez B, Subedi M, Bennet K (2018) Simultaneous detection of dopamine and serotonin—a comparative experimental and theoretical study of neurotransmitter interactions. Biosensors 9(1):3. https://doi.org/10.3390/bios9010003

Mancuso JJ, Cheng J, Yin Z, Gilliam JC, Xia X, Li X, Wong ST (2014) Integration of multiscale dendritic spine structure and function data into systems biology models. Front Neuroanat 8:130. https://doi.org/10.3389/fnana.2014.00130

Maraver MJ, Lapa A, Garcia-Marques L, Carneiro P, Raposo A (2022) Can we learn from errors? Retrieval facilitates the correction of false memories for pragmatic inferences. PLoS One 17(8):e0272427. https://doi.org/10.1371/journal.pone.0272427

Marrs TC, Maynard RL (2013) Neurotranmission systems as targets for toxicants: a review. Cell Biol Toxicol 29(6):381–396. https://doi.org/10.1007/s10565-013-9259-9

Martone ME, Gupta A, Ellisman MH (2004) e-Neuroscience: challenges and triumphs in integrating distributed data from molecules to brains. Nat Neurosci 7(5):467–472. https://doi.org/10.1038/nn1229

Matošić A (2016) Neurobiological bases of alcohol addiction. Acta Clin Croat 2016:134–150. https://doi.org/10.20471/acc.2016.55.01.19

Medaglia JD, Lynall ME, Bassett DS (2015) Cognitive network neuroscience. J Cogn Neurosci 27(8):1471–1491. https://doi.org/10.1162/jocn_a_00810

Mei H, Xia T, Feng G, Zhu J, Lin SM, Qiu Y (2012) Opportunities in systems biology to discover mechanisms and repurpose drugs for CNS diseases. Drug Discov Today 17(21-22):1208–1216

Mendis GDC, Morrisroe E, Petrou S, Halgamuge SK (2016) Use of adaptive network burst detection methods for multielectrode array data and the generation of artificial spike patterns for method evaluation. J Neural Eng 13(2):026009. https://doi.org/10.1088/1741-2560/13/2/026009

Miller JA, Ding SL, Sunkin SM, Smith KA, Ng L, Szafer A, Ebbert A, Riley ZL, Royall JJ, Aiona K, Arnold JM, Bennet C, Bertagnolli D, Brouner K, Butler S, Caldejon S, Carey A, Cuhaciyan C, Dalley RA, Dee N, Lein ES (2014) Transcriptional landscape of the prenatal human brain. Nature 508(7495):199–206. https://doi.org/10.1038/nature13185

Mittal S, Shah AK, Barkmeier DT, Loeb JA (2013) Systems biology of human epilepsy applied to patients with brain tumors. Epilepsia 54:35–39. https://doi.org/10.1111/epi.12441

Mohandas E, Rajmohan V, Raghunath B (2009) Neurobiology of Alzheimer's disease. Indian J Psychiatry 51(1):55–61. https://doi.org/10.4103/0019-5545.44908

Monroe SM, Anderson SF, Harkness KL (2019) Life stress and major depression: the mysteries of recurrences. Psychol Rev 126(6):791–816. https://doi.org/10.1037/rev0000157

Morris KF, Schwaber JS (2012) The neuroscience-systems biology disconnect: towards the NeuroPhysiome. Exp Physiol 97(4):452–454. https://doi.org/10.1113/expphysiol.2011.058297

Nelson TJ, Alkon DL (2015) Molecular regulation of synaptogenesis during associative learning and memory. Brain Res 1621:239–251. https://doi.org/10.1016/j.brainres.2014.11.054

Noble JM, Weimer LH (2014) Neurologic complications of alcoholism. Continuum 20(3):624–641. https://doi.org/10.1212/01.CON.0000450970.99322.84

Ojeda J, Ávila A (2019) Early actions of neurotransmitters during cortex development and maturation of reprogrammed neurons. Front Synaptic Neurosci 11:33. https://doi.org/10.3389/fnsyn.2019.00033

Perea G, Sur M, Araque A (2014) Neuron-glia networks: integral gear of brain function. Front Cell Neurosci 8:378. https://doi.org/10.3389/fncel.2014.00378

Picciotto M, Higley M, Mineur Y (2012) Acetylcholine as a neuromodulator: cholinergic signaling shapes nervous system function and behavior. Neuron 76(1):116–129. https://doi.org/10.1016/j.neuron.2012.08.036

Pocklington AJ, Cumiskey M, Armstrong JD, Grant SGN (2006) The proteomes of neurotransmitter receptor complexes form modular networks with distributed functionality underlying plasticity and behaviour. Mol Syst Biol 2(1):23. https://doi.org/10.1038/msb4100041

Preuss C, Pandey R, Piazza E et al (2020) A novel systems biology approach to evaluate mouse models of late-onset Alzheimer's disease. Mol Neurodegener 15:67. https://doi.org/10.1186/s13024-020-00412-5

Ramasamy A, Trabzuni D, Guelfi S, Varghese V, Smith C, Walker R, Coin L, de Silva R, Cookson MR, Singleton AB, Hardy J, Ryten M, Weale ME (2014) Genetic variability in the regulation of gene expression in ten regions of the human brain. Nat Neurosci 17(10):1418–1428. https://doi.org/10.1038/nn.3801

Ramirez-Suarez NJ, Belalcazar HM, Salazar CJ, Beyaz B, Raja B, Nguyen K, Celestrin K, Fredens J, Færgeman NJ, Hall DH, Bülow HE (2019) Axon-dependent patterning and maintenance of somatosensory dendritic arbors. Dev Cell 48(2):229–244. https://doi.org/10.1016/j.devcel.2018.12.015

Rasmusson AM, Schnurr PP, Zukowska Z, Scioli E, Forman DE (2010) Adaptation to extreme stress: post-traumatic stress disorder, neuropeptide Y and metabolic syndrome. Exp Biol Med 235(10):1150–1162. https://doi.org/10.1258/ebm.2010.009334

Rosalba S, Roberta F, Salvatore C (2019) Astrocytes: role and functions in brain pathologies. Front Pharmacol 2019:1114. https://doi.org/10.3389/fphar.2019.01114

Rudan I (2010) New technologies provide insights into genetic basis of psychiatric disorders and explain their co-morbidity. Psychiatr Danub 22(2):190–192

Russo M, Funk D, Loughlin A, Coen K, Lê AD (2018) Effects of alcohol dependence on discrete choice between alcohol and saccharin. Neuropsychopharmacology 43(9):1859–1866. https://doi.org/10.1038/s41386-018-0101-1

Sakib N, Chowdhury UN, Babul Islam M, Ahmad S, Moni MA (2020) A systems biology approach to identifying genetic factors affected by aging, lifestyle factors, and type 2 diabetes that influences Parkinson's disease progression. Inf Med 21:100448. https://doi.org/10.1016/j.imu.2020.100448

Santos JPG, Pajo K, Trpevski D, Stepaniuk A, Eriksson O, Nair AG, Keller D, Hellgren Kotaleski J, Kramer A (2021) A modular workflow for model building, analysis, and parameter estimation in systems biology and neuroscience. Neuroinformatics 20(1):241–259. https://doi.org/10.1007/s12021-021-09546-3

Sauer CS, Phetsanthad A, Riusech OL, Li L (2021) Developing mass spectrometry for the quantitative analysis of neuropeptides. Expert Rev Proteomics 18(7):607–621. https://doi.org/10.1080/14789450.2021.1967146

Scharfman HE (2007) The neurobiology of epilepsy. Curr Neurol Neurosci Rep 7(4):348–354. https://doi.org/10.1007/s11910-007-0053-z

Schena A, Johnsson K (2013) Sensing acetylcholine and anticholinesterase compounds. Angew Chem Int Ed 53(5):1302–1305. https://doi.org/10.1002/anie.201307754

Schleidgen S, Fernau S, Fleischer H, Schickhardt C, Oßa AK, Winkler EC (2017) Applying systems biology to biomedical research and health care: a précising definition of systems medicine. BMC Health Serv Res 17(1):761. https://doi.org/10.1186/s12913-017-2688-z

Schwartz TS (2020) The promises and the challenges of integrating multi-omics and systems biology in comparative stress biology. Integr Comp Biol 60(1):89–97. https://doi.org/10.1093/icb/icaa026

Shah RM, Peterson C, Strom A, Dillenburg M, Finzel B, Kitto KF, Fairbanks C, Wilcox G, Wagner CR (2019) Inhibition of HINT1 modulates spinal nociception and NMDA evoked behavior in mice. ACS Chem Neurosci 10(10):4385–4393. https://doi.org/10.1021/acschemneuro.9b00432

Sibili L, Ceccanti M, Deiana L, Coriale G, Picozzi R (2001) Maintenance of abstinence from alcohol. Lancet 358(9287):1103–1104. https://doi.org/10.1016/s0140-6736(01)06222-5

Sickmann HM, Li Y, Mørk A, Sanchez C, Gulinello M (2014) Does stress elicit depression? Evidence from clinical and preclinical studies. Curr Top Behav Neurosci 18:123–159. https://doi.org/10.1007/7854_2014_292

Skene NG, Grant SGN (2016) Identification of vulnerable cell types in major brain disorders using single cell transcriptomes and expression weighted cell type enrichment. Front Neurosci 10:16. https://doi.org/10.3389/fnins.2016.00016

Sleegers K, Lambert JC, Bertram L, Cruts M, Amouyel P, Van Broeckhoven C (2010) The pursuit of susceptibility genes for Alzheimer's disease: progress and prospects. Trends Genet 26(2):84–93. https://doi.org/10.1016/j.tig.2009.12.004

Sommer WH (2017) SyBil-AA, systems biology of alcohol addiction: modeling and validating disease state networks in human and animal brains for understanding pathophysiolgy, predicting outcomes and improving therapy, HORIZON2020. Impact 2017(10):89–91. https://doi.org/10.21820/23987073.2017.10.89

Spanagel R, Durstewitz D, Hansson A, Heinz A, Kiefer F, Köhr G, Matthäus F, Nöthen MM, Noori HR, Obermayer K, Rietschel M, Schloss P, Scholz H, Schumann G, Smolka M, Sommer W, Vengeliene V, Walter H, Wurst W, Zimmermann US, Derks EM (2013) A systems medicine research approach for studying alcohol addiction. Addict Biol 18(6):883–896. https://doi.org/10.1111/adb.12109

Südhof TC (2021) The cell biology of synapse formation. J Cell Biol 220(7):103052. https://doi.org/10.1083/jcb.202103052

Sumsion GR, Bradshaw MS, Beales JT, Ford E, Caryotakis G, Garrett DJ, LeBaron ED, Nwosu IO, Piccolo SR (2020) Diverse approaches to predicting drug-induced liver injury using gene-expression profiles. Biol Direct 15(1):1. https://doi.org/10.1186/s13062-019-0257-6

Tryon VL, Penner MR, Heide SW, King HO, Larkin J, Mizumori S (2017) Hippocampal neural activity reflects the economy of choices during goal-directed navigation. Hippocampus 27(7):743–758. https://doi.org/10.1002/hipo.22720

Turnbull L, Hütt MT, Ioannides AA et al (2018) Connectivity and complex systems: learning from a multi-disciplinary perspective. Appl Netw Sci 3:11. https://doi.org/10.1007/s41109-018-0067-2

Uffelmann E, Huang QQ, Munung NS et al (2021) Genome-wide association studies. Nat Rev Methods Primers 1:59. https://doi.org/10.1038/s43586-021-00056-9

Ung K, Tepe B, Pekarek B, Arenkiel BR, Deneen B (2020) Parallel astrocyte calcium signaling modulates olfactory bulb responses. J Neurosci Res 98(8):1605–1618. https://doi.org/10.1002/jnr.24634

Verkhratsky A, Ho MS, Zorec R, Parpura V (2019) The concept of neuroglia. Neuroglia Neurodegener Dis 2019:1–13. https://doi.org/10.1007/978-981-13-9913-8_1

Villoslada P, Steinman L, Baranzini SE (2009) Systems biology and its application to the understanding of neurological diseases. Ann Neurol 65:124–139. https://doi.org/10.1002/ana.21634

Wheeler DW, White CM, Rees CL, Komendantov AO, Hamilton DJ, Ascoli GA (2015) Hippocampome.org: a knowledge base of neuron types in the rodent hippocampus. elife 4:9960. https://doi.org/10.7554/elife.09960

Wu Z, Lin D, Li Y (2022) Pushing the frontiers: tools for monitoring neurotransmitters and neuromodulators. Nat Rev Neurosci 23(5):257–274. https://doi.org/10.1038/s41583-022-00577-6

Xavier AL, Menezes JRL, Goldman SA, Nedergaard M (2014) Fine-tuning the central nervous system: microglial modelling of cells and synapses. Philos Trans R Soc B Biol Sci 369(1654):20130593. https://doi.org/10.1098/rstb.2013.0593

Yang A, Tsai SJ (2017) New targets for schizophrenia treatment beyond the dopamine hypothesis. Int J Mol Sci 18(8):1689. https://doi.org/10.3390/ijms18081689

Yuan A, Nixon RA (2021) Neurofilament proteins as biomarkers to monitor neurological diseases and the efficacy of therapies. Front Neurosci 15:689938. https://doi.org/10.3389/fnins.2021.689938

Zhang Z, Niki K, Luo J (2020) The function of the hippocampus in bridging functional and temporal discontiguity. Neural Plast 2020:1049721. https://doi.org/10.1155/2020/1049721

Zhao Y, Zhu M, Yu Y, Qiu L, Zhang Y, He L, Zhang J (2017) Brain REST/NRSF is not only a silent repressor but also an active protector. Mol Neurobiol 54(1):541–550. https://doi.org/10.1007/s12035-015-9658-4

Computational and Experimental Approaches to Decipher the Complexity of Diseases

15

Sanket S. Rathod, Sonali S. Shinde, Prafulla B. Choudhari, Rakesh P. Dhavale, and Aniket P. Sarkate

Abstract

Systems biology, as a multidisciplinary approach, seeks to understand biological systems as a whole, and this chapter showcases the significant contributions of computational and experimental approaches within the framework of systems biology for deciphering the complexity of diseases. Systems biology adopts a holistic perspective by considering the interactions and relationships among various components within a biological system. By leveraging computational approaches such as genomics, proteomics, transcriptomics, and metabolomics, it is possible to gather comprehensive data on the molecular profiles and dynamics of diseases. These computational tools enable the identification of disease-associated biomarkers, the exploration of molecular pathways underlying disease pathogenesis, and the prediction of therapeutic responses. The integration of genomics, proteomics, transcriptomics, metabolomics, modeling and simulation, network analysis, experimental approaches, and integrative analysis provides a comprehensive toolkit for researchers to unravel the intricacies of diseases and develop effective strategies for diagnosis, treatment, and prevention.

S. S. Rathod · P. B. Choudhari
Department of Pharmaceutical Chemistry, Bharati Vidyapeeth College of Pharmacy, Kolhapur, Maharashtra, India

S. S. Shinde (✉) · A. P. Sarkate
Department of Chemical Technology, Dr. Babasaheb Ambedkar Marathwada University, Aurangabad, Maharashtra, India

R. P. Dhavale
Department of Pharmaceutics, Bharati Vidyapeeth College of Pharmacy, Kolhapur, Maharashtra, India

S. Joshi et al. (eds.), *Systems Biology Approaches: Prevention, Diagnosis, and Understanding Mechanisms of Complex Diseases*,
https://doi.org/10.1007/978-981-99-9462-5_15

Keywords

Computational biology · Diseases · Systems biology · Metabolomics · Analysis

15.1 Introduction

Systems biology is dedicated to examining biological components through systematic measurement techniques like genomics, bioinformatics, and proteomics, as well as employing mathematical and computational models to gain deeper insights into the functioning of specific biological conditions (Sordo Vieira and Laubenbacher 2022). These models play a pivotal role in the field of systems biology, serving as essential tools for analyzing complex biological systems and predicting their behavior (Meyer and Saez-Rodriguez 2021). These models also enable to simulate intricate biological processes, explore hypotheses, and uncover novel insights into the functioning of living organisms (Butcher et al. 2004; Sordo Vieira and Laubenbacher 2022). Through integrating experimental data with mathematical and computational frameworks, systems biology provides a powerful approach to comprehending the intricate dynamics and interactions within biological systems, ultimately advancing our understanding of various biological phenomena. Systems biology offers a holistic approach by integrating data from multiple levels, including genomics, proteomics, metabolomics, and clinical data. This comprehensive view enables a better understanding of the complex interactions and dynamics within biological systems. It plays a crucial role in identifying novel drug targets. By uncovering key molecular interactions, signaling pathways, and regulatory networks involved in disease processes, researchers can identify specific components or pathways that can be targeted for therapeutic intervention (Hood and Perlmutter 2004).

Systems biology aims to understand how the collective behavior of a biological system emerges from its components and interactions, encompassing various scales from subcellular to organismal levels. Moreover, computational modeling and simulation techniques in systems biology enable predictive modeling. Researchers can simulate and predict the behavior of biological systems under different conditions, facilitating the evaluation of potential drug candidates and aiding in the selection and optimization of lead compounds (Palsson 2006). Systems biology contributes to personalized medicine by considering individual genetic and molecular profiles. By integrating patient-specific data, such as genomic information and clinical data, systems biology can identify biomarkers that can be used for patient stratification and the development of targeted treatment approaches (Jain 2016).

Unlike traditional reductionist approaches that focus on individual molecules, systems biology examines the properties and interactions of multiple components. It employs a combination of experimental techniques, which measure multiple entities simultaneously, and computational approaches for analyzing complex, multivariate data and generating testable predictions (Hood and Perlmutter 2004). These experimental techniques are commonly taught in biomedical science programs at undergraduate and graduate levels. In systems biology, computational methods are applied collectively, drawing from various disciplines to analyze datasets of diverse sizes and types and construct predictive models. This interdisciplinary approach is pervasive, as computational modeling and the integration of large-scale postgenomic datasets have become integral to biological research across numerous fields. In particular, systems biology enables the study of how genotypes influence phenotypes by examining the interactions among different biological components. To implement systems biology effectively, researchers should familiarize themselves with the four key steps involved in its implementation (Sobie et al. 2011). Sometimes, systems biology is focused on unraveling the intricate interactions within complex biological systems. In the context of medicine, the potential of systems biology lies in its ability to comprehend disease processes as disturbances in normal physiological systems. This approach is essential for advancing health and well-being. However, several challenges need to be addressed within the field of systems biology. Firstly, there is a need to generate substantial amounts of data to adequately assess the variability in biological networks. Secondly, the integration of data from diverse sources into a coherent and usable knowledge base is crucial. Lastly, the acquired knowledge must be applied to construct models that optimize and elucidate the behavior of specific components within the system. These models will provide insights into system dynamics and enable the identification of strategies to improve clinical medicine (Tsongalis 2020).

In silico methods also play a vital role in systems biology by providing computational tools and techniques to analyze complex biological systems. These methods involve the use of computational models, algorithms, and simulations to gain insights into the behavior and dynamics of biological systems, ranging from molecular interactions to cellular networks and beyond. By leveraging computational power and advanced algorithms, in silico methods enable the integration of diverse omics data, prediction of system behavior, identification of key regulatory components, and exploration of emergent properties in biological systems (Palsson 2015; Schneider 2013a, b). In silico methods have become indispensable in systems biology due to the several advantages they offer. Firstly, they provide a means to efficiently process and analyze large-scale biological datasets. The rapid advancements in high-throughput technologies have generated massive amounts of biological data, and in silico methods offer efficient ways to handle, integrate, and extract meaningful information from these datasets (Dudley et al. 2011; Nesvizhskii 2010). Moreover, in silico methods enable the construction of mathematical models that simulate the behavior of biological systems. These models can capture the complex interactions and dynamics of biological components, helping to unravel the underlying mechanisms and predict system responses under different conditions (Mendes 1997).

In silico models also serve as a platform for hypothesis testing, allowing researchers to explore "what-if" scenarios and design targeted experiments to validate predictions (Bartocci and Lió 2016). Additionally, in silico methods facilitate the exploration of design principles in biological systems. By using computational algorithms and simulations, researchers can generate and optimize virtual models to study the impact of different parameters and interventions on system behavior (Voit

2018). This iterative process of simulation and analysis aids in the identification of optimal strategies for manipulating biological systems, leading to advancements in fields such as drug discovery and biotechnology (Schneider 2013b). The continuous development and improvement of in silico methods, along with the increasing availability of biological data and computational resources, open up new possibilities for understanding and harnessing the complexity of biological systems. By integrating experimental data with computational modeling, in silico methods contribute to the advancement of systems biology and provide valuable insights into the fundamental principles governing life (Hucka et al. 2003).

15.2 Computational Approaches

Computational approaches play a pivotal role in systems biology by facilitating the analysis and interpretation of complex biological data. These approaches encompass a diverse range of techniques, including mathematical modeling, simulation, network analysis, and data mining. Mathematical models are constructed to represent the dynamic behavior of biological systems, capturing the interactions among various components. These models can be deterministic or stochastic, depending on the level of detail required. Simulation techniques allow researchers to simulate the behavior of the modeled systems under different conditions, providing valuable insights into system dynamics and aiding in hypothesis generation and testing. Network analysis involves the construction and analysis of biological networks, such as protein-protein interaction (PPI) networks or gene regulatory networks (GRNs), to identify key components and their interactions. Data mining techniques are employed to extract meaningful patterns and relationships from large-scale omics datasets, facilitating the identification of biomarkers, drug targets, and potential therapeutic interventions. Additionally, computational approaches enable the integration of data from multiple sources, such as genomics, proteomics, and metabolomics, to gain a comprehensive understanding of biological systems. Computational approaches are paramount in systems biology for analyzing and integrating omics data, including genomics, proteomics, metabolomics, and other high-throughput datasets. In genomics, computational methods are utilized for genome assembly, annotation, and variant calling, enabling the identification of genetic variations associated with diseases (Li 2018). Proteomics benefits from computational tools for protein identification, quantification, and characterization, allowing the comprehensive analysis of protein expression and post-translational modifications (Nesvizhskii 2010). Metabolomics data, generated from mass spectrometry (MS) or nuclear magnetic resonance spectroscopy, are processed using computational algorithms to identify metabolites and elucidate metabolic pathways (Wishart 2016). Furthermore, computational systems biology approaches enable the integration of multi-omics data to uncover complex biological interactions. Integration of genomics and transcriptomics data provides insights into gene regulatory networks and the identification of transcriptional regulatory mechanisms (Ritchie et al. 2015). Integrative analysis of proteomics and transcriptomics data allows for the exploration of protein expression patterns in relation to gene expression profiles, aiding in the understanding of cellular processes (Cox and Mann 2008). Additionally, the integration of metabolomics data with other omics data enables the reconstruction of metabolic networks and the identification of metabolic dysregulation associated with diseases (Sumner et al. 2007). To analyze these large-scale omics datasets, computational systems biology employs various techniques such as machine learning, network analysis, and pathway enrichment analysis. Machine learning algorithms are utilized for predictive modeling, classification, and biomarker discovery (Chen and Ishwaran 2012). Network analysis methods, including graph theory and network modeling, help uncover the relationships and interactions between genes, proteins, and metabolites within biological systems (Barabási et al. 2010). Pathway enrichment analysis identifies enriched biological pathways or gene sets within omics data, shedding light on the underlying biological mechanisms (Subramanian et al. 2005).

15.2.1 Genomics

Genomics is a field of study that focuses on mapping, sequencing, and analyzing genomes, which are the complete sets of an organism's genetic material, including genes and non-coding DNA (Venter et al. 2001). It plays a crucial role in advancing our understanding of the structure, function, and evolution of genomes, as well as their impact on various biological processes. One of the pioneering achievements in genomics was the completion of the Human Genome Project (HGP) in 2003 (Collins et al. 2003b). Led by an international consortium of scientists, the HGP aimed to sequence and map the entire human genome (International Human Genome Sequencing Consortium 2004). This monumental effort provided a comprehensive reference for understanding human genetics and paved the way for subsequent research in genomics (Shendure and Ji 2008). Genomic research has broad implications across various disciplines. In healthcare, genomics has fueled the development of personalized medicine. By analyzing an individual's genomic data, healthcare professionals can gain insights into their susceptibility to certain diseases, determine optimal treatment approaches, and identify potential drug targets (Green and Guyer 2011; Strianese et al. 2020). For example, the identification of specific genetic mutations in cancer patients has led to the development of targeted therapies tailored to their unique genomic profiles (Berger and Mardis 2018).

Genomics and systems biology are closely intertwined fields that complement each other in understanding the complexity of biological systems. Genomics provides valuable insights into the structure, function, and organization of genomes, while systems biology aims to comprehensively analyze and model the interactions within biological systems (Collins et al. 2003a). Genomics provides a foundation for systems biology by providing comprehensive data on the genetic components of biological systems. Genomic data, including DNA sequences, gene expression profiles, and genetic variations, serve as fundamental building blocks for systems-level analyses (Kitano 2002b). Systems biology utilizes genomic information to construct and refine mathematical models of biological systems. These models integrate genomics data with other omics data to study the interactions and dynamics of biological processes at a system-wide level (Ideker et al. 2001a). Genomic studies have contributed to the identification of key components and networks within biological systems (Barabási and Oltvai 2004). The integration of genomics and systems biology has led to advancements in personalized medicine. Genomic data can be combined with systems-level analysis to identify disease-associated molecular pathways and predict individualized treatment responses (Hood and Friend 2011). Systems biology approaches leverage genomic data to investigate the emergent properties of biological systems. By integrating genomics data with computational modeling, systems biologists can simulate and predict system behaviors, providing insights into complex biological phenomena (Palsson 2006). Additionally, computational algorithms enable the identification of genetic variants through variant calling techniques, facilitating the discovery of disease-associated variations. Another application is the prediction of gene functions and regulatory elements based on genomic sequences, contributing to the understanding of gene regulation. Furthermore, computational systems biology techniques, such as network analysis and pathway enrichment analysis, help unravel the complex interactions between genes, identify co-expression patterns, and uncover biological pathways underlying specific phenotypes or diseases (Barabási et al. 2010; Subramanian et al. 2005). These computational advancements in genomics empower researchers to extract meaningful insights from vast genomic datasets, ultimately enhancing our understanding of genome structure, function, and its implications in health and disease.

15.2.2 Transcriptomics

Transcriptomics is a powerful scientific approach that involves the comprehensive analysis of the complete set of transcripts, including messenger RNA (mRNA), non-coding RNA, and other RNA molecules, within a specific biological entity such as a cell, tissue, organ, or organism (Yadav et al. 2018). Transcriptomics analysis is conducted to understand the dynamic nature of gene expression and regulatory mechanisms underlying various biological processes and responses (Li et al. 2015). It involves the study of relative abundances of RNA transcripts, primarily using microarray technologies. RNA microarrays, also known as chips, are specifically designed for this purpose and typically consist of a library of known transcripts, representing a vast majority of RNA species from around 20,000 human genes (Casamassimi et al. 2017; Yadav et al. 2018). This approach provides an overview of the genetic content and regulatory systems within cells. In wild-type cells, whether they are bacterial or cell lines, the genome content remains the same. However, different genes are expressed based on various applications and conditions, resulting in distinct patterns of gene expression across different organisms. Cellular regulatory mechanisms exist to switch gene expression on and off. In systems biology, transcriptomics plays a crucial role aiming to understand complex biological systems as a whole by integrating and analyzing various types of biological data (Kitano 2002b; Hood and Galas 2003).

Transcriptomic data provides essential information about gene expression levels and patterns, which serve as key components for constructing comprehensive models and networks of biological processes (Zhong et al. 2009). Transcriptomics tools enable the assessment of various aspects such as gene expression levels in a given genome, genome profiling, comparative expression levels between different experimental datasets, and the impact of different parameters on gene expression (Casamassimi et al. 2017; Conesa et al. 2016; Ideker et al. 2001a; Jiménez-Guri et al. 2013). For transcriptomic data to be widely applicable, it must be generated under diverse conditions and compared to eliminate trivial sources of variation. Transcriptomic data with integration into the other "omics" approaches may help to obtain a more comprehensive understanding of complex biological systems and unravel intricate molecular networks and interactions. This interdisciplinary approach has the potential to drive scientific discoveries, lead to the development of novel therapies, and improve our understanding of complex biological processes at the molecular level (Pazhamala et al. 2021; Yadav et al. 2018).

15.2.3 Proteomics

Proteomics, a core technology in postgenomic and systems biology approaches, plays a crucial role in understanding the molecular mechanisms underlying normal and disease phenotypes and identifying critical biomarkers (Coorssen 2022). It is a field dedicated to studying proteins, which play vital roles within cells. Proteins are subject to various post-translational modifications, making them responsive to external influences. Proteomics can be broadly categorized into four areas: sequence proteomics, structural proteomics, functional and interaction proteomics, and expression proteomics (Joshi and Patil 2017). Each area focuses on different aspects of protein properties, such as primary and three-dimensional structure, protein function, and protein abundance, respectively. For example, sequence proteomics involves determining the amino acid sequence of a given protein. Traditionally, techniques like Edman sequencing have been used to elucidate the amino acid compositions of proteins. The advancement of proteomics technologies has revolutionized our understanding of protein structure, function, and their roles in various biological processes (Aizat and Hassan 2018).

Proteomics involves the comprehensive characterization of proteins in biological systems, including their spatial distribution, temporal dynamics, response to environmental stimuli, and differences between diseased and control systems. Proteomic technologies, particularly mass spectrometry (MS), enable the analysis of proteomes, protein identification, and assessment of protein properties such as abundance, localization, post-translational modifications, and molecular interactions (Coorssen 2013, 2022; Manes and Nita-Lazar 2018). Proteome separation techniques like gel chromatography and liquid chromatography are employed, and MS analysis can be performed on intact proteins (top-down) or digested protein peptides (bottom-up). Integration of proteomics with other omics technologies enhances our understanding of complex biological systems and paves the way for personalized medicine and disease diagnosis (Hixson et al. 2017). Due to the nature of systems biology studies, not all proteomic techniques can be effectively applied in this field. For instance, the technique known as "sequence proteomics," which focuses on determining protein sequences on a small scale, may not be suitable for large-scale system studies. However, "expression proteomics" provides a promising starting point for global proteome analysis, as it allows for the cataloging of quantifiable proteins in any biological system. This approach is particularly useful for investigating perturbation experiments in samples. One notable example is the use of stable isotope labeling by/with amino acids in cell culture (SILAC), which involves labeling specific amino acids with either light or heavy isotopes to examine protein level differences between normal and treated cell lines. SILAC-based experiments provide valuable insights into the effects of perturbations on the proteome, aiding in the identification of potential drug targets or understanding disease mechanisms in a systems biology context (Aizat and Hassan 2018; Ong et al. 2002).

15.2.4 Metabolomics

Metabolomics is a cutting-edge technology that focuses on the identification and quantification of the metabolome, which refers to the dynamic collection of small molecules found in organisms or biological samples. This is a versatile field that employs various analytical tools to achieve specific research objectives (Baharum and Azizan 2018; Nicholson et al. 1999). It is important to distinguish metabolomics from metabolic profiling, as the former seeks to explore the entire repertoire of small molecules rather than specific compounds and their biochemical transformations (Weckwerth and Morgenthal 2005). Metabolomics is an innovative technology that focuses on comprehending the metabolic aspects of living organisms (Baharum and Azizan 2018). It has gained widespread acceptance in the research community as a scientific field dedicated to the analysis and interpretation of metabolite levels in biological samples. This is achieved through various approaches such as target analysis, profiling, and fingerprinting. In essence, the metabolome encompasses the entirety of metabolites present within a biological system (Oldiges et al. 2007; Putri et al. 2013). The concept of metabolomics was recently introduced to encompass two main aspects: the global analysis of all metabolites in a sample (metabolomics) and the examination of metabolic responses to drugs or diseases (Weckwerth and Morgenthal 2005). Metabolomics primarily focuses on identifying, qualitatively analyzing, and quantitatively measuring the metabolites that participate in the intricate biochemical network (Baharum and Azizan 2018; Oldiges et al. 2007). Notably, there is a growing interest in the field of metabolomics in the context of drug discovery, disease diagnostics, and treatment.

Metabolomics can be categorized into targeted analysis for quantifying specific compounds, untargeted metabolic profiling for qualitative analysis of untargeted metabolites, and metabolomics for obtaining an unbiased overview of metabolic configurations. Moreover, metabolic fingerprinting is a rapid screening technique used to identify and distinguish metabolites with biochemical relevance, reducing the need for multiple analytical platforms. This method has gained prominence in postgenomic studies and phenotypic characterization for quick identification and discrimination (Baharum and Azizan 2018). Metabolomics tools play a crucial role in the study of human diseases, particularly in the fields of oncobiology and disease biomarker research. Metabolites serve as valuable biomarkers in various conditions, including male infertility, lung cancer, Alzheimer's disease, respiratory diseases, Huntington's disease, kidney cancer, and renal cell carcinoma. These biomarkers provide essential insights into disease diagnosis, monitoring, and understanding disease mechanisms at a metabolic level. The application of metabolomics in disease research has led to significant advancements in identifying novel therapeutic targets and developing personalized treatment strategies (Yadav et al. 2018).

15.2.5 Modeling and Simulation

Modeling and simulation are powerful tools in systems biology that aid in understanding the complex behavior of biological systems. By representing biological processes mathematically and simulating their dynamics, modeling and simulation enable researchers to gain insights into the underlying mechanisms, predict system behavior, and guide experimental design. This approach helps bridge the gap between theoretical understanding and experimental observations, contributing to a deeper understanding of biological systems (Kitano 2002a). In systems biology, mathematical models are constructed to describe the interactions and dynamics of various biological components, such as genes, proteins, and metabolites. These models can be deterministic, representing average behavior, or stochastic, accounting for random fluctuations in the system (Gillespie 2007). They can take the form of ordinary differential equations, partial differential equations, or other mathematical formalisms, depending on the specific system being studied (Tyson et al. 2003). Simulation of these models allows researchers to explore how biological systems respond under different conditions and perturbations. By manipulating model parameters or simulating interventions, such as gene knockout or drug administration, simulations provide valuable insights into system dynamics and help identify key regulatory elements and pathways (Voit 2018). Simulation results can be compared with experimental data to validate the accuracy of the model and refine its parameters, leading to an iterative process of model improvement (Palsson 2015).

Moreover, modeling and simulation facilitate hypothesis testing and the generation of testable predictions. Researchers can use models to formulate hypotheses about the behavior of a biological system and simulate the expected outcomes. These predictions can guide the design of targeted experiments to validate the hypotheses and further refine the model. This iterative cycle of model development, simulation, and experimental validation accelerates scientific discovery and enhances our understanding of complex biological phenomena (Mendes 1997). The field of systems biology has witnessed significant advancements in modeling and simulation approaches, with the development of sophisticated algorithms, computational tools, and high-performance computing resources. These advancements have enabled the modeling and simulation of increasingly large-scale and complex biological systems, opening new avenues for understanding disease mechanisms, drug discovery, and personalized medicine (Le Novere 2015). The application of modeling and simulation in systems biology allows researchers to create mathematical models that capture the dynamic nature of biological processes. One of the key applications of modeling and simulation is in the study of biological network dynamics. Mathematical models can capture the interactions and regulatory processes within molecular networks, enabling the exploration of system behavior under various conditions. These models help in identifying key components and their roles in network function, predicting system responses to perturbations, and understanding how system-level properties emerge from the interactions of individual components (Alon 2020).

15.2.6 Network Analysis

Network analysis plays a pivotal role in systems biology as it provides a powerful framework to study and decipher the complex interactions within biological systems (Barabási and Oltvai 2004). Networks represent a collection of entities, such as genes, proteins, or metabolites, interconnected by edges that signify their relationships or interactions (Albert and Barabási 2002). By constructing and analyzing these networks, researchers can uncover valuable insights into the organization, dynamics, and functionality of biological systems. One prominent application of network analysis in systems biology is the construction of gene regulatory networks (GRNs). GRNs capture the intricate interactions between genes and their regulatory elements, including transcription factors and microRNAs (Davidson and Levine 2005). These networks offer a holistic perspective on how genes coordinate their activities and govern various cellular processes. Through network analysis techniques, such as Boolean modeling and differential equations, scientists can simulate and analyze the behavior of GRNs, unraveling the underlying mechanisms and regulatory dynamics (Glass and Kauffman 1973). Additionally, PPI networks are vital in elucidating the complex protein interactions that underpin cellular processes. PPI networks map the physical interactions between proteins, providing insights into protein complexes, signaling pathways, and functional modules (Vidal et al. 2011). By employing network analysis approaches, such as clustering algorithms and centrality measures, researchers can identify key proteins, predict novel protein interactions, and decipher the modular organization of PPI networks (Watts and Strogatz 1998). Metabolic networks, which depict the interconnected metabolic reactions within a cell, are another essential domain for network analysis in systems biology. These networks offer a comprehensive overview of metabolic pathways, facilitating the understanding of cellular metabolism and its regulation. Network analysis techniques, such as flux balance analysis and pathway enrichment analysis, enable researchers to identify crucial metabolic pathways, predict metabolic fluxes, and explore the effects of genetic and environmental perturbations on cellular metabolism. Furthermore, network analysis allows for the integration of multiple types of biological data, such as gene expression, proteomic data, and metabolomic data, to construct more comprehensive and accurate networks. Through data integration, researchers can refine network models, validate experimental findings, and gain a deeper understanding of the underlying biological processes.

Network analysis in systems biology offers several advantages. Firstly, it provides a holistic view of biological systems by capturing the intricate interactions between entities, such as genes, proteins, and metabolites. This allows researchers to understand the system's organization, dynamics, and functionality from a systems-level perspective (Barabási and Oltvai 2004). Additionally, network analysis enables the identification of key components within the network, such as central nodes or hubs, which play crucial roles in controlling cellular processes (Barabási and Albert 1999). Such information can guide experimental design and the development of targeted interventions. Another advantage of network analysis is its ability to predict novel interactions and relationships within biological systems. By analyzing network properties, researchers can identify potential interactions that have not yet been experimentally validated. These predictions can guide further investigation and discovery of new regulatory mechanisms or signaling pathways (Pržulj 2007). Furthermore, network analysis facilitates the integration of diverse data types, enabling researchers to merge multiple layers of biological information. This integration can provide a systems-level perspective on disease mechanisms, drug responses, and personalized medicine. However, network analysis also has limitations and challenges. One common challenge is the availability and quality of experimental data for constructing accurate and reliable networks. Insufficient data or incomplete knowledge of interactions can lead to incomplete or biased network models. Moreover, experimental noise and false positives/negatives in high-throughput datasets can affect the accuracy of network predictions (Goh et al. 2007). Another limitation is the computational complexity associated with analyzing large-scale networks. As biological systems are inherently complex, network analysis requires efficient algorithms and computational resources to handle massive datasets. Additionally, the interpretation of network analysis results may require expert knowledge in both computational biology and the specific biological domain, making it a specialized field that requires interdisciplinary collaboration (Pržulj et al. 2004). Despite these limitations, network analysis finds applications in various areas of systems biology. It aids in the discovery of gene regulatory networks, protein-protein interaction networks, and metabolic networks, shedding light on the underlying mechanisms and functional relationships within these systems. Network analysis also plays a crucial role in studying disease mechanisms, drug target identification, and biomarker discovery (Xia et al. 2010). Moreover, it contributes to the field of synthetic biology by guiding the design and engineering of synthetic genetic circuits and cellular systems (Nielsen and Keasling 2016).

15.3 Experimental Approaches

Experimental systems biology approaches encompass a range of strategies that leverage the integration of diverse data sources to gain a comprehensive understanding of complex biological systems (Schadt et al. 2009; Wang et al. 2012). These strategies include network analysis, computational modeling, omics technologies, and data-driven hypothesis generation. Network analysis involves constructing and analyzing biological networks to unravel the intricate interactions between genes, proteins, and other molecules, providing insights into disease mechanisms and potential therapeutic targets (Barabási and Oltvai 2004). Computational modeling facilitates the simulation and prediction of biological processes, enabling researchers to explore system behavior and guide experimental design (Kitano 2002b). These omics approaches enable the identification of biomarkers, elucidation of drug response mechanisms, and discovery of novel therapeutic targets (Pandey and Mann 2000). Furthermore, data-driven hypothesis generation leverages large-scale datasets to generate novel hypotheses and guide experimental investigations, accelerating the discovery of new biological insights (Dudley et al. 2011).

Experimental systems biology, which integrates various approaches such as animal models, cell culture systems, clinical trials, and high-throughput technologies, has revolutionized drug discovery and development. Systems biology provides a comprehensive framework to study complex biological systems by capturing and analyzing large-scale datasets to understand the interactions and dynamics of biological components (Butcher et al. 2004; Kitano 2002b; Westerhoff et al. 2009). Animal models, as part of experimental systems biology, allow for in vivo investigations to simulate human diseases and assess drug efficacy, safety, and pharmacokinetics (Hood et al. 2004; Hood and Friend 2011). These models provide valuable insights into disease mechanisms and therapeutic responses, aiding in the identification of potential drug targets. Cell culture systems complement animal models by offering controlled environments to examine the cellular and molecular mechanisms of drug action. They enable detailed investigations into drug-target interactions, cellular signaling pathways, and drug metabolism, facilitating the characterization of drug efficacy and toxicity. Clinical trials, another essential component, validate the therapeutic potential of drugs by evaluating their safety, efficacy, and optimal dosing in human subjects (Kola and Landis 2004). These trials provide crucial data on drug pharmacokinetics, adverse effects, and therapeutic outcomes, validating the findings from experimental systems biology and guiding the translation of discoveries into clinical practice. Through the integration of animal models, cell culture systems, clinical trials, and systems biology approaches, researchers can gain comprehensive insights into the complexity of biological systems, leading to the development of innovative and personalized therapeutic strategies. By employing these experimental systems biology approaches, it is possible to uncover complex biological relationships, enhance our understanding of diseases, and facilitate the development of innovative therapeutic interventions.

15.3.1 Experimental Strategies in Drug Discovery

In recent years, systems biology has emerged as a valuable approach to drug discovery, offering a holistic perspective by integrating experimental and computational methods to understand complex biological systems (Cho et al. 2006; Kitano 2002b). By combining high-throughput omics technologies, such as genomics, transcriptomics, proteomics, and metabolomics, with mathematical modeling and data analysis, systems biology enables the identification of key molecular networks and pathways involved in disease progression (Hood and Friend 2011). This information can guide the discovery of novel drug targets and the development of therapeutics with greater efficacy and fewer side effects (Barabási and Oltvai 2004). Furthermore, systems biology approaches facilitate the elucidation of drug mechanisms of action, prediction of drug responses, and identification of potential drug-drug interactions (Zhang and Wiemann 2009). By incorporating systems biology techniques into the drug discovery process, researchers can gain a deeper understanding of the complexities of disease biology and develop more targeted and personalized therapeutic strategies (Palsson 2006). Additionally, systems biology approaches offer the opportunity to leverage large-scale data integration and analysis, such as the integration of multi-omics data with clinical and phenotypic information, to identify biomarkers for disease diagnosis, prognosis, and patient stratification (Loscalzo and Barabasi 2011). By uncovering molecular signatures associated with specific disease subtypes or treatment responses, systems biology can guide the development of personalized medicine approaches, tailoring therapies to individual patients based on their unique biological profiles (Dudley et al. 2011). Moreover, systems biology enables the exploration of drug repurposing opportunities by uncovering connections between different diseases and identifying existing drugs that may have potential therapeutic effects in new indications (Lamb 2007). The application of systems biology in drug discovery not only enhances our understanding of disease complexity but also accelerates the translation of biological insights into actionable therapeutic strategies, fostering a more efficient and targeted approach to drug development.

Drug discovery is a complex and iterative process that involves the identification and development of novel compounds to target specific diseases. To optimize the efficiency and success rate of this process, various experimental strategies have been employed. One such strategy involves the use of high-throughput screening (HTS) techniques, which enable the rapid screening of large libraries of compounds against specific drug targets. HTS allows for the identification of lead compounds with potential therapeutic activity, which can then be further optimized through structure-activity relationship (SAR) studies (MacArron et al. 2011). SAR studies involve the systematic modification of lead compounds to explore the relationship between their chemical structure and biological activity, guiding the design of more potent and selective compounds (Oprea and Matter 2004). Another strategy is fragment-based drug discovery (FBDD), which involves the screening of small, low molecular-weight fragments that bind to a target protein, followed by the optimization of these fragments to generate lead compounds (Erlanson et al. 2016). FBDD offers advantages in terms of efficiency and the ability to explore a larger chemical space compared to traditional high-throughput screening methods (Roughley and Hubbard 2011). Additionally, computer-aided drug design (CADD) plays a crucial role in drug discovery, using computational algorithms and simulations to predict the binding affinity of small molecules to target proteins and assisting in the design of optimized drug candidates (Kitchen et al. 2004). These experimental strategies, in combination with advances in technology and our understanding of disease mechanisms, have greatly accelerated the drug discovery process and improved the likelihood of identifying effective therapeutic agents.

15.4 Integrative Analysis

Integrative analysis in systems biology is a powerful approach that combines multiple types of omics data to gain a comprehensive understanding of complex biological systems. This integrative approach allows for a holistic exploration of the interactions and relationships between various molecular components and pathways, providing insights into the underlying mechanisms of biological processes and diseases (Hood and Flores 2012). By integrating diverse omics datasets, such as genomics, transcriptomics, proteomics, and metabolomics, researchers can uncover novel associations, identify key regulatory networks, and elucidate the functional implications of molecular alterations (Lee et al. 2008). For example, the integration of genomics and transcriptomics data enables the identification of genetic variants that influence gene expression levels, providing insights into the genetic regulation of biological processes and disease phenotypes (Aguet et al. 2017). Integrative analysis also facilitates the discovery of biomarkers and therapeutic targets by identifying molecular signatures associated with specific diseases or treatment responses (Remon and Dienstmann 2018). By combining multiple omics datasets, researchers can identify common molecular features that characterize disease subtypes, allowing for personalized diagnosis and treatment strategies. To conduct integrative analysis, sophisticated computational algorithms and statistical methods are employed to integrate and analyze heterogeneous datasets (Johnson et al. 2007). These methods include dimensionality reduction techniques, network analysis, and machine learning approaches, which enable the identification of meaningful patterns and associations within the integrated data (Zhang and Horvath 2005). The application of integrative analysis in systems biology has contributed to advancements in various fields, including cancer research, neuroscience, and precision medicine. It provides a comprehensive framework for uncovering the complex interplay between molecular components and pathways, leading to a deeper understanding of biological systems and the development of novel therapeutic approaches (Ideker et al. 2001b). Integrative analysis plays a crucial role in systems biology as it offers a powerful framework to comprehensively understand the complexity of biological systems. By combining multiple omics datasets and integrating diverse biological information, integrative analysis enables the exploration of intricate interactions and relationships between molecular components and pathways, leading to novel insights into biological processes and disease mechanisms. One of the key advantages of integrative analysis is its ability to capture the multidimensional nature of biological systems. By considering multiple levels of molecular information, such as genomics, transcriptomics, proteomics, and metabolomics, researchers can obtain a more holistic view of the underlying biological processes and identify critical regulatory networks (Ideker and Krogan 2012).

Integrative analysis facilitates the identification of key molecular players and their interactions, shedding light on the mechanisms driving complex phenotypes and diseases. Furthermore, integrative analysis allows for the discovery of biomarkers and potential therapeutic targets. By integrating diverse omics data, researchers can identify molecular signatures associated with specific diseases or treatment responses. This information can aid in the development of personalized medicine approaches, where patients can receive tailored treatments based on their individual molecular profiles (Zhu et al. 2008). In addition, integrative analysis facilitates the generation of predictive models and simulations to understand system behavior. Computational techniques, combined with integrative analysis, enable the construction of mathematical models that capture the dynamics of complex biological systems. These models can be used to make testable predictions and guide experimental design and interventions. The application of integrative analysis in systems biology has already yielded significant advancements in various fields, including cancer research, drug discovery, and precision medicine. By unraveling the intricate relationships between molecular components, integrative analysis provides a deeper understanding of biological systems and opens avenues for the development of innovative strategies for disease diagnosis, treatment, and prevention (Kitano 2002b).

15.5 Future Directions and Conclusion

In conclusion, this chapter provides a comprehensive exploration of the application of systems biology and omics technologies in understanding the complexities of diseases. Through the integration of computational and experimental approaches, significant progress has been made in unraveling the molecular underpinnings of various diseases. Looking toward the future, several exciting directions emerge. First, the advancement of high-throughput technologies will continue to generate massive amounts of omics data, providing researchers with unprecedented opportunities to gain a deeper understanding of disease mechanisms. This will require the development of sophisticated computational tools and algorithms to analyze and interpret these complex datasets. The integration of multiple omics data sets, including genomics, transcriptomics, proteomics, and metabolomics, will play a pivotal role in deciphering the intricate networks and pathways involved in disease development and progression. This integrative approach will enhance our ability to identify biomarkers, discover novel therapeutic targets, and develop personalized treatment strategies. In addition, the emerging field of single-cell omics holds great promise for unraveling cellular heterogeneity and dynamics within complex diseases. By capturing the molecular profiles of individual cells, researchers can uncover unique disease subtypes and better understand cellular responses to therapeutic interventions. Furthermore, the integration of systems biology with clinical data, such as electronic health records and patient outcomes, will enable the translation of research findings into clinical practice. This integration will enhance the development of precision medicine approaches, allowing for tailored treatments that consider the individual characteristics of patients. The combination of computational and experimental approaches, along with the advancements in omics technologies, has revolutionized our understanding of disease complexity. The future of systems biology and omics research holds great promise for uncovering the intricate molecular mechanisms of diseases, paving the way for improved diagnostics, therapeutics, and patient outcomes.

References

Aguet F, Brown AA, Castel SE, Davis JR, He Y, Jo B, Mohammadi P, Park YS, Parsana P, Segrè AV, Strober BJ, Zappala Z, Cummings BB, Gelfand ET, Hadley K, Huang KH, Lek M, Li X, Nedzel JL, Zhu J (2017) Genetic effects on gene expression across human tissues. Nature 2017(550):7675. https://doi.org/10.1038/nature24277

Aizat WM, Hassan M (2018) Proteomics in systems biology. In: Advances in experimental medicine and biology, vol 1102. Springer, New York, pp 31–49. https://doi.org/10.1007/978-3-319-98758-3_3

Albert R, Barabási AL (2002) Statistical mechanics of complex networks. Rev Mod Phys 74(1):47. https://doi.org/10.1103/RevModPhys.74.47

Alon U (2020) An introduction to systems biology. Part II. In: Nowotwory, vol 62. Routledge, London. https://www.routledge.com/An-Introduction-to-Systems-Biology-Design-Principles-of-Biological-Circuits/Alon/p/book/9781439837177

Baharum SN, Azizan KA (2018) Metabolomics in systems biology. In: Advances in experimental medicine and biology, vol 1102. Springer, New York, pp 51–68. https://doi.org/10.1007/978-3-319-98758-3_4

Barabási AL, Albert R (1999) Emergence of scaling in random networks. Science 286(5439):509–512. https://doi.org/10.1126/SCIENCE.286.5439.509

Barabási AL, Oltvai ZN (2004) Network biology: understanding the cell's functional organization. Nat Rev Genet 5(2):101–113. https://doi.org/10.1038/nrg1272

Barabási AL, Gulbahce N, Loscalzo J (2010) Network medicine: a network-based approach to human disease. Nat Rev Genet 12(1):56–68. https://doi.org/10.1038/nrg2918

Bartocci E, Lió P (2016) Computational modeling, formal analysis, and tools for systems biology. PLoS Comput Biol 12(1):1004591. https://doi.org/10.1371/journal.pcbi.1004591

Berger MF, Mardis ER (2018) The emerging clinical relevance of genomics in cancer medicine. Nat Rev Clin Oncol 15(6):353–365. https://doi.org/10.1038/s41571-018-0002-6

Butcher EC, Berg EL, Kunkel EJ (2004) Systems biology in drug discovery. Nat Biotechnol 22(10):1253–1259. https://doi.org/10.1038/nbt1017

Casamassimi A, Federico A, Rienzo M, Esposito S, Ciccodicola A (2017) Transcriptome profiling in human diseases: new advances and perspectives. Int J Mol Sci 18(8):1652. https://doi.org/10.3390/ijms18081652

Chen X, Ishwaran H (2012) Random forests for genomic data analysis. Genomics 99(6):323–329. https://doi.org/10.1016/J.YGENO.2012.04.003

Cho CR, Labow M, Reinhardt M, van Oostrum J, Peitsch MC (2006) The application of systems biology to drug discovery. Curr Opin Chem Biol 10(4):294–302. https://doi.org/10.1016/j.cbpa.2006.06.025

Collins FS, Green ED, Guttmacher AE, Guyer MS, US National Human Genome Research Institute (2003a) A vision for the future of genomics research. Nature 422(6934):835–847
Collins FS, Morgan M, Patrinos A (2003b) The human genome project: lessons from large-scale biology. Science 300:286–290. www.sciencemag.org
Conesa A, Madrigal P, Tarazona S, Gomez-Cabrero D, Cervera A, McPherson A, Szcześniak MW, Gaffney DJ, Elo LL, Zhang X, Mortazavi A (2016) A survey of best practices for RNA-seq data analysis. Genome Biol 17(1):13. https://doi.org/10.1186/s13059-016-0881-8
Coorssen JR (2013) Proteomics. In: Brenner's encyclopedia of genetics, 2nd edn. Elsevier, Amsterdam, pp 508–510. https://doi.org/10.1016/B978-0-12-374984-0.01231-6
Coorssen JR (2022) Proteomics. In: Reference module in life sciences. Elsevier, Amsterdam. https://doi.org/10.1016/B978-0-12-822563-9.00058-5
Cox J, Mann M (2008) MaxQuant enables high peptide identification rates, individualized p.p.b.-range mass accuracies and proteome-wide protein quantification. Nat Biotechnol 26(12):1367–1372. https://doi.org/10.1038/nbt.1511
Davidson E, Levine M (2005) Gene regulatory networks. Proc Natl Acad Sci U S A 102(14):4935. https://doi.org/10.1073/PNAS.0502024102
Dudley JT, Sirota M, Shenoy M, Pai RK, Roedder S, Chiang AP, Morgan AA, Sarwal MM, Pasricha PJ, Butte AJ (2011) Computational repositioning of the anticonvulsant topiramate for inflammatory bowel disease. Sci Transl Med 3(96):96. https://doi.org/10.1126/SCITRANSLMED.3002648
Erlanson DA, Fesik SW, Hubbard RE, Jahnke W, Jhoti H (2016) Twenty years on: the impact of fragments on drug discovery. Nat Rev Drug Discovery 15(9):605–619. https://doi.org/10.1038/nrd.2016.109
Gillespie DT (2007) Stochastic simulation of chemical kinetics. Annu Rev Phys Chem 58:35–55. https://doi.org/10.1146/ANNUREV.PHYSCHEM.58.032806.104637
Glass L, Kauffman SA (1973) The logical analysis of continuous, non-linear biochemical control networks. J Theor Biol 39(1):103–129. https://doi.org/10.1016/0022-5193(73)90208-7
Goh KI, Cusick ME, Valle D, Childs B, Vidal M, Barabási AL (2007) The human disease network. Proc Natl Acad Sci U S A 104(21):8685–8690. https://doi.org/10.1073/PNAS.0701361104/SUPPL_FILE/01361FIG13.PDF
Green ED, Guyer MS (2011) Charting a course for genomic medicine from base pairs to bedside. Nature 470(7333):204–213. https://doi.org/10.1038/nature09764
Hixson KK, Lopez-Ferrer D, Robinson EW, Paša-Tolić L (2017) Proteomics. In: Encyclopedia of spectroscopy and spectrometry. Elsevier, Amsterdam, pp 766–773. https://doi.org/10.1016/B978-0-12-803224-4.00061-3
Hood L, Flores M (2012) A personal view on systems medicine and the emergence of proactive P4 medicine: predictive, preventive, personalized and participatory. New Biotechnol 29(6):613–624. https://doi.org/10.1016/J.NBT.2012.03.004
Hood L, Friend SH (2011) Predictive, personalized, preventive, participatory (P4) cancer medicine. Nat Rev Clin Oncol 8(3):184–187. https://doi.org/10.1038/NRCLINONC.2010.227
Hood L, Galas D (2003) The digital code of DNA. Nature 421:444–448. https://doi.org/10.1038/nature01409
Hood L, Perlmutter R (2004) The impact of systems approaches on biological problems in drug discovery. Nat Biotechnol 22:1215–1217
Hood L, Heath JR, Phelps ME, Lin B (2004) Systems biology and new technologies enable predictive and preventative medicine. Science 306(5696):640–643. https://doi.org/10.1126/SCIENCE.1104635
Hucka M, Finney A, Sauro HM, Bolouri H, Doyle JC, Kitano H, Arkin AP, Bornstein BJ, Bray D, Cornish-Bowden A, Cuellar AA, Dronov S, Gilles ED, Ginkel M, Gor V, Goryanin II, Hedley WJ, Hodgman TC, Hofmeyr J-H, Hunter PJ, Juty NS, Kasberger JL et al (2003) The systems biology markup language (SBML): a medium for representation and exchange of biochemical network models. Bioinformatics 19(4):524–531. https://doi.org/10.1093/BIOINFORMATICS/BTG015

Ideker T, Krogan NJ (2012) Differential network biology. Mol Syst Biol 8:565. https://doi.org/10.1038/MSB.2011.99

Ideker T, Thorsson V, Ranish JA, Christmas R, Buhler J, Eng JK, Bumgarner R, Goodlett DR, Aebersold R, Hood L (2001a) Integrated genomic and proteomic analyses of a systematically perturbed metabolic network. Science 292(5518):929–934. https://doi.org/10.1126/SCIENCE.292.5518.929

Ideker T, Galitski T, Hood L (2001b) A new approach to decoding life: systems biology. Annu Rev Genomics Hum Genet 2:343–372

International Human Genome Sequencing Consortium (2004) Finishing the euchromatic sequence of the human genome. Nature 431(7011):931–945. http://www.genome.gov/10000923

Jain KK (2016) Role of proteomics in the development of personalized medicine. Adv Protein Chem Struct Biol 102:41–52. https://doi.org/10.1016/bs.apcsb.2015.09.002

Jiménez-Guri E, Huerta-Cepas J, Cozzuto L, Wotton KR, Kang H, Himmelbauer H, Roma G, Gabaldón T, Jaeger J (2013) Comparative transcriptomics of early dipteran development. BMC Genomics 14(1):123. https://doi.org/10.1186/1471-2164-14-123

Johnson WE, Li C, Rabinovic A (2007) Adjusting batch effects in microarray expression data using empirical Bayes methods. Biostatistics 8(1):118–127. https://doi.org/10.1093/BIOSTATISTICS/KXJ037

Joshi K, Patil D (2017) Proteomics. In: Innovative approaches in drug discovery: ethnopharmacology, systems biology and holistic targeting. Elsevier, Amsterdam, pp 273–294. https://doi.org/10.1016/B978-0-12-801814-9.00009-X

Kitano H (2002a) Computational systems biology. Nature 420(6912):206–210. https://doi.org/10.1038/nature01254

Kitano H (2002b) Systems biology: a brief overview. Science 295(5560):1662–1664. https://doi.org/10.1126/SCIENCE.1069492

Kitchen DB, Decornez H, Furr JR, Bajorath J (2004) Docking and scoring in virtual screening for drug discovery: methods and applications. Nat Rev Drug Discov 3(11):935–949. https://doi.org/10.1038/nrd1549

Kola I, Landis J (2004) Can the pharmaceutical industry reduce attrition rates? Nat Rev Drug Discov 3(8):711–716

Lamb J (2007) The connectivity map: a new tool for biomedical research. Nat Rev Cancer 7(1):54–60. https://doi.org/10.1038/nrc2044

Le Novere N (2015) Quantitative and logic modelling of molecular and gene networks. Nat Rev Genet 16(3):146–158. https://doi.org/10.1038/nrg3885

Lee H, Kong SW, Park PJ (2008) Integrative analysis reveals the direct and indirect interactions between DNA copy number aberrations and gene expression changes. Bioinformatics 24(7):889–896. https://doi.org/10.1093/BIOINFORMATICS/BTN034

Li H (2018) Minimap2: pairwise alignment for nucleotide sequences. Bioinformatics 34(18):3094–3100. https://doi.org/10.1093/BIOINFORMATICS/BTY191

Li CF, Zhu Y, Yu Y, Zhao QY, Wang SJ, Wang XC, Yao MZ, Luo D, Li X, Chen L, Yang YJ (2015) Global transcriptome and gene regulation network for secondary metabolite biosynthesis of tea plant (Camellia sinensis). BMC Genomics 16(1):560. https://doi.org/10.1186/s12864-015-1773-0

Loscalzo J, Barabasi AL (2011) Systems biology and the future of medicine. Wiley Interdiscip Rev Syst Biol Med 3(6):619–627. https://doi.org/10.1002/WSBM.144

MacArron R, Banks MN, Bojanic D, Burns DJ, Cirovic DA, Garyantes T, Green DVS, Hertzberg RP, Janzen WP, Paslay JW, Schopfer U, Sittampalam GS (2011) Impact of high-throughput screening in biomedical research. Nat Rev Drug Discov 10(3):188–195. https://doi.org/10.1038/nrd3368

Manes NP, Nita-Lazar A (2018) Application of targeted mass spectrometry in bottom-up proteomics for systems biology research. J Proteome 189:75–90. https://doi.org/10.1016/J.JPROT.2018.02.008

Mendes P (1997) Biochemistry by numbers: simulation of biochemical pathways with Gepasi 3. Trends Biochem Sci 22(9):361–363. https://doi.org/10.1016/S0968-0004(97)01103-1

Meyer P, Saez-Rodriguez J (2021) Advances in systems biology modeling: 10 years of crowdsourcing DREAM challenges. Cell Syst 12:636–653. https://doi.org/10.1016/j.cels.2021.05.015
Nesvizhskii AI (2010) A survey of computational methods and error rate estimation procedures for peptide and protein identification in shotgun proteomics. J Proteome 73(11):2092–2123. https://doi.org/10.1016/J.JPROT.2010.08.009
Nicholson JK, Lindon JC, Holmes E (1999) "Metabonomics": understanding the metabolic responses of living systems to pathophysiological stimuli via multivariate statistical analysis of biological NMR spectroscopic data. Xenobiotica 29(11):1181–1189. www.taylorandfrancis.com
Nielsen J, Keasling JD (2016) Engineering cellular metabolism. Cell 164(6):1185–1197. https://doi.org/10.1016/J.CELL.2016.02.004
Oldiges M, Lütz S, Pflug S, Schroer K, Stein N, Wiendahl C (2007) Metabolomics: current state and evolving methodologies and tools. Appl Microbiol Biotechnol 76(3):495–511. https://doi.org/10.1007/s00253-007-1029-2
Ong SE, Blagoev B, Kratchmarova I, Kristensen DB, Steen H, Pandey A, Mann M (2002) Stable isotope labeling by amino acids in cell culture, SILAC, as a simple and accurate approach to expression proteomics. Mol Cell Proteomics 1(5):376–386. https://doi.org/10.1074/mcp.M200025-MCP200
Oprea TI, Matter H (2004) Integrating virtual screening in lead discovery. Curr Opin Chem Biol 8(4):349–358. https://doi.org/10.1016/J.CBPA.2004.06.008
Palsson BØ (2006) Properties of reconstructed networks. Cambridge University Press, Cambridge, pp 89–107. https://www.cambridge.org/core/books/systems-biology/F6C21EFCF03BFF6784 1CABF22B830106%0A
Palsson BO (2015) Systems biology: constraint-based reconstruction and analysis. Cambridge University Press, Cambridge, pp 1–531. https://doi.org/10.1017/CBO9781139854610
Pandey A, Mann M (2000) Proteomics to study genes and genomes. Nature 405(6788):837–846. https://doi.org/10.1038/35015709
Pazhamala LT, Kudapa H, Weckwerth W, Millar AH, Varshney RK (2021) Systems biology for crop improvement. Plant Genome 14(2):e20098. https://doi.org/10.1002/tpg2.20098
Pržulj N (2007) Biological network comparison using graphlet degree distribution. Bioinformatics 23(2):e177–e183. https://doi.org/10.1093/BIOINFORMATICS/BTL301
Pržulj N, Corneil DG, Jurisica I (2004) Modeling interactome: scale-free or geometric? Bioinformatics 20(18):3508–3515. https://doi.org/10.1093/BIOINFORMATICS/BTH436
Putri SP, Nakayama Y, Matsuda F, Uchikata T, Kobayashi S, Matsubara A, Fukusaki E (2013) Current metabolomics: practical applications. J Biosci Bioeng 115(6):579–589. https://doi.org/10.1016/j.jbiosc.2012.12.007
Remon J, Dienstmann R (2018) Precision oncology: separating the wheat from the chaff. ESMO Open 3(6):e000446. https://doi.org/10.1136/ESMOOPEN-2018-000446
Ritchie ME, Phipson B, Wu D, Hu Y, Law CW, Shi W, Smyth GK (2015) Limma powers differential expression analyses for RNA-sequencing and microarray studies. Nucleic Acids Res 43(7):e47. https://doi.org/10.1093/NAR/GKV007
Roughley SD, Hubbard RE (2011) How well can fragments explore accessed chemical space? A case study from heat shock protein 90. J Med Chem 54(12):3989–4005. https://doi.org/10.1021/JM200350G/SUPPL_FILE/JM200350G_SI_001.PDF
Schadt EE, Friend SH, Shaywitz DA (2009) A network view of disease and compound screening. Nat Rev Drug Discov 8(4):286–295. https://doi.org/10.1038/nrd2826
Schneider MV (2013a) Defining systems biology: a brief overview of the term and field. Methods Mol Biol 1021:1–11. https://doi.org/10.1007/978-1-62703-450-0_1
Schneider MV (ed) (2013b) In silico systems biology, vol 1021. Humana Press, Totowa. https://doi.org/10.1007/978-1-62703-450-0
Shendure J, Ji H (2008) Next-generation DNA sequencing. Nat Biotechnol 26(10):1135–1145. https://doi.org/10.1038/nbt1486
Sobie EA, Lee Y-S, Jenkins SL, Iyengar R (2011) Systems biology—biomedical modeling. Sci Signal 4:190

Sordo Vieira L, Laubenbacher RC (2022) Computational models in systems biology: standards, dissemination, and best practices. Curr Opin Biotechnol 75:102702. https://doi.org/10.1016/j.copbio.2022.102702

Strianese O, Rizzo F, Ciccarelli M, Galasso G, D'agostino Y, Salvati A, Del Giudice C, Tesorio P, Rusciano MR (2020) Precision and personalized medicine: how genomic approach improves the management of cardiovascular and neurodegenerative disease. Gene 11(7):1–24. https://doi.org/10.3390/genes11070747

Subramanian A, Tamayo P, Mootha VK, Mukherjee S, Ebert BL, Gillette MA, Paulovich A, Pomeroy SL, Golub TR, Lander ES, Mesirov JP (2005) Gene set enrichment analysis: a knowledge-based approach for interpreting genome-wide expression profiles. Proc Natl Acad Sci U S A 102(43):15545–15550. https://doi.org/10.1073/PNAS.0506580102/SUPPL_FILE/06580FIG7.JPG

Sumner LW, Amberg A, Barrett D, Beale MH, Beger R, Daykin CA, Fan TWM, Fiehn O, Goodacre R, Griffin JL, Hankemeier T, Hardy N, Harnly J, Higashi R, Kopka J, Lane AN, Lindon JC, Marriott P, Nicholls AW, Viant MR (2007) Proposed minimum reporting standards for chemical analysis Chemical Analysis Working Group (CAWG) Metabolomics Standards Initiative (MSI). Metabolomics 3(3):211–221. https://doi.org/10.1007/S11306-007-0082-2

Tsongalis GJ (2020) Integrative systems biology. In: Essential concepts in molecular pathology. Elsevier, Amsterdam, pp 165–175. https://doi.org/10.1016/B978-0-12-813257-9.00010-3

Tyson JJ, Chen KC, Novak B (2003) Sniffers, buzzers, toggles and blinkers: dynamics of regulatory and signaling pathways in the cell. Curr Opin Cell Biol 15(2):221–231. https://doi.org/10.1016/S0955-0674(03)00017-6

Venter JC, Adams MD, Myers EW, Li PW, Mural RJ, Sutton GG, Smith HO, Yandell M, Evans CA, Holt RA, Gocayne JD, Amanatides P, Ballew RM, Huson DH, Wortman JR, Zhang Q, Kodira CD, Zheng XH, Chen L, Zhu X (2001) The sequence of the human genome. Science 291:1304–1351. http://science.sciencemag.org/

Vidal M, Cusick ME, Barabási AL (2011) Interactome networks and human disease. Cell 144(6):986–998. https://doi.org/10.1016/J.CELL.2011.02.016

Voit EO (2018) A first course in systems biology. Routledge, London. https://www.routledge.com/A-First-Course-in-Systems-Biology/Voit/p/book/9780815345688

Wang RS, Saadatpour A, Albert R (2012) Boolean modeling in systems biology: an overview of methodology and applications. Phys Biol 9(5):e055001. https://doi.org/10.1088/1478-3975/9/5/055001

Watts DJ, Strogatz SH (1998) Collective dynamics of 'small-world' networks. Nature 393(6684):440–442. https://doi.org/10.1038/30918

Weckwerth W, Morgenthal K (2005) Metabolomics: from pattern recognition to biological interpretation. Drug Discov Today 10:1551–1558

Westerhoff HV, Winder C, Messiha H, Simeonidis E, Adamczyk M, Verma M, Bruggeman FJ, Dunn W (2009) Systems biology: the elements and principles of Life. FEBS Lett 583(24):3882–3890. https://doi.org/10.1016/J.FEBSLET.2009.11.018

Wishart DS (2016) Emerging applications of metabolomics in drug discovery and precision medicine. Nat Rev Drug Discov 15(7):473–484. https://doi.org/10.1038/nrd.2016.32

Xia J, Wishart DS, Valencia A (2010) MetPA: a web-based metabolomics tool for pathway analysis and visualization. Bioinformatics 26(18):2342–2344. https://doi.org/10.1093/BIOINFORMATICS/BTQ418

Yadav D, Tanveer A, Malviya N, Yadav S (2018) Overview and principles of bioengineering: the drivers of omics technologies. In: Omics technologies and bio-engineering: towards improving quality of life, vol 1. Elsevier, Amsterdam, pp 3–23. https://doi.org/10.1016/B978-0-12-804659-3.00001-4

Zhang B, Horvath S (2005) A general framework for weighted gene co-expression network analysis. Stat Appl Genet Mol Biol 4(1):17. https://doi.org/10.2202/1544-6115.1128/MACHINEREADABLECITATION/RIS

Zhang JD, Wiemann S (2009) KEGGgraph: a graph approach to KEGG PATHWAY in R and bioconductor. Bioinformatics 25(11):1470. https://doi.org/10.1093/BIOINFORMATICS/BTP167

Zhong W, Gerstein M, Snyder M (2009) RNA-Seq: a revolutionary tool for transcriptomics. Nat Rev Genet 10:57–63

Zhu J, Zhang B, Smith EN, Drees B, Brem RB, Kruglyak L, Bumgarner RE, Schadt EE (2008) Integrating large-scale functional genomic data to dissect the complexity of yeast regulatory networks. Nat Genet 40(7):854–861. https://doi.org/10.1038/ng.167

Complex Disease Prediction Using Systems Biology Approach

16

Meghashree H. R., S. Yuvaraj, Sunitha C. Mesta, J. Satish Kumar, and E. Sumitha

Abstract

Imagine a mirror maze this complex breakthrough can be analogous to what many researchers face when trying to decipher the intricacy and uncertainties of biological systems as well as genetics linked to the 'complexity of human diseases'. Recent innovations in sequencing are characterised by humongous amounts of data from contrasting sources, creating demand for the development of novel tools to extract biologically meaningful data. The starkness of genomic and biomedical data associated with complex diseases creates a new challenge to understand the interconnection between genes and proteins, which therefore requires the integration of data from a wide range of protein structures to decipher the complexity of diseases. Thus, computational tools have become critical for the integration, representation, and visualisation of heterogeneous biological data.

M. H. R. · E. Sumitha (✉)
Department of Biotechnology and Bioinformatics, JSS Academy of Higher Education and Research, Mysore, Karnataka, India

S. Yuvaraj
Department of Pharmaceutical Chemistry, Cauvery College of Pharmacy, Mysore, Karnataka, India

S. C. Mesta
Department of Microbiology and Tissue Culture, JSS Academy of Higher Education and Research, Mysore, Karnataka, India

J. Satish Kumar
Department of Environmental Science, JSS Academy of Higher Education and Research, Mysore, Karnataka, India

S. Joshi et al. (eds.), *Systems Biology Approaches: Prevention, Diagnosis, and Understanding Mechanisms of Complex Diseases*,
https://doi.org/10.1007/978-981-99-9462-5_16

Keywords

Single-nucleotide polymorphism · Glioblastoma · Machine learning · Deep learning · Random forest · Digenic inheritance

16.1 Introduction

In the twenty-first century, mankind has seen vast advancements in most of the fields, which include technology, health care, and electronics, they have made it easy for mankind to do everyday tasks (Bohr and Memarzadeh 2020). Contrastingly, one's lifestyle is more prone to complex diseases such as cancer, diabetes, genetic disorders, cardiovascular diseases, etc. The major cause of this disease is a combination of genetic variations, lifestyle, and other environmental factors (Roth et al. 2020). In modern times, current researchers are supported by wet lab experiments along with soft algorithms to understand the basics and complexities of biological systems and challenges that can be rectified using computational tools and techniques, which is leading to the production of tremendous data and encouraging advancements in 'computational biology'. In this chapter, we will be discussing tools and techniques such as multiple sequence alignment, next-generation sequencing, and upcoming branch pharmacogenomics.

Complex diseases are often difficult to decipher because almost all complex diseases involve more than one gene that interacts with each other, or sometimes gene interaction with outside aspects such as viruses or components of the diet. Mutations in DNA are a major cause of genetic diseases. Surprisingly, mutations that occur naturally differ among individuals because mutations affect the structural interactions, functional properties of other DNA, and the expression of its proteins. The ability to determine whether a mutation is harmless or not in complex diseases would pave the way for predictive diagnosis and personalized medicine (Anwaar et al. 2022). The vast generality of complex diseases, most of which still need significant improvement with respect to diagnostics, includes digenetic inheritance, multiple sclerosis, and autoimmune diseases (Parkinson's disease, Alzheimer's, etc.). This chapter reveals the capacity and challenges of computational tools and techniques used for decoding the complexity of three diseases, focusing on the computational approach to deciphering glioblastoma, role of single-nucleotide polymorphisms (SNPs) in disease diagnosis, and machine learning approaches to understanding rare genetic disorders and their significance. In order to understand computational methods, a brief introduction is discussed at the beginning of the paper as follows.

16.2 Computational Techniques and Tools

Sequence alignment is a way of arranging the two sequences of DNA, RNA, or protein in order to identify regions of similarity between the sequences and annotate the function of the gene or protein. Generally, a basic alignment local search tool is used, popularly called 'Google' of bioinformatics BLAST (Ejigu and Jung 2020). BLAST has many types and inputs, the data in FASTA format prior to BLAST is used to compare nucleotide queries against nucleotide sequence databases, and P Blast is used to compare protein queries against protein sequences. Understanding the protein sequences gives insight into protein subcellular location, creates 3D protein map to analyse for protein regulations and interactions with small molecules, and aids in deciphering cascades of signals (Lu et al. 2021). Based on the signalling cascade, one can precisely modulate the machinery inside the cell. Blastx gives results against translated nucleotides in protein sequences. tBlastn gives results against proteins and translated nucleotides.

16.2.1 Multiple Sequence Alignment (MSA)

MSA may refer to the process of aligning two or more sequences of DNA, RNA, and proteins of similar region and length. MSA gives a brief idea about conserved regions among the sequences, which makes it easy to detect homology between newly sequenced genes and existing gene families. Major tools are used for MSA CLUSTAL. Majorly used for phylogenetic analysis, K-align sets the unaligned sequences, and MUSCLE (multiple sequence comparison by log-expectation) is a computer software vastly used to align protein and nucleotide sequences. This tool is good at handling small sets of data but shows redundancy when handling large datasets to upgrade the ability to solve with faster accuracy and efficiency. The MSA image is illustrated with the example given below. Figure 16.1 multiple sequence alignment done for von Willebrand disease, were failure of platelets to perform initial adhesion and aggregation which affects the primary haemostasis. As decrease in coagulation factor FVIII is observed, it also affects secondary haemostasis. Sequence for Von Willebrand disease was derived from 10 different species including (accession numbers: Mus musculus, Q8CIZ8; Canis familiaris, Q28295; Sus scrofa, Q28833; Homo sapiens, P04275; Macaca mulatta, F5XVB6; Oryctolagus cuniculus, F5XVB8; Xenopus tropicalis, F5XVC3; Danio rerio, E7F9Z1; Bos taurus, F1N0R5; and Gallus gallus, F5XVB5) were obtained from SwissProt and

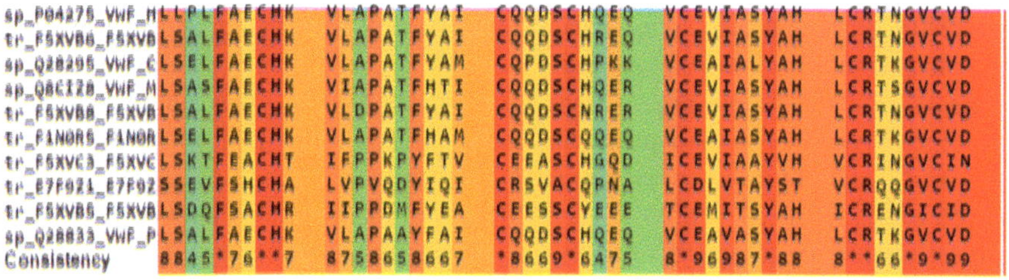

Fig. 16.1 Schematic representation of multiple alignment of Von Willebrand factor (VWF) sequences. As per sequence residues are: CHQCECR is shown in red, QDEASVES is yellow, and all other residues are coloured according to conservation status. The residue range for each sequence is shown after the sequence name

Trembl databases using PSI-BLAST and multiple sequences in terms of local alignment was done using multiple sequence alignment was performed with PRALINE (http://www.ibi.vu.nl/programs/praline).

16.2.2 Assembly and Annotations

The process of identifying and arranging overlapping sequences is of two types, firstly De novo assembly is used for novel sequences and then comparative assembly, where the sequence is compared to a standard library. MSA tools based on de novo assembly pair-wise tree construction can be used (Dida and Yi 2021). The process of assigning biological functions to the sequences involves mainly three steps, as follows:

1. Identification of portions of proteins that do not code for proteins.
2. Identifying elements on the genome for gene prediction.
3. Attaching biological information to these elements (Raghavan et al. 2022).

MSA is the initial step for functional annotations of genes, and as mentioned earlier in this chapter, MSA develops a draught for phylogenetic studies and the necessity for comparative genomics and metagenomics (Martin et al. 2021). With advances in technology and the internet, MSA can handle large datasets using PASTA format (a multiple sequence method that uses divide and conquer plus iteration base alignment to scale with high accuracy to large sequence datasets). Unreliable alignment regions in MSA are detected with modern tools like GUDIANCE2, and hardware driver problems are resolved with modern usage of GPU (graphics processing unit) and eminent processors like Cuda Core. Along with MSA, we have CUDA ClustalW.

16.3 Next Generation Sequencing (NGS)

The Human Sequencing Genome Project was a landmark effort made at the global level that started in the year 1990 to uncover a set of 3.2 billion DNA bases of the human genome (Craig Venter et al. 2001). On a long run till 2003, 85% of the project was completed, but some ongoing research projects are going to fill some of the gaps in the human genome project. In total, it took 32 years to have a broad idea, but interestingly, now it takes only 1 day to sequence with tremendous application of the next-generation sequencing (Schloss et al. 2020). Sequencing the genome of patients serves as a preliminary source for doctors to rectify any mutations in the genome sequence. The main principle involves DNA being cut into small fragments at specific sites by high-frequency sound waves or restriction enzymes and subjected to sequencing (Schloss et al. 2020). Later, these small-sequenced segments are assembled for comparative analysis with the human genome sequence and the data interpreted. NGS is used for sequencing both DNA and RNA. Initial isolation of samples is done by wet lab experiments before subjecting them to sequencing. Purification of the sample is a must, and degradation should be avoided, especially when RNA is reverse transcribed to DNA before sequencing and a standard library is prepared (a collection of shot fragments of DNA). Short sequences of DNA called adapters (adapters contain an index) get added to each end of small fragments of DNA. A successful library is created by removing all non-bound adapters, and based on the requirements, the library is subjected to a polymerase chain reaction to yield a high amount of library with the correct size. During the human genome project, Sanger sequencing was used due to its limitations, which gave rise to many NGS platforms (Terekhov et al. 2020).

The NGS is performed by instrument Illumina, where sequencing by synthesis reaction is observed. This method involves flow cell on glass surface sequencing, which occurs such that oligonucleotides are made to bound the surface of the flow cell (Raghavan et al. 2022). The initial adapter sequence of the library should be an oligonucleotide. First, the library gets denatured, hence single strands of DNA are formed. The denatured library is added to the flow cell, which attaches to the oligonucleotide. Usually, the attached strand is forward in nature; similarly, the reverse strand is made as the forward strand gets washed away. As a result, the library is attached to the flow cell in case sequencing is initiated. If the fluorescent signal is too weak for detection, clonal amplification is inferres an unique library fragment cluster generation is done. Along with Illumina, other NGS platforms like Ion Torrent Personal Genome Machine (PGM) sequence templates are generated on a bead via emulsion PCR; generally, clonal amplification is observed in advance of sequencing, but certain platforms like Pac Biosciences, to obtain high-sensitivity single-molecule real-time sequencing, are done with fluorescence detection. Illumina NGS platforms can sequence DNA of approximately 5000 MB at a cost of $0.50/MB/day. Though NGS is an emerging industry, it still faces certain challenges to perform NGS, ultimately. Infrastructure with minimal machinery costs, well-maintained labs with standard protocols, and hands-on experience professionals with adequate knowledge from both streams of life sciences and computer science to understand how to integrate wet lab experiments with soft algorithms and a wide idea about data sampling and storage is necessary, and major understanding of the current research scenario on how to interconnect the NGS data for other pathological studies is missing. These need to be addressed by researchers and technology developers.

NGS is a revolutionary sequencing technology that has taken computational biology to the next level as databases of m-RNA, DNA, and proteins are produced parallelly, and wet lab experiments produce data that can be converted to soft algorithms with cost effectiveness and less time to diagnose. NGS has shown great significance in understanding the human genetics of health and disease. Metagenomics, which produces huge datasets, needs computational tools and techniques to handle huge datasets so that accuracy and faster results can be achieved. Thus, current diagnosis demands the development of new tools for accuracy. In the era of machine learning and deep learning models, researchers are able to interpret this data rather than understand its functional or taxonomical information. The holistic approach to data has connected many interdisciplinary concepts to decipher the complexity of diseases at ground level.

A set of instructions written in the form of code using object-oriented programming languages like Python and R (for computing and graphical representation) are often used to investigate NGS data. To decipher complex diseases, transcriptomic analysis (very important as the transcriptome contains all sets of m-RNA) can be done by NGS. Other studies of noncoding RNA and chromatin immunoprecipitation can also be achieved. Next-generation sequencing is a powerful tool as it analyses long noncoding m-RNAs called IncRNAs, which play a major role in cancer, which is difficult to diagnose at the initial stages, or small noncoding RNAs called microRNAs. In this chapter, we further discuss NGS and its major significance in disease diagnosis.

16.4 Glioblastoma and Computational Approach for Diagnosis

Glioblastoma is a type of brain cancer considered most lethal due to its invasiveness, unspecific symptoms, and fast tumour growth, as per studies. If 60% of gliomas are present, then the surveillance of patients is expected to take only a year to 18 months; hence, the major challenge of this disease is that it cannot rectify the tumour and other symptoms in the early stages of life to decipher the pattern of genomic instability and activation of oncogenes. Next-generation sequencing is functional on the basis of the patient's genetics and molecular subtypes subjected to sequencing to enhance glioblastoma diagnosis. Based on epigenetic modifications such as methylation and histone acetylation, sample classification is done for novel diagnosis development. The idea of epigenetic profiling in glioblastoma provides a brief about sample classification, which can differ in individuals, and the obtained data are subjected to computational analysis like machine learning and deep learning using algorithms to understand the patterns among individual patients, bringing them closer to precision medicine.

16.4.1 Classical Outlook of Glioblastoma

The Cancer Genome Atlas (TCGA) for glioblastoma project has revealed significant changes occur majorly in 3 pathways: RAT sarcoma/receptor tyrosine kinase, retinoblastoma (RB) p53, with gene alterations on 88%, 87%, and 78%, respectively (Fig. 16.2). According to the World Health Organization, the fourth grade of glioblastoma accounts for 70% of the formation of gliomas. Glioblastoma is genetically classified into two types.

Inquisitively, most of the gene alteration observed was on epidermal growth factor receptor (EGFR, approximately 45%). Based on molecular profiling studies of glioblastoma, four types of glioblastomas are defined as follows: classical glioblastoma, mesenchymal glioblastoma, and neural and pro-neural glioblastoma; hence, keen observation of their studies shows the heterogeneity of tumours present both

Fig. 16.2 Schematic representation of genomic root cause leading to cancer

inter and intra at the molecular level (Verhaak et al. 2010). Based on tumour location, anaplasia (poor cellular differentiation), differentiation pattern, necrosis, and mitotic activity, genetically glioblastoma can be classified into 2 subtypes.

16.4.2 Isocitrate Dehydrogenase (IDH) Mutant Type

Histopathologically, isocitrate dehydrogenase (IDH) wildtype and isocitrate dehydrogenase (IDH) mutant type are the same, but they differ based on the clinical progression of tumours. In comparison, the surveillance rate is slightly higher in the isocitrate dehydrogenase (IDH) mutant type than in the isocitrate dehydrogenase (IDH) wildtype.

The ambiguous symptoms like mood swings, fatigue, numbness, and slight memory loss observed in gliomas patients at the advanced level of tumour progression require conclusive diagnosis histopathologically, but in recent times, the WHO has classified brain tumours based on the amalgamation of both molecular parameters and phenotypes, which is showing better diagnostic accuracy. In general, cancer diagnosis is followed by surgery to remove tumours, followed by chemotherapy and temozolomide (approved by the FDA, United States of America) supplied orally, which is a lipophilic molecule with the ability to bind to plasma proteins and penetrate through the blood–brain barrier. Unfortunately, after this belligerent treatment, patient life expectancy is still 18 months (Thomas et al. 2017).

Single-cell RNA sequencing (scRNA-Seq) technology has become researchers art approach as it can capture resolution of individual cells of both tumour and non-tumour cells, and in this glioblastoma, it is uncomplicated to examine the intratumour heterogeneity precisely (Pang et al. 2019). This chapter discusses about NGS and its major significance in disease diagnosis earlier can be connected to understand the molecular heterogeneity of the glioblastoma by constructing next-generation sequence panels containing specific set of glioblastoma gens from patients (Synhaeve et al. 2018).

16.5 Novel Diagnostic Computational Approaches

To decipher the complexity of disease such as cancer and its subtypes, it requires huge data samples to frame suitable algorithms and also reference datasets from healthy donors to understand the molecular background of disease and have a better comparison between patients and healthy donors.

In the case of glioblastoma, the availability of heterogeneity tumour patterns can be analysed using computational tools and techniques; therefore, advancing fields such as artificial intelligence and its application field 'machine learning' (algorithmically built mathematical models) have gained much attention in biomedical sciences for eminent diagnosis from large datasets and therapeutic motives as machines learn from datasets by building statistical models from huge datasets; hence, the prediction of disease patterns is obtained. Therefore, artificial intelligence and machine learning have vital roles to play in novel diagnostics. In contrast to disease diagnosis, classical statistics methods and models were used to relate the correspondence between covariates (clinical factors) and outcome as a single dependent variable obtained from a sample population and protruded to a large population. In brief, statistical models are generally good at deciding on treatment and approach based on the survival of the patients, whereas machine learning models seek to predict outcomes using clinical factors as input attributes. In detail, machine learning and its diagnostic approach to various diseases are explained.

16.6 Deep Learning and Medicine

Deep learning is a subtype of machine learning where raw data is constructed as structured models to establish neural models. The main significance of deep learning models training set of data uses labelled data, which is fruitful to optimise any research parameter-based prediction for prognosis gene discovery with respect to glioblastoma, while machine learning has shown its capacity in gene expression profiles of glioblastoma patients and their survivors (Verhaak et al. 2010).

16.6.1 SNPS and Its Importance in Disease Diagnosis with Machine Learning Methods

To decipher genetically complex diseases at ground level, identification of pathogenic mutations is important, and single-nucleotide polymorphisms play a vital role in diagnosis. The ambiguity of diseases can be reduced with plausible combined computational and experimental methods. Single-nucleotide polymorphisms are random point mutations that occur in DNA where single nucleotides are altered individually based on this genetic difference. Types of SNPs (Fig. 16.3a) play a vital role in understanding both the coding and noncoding regions if they have point mutations that can impact specific amino acids that encode specific proteins; hence, SNPs are initial hints to diagnose particular root causes of complex genetic diseases (Fig. 16.3b).

Macromolecules such as DNA, RNA, and proteins can have changes in their biophysical properties with respect to genetic differences, which include electrostatic properties, protein–DNA interactions, protein–RNA interactions, protein–protein interactions, and protein membrane interactions. These changes correspond to changes in the hydrogen bond network, hydrophobicity, Pka, and flexibility, leading to structural disorder. To understand disease-causing mutations, particular

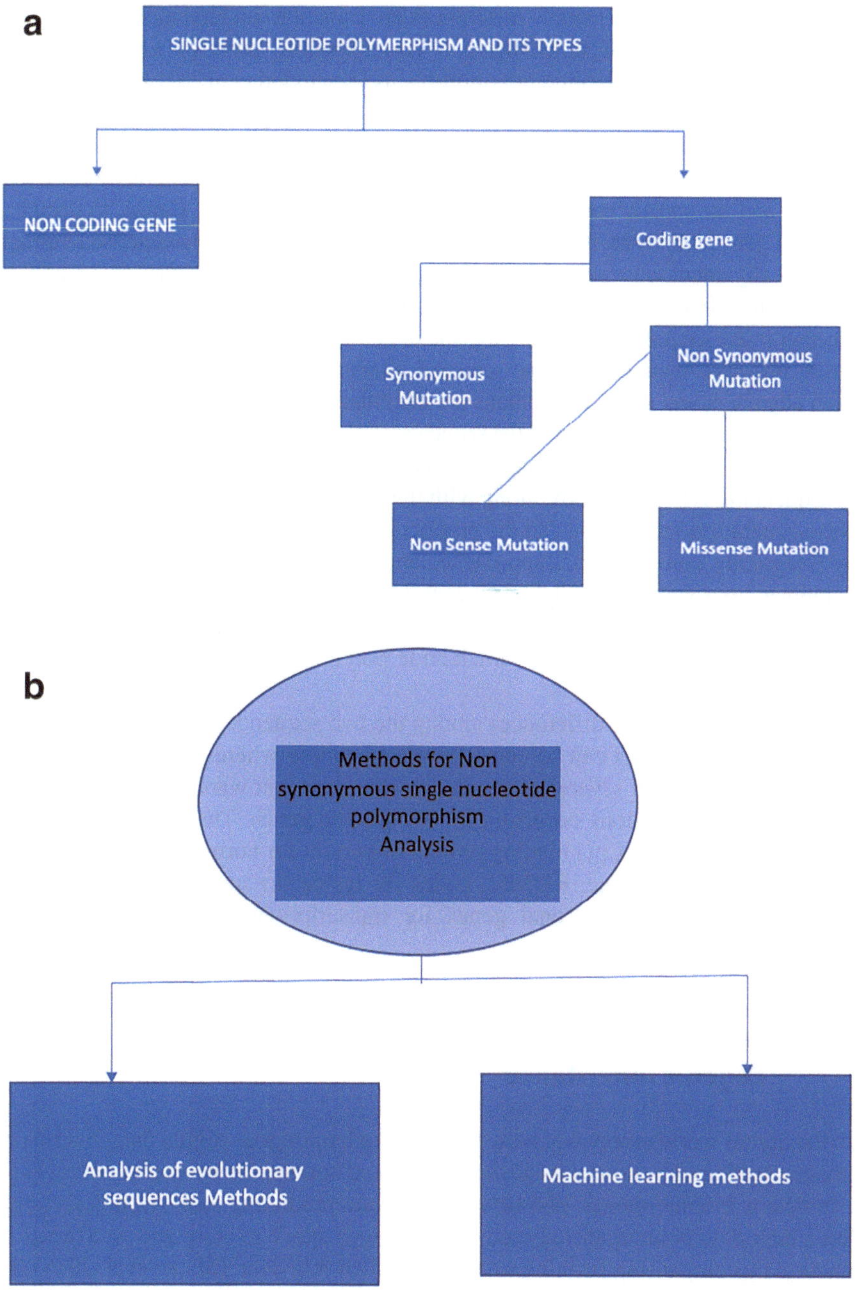

Fig. 16.3 (**a**) Illustration of types and subtypes of SNPS. (**b**) Current research approaches towards SNPS

nucleotide sequences of genetic differences are critical to understanding changes in amino acids that result in changes in corresponding proteins known as non-synonymous single-nucleotide polymorphisms (Kucukkal et al. 2014).

In genetics, missense mutation is the type of non-synonymous mutation where a single nucleotide change results in a codon that codes for different amino acids responsible for producing different proteins. Ultimately, these missense mutations are disease-causing and are mostly seen at evolutionary conserved positions. Hence, to decipher complex genetic diseases, missense mutation sequences are often characterised through multiple alignment sequences. In most cases, structural similarity implies the identity of the sequence. In recent years, computational methods have been developed to identify whether missense mutations are pathogenic or not. Due to the availability of large databases, missense mutations can be screened efficiently, where results are normalised from alignment and specific mutations are identified by applying a degree of tolerance based on the data conservation of the sequence (Kucukkal et al. 2014).

In addition to the experimental approach, the last decade has seen advancements in sequencing DNA and RNA, along with the prediction of proteins in 3D structure. Next-generation sequencing and the production of huge amounts of data are utilised by integrating computational approaches and machine learning methods to create combined approaches to deciphering complex diseases. Machine learning methods such as neural networks, decision trees, random forests, and support vector machines have been used to predict single-nucleotide polymorphism, as discussed with the following examples:

Based on the genetic differences among the two sequences involved in X-linked mental retardation, this can be considered an example where many methods were applied to understand several molecular mechanisms that were revealed in studies with respect to mutations occurring in two specific genes. The 2 genes, MEPCP2 gene, code for methyl CpG binding protein important for normal functioning of the cell, and the second gene, KDM5c, codes for lysine (k) specific demethylase 5c, which participates in neuronal genes for transcriptional repression (Kucukkal et al. 2014).

16.7 Machine Learning Approaches to Explore Digenic Inheritance

In the digital world, every day we are producing humongous digital data in terms of images, bank transactions, spreadsheets, etc., and not only by people's actions, even computers, mobile phones, and smart devices are producing data. All this accumulated bulk data is unorganised (and requires huge space to store and is difficult to manage by companies), forming a new field called 'big data' (Slatko et al. 2018). To simplify and understand big data, a new branch has emerged called 'machine learning' (ML). ML is basically making machines to learn from the data (algorithms) by building projection models to improve the performance of any set of tasks. The ultimate notion of ML is to make computers understand data and its patterns and learn how to solve and predict complex problems rather than give solutions. Recent breakthroughs in the biological sciences have given many soft tools and approaches to decipher uncertainties and complex disease patterns in biological systems using machine learning approaches (Shoaib et al. 2023). Arthur C. Clarke famously once said, 'Any sufficiently advanced technology is indistinguishable from magic. I found machine learning to be not magic but current innovations in soft algorithms that can characterize data from various sources and understand the interplay between gene expression and proteins, which is explained further with an example of a complex rare disease'.

16.7.1 Digenic Inheritance: Beyond Mendelian Genetics

In classical genetics, human diseases have heritable traits passed from parents to offspring; generally, variation occurs at a single locus or gene. A recent breakthrough in NGS revealed genes causing rare diseases that do not obey Mendelian laws, so both genotypic and phenotypic ratios will be different. Despite the increasing number of rare genetic diseases, it is crucial to diagnose complex genetic diseases (Okazaki and Ott 2022). Digenic inheritance is a type of coinheritance where DNA variants or mutations can be seen at two different genetic loci; hence, DI is popularly known as the simplest form of inheritance, causing some of the rare disorders such as retinitis pigmentosa (Fig. 16.4).

16.7.2 Beyond Mendelian Genetics

Traditional human genetic methods have looked at one DNA variant at a time for linkage or correlation with disease for heritable traits, which has been highly successful for Mendelian traits, that is, heritable diseases that are due to a single variant (Fig. 16.5).

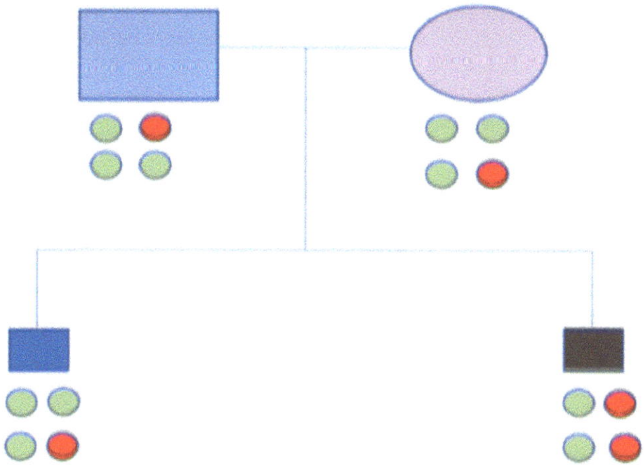

Fig. 16.4 Schematic representation of digenic inheritances from parents to offspring

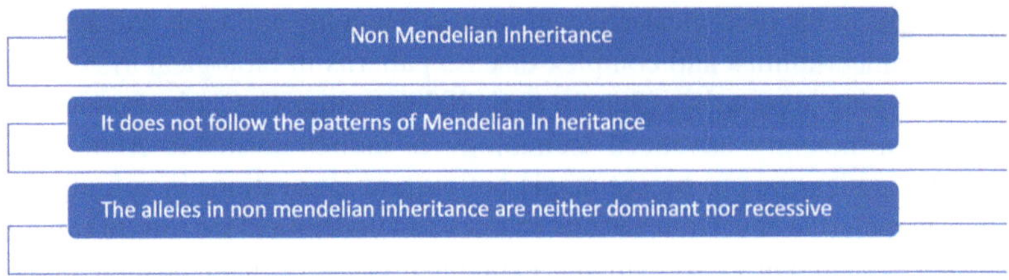

Fig. 16.5 A general overview of non-Mendelian inheritance

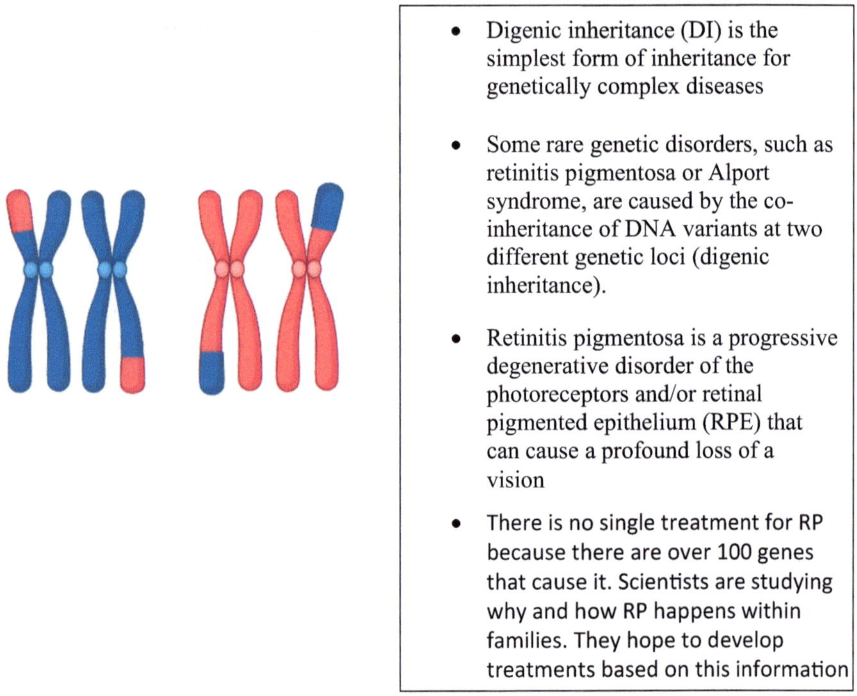

Fig. 16.6 Chromosomes having digenic inheritance (Yoshida et al. 2023)

16.7.3 Machine Learning Approaches to Explore Digenic Inheritance

Digenic inheritance refers to a type of inheritance pattern where the expression of a trait or disease is determined by the interaction of two different genes (Schäffer 2013) (Fig. 16.6).

16.8 Machine Learning Approaches

In the digital world, every day we are producing humongous digital data in terms of images, bank transactions, spreadsheets, etc., and not only by people's actions, even computers, mobile phones, and smart devices are producing data. All this accumulated bulk data is unorganised (and requires huge space to store and is difficult to manage by companies), forming a new field called 'big data'. Machine learning is derived from mathematical models that are built algorithmically, where the machine learns from the data provided and improves its efficiency and prediction based on data analysis and its algorithms. Modern-day machine learning is depicted in Fig. 16.7a.

To simplify and understand big data and its pattern, a new branch has emerged called 'machine learning' (ML). ML is basically making machines learn from the data (algorithms) by building projection models to improve the performance of any set of tasks. The ultimate notion of ML is to make computers understand data and its patterns and learn how to solve and predict complex problems rather than give solutions. Recent breakthroughs in the biological sciences have given many soft tools and approaches to decipher uncertainties and complex disease patterns in biological systems using machine learning approaches (Yoshida et al. 2023).

Various machine learning models are developed to statistically classify biological data and understand its significance analysis. Based on learning capability, popular machine learning models, decision tree (DT), random forest (RF), K nearest neighbours (KNN), logistic regression (LR), Gaussian Naïve Bayesian (GNB), XG Boost (XGB), are extensively used for biological data. General steps followed for machine learning approach are represented in Fig. 16.7a.

Decision tree majorly used for supervised learning this method meant for classification and regression data it divides data into trees and node where nodes represents descriptors or features and leaves represents its outcomes, random forest is the extended part of Decision tree its an ensemble method used for classification and regression of biological data with effective accuracy.one such example is well explained how to build immunological dataset using machine learning models to as part of bioinformatics approach to predict IL-6 inducing epitopes for vaccine design (Dhall et al. 2021).

Interleukin-6 (IL-6) is a multifunctional cytokine that plays a significant role in the immune response and inflammation. It is involved in various cellular processes, including the immune system, haematopoiesis, and nervous system. IL-6 is released by several cell types, such as T cells, B cells, monocytes, macrophages, and dendritic cells, and it can also be produced by non-hematopoietic cells like endothelial cells, keratinocytes, and smooth muscle cells. The IL-6 gene is located on chromosome 7q22-q23 and consists of 6 exons spread over approximately 600 kb. The IL-6 protein exists as a 262-kDa monomer and as a 230-kDa homodimer. There are two main IL-6 receptors: IL-6Rα and IL-6Rβ. IL-6Rα is the signal-transducing receptor, and IL-6Rβ is the accessory receptor that modulates the signalling process. IL-6 binds to IL-6Rα, and the complex of IL-6 with IL-6Rα forms a dimer that initiates intracellular signalling through the Janus kinase (JAK) family and the STAT family. IL-6 has various biological effects, including the promotion of B-cell growth and differentiation, enabling T-cell activation and modulating the immune response. It also plays a role in haematopoiesis, as it stimulates the production of red blood cells, white blood cells, and platelets. IL-6 is involved in the regulation of the immune response to infection, inflammation, and malignancy. Elevated IL-6 levels have been observed in patients with certain types of cancer, such as B-cell lymphomas, Hodgkin's lymphoma, and breast cancer. The IL-6/IL-6R system is a potential

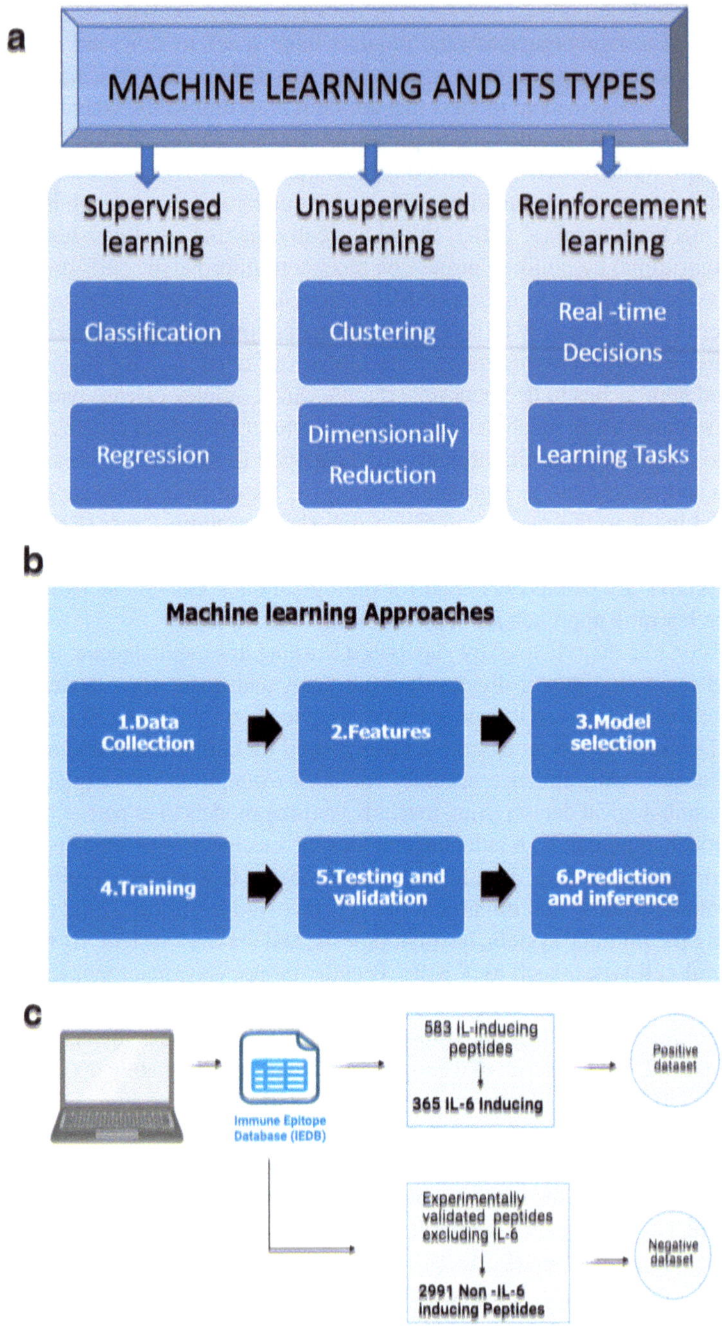

Fig. 16.7 (**a**) Modern-day machine learning. (**b**) Outline for machine learning approach. (**c**) Schematic overview of the mixed random forest approach

target for new therapeutic strategies in cancer treatment, as it is involved in tumour growth, angiogenesis, and the invasion of tumour cells. IL-6 inhibitors have shown antitumour effects, and clinical trials have demonstrated the safety and efficacy of these drugs.

The workflow of 'computer-aided prediction and design of IL-6 inducing peptides where IL-6 plays a crucial role in COVID-19' by Dhall et al. (2021) aimed to develop a method for predicting IL-6 inducing peptides, which play a vital role in the progression of COVID-19. The authors have constructed a dataset from the Immune Epitope Database (IEDB) and used machine learning techniques to predict IL-6 inducing peptides. The study achieved a maximum AUROC of 0.84 and 0.83 on training and independent validation datasets, respectively, using the random forest-based model. Other classifiers, such as decision tree, logistic regression, K-nearest neighbours, and Gaussian Naive Bayes, performed poorly on the training and validation dataset. The authors also developed a web server named IL-6Pred and a standalone package for predicting, designing, and screening IL-6 inducing peptides.

Among the six machine learning models, random forest (RF) demonstrated the best performance with an AUCROC of 0.893 and 0.863 and accuracy of 75.79 and 73.32 on training and validation datasets, respectively. XGB also performed well on the training and validation datasets, with an AUCROC of 0.87 and 0.82 and accuracy of 86.29 and 84.65, respectively. However, there was a considerable difference in sensitivity and specificity between the two models. Other classifiers, such as DT, LR, KNN, and GNB, performed poorly on the training and validation datasets. In conclusion, the study by Dhall et al. (2021) aimed to emphasize the nature of IL-6 inducing peptides and develop models to identify their IL-6 inducing potential. The dataset used in the study was constructed from IEDB, and the research provides valuable insights into the potential use of IL-6 inducing peptides in the development of vaccines against COVID-19.

16.8.1 Machine Learning Approaches in Digenic Inheritance

Machine learning approaches can be used to understand and analyse digenic inheritance by examining large datasets of genetic information and identifying patterns and correlations between different genes (Okazaki and Ott 2022). One approach to analysing digenic inheritance using machine learning is to use classification algorithms to identify groups of genes that are likely to interact with each other. For example, a supervised learning algorithm such as a decision tree or random forest could be trained on a dataset of genetic information (see Fig. 16.8) to predict the likelihood of a specific trait or disease based on the presence or absence of certain gene combinations.

Another approach is to use unsupervised learning techniques such as clustering or dimensionality reduction to identify patterns or relationships between different genes that may be involved in digenic inheritance (Okazaki and Ott 2022). This could involve using techniques such as principal component analysis to visualise genetic data in a lower-dimensional space and identify clusters of genes that are

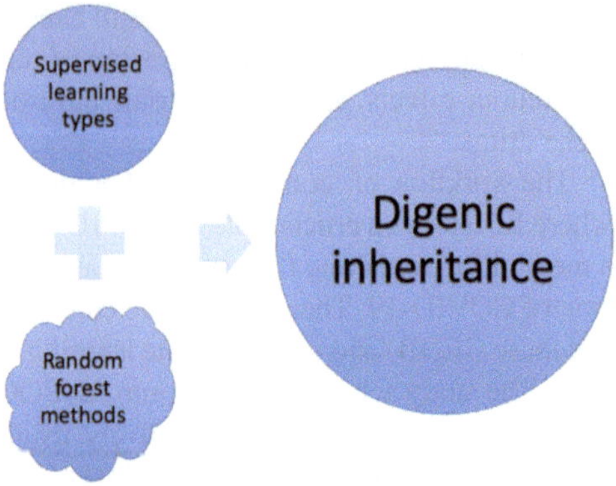

Fig. 16.8 Schematic overview of the mixed random forest approach

co-expressed or co-regulated. Ultimately, the goal of using machine learning techniques to understand and analyse digenic inheritance is to identify specific gene combinations that are associated with particular traits or diseases and to gain insights into the underlying biological mechanisms that govern these complex interactions (Nachtegael et al. 2022). This knowledge could ultimately lead to more effective treatments and interventions for a wide range of genetic disorders. Machine learning techniques can be applied to understand and analyse digenic inheritance by examining the interactions between two or more genes that contribute to a particular trait or disease (De La Vega et al. 2021). This approach can be used to identify complex patterns and relationships between genes, as well as to predict the effects of different gene combinations on the phenotype. Digenic interaction effect predictor (DIEP) is to use supervised learning techniques, such as regression or classification, to build predictive models of gene interactions.

16.8.2 Digenic Interaction Effect Predictor (DIEP)

An accurate machine learning approach to identify the genome-wide pathogenic coding gene pairs with digenic interaction effects (Yuan et al. 2022). This approach achieved high accuracy and sensitivity in independent testing datasets which is generally depicted in Fig. 16.9. This model uses five features to make its predictions; they are gene-based, protein-based, structure-based, expression-based, and phenotype-based. It should be noted that the digenic interaction estimated by DIEP is based on the gene-level functional relation per se (Rao et al. 2014). A positive prediction by DIEP depicts that the two genes in a pair may be functionally alternative, and distortions at both genes are needed to distort a biological function and subsequently cause a disease. Two genes with a digenic interaction effect may not necessarily be pathogenic in a person unless both genes had deleterious mutations. Conversely, deleterious mutations will not manifest interactive pathogenic effects unless they occur in genes with digenic effects (Yuan et al. 2022).

These models could be trained on datasets that include information on both the genotypes and phenotypes of individuals with different combinations of digenic

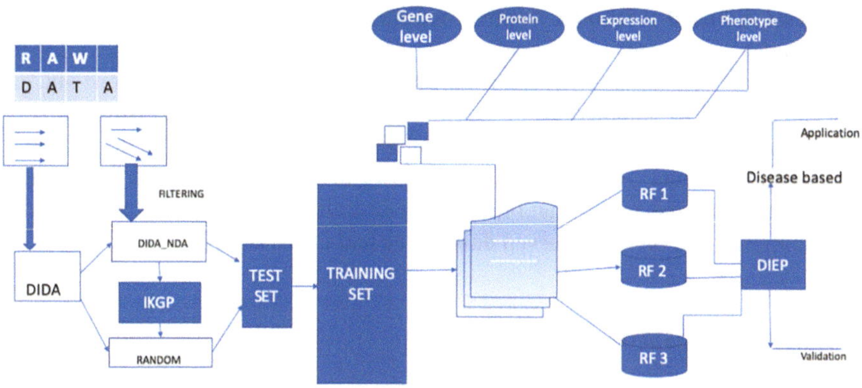

Fig. 16.9 General overview of digenic interaction effect predictor (DIEP)

variants. By analysing these datasets, machine learning algorithms could identify which gene combinations are associated with specific phenotypic outcomes and potentially identify novel genetic interactions that were not previously known (Yuan et al. 2022). Another potential approach is to use unsupervised learning techniques, such as clustering or dimensionality reduction, to identify patterns and relationships between digenic variants in large datasets. These techniques could be used to identify groups of genes that are co-expressed or functionally related, which could provide insight into the underlying mechanisms of digenic inheritance. Overall, machine learning techniques have the potential to be a powerful tool for understanding and analysing digenic inheritance and could lead to new insights into the complex interactions between genes that underlie many genetic diseases and traits (Yuan et al. 2022). However, it is important to note that these techniques must be used carefully and rigorously, and that the results of machine learning analyses must be validated and confirmed using experimental methods.

16.9 Combining Disease Association Effects Over Multiple Variants

When studying the association of genetic variants with a disease or trait, it is common to analyse multiple variants at once. To combine the effects of multiple variants, there are several methods that can be used, including:

Summing the effects: One way to combine the effects of multiple variants is to simply sum the effect sizes (e.g., odds ratios, beta coefficients) across all variants. This assumes that the effects of each variant are independent and additive, which may not always be true (Holcomb et al. 2021).

Weighted sum: A more sophisticated approach is to weight the effect sizes of each variant by their estimated standard errors. This method is known as a meta-analysis and takes into account the uncertainty of each variant's effect estimate (Nussinov 2015).

Regression-based methods: Another approach is to use a regression model that includes all variants simultaneously as predictors. This method allows for the effects of each variant to be adjusted for the effects of the other variants and can also account for interactions between variants (Mahbobi and Tiemann 2015).

Bayesian methods: Bayesian methods can also be used to combine the effects of multiple variants. These methods use a prior distribution to model the relationship between the variants and the disease or trait and then update this prior distribution based on the observed data (Wolc and Dekkers 2022).

16.10 Conclusion

Increased significance has been given to the prediction of causative mutations, or SNPs, due to the rapid advances in genome sequencing techniques. The key steps include the construction of a library with respect to the gene defect associated with the disease-causing mutation, where individual databases are maintained with the disease-causing mutation or where the SNPs have increased in vitality. For quick and accurate diagnostics, these databases' growing size and number are crucial since the individual's genome is the only data needed. The disease predisposal prediction can be done quickly when the individual genome has been mapped onto the disease database. Thousand genome database results revealed that in the annotated genes, nearly 250–300 variants causing loss of function were observed, and variants spanning 50–100 were found to be associated with the genetic disorders. In addition to these key observations, the severity of the disease was found to be associated with multiple factors, and the same mutation exhibited different phenotypes in a series of patients. This puts forth the notion of modifying the approach to the prediction of disease-causing mutations without the aid of databases and comparison. The best strategy, as described in this review, may be to rely on group efforts that make use of broader perspectives from both computational and experimental approaches. It remains challenging to probe each genetic variation by in silico analysis (i.e., molecular modelling studies where characteristics including stability and interactions with neighbouring atoms are compared in the presence of the wildtype and in vitro studies where the plasmid carrying the desired mutation is expressed in a mammalian cell line). However, to understand the overall impact of the mutations, in vivo approaches are needed (animal models: rat, zebrafish, canine, etc.). These experiments are time-consuming because they require careful selection of the target genes and the key mutations. To access the neuropathology of the disease, if the disease follows an X-linked pattern of inheritance, the characteristics, including phenotype, brain morphology, and function, need to be studied to understand the nature of nucleotide change. Zebrafish are ideally suited to carry out these initial studies because they are genetically tractable, plentiful, develop quickly, and are not dependent on their mothers. Importantly, they share significant genetic sequence identity with humans, have organ systems that are similar to those of mammals, and frequently start showing signs of disease phenotypes five days after development starts. Genetic make-up and phenotypic similarities to humans make the zebrafish ideal in comparison to other model organisms. The presence of fluorescent reporter genes allows for live imaging of the whole organism. This provides a real-time readout of affected cells, tissues, and organs, which is not possible in mammalian models. Hence, the zebrafish acts as an ideal model to screen a large number of variants to analyse (1) behaviour, (2) physiological, and (3) morphological outcomes, including epigenetic and genetic outcomes. These studies can be extended to higher mammalian animal models. Although lengthy, expensive, and time-consuming, these studies are essential for the subsequent stage of disease modelling. Mammalian studies are more effective and fruitful when they are highly focused on specific variants of interest. Combining in silico, in vitro, and in vivo techniques can help identify and understand the mechanisms underlying disease variants. The complexity of the diseases can be understood using tools, i.e., network biology. Network-based approaches make use of the theory that dysregulated modules rather than individual genes can provide a more comprehensive understanding of complex diseases. The idea of modularity in molecular networks is widely accepted, and module-based methods have a number of benefits, such as improved disease classification and robustness in the detection of dysregulated pathways. Additionally, network-based formulations enable the use of a wide range of graph theory-developed techniques, including shortest paths, network flow, and Steiner trees. The absence of mechanistic explanations is one of the limitations of network-based methods. Despite these drawbacks, network analysis has been successfully used to generate testable hypotheses in numerous disease studies.

References

Anwaar A, Varma AK, Baruah R (2022) In silico-based structural evaluation to categorize the pathogenicity of mutations identified in the RAD class of proteins. ACS Omega 8(11):10266–10277. https://doi.org/10.1021/acsomega.2c07802

Bohr A, Memarzadeh K (2020) The rise of artificial intelligence in healthcare applications. Artif Intell Healthc 25:1391–1396. https://doi.org/10.1016/B978-0-12-818438-7.00002-2

Craig Venter J, Adams MD, Myers EW, Li PW, Mural RJ, Sutton GG, Smith HO, Yandell M, Evans CA, Holt RA, Gocayne JD, Amanatides P, Ballew RM, Huson DH, Wortman JR, Zhang Q, Kodira CD, Zheng XH, Chen L, Zhu X (2001) The sequence of the human genome. Science 291(5507):1304–1351. https://doi.org/10.1126/SCIENCE.1058040

De La Vega FM, Chowdhury S, Moore B, Frise E, McCarthy J, Hernandez EJ, Wong T, James K, Guidugli L, Agrawal PB, Genetti CA, Brownstein CA, Beggs AH, Löscher BS, Franke A, Boone B, Levy SE, Õunap K, Pajusalu S, Kingsmore SF (2021) Artificial intelligence enables comprehensive genome interpretation and nomination of candidate diagnoses for rare genetic diseases. Genome Med 13(1):1–19. https://doi.org/10.1186/S13073-021-00965-0/FIGURES/5

Dhall A, Patiyal S, Sharma N, Usmani SS, Raghava GPS (2021) Computer-aided prediction and design of IL-6 inducing peptides: IL-6 plays a crucial role in COVID-19. Brief Bioinform 22(2):259. https://doi.org/10.1093/bib/bbaa259

Dida F, Yi G (2021) Empirical evaluation of methods for de novo genome assembly. PeerJ Comp Sci 7:1–31. https://doi.org/10.7717/PEERJ-CS.636

Ejigu GF, Jung J (2020) Review on the computational genome annotation of sequences obtained by next-generation sequencing. Biology 9(9):295. https://doi.org/10.3390/BIOLOGY9090295

Holcomb D, Hamasaki-Katagiri N, Laurie K, Katneni U, Kames J, Alexaki A, Bar H, Kimchi-Sarfaty C (2021) New approaches to predict the effect of co-occurring variants on protein characteristics. Am J Hum Genet 108(8):1502. https://doi.org/10.1016/J.AJHG.2021.06.011

Kucukkal TG, Yang Y, Chapman SC, Cao W, Alexov E (2014) Computational and experimental approaches to reveal the effects of single nucleotide polymorphisms with respect to disease diagnostics. Int J Mol Sci 15(6):9670. https://doi.org/10.3390/ijms15069670
Lu J, Wu T, Zhang B, Liu S, Song W, Qiao J, Ruan H (2021) Types of nuclear localization signals and mechanisms of protein import into the nucleus. Cell Commun Signal 19(1):1–10. https://doi.org/10.1186/S12964-021-00741-Y/FIGURES/1
Mahbobi M, Tiemann TK (2015) Regression basics. BCcampus, Victoria
Martin R, Dreßler H, Hattab G, Hackl T, Fischer MG, Heider D (2021) MOSGA 2: comparative genomics and validation tools. Comput Struct Biotechnol J 19:5504. https://doi.org/10.1016/J.CSBJ.2021.09.024
Nachtegael C, Gravel B, Dillen A, Smits G, Nowé A, Papadimitriou S, Lenaerts T (2022) Scaling up oligogenic diseases research with OLIDA: the oligogenic diseases database. Database 2022:23. https://doi.org/10.1093/DATABASE/BAAC023
Nussinov R (2015) Advancements and challenges in computational biology. PLoS Comput Biol 11(1):1004053. https://doi.org/10.1371/JOURNAL.PCBI.1004053
Okazaki A, Ott J (2022) Machine learning approaches to explore digenic inheritance. Trends Genet 38(10):1013–1018. https://doi.org/10.1016/J.TIG.2022.04.009
Pang L, Hu J, Li F, Yuan H, Yan M, Liao G, Xu L, Pang B, Ping Y, Xiao Y, Li X (2019) Discovering rare genes contributing to cancer stemness and invasive potential by GBM single-cell transcriptional analysis. Cancer 11(12):2025. https://doi.org/10.3390/cancers11122025
Raghavan V, Kraft L, Mesny F, Rigerte L (2022) A simple guide to de novo transcriptome assembly and annotation. Brief Bioinform 23(2):1–30. https://doi.org/10.1093/BIB/BBAB563
Rao VS, Srinivas K, Sujini GN, Kumar GNS (2014) Protein-protein interaction detection: methods and analysis. Int J Proteom 2014:1–12. https://doi.org/10.1155/2014/147648
Roth GA, Mensah GA, Johnson CO, Addolorato G, Ammirati E, Baddour LM, Barengo NC, Beaton A, Benjamin EJ, Benziger CP, Bonny A, Brauer M, Brodmann M, Cahill TJ, Carapetis JR, Catapano AL, Chugh S, Cooper LT, Coresh J, Fuster V (2020) Global burden of cardiovascular diseases and risk factors, 1990-2019: update from the GBD 2019 study. J Am Coll Cardiol 76(25):2982–3021. https://doi.org/10.1016/J.JACC.2020.11.010
Schäffer AA (2013) Digenic inheritance in medical genetics. J Med Genet 50(10):101713. https://doi.org/10.1136/jmedgenet-2013-101713
Schloss JA, Gibbs RA, Makhijani VB, Marziali A (2020) Cultivating DNA sequencing technology after the human genome project. Annu Rev Genomics Hum Genet 21:117–138. https://doi.org/10.1146/Annurev-Genom-111919-082433
Shoaib M, Shah B, El-Sappagh S, Ali A, Ullah A, Alenezi F, Gechev T, Hussain T, Ali F (2023) An advanced deep learning models-based plant disease detection: a review of recent research. Front Plant Sci 14:1158933. https://doi.org/10.3389/FPLS.2023.1158933/BIBTEX
Slatko BE, Gardner AF, Ausubel FM (2018) Overview of next generation sequencing technologies. Curr Protoc Mol Biol 122(1):e59. https://doi.org/10.1002/CPMB.59
Synhaeve NE, van den Bent MJ, French PJ, Dinjens WNM, Atmodimedjo PN, Kros JM, Verdijk R, Dirven CMF, Dubbink HJ (2018) Clinical evaluation of a dedicated next generation sequencing panel for routine glioma diagnostics. Acta Neuropathol Commun 6(1):126. https://doi.org/10.1186/s40478-018-0633-y
Terekhov SS, Eliseev IE, Ovchinnikova LA, Kabilov MR, Prjibelski AD, Tupikin AE, Smirnov IV, Belogurov AA, Severinov KV, Lomakin YA, Altman S, Gabibov AG (2020) Liquid drop of DNA libraries reveals total genome information. Proc Natl Acad Sci U S A 117(44):27300–27306. https://doi.org/10.1073/PNAS.2017138117
Thomas A, Tanaka M, Trepel J, Reinhold WC, Rajapakse VN, Pommier Y (2017) Temozolomide in the era of precision medicine. Cancer Res 77(4):823–826. https://doi.org/10.1158/0008-5472.CAN-16-2983
Verhaak RGW, Hoadley KA, Purdom E, Wang V, Qi Y, Wilkerson MD, Miller CR, Ding L, Golub T, Mesirov JP, Alexe G, Lawrence M, O'Kelly M, Tamayo P, Weir BA, Gabriel S, Winckler W, Gupta S, Jakkula L, Hayes DN (2010) Integrated genomic analysis identifies clinically relevant

subtypes of glioblastoma characterized by abnormalities in PDGFRA, IDH1, EGFR, and NF1. Cancer Cell 17(1):20. https://doi.org/10.1016/j.ccr.2009.12.020

Wolc A, Dekkers JCM (2022) Application of Bayesian genomic prediction methods to genome-wide association analyses. Genet Sel Evol 54(1):1–12. https://doi.org/10.1186/S12711-022-00724-8

Yoshida Y, Uchida K, Kodo K, Ishizaki-Asami R, Maeda J, Katsumata Y, Yuasa S, Fukuda K, Kosaki K, Watanabe Y, Nakagawa O, Yamagishi H (2023) A genetic and developmental biological approach for a family with complex congenital heart diseases—evidence of digenic inheritance. Front Cardiovasc Med 10:1135141. https://doi.org/10.3389/FCVM.2023.1135141/BIBTEX

Yuan Y, Zhang L, Long Q, Jiang H, Li M (2022) An accurate prediction model of digenic interaction for estimating pathogenic gene pairs of human diseases. Comput Struct Biotechnol J 20:3639–3652. https://doi.org/10.1016/J.CSBJ.2022.07.011

Systems Biology in Understanding the Human Gut Microbiome and Related Diseases Highlighting Metabolic Modeling and Analysis

17

Priyanka Narad, Romasha Gupta, Sudeepti Kulshrestha, and Abhishek Sengupta

Abstract

The human microbiome has an impact on human health. In physiological settings, genome-scale metabolic modeling (GSMM) is a comparatively modern method to predict diet-microbiome, microbe-microbe, and host-microbe interactions. Recent discoveries have led to a completely new outlook on the subject that suggests our microbiota may play a role in the onset of many different diseases, such as cancer, gastrointestinal problems, and metabolic disorders. Due to the complicated etiology of various disorders, the precise mechanisms by which certain alterations in the makeup of the gut microbiota are related to the emergence of such diseases in humans are yet unknown. Multiple metagenomics and metabolomics studies have shown that microbial dysbiosis increases cancer risk through a variety of mechanisms. Using a systems biology strategy for facilitating learning of the underlying variations of the microbiome between physiological and pathological phases can ultimately help to uncover potential innovative therapeutic and diagnostic approaches.

Keywords

Systems biology · Microbiome · Gut microbiota-related diseases · GSMM

P. Narad
Division of Biomedical Informatics (BMI), Indian Council of Medical Research, Ansari Nagar, New Delhi, India

R. Gupta · S. Kulshrestha · A. Sengupta (✉)
Amity Institute of Biotechnology, Amity University, Noida, Uttar Pradesh, India
e-mail: asengupta@amity.edu

S. Joshi et al. (eds.), *Systems Biology Approaches: Prevention, Diagnosis, and Understanding Mechanisms of Complex Diseases*,
https://doi.org/10.1007/978-981-99-9462-5_17

17.1 Introduction

The microbial communities that inhabit the human body are crucial to human health. Due to the advancements in next-generation sequencing (NGS), microbial communities in health and disease may be quickly and comprehensively assessed. By incorporating biological networks and employing mathematical models, complex cellular or tissue interactions are attempted to be characterized using systems biology methodologies (Malla et al. 2019). The health of the host depends on the human microbiome, adding to the innate immunity, energy harvest, infection resistance, and microbiome dysbiosis that have been implicated in a variety of disorders (Althani et al. 2016). Comparative metagenomics techniques are frequently used to examine variations between individuals' health and disease. The human microbiome's co-occurrence patterns are substantially restricted by anatomical position, implying that organisms are conducive to specific environments. On a wider level, literary works' co-occurrence analysis has revealed the impact of competing metabolic interactions on the way of life and ecological strategy on a global scale. The human microbiota is engrossed in a plethora of physiological tasks, including immune system development, drug metabolism, food breakdown, pathogen protection, and vitamin generation (Sekirov et al. 2010). Furthermore, dysbiosis (abnormalities in the homeostatic balance of gut microbial communities) has an association with several disorders, including viral diseases, obesity, metabolic diseases, diabetes, numerous types of liver disease, and inflammatory bowel disease (IBD), according to research. Integrating our present understanding of systems biology methodologies with synthetic biology gives a means for manipulating and designing microbial systems to develop better diagnosis and treatment. The human microbiome seems to be a prominent topic of research due to its enormous health effects and has ten times the number of cells and a further 150 times the number of genes as the human host (Blaser et al. 2013). Several complex diseases are found to be connected to the microbiome's taxonomic and genetic diversity, according to comparative metagenomics studies. Extending statistical and quantitative studies and using systems biology methodologies to explore the microbiome at numerous levels would be extremely beneficial to microbiome research. Systems biology's holistic approaches to promoting perfect health and preventing diseases are gaining traction in medical care (Fiandaca et al. 2017).

In many bacterial systems, genome-scale system-level modeling has predicted correctly cellular behavior and supplied new design principles when exploring microbial species. A scalable and extendable computer technique for evaluating and forecasting biological function is genome-scale metabolic modeling (Chowdhury and Fong 2020). In a constraint-based reconstruction and analysis approach, organisms' genome-scale metabolic systems are used to emulate metabolism across different species within a specific microbiome and aid in the generation of original hypotheses about the interactions between microbes. Genome-scale models (GEMs), which use techniques like flux balance analysis (FBA) to simulate the organisms' growth in different situations, have evolved into important parameters for producing testable hypotheses on their metabolic functions. GEMs (genome-scale models) are used in most microbiome investigations to formally model a microbiome. Genome-scale metabolic modeling (GSMM), a constraint-based mathematical modeling approach, is being utilized to investigate gut ecosystems, elucidating microbial metabolic relationships with their host (Sen and Orešič 2019). GEMs of cataloged human gut microorganisms depending on metabolic processes have recently been constructed. Within a computational framework, GEMs can combine numerous biological data. Various omics datasets may all be integrated using the complex interaction of metabolites, genes, and enzymes as a scaffold. Integration of proteomic, transcriptomic, or regulatory data could enhance both the reconstruction and validation of these models even more.

17.2 Comprehending the Intricacy of Biological Networks at Systems Level

A systems-level approach is necessary to better understand these dynamics and characterize the microbiome alterations underlying health and disease mechanisms since the interactions between the microbiome, host, and microbes are complicated. By incorporating biological networks and employing mathematical models, complex cellular relations are attempted to be characterized using systems biology methodologies. Furthermore, biological network modeling could be utilized to incorporate and explore multi-omics data for a much more comprehensive view of microbial communities. Current molecular procedures are primarily concerned with biomolecules, such as proteins and nucleic acids, as well as their sequences, functions, and structures within the cell. Systems biology is an effort to comprehend a system made up of all these different molecules. If one wants to determine the connections within the system, one will observe that interactions concerning molecules change. As a result, a biological role emerges from the interrelatedness of the molecules within the system, instead of from the isolated molecules. It mainly focuses on understanding the structure of the system, the functionality of the constituent components, the network topologies, and the dynamics that constitute the synergistic system. In simple language, this statement means that systems biology attempts to understand the dynamics and structure of a system as a whole rather than understanding its genes and proteins in isolation (Kitano 2000). A system is coherently made up of a network of codependent and consequently interrelated components. Every system produces emergent behavior, which is a unique trait shared solely by the entire system and not to a significant extent by the individual components. Systems biology is a multidisciplinary approach to understanding life's dynamics, aiming to establish how the elements of a biological system interact with one another in space and time to create useful networks. Control and functional processes occur as emergent qualities of networks, not of their components, in biological systems across all levels of complexity, from the subcellular to the cell, tissue, and organ levels, as well as in human behavior. These biological networks self-organize, with somewhat autonomously operating sub-systems that have their network architecture, which is frequently made up of modular pieces found in many locations.

The main objective is to integrate data from multi-omics technologies to disclose the mechanisms involving the microbiome from a systematic perspective, as well as illuminate complex relationships among the microbial population, host, and the environment. For interpreting microbiome omics data, many bioinformatic pipelines are available. The outcomes and biological interpretation of a microbiome study can be influenced at any level, from experimental designs to data processing. As a result, at the data-analysis phase of the process, standardization is necessary, which will enhance the microbiome field's uniformity and comparability. With its comprehensive perspective, systems biology can provide an integrative framework for linking multi-omics data and elucidating intricate connections between the various components (Dahal et al. 2020). This discipline combines experimental methods with computational methods to find molecular processes that take place in intricate networks, such as those seen in biological systems (metagenomics, metaproteomics, metatranscriptomics, and metabolomics). The basic purpose of systems biology is to use mathematical representations to show the connections that actually exist inside a cell or tissue in order to appreciate the complexity of biological networks at the systems level. As massive omics data become more widely available, data integration has become much tougher due to various forms of data processing. Comparing and integrating diverse omics datasets is challenging since each omics discipline has its own analytical techniques and experiment designs.

17.3 Understanding the Human Gut Microbiome

17.3.1 Role of Gut Microbiota in the Human Body

The human microbiome, primarily the gut microbiome, is of primary purpose and focus. The biological and functional investigation of the microbiome of the gut through next-generation sequencing (NGS) has exposed several rich diversities (Wensel et al. 2022). On the other hand, it is uncertain what mechanisms govern interactions involving diet, gut microbiota, and host metabolism. The gut microbiota also produces gases while metabolizing short-chain fatty acids and other biological substances. Owing to the development of high-throughput meta-omics technology, there has been significant progress in understanding how the gut microbiota affects both health and disease (Lepage et al. 2013). The creation of novel computational tools and statistical models is required due to the enormous amount of biological data generated by gut microbiome research. Researchers are now starting to look at the boundaries of the gut microbiota using such instruments.

17.3.1.1 Nutrient Metabolism and Energy Storage

The human body's gut microbiota contributes to significant biochemical activity by performing additional metabolic tasks and controlling several elements of mediated activation in cellular growth and gene expression. The intestinal microbiota supplies enzymes for the utilization of nondigestible polysaccharides, host-derived glycoconjugates, deconjugation, and dehydroxylation of bile acids, reducing cholesterol, and synthesis of vitamins, isoprenoids, and amino acids (Vernocchi et al. 2020). The commensal microbiota's ability to consume essential nutritional polysaccharides that humans would otherwise be unable to use and to make short-chain fatty acids (SCFA) appears to have an impact on the host's ability to get energy from the diet (Cockburn and Koropatkin 2016).

17.3.1.2 Neurohormonal Function

Hormonal and neuropeptide production and function in peripheral organs, enteroendocrine cells of the digestive tract mucosa, and the nervous system, which are essential for the control of energy balance, may also be affected by the bacteria in the gut. Moreover, it has been found that the composition of the gut microbiota and stress-related chemicals, which affect energy balance, interact. Infant monkeys with low fecal Bifidobacterium counts were shown to have higher cortisol plasma levels and late-stage pregnant stress (Chung et al. 2019). Early maternal separation induced stress, elevated plasma corticosterone, triggered the systemic immunological response, and altered the fecal microbiota in male rat pups when compared to the control group.

17.3.1.3 Immune Function

Obesity caused by high-fat diets (HFDs) and associated metabolic disorders is defined by a low-grade inflammatory state that has been linked to altered gut microbiota composition and elevated plasma levels of lipopolysaccharide (Kang et al. 2017). This immune system-microbiota partnership, when functioning at its best, ensures the maintenance of regulatory pathways and the generation of defensive responses against infections that are responsible for maintaining tolerance to harmless antigens. The microbiota guards the intestine from colonization by foreign pathogens and potentially deadly resident microorganisms through a number of mechanisms, such as competing directly for limiting resources and the modulation of host immune responses.

17.3.1.4 Gut Microbiota Regulation in the Treatment of Metabolic Disorders

In view of research connecting changes in the diversity and composition of the gut microbiota to the formation of metabolic diseases, strategies aimed at restoring the gut microbiota are appearing as promising and interesting therapeutics. Cardiovascular disease (CVD) is more common due in large part to metabolic diseases such as insulin resistance, obesity, and diabetes mellitus (DM), which are linked to higher risk factors for other diseases (Gable et al. 2006). Only a small percentage of treated patients benefit significantly from treatment regimens, even though a variety of pharmacological medications are available for the management of metabolic illnesses. Because of the metabolic abnormalities that frequently accompany these disorders, patients are more likely to develop cardiovascular disease and die early. Only a small percentage of treated patients benefit significantly from treatment regimens, even though a variety of pharmacological medications are available for the management of metabolic illnesses. The involvement of gut microorganisms has been expanded to numerous other diseases, with new research and high-throughput technologies. Although modifying gut microbiota seems to be an interesting strategy in the management of metabolic diseases, the etiology of metabolic disorders is not solely dependent on the participation of gut microbiota. Numerous factors, including genetics, increased consumption of meals high in energy, and decreased or no physical exercise, contribute to the development of metabolic illnesses. Therefore, if successful, the modification of gut microbiota may also be used as a supplement to the present therapies for metabolic diseases (Boulangé et al. 2016).

17.3.1.4.1 Diet Impacts the Gut Microbiome

Intestinal bacteria composition and function appear to be quickly regulated by dietary composition and calorie intake. Fecal transplants were used to create a metabolic syndrome profile, demonstrating the crucial part the microbiota plays in this condition. Obesity and metabolic syndrome may indeed be associated with significant microbiome alterations, according to studies. Numerous sources of proof imply that dietary patterns may have a significant impact on the microbiota's composition (Graf et al. 2015). A high-fiber, low-fat diet preserves microbial diversity, allowing people to get the most energy from fiber while also preventing inflammation and noninfectious colon illnesses. Even though it is not apparent whether this link results in protection from metabolic diseases, a number of metagenomics studies imply that promising results in metabolic disorders are caused by higher microbial diversity. Despite these rapid interactions, long-term dietary patterns are nevertheless crucial for determining a person's gut composition, and the effectiveness of a particular diet is greatly influenced by the level of implementation of the gut microbiota.

17.3.1.4.2 Metabolite Control by Microorganisms

Diverse amounts of metabolites are produced by intestinal bacteria. Studies have shown that SCFAs directly support the host's defense against metabolic disorders (van der Hee and Wells 2021). For instance, SCFAs represent a key energy supplier for gut epithelial cells through stimulating G-protein-coupled receptors that influence enteroendocrine regulation (Samuel et al. 2008). Propionate can trigger intestinal gluconeogenesis and butyrate can operate as a major source of energy for colonocytes, using the gut-brain neural axis to defend the host from diet-induced obesity. It is interesting to note that eating a complex diet led to higher SCFA levels and greater diversity in the gut flora. Although it has become clear that bile acids function as signaling molecules for metabolic pathways, there are still significant concerns regarding whether supplementary administration of specific commensal bacteria can manage and control bile acid metabolic activity and the key element of these amended bile acids in metabolic disorders.

17.3.1.4.3 Metabolic Disease Symbionts

A single species of gut microbiota can control metabolic diseases in many different ways. Consuming probiotics or prebiotics beneficial bacteria is one such tactic

(Parkes 2007). For instance, metformin, a medication frequently used to treat type 2 diabetes, enhanced the glycemic index and increased the prevalence of A. muciniphila in mice fed a high-fat diet (HFD) (Foretz et al. 2019). Additionally, oral administration of A. muciniphila to mice on the HFD diet had antidiabetic benefits without the use of metformin (Pascale et al. 2019). A follow-up investigation to determine the underlying mechanisms causing the interactions between those bacteria's soluble products and their beneficial symbionts would be fascinating.

17.4 Database, Tools, and Repositories for Gut Microbiota

The gut microbiome is among the most researched microbiome regions in humans, making it an excellent target for GEM analysis. The AGORA set of semiautomated reconstruction of 773 human gut bacteria has formed the most extensive models of the gut microbiome's metabolism at the genome scale to date (Magnúsdóttir et al. 2017). The models contained in the AGORA collection were created using ModelSEED and Kbase semiautomated procedures. The models created can be used as a springboard for more thorough analysis (Sieow et al. 2019). GEMs have been used to explore the human gut microbiome's system based on the successful rebuilding of bacteria within the gut microbiome. The capability to simulate and examine the human gut microbiota may be utilized in order to investigate particular elements of digestive problems. Bile acid metabolism was examined in one study as only a host-microbiome integrated activity, in which the host is in charge of primary bile production, but the gut microbiome is in charge of bile acid deconjugation and biotransformation. It was asserted that the deconjugation and biotransformation of 693 human gut microorganisms are carried out by a consortium rather than by a single type of species and that this microbiome metabolic competency is severely reduced in juvenile patients who were suffering from inflammatory bowel disease (Heinken et al. 2019). GEMs were utilized to develop customized models to examine the synthesis of short-chain fatty acids (SCFAs) in Crohn's disease. Patient-specific models were used to predict specific deficiencies and recommend dietary changes that could be made to make up for them (Bauer and Thiele 2018).

The Human Microbiome Project (HMP) (Peterson et al. 2009), MetaHIT (Li et al. 2014), and American Gut (McDonald et al. 2018) are only a few of the collaborative research and major consortia that have taxonomically and functionally examined the gut microbiome in both healthy and disease states. MEtaGenome ANalyzer (MEGAN) is a user-friendly microbiome analysis toolset that enables the examination and evaluation of large-scale metagenomics datasets from both functional and taxonomical standpoints (Huson et al. 2007). A web-based, semiautomated reconstruction tool for prokaryotes was released, enabling the processing of draught reconstructions that are suited to the COBRA approach (Henry et al. 2010). Many mathematical modeling methods rely on linear programming (LP) to address one or even more flux balance analysis (FBA) challenges for computing functional states of metabolic pathway features in silico (Lewis et al. 2012).

17.4.1 Databases and Model Repositories Available for Gut Microbiota Modeling

The human gut's microbial population is humming with metabolic activity and is vital to the physiology and well-being of the host. The host's metabolic abilities are complemented by the vast and varied repertoire of bacterial metabolic processes, enabling it to, for instance, digest otherwise indigestible carbohydrates and produce healthy vitamins. Acquiring, analyzing, and evaluating microbiome-metabolome datasets from various studies, however, is often a laborious, difficult, and time-consuming procedure. Here we list the databases and model repositories that are accessible for modeling the gut microbiota (Table 17.1).

17.4.2 Databases of Metabolic Pathways and Enzymes Available for Gut Microbiota Modeling

With hundreds of species, the human gut microbiome is a sophisticated ecology. Metabolic cross talk between gut bacteria can be found with the aid of metabolic modeling. It is possible to study various food elements and microbiota makeup. By combining various datasets, it is possible to develop models of each individual's gut microbiota. Below is the list of databases of metabolic pathways and enzymes available for gut microbiota modeling (Table 17.2).

17.4.3 Modeling Tools Available for Gut Microbiota Modeling

Human health and disease are strongly influenced by the gut microbiota and its metabolites. With the help of developments in mass spectrometry, high-throughput sequencing, and other technologies for omics assays, we are now better able to produce massive amounts of data that are used to study temporal fluctuations in the makeup and capabilities of microbial communities. Tools and techniques that can accurately characterize the dependencies included in time-series data are required to understand mechanisms (Table 17.3).

Table 17.1 Databases and model repositories available for gut microbiota modeling

Tools	Description	References
ModelSEED	A web-based metabolic modelling library	Henry et al. (2010)
BiGG	A database of gold standard GEMs that is open to the public	King et al. (2016)
Human Metabolic Atlas (HMA)	A web-based resource for human metabolism that is free to the public	Pornputtapong et al. (2015)
Virtual Metabolic Human (VMH)	A database for human and gut microbial metabolism that is open to the public (GEMs)	Noronha et al. (2019)

Table 17.2 Databases of metabolic pathways and enzymes available for gut microbiota modeling

Tools	Description	References
KEGG	A database that contains large-scale molecular datasets as well as specific pathway information	Lewis et al. (2012)
UniProt.	A curated database of protein information that is open to the public	UniProt Consortium (2018)
REACTOME	A repository of biological pathways that is open to the public	D'Eustachio (2011)
BRENDA	An information retrieval system about enzymes and their liquids	Schomburg et al. (2017)
MetaCyc/ HumanCyc	A curated collection of metabolic pathways that have been empirically validated. HumanCyc is a curated library of metabolic pathways in humans	Chen et al. (2021)

Table 17.3 Modeling tools available for gut microbiota modeling

Tools	Description	References
COBRA (microbiome modeling toolbox)	Tools and techniques for pairwise and community modelling of microbiota are included in a MATLAB suite for constraint-based modelling (CBM). COBRA can also be used to rebuild and analyse GEM data	Schellenberger et al. (2011)
Kbase	A web-based platform for metabolic modelling and systems biology. It may be utilized to perform GEM reconstruction and analyses automatically	Arkin et al. (2016)
RAVEN	Tools for modelling diet-microbiota interactions are included in a MATLAB suite for CBM. This helps to recreate and analyse GEM data	Agren et al. (2013)
MetExplore	An open-access web-server for metabolomic dataset and genome-scale metabolic networks integrative analysis	Cottret et al. (2010)
jQMM library	Modeling internal metabolic fluxes with an open-source, Python-based framework. FBA and 13C Metabolic Flux Analysis (MFA) can both benefit from the toolset	Birkel et al. (2017)
MCM	Based on traditional FBA, a tool for CBM of microbial community models	Louca and Doebeli (2015)
OptCom	FBA of microbial communities using a modelling framework	Zomorrodi et al. (2014)
BacArena	An R programme for individual-based and community-based microorganism analysis in the gut	Bauer et al. (2017)

17.5 Systems Biology and Metabolic Modeling

17.5.1 Application of GEMs and GSMM in Human Microbiome

GEMs are linked lists of biochemical variables from a recipient bacterium that encompass information on reaction directionality, stoichiometry, compartmentalization, and gene and protein relationships. As a result, these models may generate gene-protein-reaction linkages, making them useful for integrating various omics data, to create a genotype-phenotype relationship. GEMs are used in constraint-based modeling by applying sets of constraints to consider certain objective functions like growth or substrates (Gu et al. 2019). Relying on the constraints and objective function, multiple optimization algorithms, such as linear programming, can be used to determine the best applicable strategy. GEMs have proven to be effective in generating tissue or cell and microbiological-specific models using these applications. GEMs are commonly used in human microbiome research to enhance one's knowledge of the connection between both the host along the microbiota and the consequences of the host's reaction to the microbial population. Transcriptional information, the rate at which molecules regenerate via a metabolic pathway, referred to as flux, and the status (on or off) of gene expression determined by high-throughput data can all constrain GEMs. These restrictions may specify a certain condition or setting under which data on the microorganisms' basic metabolic capacity may be gathered. Several GEMS are being joined along their extracellular divisions to form a community model. This community model has a regular compartment that looks like a particular body cavity, like the human intestine. This method can be employed to identify beneficial microorganisms for human health and, as a result, can be utilized to treat illnesses linked to the human microbiota. Flux variability analysis as well as other simulations are leveraged to evaluate the metabolic exchange reaction's flux span among the strains of bacteria in the microbial community. Using GEMs and combining high-throughput data to understand the microbiome necessitates gaining knowledge about microbial populations. This comprehension has the potential to alter the microbiome in ways that benefit the medical business, either directly or indirectly. In 2016, the NIH Human Microbiome Project published a set of papers aimed at classifying and sequencing bacteria related to various human microbiome locales (Peterson et al. 2009).

In contrast to primarily genome-centric techniques, mechanistic approaches to investigations of gut microbial populations are gaining traction. As a result, the popular computational tool GSMM is used to study the metabolism of gut microbes and how they interact with their hosts. Additionally, longitudinal meta-omics information is beneficial for contextualizing and personalizing GEMs, offering a picture of metabolic processes throughout time. Personalized microbiome models may serve to lower the expenses of clinical research, and contribute to the growth of generating novel medicines, either at the individual level or for a specific patient group. Many initiatives are being made to look at the interactions between drugs, microbes, and nutrition using constraint-based models and pharmacokinetics. GEMs cannot include enzymatic rates or metabolite quantities because they are stoichiometric models, which is a drawback of the GSMM approach (Price et al. 2004). Combining computational and experimental methods may help to show the diverse nature of the human gut microbiome as well as its essential linkages with the host metabolic apparatus. It might fill in research gaps in the gut microbiome and help us understand the human gut microbiota in both healthy and diseased states.

17.6 Metabolic Disorders and Human Gut Microbiota

17.6.1 Gut Microbiota Association with Obesity and Metabolic Disorders

The term "gut microflora" or "microbiota" refers to the tremendously large, diverse, and complicated microbial population that lives in the human intestine. In addition to genetic and environmental variables, the onset of obesity as well as metabolic syndrome is linked to pathways that connect the immune system with metabolism and vice versa. When obesity first develops, which is a complex process involving genetic as well as environmental variables, a number of chronic problems, such as high blood pressure, elevated triglyceride levels, decreased rates of high-density lipoprotein, and high rising glucose levels, commonly follow. Those who meet at least three of these criteria are clinically diagnosed with the metabolic syndrome, which puts them at greater risk for specific diseases like type 2 diabetes and cardiovascular diseases (Meigs et al. 2007). Consider the metabolic syndrome as a clinical diagnostic that is mechanistically triggered by a complex confluence of elements, such as poor fat storage, insulin action, and immunity. Obese individuals' gut microbiota's metabolic performance and composition have experienced considerable modifications, which seems to allow the microbiota to obtain more energy from the diet (Murphy et al. 2015). These studies on the relationship between intestinal microbial flora and obesity have revealed these profound changes. Furthermore, these investigations have shown that the host epithelial cells and the gut microbiota collaborate to indirectly regulate energy production and storage.

17.6.1.1 Physiopathology of Obesity and the Metabolic Syndrome

Although some genetic variants have been associated with a higher likelihood of developing type 2 diabetes and obesity, their heritability is very low. Recently, it has been evident that the gut microbiota is one of the primary environmental factors driving metabolic disorders (Ghaisas et al. 2016). The host's immune system and energy balance are supported by the host's gut microbiota, which is increasingly recognized as a unique endocrine organ. Changes in the gut microbial composition brought on by external stressors can radically alter the symbiotic relationship between the gut bacteria and the host, which promotes the emergence of metabolic diseases.

17.6.1.2 Gut microbiota Mechanism Affecting the Obesity-related Metabolic Disorders

The gut microbiota and its host have a symbiotic connection, but dysbiosis can arise when mutually supportive or commensal bacteria are changed or outcompeted by less beneficial or pathogenic bacteria. Different perspectives of the gut microbiota are identified to be involved in metabolic diseases and obesity, according to a stream of research (Aron-Wisnewsky et al. 2020). Since the methods used usually yielded coarse taxonomic categorization or intricate relationships among the diet, gut microbiota, and host, the early results were sometimes inconsistent. Evidence in animal experiments indicates that disturbed microbiota can lead to weight gain and an adiposity-related metabolic profile, and preliminary research in people suggests the same causal link (Zhao et al. 2022) (Fig. 17.1).

Advancements in omics technology allowed for the identification of major microbiome features linked to food and obesity. Weight fluctuations displayed broad molecular fingerprints of chronic disorders, including hypertrophic cardiomyopathy and insulin resistance, based on a comprehensive examination of many omics data over a long-term weight fluctuation research (Deidda et al. 2015). The role of adipose tissue in the pathophysiology of obesity is well known, and thus obesity and related health disorders could be prevented by targeting white adipose tissue. The advent of GEMs for adipocytes allows for mechanistic explanations of metabolic differences between lean and obese individuals, taking into account the inclusion of gene expression and plasma metabolomics datasets into the model (Mardinoglu et al. 2013).

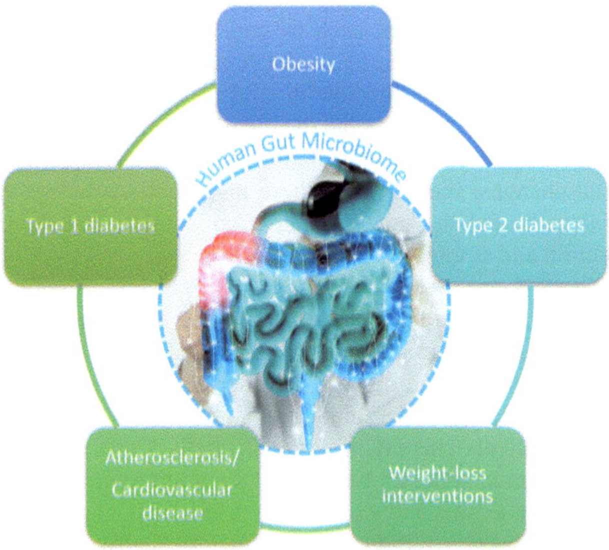

Fig. 17.1 Various metabolic diseases as well as weight-loss interventions are found to be associated with the human gut microbiome

17.6.2 Microbiota Related to Diabetes and Cardiovascular Diseases

Diabetes mellitus (DM) affects about 425 million people globally and is a significant public health concern (Glovaci et al. 2019). More and more evidence reveals that perhaps gut bacteria may contribute to diabetes etiology. An integrated taxonomic and functional analysis of multiple meta-omics data revealed that family membership had a significant impact on the inter- and intraindividual variance of the microbial community in the locale of type 1 DM. Furthermore, two large populations of healthy people found that environmental factors influenced gut microbiota composition more than host genetics. Obesity is a significant cause of type 2 diabetes (T2D), and the two are linked (Van Greevenbroek et al. 2013). T2D was also affected by the linkages and mechanisms that link obesity with the gut flora, but it is also crucial to look into the specific linkages and mechanisms that could trigger T2D. Insulin resistance and occasionally impaired insulin synthesis characterize T2D, resulting in poor glucose absorption by cells and high blood glucose levels.

The most prevalent cause of mortality worldwide is cardiovascular disease (CVD), which mostly includes heart attacks and strokes and accounts for almost 30% of all fatalities (Reddy et al. 2022). Obesity and diabetes are substantial risk factors for heart disease. Plaques formed by cholesterol or macrophages in the artery wall may burst and restrict blood flow downstream, leading to a heart attack or a stroke. The gut microbiota's metabolism of bile acids is of particular importance in the setting of atherosclerosis (Li and Tang 2017). After being created from cholesterol within the liver and discharged in the duodenum, bile acids are detergents that thin out lipids and fat-soluble substances and are extensively entirely absorbed in the distal ileum. In the small intestine and colon, microorganisms deconjugate and dehydroxylate bile acids to form secondary bile acids. They drain cholesterol from the enterohepatic circulation when the secondary bile acid lithocholic acid is secreted in the stool. The significance of microbial bile acid metabolism in atherosclerosis is unclear. Several studies conducted on humans and animals have lately revealed that the gut microbiome is an important aspect of the environmental component that contributes to the progression of a variety of liver illnesses, including end-stage cirrhosis and cancer (Schnabl and Brenner 2014). Evidence suggests that this may perhaps be a primary target for disease treatments, along with keeping an eye on how microbiome-based approaches are being used to treat liver diseases.

17.6.3 The Role of the Gut Microbiota in Inflammatory Bowel Diseases

The gut microbiota's bacterial component has received the most research attention, although more work is still needed on its function and strain-level resolution. A substantial impact on IBD and human health may also be caused by unstudied components of the gut microbiota, such as viruses and fungi, and the interactions they have with the gut bacteria. The intestinal system is affected by a chronic inflammatory ailment called the inflammatory bowel disease (IBD), which has no known etiology (Naser et al. 2014). Studies on the etiology of inflammatory bowel disease (IBD) have focused on a number of variables, such as the host genetics and immunological reactions, the gut microbiota, and the significance of environmental triggers in disease pathogenesis (Knights et al. 2014). The link between IBD and gut dysbiosis has been shown time and time again. Due to the rise in the use of high-throughput deep sequencing technologies over the past 10 years, we have been able to gradually disclose the role of the microbiome in the emergence of IBD. IBD is one of the most thoroughly researched inflammatory illnesses, and it is intimately linked to the gut flora.

17.6.4 Gut Microbiota and Osteoporosis

A tremendous economic load on healthcare systems could be caused by osteoporosis, a serious issue with bone health. The variations in peak bone mass and age-related bone loss are primarily influenced by inherited and environmental variables. Research on how gut bacteria may impact bone health is still relatively recent. Recent research has shown that gut bacteria regulate bone mass by influencing the immune system (Hernandez et al. 2016). An imbalanced microbial composition of the gut microbiota has been associated with a variety of inflammatory conditions, both inside and outside the GI tract, including inflammatory bowel diseases, allergies, rheumatoid arthritis, obesity, and metabolic syndrome. Additionally, autoimmune and inflammatory disorders have been linked to reduced bone mass, indicating a connection between the immune system and bone metabolism. Even though everyone's gut microbiota is unique, older persons' gut microbial composition has undergone considerable alterations. Inflammation results from a change in the gut microbiota in the aged from obligate to facultative anaerobes. Research revealed that as older persons move from living in the community to long-term care facilities, significant microbial changes occur in their gut microbiome. These changes in bone and body composition increase the risk of osteoporosis (Rosen and Klibanski 2009).

17.6.5 Weight-loss Interventions

The gut flora may considerably contribute to this variation in losing weight responsiveness and appears to significantly contribute to the development of obesity (Dao et al. 2016). Since the human gut microbiota may be changed by diet, it is possible to target specific microorganisms through well-defined dietary methods to find affordable treatments for noncommunicable diseases such as pandemic obesity, overweight, and associated metabolic comorbidities. However, low-to-moderate success rates in weight control are frequently seen with randomly controlled dietary regimens, making clinical practice application challenging. Changes in the composition or diversity of the gut microbiota have been linked to the pathophysiology of obesity in both animal models and humans through a variety of mechanisms, including energy consistency uptake, inflammatory responses, appetite regulation, and the generation of molecules that alter the metabolism. The gut microbiota has been demonstrated to shift in response to weight loss, and there is evidence to suggest that the gut microbiota and metabolites produced from it may serve as key mediators of the body's reaction to dietary energy restriction (Remely et al. 2015).

17.6.6 Intestinal Microbiota Dysbiosis of the Liver

The prevalence rate of nonalcoholic fatty liver disease (NAFLD), which is commonly thought of as the hepatic component of the metabolic syndrome, ranges from 20% to 40% in adult populations worldwide (Moore 2010). NAFLD encompasses a wide range of diseases, from simple steatosis to nonalcoholic steatohepatitis (NASH), an inflammatory, advanced form of NAFLD (Hübscher 2006). Disturbed gut microbiota might be one of the numerous important aspects in the pathogenesis of NAFLD and NASH, which is something that is becoming increasingly acknowledged. People with liver cirrhosis and liver failure have significantly different functional capabilities of the gut microbiota than those who are healthy. As hallmarks of hepatic encephalopathy, the changes include a high enrichment of nitrate absorption into or dissimilation from ammonia. As a result of protoporphyrin toxicity, an elevation in phosphotransferase pathways may indicate a greater need for protein following liver tissue injury, and an increase in microbial hemoglobin synthesis may cause liver damage (Sarkany et al. 2010).

17.6.7 Gut Microbiota Disruption Due to Malnutrition

The primary reason for mortality in children under 5 years of age worldwide is malnutrition, affecting 160 million people. The development of a healthy gut microbiota, defined by a brief bifidobacterial bloom before an increase in anaerobic bacteria during late infancy, depends on breastfeeding, water, and food security, and these are all important preventive factors against malnutrition (Fan and Pedersen 2021). Chronic malabsorption, diarrhea, and systemic invasion from pathogenic bacteria are all associated with persistent malnutrition, which causes damage to the healthy, mature anaerobic gut microbiota to gradually result in a deficit in the production of energy, immune responses, and vitamin synthesis.

17.6.8 Human Gut Microbiota and Carcinogenesis

17.6.8.1 Cancer of the Digestive System

The human microbiota, specifically found in the gut, may influence health favorably and contribute to serious metabolic, immune, and neurological problems. The formation and progression of the cancer disease may also be significantly influenced by the gut flora. The gut microbiome, which comprises the entire genome of the gut microbiota of the host, has 100 times as many genes as the human genome. By combining next-generation sequencing with computational analysis of the 16S rRNA amplicons, metagenomics has made it possible to assess the diversity and abundance of the gut microbiome within the last ten years. Together with improvements in transcriptomics and metabolomics, metagenomics investigations have made it possible to explain how some bacterial species affect the health of hosts. This is an advancement from descriptive investigations of microbiome composition to functional annotation, which is now assisting in determining the true influence of microbial architecture on human health. One of the conditions linked to the gut microbiota that has been studied the most is tumorigenesis. Both local gastrointestinal malignancies and additional distant tumors have been linked to this association (Cani and Jordan 2018). Metabolomics and metagenomics studies brought to light the gut microbiome's diverse role in cancer prevention, carcinogenesis, and anticancer therapy. Actually, the gut microbiota might also prevent or encourage the growth of tumors. Metabolomics and metagenomics studies brought to light the gut microbiome's diverse role in cancer prevention, carcinogenesis, and anticancer therapy. Actually, the gut microbiota might also prevent or encourage the growth of tumors. Although this connection has long been researched, it is only poorly understood. The intricacy and bi-directionality of the relationship between the microbiome and cancer are shown by all of the present research. As a result, the development of cancer may change the microbiome, and alterations to the microbiome may then affect the advancement of cancer.

17.6.8.1.1 Oral Cavity Cancer

Oral mucosa is the origin of oral cavity cancer, most notably oral squamous cell carcinoma (OSCC). It ranks as the 14 most prevalent kind of cancer worldwide. Almost 90% of all oral malignancies are squamous cell carcinomas (Feller and Lemmer 2012). Oral squamous cell carcinoma may occur due to a variety of environmental variables, including changes in the microbiota (OSCC). According to studies, the microbiota in the mouth and stomach could be used as cancer indicators. Several oral microbiota species have been linked to the onset and spread of cancer, both in the oral cavity and in other distant anatomical sites. In particular, *F. nucleatum* and *P. gingivalis* carriage are linked to the occurrence of cancer (Stasiewicz and Karpiński 2022). The development of pro-inflammatory conditions, immunological suppression, and the suppression of apoptosis are some of the processes of bacterial-mediated carcinogenesis. In this way, oral bacteria can cause a number of cancer's telltale signs.

17.6.8.1.2 Esophageal Cancer

Throughout the world, esophageal cancer ranks as the sixth most common cause of cancer-related mortality and the eighth most frequently diagnosed malignancy. Adenocarcinoma and squamous cell carcinoma are the two main histological forms of esophageal cancer (Greenawalt et al. 2007). According to experimental data, the human intestinal microbiome can have an impact on the growth and spread of tumors in the gastrointestinal tract by destroying DNA, turning on oncogenic signaling pathways, generating metabolites that support tumor growth, and preventing the immune system from fighting them. Esophageal cancer can be treated or prevented by manipulating the microbiota, and new therapies and strategies can be developed by studying the pathways and microbiome components to its development and progression.

17.6.8.1.3 Gastric Cancer

As the fourth most common cancer-related fatality worldwide, stomach cancer remains a serious health concern despite its decreasing frequency (Engstrand and Graham 2020). Chronic inflammation frequently occurs before the onset of stomach cancer, especially in cases of intestinal cancer, as is generally accepted. The microbiome can distinguish between different tumor types, including stomach cancer. More people are becoming aware of the significance that the gut microbiota plays in the development of cancer. While 15–20% of cancers are known to be triggered by distinct infectious agents, additional cancers are linked to the overall gut microbiota with or without the presence of specific trigger species. The microbiome has a complex effect on the chance of developing cancer and includes effects on the host metabolism, immunological response, host/microbial sensing networks, and cell division. Recent developments in sequencing technologies have revealed important discrepancies between the oral and/or esophageal microbiota and the gastric microbiota.

17.6.8.1.4 Colorectal Cancer

The third highest incidence of cancer in the world and one of the most often diagnosed malignancies is colorectal cancer. The strong connection between its onset and diet raises the possibility that the gut microbiota have an effect on the initiation and spread of colorectal cancer (Saus et al. 2019). The fine line separating health from disease is determined by the presence of a homeostatic gut microbiota, which is essential for maintaining host homeostasis. The gut microbiota has been identified as a key environmental component in the etiology of colorectal cancer. For screening, early detection, and treatment outcome prediction in colorectal cancer, it is crucial to know the makeup of bacterial strains and how geography, race, sex, and nutrition affect the microbiome.

17.6.8.1.5 Pancreatic Cancer

An aggressive and fatal cancer is pancreatic ductal adenocarcinoma (PDAC) (Sarantis et al. 2020). The gut microbiome may lead to PDAC susceptibility, onset, and development. Over half of the PDAC sufferers have advanced or metastatic illnesses that make them inoperable. Even after resection, PDAC frequently returns, and chemotherapy is frequently ineffective. There is a need for novel therapeutic approaches and early diagnosis techniques. Some types of hepatotropic viruses, well-studied oral, gastrointestinal, and intrapancreatic microorganisms, as well as bactibilia, may have etiological or modulatory functions in the emergence of pancreatic cancer (Wei et al. 2019). A deeper comprehension of the microbiota in

PDAC may help develop approaches for early-stage cancer diagnostics and cancer bacteriotherapy, adjusting therapeutic efficacy, even reducing side effects, opening up new possibilities, and giving hope to PDAC patients in need. The variables influencing the various survival outcomes are still mysterious and complicated as of now. Therefore, there is an urgent need for novel methods of early detection, prevention, and treatment of pancreatic cancer.

17.6.8.1.6 Liver Cancer

Disturbances in the human gut flora are associated with a pattern of hepatocyte destruction and regeneration that is typical of chronic liver disease (CLD) (Temraz et al. 2021). The persistence of chronic inflammation has been shown to result in an aggregation of epigenetic and genetic modifications that lead to hepatocellular carcinoma (HCC). The significance and prognostic impact of intestinal bacteria on hepatocarcinogenesis has thus attracted more research in past years (Rattan et al. 2020). The liver is directly exposed to gut microbial components and metabolites through the liver portal vein. Metabolites and other components of the microbiota, which are important in this process and play a major role in the development of hepatocellular carcinoma (HCC), are primarily responsible for this. Chronic liver disease (CLD), which originates from the vicious cycle of liver injury, inflammation, and regeneration that may extend for dozens of years, is the only condition in which HCC occurs almost exclusively. Recent research on the microbiome-gut-liver axis has improved our knowledge of how the development of HCC and liver disease is influenced by gut bacteria (Moreno-Gonzalez and Beraza 2021).

17.6.8.2 Cancer of Urogenital System

Different urological diseases have been linked to urine microbiome dysbiosis (Markowski et al. 2019). Due to the high rates of morbidity and mortality, bladder cancer has caught the attention of scientists in recent years. Regrettably, bladder cancer's origin and mechanism are yet unknown. Both environmental risk factors and genetic abnormalities could be to blame. While the healthy urinary tract is believed to be sterile, the presence of bacteria has historically been regarded as a urinary tract infection. It is commonly acknowledged that urine is antagonistic to germs' ability to survive. Men and women have different urine microbiota compositions. Recent data is compelling evidence that the human microbiome contributes in a variety of ways to the etiology and treatment of genitourinary cancers (Brubaker and Wolfe 2015). Preclinical and clinical studies demonstrate that microbial organisms are inextricably linked to many disease processes despite the complexity and incomplete definition of these relationships. With more research, both the urine microbiome and the gut microbiome may eventually serve as trustworthy predictors of disease onset and prognostic indications.

17.6.8.2.1 Cervical Cancer

Cervical cancer is one of the most dangerous malignancies that strike women and is the third leading cause of cancer-related fatalities worldwide. Infection with the human papillomavirus (HPV) increases the risk of developing cervical cancer. The microbiota can alter human physiology and immunological responses, increasing the risk of infection and predisposition to neoplastic disease (Castanheira et al. 2021). Due to its subtle signs, cervical cancer is sometimes overlooked in its early stages. However, as cervical cancer progresses to the final stage, a number of typical symptoms, including abnormal vaginal discharge, vaginal bleeding, and dyspareunia, manifest (Curty et al. 2019). Cervical cancer mortality rates do not dramatically reduce despite the fact that routine screening and various forms of treatment can stop the disease's onset. Examining common treatments for cervical cancer is critically needed.

17.6.8.2.2 Bladder Cancer

Bladder cancer is the most commonly encountered type of urinary system cancer. Males are three times more inclined than females to get this condition, and the risk of having it rises with age. According to studies examining the spectrum of health and disease states, microbial populations can have an impact on urological disorders (Bučević Popović et al. 2018). The conventional wisdom that healthy people have sterile pee and bladders has been refuted by new findings. Microbes in the urinary system can now be found thanks to modern culture and sequencing techniques. The present focus of bladder cancer research is on determining how commensal urinary microbiota may alter bladder cancer vulnerability to progression and have an effect on the efficacy of treatment via manipulation of the anticancer immune response (Annels et al. 2020). Every stage of bladder cancer's growth is likely to be affected differently by the microbiome, and it also likely influences how the body's natural immune system reacts to tumors. Therefore, it has the potential to be useful as a biomarker and treatment for bladder cancer, particularly because immune checkpoint therapy is now authorized for both early and late-stage disease.

17.6.8.2.3 Prostate Cancer

Prostate cancer is the most prevalent tumor of the male urinary system (Ohadian Moghadam and Momeni 2021). Prostate cancer may be caused by several biological processes mediated by different bacteria and their metabolites. Tumor occurrence and development may be affected by modifications to the gut microflora's makeup and biochemical imbalances. As a result of the development of new therapeutic options like immunotherapeutics and sequencing technologies like next-generation sequencing and metagenomics, there is optimism that the microbiome, a key factor in the immune system's regulation, might prove to be a promising new therapeutic option in the management of cancers, such as prostate cancer.

17.6.8.3 Lung Cancer

Lung cancer is a serious threat to people's health because it is the most common cancer diagnosis worldwide. Healthy human lungs have a particular pattern of microbiota and are frequently exposed to the outside environment. They also share many important pathological and physiological traits with the intestinal system

(Halley et al. 2020). Lung cancer development and progression are influenced by gut dysbiosis through systemic inflammation, genotoxicity, and compromised immune surveillance. A unique biomarker for predicting immunotherapy sensitivity and side effects in lung cancer patients may also be found in the gut flora (Wu et al. 2021). The crucial functions that the microbiota plays in maintaining lung homeostasis need not be disregarded, as microbiome dysbiosis may promote tumor development and progression by the production of cytokines and toxins, among other mechanisms.

17.6.8.4 Melanoma

Despite being the least common form of skin cancer, melanoma causes the most fatalities. Skin cancer has a complex etiology, and while certain environmental risk factors are well understood, many are still unknown (Makaranka et al. 2022). Skin cancer's onset, progression, and response to treatment may all be influenced by changes in the microbiota of the skin. These factors may include activation of the skin's immune system, generation of microbial metabolites and toxins, barrier disruption, and UV radiation. Additionally, melanoma patients' gut bacterial and fungal profiles were found to be considerably different from controls. The commensal microbiota may affect anti-tumor immunity and, consequently, the response to immune checkpoint inhibitor (ICI) therapy in cancer patients, according to numerous studies of ICI therapies. Studies show that gut microbiota are essential for repairing chemotherapy- and radiation-induced damage during chemotherapy treatments, and therapeutic modification of gut microbiota can be a successful approach to managing adverse effects. Also, the gut microbiome has both tumor-promoting and tumor-suppressive characteristics, which may assist in melanoma formation (Bhatt et al. 2017).

17.6.8.5 Breast Cancer

Breast cancer (BC) is the second leading cause of death for women worldwide and the most prevalent malignancy in women. Some breast cancer cases' onset is thought to be impacted by gut microbial dysbiosis (Chen et al. 2019), which presents an unmet opportunity for intervention and the chance to save lives. Homeostasis and gut normobiosis are disrupted by several circumstances, which favor the emergence of disease. The most significant estrogen-related risk factors for BC are endogenous and exogenous estrogen exposure, high breast density, a history of atypical hyperplasia, lifestyle variables such as a Western diet, obesity, and alcohol usage, and genetic vulnerability. In recent years, it has been proven that gut bacteria may regulate the effectiveness and side effects of cancer therapy. Yet it is also evident that the gut microbiota and both cancer and anticancer medicines interact in a two-way manner. Microbiota may soon be recognized as a nonheritable risk factor for BC, which may be influenced by a customized preventive and therapeutic strategy. In situations of established BC, the microbiota may potentially be a prognostic and predictive factor of therapeutic response or have negative side effects. Additionally, altering the microbiome can help BC patients have better results.

17.7 Strategies to Build up a Healthy Gut

17.7.1 Therapeutic Approaches for Enhancing Human Gut Microbiome

Research on the metabolic, signaling, and immune connections between the gut bacteria and the host, and how these interactions control the host brain, muscle, liver, and gut functioning, has led to the idea of therapeutic microbial alteration to treat or prevent diseases. The selection of specific gut bacterial strains and the augmentation of the gut microbial ecology represent prospective therapeutic approaches to control energy intake and reduce the prevalence of obesity and metabolic syndrome. The therapy of metabolic illnesses using a novel prospective therapeutic strategy involves the gut flora. A unique treatment approach for glycemic management involves the regulation of the gut microbiota, which is carried out by well-known substances like prebiotics, probiotics, and some medications like metformin. Considering the therapeutic and financial promise, microbiome therapeutics remain in the early stages of development and are dealing with several logistical and organizational problems that call for further research (Bajaj et al. 2022). Even though research has demonstrated the effectiveness of microbiome therapeutics, more research is necessary to completely understand the microbiome and the manner in which it interacts with the host before this idea of microbiome therapeutics can be implemented in clinical trials to produce a manual for effective treatment. This era of microbiome therapeutics and the combined efforts could aid in the clinical application of disease treatment.

17.7.1.1 Prebiotics

Preoccupations about the connections between prebiotics, the typical diet, local flora, and host health have led to a regular updating of the criteria used to classify prebiotics in recent years. Prebiotics are purposely fermented meals that alter the makeup and/or function of the gut microbiota to benefit the host's well-being (Gibson 2008). High-fiber foods provide a lot of prebiotics that pass through the human digestive system undigested and when they get to the large intestine, gut microbes start fermenting them there. Prebiotics are essential nourishment for the bacteria in our stomach.

17.7.1.2 Probiotics

Probiotics are live bacteria that, when consumed in the appropriate amounts, have a favorable effect on the health of the host. The use of probiotics or next-generation beneficial bacteria to alter the microbiota represents a potential future direction for the creation of dietary or pharmacological aids for maintaining health. Probiotics are widely marketed and consumed, typically as functional foods or nutritional supplements. A few of the processes of probiotics include the control of intestinal microbial communities, the elimination of pathogens, immunomodulation, encouragement of epithelial cell proliferation and differentiation, and reinforcement of the intestinal barrier. The use of probiotics has been advocated as a therapeutic and preventive strategy to assist the gut microbiome in regaining its normal makeup and functionality. However, information from research on the human microbiome may help identify new native microbial species and develop techniques to positively affect alterations in the gut's microbial population (Sanders et al. 2019).

17.7.1.3 Synbiotics

Probiotics and prebiotics together in combination are known as synbiotics. Due in part to the lack of logical methods to evaluate these items for synbiotic activity before clinical trials, the formulation of synbiotics remains difficult. Synbiotics are now formulated as either complementary or synergistic substances. Probiotics and prebiotics are combined to create complementary synbiotics, each of which acts independently and has been proven to have clinically significant benefits to health. Synbiotics have a substantial impact on the gut microbiome and improve both human and animal health. Consumers would benefit from improved health outcomes beyond those provided by the probiotic microbe or the prebiotic substrate, both from a commercial and clinical standpoint. In comparison to probiotics or prebiotics alone, synergistic synbiotics may improve clinical results by enhancing the formation and/or retention of the companion bacteria (Markowiak and Śliżewska 2017).

17.7.1.4 Antibiotics

For many years, antibiotics have been used to treat bacterial diseases by limiting the spread of bacterial pathogens. They are destabilizing substances that put the balance of the gut's microbial species under stress. This imbalance is exhibited by a decline in species diversity and a concurrent overpopulation of pathogenic species, often known as pathobionts. Despite the fact that antibiotics save millions of lives, they also reduce household bacteria, which are essential for a healthy gut. The amount and length of antibiotic treatment, the mode of administration, the pharmacodynamic and pharmacokinetic qualities, and the range of the antibiotic agent are all factors that can modify the gut microbiota in diverse ways (Schwartz et al. 2020).

17.7.1.5 Drugs

Pharmaceuticals can change the populations of gut bacteria, and the variety of gut microbes can influence therapeutic effectiveness by changing the bioavailability and bioactivity of drugs. Drugs can be metabolized into a variety of metabolites by gut microbial activity or microbiota-host co-metabolism. Physicians need to be aware that nonantibiotic medications might also alter the gut microbiota, which could ultimately result in subpar health outcomes. This is because antibiotic usage is not the only factor that affects the gut microbiome. A thorough understanding of how gut microorganisms metabolize, consume, and bio-transform pharmaceuticals will open new potential channels for controlling the gut microbiome to enhance the efficacy of medications and biologics, as the pharmaco-microbiomes area is emerging and garnering attention (Nichols et al. 2019). Understanding pharmacological processes and how specific drug side effects occur depends on how pharmaceuticals interact with the makeup of the gut microbiome.

17.8 Microbiota Interventions to Improve Metabolic Health

17.8.1 Targeted Interventions

Intestinal microbiome's impact on metabolic health is demonstrated through targeted therapies including CRISPR-Cas9 editing, Bacteriophage therapy, and bioengineered commensals and drugs. The particular pharmacological inhibition of the microbial synthesis of trimethylamine is a classic example of how a new avenue for improving metabolism involves offering drugs that specifically target certain microbially generated compounds. Future research could use bacteriophages and archaeal viruses to focus on dysbiotic microbiome regions in individuals with metabolic issues because most of them are strain-specific (Ramezani and Raj 2014).

17.8.2 Untargeted Interventions

Microbiome-directed therapies that are untargeted in people include dietary or activity changes, probiotic, prebiotic, or postbiotic trials, and heterologous as well as autologous fecal microbiota transplantation. Interventions that involve more physical activity have been linked to improvements in cardiorespiratory fitness and insulin sensitivity as well as alterations in the gut microbiota that are adaptable and occasionally transmissible. These modifications include a better capacity to metabolize short-chain fatty acids, branched-chain amino acids, and lactate. Probiotics do not consistently have an impact on the microbial makeup of the gut, according to systematic analyses of brief randomized controlled studies in healthy individuals. However, certain probiotics might not even require colonization of the intestine to have a positive impact on metabolic health. It is yet unknown if heterologous fecal microbiota transplantation will be an efficient way to avoid or cure metabolic dysregulation with etiologies that are significantly more complicated than chronic infections. The efficiency and viability of the much simpler autologous fecal microbiota transplantation saved before the onset of a metabolic disorder need to be investigated.

17.9 Challenges

In spite of the abundance of published research linking the gut microbiota to metabolic diseases, there are still a number of challenges. These issues, which require attention, have to do with how the host's systems are modified, influenced, and interfered with by the gut microorganisms to produce systemic metabolic effects. Prebiotics, probiotics, and antibiotics are examples of potential modulation techniques for the gut microbiota, although they come with significant limitations. The viability and durability of ingested microorganisms in the gut are among these restrictions. The host-specificity of probiotics is another factor that prevents the health effects associated with probiotics from being generalized. However, the positive benefits of prebiotics are frequently transient and can be overridden or counteracted by diets. Studies comparing the various microorganisms frequently for use as probiotics are essential, both separately and collectively. Interventions that weigh the effectiveness of a probiotic-prebiotic combination versus either medication are also essential. Antimicrobial medicines' lack of bacterial species specificity is one of the key barriers to their capacity to modify the gut microbiota. Future studies should focus on individualized gut microbiota modulation. This will need tracking down food ingredients that can respectively prevent and promote the growth of hazardous and advantageous bacteria. Also, it is recognized that the type of food consumed affects the diversity and composition of the gut microbiota.

17.10 Conclusion

Systems biology studies have indeed transformed genomics, which has the power to revolutionize metagenomics and, in particular, human microbiome research. Modeling at the system level frequently relies on a huge amount of data in various forms. In the future, genomic and metagenomic sequencing will be critical for developing more accurate and reliable models. For appropriately modeling the microbiome ecosystem, precise identification of species composition and genetic data on individual species are particularly necessary. The Human Microbiome Project has been primarily responsible for sequencing over a thousand human-associated genomes, but hundreds more microbiome species will likely be sequenced in the near future. The total number of species that inhabit the human body may not include these species. However, advances in next-generation sequencing technology are anticipated to soon make a large number of additional microbial genomes accessible. It is possible to scale up ecosystem models to include the majority of the major microbiome species when several human-associated reference sequences become available. Likewise, high-coverage shotgun metagenomics information will aid in the generation of models for the superior supraorganisms that accurately depict the metabolic potential of the community. Data gathering should be informed by modeling efforts, which should provide viable hypotheses that point to important genes and species. When advances in next-generation sequencing and new technologies are made, the pace at which genomic or metagenomic data is gathered will accelerate. These technologies face significant informatics issues, many of which stem from platforms like Illumina and SOLiD's short read length. For the reconstruction of both multispecies and supraorganism models in metabolic modeling, precise annotation is essential. The development of novel algorithms will be necessary for billions of short reads in order to do reads assembly, mapping, and annotation to advance these technologies further and enable the assembly of full microbial genomes explicitly from data of shotgun metagenomic as well as high-resolution functional annotation of microbiomes. Systems-based analyses along with in silico modeling of the microbiome are not limited to genomic or metagenomics data. Other sorts of data could also aid in the experimental verification of such models. Metatranscriptomic and metametabolomic data, in particular, will improve modeling frameworks by allowing for a more accurate description of the metabolic activity of the microbiome and the accessibility of metabolites in the environment. Additional evidence is also necessary for using specialized modeling frameworks. A new constraint-based model for understanding the relationship between a host and a typical gut bacteria was recently developed. By quantifying nutrition changes in absorption and excretion rates, the model was said to be helpful in forecasting growth dependence and cross-feeding. There are currently multi-scale theories of metabolic reactions of humans that integrate whole-body physiology with cellular size models. Yet a thorough framework is required that takes into account interactions between microbes and their hosts, as well as the microbiome's overall interaction with the host, and can be increased in size to precisely represent the many species that comprise the human microbiome. Diagnostics and intervention in medically critical situations would be enhanced by the ability to foresee the precise impact of dietary supplements, microbial supplements, or pharmacological therapies on the metabolic activity of the microbiome. Reports of disease eradication through total microbiota transfer sparked initial interest, although more tailored medicines have lately been proposed. Such methods would be safer, more cost-effective, and eventually more effective if scientists had better insight into microbial interactions.

Because of advancements in high-throughput technologies and systems based on biology research techniques, it may be possible to better comprehend how the gut microbiota impacts health in humans, which could influence the formation of preventive and tailored treatment strategies. We also expect that over the next few years, these types of models will be able to identify novel blood biomarkers linked to the gut microbiota, which are known to induce illness development and/or progression, allowing for early diagnosis.

References

Agren R, Liu L, Shoaie S, Vongsangnak W, Nookaew I, Nielsen J (2013) The RAVEN toolbox and its use for generating a genome-scale metabolic model for Penicillium chrysogenum. PLoS Comput Biol 9(3):e1002980

Althani AA, Marei HE, Hamdi WS, Nasrallah GK, El Zowalaty ME, Al Khodor S, Cenciarelli C (2016) Human microbiome and its association with health and diseases. J Cell Physiol 231(8):1688–1694

Annels NE, Simpson GR, Pandha H (2020) Modifying the non-muscle invasive bladder cancer immune microenvironment for optimal therapeutic response. Front Oncol 10:175

Arkin AP, Stevens RL, Cottingham RW, Maslov S, Henry CS, Dehal P, Yoo S (2016) The DOE systems biology knowledgebase (KBase). BioRxiv, 096354

Aron-Wisnewsky J, Vigliotti C, Witjes J, Le P, Holleboom AG, Verheij J, Clément K (2020) Gut microbiota and human NAFLD: disentangling microbial signatures from metabolic disorders. Nat Rev Gastroenterol Hepatol 17(5):279–297

Bajaj JS, Ng SC, Schnabl B (2022) Promises of microbiome-based therapies. J Hepatol 76(6):1379–1391

Bauer E, Thiele I (2018) From metagenomic data to personalized in silico microbiotas: predicting dietary supplements for Crohn's disease. NPJ Syst Biol Appl 4(1):27

Bauer E, Zimmermann J, Baldini F, Thiele I, Kaleta C (2017) BacArena: individual based metabolic modeling of heterogeneous microbes in complex communities. PLoS Comput Biol 13(5):e1005544

Bhatt AP, Redinbo MR, Bultman SJ (2017) The role of the microbiome in cancer development and therapy. CA Cancer J Clin 67(4):326–344
Birkel GW, Ghosh A, Kumar VS, Weaver D, Ando D, Backman TW, Martín HG (2017) The JBEI quantitative metabolic modeling library (jQMM): a python library for modeling microbial metabolism. BMC Bioinformatics 18(1):1–11
Blaser M, Bork P, Fraser C, Knight R, Wang J (2013) The microbiome explored: recent insights and future challenges. Nat Rev Microbiol 11(3):213–217
Boulangé CL, Neves AL, Chilloux J, Nicholson JK, Dumas ME (2016) Impact of the gut microbiota on inflammation, obesity, and metabolic disease. Genome Med 8:1–12
Brubaker L, Wolfe AJ (2015) The new world of the urinary microbiota in women. Am J Obstet Gynecol 213(5):644–649
Bučević Popović V, Šitum M, Chow CET, Chan LS, Roje B, Terzić J (2018) The urinary microbiome associated with bladder cancer. Sci Rep 8(1):12157
Cani PD, Jordan BF (2018) Gut microbiota-mediated inflammation in obesity: a link with gastrointestinal cancer. Nat Rev Gastroenterol Hepatol 15(11):671–682
Castanheira CP, Sallas ML, Nunes RAL, Lorenzi NPC, Termini L (2021) Microbiome and cervical cancer. Pathobiology 88(2):187–197
Chen J, Douglass J, Prasath V, Neace M, Atrchian S, Manjili MH, Habibi M (2019) The microbiome and breast cancer: a review. Breast Cancer Res Treat 178:493–496
Chen IMA, Chu K, Palaniappan K, Ratner A, Huang J, Huntemann M, Kyrpides NC (2021) The IMG/M data management and analysis system v. 6.0: new tools and advanced capabilities. Nucleic Acids Res 49(D1):D751–D763
Chowdhury S, Fong SS (2020) Leveraging genome-scale metabolic models for human health applications. Curr Opin Biotechnol 66:267–276
Chung E, Grue KA, Kaur G, Mallory B, Serrano CR, Ullevig SL, Umeda M (2019) Maternal exercise before and during pregnancy alleviates metabolic dysfunction associated with high-fat diet in pregnant mice, without significant changes in gut microbiota. Nutr Res 69:42–57
Cockburn DW, Koropatkin NM (2016) Polysaccharide degradation by the intestinal microbiota and its influence on human health and disease. J Mol Biol 428(16):3230–3252
Cottret L, Wildridge D, Vinson F, Barrett MP, Charles H, Sagot MF, Jourdan F (2010) MetExplore: a web server to link metabolomic experiments and genome-scale metabolic networks. Nucleic Acids Res 38(2):W132–W137
Curty G, de Carvalho PS, Soares MA (2019) The role of the cervicovaginal microbiome on the genesis and as a biomarker of premalignant cervical intraepithelial neoplasia and invasive cervical cancer. Int J Mol Sci 21(1):222
D'Eustachio P (2011) Reactome knowledgebase of human biological pathways and processes. Bioinf Comparative Proteomics 2011:49–61
Dahal S, Yurkovich JT, Xu H, Palsson BO, Yang L (2020) Synthesizing systems biology knowledge from omics using genome-scale models. Proteomics 20(17-18):1900282
Dao MC, Everard A, Clément K, Cani PD (2016) Losing weight for a better health: role for the gut microbiota. Clin Nutr Exp 6:39–58
Deidda M, Piras C, Bassareo PP, Dessalvi CC, Mercuro G (2015) Metabolomics, a promising approach to translational research in cardiology. IJC Metabol Endocr 9:31–38
Engstrand L, Graham DY (2020) Microbiome and gastric cancer. Dig Dis Sci 65:865–873
Fan Y, Pedersen O (2021) Gut microbiota in human metabolic health and disease. Nat Rev Microbiol 19(1):55–71
Feller L, Lemmer J (2012) Oral squamous cell carcinoma: epidemiology, clinical presentation and treatment. J Cancer Ther 3(4):263–268
Fiandaca MS, Mapstone M, Connors E, Jacobson M, Monuki ES, Malik S, Federoff HJ (2017) Systems healthcare: a holistic paradigm for tomorrow. BMC Syst Biol 11(1):1–17
Foretz M, Guigas B, Viollet B (2019) Understanding the glucoregulatory mechanisms of metformin in type 2 diabetes mellitus. Nat Rev Endocrinol 15(10):569–589

Gable DR, Hurel SJ, Humphries SE (2006) Adiponectin and its gene variants as risk factors for insulin resistance, the metabolic syndrome and cardiovascular disease. Atherosclerosis 188(2):231–244

Ghaisas S, Maher J, Kanthasamy A (2016) Gut microbiome in health and disease: linking the microbiome–gut–brain axis and environmental factors in the pathogenesis of systemic and neurodegenerative diseases. Pharmacol Ther 158:52–62

Gibson GR (2008) Prebiotics as gut microflora management tools. J Clin Gastroenterol 42:S75–S79

Glovaci D, Fan W, Wong ND (2019) Epidemiology of diabetes mellitus and cardiovascular disease. Curr Cardiol Rep 21:1–8

Graf D, Di Cagno R, Fåk F, Flint HJ, Nyman M, Saarela M, Watzl B (2015) Contribution of diet to the composition of the human gut microbiota. Microb Ecol Health Dis 26(1):26164

Greenawalt DM, Duong C, Smyth GK, Ciavarella ML, Thompson NJ, Tiang T, Phillips WA (2007) Gene expression profiling of esophageal cancer: comparative analysis of Barrett's esophagus, adenocarcinoma, and squamous cell carcinoma. Int J Cancer 120(9):1914–1921

Gu C, Kim GB, Kim WJ, Kim HU, Lee SY (2019) Current status and applications of genome-scale metabolic models. Genome Biol 20:1–18

Halley A, Leonetti A, Gregori A, Tiseo M, Deng DM, Giovannetti E, Peters GJ (2020) The role of the microbiome in cancer and therapy efficacy: focus on lung cancer. Anticancer Res 40(9):4807–4818

Heinken A, Ravcheev DA, Baldini F, Heirendt L, Fleming RM, Thiele I (2019) Systematic assessment of secondary bile acid metabolism in gut microbes reveals distinct metabolic capabilities in inflammatory bowel disease. Microbiome 7:1–18

Henry CS, DeJongh M, Best AA, Frybarger PM, Linsay B, Stevens RL (2010) High-throughput generation, optimization and analysis of genome-scale metabolic models. Nat Biotechnol 28(9):977–982

Hernandez CJ, Guss JD, Luna M, Goldring SR (2016) Links between the microbiome and bone. J Bone Miner Res 31(9):1638–1646

Hübscher SG (2006) Histological assessment of non-alcoholic fatty liver disease. Histopathology 49(5):450–465

Huson DH, Auch AF, Qi J, Schuster SC (2007) MEGAN analysis of metagenomic data. Genome Res 17(3):377–386

Kang C, Wang B, Kaliannan K, Wang X, Lang H, Hui S, Mi M (2017) Gut microbiota mediates the protective effects of dietary capsaicin against chronic low grade inflammation and associated obesity induced by high-fat diet. MBio 8(3):e00470

King ZA, Lu J, Dräger A, Miller P, Federowicz S, Lerman JA, Lewis NE (2016) BiGG models: a platform for integrating, standardizing and sharing genome scale models. Nucleic Acids Res 44(D1):D515–D522

Kitano H (2000) Perspectives on systems biology. N Gener Comput 18:199–216

Knights D, Silverberg MS, Weersma RK, Gevers D, Dijkstra G, Huang H, Xavier RJ (2014) Complex host genetics influence the microbiome in inflammatory bowel disease. Genome Med 6:1–11

Lepage P, Leclerc MC, Joossens M, Mondot S, Blottière HM, Raes J, Doré J (2013) A metagenomic insight into our gut's microbiome. Gut 62(1):146–158

Lewis NE, Nagarajan H, Palsson BO (2012) Constraining the metabolic genotype-phenotype relationship using a phylogeny of in silico methods. Nat Rev Microbiol 10(4):291–305

Li DY, Tang WW (2017) Gut microbiota and atherosclerosis. Curr Atheroscler Rep 19:1–12

Li J, Jia H, Cai X, Zhong H, Feng Q, Sunagawa S, Wang J (2014) An integrated catalog of reference genes in the human gut microbiome. Nat Biotechnol 32(8):834–841

Louca S, Doebeli M (2015) Calibration and analysis of genome-based models for microbial ecology. elife 4:e08208

Magnúsdóttir S, Heinken A, Kutt L, Ravcheev DA, Bauer E, Noronha A, Thiele I (2017) Generation of genome-scale metabolic reconstructions for 773 members of the human gut microbiota. Nat Biotechnol 35(1):81–89

Makaranka S, Scutt F, Frixou M, Wensley KE, Sharma R, Greenhowe J (2022) The gut microbiome and melanoma: a review. Exp Dermatol 31(9):1292–1301
Malla MA, Dubey A, Kumar A, Yadav S, Hashem A, Abd Allah EF (2019) Exploring the human microbiome: the potential future role of next-generation sequencing in disease diagnosis and treatment. Front Immunol 9:2868
Mardinoglu A, Agren R, Kampf C, Asplund A, Nookaew I, Jacobson P, Nielsen J (2013) Integration of clinical data with a genome-scale metabolic model of the human adipocyte. Mol Syst Biol 9(1):649
Markowiak P, Śliżewska K (2017) Effects of probiotics, prebiotics, and synbiotics on human health. Nutrients 9(9):1021
Markowski MC, Boorjian SA, Burton JP, Hahn NM, Ingersoll MA, Vareki SM, Sfanos KS (2019) The microbiome and genitourinary cancer: a collaborative review. Eur Urol 75(4):637–646
McDonald D, Hyde E, Debelius JW, Morton JT, Gonzalez A, Ackermann G, Knight R (2018) American gut: an open platform for citizen science microbiome research. Msystems 3(3):e00031
Meigs JB, Rutter MK, Sullivan LM, Fox CS, D'Agostino RB Sr, Wilson PW (2007) Impact of insulin resistance on risk of type 2 diabetes and cardiovascular disease in people with metabolic syndrome. Diabetes Care 30(5):1219–1225
Moore JB (2010) Non-alcoholic fatty liver disease: the hepatic consequence of obesity and the metabolic syndrome. Proc Nutr Soc 69(2):211–220
Moreno-Gonzalez M, Beraza N (2021) The role of the microbiome in liver cancer. Cancer 13(10):2330
Murphy EA, Velazquez KT, Herbert KM (2015) Influence of high-fat-diet on gut microbiota: a driving force for chronic disease risk. Curr Opin Clin Nutr Metab Care 18(5):515
Naser SA, Sagramsingh SR, Naser AS, Thanigachalam S (2014) Mycobacterium avium subspecies paratuberculosis causes Crohn's disease in some inflammatory bowel disease patients. World J Gastroenterol 20(23):7403
Nichols RG, Peters JM, Patterson AD (2019) Interplay between the host, the human microbiome, and drug metabolism. Hum Genomics 13(1):1–10
Noronha A, Modamio J, Jarosz Y, Guerard E, Sompairac N, Preciat G, Thiele I (2019) The virtual metabolic human database: integrating human and gut microbiome metabolism with nutrition and disease. Nucleic Acids Res 47(D1):D614–D624
Ohadian Moghadam S, Momeni SA (2021) Human microbiome and prostate cancer development: current insights into the prevention and treatment. Front Med 15:11–32
Parkes GC (2007) An overview of probiotics and prebiotics. Nurs Stand 21(20):43
Pascale A, Marchesi N, Govoni S, Coppola A, Gazzaruso C (2019) The role of gut microbiota in obesity, diabetes mellitus, and effect of metformin: new insights into old diseases. Curr Opin Pharmacol 49:1–5
Peterson J, Garges S, Giovanni M, McInnes P, Wang L, Schloss JA (2009) The NIH human microbiome project. Genome Res 19(12):2317–2323
Pornputtapong N, Nookaew I, Nielsen J (2015) Human metabolic atlas: an online resource for human metabolism. Database 2015:68
Price ND, Reed JL, Palsson BØ (2004) Genome-scale models of microbial cells: evaluating the consequences of constraints. Nat Rev Microbiol 2(11):886–897
Ramezani A, Raj DS (2014) The gut microbiome, kidney disease, and targeted interventions. J Am Soc Nephrol 25(4):657–670
Rattan P, Minacapelli CD, Rustgi V (2020) The microbiome and hepatocellular carcinoma. Liver Transpl 26(10):1316–1327
Reddy TSE, Sripathi SR, Akula D, Palaniswamy S, Subramani R (2022) Cardiovascular disease prediction using machine learning and deep learning. In: 2022 6th international conference on computation system and information technology for sustainable solutions (CSITSS). IEEE, Piscataway, pp 1–5
Remely M, Tesar I, Hippe B, Gnauer S, Rust P, Haslberger AG (2015) Gut microbiota composition correlates with changes in body fat content due to weight loss. Benefic Microbes 6(4):431–439

Rosen CJ, Klibanski A (2009) Bone, fat, and body composition: evolving concepts in the pathogenesis of osteoporosis. Am J Med 122(5):409–414

Samuel BS, Shaito A, Motoike T, Rey FE, Backhed F, Manchester JK, Gordon JI (2008) Effects of the gut microbiota on host adiposity are modulated by the short chain fatty-acid binding G protein-coupled receptor, Gpr41. Proc Natl Acad Sci 105(43):16767–16772

Sanders ME, Merenstein DJ, Reid G, Gibson GR, Rastall RA (2019) Probiotics and prebiotics in intestinal health and disease: from biology to the clinic. Nat Rev Gastroenterol Hepatol 16(10):605–616

Sarantis P, Koustas E, Papadimitropoulou A, Papavassiliou AG, Karamouzis MV (2020) Pancreatic ductal adenocarcinoma: treatment hurdles, tumor microenvironment and immunotherapy. World J Gastrointest Oncol 12(2):173

Sarkany RPE, Breathnach SM, Morris AAM, Weismann K, Flynn PD (2010) Metabolic and nutritional disorders. Rook's Textb Dermatol 1:1–103

Saus E, Iraola-Guzmán S, Willis JR, Brunet-Vega A, Gabaldón T (2019) Microbiome and colorectal cancer: roles in carcinogenesis and clinical potential. Mol Asp Med 69:93–106

Schellenberger J, Que R, Fleming RM, Thiele I, Orth JD, Feist AM, Palsson BØ (2011) Quantitative prediction of cellular metabolism with constraint-based models: the COBRA Toolbox v2. 0. Nat Protoc 6(9):1290–1307

Schnabl B, Brenner DA (2014) Interactions between the intestinal microbiome and liver diseases. Gastroenterology 146(6):1513–1524

Schomburg I, Jeske L, Ulbrich M, Placzek S, Chang A, Schomburg D (2017) The BRENDA enzyme information system–from a database to an expert system. J Biotechnol 261:194–206

Schwartz DJ, Langdon AE, Dantas G (2020) Understanding the impact of antibiotic perturbation on the human microbiome. Genome Med 12(1):1–12

Sekirov I, Russell SL, Antunes LCM, Finlay BB (2010) Gut microbiota in health and disease. Physiol Rev 90(3):859–904

Sen P, Orešič M (2019) Metabolic modeling of human gut microbiota on a genome scale: an overview. Meta 9(2):22

Sieow BFL, Nurminen TJ, Ling H, Chang MW (2019) Meta-omics-and metabolic modeling-assisted deciphering of human microbiota metabolism. Biotechnol J 14(9):1800445

Stasiewicz M, Karpiński TM (2022) The oral microbiota and its role in carcinogenesis. In: Seminars in cancer biology, vol 86. Academic, Cambridge, pp 633–642

Temraz S, Nassar F, Kreidieh F, Mukherji D, Shamseddine A, Nasr R (2021) Hepatocellular carcinoma immunotherapy and the potential influence of gut microbiome. Int J Mol Sci 22(15):7800

UniProt Consortium (2018) UniProt: the universal protein knowledgebase. Nucleic Acids Res 46(5):2699

van der Hee B, Wells JM (2021) Microbial regulation of host physiology by short-chain fatty acids. Trends Microbiol 29(8):700–712

Van Greevenbroek MM, Schalkwijk CG, Stehouwer CD (2013) Obesity-associated low-grade inflammation in type 2 diabetes mellitus: causes and consequences. Neth J Med 71(4):174–187

Vernocchi P, Del Chierico F, Putignani L (2020) Gut microbiota metabolism and interaction with food components. Int J Mol Sci 21(10):3688

Wei MY, Shi S, Liang C, Meng QC, Hua J, Zhang YY, Yu XJ (2019) The microbiota and microbiome in pancreatic cancer: more influential than expected. Mol Cancer 18(1):1–15

Wensel CR, Pluznick JL, Salzberg SL, Sears CL (2022) Next-generation sequencing: insights to advance clinical investigations of the microbiome. J Clin Investig 132(7):e154944

Wu M, Bai J, Ma C, Wei J, Du X (2021) The role of gut microbiota in tumor immunotherapy. J Immunol Res 2021:5061570

Zhao P, Xu A, Leung WK (2022) Obesity, bone loss, and periodontitis: the interlink. Biomolecules 12(7):865

Zomorrodi AR, Islam MM, Maranas CD (2014) d-OptCom: dynamic multi-level and multi-objective metabolic modeling of microbial communities. ACS Synth Biol 3(4):247–257

Understanding Colon Cancer: Causes, Prevention, and Complementary Therapies Including Therapeutic Foods, Beverages, and Meditation

Mahalakshmi S. Biradar, Shankar Thapa, Sonali S. Shinde, Aniket P. Sarkate, and Kalusing S. Padvi

Abstract

Colon cancer is a significant health concern globally, and understanding its biology, causes, prevention, and complementary therapies is crucial for effective management. This chapter provides an overview of colon cancer, exploring its biological aspects, common causes, preventive measures, and various complementary therapies, including therapeutic foods, beverages, and meditation. Colon cancer development is influenced by various variables, including family history, age group, inflammatory bowel diseases, genetic conditions, sedentary lifestyle, obesity, certain dietary habits, smoking, and excessive alcohol consumption. Understanding these factors helps people to take preventative actions, such as leading an active lifestyle, eating a balanced diet, and taking part in regular checkups for early diagnosis. Complementary therapies play a valuable role in the management of colon cancer, either as supportive measures alongside conventional treatments or as preventive approaches.

Keywords

Colon cancer · Diet · Meditation · Fibers · Green tea · Smoking

M. S. Biradar
Department of Pharmaceutical Chemistry, Hillside College of Pharmacy, Rajiv Gandhi University of Health Science, Bengaluru, India

S. Thapa
Department of Pharmacy, Universal College of Medical Sciences, Bhairahawa, Nepal

S. S. Shinde · A. P. Sarkate (✉) · K. S. Padvi
Department of Chemical Technology, Dr. Babasaheb Ambedkar Marathwada University, Aurangabad, Maharashtra, India

S. Joshi et al. (eds.), *Systems Biology Approaches: Prevention, Diagnosis, and Understanding Mechanisms of Complex Diseases*,
https://doi.org/10.1007/978-981-99-9462-5_18

18.1 Introduction

Colon cancer, often referred to as colorectal cancer, is a serious public health issue and the main reason for cancer-related deaths globally (Xi and Xu 2021). It arises from the lining of the colon or rectum and starts as little growths called polyps that may later develop into cancer (Marley and Nan 2016). Despite advancements in early identification and treatment, colon cancer continues to be a significant health threat, particularly in developing countries where lifestyle factors such as unhealthy diet, sedentary behaviour, and smoking are prevalent. Understanding the biology of colon cancer and developing innovative therapeutics have improved greatly in recent years, but there is still much to learn about the disease (Keum and Giovannucci 2019; Biradar et al. 2023).

Despite early detection and treatment advances, many patients still present with advanced disease and poor prognosis. Furthermore, there are disparities in screening and treatment access for certain populations, leading to unequal outcomes. Therefore, there is a pressing need for further research to improve the understanding of the biology, causes, treatment, and prevention of colorectal cancer and address issues of equity in access to care (Alcaraz et al. 2020). Understanding the molecular and genetic factors underlying the onset and spread of colon cancer is one significant research gap. While several genetic mutations that increase the risk of colon cancer have been identified, the full spectrum of genetic alterations and their interactions with environmental factors is not yet fully understood. Understanding and addressing the underlying causes of these disparities is essential to improving outcomes and reducing the burden of colon cancer on these populations (Elmore et al. 2021).

In recent years, complementary therapies, including therapeutic foods, beverages, and meditation, have garnered attention as potential supportive approaches in the management of colon cancer. Therapeutic foods and beverages, such as certain fruits (Berries and Oranges), vegetables, herbs, spices, and teas, possess bioactive compounds with potential anti-cancer properties. Incorporating these approaches into a well-balanced diet may help enhance the body's natural defence mechanism against colon cancer (Tabriz et al. 2021). Similarly, meditation and mindfulness practices have shown promise in reducing stress, improving overall well-being, and potentially supporting conventional cancer treatments (Mao et al. 2022).

18.2 Biology

Colon cancer typically starts as noncancerous growths called polyps on the inner lining of the colon or rectum. Over time, these polyps can develop into cancerous tumours. Although the specific cause of colon cancer is unknown, there are several risk factors that raise the possibility to develop the illness. These genetic mutations may be inherited from one's parents (germline mutations) or acquired throughout one's lifetime (somatic mutations) (Markowitz and Bertagnolli 2009).

18.2.1 Tumour Microenvironment

The surrounding cellular and non-cellular elements that interact with tumour cells in the colon are referred to as the tumour microenvironment in colon cancer. It consists of various cell types, extracellular matrix components, blood vessels, and immune cells, among others. The tumour microenvironment is essential for the growth, development, immune response, and response to therapy of tumours (Arneth 2020).

(a) Stromal cells: The tumour stroma consists of fibroblasts, myofibroblasts, and other connective tissue cells. These cells create extracellular matrix substances like collagen and fibronectin that support the tumour structurally and enhance the survival, migration, and invasion of tumour cells (Shiga et al. 2015).
(b) Immune cells: The immune cells within the tumour microenvironment include tumour-infiltrating lymphocytes (TILs), dendritic cells, myeloid-derived suppressor cells (MDSCs), and macrophages. The immune response in colon cancer is complex, with a balance between anti-tumor immune responses and immunosuppressive mechanisms. Immune cells can have both tumour-promoting and tumour-suppressing functions, and their interactions with tumour cells influence disease progression and treatment response (Hanus et al. 2021).
(c) Angiogenesis: Angiogenesis is a critical mechanism in tumour development and progression. In colon cancer, tumours promote the recruitment of endothelial cells and the development of new blood vessels to provide a sufficient blood supply for tumour growth. A key signalling molecule in colon cancer angiogenesis is vascular endothelial growth factor (VEGF) (Lugano et al. 2020).
(d) Extracellular matrix (ECM): The ECM provides structural support to the tumour and influences tumour cell behaviour. Alterations in the composition and organization of the ECM can affect tumour invasion, metastasis, and response to therapy (Arneth 2020).

18.2.2 Genetic Mutation

In the onset and spread of colon cancer, genetic abnormalities are a major factor. Some key genetic mutations associated with colon cancer are:

(a) Adenomatous Polyposis Coli (APC) gene mutation: APC is a gene that suppresses cancer growth and helps to limit cell division and proliferation. Defects in the APC gene can cause colon polyps to develop, which are benign growths that can later turn cancerous (Aghabozorgi et al. 2019).
(b) KRAS gene mutation: The KRAS gene is involved in cell signalling systems that regulate cell division and proliferation. KRAS gene mutations are often identified in colon cancer and it can result in uncontrolled cell proliferation and contribute to tumour development (Cicenas et al. 2017).

(c) TP53 gene mutation: The TP53 gene, also known as the p53 gene, is a tumour-suppressing gene that is essential for regulating DNA repair and cell cycle progression. A more severe type of colon cancer and a worse prognosis are connected with TP53 gene mutations (Nakayama and Oshima 2019).
(d) BRAF gene mutation: The BRAF gene participates in cell signalling pathways that are connected to cell survival and proliferation. A subtype of colon malignancies known as microsatellite stable (MSS) tumours frequently exhibits mutations in the BRAF gene, notably the V600E mutant. A worse prognosis and resistance to several therapies are linked to BRAF mutations (Hideshima and Anderson 2021).
(e) DNA mismatch repair (MMR) gene mutations: Correcting errors created during DNA replication requires the DNA MMR genes, including PMS2, MLH1, MSH6, and MSH2. Microsatellite instability (MSI), a condition that involves frequent DNA replication mistakes, can result from mutations in MMR genes. MSI is commonly observed in a subset of colon cancers called microsatellite instability-high (MSI-H) tumours (Li and Martin 2016).

18.2.3 Pathway and Signalling Molecules

Colon cancer involves alterations in various signalling pathways and molecules that regulate cell growth, division, apoptosis (cell death), and tumour progression.

(a) β-catenin/Wnt pathway: Controlling cell proliferation and differentiation is mostly governed by the Wnt/-catenin pathway. APC gene mutations or β-catenin activation can result in catenin stabilisation and nuclear translocation, which makes specific genes active that are important in cell cycle progression and cancer development (Kim et al. 2019).
(b) Ras/Raf/MEK/ERK pathway: The nucleus receives signals from receptors on the cell surface via this route, which helps cells develop and survive. Constitutive activation of this pathway can be triggered by mutations in the genes of KRAS or BRAF, which can result in uncontrolled cell growth and resistance to certain treatments (Fang and Richardson 2005).
(c) PI3K/AKT/mTOR pathway: This pathway regulates cell survival, growth, and metabolism. Activation of this pathway, often carried on by PIK3CA mutations or PTEN function loss, can promote tumour cell survival, proliferation, and resistance to apoptosis (Stefani et al. 2021).
(d) TGF-β signalling pathway: In early stages of cancer, the transforming growth factor-beta (TGF-) pathway suppresses tumour development, but as the disease progresses, it promotes tumour growth. In colon cancer, alterations in TGF-β signalling components can disrupt its tumour-suppressive effects and facilitate tumour growth and invasion (Itatani et al. 2019).
(e) Notch signalling pathway: Cell fate determination, differentiation, and self-renewal are all influenced by notch signalling. The notch pathway has been

found to be abnormally activated in colon cancer and is linked to tumour cell survival, proliferation, and resistance to apoptosis (Tyagi et al. 2020).

(f) MAPK/ERK pathway: Numerous biological processes, including cell division, proliferation, and survival, are regulated by the MAPK/ERK pathway. Dysregulation of this pathway, often through mutations or amplifications of genes such as EGFR, can contribute to tumour growth and progression (Tyagi et al. 2020).

These are just a few examples of the numerous pathways and molecules involved in colon cancer. The signalling pathways in cancer are complex and interrelated, and changes in several pathways frequently occur in colon cancer tumours. It is vital to remember this. Understanding the specific alterations in these pathways and molecules is important for developing targeted treatments and personalized therapies approaches for colon cancer patients.

18.3 Causes of CC

Colon cancer development can be influenced by several variables.

18.3.1 Age

Age affects colon cancer incidence, and those over 50 are more likely to be diagnosed with the disease. According to the American Cancer Society, persons over the age of 50 account for more than 90% of cases of colon cancer (Title et al. 2018). It is believed that age-related changes in the cells lining the colon, as well as accumulated damage to DNA over time, may contribute to the development of cancerous cells (Khansari et al. 2009). However, it should be noted that the findings of the most recent research showed that colorectal cancer occurrences are rising among young individuals (20–49 years old) in both the United States and Europe (Peterse et al. 2018).

18.3.2 Hereditary

Colon cancer development may be significantly influenced by hereditary factors. The chance of getting colon cancer can be raised by several inherited diseases, such as:

(a) Lynch syndrome (hereditary non-polyposis colorectal cancer): This is brought on by mutations in certain DNA repair genes (Carethers and Stoffel 2015). Lynch syndrome patients are far more likely to develop colon cancer, as well as endometrial, ovarian, stomach, and pancreatic cancer (Bucksch et al. 2020).

(b) Familial adenomatous polyposis (FAP): This disorder causes the colon and rectum to produce numerous polyps or thousands of them (Rengifo-Cam et al. 2016). People with FAP nearly invariably acquire colon cancer if they are not treated (Waller and Sarah Findeis 2016).
(c) MUTYH-associated polyposis (MAP): Mutations in the MUTYH gene enhance the likelihood of developing colon cancer in MAP patients (Kanth et al. 2017).
(d) Juvenile polyposis syndrome (JPS): In this genetic disorder, the colon and rectum frequently acquire many polyps at an early age (Kidambi et al. 2019).
(e) Peutz-Jeghers syndrome (PJS): In this health condition, pigmented patches on the skin and mucous membranes as well as intestinal polyps occur (Latchford and Clark 2022). Colon cancer risk is higher for those with PJS (Campos et al. 2015).

18.3.3 Diet

Diet has a significant impact on colon cancer development. A recent study has found that a high diet of red and processed meat raises the chance of developing colon cancer (Doaei et al. 2018). This may be due to the presence of compounds such as heme iron and nitrites/nitrates in these meats, which can damage the lining of the colon (Hammerling et al. 2016). A diet poor in fruits and vegetables, which are high in fibre, antioxidants, and vitamins, has been related to an increased risk of colon cancer (Song et al. 2015).

18.3.4 Sedentary Lifestyle

Colon cancer risk has been associated with a sedentary lifestyle and irregular exercise. Regular physical activity reduces inflammation and oxidative stress, which both have been associated with the occurrence of colon cancer, according to a number of studies (Sanchez et al. 2012; Perše 2013). In addition to lowering insulin levels and promoting regular bowel movements, exercise may also assist to enhance insulin sensitivity, which lowers the colon's exposure to potentially cancer-causing substances in waste products (Boyd 2003; American Cancer Society 2019).

18.3.5 Smoking and Alcoholism

Many malignancies, including colon cancer, are more likely to occur in persons who smoke or drink alcohol (Knekt et al. 1998). Smoking has been proven to promote inflammation and oxidative stress in the body, which may help malignant cells proliferate (Forman and Zhang 2021). The DNA in the cells lining the colon can be damaged by the carcinogens in cigarette smoke, and alcohol, which can result in the growth of cancerous cells (Li et al. 2014).

18.3.6 Inflammatory Bowel Movements

Inflammatory bowel disease (IBD), which includes ulcerative colitis and Crohn's disease, is characterized by chronic intestinal inflammation. IBD is a recognised risk factor for colon cancer because persistent inflammation can result in DNA damage and the growth of malignant cells (Ullman and Itzkowitz 2011). People with chronic and severe colitis have a higher chance of developing colon cancer (Herszényi et al. 2014). The likelihood of getting colon cancer is closely correlated with the length and severity of colon inflammation (Gupta et al. 2007) (Fig. 18.1).

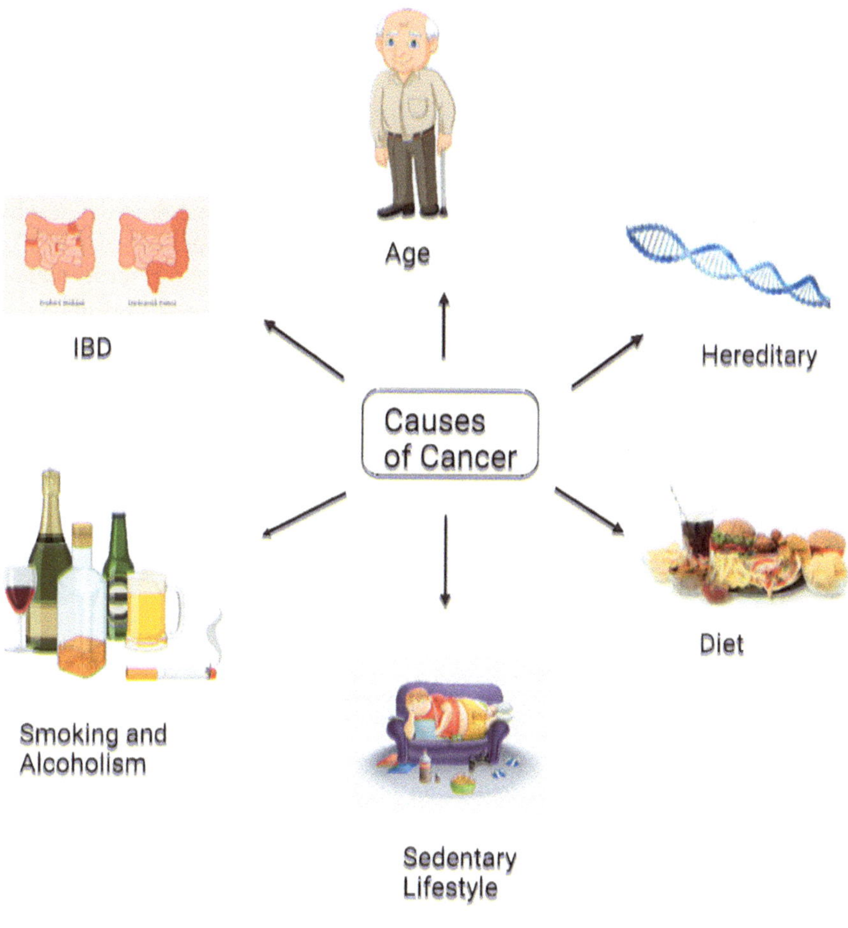

Fig. 18.1 Various causes of colon cancer

18.4 Prevention of CC

18.4.1 Screening

Screening for colon cancer can help to detect the disease at early stage when it is most treatable and can even prevent it from developing in the first place. The guidelines for screening vary depending on a person's age and risk factors for colon cancer. According to the American Cancer Society, people should begin routine colon cancer screening at age of 45 years (Wolf et al. 2018). The need for early screening or more frequent screening may be necessary for those with associated risks, such as a genetic background of colon cancer (Lowery et al. 2016).

18.4.2 Healthy Diet

A nutritious diet can be very helpful in preventing colon cancer. Include a variety of fruits, vegetables, whole grains, legumes, and nuts in diet as these foods are full of fibre, which helps in maintaining regular bowel movements and promote healthy colon (Baena and Salinas 2015; Slavin 2013). Choosing lean protein sources like chicken, fish, and plant-based proteins instead of red and processed meats, which have been associated with an elevated risk of colon cancer, is a good way to reduce the consumption of these foods (Baena and Salinas 2015).

Incorporating low-fat dairy products, leafy green vegetables, fortified plant-based milk, and calcium-rich foods into diet will reduce the possibility of colon cancer (Baron et al. 2015). Including sources of vitamin D containing foods in the diet, such as fatty fish, fortified dairy products, and sunlight exposure will help for maintaining a healthy immune system and may have a protective effect against colon cancer (Meeker et al. 2016).

18.4.3 Regular Exercise

Regular exercise has been shown to decrease inflammation in the body, including the colon, which may help lower the risk of colon cancer (Chan and Giovannucci 2010). Physical activity helps improve insulin sensitivity; high insulin levels and insulin resistance have been linked to an increased occurrence of colon cancer (Tsugane and Inoue 2010). By improving insulin sensitivity, exercise may help lower the risk of colon cancer (Komninou et al. 2003).

Regular exercise can speed up waste removal from the colon and help maintain regular bowel movements. This may assist to decrease the colon's exposure to possible carcinogens found in the faeces (Citronberg et al. 2014). By exercising regularly, one can maintain a healthy weight or reduce weight, which lowers the risk of colon cancer as obesity and excess body weight have been linked to an increase in the chance of colon cancer (Rock et al. 2020). Exercise contains immune-boosting properties that can aid the body's defences in identifying and eliminating malignant cells (Gleeson et al. 2004).

The American Cancer Society recommends adding 150 min of aerobics with moderate intensity or 75 min of intense aerobic exercise per week to strengthen muscles performed at least 2 days a week (Piercy et al. 2018). Incorporating regular exercise into lifestyle and maintaining an active routine is key for reaping the benefits and reducing the possibility of colon cancer.

18.4.4 Avoiding Usage of Tobacco and Alcohol

Limiting the use of tobacco and alcohol are crucial steps in the prevention of colon cancer as they are known for risk factor for colon cancer. The toxic substance in tobacco can enter the bloodstream by smoking and affect the colon directly, increasing the risk of DNA damage and the development of cancerous cells (Cao et al. 2015). Alcohol can cause oxidative stress and damage DNA, leading to the formation of cancer cells. It can also interfere with the body's ability to absorb certain nutrients that are essential for colon health (Klarich et al. 2015). Quitting smoking and limiting alcohol intake or avoiding it altogether significantly reduces the risk of colon cancer, as well as other smoking-related health issues (Tseng et al. 2010).

18.4.5 Treating Inflammatory Bowel Disease

Effective management of inflammatory bowel disease (IBD) is crucial for lowering the chance of colon cancer development in those who have this illness (Borody et al. 2019). While colon cancer incidence is increased by IBD, appropriate management and treatment can help minimize the associated risks. Various medications are available to manage IBD and reduce inflammation in the gastrointestinal tract (Troncone and Monteleone 2017) (Fig. 18.2).

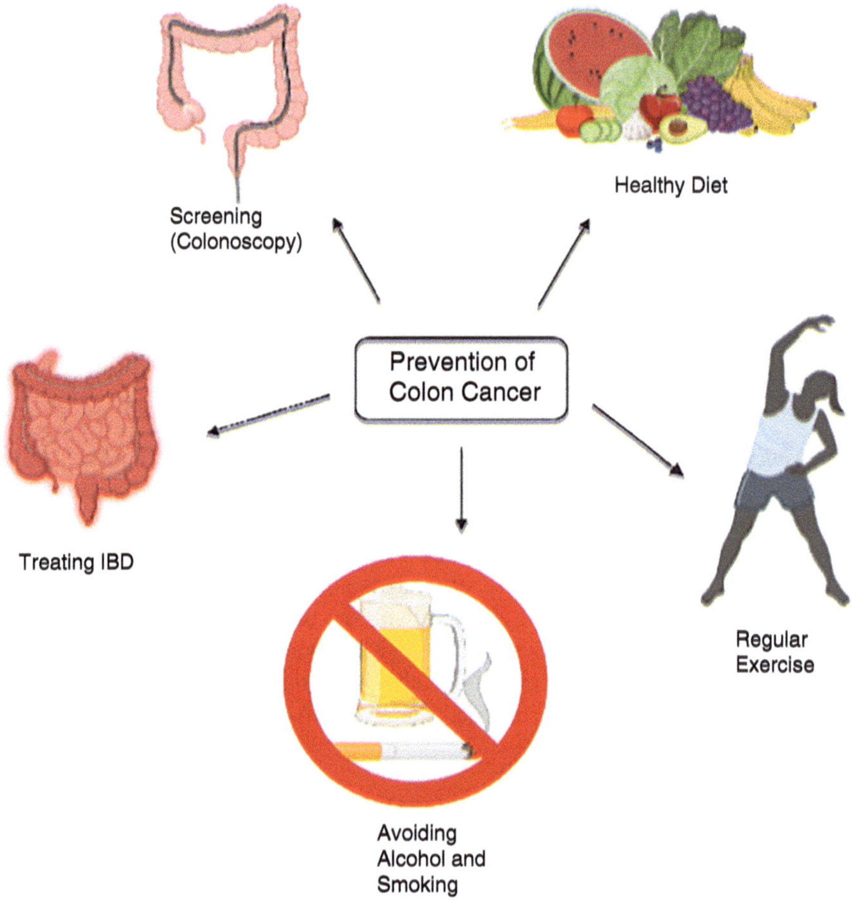

Fig. 18.2 Preventive measures for avoiding colon cancer risk

18.5 Therapeutic Food and Beverages

18.5.1 Fiber-Rich Foods

Fruits, vegetables, legumes, and whole grains, are examples of fibre-rich diets that can be helpful in the treatment and prevention of colon cancer. They are rich in antioxidants and phytochemicals, and they possess anti-inflammatory and anti-cancer properties, which helps to protect cells from damage and potentially reduce the possibility of colon cancer (Samtiya et al. 2021).

Adequate fibre intake can help maintain regular bowel movements, preventing constipation and promoting healthy digestion. This is important for reducing the exposure of the colon to potential carcinogens present in the stool and minimizing the time that waste products spend in contact with the colon lining (Gianfredi et al.

2018). Fiber serves as a prebiotic, providing nourishment and helps to encourage the development of good bacteria in the gut. These beneficial bacteria generate short-chain fatty acids (SCFAs) that have anti-inflammatory effects and support a healthy colon environment (Xu et al. 2022).

A variety of fibre-rich foods, like fruits (berries, apples, oranges, and bananas), vegetables (leafy greens, broccoli, Brussels sprouts, carrots, and bell peppers), whole grains (oats, brown rice, and quinoa), legumes (lentils, beans, and split peas), and nuts and seeds (almonds, flax seeds, chia seeds, and walnuts) incorporating these foods into the diet will decrease the possibilities of colon cancer.

18.5.2 Cruciferous Vegetables

The Brassicaceae family of vegetables, which includes cauliflower, broccoli, cabbage, Brussels sprouts, bok choy, and kale is known as the cruciferous vegetable group. These vegetables are rich in nutrients, good source of dietary fibre, and contain bioactive compounds such as Sulforaphane and Indole-3-Carbinol (I3C) in high amount that have been associated with potential therapeutic effects against colon cancer (Tao et al. 2018; Popolo et al. 2017). These compounds have been shown to help inhibit the growth of colon cancer cells, induce apoptosis in cancer cells, and reduce the risk of tumour formation in animal studies (Feitelson et al. 2015).

Cruciferous vegetables may help lessen the chance of colon cancer development because they reduce inflammation, which is a possibility for colon cancer (Ekenci et al. 2022). These vegetables compound that help to eliminate harmful toxins and free radicals that can damage cells and promote cancer development. This detoxification and antioxidant activity may contribute to the prevention of colon cancer (Pierro 2015).

18.5.3 Fruits and Berries

Fruits and berries are excellent sources of dietary fibre, which is beneficial for promoting regular bowel movements and maintaining a healthy digestive system. Adequate fibre intake is related with a lower risk of colon cancer and helps to nourish beneficial gut bacteria, supporting a healthy gut microbiota (Host 2018). Fruits and berries are packed with antioxidants, including vitamins C and E, flavonoids, and other phytochemicals. These antioxidants help neutralize harmful free radicals in the body, protecting cells from damage (Borek 2004).

Fruits and berries are rich sources of vitamins, minerals, and other phytochemicals. The phytochemicals like resveratrol has shown anti-inflammatory, anti-cancer effects, and ellagic acid in berries has demonstrated anti-cancer activity in laboratory studies (Ko et al. 2017). Adequate vitamin and mineral intake are essential for ensuring a healthy colon and minimising the risk of colon cancer (Ames and Wakimoto 2002).

18.5.4 Omega-3 Fatty Acids

Researchers have studied omega-3 fatty acids for their potential therapeutic effects in colon cancer prevention and therapy. These acids are mostly found in fatty fish, flaxseeds, chia seeds, and walnuts. They are most effective when part of a balanced diet and healthy lifestyle. Incorporating these foods into diet can be a good way to increase omega-3 fatty acid intake (Cockbain et al. 2012).

These Omega-3 fatty acids have various activity like anti-inflammatory which reduces the risk of chronic inflammation and it has an impact with a reduced risk of developing polyps, particularly adenomatous polyps, which have the potential to become cancerous over time. By reducing inflammation and inhibiting polyps formation, they are contributing to colon cancer prevention (Nabavi et al. 2015; Cockbain et al. 2012). It can influence gene expression and signalling pathways involved in apoptosis, cell growth, and differentiation (Borsini et al. 2020). Studies have shown that omega-3 fatty acids may help regulate gene expression and signalling pathways involved in cell growth, differentiation, and apoptosis, potentially it inhibit the growth of cancer cells (Chamberland and Moon 2015).

18.5.5 Garlic

A common culinary item, garlic, has been acknowledged for its potential medicinal benefits, particularly its involvement in the management and prevention of colon cancer (Hossain et al. 2022). Garlic contains bioactive compounds, particularly organosulfur compounds such as allicin, which have been shown to possess anticancer properties. In laboratory studies, these substances have shown to have a variety of effects, such as reducing the growth and proliferation of colon cancer cells, causing apoptosis in cancer cells, and delaying the development of tumours (Asemani et al. 2019; Gupta et al. 2010).

Garlic has been demonstrated anti-inflammatory properties by lowering the generation of inflammatory agents' substances in the body. By mitigating inflammation, garlic may help lower the risk of colon cancer development (Rayburn et al. 2009). It has immune-modulating properties, meaning it can help enhance immune function. A robust immune system is crucial for identifying and eliminating abnormal cells, including potentially cancerous cells. By supporting immune function, garlic may participate in the prevention and management of colon cancer (Song et al. 2015).

18.5.6 Turmeric

Turmeric, a vibrant yellow spice commonly used in Indian cuisine, has gained attention for its potential therapeutic properties, including its role in colon cancer prevention and management. The active component in turmeric, known as curcumin, is responsible for many of its health benefits. By blocking several inflammatory pathways in the body, curcumin has been found to have strong anti-inflammatory properties. Curcumin is also a powerful antioxidant that aids in the neutralisation of free radicals, which can result in cellular damage. By reducing inflammation, reducing oxidative stress, and protecting cells from damage, turmeric may contribute to colon cancer prevention (Boroumand et al. 2018).

Curcumin has shown anti-cancer benefits, such as preventing the development and division of colon cancer cells, triggering apoptosis in cancer cells by modifying several signalling pathways, and preventing the creation of blood vessels that feed tumours (Mansouri et al. 2020; Liczbiński et al. 2020).

18.5.7 Green Tea

Green tea is often considered a therapeutic beverage due to its potential benefits in preventing and managing various health conditions, including colon cancer. Green tea is rich in polyphenols, particularly catechins such as epigallocatechin gallate (EGCG) (Musial and Alicja Kuban-Jankowska 2020). Strong antioxidant and anti-inflammatory capabilities of these substances help to protect cells from free radical damage and decrease inflammation, respectively. By neutralizing harmful oxidative stress and reducing inflammation, green tea may help lower the chance of developing colon cancer and inhibit tumour growth and progression (Xu et al. 2010; Pisoschi et al. 2021).

Green tea polyphenols have been shown potential action to protect DNA from damage caused by environmental factors, carcinogens, and oxidative stress. Maintaining the integrity of DNA is essential for avoiding genetic changes that might lead to colon cancer (Navarro-Perán et al. 2007). Multiple signalling pathways involved in cell growth, inflammation, and apoptosis can be affected by green tea polyphenols. By modulating these pathways, green tea may help regulate the behaviour of colon cancer cells and inhibit their growth (Almatrood et al. 2020) (Fig. 18.3).

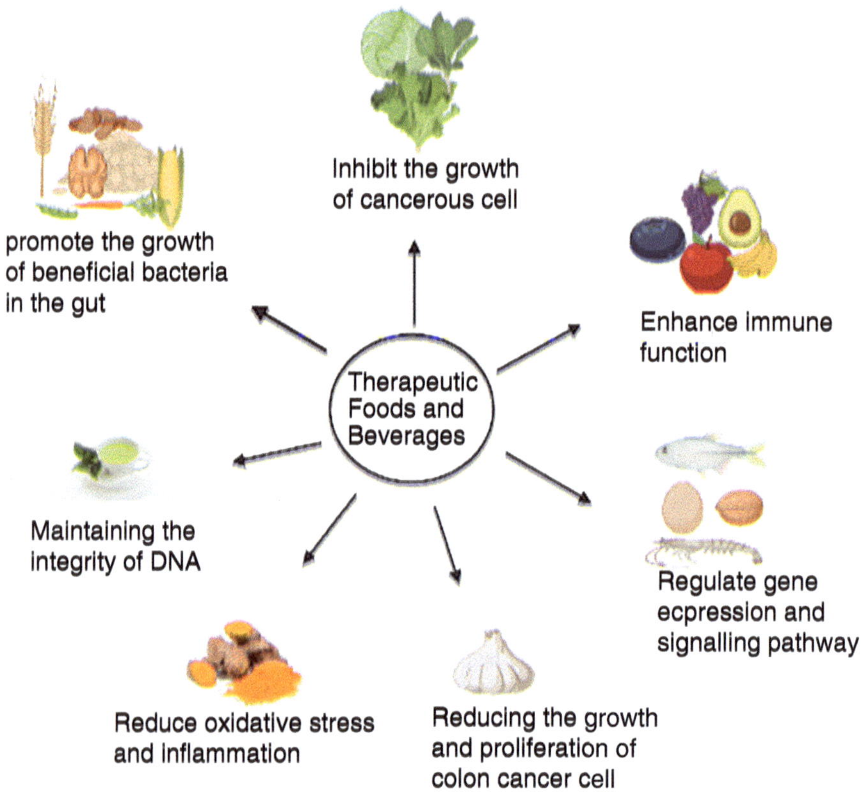

Fig. 18.3 Therapeutic foods, beverages, and their uses

18.6 Meditation in Colon Cancer

Meditation has shown to be a helpful practice in lowering stress for individuals with colon cancer. Meditation techniques, such as mindfulness meditation, promote the relaxation response in the body (Johns et al. 2016). By managing stress through meditation, individuals may experience improved sleep quality. This response is characterized by a state of deep relaxation, decreased stress hormone levels, lowered blood pressure, and reduced heart rate. The relationship between sleep and colon cancer is complex. Disrupted sleep patterns, sleep disorders, or poor sleep quality may affect the immune system, hormone regulation, and overall health, potentially impacting cancer development, progression, and treatment outcomes. Additionally, poor sleep can contribute to increased inflammation, which is linked to increase risk of colon cancer. By improving sleep quality, meditation may indirectly support overall health and potentially influence the underlying factors related to colon cancer development (Medic et al. 2017).

Meditation can be a beneficial practice for managing symptoms in individuals with colon cancer. While it may not directly treat the cancer itself, it can provide various physical and psychological benefits that can support symptom management and overall well-being. Meditation may have indirect effects on boosting the immune system in individuals with colon cancer. Chronic stress may harm the immune system by reducing immunological response. By reducing stress levels, meditation may help mitigate the impact of chronic stress on the immune system. Meditation has been found to have anti-inflammatory qualities and the capacity to modify inflammatory indicators in the body. By reducing inflammation, meditation may indirectly support immune health (Ali et al. 2013). Meditation can help people with colon cancer improve their quality of life. While it may not directly treat cancer itself, it can provide various physical, emotional, and psychological benefits that contribute to overall well-being (Kieviet-Stijnen et al. 2008).

18.7 Conclusion

Understanding colon cancer is a multifaceted endeavour encompassing its biology, causes, prevention strategies, and the potential role of complementary therapies in its management. The biology of colon cancer highlights the development and progression of abnormal growths in the colon or rectum, emphasizing the importance of unravelling cellular and genetic mechanisms for targeted interventions. Identifying the causes of colon cancer enables individuals and healthcare professionals to implement preventive measures effectively. Age, family history, lifestyle factors, and dietary habits increase the possibility of acquiring colon cancer. A healthy lifestyle, which includes frequent exercise, eating a balanced diet, and avoiding identified risk factors, can greatly minimise an individual's vulnerability to this illness.

In recent years, complementary therapies have gained recognition as supportive approaches in the management of colon cancer. Therapeutic foods, beverages, and mindfulness practices such as meditation offer potential benefits in enhancing the body's natural defences, reducing stress, and complementing conventional treatments. Incorporating these strategies into comprehensive cancer care may improve treatment outcomes and overall well-being. By integrating knowledge of the biology, causes, prevention strategies, and complementary therapies, individuals and healthcare professionals can take a holistic approach to combat colon cancer. This approach emphasizes the importance of early detection through regular screenings and emphasizes the adoption of healthy lifestyle choices. Additionally, complementary therapies provide opportunities to enhance conventional treatment outcomes and improve patients' quality of life.

Continued research and advancements in understanding colon cancer will pave the way for more targeted treatments and improved preventive strategies. Collaborative efforts between medical professionals, researchers, and individuals impacted by colon cancer are essential to furthering our understanding of this disease and reducing its global burden. In conclusion, by fostering a comprehensive understanding of colon cancer and implementing a multifaceted approach encompassing biology, causes, prevention, and complementary therapies, we can make significant strides in preventing and managing this disease. Through education, awareness, and collaborative efforts, we can improve outcomes for individuals affected by colon cancer and work towards a future with reduced incidence and improved quality of life.

References

Aghabozorgi AS, Bahreyni A, Soleimani A, Bahrami A, Khazaei M, Ferns GA, Avan A, Hassanian SM (2019) Role of adenomatous polyposis coli (APC) gene mutations in the pathogenesis of colorectal cancer; current status and perspectives. Biochimie 157:64–71. https://doi.org/10.1016/j.biochi.2018.11.003

Alcaraz KI, Wiedt TL, Daniels EC, Yabroff KR, Guerra CE, Wender RC (2020) Understanding and addressing social determinants to advance cancer health equity in the United States: a blueprint for practice, research, and policy. CA Cancer J Clin 70(1):31–46. https://doi.org/10.3322/caac.21586

Ali T, Choe J, Awab A, Wagener TL, Orr WC (2013) Sleep, immunity and inflammation in gastrointestinal disorders. World J Gastroenterol 19(48):9231. https://doi.org/10.3748/wjg.v19.i48.9231

Almatrood SA, Almatroudi A, Khan AA, Alhumaydh FA, Alsahl MA, Rahmani AH (2020) Potential therapeutic targets of epigallocatechin gallate (EGCG), the most abundant catechin in green tea, and its role in the therapy of various types of cancer. Molecules 25(14):3146. https://doi.org/10.3390/molecules25143146

American Cancer Society (2019) American Cancer Society guidelines on nutrition and physical activity for cancer prevention. Am Cancer Soc 62(1):30–67. https://doi.org/10.3322/caac.20140.Available

Ames BN, Wakimoto P (2002) Are vitamin and mineral deficiencies a major cancer risk? Nat Rev Cancer 2(9):694–704. https://doi.org/10.1038/nrc886

Arneth B (2020) Tumor microenvironment. Medicina 56(1):15. https://doi.org/10.3390/medicina56010015

Asemani Y, Zamani N, Bayat M, Amirghofran Z (2019) Allium vegetables for possible future of cancer treatment. Phytother Res 33(12):3019–3039. https://doi.org/10.1002/ptr.6490

Baena R, Salinas P (2015) Diet and colorectal cancer. Maturitas 80(3):258–264. https://doi.org/10.1016/j.maturitas.2014.12.017

Baron JA, Barry EL, Mott LA, Rees JR, Sandler RS, Snover DC, Bostick RM, Ivanova A, Cole BF, Ahnen DJ, Beck GJ, Bresalier RS, Burke CA, Church TR, Cruz-Correa M, Figueiredo JC, Goodman M, Kim AS, Robertson DJ, Summers RW (2015) A trial of calcium and vitamin D for the prevention of colorectal adenomas. N Engl J Med 373(16):1519–1530. https://doi.org/10.1056/nejmoa1500409

Biradar MS, Nargund SL, Thapa S (2023) Synthesis and cytotoxicity assay of aniline substituted thienopyrimidines for anti-colorectal cancer activity. Results Chem 5:100926. https://doi.org/10.1016/j.rechem.2023.100926

Borek C (2004) Dietary antioxidants and human cancer. Integr Cancer Ther 3(4):333–341. https://doi.org/10.1177/1534735404270578

Borody TJ, Eslick GD, Clancy RL (2019) Fecal microbiota transplantation as a new therapy: from Clostridioides difficile infection to inflammatory bowel disease, irritable bowel syndrome, and colon cancer. Curr Opin Pharmacol 49:43–51. https://doi.org/10.1016/j.coph.2019.04.017

Boroumand N, Samarghandian S, Hashemy SI (2018) Immunomodulatory, anti-inflammatory, and antioxidant effects of curcumin. J Herb Med Pharmacol 7(4):211–219. https://doi.org/10.15171/jhp.2018.33

Borsini A, Stangl D, Jeffries AR, Pariante CM, Thuret S (2020) The role of omega-3 fatty acids in preventing glucocorticoid-induced reduction in human hippocampal neurogenesis and increase in apoptosis. Transl Psychiatry 10(1):219. https://doi.org/10.1038/s41398-020-00908-0
Boyd DB (2003) Insulin and cancer. Integr Cancer Ther 2(4):315–329. https://doi.org/10.1177/1534735403259152
Bucksch K, Zachariae S, Aretz S, Büttner R, Holinski-Feder E, Holzapfel S, Hüneburg R, Kloor M, Von Knebel Doeberitz M, Morak M, Möslein G, Nattermann J, Perne C, Rahner N, Schmiegel W, Schulmann K, Steinke-Lange V, Strassburg CP, Vangala DB, Engel C (2020) Cancer risks in lynch syndrome, lynch-like syndrome, and familial colorectal cancer type X: a prospective cohort study. BMC Cancer 20(1):1–11. https://doi.org/10.1186/s12885-020-06926-x
Campos FG, Figueiredo MN, Martinez CAR (2015) Colorectal cancer risk in hamartomatous polyposis syndromes. World J Gastrointest Surg 7(3):25
Cao Y, Willett WC, Rimm EB, Stampfer MJ, Giovannucci EL (2015) Light to moderate intake of alcohol, drinking patterns, and risk of cancer: results from two prospective US cohort studies. BMJ 351:1–8. https://doi.org/10.1136/bmj.h4238
Carethers JM, Stoffel EM (2015) Lynch syndrome and Lynch syndrome mimics: the growing complex landscape of hereditary colon cancer. World J Gastroenterol 21(31):9253. https://doi.org/10.3748/wjg.v21.i31.9253
Chamberland JP, Moon HS (2015) Down-regulation of malignant potential by alpha linolenic acid in human and mouse colon cancer cells. Familial Cancer 14(1):25–30. https://doi.org/10.1007/s10689-014-9762-z
Chan AT, Giovannucci EL (2010) Primary prevention of colorectal cancer. Gastroenterology 138(6):2029–2043. https://doi.org/10.1053/j.gastro.2010.01.057
Cicenas J, Tamosaitis L, Kvederaviciute K, Tarvydas R, Staniute G, Kalyan K, Meskinyte-Kausiliene E, Stankevicius V, Valius M (2017) KRAS, NRAS and BRAF mutations in colorectal cancer and melanoma. Med Oncol 34:1–11. https://doi.org/10.1007/s12032-016-0879-9
Citronberg J, Kantor ED, Potter JD, White E (2014) A prospective study of bowel movement frequency, constipation, and laxative use on colorectal cancer risk. Am J Gastroenterol 109(10):1640–1649. https://doi.org/10.1038/ajg.2014.233
Cockbain AJ, Toogood GJ, Hull MA (2012) Omega-3 polyunsaturated fatty acids for the treatment and prevention of colorectal cancer. Gut 61(1):135–149. https://doi.org/10.1136/gut.2010.233718
Doaei S, Hajiesmaeil M, Aminifard A, Mosavi-Jarrahi SA, Akbari ME, Gholamalizadeh M (2018) Effects of gene polymorphisms of metabolic enzymes on the association between red and processed meat consumption and the development of colon cancer; a literature review. J Nutr Sci 7:e26. https://doi.org/10.1017/jns.2018.17
Ekenci D, Yılmaz B, Capasso R, Ozer D (2022) Cruciferous vegetables and their bioactive metabolites: from prevention to novel therapies of colorectal cancer. Evid Based Complement Alternat Med 2022:1534083. https://doi.org/10.1155/2022/1534083
Elmore LW, Greer SF, Daniels EC, Saxe CC, Melner MH, Krawiec GM, Cance WG, Phelps WC (2021) Blueprint for cancer research: critical gaps and opportunities. CA Cancer J Clin 71(2):107–139. https://doi.org/10.3322/caac.21652
Fang JY, Richardson BC (2005) The MAPK signalling pathways and colorectal cancer. Lancet Oncol 6(5):322–327. https://doi.org/10.1016/S1470-2045(05)70168-6
Feitelson MA, Arzumanyan A, Kulathinal RJ, Blain SW, Holcombe RF, Mahajna J, Marino M, Martinez-chantar ML, Nawroth R, Sanchez-garcia I, Sharma D, Saxena NK, Singh N, Vlachostergios PJ, Guo S, Honoki K, Fujii H, Georgakilas AG, Amedei A, Bhakta D (2015) Sustained proliferation in cancer: mechanisms and novel therapeutic targets. Semin Cancer Biol 35:S25–S54. https://doi.org/10.1016/j.semcancer.2015.02.006
Forman HJ, Zhang H (2021) Targeting oxidative stress in disease: promise and limitations of antioxidant therapy. Nat Rev Drug Discov 20(9):689–709. https://doi.org/10.1038/s41573-021-00233-1

Gianfredi V, Salvatori T, Villarini M, Moretti M, Nucci D, Realdon S (2018) Is dietary fibre truly protective against colon cancer? A systematic review and meta-analysis. Int J Food Sci Nutr 69(8):904–915. https://doi.org/10.1080/09637486.2018.1446917

Gleeson M, Nieman DC, Pedersen BK (2004) Exercise, nutrition and immune function. J Sports Sci 22(1):115–125. https://doi.org/10.1080/0264041031000140590

Gupta RB, Harpaz N, Itzkowitz S, Hossain S, Matula S, Kornbluth A, Bodian C, Ullman T (2007) Histologic inflammation is a risk factor for progression to colorectal neoplasia in ulcerative colitis: a cohort study. Gastroenterology 133(4):1099–1105. https://doi.org/10.1053/j.gastro.2007.08.001

Gupta SC, Kim JH, Prasad S, Aggarwal BB (2010) Regulation of survival, proliferation, invasion, angiogenesis, and metastasis of tumor cells through modulation of inflammatory pathways by nutraceuticals. Cancer Metastasis Rev 29(3):405–434. https://doi.org/10.1007/s10555-010-9235-2

Hammerling U, Bergman Laurila J, Grafström R, Ilbäck NG (2016) Consumption of red/processed meat and colorectal carcinoma: possible mechanisms underlying the significant association. Crit Rev Food Sci Nutr 56(4):614–634. https://doi.org/10.1080/10408398.2014.972498

Hanus M, Parada-Venegas D, Landskron G, Wielandt AM, Hurtado C, Alvarez K, Hermoso MA, López-Köstner F, De la Fuente M (2021) Immune system, microbiota, and microbial metabolites: the unresolved triad in colorectal cancer microenvironment. Front Immunol 12:612826. https://doi.org/10.3389/fimmu.2021.612826

Herszényi L, Barabás L, Miheller P, Tulassay Z (2014) Colorectal cancer in patients with inflammatory bowel disease: the true impact of the risk. Dig Dis 33(1):52–57. https://doi.org/10.1159/000368447

Hideshima T, Anderson KC (2021) Signaling pathway mediating myeloma cell growth and survival. Cancer 13(2):216. https://doi.org/10.3390/cancers13020216

Hossain MS, Kader MA, Goh KW, Islam M, Khan MS, Harun-Ar Rashid M, Ooi DJ, Melo Coutinho HD, Al-Worafi YM, Moshawih S, Lim YC, Kibria KMK, Ming LC (2022) Herb and spices in colorectal cancer prevention and treatment: a narrative review. Front Pharmacol 13(June):1–23. https://doi.org/10.3389/fphar.2022.865801

Host C (2018) The impact of dietary fiber on gut microbiota in host health and disease. Cell Host Microbe 23(6):705–715. https://doi.org/10.1016/j.chom.2018.05.012

Itatani Y, Kawada K, Sakai Y (2019) Transforming growth factor-β signaling pathway in colorectal cancer and its tumor microenvironment. Int J Mol Sci 20(23):5822. https://doi.org/10.3390/ijms20235822

Johns SA, Von Ah D, Brown LF, Beck-Coon K, Talib TL, Alyea JM, Monahan PO, Tong Y, Wilhelm L, Giesler RB (2016) Randomized controlled pilot trial of mindfulness-based stress reduction for breast and colorectal cancer survivors: effects on cancer-related cognitive impairment. J Cancer Surviv 10:437–448. https://doi.org/10.1007/s11764-015-0494-3

Kanth P, Grimmett J, Champine M, Burt R, Samadder NJ (2017) Hereditary colorectal polyposis and cancer syndromes: a primer on diagnosis and management. Am J Gastroenterol 112(10):1509–1525. https://doi.org/10.1038/ajg.2017.212

Keum NN, Giovannucci E (2019) Global burden of colorectal cancer: emerging trends, risk factors and prevention strategies. Nat Rev Gastroenterol Hepatol 16(12):713–732. https://doi.org/10.1038/s41575-019-0189-8

Khansari N, Shakiba Y, Mahmoudi M (2009) Chronic inflammation and oxidative stress as a major cause of age- related diseases and cancer. Recent Patents Inflamm Allergy Drug Discov 3(1):73–80. https://doi.org/10.2174/187221309787158371

Kidambi TD, Kohli DR, Samadder NJ, Singh A (2019) Hereditary polyposis syndromes. Curr Treat Options Gastroenterol 17:650–665. https://doi.org/10.1007/s11938-019-00251-4

Kieviet-Stijnen A, Visser A, Garssen B, Hudig W (2008) Mindfulness-based stress reduction training for oncology patients: patients' appraisal and changes in well-being. Patient Educ Couns 72(3):436–442. https://doi.org/10.1016/j.pec.2008.05.015

Kim WK, Kwon Y, Jang M, Park M, Kim J, Cho S, Jang DG, Lee WB, Jung SH, Choi HJ, Min BS, Il Kim T, Hong SP, Paik YK, Kim H (2019) β-Catenin activation down-regulates cell-cell junction-related genes and induces epithelial-to-mesenchymal transition in colorectal cancers. Sci Rep 9(1):1–15. https://doi.org/10.1038/s41598-019-54890-9

Klarich DS, Brasser SM, Hong MY (2015) Moderate alcohol consumption and colorectal cancer risk. Alcohol Clin Exp Res 39(8):1280–1291. https://doi.org/10.1111/acer.12778

Knekt P, Hakama M, Järvinen R, Pukkala E, Heliövaara M (1998) Smoking and risk of colorectal cancer. Br J Cancer 78(1):136–139. https://doi.org/10.1038/bjc.1998.455

Ko JH, Sethi G, Um JY, Shanmugam MK, Arfuso F, Kumar AP, Bishayee A, Ahn KS (2017) The role of resveratrol in cancer therapy. Int J Mol Sci 18(12):2589. https://doi.org/10.3390/ijms18122589

Komninou D, Ayonote A, Richie JP, Rigas B (2003) Insulin resistance and its contribution to colon carcinogenesis. Exp Biol Med 228(4):396–405. https://doi.org/10.1177/153537020322800410

Latchford AR, Clark SK (2022) Gastrointestinal aspects of Peutz-Jeghers syndrome. Best Pract Res Clin Gastroenterol 2022:101789. https://doi.org/10.1016/j.bpg.2022.101789

Li SKH, Martin A (2016) Mismatch repair and colon cancer: mechanisms and therapies explored. Trends Mol Med 22(4):274–289. https://doi.org/10.1016/j.molmed.2016.02.003

Li LF, Chan RLY, Lu L, Shen J, Zhang L, Wu WKK, Wang L, Hu T, Li MX, Cho CH (2014) Cigarette smoking and gastrointestinal diseases: the causal relationship and underlying molecular mechanisms (review). Int J Mol Med 34(2):372–380. https://doi.org/10.3892/ijmm.2014.1786

Liczbiński P, Michałowicz J, Bukowska B (2020) Molecular mechanism of curcumin action in signaling pathways: review of the latest research. Phytother Res 34(8):1992–2005. https://doi.org/10.1002/ptr.6663

Lowery JT, Ahnen DJ, Schroy PC, Hampel H, Baxter N, Boland CR, Burt RW, Butterly L, Doerr M, Doroshenk M, Feero WG, Henrikson N, Ladabaum U, Lieberman D, McFarland EG, Peterson SK, Raymond M, Samadder NJ, Syngal S, Smith R (2016) Understanding the contribution of family history to colorectal cancer risk and its clinical implications: a state-of-the-science review. Cancer 122(17):2633–2645. https://doi.org/10.1002/cncr.30080

Lugano R, Ramachandran M, Dimberg A (2020) Tumor angiogenesis: causes, consequences, challenges and opportunities. Cell Mol Life Sci 77:1745–1770. https://doi.org/10.1007/s00018-019-03351-7

Mansouri K, Rasoulpoor S, Daneshkhah A, Abolfathi S, Salari N, Mohammadi M, Rasoulpoor S, Shabani S (2020) Clinical effects of curcumin in enhancing cancer therapy: a systematic review. BMC Cancer 20(1):1–11. https://doi.org/10.1186/s12885-020-07256-8

Mao JJ, Pillai GG, Andrade CJ, Ligibel JA, Basu P, Cohen L, Khan IA, Mustian KM, Puthiyedath R, Dhiman KS (2022) Integrative oncology: addressing the global challenges of cancer prevention and treatment. CA Cancer J Clin 72(2):144–164. https://doi.org/10.3322/caac.21706

Markowitz SD, Bertagnolli MM (2009) Molecular basis of colorectal cancer. N Engl J Med 361(25):2449–2460. https://doi.org/10.1056/NEJMra0804588

Marley AR, Nan H (2016) Epidemiology of colorectal cancer. Int J Mol Epidemiol Genet 7(3):105

Medic G, Wille M, Hemels MEH (2017) Short- and long-term health consequences of sleep disruption. Nat Sci Sleep 9:151–161. https://doi.org/10.2147/NSS.S134864

Meeker S, Seamons A, Maggio-price L, Paik J, Meeker S, Seamons A, Maggio-price L (2016) Protective links between vitamin D, inflammatory bowel disease and colon cancer. World J Gastroenterol 22(3):933–948. https://doi.org/10.3748/wjg.v22.i3.933

Musial C, Alicja Kuban-Jankowska MG-P (2020) Beneficial properties of green tea catechins. Int J Mol Sci 21(5):1744. https://doi.org/10.3390/ijms21051744

Nabavi SF, Bilotto S, Russo GL, Orhan IE, Habtemariam S, Daglia M, Devi KP, Loizzo MR, Tundis R, Nabavi SM (2015) Omega-3 polyunsaturated fatty acids and cancer: lessons learned from clinical trials. Cancer Metastasis Rev 34:359–380. https://doi.org/10.1007/s10555-015-9572-2

Nakayama M, Oshima M (2019) Mutant p53 in colon cancer. J Mol Cell Biol 11(4):267–276. https://doi.org/10.1093/jmcb/mjy075

Navarro-Perán E, Cabezas-Herrera J, del Campo LS, Rodríguez-López JN (2007) Effects of folate cycle disruption by the green tea polyphenol epigallocatechin-3-gallate. Int J Biochem Cell Biol 39(12):2215–2225. https://doi.org/10.1016/j.biocel.2007.06.005

Perše M (2013) Oxidative stress in the pathogenesis of colorectal cancer: cause or consequence? Biomed Res Int 2013:725710. https://doi.org/10.1155/2013/725710

Peterse EFP, Meester RGS, Siegel RL, Chen JC, Dwyer A, Ahnen DJ, Smith RA, Zauber AG, Lansdorp-Vogelaar I (2018) The impact of the rising colorectal cancer incidence in young adults on the optimal age to start screening: Microsimulation analysis I to inform the American Cancer Society colorectal cancer screening guideline. Cancer 124(14):2964–2973. https://doi.org/10.1002/cncr.31543

Piercy KL, Troiano RP, Ballard RM, Carlson SA, Fulton JE, Galuska DA, George SM, Olson RD (2018) The physical activity guidelines for Americans. JAMA 320(19):2020–2028. https://doi.org/10.1001/jama.2018.14854

Pierro FD (2015) Antioxidants and cancer: a debate on prevention, progression, hormesis, and cruciferous vegetables. Forum Nutr 14(4):175–179. https://doi.org/10.1007/s13749-015-0064-3

Pisoschi AM, Pop A, Iordache F, Stanca L, Predoi G, Serban AI (2021) Oxidative stress mitigation by antioxidants - an overview on their chemistry and influences on health status. Eur J Med Chem 209:112891. https://doi.org/10.1016/j.ejmech.2020.112891

Popolo A, Pinto A, Daglia M, Fazel S, Ahmad A, Rastrelli L (2017) Two likely targets for the anti-cancer effect of indole derivatives from cruciferous vegetables: PI3K/Akt/mTOR signalling pathway and the aryl hydrocarbon receptor. Semin Cancer Biol March:1–6. https://doi.org/10.1016/j.semcancer.2017.06.002

Rayburn ER, Ezell SJ, Zhang R (2009) Anti-inflammatory agents for cancer therapy. Mol Cell Pharmacol 1(1):29–43. https://doi.org/10.4255/mcpharmacol.09.05

Rengifo-Cam W, Jasperson KW, Burt RW (2016) Familial adenomatous polyposis. In: Intestinal polyposis syndromes: diagnosis and management. Springer, Cham, pp 173–195. https://doi.org/10.1007/978-3-319-28103-2

Rock CL, Thomson C, Gansler T, Gapstur SM, McCullough ML, Patel AV, Andrews KS, Bandera EV, Spees CK, Robien K, Hartman S, Sullivan K, Grant BL, Hamilton KK, Kushi LH, Caan BJ, Kibbe D, Black JD, Wiedt TL, Doyle C (2020) American Cancer Society guideline for diet and physical activity for cancer prevention. CA Cancer J Clin 70(4):245–271. https://doi.org/10.3322/caac.21591

Samtiya M, Aluko RE, Dhewa T, Moreno-Rojas JM (2021) Potential health benefits of plant food-derived bioactive components: an overview. Foods 10(4):839. https://doi.org/10.3390/foods10040839

Sanchez NF, Stierman B, Saab S, Mahajan D, Yeung H, Francois F (2012) Physical activity reduces risk for colon polyps in a multiethnic colorectal cancer screening population. BMC Res Notes 5:1–8. https://doi.org/10.1186/1756-0500-5-312

Shiga K, Hara M, Nagasaki T, Sato T, Takahashi H, Takeyama H (2015) Cancer-associated fibroblasts: their characteristics and their roles in tumor growth. Cancer 7(4):2443–2458. https://doi.org/10.3390/cancers7040902

Slavin J (2013) Fiber and prebiotics: mechanisms and health benefits. Nutrients 5(4):1417–1435. https://doi.org/10.3390/nu5041417

Song M, Garrett WS, Chan AT (2015) Nutrients, foods, and colorectal cancer prevention. Gastroenterology 148(6):1244–1260. https://doi.org/10.1053/j.gastro.2014.12.035

Stefani C, Miricescu D, Stanescu-Spinu I-I, Nica RI, Greabu M, Totan AR, Jinga M (2021) Growth factors, PI3K/AKT/mTOR and MAPK signaling pathways in colorectal cancer pathogenesis: where are we now? Int J Mol Sci 22(19):10260. https://doi.org/10.3390/ijms221910260

Tabriz ER, Ramezani M, Heydari A, Aledavood SA (2021) Health-promoting lifestyle in colorectal cancer survivors: a qualitative study on the experiences and perspectives of colorectal cancer survivors and healthcare providers. Asia Pac J Oncol Nurs 8(6):696–710. https://doi.org/10.4103/apjon.apjon-2132

Tao J, Li S, Li H (2018) Plant foods for the prevention and management of colon cancer. J Funct Foods 42:95–110. https://doi.org/10.1016/j.jff.2017.12.064

Title S, Missing T, Facts CC (2018) Screening for colon and rectal cancer in average-risk adults. CA Cancer J Clin 68(4):282–283. https://doi.org/10.3322/caac.21458

Troncone E, Monteleone G (2017) The safety of non-biological treatments in ulcerative colitis. Expert Opin Drug Saf 16(7):779–789. https://doi.org/10.1080/14740338.2017.1340936

Tseng TS, Lin HY, Martin MY, Chen T, Partridge EE (2010) Disparities in smoking and cessation status among cancer survivors and non-cancer individuals: a population-based study from National Health and Nutrition Examination Survey. J Cancer Surviv 4(4):313–321. https://doi.org/10.1007/s11764-010-0127-9

Tsugane S, Inoue M (2010) Insulin resistance and cancer: epidemiological evidence. Cancer Sci 101(5):1073–1079. https://doi.org/10.1111/j.1349-7006.2010.01521.x

Tyagi A, Sharma AK, Damodaran C (2020) A review on notch signaling and colorectal cancer. Cell 9(6):1549. https://doi.org/10.3390/cells9061549

Ullman TA, Itzkowitz SH (2011) Intestinal inflammation and cancer. Gastroenterology 140(6):1807–1816. https://doi.org/10.1053/j.gastro.2011.01.057

Waller A, Sarah Findeis MJL (2016) Familial adenomatous polyposis. J Pediatric Genet 5(2):78–83. https://doi.org/10.1055/s-0036-1579760

Wolf AMD, Fontham ETH, Church TR, Flowers CR, Guerra CE, LaMonte SJ, Etzioni R, McKenna MT, Oeffinger KC, Shih Y-CT, Walter LC, Andrews KS, Brawley OW, Brooks D, Fedewa SA, Manassaram-Baptiste D, Siegel RL, Wender RC, Smith RA (2018) Colorectal cancer screening for average-risk adults: 2018 guideline update from the American Cancer Society. CA Cancer J Clin 68(4):250–281. https://doi.org/10.3322/caac.21457

Xi Y, Xu P (2021) Global colorectal cancer burden in 2020 and projections to 2040. Transl Oncol 14(10):101174. https://doi.org/10.1016/j.tranon.2021.101174

Xu Y, Zhang J, Xiong L, Zhang L, Sun D, Liu H (2010) Green tea polyphenols inhibit cognitive impairment induced by chronic cerebral hypoperfusion via modulating oxidative stress. J Nutr Biochem 21(8):741–748. https://doi.org/10.1016/j.jnutbio.2009.05.002

Xu T, Wu X, Liu J, Sun J, Wang X, Fan G, Meng X, Zhang J, Zhang Y (2022) The regulatory roles of dietary fibers on host health via gut microbiota-derived short chain fatty acids. Curr Opin Pharmacol 62:36–42. https://doi.org/10.1016/j.coph.2021.11.001

The Potential of Systems Biology to Understand the Tenets of Host-pathogen Interaction, Toxicology, and Aging

19

Koushik Nayak, Sumanta Daw, and Pratick Singha

Abstract

This chapter highlights the potential of systems biology in elucidating the fundamental principles underlying host-pathogen interaction, toxicology, and aging. Systems biology, an interdisciplinary field, integrates computational modelling, data analysis, and experimental techniques to provide a holistic understanding of complex biological systems. By integrating various layers of biological information, including genomics, proteomics, metabolomics, and interactomics, systems biology offers a comprehensive framework to decipher the intricate mechanisms involved in these processes. Host-pathogen interactions are dynamic and influenced by a myriad of factors. Systems biology enables the integration of large-scale omics data, facilitating the identification of key molecular players and regulatory networks. By employing computational modelling, systems biology can simulate and predict the behaviour of complex host-pathogen systems, aiding in the development of novel therapeutic strategies and effective vaccines. Toxicology, which investigates the adverse effects of chemicals on living organisms, plays a crucial role in risk assessment and the development of safer substances. Systems biology approaches enable the integration of diverse data types, such as gene expression profiles, signalling pathways, and toxicokinetic infor-

K. Nayak (✉)
Department of Medical Biotechnology, Ramakrishna Mission Vivekananda Educational and Research Institute, Kolkata, West Bengal, India

S. Daw
Department of Microbiology, Maulana Abul Kalam Azad University of Technology, Nadia, West Bengal, India

P. Singha
Department of Marine Biotechnology, Bharathidasan University, Tiruchirappalli, Tamil Nadu, India

S. Joshi et al. (eds.), *Systems Biology Approaches: Prevention, Diagnosis, and Understanding Mechanisms of Complex Diseases*,
https://doi.org/10.1007/978-981-99-9462-5_19

mation. This integration allows for a comprehensive understanding of toxicological processes, aiding in the assessment of chemical toxicity, prediction of adverse effects, and identification of early detection biomarkers. Aging is a complex biological process influenced by various genetic, environmental, and lifestyle factors. Systems biology provides a systems-level understanding of aging by integrating multi-omics data and considering the interplay between different cellular processes. By identifying molecular signatures associated with aging, systems biology unravels key mechanisms and pathways involved in the aging process. This knowledge has implications for developing interventions to delay aging and age-related diseases.

Keywords

System biology · Aging · Host-pathogen interaction · Toxicology

19.1 Introduction

In recent years, the field of biology has witnessed a remarkable shift towards a more comprehensive and integrative approach known as systems biology. This paradigm shift recognizes the intricate interconnectedness of biological systems and aims to understand their behaviour and functionality as a whole, rather than studying individual components in isolation (Dix et al. 2016). Systems biology has proven to be particularly valuable in unravelling the complexities of host-pathogen interactions, toxicology, and aging.

Host-pathogen interactions lie at the heart of infectious diseases, shaping the outcomes of infection and influencing the strategies pathogens employ to evade the host's immune system. Traditional reductionist approaches have yielded valuable insights into specific aspects of these interactions. However, they often fail to capture the intricate network of molecular events that occur across multiple scales and dynamic time frames. Systems biology, with its ability to integrate diverse data types and construct predictive models, offers a powerful framework to decipher the underlying principles governing host-pathogen dynamics. By combining genomics, transcriptomics, proteomics, and metabolomics data with computational modelling and network analysis, systems biology enables researchers to gain a holistic understanding of these interactions (Pinu et al. 2019). This knowledge can pave the way for the development of more effective therapeutic interventions and preventive strategies against infectious diseases.

Toxicology, on the other hand, deals with the study of the adverse effects of chemicals and substances on living organisms. Traditional toxicological approaches primarily focus on understanding the effects of individual toxicants and their mechanisms of action. However, organisms are exposed to numerous chemicals simultaneously, leading to complex interactions and potential synergistic or antagonistic effects. Systems biology offers a framework to comprehensively assess the impact of multiple toxicants on biological systems (Aguayo-Orozco et al. 2019). By integrating high-throughput data, such as transcriptomics, proteomics, and metabolomics, with computational modelling and advanced statistical analyses, systems toxicology provides a holistic understanding of the complex interactions between toxicants and biological pathways (Sturla et al. 2014). This approach has the potential to identify biomarkers of toxicity, predict adverse outcomes, and support the development of safer chemicals and pharmaceuticals.

Aging is a complex biological process characterized by a progressive decline in physiological function and an increased susceptibility to diseases. Traditional approaches to studying aging have often focused on individual molecular pathways or specific age-related diseases (Guo and Chiang 2022). However, aging is a multifaceted phenomenon influenced by a wide array of interconnected genetic, epigenetic, and environmental factors. Systems biology provides a powerful toolset to unravel the underlying mechanisms and dynamics of aging. By integrating data from genomics, transcriptomics, proteomics, and metabolomics, along with computational modelling and network analysis, systems biology can help identify key molecular drivers of aging and understand how they interact within complex biological networks (Dhillon et al. 2020). Furthermore, systems biology approaches enable the identification of potential interventions to promote healthy aging and delay the onset of age-related diseases.

In this chapter, we will explore the potential of systems biology to unravel the tenets of host-pathogen interactions, toxicology, and aging. We will discuss the conceptual framework of systems biology, highlight its key methodologies and computational tools, and provide examples of its applications in these fields. By embracing the systems biology approach, researchers can gain deeper insights into the intricate dynamics of host-pathogen interactions, toxicological responses, and the aging process, ultimately leading to improved disease prevention, therapeutic strategies, and interventions to promote healthy aging.

19.2 Doctrine of Host-Pathogen Interaction in View of System Biology

As the human civilization is updating day by day, there are several threats that are also increasing with it. One of the common threats is related to pathogens. Understanding the defence mechanism of the host system is very essential to know more about how it works against several pathogens. The host system might consist of several mechanism, by which it can eradicate the pathogen from its body. Sometimes this mechanism also might not work. These all possibilities along with the response towards pathogen makes the 'Host-Pathogen Interaction'. Infectious diseases are one of the major causes of majority of deaths worldwide every year. The changes in pathogen's mechanism makes them way stronger to fight against several drugs. Therefore, the need of several novel therapeutic techniques is also increasing and they are known as 'theranostics'. This technique relies on systems biological approach to understand the underlying mechanism between the interaction of pathogen and its targeted host (Durmuş et al. 2015). Systems biology is a part of interdisciplinary science that mainly focused on non-linear interaction between the biological entities with the help of biomedical, computational, mathematical, and engineering approaches (Fig. 19.1).

It focuses on studying the non-linear interactions among biological entities by using omics data, which includes genomics, transcriptomics, proteomics, and metabolomics. The traditional approach to studying biological systems is reductionism, which focuses on studying a few molecules and interactions. In contrast, systems biology takes a more holistic approach by modelling biological phenomena using large-scale data. This approach allows for a better understanding of the complex interactions between molecules and systems in living organisms. One of the major ideas is to study about the protein-protein interaction between the host and the pathogen to understand the phenomena better (Stebbins 2005; Durmuş et al. 2015). Other than the protein data, several ncRNA and other metabolites also plays a crucial role in host-pathogen interaction (Gottwein and Cullen 2008; Skalsky and Cullen 2010; Eisenreich et al. 2013; Saayman et al. 2014).

From future perspective, it is important to collect various levels of omics data from the pathogen and its targeted host and by using several bioinformatic tool it is possible to construct a brief idea and analysis of different levels of interaction that is happening inside of the host system. These may include gene-gene interaction,

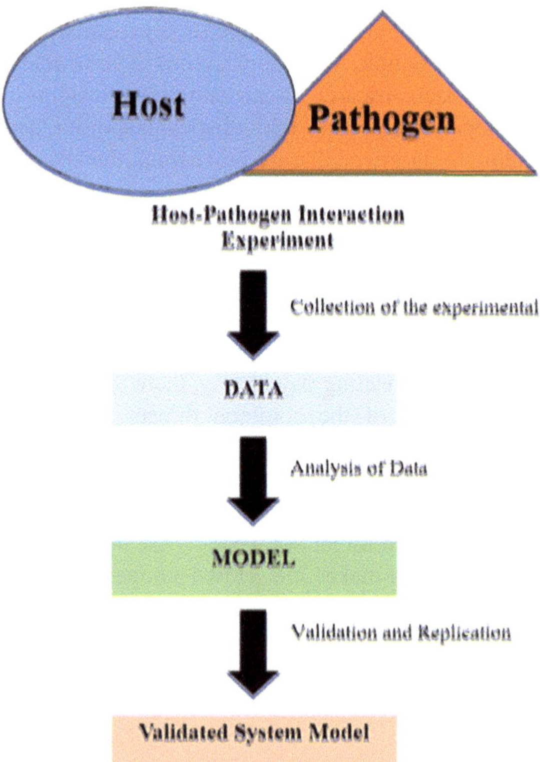

Fig. 19.1 Schematic modelling system for pathogen-host interaction (PHI)

protein-protein interaction, and different levels of gene expression (Westermann et al. 2012; Schulze et al. 2015).

19.2.1 Influence of Whole Genome and Genome Sequencing in Pathogen-Host Interaction (phi)

In the field of genetical study, it is very important to know the whole genome preposition of an organism. This data is very useful for further investigation related to phi. This could be helpful to identify the exact gene and the exact changes related to gene that is happening during the phi. Until a time period the data about few genes were known but it was very important to mine the whole genome data. It was then further done by shotgun sequencing in *Haemophilus influenzae* Rd (Fleischmann et al. 1995). This discovery opened a new horizon in the area of genomics and related studies. Later a numerous number of genome data has been extracted from various organisms. These data sets also allowed to dig more in genome-wide study and later various studies like targeted mutagenesis, genome signature, microarray technology, and various sequencing methods includes RNA sequencing (Leonard et al. 2003; Opijnen et al. 2009; Perkins et al. 2013). From the host point of view, the first breakthrough was the development of human genome sequencing. At first it was not enough to describe the whole structure of host-pathogen interaction but later studies on knock-out mouse, zebrafish, and generation of transgenic mice, these all experiments were significant landmarks in host-genome interaction mechanism (Adams et al. 2015; Howe et al. 2013; White et al. 2013).

As the researches on whole genome was going on, it opened a new era in the field of genome sequencing. Within a short span of time the different genome sequencing came to an affordable level. Among all the type of sequencing method, the next-generation sequencing probably have the more impact than others. As the cost of next-generation sequencing dropped by time and its efficiency increased, it was more accessible to the broader research community. These inventions were not only helpful for genome study but are also very useful in construct of phylogenetic tree, matching the whole genome of different organisms and others (Holt et al. 2008; Chan and Ragan 2013; Klemm and Dougan 2016). In addition, later genome-wide association studies (GWAS) are now also possible in bacteria (Chewapreecha et al. 2014; Sheppard et al. 2013). The whole genome-based studies of organisms are also very helpful in the study of phylogeographical analysis. It allows to monitor the availability of different pathogens in the given geographical area in real time. Phylogeographical analysis is used to monitor the spread of different pathogens in different parts of the world including different health sector and different habitats. This technology is a breakthrough in the field of prophylaxis (Klemm and Dougan 2016; Wong et al. 2015; Nicolas-Chanoine et al. 2014; Okoro et al. 2012; Kaur et al. 2019; Chewapreecha et al. 2014; Sheppard et al. 2013; Alam et al. 2014).

19.2.2 Screening of Genome

In Host The most common host system is the human system; thus, it is very important to know all the preposition of human genome. The availability of data makes it easier to find out the most valuable genes that is linked with pathogen interaction. The identification of genes and particular loci is the main purpose of the genome screening in human and it helps to identify the susceptibility of the human body towards any pathogen-borne disease (Huang et al. 2017). The GWAS is based in a way that it uses the chromosome mapping and genotyping of populations to identify whether there is a link present between a particular disease in a particular area and its host's response. By this practice and in a minimal time with a low cost, it is possible to target a wide range of diseases ranging from diabetes to IBD and their association with microbial world (Walker et al. 2011). In such diseases a variety of genetic changes are also present. GWAS is used to link between the genetic changes with particular diseases that is caused by any pathogens. GWAS is a useful weapon to link between the host and the pathogen but in other hand it also has some limitations. The poor diagnosis or incomplete diagnosis can complicate the function of GWAS, some common diseases can be linked with multiple pathogens and our body respond differently to the pathogens. So, it is obvious that the gene expression will also be different in these cases (Goh and Knight 2017).

In Pathogens The analysis of pathogen genomes has emerged as a critical aspect of contemporary research. By examining the genetic information of pathogens, researchers aim to identify and comprehend various elements such as virulence factors and potential drug targets. Genome-wide screening provides a comprehensive exploration of the entire genetic composition of pathogens, facilitating the identification of crucial genes and their functions. The insights obtained from genome screening significantly contribute to the development of targeted interventions, encompassing vaccines and strategies to combat infectious diseases effectively. Following the acquisition of pathogen genome data, our research focus shifted towards genome-wide association screening in microbes to pinpoint virulent strains. In earlier methodologies, this process involved constructing a library of knockout mutations within the bacterial genome. Notably, we successfully generated libraries for *Mycobacterium tuberculosis* and *Salmonella typhimurium*, as detailed by Porwollik et al. in 2014. These libraries now stand as valuable tools for a thorough examination of the entire genome or a focused investigation into specific genes and their expressions. This research enhances our understanding and contributes to ongoing efforts in combating infectious diseases.

Generating a mobile element such as transposons are another kind of genome-wide screening technique in pathogens. Transposon is generated within a bacterial cell and it contains the whole genome data. After the library generation is completed, it contains numerous numbers of transposon that contain the genome data and also, they can be used for further experiments. These sequence then further gives data about the insertion sequence and adjacent sequence. Some special terms such as 'Tn-Seq' and 'TraDIS' are used to describe these processes (Langridge et al. 2009; Troy et al. 2016; Pickard et al. 2013).

19.2.3 Screening of Host Cell

As of now, several types of model organisms such as mice, flies, and worms are being used for the human cell screening because linking genotype to phenotype in human itself is a challenging work. To do the screening in model organisms, human cell lines are assembling to broadly represent human genetic diversity. One of these collections are HapMap lymphoblastoid of collection of B cell lines that represent the MHC diversity in humans. In a laboratory study where the HapMap approach was fused with *S. typhimurium* cell invasion assays, it shows that lack of function allele of a gene CARD8 and also associated with increased cell death (Ko et al. 2009, 2012; Miller and Chaudhary 2016; Salinas et al. 2014).

RNAi or RNA interference technology is a key technique to study the genetic screening of human cell. RNAi is linked with mRNA thus it helps to study the biological functions of human genes by mRNA (Fire et al. 1998). The combination of portable gene delivery, known gene knockdown leads to study towards cancer biology, cell biology, signal transduction, and cellular response to pathogen. One of the major limitations in RNAi technology is that sometimes it leads to partial silencing of the mRNA and sometimes it can also target the non-targeted gene (Sasaki et al. 2017). Apart from RNAi technology, one of the most recent techniques are CRISPR-Cas9 mediated gene manipulation. It provides a high amount of efficacy in case of gene knockout (Sun et al. 2017). This technique can be used in gene manipulation by targeting the particular protein coding sequence, lncRNA, mRNA, and so on.

19.2.4 Study of Challenges

An interesting area where systems biology could have an impact is that of controlled human infection models (CHIMs) with pathogens and vaccines. As technology has developed in such a way that nowadays a small amount of sample can be used to diagnose the patient. In addition, it is nowadays possible to screen the DNA or the RNA from a small amount of sample along with several other screening that are related to the diagnosis process. Collection data from an infected human sample is the ultimatum of host-pathogen interaction in CHIM. Samples can be taken directly from body fluid and can be processed further by genome sequencing, ELISA, RIA, several protein assays, proteomics, metabolomics, and the data can be stored for further studies. This approach is nowadays using in production of vaccines (Akondy et al. 2017; Kazmin et al. 2017).

19.3 Postulation of Toxicology in View of System Biology

Toxicology is the scientific study of adverse effects that substances or agents can have on living organisms. Traditional approaches in toxicology focus on understanding the mechanisms by which individual chemicals exert their toxic effects. However, the complexity of biological systems and the interactions between chemicals and organisms present challenges for conventional toxicology methods. Systems biology, an interdisciplinary field that integrates biology, computational modelling, and data analysis, offers new opportunities to comprehensively understand toxicological processes (Kitano 2002; van der Greef et al. 2010). This chapter explores the potential of systems biology in unravelling the intricacies of toxicology.

19.3.1 Network-Based Approaches in Toxicology

Systems biology employs network-based approaches to understand the behaviour of biological systems as a whole. Biological systems consist of interconnected networks of molecules and genes, and toxic agents can perturb these networks, leading to adverse effects (Barabási et al. 2011). By constructing and analysing these networks, systems biologists can identify key nodes, interactions, and pathways that play a crucial role in toxicological responses.

For example, the reconstruction of gene regulatory networks can elucidate how toxicants alter gene expression patterns and disrupt normal cellular processes (Butcher et al. 2004). Integration of these networks with toxicological data, such as dose-response relationships and toxicokinetic parameters, enhances our understanding of the underlying mechanisms of toxicity.

Network Construction in Toxicology Networks are graphical representations of interconnected elements, where nodes represent entities such as genes, proteins, or chemicals, and edges denote the relationships or interactions between them. In toxicology, networks can be constructed using various data sources, including gene expression profiles, protein-protein interactions, chemical-protein interactions, and adverse drug reaction databases. Integration of these diverse data types enables the construction of comprehensive networks that capture the intricate interactions underlying toxicity.

One commonly used approach for network construction is based on gene expression data. By analysing transcriptomic profiles of cells or tissues exposed to toxicants, differentially expressed genes can be identified and used to construct toxicological gene networks. For instance, Smith et al. (2019) used this approach to construct a gene co-expression network in rat liver exposed to the hepatotoxicant acetaminophen. The network revealed modules of functionally related genes, providing insights into the molecular mechanisms underlying acetaminophen-induced liver toxicity.

Another approach involves the construction of protein-protein interaction (PPI) networks. PPI networks provide a global view of the functional relationships between proteins and can be utilized to explore the toxicological effects of chemicals on cellular processes. Wang et al. (2018) constructed a PPI network to investigate the mechanisms of action of environmental toxicants. The network analysis revealed key protein modules associated with toxicity pathways, shedding light on the potential targets of these toxicants.

Network Analysis and Visualization Once the networks are constructed, various computational techniques can be applied to analyse and interpret the toxicological information embedded within them. Network analysis methods, such as centrality measures, clustering algorithms, and functional enrichment analysis, help identify important nodes and modules within the network and uncover underlying toxicity mechanisms.

Centrality measures, such as degree centrality and betweenness centrality, quantify the importance of nodes within a network. For example, Li et al. (2020) utilized degree centrality to identify key genes involved in liver fibrosis induced by carbon tetrachloride. The highly connected nodes in the network were found to be associated with crucial molecular pathways implicated in liver fibrosis progression.

Clustering algorithms, such as the Markov Cluster Algorithm (MCL) or the Louvain algorithm, group nodes with similar properties into modules or clusters. These modules often represent functional units or pathways affected by toxicants. Huang et al. (2021) applied the MCL algorithm to identify toxicant-specific modules in a chemical-protein interaction network. The resulting modules revealed potential targets and pathways associated with specific toxicological effects.

Visualization of networks plays a crucial role in understanding their structure and highlighting key elements. Various network visualization tools, such as Cytoscape (Shannon et al. 2003) and Gephi (Bastian et al. 2009), enable the exploration and interpretation of complex toxicological networks. By employing visual representations, researchers can easily identify densely connected regions, hub nodes, and patterns within the networks, aiding in hypothesis generation and experimental design.

Network-based Toxicity Prediction and Risk Assessment Network-based approaches offer great potential in predicting and assessing the toxicity of chemicals. By integrating toxicological networks with additional information, such as chemical structure or functional annotations, predictive models can be developed to estimate the toxicity of untested compounds.

One application is the prediction of drug-induced toxicities. Jin et al. (2022) proposed a network-based method that integrated chemical-protein interaction networks with drug side-effect data. By analysing the connectivity patterns of drugs and side-effect associations, the model accurately predicted novel drug-toxicity relationships, facilitating early identification of potential adverse effects.

Furthermore, network-based approaches can aid in risk assessment by assessing the potential impact of chemical mixtures on biological systems. Wang et al. (2021) developed a network-based framework that integrated chemical exposure data with toxicological networks. The framework quantified the perturbation of key network modules by chemical mixtures, providing insights into the combined effects of multiple toxicants and assisting in the identification of high-risk mixtures.

Network-based approaches have revolutionized toxicology by providing a systems-level understanding of toxicity mechanisms. By integrating diverse data sources, constructing complex networks, and employing advanced computational techniques, these approaches have the potential to uncover hidden relationships, predict toxicity, and aid in risk assessment. As technology continues to advance, network-based approaches will undoubtedly play a pivotal role in the future of toxicological research.

19.3.2 High-Throughput Technologies and Omics Data Integration

Systems biology approaches enable the integration of omics data to identify key molecular signatures associated with toxicity (Patterson et al. 2006). Comparative analysis of toxicological responses across different compounds and dose levels helps uncover common patterns and identify biomarkers that can predict toxic outcomes (Wang and Zhang 2013). Integration of multiple omics datasets provides a more comprehensive view of the toxicological landscape, capturing interactions between different biological layers and revealing hidden relationships.

High-Throughput Technologies in Toxicology High-throughput technologies enable the simultaneous measurement of thousands or even millions of molecular features, providing a wealth of data for toxicological investigations. Several key high-throughput technologies have significantly advanced toxicology research:

- *Genomics*: Genomic technologies, such as DNA microarrays and next-generation sequencing (NGS), enable the comprehensive analysis of gene expression, DNA sequence variations, and epigenetic modifications. These technologies have been instrumental in identifying toxicant-responsive genes and understanding the molecular basis of toxicity (Grafström et al. 2020).
- *Transcriptomics*: Transcriptomics focuses on the measurement of gene expression levels. Techniques like RNA-seq have revolutionized toxicology by enabling the identification of differentially expressed genes and the exploration of toxicological pathways (Kamp et al. 2018).
- *Proteomics*: Proteomics involves the large-scale study of proteins and their modifications. Mass spectrometry-based proteomics allows for the identification and quantification of proteins, providing insights into toxicant-induced changes in protein expression and post-translational modifications (Wei et al. 2021).

- *Metabolomics*: Metabolomics analyses the comprehensive profile of small molecules, metabolites, in biological samples. Mass spectrometry and nuclear magnetic resonance spectroscopy are commonly used techniques to identify and quantify metabolites, aiding in the understanding of toxicant-induced metabolic perturbations (Jones et al. 2020).

Integration of Omics Data Integrating multiple omics datasets is crucial for obtaining a holistic understanding of toxicity. Integration can occur at various levels, including data fusion, pathway analysis, and network-based approaches.

- *Data fusion*: Data fusion involves integrating multiple omics datasets, such as genomics, transcriptomics, proteomics, and metabolomics, to capture a more comprehensive picture of toxicity. For example, Lai et al. (2021) integrated transcriptomic and metabolomic data to identify key metabolic pathways and potential biomarkers associated with acetaminophen-induced liver injury.
- *Pathway analysis*: Pathway analysis aims to identify the biological pathways that are perturbed by toxicants. By integrating omics data with pathway databases, such as Gene Ontology and Kyoto Encyclopedia of Genes and Genomes (KEGG), researchers can identify enriched pathways and gain insights into toxicological mechanisms (Yang et al. 2019).
- *Network-based approaches*: Network-based approaches integrate omics data to construct molecular interaction networks, such as gene regulatory networks and protein-protein interaction networks. These networks capture the complex interactions between molecules and can provide valuable insights into toxicity mechanisms (Calura et al. 2019).

Applications in Toxicology Integration of high-throughput technologies and omics data has led to several important applications in toxicology research:

- *Identification of biomarkers*: Omics data integration facilitates the identification of biomarkers that can predict toxicity or assess the efficacy of therapeutic interventions. For example, Nambiar et al. (2022) integrated transcriptomic and proteomic data to identify potential biomarkers associated with neurotoxicity, enabling the early detection of adverse effects.
- *Mode of action elucidation*: Omics data integration helps unravel the mode of action of toxicants by identifying key molecular pathways and targets. This information is critical for understanding toxicity mechanisms and developing strategies for intervention or prevention (Wilmes et al. 2020).
- *Risk assessment*: Integration of high-throughput data enhances the accuracy of risk assessment models by incorporating comprehensive molecular information. By integrating exposure data with omics profiles, researchers can develop pre-

dictive models for assessing the potential risks associated with chemical exposure (Yauk et al. 2019).

The integration of high-throughput technologies and omics data has revolutionized toxicology research. By leveraging the power of large-scale data analysis, researchers can gain a deeper understanding of toxicological pathways, identify biomarkers, and improve risk assessment strategies. Continued advancements in high-throughput technologies and data integration techniques will undoubtedly drive further progress in the field of toxicology.

19.3.3 Modelling and Simulation of Toxicological Processes

Computational modelling and simulation are essential components of systems biology approaches. Mathematical models capturing the dynamics of biological systems allow researchers to simulate and predict the effects of toxicants (Chakravarti and Guha 2011). These models can range from simple kinetic models of biochemical reactions to complex multi-scale models integrating cellular, tissue, and organ-level processes.

Modelling toxicological processes facilitates exploration of "what-if" scenarios, aiding in the identification of critical factors and potential intervention strategies (Kholodenko 2006). Moreover, models bridge the gap between in-vitro and in-vivo experiments by extrapolating and scaling biological responses across different levels of organization whereas toxicological processes involve complex interactions between chemicals and biological systems, making their study challenging. Traditional experimental approaches provide valuable insights, but they are often time-consuming and costly. Modelling and simulation techniques, combined with systems biology approaches, offer powerful tools to understand toxicological processes comprehensively. This chapter explores the application of modelling and simulation in toxicology, specifically focusing on the integration of systems biology, and highlights their potential in elucidating toxicity mechanisms, predicting adverse outcomes, and guiding risk assessment.

Modelling Approaches in Toxicology

- *Physiologically Based Pharmacokinetic (PBPK) Models*: PBPK models simulate the absorption, distribution, metabolism, and excretion (ADME) of chemicals in the body. By incorporating physiological parameters, such as blood flow rates and tissue volumes, these models predict the concentration of chemicals in different organs and tissues over time (Rowland et al. 2017).
- *Pharmacodynamic models*: Pharmacodynamic models describe the relationship between drug concentrations and their effects on biological systems. These models capture the dose-response relationships, target interactions, and downstream signalling pathways involved in toxicological processes (Lipnik et al. 2021).

- *Systems biology models*: Systems biology models integrate diverse biological components, such as genes, proteins, and metabolites, to simulate complex biological systems. These models capture the dynamic interactions and regulatory mechanisms underlying toxicological processes, providing insights into toxicity mechanisms and predicting system-level responses (Lehár et al. 2008).

Applications of Modelling and Simulation in Toxicology

- *Mechanistic understanding*: Modelling and simulation techniques facilitate the exploration of toxicity mechanisms by integrating experimental data and existing knowledge. For example, a systems biology model developed by Zhou et al. (2020) incorporated molecular interactions and metabolic pathways to simulate the disposition of toxic compounds in the liver and predict hepatotoxicity.
- *Adverse outcome prediction*: Models that simulate the interactions between chemicals and biological systems can predict adverse outcomes and guide risk assessment. Zhang et al. (2019) developed a computational model that predicted hepatotoxicity of drugs based on chemical structures and known toxicity mechanisms, aiding in early identification of potential toxicants.
- *Virtual testing*: Modelling and simulation enable virtual testing of chemicals, reducing reliance on animal experiments. Virtual models can simulate the effects of toxicants on biological systems, predicting toxicity outcomes and guiding decision-making in chemical development and safety evaluation (Nelms et al. 2019).
- *Integration of systems biology in toxicological modelling*: Systems biology approaches enhance toxicological modelling by integrating multidimensional data, such as genomics, transcriptomics, proteomics, and metabolomics, to capture the complexity of biological systems. These integrated models enable a more comprehensive understanding of toxicological processes by incorporating molecular interactions, regulatory networks, and system-wide responses (Lehár et al. 2008).

Challenges and Future Directions While modelling and simulation, coupled with systems biology, offer significant advantages in toxicology research, several challenges need to be addressed:

- *Data integration*: Integrating diverse omics data from multiple sources remains a challenge due to differences in experimental protocols and data formats. Efforts are needed to standardize data collection, sharing, and integration to improve model accuracy and reproducibility (Sturla et al. 2014).
- *Model complexity*: Developing accurate and interpretable models for complex biological systems is challenging. Advancements in computational techniques, such as machine learning and artificial intelligence, can aid in handling the complexity and heterogeneity of toxicological data (Rodgers et al. 2018).

- *Experimental validation*: Validating systems biology models with experimental data is crucial for their reliability and acceptance. Integrating experimental data into model development and calibration is necessary to ensure accurate predictions and improve model performance (Bray 2019).

Modelling and simulation techniques, integrated with systems biology approaches, provide valuable tools for understanding toxicological processes comprehensively. By combining multidimensional data and mathematical modelling, these approaches enable mechanistic insights, adverse outcome prediction, and virtual testing. Overcoming challenges related to data integration, model complexity, and experimental validation will further enhance the application of modelling and simulation, coupled with systems biology, in toxicology, ultimately improving risk assessment and ensuring safer practices.

19.3.4 Challenges and Future Directions

While systems biology holds great promise for advancing toxicology, several challenges need to be addressed. Integration of diverse data types, standardization of experimental protocols, and development of comprehensive computational models pose technical and logistical challenges (Thomas et al. 2013). Interpretation and validation of complex models require robust experimental data and collaborations among toxicologists, biologists, and computational scientists (Hartung 2009).

Future advancements in high-throughput technologies, data integration methods, and computational tools will further enhance the application of systems biology in toxicology. Continued efforts towards more accurate and predictive models, coupled with advancements in experimental techniques, will contribute to a deeper understanding of the complex interactions between toxicants and biological systems.

19.3.5 Summarization

Systems biology offers a powerful framework for understanding the tenets of toxicology by integrating experimental data, computational modelling, and data analysis. Network-based approaches, omics data integration, and computational modelling provide a holistic view of toxicological processes. Despite challenges, systems biology holds tremendous potential to unravel the complexities of toxicology and facilitate the development of safer chemicals and improved risk assessment strategies.

19.4 About Aging in View of System Biology

Systems biology is an interdisciplinary field that integrates biological experimentation, computational modelling, and data analysis to study complex biological systems as a whole (Ideker et al. 2001). It aims to understand the dynamic interactions and behaviours of biological components, such as genes, proteins, and metabolites, and how they collectively contribute to the functioning of living organisms. In recent years, systems biology has emerged as a powerful framework for investigating the intricate processes of aging. Aging is a complex biological process characterized by the progressive decline in physiological function and increased vulnerability to disease. Understanding the underlying mechanisms of aging has been a long-standing challenge in biology. Traditional reductionist approaches have provided valuable insights into specific aspects of aging but fail to capture the interconnectedness and emergent properties of biological systems. In recent years, systems biology has emerged as a powerful framework to study aging, integrating high-throughput data, computational modelling, and network analysis to unravel the intricate dynamics of aging processes. This chapter explores the potential of systems biology in advancing our understanding of the tenets of aging.

Systems biology offers a holistic approach to understanding aging by considering the interplay and dependencies among the various components of a biological system. By integrating high-throughput experimental techniques, such as genomics, transcriptomics, proteomics, and metabolomics, with computational modelling and network analysis, systems biology can provide a deeper understanding of the underlying mechanisms and processes driving aging.

Through the examination of dynamic biological networks and the complex interactions between molecules and pathways, systems biology can uncover key molecular players, signalling pathways, and regulatory mechanisms involved in aging (Barabási et al. 2011). It sheds light on the interconnectedness of cellular processes and their contributions to the overall aging phenotype. Furthermore, systems biology approaches enable the exploration of interventions, such as caloric restriction or pharmacological agents, to modulate aging-related processes and potentially extend healthy lifespan.

Systems biology offers a comprehensive and integrative framework to comprehend the complex nature of aging. By combining experimental data, computational modelling, and network analysis, it holds great promise in unravelling the tenets of aging and identifying novel strategies for promoting healthy aging and mitigating age-related diseases.

19.4.1 Systems Biology: A Holistic Approach

Systems biology seeks to understand biological systems as a whole, considering the interactions and interdependencies among their components. It combines experimental techniques from molecular biology, genomics, proteomics, and metabolomics with computational tools to generate comprehensive and quantitative models of biological processes (Ideker et al. 2001). By examining the dynamic behaviour of biological networks, systems biology can elucidate the underlying mechanisms of complex biological phenomena.

19.4.2 Network Analysis in Aging Research

One of the key methodologies in systems biology is network analysis, which provides a framework for understanding the organization and behaviour of biological systems. Aging is a multifactorial process influenced by a myriad of molecular interactions and cellular pathways. Network analysis allows researchers to identify crucial nodes, or molecules, and their connections within the aging network. For example, a study by Barabási et al. (2011) utilized network analysis to identify highly connected "hub" genes involved in aging processes, revealing their central role in regulating age-related phenotypes.

19.4.3 Integrative Omics Approaches

Systems biology also leverages integrative omics approaches to study aging comprehensively. Integrative omics involves the integration of multiple levels of molecular data, including genomics, transcriptomics, proteomics, and metabolomics. This integration allows researchers to capture the complex interactions between genes, proteins, and metabolites that drive aging processes. For instance, a recent study by used integrative omics to identify key molecular pathways involved in the aging of the immune system, highlighting potential targets for therapeutic interventions.

19.4.4 Computational Modelling in Aging Research

Computational modelling plays a critical role in systems biology by simulating and predicting the behaviour of biological systems. In aging research, computational models can help unravel the intricate dynamics of aging processes and make predictions about the effects of interventions. For example, a study by Chen et al. (2019) developed a computational model to investigate the impact of caloric restriction on aging. The model predicted that caloric restriction activates specific signalling pathways, leading to increased lifespan in various organisms.

19.4.5 Challenges and Future Directions

Despite its promises, systems biology also faces several challenges in the study of aging. The complexity and heterogeneity of aging processes make it difficult to capture the full scope of aging in a single model. Additionally, the integration of diverse omics data and the development of accurate computational models pose technical and computational challenges. Nevertheless, continued advancements in high-throughput technologies, data integration techniques, and computational methods hold great potential for addressing these challenges.

19.4.6 Conclusion

Systems biology offers a holistic and integrative approach to unravel the tenets of aging. By examining biological systems as a whole and employing network analysis, integrative omics approaches, and computational modelling, systems biology provides valuable insights into the intricate dynamics of aging processes. It holds the potential to identify novel therapeutic targets and interventions to promote healthy aging.

Author's Contribution There has been no conflict of interest for preparation of this manuscript. All authors have equally contributed.

References

Adams DJ, Doran AG, Lilue J, Keane TM (2015) The mouse genomes project: a repository of inbred laboratory mouse strain genomes. Mamm Genome 26(9):403–412

Aguayo-Orozco A, Taboureau O, Brunak S (2019) The use of systems biology in chemical risk assessment. Curr Opin Toxicol 15:48–54

Akondy RS, Fitch M, Edupuganti S, Yang S, Kissick HT, Li KW, Youngblood BA et al (2017) Origin and differentiation of human memory CD8 T cells after vaccination. Nature 552(7685):362–367

Alam T, Petit III RA, Crispell EK, Thornton TA, Conneely KN, Jiang Y, Satola SW, Read TD (2014) Dissecting vancomycin-intermediate resistance in Staphylococcus aureus using genome-wide association. Genome Biol Evol 6(5):1174–1185

Barabási AL, Gulbahce N, Loscalzo J (2011) Network medicine: a network-based approach to human disease. Nat Rev Genet 12(1):56–68

Bastian M, Heymann S, Jacomy M (2009) Gephi: an open source software for exploring and manipulating networks. International AAAI Conference on Weblogs and Social Media, Washington

Bray MA (2019) The challenges of model validation in systems biology. Drug Discov Today 24(11):2214–2220

Butcher EC, Berg EL, Kunkel EJ (2004) Systems biology in drug discovery. Nat Biotechnol 22(10):1253–1259

Calura E, Martini P, Sales G, Belcastro V, Romualdi C, Lanfranchi G (2019) Network-based approaches to explore complex biological systems towards network medicine. Gene 10(1):29

Chakravarti A, Guha R (2011) A perspective on toxicity profiling. Toxicol Mech Methods 21(3):202–217

Chan CX, Ragan MA (2013) Next-generation phylogenomics. Biol Direct 8(1):3. https://doi.org/10.1186/1745-6150-8-3

Chen K-H, Wang T-F, Hu Y-J (2019) Protein-protein interaction prediction using a hybrid feature representation and a stacked generalization scheme. BMC Bioinformatics 20(1). https://doi.org/10.1186/s12859-019-2907-1

Chewapreecha C, Marttinen P, Croucher NJ, Salter SJ, Harris SR, Mather AE, Hanage WP et al (2014) Comprehensive identification of single nucleotide polymorphisms associated with beta-lactam resistance within pneumococcal mosaic genes. PLoS Genet 10(8):e1004547. https://doi.org/10.1371/journal.pgen.1004547

Dhillon BK, Smith M, Baghela A, Lee AHY, Hancock REW (2020) Systems biology approaches to understanding the human immune system. Front Immunol 11(July):1–8. https://doi.org/10.3389/fimmu.2020.01683

Dix A, Vlaic S, Guthke R, Linde J (2016) Use of systems biology to decipher host–pathogen interaction networks and predict biomarkers. Clin Microbiol Infect 22(7):600–606. https://doi.org/10.1016/j.cmi.2016.04.014

Durmuş S, Çakır T, Özgür A, Guthke R (2015) A review on computational systems biology of pathogen–host interactions. Front Microbiol 6:235. https://www.frontiersin.org/articles/10.3389/fmicb.2015.00235

Eisenreich W, Heesemann J, Rudel T, Goebel W (2013) Metabolic host responses to infection by intracellular bacterial pathogens. Front Cell Infect Microbiol 3:24. https://www.frontiersin.org/articles/10.3389/fcimb.2013.00024

Fire A, Xu S, Montgomery MK, Kostas SA, Driver SE, Mello CC (1998) Potent and specific genetic interference by double-stranded RNA in caenorhabditis elegans. Nature 391(6669):806–811. https://doi.org/10.1038/35888

Fleischmann RD, Adams MD, White O, Clayton RA, Kirkness EF, Kerlavage AR, Bult CJ et al (1995) Whole-genome random sequencing and assembly of haemophilus influenzae Rd. Science 269(5223):496–512. https://doi.org/10.1126/science.7542800

Goh C, Knight JC (2017) Enhanced understanding of the host–pathogen interaction in sepsis: new opportunities for omic approaches. Lancet Respir Med 5(3):212–223. https://doi.org/10.1016/S2213-2600(17)30045-0

Gottwein E, Cullen BR (2008) Viral and cellular MicroRNAs as determinants of viral pathogenesis and immunity. Cell Host Microbe 3(6):375–387. https://doi.org/10.1016/j.chom.2008.05.002

Grafström RC, Panchapakesan K, Högberg J (2020) High-throughput genomics: from raw data to toxicology and safety evaluation of drugs. Annu Rev Pharmacol Toxicol 60:235–256

Guo J, Chiang W-C (2022) Mitophagy in aging and longevity Abstract. IUBMB Life 74(4):296–316. 10.1002/iub.v74.4. https://doi.org/10.1002/iub.2585

Hartung T (2009) Toxicology for the twenty-first century. Nature 459(7245):906–913

Holt KE, Parkhill J, Mazzoni CJ, Roumagnac P, Weill F-X, Goodhead I, Rance R et al (2008) High-throughput sequencing provides insights into genome variation and evolution in salmonella typhi. Nat Genet 40(8):987–993. https://doi.org/10.1038/ng.195

Howe K, Clark MD, Torroja CF, Torrance J, Berthelot C, Muffato M, Collins JE et al (2013) The Zebrafish reference genome sequence and its relationship to the human genome. Nature 496(7446):498–503. https://doi.org/10.1038/nature12111

Huang H, Fang M, Jostins L, Mirkov MU, Boucher G, Anderson CA, Andersen V et al (2017) Fine-mapping inflammatory bowel disease loci to single-variant resolution. Nature 547(7662):173–178

Huang T, Xu Z, Liu X (2021) Identifying toxicant-specific modules in chemical-protein interaction network by clustering analysis. Ecotoxicol Environ Saf 208:111726

Ideker T, Thorsson V, Ranish JA, Christmas R, Buhler J, Eng JK, Bumgarner R, Goodlett DR, Aebersold R, Hood L (2001) Integrated genomic and proteomic analyses of a systematically perturbed metabolic network. Science 292(5518):929–934. https://doi.org/10.1126/science.292.5518.929

Jin M, Yang J, Lin W, Tian Y (2022) Network-based prediction of drug-induced toxicity using drug side-effect data and chemical-protein interactome. Environ Health Perspect 130(2):027001

Jones DP, Park Y, Ziegler TR (2020) Nutritional metabolomics: progress in addressing complexity in diet and health. Annu Rev Nutr 40:169–196

Kamp HG, van Ommen B, Romijn JA (2018) A digital biomarker to assess environmental exposure using gene expression in peripheral blood cells. Environ Health Perspect 126(4):047007

Kaur S, Ali A, Ahmad U, Siahbalaei Y, Pandey AK, Singh B (2019) Role of single nucleotide polymorphisms (SNPs) in common migraine. Egypt J Neurol Psychiatry Neurosurg 55(1):47

Kazmin D, Nakaya HI, Lee EK, Johnson MJ, van der Most R, van den Berg RA, Ballou WR et al (2017) Systems analysis of protective immune responses to RTS,S malaria vaccination in humans. Proc Natl Acad Sci 114(9):2425–2430. https://doi.org/10.1073/pnas.1621489114

Kholodenko BN (2006) Cell-signalling dynamics in time and space. Nat Rev Mol Cell Biol 7(3):165–176. https://doi.org/10.1038/nrm1838

Kitano H (2002) Systems biology: a brief overview. Science 295(5560):1662–1664. https://doi.org/10.1126/science.1069492

Klemm E, Dougan G (2016) Advances in understanding bacterial pathogenesis gained from whole-genome sequencing and phylogenetics. Cell Host Microbe 19(5):599–610. https://doi.org/10.1016/j.chom.2016.04.015

Ko DC, Shukla KP, Fong C, Wasnick M, Brittnacher MJ, Wurfel MM, Holden TD et al (2009) A genome-wide in vitro bacterial-infection screen reveals human variation in the host response associated with inflammatory disease. Am J Hum Genet 85(2):214–227

Ko DC, Gamazon ER, Shukla KP, Pfuetzner RA, Whittington D, Holden TD, Brittnacher MJ et al (2012) Functional genetic screen of human diversity reveals that a methionine salvage enzyme regulates inflammatory cell death. Proc Natl Acad Sci U S A 109(35):2343–2352. https://doi.org/10.1073/pnas.1206701109

Lai X, Hou Q, Zheng J, Li M, Wu J (2021) Integration of transcriptomic and metabolomic data reveals key metabolic pathways in acetaminophen-induced liver injury. Toxicol Sci 181(2):247–257

Langridge GC, Phan M-D, Turner DJ, Perkins TT, Parts L, Haase J, Charles I et al (2009) Simultaneous assay of every salmonella typhi gene using one million transposon mutants. Genome Res 19(12):2308–2316. https://doi.org/10.1101/gr.097097.109

Lehár J, Krueger AS, Avery W, Heilbut AM, Johansen LM, Price ER, Lee MS (2008) Synergistic drug combinations tend to improve therapeutically relevant selectivity. Nat Biotechnol 27(7):659–666

Leonard II, Edward E, Takata T, Blaser MJ, Falkow S, Tompkins LS, Gaynor EC (2003) Use of an open-reading frame–specific campylobacter Jejuni DNA microarray as a new genotyping tool for studying epidemiologically related isolates. J Infect Dis 187(4):691–694. https://doi.org/10.1086/368268

Li X, Gao X, Liu F, Li C (2020) Identifying key genes and miRNAs in liver fibrosis induced by carbon tetrachloride through a network-based approach. Front Genet 11:573

Lipnik K, Khan S, Ammar A, Vishwakarma V (2021) In silico methods in the prediction of chemical toxicity. In: Computational approaches for toxicity prediction. Elsevier, Amsterdam, pp 25–57

Miller SI, Chaudhary A (2016) A cellular GWAS approach to define human variation in cellular pathways important to inflammation. Pathogens 5(2):39. https://doi.org/10.3390/pathogens5020039

Nambiar S, Parmar VS, Pujari MP, Parmar PP, Suryavanshi PM (2022) Integrated transcriptomic and proteomic analysis to identify potential biomarkers of neurotoxicity in zebrafish. Ecotoxicol Environ Saf 230:113002

Nelms MD, Byler KG, Groth T, Li Z (2019) The evolution of predictive toxicology: from reactive to proactive toxicology and risk assessment. Toxicol Sci 167(2):298–309

Nicolas-Chanoine M-H, Bertrand X, Madec J-Y (2014) Escherichia coli ST131, an Intriguing Clonal Group. Clin Microbiol Rev 27(3):543–574. https://doi.org/10.1128/cmr.00125-13

Okoro CK, Kingsley RA, Connor TR, Harris SR, Parry CM, Al-Mashhadani MN, Kariuki S et al (2012) Intracontinental spread of human invasive salmonella typhimurium pathovariants in Sub-Saharan Africa. Nat Genet 44(11):1215–1221. https://doi.org/10.1038/ng.2423

Opijnen T, Bodi KL, Camilli A (2009) Tn-Seq: high-throughput parallel sequencing for fitness and genetic interaction studies in microorganisms. Nat Methods 6(10):767–772. https://doi.org/10.1038/nmeth.1377

Patterson TA, Lobenhofer EK, Fulmer-Smentek SB et al (2006) Performance comparison of one-color and two-color platforms within the MicroArray Quality Control (MAQC) project. Nat Biotechnol 24(9):1140–1150

Perkins TT, Davies MR, Klemm EJ, Rowley G, Wileman T, James K, Keane T et al (2013) ChIP-Seq and transcriptome analysis of the OmpR regulon of salmonella enterica serovars typhi and typhimurium reveals accessory genes implicated in host colonization. Mol Microbiol 87(3):526–538. https://doi.org/10.1111/mmi.12111

Pickard D, Kingsley RA, Hale C, Turner K, Sivaraman K, Wetter M, Langridge G, Dougan G (2013) A genome wide mutagenesis screen identifies multiple genes contributing to Vi capsular expression in salmonella enterica serovar typhi. J Bacteriol 195(6):1320–1326. https://doi.org/10.1128/jb.01632-12

Pinu FR, Beale DJ, Paten AM, Kouremenos K, Swarup S, Schirra HJ, Wishart D (2019) Systems biology and multi-omics integration: viewpoints from the metabolomics research community. Meta 9(4):1–31. https://doi.org/10.3390/metabo9040076

Porwollik S, Santiviago CA, Cheng P, Long F, Desai P, Fredlund J, Srikumar S et al (2014) Defined single-gene and multi-gene deletion mutant collections in Salmonella Enterica Sv typhimurium. PLoS ONE 9(7):e99820. https://doi.org/10.1371/journal.pone.0099820

Rodgers JL, Jones J, Bolleddula J, Thakrar S, Vakil H (2018) Machine learning approaches in precision medicine in solid cancers. J Clin Med 7(12):471

Rowland M, Tozer TN, Riegelman S (2017) Clinical pharmacokinetics and pharmacodynamics: concepts and applications, 5th edn. Wolters Kluwer, Alphen aan den Rijn

Saayman S, Ackley A, Turner A-MW, Famiglietti M, Bosque A, Clemson M, Planelles V, Morris KV (2014) An HIV-encoded antisense long noncoding RNA epigenetically regulates viral transcription. Mol Ther 22(6):1164–1175. https://doi.org/10.1038/mt.2014.29

Salinas RE, Ogohara C, Thomas MI, Shukla KP, Miller SI, Dennis CK (2014) A cellular genome-wide association study reveals human variation in microtubule stability and a role in inflammatory cell death. Mol Biol Cell 25(1):76–86. https://doi.org/10.1091/mbc.E13-06-0294

Sasaki K, Kurahara H, Young ED, Natsugoe S, Ijichi A, Iwakuma T, Welch DR (2017) Genome-wide in vivo RNAi screen identifies ITIH5 as a metastasis suppressor in pancreatic cancer. Clin Exp Metastasis 34(3–4):229–239. https://doi.org/10.1007/s10585-017-9840-3

Schulze S, Henkel SG, Driesch D, Guthke R, Linde J (2015) Computational prediction of molecular pathogen-host interactions based on dual transcriptome data. Front Microbiol 6:65

Shannon P, Markiel A, Ozier O, Baliga NS, Wang JT, Ramage D, Ideker T (2003) Cytoscape: a software environment for integrated models of biomolecular interaction networks. Genome Res 13(11):2498–2504

Sheppard SK, Didelot X, Meric G, Torralbo A, Jolley KA, Kelly DJ, Bentley SD, Maiden MCJ, Parkhill J, Falush D (2013) Genome-wide association study identifies vitamin B5 biosynthesis as a host specificity factor in campylobacter. Proc Natl Acad Sci 110(29):11923–11927. https://doi.org/10.1073/pnas.1305559110

Skalsky RL, Cullen BR (2010) Viruses, microRNAs, and host interactions. Annu Rev Microbiol 64(1):123–141. https://doi.org/10.1146/annurev.micro.112408.134243

Smith CA, O'Maille G, Want EJ, Qin C, Trauger SA, Brandon TR, Fiehn O (2019) METLIN: a metabolite mass spectral database. Ther Drug Monit 27(6):747–751

Stebbins CE (2005) Structural microbiology at the pathogen-host interface. Cell Microbiol 7(9):1227–1236. https://doi.org/10.1111/j.1462-5822.2005.00564.x

Sturla SJ, Boobis AR, Fitzgerald RE, Hoeng J, Kavlock RJ, Schirmer K, Whelan M, Wilks MF, Peitsch MC (2014) Systems toxicology: from basic research to risk assessment. Chem Res Toxicol 27(3):314–329

Sun J, Katz S, Dutta B, Wang Z, Fraser IDC (2017) Genome-wide SiRNA screen of genes regulating the LPS-induced TNF-α response in human macrophages. Sci Data 4(March):170007. https://doi.org/10.1038/sdata.2017.7

Thomas RS, Philbert MA, Auerbach SS et al (2013) Incorporating new technologies into toxicity testing and risk assessment: moving from 21st century vision to a data-driven framework. Toxicol Sci 136(1):4–18

Troy EB, Lin T, Gao L, Lazinski DW, Lundt M, Camilli A, Norris SJ, Linden TH (2016) Global Tn-Seq analysis of carbohydrate utilization and vertebrate infectivity of Borrelia Burgdorferi. Mol Microbiol 101(6):1003–1023. https://doi.org/10.1111/mmi.13437

van der Greef J, Stroobant P, van der Heijden R, Hankemeier T (2010) Systems toxicology: an emerging discipline. WIREs Syst Biol Med 2(5):532–549

Walker AW, Sanderson JD, Churcher C, Parkes GC, Hudspith BN, Rayment N, Brostoff J, Parkhill J, Dougan G, Petrovska L (2011) High-throughput clone library analysis of the mucosa-

associated microbiota reveals dysbiosis and differences between inflamed and non-inflamed regions of the intestine in inflammatory bowel disease. BMC Microbiol 11(1):7
Wang J, Zhang XS (2013) Integration of multiple biological data for prediction of toxicological effects of environmental chemicals: a machine learning approach. J Chem Inf Model 53(10):2621–2629
Wang X, Gao X, Sun X, Xu X (2018) Construction of a protein-protein interaction network and its applications in environmental toxicology. Environ Toxicol Chem 37(7):1783–1793
Wang J, Li M, Wang H, Pan Y, Dai W (2021) A network-based framework for assessing the combined toxicity of chemical mixtures. Environ Pollut 272:116053
Wei Y, Chen Y, Chen L, Chen B, He Y (2021) Current trends and applications of mass spectrometry-based proteomics in environmental toxicology. J Hazard Mater 404:124073
Westermann AJ, Gorski SA, Vogel J (2012) Dual RNA-Seq of pathogen and host. Nat Rev Microbiol 10(9):618–630
White JK, Gerdin A-K, Karp NA, Ryder E, Buljan M, Bussell JN, Salisbury J et al (2013) Genome-wide generation and systematic phenotyping of knockout mice reveals new roles for many genes. Cell 154(2):452–464
Wilmes A, Limonciel A, Aschauer L, Moenks K, Bielow C, Leonard MO, Dekant W (2020) Application of integrated transcriptomic, proteomic and metabolomic profiling for the delineation of mechanisms of drug-induced cell stress. J Proteome 225:103877
Wong VK, Baker S, Pickard DJ, Parkhill J, Page AJ, Feasey NA, Kingsley RA et al (2015) Phylogeographical analysis of the dominant multidrug-resistant H58 clade of salmonella typhi identifies inter- and intracontinental transmission events. Nat Genet 47(6):632–639
Yang C, Huang Y, Li J, Feng X, Huang Q (2019) Integration of transcriptomic, proteomic, and metabolomic data reveals the toxicological effect of bisphenol S exposure on zebrafish larvae. Environ Sci Technol 53(15):9345–9356
Yauk CL, Buick JK, Williams A, Swartz CD, Recio L, Li HH, Marchetti F (2019) Integrating human clinical and environmental data to advance predictive toxicology. Toxicol Sci 169(1):32–47
Zhang W, Xu Y, Chen S, Chen X, Zhang J, Xu JJ (2019) Predicting hepatotoxicity using Tox21 data and diverse drug chemical structures. Chem Res Toxicol 32(8):1585–1596
Zhou J, Sun P, Cheng F, Zhang X (2020) Predicting drug hepatotoxicity using toxicogenomics signature and structure-based similarity. Toxicol Sci 175(2):293–302

Systems Biology Approaches as a Tool in Understanding Neurodevelopmental Disorders: Some Case Studies

20

Anuvab Dey, Subhrojyoti Ghosh, Sourav Majhi, Aishee Nishat Hoque, and Aneshwa Chakrabarti

Abstract

Systems biology has emerged as a powerful approach in neurology and neurosciences, offering novel insights into the complex mechanisms underlying various neurological diseases. The intricate nature of the nervous system necessitates a comprehensive approach to comprehending neurological disorders. Systems biology integrates multidisciplinary techniques, including genomics, proteomics, metabolomics, and computational modelling, to unravel the intricate networks of

Subhrojyoti Ghosh, Sourav Majhi, Aishee Nishat Hoque and Aneshwa Chakrabarti contributed equally with all other contributors.

A. Dey
Department of Biological Sciences and Biological Engineering, Indian Institute of Technology Guwahati, Guwahati, Assam, India

Department of Biotechnology, Heritage Institute of Technology, Kolkata, West Bengal, India
e-mail: anuvab2000dey@gmail.com

S. Majhi · A. N. Hoque
Department of Biological Sciences and Biological Engineering, Indian Institute of Technology Guwahati, Guwahati, Assam, India
e-mail: srvmajhi1@gmail.com; aisheenishat99@gmail.com

S. Ghosh (✉)
Department of Biotechnology, IIT Madras, Chennai, Tamil Nadu, India

Department of Biotechnology, Heritage Institute of Technology, Kolkata, West Bengal, India
e-mail: subhrojyotighosh8@gmail.com

A. Chakrabarti
Department of Chemistry and Chemical Biology, IIT ISM Dhanbad, Dhanbad, India
e-mail: aneshwa.chakrabarti.bt22@heritageit.edu.in

S. Joshi et al. (eds.), *Systems Biology Approaches: Prevention, Diagnosis, and Understanding Mechanisms of Complex Diseases*,
https://doi.org/10.1007/978-981-99-9462-5_20

molecular interactions within neurons, glia, and other central and peripheral nervous system cells. Recently, systems biology has made significant strides in elucidating the pathogenesis of neurodegenerative disorders like Parkinson's disease, Alzheimer's disease, and amyotrophic lateral sclerosis. Researchers have identified key molecular players and dysregulated pathways by integrating large-scale omics data with advanced computational methods, shedding light on disease progression, and potential therapeutic targets. Furthermore, systems biology has improved our knowledge of neurodevelopmental disorders, which includes disorders in the autism spectrum and intellectual disabilities. Researchers have revealed disrupted molecular pathways underlying these conditions by deciphering gene regulatory networks and functional modules, offering new avenues for targeted interventions, and personalized medicine. Moreover, systems biology has facilitated the exploration of complex brain functions, including learning, memory, and behaviour. Integration of multi-level data, such as transcriptomics and neuroimaging, has unravelled the intricate interplay between genes, neural circuits, and cognitive processes. This book chapter presents a comprehensive overview of the usage of systems biology in neurology and neurosciences. It demonstrates how the integration of high-throughput technologies, computational modelling, and multi-omics data has widened our understanding of neurological diseases, paving the way for personalized medicine and improved patient outcomes. The chapter also highlights the potential for further advancements in the field and the importance of interdisciplinary collaborations to tackle the difficulties of studying complex neurological systems.

Keywords

Systems biology · Neurological disorders · High-throughput technologies · Personalized medicine · Biomarkers

20.1 Introduction to Systems Biology in Neurology and Neurosciences

20.1.1 Definition and Scope of Systems Biology

Systems biology involves studying of biological systems through systematic genetic, biological, or chemical disturbances, observation of the responses of proteins, genes, and informational pathways, assimilating these data, and finally, developing mathematical models that define the configuration of the system and its response to specific perturbations. It is an interdisciplinary field that combines the principles from biology, physics, mathematics, and computer science to study entire biological systems instead of studying individual components in isolation (Ideker et al. 2001). Systems biology is an approach that is driven by hypothesis, broad, iterative, quantitative, integrative, and dynamic for studying biological systems. It seeks to understand emergent properties, maintain robustness, and leverage modularity to get insights into intricate biological phenomena and enable predictions and interventions in biological systems.

The scope of systems biology in the future is vast and has a lot of potential for advancing our knowledge of life at the systems level. Systems biology will revolutionize the practice of medicine. Researchers could identify vital biological processes and pathways involved in diseases by integrating large-scale data from genomics, proteomics, and other omics technologies. Drug development will become easier with knowledge of disease-perturbed networks, and pharmacological approaches will concentrate on reducing disease-mediated changes. This will result in the development of personalized treatment regimens and the discovery of new therapeutic targets (Aderem 2005). Systems biology gives the foundation for designing and engineering biological systems with desired functions. By applying mathematical modelling and computational simulations, researchers will determine the behaviour of sophisticated genetic circuits and optimize the design of synthetic organisms. Systems biology approaches enable the redesign and optimization of metabolic pathways in microorganisms to produce valuable compounds like biofuels, pharmaceuticals, and industrial chemicals. Researchers will be able to control metabolic networks to increase production efficiency and yield by understanding the relationships between genes, enzymes, and metabolites. Systems biology also helps in understanding complicated ecological interactions such as those in microbial communities or ecosystems. The dynamics of species interactions, nutrient cycling, and ecosystem stability can be understood by analyzing genomic, transcriptomic, and environmental data. This information can help with the mitigation of several environmental problems, ecological conservation, sustainable agriculture, and bioremediation (Toscano and Oehlke 2005).

20.1.2 The Relevance of Systems Biology in Neurology and Neurosciences

Systems biology provides a comprehensive approach to understanding the intricate biological mechanisms that govern the functioning and organization of the nervous system. Brain functions and neurological disorders are the results of intricate interactions among genes, proteins, cells, circuits, and systems, which can be effectively studied using systems biology principles. Systems biology attempts to comprehend the entire system instead of focusing on individual components. This approach is useful in neurology and neuroscience for understanding the complex interactions and feedback loops between neurons, glial cells, and neurotransmitters, among other nervous system components. It provides insights into how the system functions as a whole and how malfunction or dysregulation in particular parts might result in neurological disorders. The brain operates at multiple scales ranging from molecules and cells to circuits and networks. Systems biology integrates data from several sources, such as genomes, proteomics, transcriptomics, and imaging techniques, to comprehensively view the nervous system. By combining information from multiple scales, researchers can identify emergent features and comprehend how changes at one level affect the system as a whole (Diaz-Beltran et al. 2013; Barabási and Albert 1999). Systems biology uses network analysis techniques to decipher the complex connection patterns in the brain. Researchers can identify important hubs, routes, and modules essential for both healthy brain function and neurological disorders by building and analyzing complex networks of interactions between genes, proteins, and various brain regions. This approach aids in understanding the neural circuits' dynamics and how perturbations can lead to diseases (Barabási and Oltvai 2004; Grant 2003). Systems biology uses computational models to analyze information processing, stimulate neural networks, and determine how alterations in molecular and cellular components affect the system. These models help in generating testable hypotheses and offer a platform for examining the effects of interventions and treatments in neuroscience (Diaz-Beltran et al. 2013). The systems biology approach has made it possible to enable personalized medicine in neurology. By integrating individual patient data, including genetic information, clinical profiles, and imaging data, with large-scale datasets, researchers can develop personalized models of brain function and disease. This allows for tailoring treatments to individual patients based on their specific molecular and genetic profiles, improving the effectiveness of interventions.

Overall, it can be said that systems biology is highly relevant as it provides a holistic and integrative approach to neurology and neuroscience. It offers insights into typical brain function, mechanisms behind neurological disorders, and opportunities for designing personalized therapeutic methods by taking into consideration the complex interactions and fluctuations of the nervous system.

20.1.3 Overview of Associated Neurological Diseases

Neurological disorders are complex diseases affected by environmental as well as genetic factors. Multiple pathogenic determinants are associated with neurological diseases, which exhibit a progressive loss of neuronal function and survival. Classical reductionist approaches that are focused on individual genes have only partially revealed the underlying complexities of these disorders. Therefore, although these studies have shed light on the nature of these disorders, still a complete knowledge of their etiopathogenesis is still lacking. Conversely, the perspective of systems biology seeks to incorporate the cellular and molecular pathways involved in the progression of these disorders. This approach considers how various biological systems interact with one another and their surrounding environment. Structural and dynamic analysis of molecular networks provides insights into their healthy and diseased states, providing possibilities for therapeutic interventions. Thus, a systems biology approach appears to be more effective in understanding these multifactorial diseases' molecular complexities (Diaz-Beltran et al. 2013).

A brief overview of some neurological disorders to lend insight into how systems biology helps to understand the underlying complexity of these disorders:

Alzheimer's disease (AD): The biochemical processes that contribute to AD, such as amyloid-beta buildup, tau protein phosphorylation, inflammation, and synaptic dysfunction, have been uncovered with the help of systems biology. By integrating genetic, genomic, and proteomic data, researchers have identified essential genes and proteins associated with AD pathology (Hampel et al. 2020).

Parkinson's disease (PD): Systems biology has provided information on the complex molecular networks underlying PD. It has discovered genes and proteins related to oxidative stress, neuroinflammation, alpha-synuclein, protein misfolding and aggregation, and mitochondrial dysfunction. By analyzing large-scale datasets, systems biology has provided insights into disease progression (Wood et al. 2015).

Autism spectrum disorders (ASD): Systems biology approaches have helped unravel the genetic and molecular factors that cause ASD. Researchers have discovered altered biochemical pathways that contribute to the development of the brain, synapse function, as well as neural connection in people with ASD by fusing genetic and gene expression data. This has improved our understanding of the biology behind ASD (Diaz-Beltran et al. 2013).

Multiple sclerosis (MS): Bayesian inference provided valuable insights into the immune dysregulation and neurodegeneration associated with MS. It also helped to identify novel therapeutic targets in MS (Palacios et al. 2007).

20.2 Principles of Systems Biology

Systems biology is an interdisciplinary discipline that tries to comprehend living systems as a dynamic and integrated networks of molecules, cells, and organisms. In order to gather insights into the complicated behaviour of biological systems, it tries to integrate experimental methods with computer and mathematical modelling.

20.2.1 Systems Thinking and Holistic Approaches

Systems thinking is an approach to problem-solving that emphasizes the connectivity and interdependence of a system's diverse parts. This means that systems thinking requires considering the system as a whole and the connections among its components while analyzing a biological system. This approach makes it possible to recognize the intricate feedback loops and nonlinear interactions that define biological systems.

Systems biologists often use the term "holistic approaches" to describe the comprehensive study of biological systems that takes into account the interactions and integration of different parts at various levels, from molecules and cells to tissues, organs, and organisms. In systems biology experiments, which use holistic approaches rather than predetermined or known hypotheses, all data are collected and evaluated to produce a hypothesis that may be further explored (Horgan and Kenny 2011). The exclusively effective way to comprehend the complexity involved in figuring out the biology of various diseases is to use a holistic approach as an in-silico modelling strategy by networking the pathways for multiple diseases, including both environmental and genetic factors (Gupta and Misra 2016).

20.2.2 Computational Modelling and Simulation

Systems biology, a discipline that tries to comprehend complex biological systems by combining experimental data (from various sources, like genomics, proteomics, metabolomics, and high-throughput techniques) with mathematical models, depends extensively on computational modelling and simulation. Computational models combining numerous interactions on multiple spatial and temporal scales are developed to describe biological systems, including molecular interactions, cellular networks, and even entire organisms. Model reduction, model integration, and modelling at many scales are necessary for this. Executing simulations of extensive integrated models that encompass these various scales requires significant computational resources, except when the multiscale nature of biological problems is systematically leveraged (Dada and Mendes 2011).

Depending on the system being studied and the level of detail necessary, these models may be based on differential equations, stochastic processes, Boolean logic, or other mathematical frameworks (Sordo Vieira and Laubenbacher 2022; Hemedan et al. 2022; Kohl et al. 2010). The only significant distinction between a biologist who utilizes mathematical modelling compared to someone who does not is that the former examines the quantitative aspects of his theories, involving the use of computational experiments to evaluate their reliability. Because quantitatively reasonable predictions enhance later hypothesis-driven experimental research, the potential benefits of doing so are evident (Dada and Mendes 2011; Kohl et al. 2010).

Thus, computational models are used in order to develop drugs that are safe for patients and have minimal side effects, to track infectious diseases among populations, to effectively collect, filter, evaluate, and present information about health, and to give doctors recommendations for disease treatment based on specific patient data. Hence, systems biology can benefit significantly from computer modelling and simulation's sophisticated tools for comprehending and analyzing complex biological systems. They let researchers make predictions, test hypotheses, and gain knowledge of the dynamics and behaviour of biological systems by bridging the gap between experimental data and theoretical understanding.

20.2.3 Network Analysis and Dynamics

A significant aspect of network analysis is the study of connections or interactions between diverse constituents of a living system. In systems biology, these components can represent molecules like genes, proteins, or metabolites, and the interactions between them can take the form of physical interaction, regulatory relationships, signalling pathways, or metabolic processes. Network analysis approaches aim to identify these networks' patterns, properties, and organizational principles. Using graph theory, one common network analysis approach is constructing and analyzing biological networks. Molecular components found inside a cell are frequently regarded as nodes in a network representation of intracellular biological networks. At the same time, the direct or indirect interactions between them are links or edges.

Different kinds of mathematical structures known as graphs can be used to depict various intracellular molecular biological networks. Examples include (1) metabolic networks, which are accessible through the EcoCyc database and include networks from various organisms; (2) cell signaling networks, including a network that represents the pathways in mammalian neurons and many more. Using graph theory approaches, researchers can examine several network properties, such as connectivity, centrality, modularity, and clustering coefficients. These properties aid in understanding the organization and significance of individual network components (Wynn et al. 2012).

On the other side, network dynamics focuses on researching how biological networks evolve and change over time. Because biological systems are dynamic, interactions and feedback mechanisms between the various parts of the system also have an impact on how they behave. The (1) structural network analysis methodologies and (2) computational and mathematical models founded on ordinary differential equations, Boolean networks, agent-based models, and stochastic models are two distinct paradigms of systems biology that are routinely used to build and analyze the dynamic network models of representing molecular interactions deduced from experimental data. The usage of logic-based methods to get the anticipated outcomes is a third, increasingly significant network analysis paradigm in systems biology. Despite their qualitative nature, logic-based approaches can shed light on the dynamics of highly complex gene regulation and signal transduction networks without the burden of vast parameter space (Wynn et al. 2012).

In summary, network analysis and dynamics provide valuable tools for studying the structure and behaviour of living systems in systems biology. These approaches help uncover the underlying principles of complex biological networks, enabling a better comprehension of the functions and mechanisms of living systems.

20.2.4 Data Integration and Omics Technologies

Data integration involves combining data from several sources, including genomic, transcriptomic, proteomic, and metabolomic data. This allows researchers to spot patterns and correlations that might not be visible when looking at each dataset separately. A broader perspective of biological systems and their functions may result from this integration. On the other hand, omics technologies pertain to a group of techniques used to analyze vast amounts of biological data and embrace a comprehensive perspective of the molecules comprising a cell, tissue, or organism. Genomic, transcriptomic, proteomic, and metabolomic techniques are some of the methods used. Different types of biological data, such as DNA sequences, levels of gene expression, protein structures, and metabolite concentrations, are the focus of each omics technique. Several omics approaches provide enormous volumes of data at various levels with extraordinary precision, leading to the creation of databases and resources. Due to the enormous number and volume of data, the problem of continually elaborating information continues. A thorough grasp of biological activities' complexity is also necessary because biomolecule identification alone will not provide much information. As a result, systems biology uses computational methods and multi-omics platforms to integrate data in many contexts (Gomez-Cabrero et al. 2014; Maghuly and Marzban 2021). Two categories of existing omics data visualization technologies for integrating omics data are tools based on biological networks and tools based on pathways (Kuo et al. 2013).

20.3 Analytical Techniques in Systems Biology

Systems biology is a multidisciplinary field that seeks to comprehend biological systems as a whole while considering the interactions and behaviours of each constituent part. Analytical techniques play a significant role in systems biology by providing methods to measure, quantify, and analyze the many elements and processes within biological systems.

20.3.1 Genomics and Transcriptomics

Genomics and transcriptomics are two closely related fields in systems biology that involve the study of an organism's genetic information and gene expression patterns, respectively. Genomics encompasses the study of an organism's entire genome, including the sequencing, mapping, and analysis of its genes and their functions. The ability to rapidly and affordably sequence whole genomes and transcriptomes has revolutionized genomics through the use of high-throughput DNA sequencing technologies like next-generation sequencing (Khodadadian et al. 2020). According to Hieter et al., it was stated that a powerful novel viewpoint on the comprehensive operation of biological systems would be provided by functional genomics to enhance the specific comprehension of gene function provided by conventional techniques (Hieter and Boguski 1997). The concept of a "holistic understanding" was further developed using whole-genome datasets, mathematical models, and the knowledge gained from complex system studies (Likić et al. 2010).

On the other hand, the study of an organism's entire collection of RNA transcripts, or transcriptome, is the subject of transcriptomics. Researchers can evaluate gene expression levels, identify alternative splicing events, and investigate the dynamics of transcriptional regulation using techniques like microarrays, RNA sequencing, differential gene expression analysis, and so on (Khodadadian et al. 2020).

Understanding the molecular mechanisms underlying biological processes, development, diseases, and drug responses requires knowledge from both genomics and transcriptomics. Researchers can learn a great deal about how genes are regulated, find novel therapeutic targets, and investigate the complexity of biological systems by integrating data from these two domains.

20.3.2 Proteomics and Metabolomics

Proteomics and metabolomics, two branches of the omics sciences, play a crucial role in advancing the field of systems biology. These disciplines are essential for understanding the intricate interactions and molecular dynamics underlying biological processes, necessitating comprehensive investigations of proteins and metabolites within biological systems. The study of the complete set of proteins (proteome) present in a cell, tissue, or organism is referred to as proteomics. It includes several techniques for identifying, quantifying, and characterizing proteins. Identifying and quantifying proteins, studying post-translational modifications and protein expression patterns, and investigating protein-protein interactions are all widely accomplished using methods like mass spectrometry (MS) and two-dimensional gel electrophoresis. In systems biology, proteomics is essential to understand how cellular processes are regulated. It allows the identification of alterations in protein expression or modifications under different conditions and the identification of essential proteins associated with specific pathways or disorders. Advancements in quantitative proteomics MS are generating a growing collection of datasets suitable for applications in systems biology. These recent and ongoing developments enable the utilization of large-scale proteomic data for studying complex biological systems. In contrast, microarray data was initially most suitable for systems modelling. Proteomic datasets are undoubtedly essential for creating network models with reliable predictive potential, given the established lack of correlations between mRNA and protein expression levels (Horgan and Kenny 2011; Moore and Weeks 2011).

The investigation of comprehensive metabolite profiles within a biological system (cell, tissue, or organism) under specific circumstances is known as metabolomics. Metabolites are the end product of gene expression. However, they also function as an integral part of the regulatory system because of how the genome of the system interacts with its environment. Their measurement and analysis provide insights into the metabolic state of an organism. Metabolomics techniques involve identifying and quantifying metabolites using various analytical approaches, such as MS and nuclear magnetic resonance spectroscopy (NMR) (Rochfort 2005).

Systems biology aims to integrate genomes, metabolomics, transcriptomics, and proteomics; this topic has been extensively discussed (Rochfort 2005). Proteomics and metabolomics are often integrated with systems biology studies to obtain a more comprehensive view of biological systems. By combining proteomic and metabolomic data, researchers can map out the connections between proteins and metabolites and gain insights into their coordinated behaviour and functional relationships. Integrated proteomics and metabolomics approaches enable the construction of dynamic models that capture the complex interactions between genes, proteins, and metabolites. These models can be used to simulate and predict the behaviour of living systems under different conditions or perturbations, aiding in the identification of potential drug targets, understanding disease mechanisms, and optimizing biotechnological processes (Costanzo et al. 2017).

20.3.3 Imaging Techniques in Neurosciences

Neuroscience research is greatly aided by imaging tools, which make it possible to visualize and examine the structure and operation of the brain. The mechanisms behind many neurological functions and conditions can be better understood because of these imaging techniques. Many brain imaging tools are obtainable to cognitive neuroscientists, including Electroencephalography (EEG), positron emission tomography (PET), Magnetic Resonance Imaging (MRI), Functional MRI (fMRI), and so on (Xue et al. 2010).

- MRI employs powerful magnetic fields and radio waves to produce detailed brain anatomy images. It gives high-resolution structural information. MRI can also be used for functional imaging by measuring blood flow variations, which enables researchers to examine how the brain responds to various tasks or stimuli.
- One of the most recent advancements in neuroimaging technology is fMRI. fMRI is a specific application of MRI that measures changes in blood oxygenation levels as a proxy for neural activity. By detecting regional changes in blood flow, fMRI can identify brain areas associated with specific cognitive processes or tasks, offering insights into brain function and connectivity (Xue et al. 2010).
- EEG, first developed almost a century ago, uses electrodes positioned on the scalp to measure the brain's electrical activities. It is frequently used in studies of sleep, epilepsy, and cognitive functions and gives real-time information about brain waves. EEG can also be integrated with imaging methods like fMRI (also referred to as EEG-fMRI) to attain significant temporal and spatial resolution (Xue et al. 2010).
- PET imaging involves injecting a small amount of a radioactive substance (tracer) into the body, which emits positrons. PET scans can reveal brain activity, metabolism, receptor binding, and neurochemical processes by detecting the gamma rays created when positrons interact with electrons.

20.3.4 High-Throughput Technologies

In systems biology, high-throughput technologies are essential because they let researchers create large-scale datasets and learn more about complex processes in biology. Researchers may explore and comprehend biological processes at a systems level due to these technologies' fast analysis and processing of large amounts of biological data.

Genomic research has been transformed by high-throughput sequencing methods like next-generation sequencing (NGS). A thorough understanding of genetic and gene expression profiles is made possible by the quick sequencing of complete genomes, transcriptomes, and epigenomes made possible by NGS (Ambardar et al. 2016). High-throughput proteomics technologies involve identifying and quantifying proteins on a large scale and as fast as possible. So mass spectrometry-based proteomics is widely used, and HT proteomics focuses on data-independent acquisition (DIA) techniques like Sequential Windowed Acquisition of all Theoretical Mass Spectra (SWATH-MS). Specifically in proteomics investigations using rapid chromatographic gradients, DIA has been shown to increase proteome depth, data completeness, robustness, and quantification accuracy over time (Messner et al. 2023). High-throughput metabolomics technologies use methods like gas chromatography-mass spectrometry (GC-MS) and liquid chromatography-mass spectrometry (LC-MS) to identify as well as to quantify compounds. High-throughput LC-MS-based metabolomics can offer an in-depth examining the low molecular weight naturally occurring metabolites within cells, which indicate variations in cellular regulation and metabolic pathways (Wang et al. 2020). High-throughput technologies for interactomics include yeast two-hybrid assays, protein microarrays, and mass spectrometry-based approaches like affinity purification coupled with mass spectrometry (AP-MS). These methodologies allow the detection and characterization of protein-protein interactions, protein-DNA interactions, and protein complex compositions on a large scale (Blikstad and Ivarsson 2016). The large-scale analysis of cellular and molecular events is made possible by high-throughput imaging technology. Automated microscopy systems with robotic sample handling, image acquisition, and image analysis capabilities can produce large volumes of imaging data. These techniques have been used for studying protein localization, cellular phenotypes, and dynamic processes inside cells and tissues (Perlman et al. 2004).

When coupled with cutting-edge computational and statistical techniques, these high-throughput technologies make it easier to explore complex biological systems and provide large-scale datasets that enhance systems biology research.

20.4 Application of Systems Biology in Neurodegenerative Diseases

20.4.1 Alzheimer's Disease: Insights from Systems Biology

Alzheimer's disease (AD) is the most prevalent form of dementia that is characterised by progressive deterioration of cognitive functions, memory loss, and behavioural changes. Although extensive studies have been conducted to uncover the underlying mechanisms of AD, still a lot remains unknown. Systems biology has been used in AD research, and it provided valuable insights into the pathogenesis of the disease.

Network analysis of AD revealed the accumulation of insoluble deposits of amyloid-β (Aβ) protein outside of brain cells and hyperphosphorylated tau protein inside brain cells, forming neurotoxic aggregates (Lee et al. 2017). The systems biology approach has helped elucidate the intricate molecular pathways involved in producing, clearing, and aggregating Aβ and tau proteins. It has provided insights into AD's complex interplay between genetic and environmental factors. To thoroughly understand the pathways and networks relating to AD, systems biology techniques have been used to analyze GWAS (genome-wide association studies) data and combine it with other omics data, including transcriptomics and proteomics. These approaches have also highlighted the significance of non-neuronal cells like astrocytes and microglia in AD pathogenesis. These cells play crucial roles in neuroinflammation, synaptic function, and the clearance of Aβ. Systems biology models can simulate the interactions between neurons and these supporting cells, providing insights into the complex cellular crosstalk that occurs in AD (Wood et al. 2015). Additionally, computational models of AD progression have been created using systems biology. These models integrate multiple data types, such as clinical data, imaging data, and molecular profiles, to simulate disease progression over time. By capturing the dynamics of AD, these models can help with disease outcome prediction, the assessment of prospective treatment strategies, and the discovery of new biomarkers. Overall, systems biology approaches have provided valuable insights into the pathophysiology of AD, identified potential biomarkers, and opened up new possibilities for therapeutic interventions in the future (Wang et al. 2019; Savelieff et al. 2022).

20.4.2 Parkinson's Disease (PD): Systems Approaches for Mechanistic Understanding

PD is considered the second most prevalent neurodegenerative condition, with AD being the only disorder that surpasses it in frequency. PD is distinguished by motor difficulties, including muscle stiffness, tremors, and slowness of movement (bradykinesia). These symptoms arise from the deterioration of dopaminergic neurons in the substantia nigra area of the brain. A combination of genetic and environmental factors is believed to cause PD. To gain a mechanistic comprehension of the disease, researchers have taken a systems approach in their studies (Wood et al. 2015; Savelieff et al. 2022).

High-throughput technologies such as genome-wide association studies (GWAS) and transcriptomic profiling identify genetic variants and gene expression changes associated with PD. Various modifiable risk factors and exposure to potentially toxic substances can also be said to contribute to its pathogenesis. Proteomics and metabolomics profiling techniques identify alteration in protein expression, post-transcriptional modification, and metabolic pathways in PD. Histologically, the accumulation of protein aggregates called Lewy bodies, which contain α-synuclein, are found in different regions of the nervous system (Braak et al. 2013). Moreover, PD progression is not solely dependent on cell-autonomous processes, as astroglia, a type of brain cell, also contribute to the disease progression in a non-cell autonomous manner. Lipidomics has also revealed dysregulation of lipid metabolisms due to mutation of related genes such as glucosylceramidase beta (GBA), sphingomyelin phosphodiesterase 1 (SMPD1), galactosylceramidase (GALC), phospholipase A2 group VI (PLA2G6), and sterol regulatory element-binding transcription factor 1 (SREBF1) (Alecu and Bennett 2019). Furthermore, metagenomic studies have been conducted in PD due to the accumulation of α-synuclein fibrils in the gastrointestinal tract. These studies explore the possible link between PD and dysbiosis (imbalance) in the gut microbiome. Computational models, such as mathematical models and simulations, help investigate how cellular and molecular changes contribute to neuronal dysfunction and degeneration. It also helps develop and optimize deep brain stimulation (DBS) as a proficient approach for PD therapy. Neuroimaging techniques such as MRI and single-photon emission computed tomography (SPECT) provide valuable information about disease progression, neurodegeneration patterns, and potential therapies' effects (Savelieff et al. 2022; Yu et al. 2020).

By employing these systems approaches, researchers aim to unravel the complex mechanisms underlying PD, which will help in the development of targeted therapies and personalized medicines.

20.4.3 Huntington's Disease (HD): Network Analysis and Therapeutic Strategies

HD is an autosomal dominant progressive neurodegenerative disorder delineated by motor dysfunction, cognitive decline, and psychiatric abnormalities. The condition arises from an unsteady increase in the repetition of a trinucleotide sequence (CAG) within the coding section of the IT-15 gene. In HD, this gene results in the generation of an abnormal form of the huntingtin (Htt) protein, which contains an elongated polyglutamine (polyQ) tract (Giorgini and Muchowski 2005). The mutant Htt protein aggregates into insoluble inclusion bodies within neurons when the polyQ region in the protein crosses a critical threshold of roughly 37 glutamines. This leads to the degeneration of specific brain regions, including the cerebral cortex and striatum. It is widely believed that the mutant Htt protein exerts its pathological effects in a dominant manner, causing cytotoxicity. Since most HD patients have one standard copy and one mutant copy of the IT-15 gene, the toxic effects of mutant Htt are thought to result from abnormal protein interactions involving the mutant protein. However, recent studies suggest that HD pathogenesis may also be influenced by the depletion of normal Htt protein and the loss of its regular interactions. In order to better comprehend the pathological mechanism of HD, researchers have generated a protein-protein interaction network involving the Htt protein, which is implicated in HD. This identified novel interaction partners of Htt, including G-protein-coupled receptor kinase interactor 1 (GIT1), which was found to be involved in Htt aggregation. Other proteins that interacted with GIT1, such as BARD1 and HIP5, were also recognized, which provide further avenues for investigating the role of GIT1 in Htt aggregation and HD progression (Wood et al. 2015; Giorgini and Muchowski 2005).

Currently, there is no cure for HD, but several therapeutic strategies are being investigated that aim to alleviate symptoms, slow disease progression, or target-specific aspects of HD pathology. New medications are explored that target specific HD-related pathways, such as reducing mutant huntingtin production, enhancing the removal of mHTT aggregates, or modulating the neurotransmitter systems affected by the disease. RNA interference (RNAi) and antisense oligonucleotides (ASOs) are also being investigated as potential treatments for HD. These methods aim to suppress the expression of mutant huntingtin by targeting the HTT mRNA. Stem cell-based therapies show promise in replacing damaged and affected neurons. Deep Brain Stimulation entails the placement of electrodes in precise areas of the brain to regulate neuronal activity. It is being investigated as a symptomatic treatment for HD-related movement disorders. Finally, non-pharmacological interventions, such as physical, occupational, and speech therapy, can help manage motor and cognitive symptoms and enhance the quality of life for individuals with HD (Giorgini and Muchowski 2005).

20.4.4 Amyotrophic Lateral Sclerosis (ALS): Systems Perspectives

ALS, also known as Lou Gehrig's disease, is a progressive neurological disorder that affects neurons in the motor cortex and spinal cord, which causes muscle stiffness and weakness, loss of motor function, and eventually paralysis, leading to death within 5–8 years of onset of the symptoms (Wood et al. 2015). Looking at ALS from a systems perspective involves examining the various interconnected systems and processes that play a vital role in the initiation and advancement of the disease. Extensive molecular studies have revealed several significant pathological features of ALS that are common to other neurodegenerative diseases. Some of the factors involved in the development and progression of the disease are abnormalities in TDP-43 protein accumulations, disturbances in autophagy, excitotoxicity, impaired metabolism, malfunctioning mitochondria, inflammation, and oxidative stress. In the histopathology of ALS, TAR DNA-binding protein 43 (TDP-43) inclusion bodies are frequently observed. TDP-43 is responsible for regulating processes such as transcription, pre-mRNA splicing, mRNA translation, and microRNA biosynthesis. Therefore, numerous studies have concentrated on investigating the disruption of the epigenome and transcriptome in ALS, including the role of microRNAs (miRNAs). miRNAs are short non-coding RNAs that degrade specific target mRNAs, thus preventing their expression and influencing various aspects such as neurogenesis, neuromuscular junction structure and function, and inflammation in ALS (Rinchetti et al. 2018).

Environmental exposures can also influence the epigenome through DNA methylation and histone acetylation alterations. In post-mortem spinal cord tissue from sporadic ALS patients, epigenetic dysregulation and higher levels of overall DNA methylation and hydroxymethylation have been noted. A collection of dysregulated genes that displayed changes in DNA methylation were identified through analysis of differentially expressed genes (DEGs) and differentially methylated genes (DMGs). These genes were associated with biological pathways involved in neuron adhesion, immunological response, plasma membrane function, and defense response. Transcriptomic profiling in ALS has provided insights into the spatiotemporal progression of the disease. Studies on mice and human spinal cord tissue revealed dysregulation of various processes like extracellular matrix-receptor interaction, immune signalling pathways, axon guidance, and metabolic pathways. Impaired metabolism, particularly lipid metabolism, emerged as a hallmark of ALS (Maniatis et al. 2019). Additionally, metabolomics analysis of ALS patient plasma revealed changes in lipid metabolism as well as disruption in creatine metabolism, which is likely associated with muscle wasting in ALS (Savelieff et al. 2022).

Thus, taking a systems perspective helps researchers and doctors understand the complex interactions and nature of the various systems involved in ALS, which may offer valuable insights for the advancement of novel therapeutic strategies.

20.5 Systems Biology and Neurodevelopmental Disorders

20.5.1 Autism Spectrum Disorders (ASD): Unraveling Complexity Through Systems Biology

ASD represent a collection of neurodevelopmental conditions depicted by challenges in social interaction, communication difficulties, and repetitive and restricted behaviours. ASD, being a multifaceted disorder with a wide array of symptoms and varying degrees of severity, still holds elusive causes that have yet to be fully comprehended. Nonetheless, the realm of systems biology has emerged as a potent instrument in unravelling the intricacies of ASD and shedding light on its underlying biological mechanisms (Diaz-Beltran et al. 2013).

Systems biology, an interdisciplinary field that amalgamates data from various biological levels encompassing genes, proteins, cells, and organs, aims to fathom their collective interactions and holistic functionality. Within the context of ASD, systems biology endeavours to identify the molecular and cellular networks involved in brain development and functioning while scrutinizing perturbations that potentially contribute to the pathology of ASD.

An inherent challenge in ASD research stems from its heterogeneity. ASD encompasses a spectrum of disorders wherein a multitude of genetic, environmental, and epigenetic factors contribute to its etiology. Systems biology methodologies prove instrumental in deciphering this complexity by analyzing expansive datasets and integrating information from diverse sources like genomics, transcriptomics, proteomics, and metabolomics. By implementing systems biology techniques, researchers have successfully detected numerous genetic risk factors associated with ASD. GWAS have uncovered common genetic variations that confer susceptibility to ASD. Furthermore, studies utilizing whole-exome sequencing have detected rare de novo mutations in genes implicated in synaptic function and neuronal development. The integration of these genetic findings with functional genomic data has culminated in the identification of converging molecular pathways and disrupted biological processes in ASD. Systems biology approaches extend their scope to encompass environmental factors that may contribute to ASD. Exposure to specific toxins or maternal immune activation during pregnancy has been associated with an elevated risk of ASD. By integrating environmental data with genetic and molecular information, researchers can glean insights into gene-environment interactions and their implications for ASD susceptibility (Bölte et al. 2019). Moreover, systems biology aids in unravelling the intricate interplay between different biological systems. For instance, the examination of the interactions between the immune system and the brain holds promise in elucidating the role of neuroinflammation in ASD. Researchers can discern potential therapeutic targets for intervention by constructing models to depict these interactions (Dhillon et al. 2020; Noris García et al. 2022).

20.5.2 Schizophrenia: Integrative Approaches for a Systems Understanding

Schizophrenia, a complex and chronic mental disorder, affects around 1% of the global population. It encompasses a range of symptoms, including hallucinations, delusions, social withdrawal, and reduced emotional expression. Unravelling the underlying causes of this condition necessitates a comprehensive exploration that considers the intricate interplay between genetic, environmental, and neurobiological factors. Researchers have embraced integrative approaches grounded in systems biology to delve deep into the complexities of this disorder (Desai et al. 2013).

In the realm of schizophrenia research, integrative approaches synthesize data from various biological levels, such as genetics, epigenetics, neuroimaging, and clinical assessments. The amalgamation of these diverse sources aims to unveil the intricate network of interactions and dysregulations that drive the development and progression of schizophrenia.

From a genetic perspective, GWAS have detected certain common genetic variants that increase the risk of schizophrenia. However, these variants only explain a fraction of the disorder's heritability. Integrative approaches merge genetic data with other layers of information, like gene expression profiles, to identify disrupted molecular pathways and biological processes involved in schizophrenia. This holistic perspective enables a more comprehensive understanding of the disorder's genetic foundations (Wahbeh and Avramopoulos 2021).

In the domain of neuroimaging, studies utilizing techniques like fMRI and diffusion tensor imaging (DTI) play a pivotal role in unravelling the neurobiological alterations associated with schizophrenia. Integrative approaches integrate these neuroimaging findings with genetic and clinical data to reveal aberrant brain networks and their intricate relationship with symptoms and cognitive impairments. Environmental factors, including prenatal infections, early life stress, and substance abuse, have also been implicated in the development of schizophrenia. Integrative approaches incorporate these environmental factors into the analysis, along with genetic and neurobiological data. This comprehensive perspective allows researchers to investigate gene-environment interactions and their impact on the disorder's development. By employing integrative approaches, researchers strive to achieve a systems-level understanding of schizophrenia. This pursuit involves unravelling the intricate interactions among genetic, environmental, and neurobiological factors and their collective contribution to the diverse clinical manifestations of the disorder. Such an understanding holds the potential to identify novel biomarkers, therapeutic targets, and personalized treatment strategies for individuals grappling with schizophrenia (Wahbeh and Avramopoulos 2021).

20.6 Systems Medicine in Neurology and Neurosciences

20.6.1 Personalized Medicine: Systems Approaches to Diagnosis and Treatment

Systems biology is the computational and mathematically simulating of complicated biological systems to research biological component systems (such as molecules, cells, animals, or whole species). This is turning out to be essential for all branches of biology and medicine. Thus, systems biology may be viewed as a "holistic" method for examining the complexity of biological systems (Loscalzo and Barabasi 2011). From a biological perspective, organisms are recognized as dynamic and intricate entities due to their composition of multiple interconnected networks that engage in interactions across various levels, encompassing the genome, molecules, cells, and, ultimately, the entire organism. This holistic organism interacts with its surrounding environment. As a result, it can be challenging to predict an organism's behaviour solely by looking at how each of its parts works. To put it another way, a living thing might be thought of as a "network of networks," hence systems biology concentrates on the comprehensive examination of these networks' dynamic behaviours at many scales in order to create theories for biological action and dynamical shifts. By pre-identifying patients who are likely to derive the greatest benefit from a particular treatment and distinguishing them from those who might experience adverse effects, this approach can contribute to comprehending the factors that influence treatment responses and enable the customization of targeted therapies. Through the integration of individualized genetic, molecular, and environmental factors using diagnostic tools or imaging technologies, personalized medicine can be seen as an expansion of traditional methods for comprehending and managing diseases. The comprehensive examination of these networks' dynamic behaviours at many scales in order to create theories for biological action and dynamical shifts. In order to analyze the patient's unique traits, this technique may thus make it easier to determine the factors that influence how well a patient responds to therapy. As a result, the focus of the new paradigm is moving from studying the "average" cell to the "discrete" cell subtype in a population of cells, much as it is doing with studying "the average" patient in a large population to evaluating the distinctive traits of a particular patient. In order to create markers that may identify the illness at an early stage of its development and to enable targeted treatments to avert pathological consequences, several molecular screening techniques have already been created for the risk stratification of patients. For instance, a lot of work is presently put into finding the precise molecular markers in a collection of cancerous cells in order to build precise therapies. Although such markers are not highly personalized, current cardiology practice also depends on a number of biomarkers, imaging, and clinical data for the primary and secondary prevention of neurological disorders. However, genetic testing and contemporary transcriptomic, proteomic, and metabolomic technologies are able to collect and provide information about the interactions between an individual and their environment (Gotovac et al. 2014).

20.6.2 Biomarkers and Diagnostic Tools in Systems Neurology

Table 20.1 provides a brief idea about the different biomarkers that can be used for diagnostics purposes in neurological disorders.

Table 20.1 Details of identifiable biomarkers

Sl. no	Biomarker name	Source	Type	Detection strategy	Reference
1.	Circulating nucleic acids	Body fluid of the patient	Cell-free DNA and RNA	This could be a reflection of distinctive features of the biological processes that are involved in neurological diseases	Bruno et al. (2020)
2.	Neurofilament protein	Cerebrospinal fluid of patient	Protein	Major neuron-specific components that are sensitive to neurodegeneration and neuronal injury in a wide range of neurologic diseases and maintain structural integrity	Housley et al. (2015)
3.	Oligoclonal bands	Cerebrospinal fluid of patient	Protein	These proteins indicate multiple sclerosis characterized by central nervous system inflammation	Mayeux (2004)
4.	Apolipoprotein A1	Body fluid of the patient	Protein	Providing information regarding the pathogenesis of neurological disorders, with which information researchers can determine whether other genes or environmental risk factors altered disease risk	Chmielewska et al. (2018)
5.	Rno-miR-9a-3p	Serum of patient	miRNA	Potential marker for increased epileptic-to-normal neuronal differentiation	Zhang et al. (2010)

20.6.3 Translational Challenges and Future Directions

Translational systems biology (TSB) involves the utilization of dynamic mathematical modelling and specific design principles in biological systems to establish connections between mechanisms and phenomena, with a primary focus on influencing clinical practice. This approach has broad applicability across various domains, including inflammation and wound healing, by integrating and interlinking them (Nord et al. 2015).

In the field of genetics, there have been endeavours to establish connections between genetic variations and gene expression, as well as protein-protein interaction (PPI) networks, resulting in valuable insights. However, in the context of complex diseases, the majority of genetic variations associated with the disease are located in noncoding regulatory regions of the genome. To further advance high-throughput molecular investigations in the brain, the next crucial step is to understand regulatory changes and interactions during development. Techniques like histone mark profiling and chromosome conformation capture can aid in achieving this objective. System biology has given a perspective to this question of high-throughput studies (Zhang et al. 2013; Chen and Geschwind 2022).

Moreover, attaining comprehension of the transcriptomic and epigenetic alterations at a finer scale, such as within uniform cellular populations or at a single-cell resolution, will substantially augment our knowledge of the intricate mechanisms underlying normal human brain development. The PsychENCODE consortium is currently constructing initial maps of these neurobiologically relevant epigenetic landscapes and cell type differences, primarily at the tissue level (Parikshak et al. 2015).

Genome-wide transcriptomic approaches offer a quantitative endophenotype, which can serve as a biomarker for disease states and be utilized to improve the measurement of disease states in genetic association studies. Additionally, correlations between transcriptomic and other molecular systems measurements and phenotypes from systems neuroscience, such as MRI and fMRI, as well as behavioural phenotypes, can help identify non-invasive indicators of disease states (Chen et al. 2014; Civelek and Lusis 2013).

20.7 Case Studies and Success Stories

20.7.1 Breakthroughs in Systems Biology for Neurological Disorders

In recent years, significant breakthroughs have been made in the field of systems biology, particularly in relation to neurological disorders. Systems biology is an interdisciplinary approach combining various scientific disciplines, including biology, mathematics, computer science, and engineering, to understand complex biological systems holistically. This approach has provided new insights into the underlying mechanisms of neurological disorders and has the potential to revolutionize the way we diagnose, treat, and prevent these conditions (Lozano et al. 2019).

One of the critical breakthroughs in systems biology for neurological disorders is the development of high-throughput technologies for large-scale data collection. These technologies enable researchers to gather massive amounts of data on genes, proteins, and other molecular components involved in neurological disorders. By integrating and analyzing this vast dataset, scientists can identify critical molecular pathways and networks that contribute to the development and progression of these disorders. This comprehensive understanding of the intricate biological systems involved allows for the identification of novel therapeutic targets. Furthermore, systems biology has contributed to the development of computational models and simulations that can mimic the behaviour of the brain and nervous system. These models provide a platform for studying neurological disorders' dynamics and testing potential drug candidates in a virtual setting. By leveraging computational power, researchers can simulate the effects of various interventions and optimize treatment strategies for individual patients (Paul et al. 2021).

20.7.2 Impact of Systems Biology on Clinical Practice and Patient Outcomes

The impact of systems biology on clinical practice and patient outcomes is profound and far-reaching. By providing a holistic understanding of biological systems and their dysregulation in diseases, systems biology has the potential to transform the way we approach diagnosis, treatment, and patient care (Naylor and Chen 2010).

One of the key impacts of systems biology on clinical practice is the development of personalized medicine. Integrating patient-specific data, including genomic information, molecular profiles, and clinical data, systems biology allows for tailoring treatment plans to individual patients. This personalized approach considers the unique molecular characteristics of each patient's disease, enabling targeted therapies and reducing the likelihood of adverse effects. Systems biology can improve treatment efficacy and patient outcomes by optimizing treatment strategies based on a patient's specific profile (Strianese et al. 2020).

Additionally, systems biology contributes to the identification of novel biomarkers for early disease detection and prognosis. Researchers can uncover molecular signatures indicating diseases at their earliest stages by analyzing multi-omics data and utilizing advanced computational algorithms. This early detection facilitates timely interventions, leading to more successful treatment outcomes and improved patient survival rates. Furthermore, using biomarkers to monitor disease progression allows for the adjustment of treatment plans as needed, ensuring that patients receive the most appropriate care throughout their disease journey (Yan et al. 2018).

Systems biology also plays a crucial role in drug discovery and development. By utilizing computational models and simulations, researchers can predict the effects of potential drug candidates on complex biological systems. This approach enables the identification of promising drug targets and the optimization of treatment regimens before embarking on costly and time-consuming clinical trials. By streamlining the drug development process, systems biology expedites the translation of scientific discoveries into effective therapeutics, benefitting patients by providing them with more treatment options and accelerating the availability of innovative therapies (Kraljevic et al. 2004).

Moreover, systems biology contributes to a deeper understanding of disease mechanisms, allowing clinicians to make more informed decisions regarding treatment options. Systems biology uncovers potential off-target effects and drug interactions by elucidating diseases' intricate molecular networks and pathways, thereby minimizing adverse events and optimizing treatment outcomes (Zou et al. 2013).

The impact of systems biology on clinical practice and patient outcomes is transformative. Systems biology can revolutionize patient care by enabling personalized medicine, facilitating early disease detection, accelerating drug discovery, and enhancing treatment decision-making. Integrating systems biology approaches into clinical practice can improve treatment efficacy, reduce adverse effects, and ultimately enhance patients' overall well-being and outcomes (Flores et al. 2013).

20.8 Ethical and Social Implications of Systems Biology in Neurology

20.8.1 Privacy and Data Protection in Systems Neuroscience

Researchers have been able to collect a lot of data from the brains of individual participants due to improvements in techniques for monitoring and recording brain activity. The speed and scope of data processing and analysis have increased because of breakthroughs in systems neuroscience. Neuroscience research has joined the considerable data age with the help of these cutting-edge software tools and algorithms to analyze enormous datasets. Increased rigor and reproducibility in neuroscience research resulted from the availability and volume of neuroscience data. This may also lead to new understandings of the processes underlying brain function in both healthy and diseased states. However, technology also brings up critical ethical challenges like data privacy (Eberwine and Kahn 2020). The sharing and going back over of neuroscience information could bring about the unexpected openness of this touchy data since individual sensitive information is frequently paired with neuroscience data (such as a person's diagnosis of a neurological condition or their genetic makeup). Reverse inference from neuroimaging or brain recording data may reveal the cognitive and emotional states of the participants. To address these issues, a number of legislative and policy proposals have been made to safeguard neuroscience data. The majority of these strategies have concentrated on limiting who may access and disclose neuroscience data, placing stringent restrictions on who can access data for what objectives, enforcing severe data safeguards (like deidentification or encryption of information), or making novel common liberties to safeguard mental security (Ienca et al. 2022).

20.8.2 Ethical Considerations in Personalized Medicine

Personalized medicine is frequently viewed favourably. Knowing more about a person's genetic makeup might be helpful when developing a treatment plan because new health technologies are being created to treat health concerns. The academic literature has discussed a number of ethical issues in addition to the numerous benefits of personalized medicine. Individuals having the option to get treatments that are bound to help them is, obviously, one of the super moral benefits of customized medication. This fulfills the clinical commitment to do the greatest good (and the least bad). Another difficulty with personalized medicine is the morality of inadvertent discoveries. What are our ethical responsibilities in the event that someone undergoes a medical examination with the intention of receiving individualized care and unexpected information about their health condition or overall health is discovered along with its confidentiality and privacy (Larijani et al. 2017)!

20.8.3 Societal Implications and Public Engagement

There are growing worries about the societal repercussions of the numerous systems biology applications that are being developed. Systems biology is sometimes referred to as a "wicked problem," necessitating policy or regulatory judgments that balance technological and ethical issues. To bargain the proper role of these applications in society, many of the policy issues surrounding systems biology will rely on input from scientists, legislators, public stakeholders, and industry—and all of their various value systems, policy preferences, and understandings (Bedau et al. 2009). Additional studies on scientific experts and choices made on the application of science in society have concentrated on the factors that these choices are based on. The public at least implicitly understands that many of the consequences of systems biology cannot be resolved based only on science and that scientists and specialists may have quite different values and perspectives than other stakeholders regarding the ramifications. As a result, both the general public and specialist groups are increasingly demanding public involvement (Rose et al. 2018).

20.9 Conclusion

The application of systems biology in neurology and neurosciences has significantly advanced our understanding of associated diseases. Through the integration of high-throughput technologies, computational modelling, and multi-omics data, researchers have gained valuable insights into the complex molecular pathways and networks underlying neurological disorders. This holistic approach has led to breakthroughs in identifying potential biomarkers for early diagnosis, discovering novel therapeutic targets, and developing personalized treatment strategies. Furthermore, systems biology has paved the way for virtual simulations, enabling the study of disease progression, and testing interventions in a virtual setting. The continued collaboration and integration of diverse disciplines will further enhance our understanding and management of neurological diseases.

References

Aderem A (2005) Systems biology: its practice and challenges. Cell 121(4):511–513. https://doi.org/10.1016/J.CELL.2005.04.020

Alecu I, Bennett SAL (2019) Dysregulated lipid metabolism and its role in α-synucleinopathy in Parkinson's disease. Front Neurosci 13:328. https://doi.org/10.3389/FNINS.2019.00328

Ambardar S, Gupta R, Trakroo D, Lal R, Vakhlu J (2016) High throughput sequencing: an overview of sequencing chemistry. Indian J Microbiol 56(4):394–404. https://doi.org/10.1007/S12088-016-0606-4/TABLES/1

Barabási AL, Albert R (1999) Emergence of scaling in random networks. Science 286(5439):509–512. https://doi.org/10.1126/SCIENCE.286.5439.509/ASSET/72325557-9D24-4C96-BEEB-EAF2CA37B5FF/ASSETS/GRAPHIC/SE4097898002.JPEG

Barabási AL, Oltvai ZN (2004) Network biology: understanding the cell's functional organization. Nat Rev Genet 5(2):101–113. https://doi.org/10.1038/NRG1272

Bedau MA, Parke EC, Tangen U, Hantsche-Tangen B (2009) Social and ethical checkpoints for bottom-up synthetic biology, or protocells. Syst Synth Biol 3(1-4):65. https://doi.org/10.1007/S11693-009-9039-2

Blikstad C, Ivarsson Y (2016) High-throughput methods for identification of protein-protein interactions involving short linear motifs. Cell Commun Signal 13(1):1–9. https://doi.org/10.1186/S12964-015-0116-8

Bölte S, Girdler S, Marschik PB (2019) The contribution of environmental exposure to the etiology of autism spectrum disorder. Cell Mol Life Sci 76(7):1275. https://doi.org/10.1007/S00018-018-2988-4

Braak H, Brettschneider J, Ludolph AC, Lee VM, Trojanowski JQ, Del TK (2013) Amyotrophic lateral sclerosis–a model of corticofugal axonal spread. Nat Rev Neurol 9(12):708–714. https://doi.org/10.1038/NRNEUROL.2013.221

Bruno DCF, Donatti A, Martin M et al (2020) Circulating nucleic acids in the plasma and serum as potential biomarkers in neurological disorders. Braz J Med Biol Res 53(10):1–15. https://doi.org/10.1590/1414-431X20209881

Chen GT, Geschwind DH (2022) Challenges and opportunities for precision medicine in neurodevelopmental disorders. Adv Drug Deliv Rev 191:114564. https://doi.org/10.1016/J.ADDR.2022.114564

Chen JC, Alvarez MJ, Talos F et al (2014) Identification of causal genetic drivers of human disease through systems-level analysis of regulatory networks. Cell 159(2):402–414. https://doi.org/10.1016/J.CELL.2014.09.021

Chmielewska N, Szyndler J, Makowska K, Wojtyna D, Maciejak P, Płaźnik A (2018) Looking for novel, brain-derived, peripheral biomarkers of neurological disorders. Neurol Neurochir Pol 52(3):318–325. https://doi.org/10.1016/j.pjnns.2018.02.002

Civelek M, Lusis AJ (2013) Systems genetics approaches to understand complex traits. Nat Rev Genet 15(1):34–48. https://doi.org/10.1038/nrg3575

Costanzo M, Zacchia M, Bruno G et al (2017) Integration of proteomics and metabolomics in exploring genetic and rare metabolic diseases. Kidney Dis 3(2):66–77. https://doi.org/10.1159/000477493

Dada JO, Mendes P (2011) Multi-scale modelling and simulation in systems biology. Integr Biol 3(2):86–96. https://doi.org/10.1039/C0IB00075B

Desai PR, Lawson KA, Barner JC, Rascati KL (2013) Estimating the direct and indirect costs for community-dwelling patients with schizophrenia. J Pharm Health Serv Res 4(4):187–194. https://doi.org/10.1111/JPHS.12027/EPDF

Dhillon BK, Smith M, Baghela A, Lee AHY, Hancock REW (2020) Systems biology approaches to understanding the human immune system. Front Immunol 11:1683. https://doi.org/10.3389/FIMMU.2020.01683

Diaz-Beltran L, Cano C, Wall DP, Esteban FJ (2013) Systems biology as a comparative approach to understand complex gene expression in neurological diseases. Behav Sci 3(2):253–273. https://doi.org/10.3390/BS3020253

Eberwine J, Kahn J (2020) The BRAIN initiative and neuroethics: Enabling and Enhancing Neuroscience Advances for Society. AJOB Neurosci 11(3):135–139. https://doi.org/10.1080/21507740.2020.1778121

Flores M, Glusman G, Brogaard K, Price ND, Hood L (2013) P4 medicine: how systems medicine will transform the healthcare sector and society. Pers Med 10(6):565. https://doi.org/10.2217/PME.13.57

Giorgini F, Muchowski PJ (2005) Connecting the dots in Huntington's disease with protein interaction networks. Genome Biol 6(3):1–5. https://doi.org/10.1186/GB-2005-6-3-210/FIGURES/2

Gomez-Cabrero D, Abugessaisa I, Maier D et al (2014) Data integration in the era of omics: current and future challenges. BMC Syst Biol 8(2):I1. https://doi.org/10.1186/1752-0509-8-S2-I1/FIGURES/2

Gotovac K, Hajnšek S, Pašić MB, Pivac N, Borovečki F (2014) Personalized medicine in neurodegenerative diseases: how far away? Mol Diagn Ther 18(1):17–24. https://doi.org/10.1007/S40291-013-0058-Z/METRICS

Grant SGN (2003) Systems biology in neuroscience: bridging genes to cognition. Curr Opin Neurobiol 13(5):577–582. https://doi.org/10.1016/j.conb.2003.09.016

Gupta MK, Misra K (2016) A holistic approach for integration of biological systems and usage in drug discovery. In: Network modeling analysis in health informatics and bioinformatics, vol 5. Springer, Cham. https://doi.org/10.1007/S13721-015-0111-4

Hampel H, Caraci F, Cuello AC et al (2020) A path toward precision medicine for neuroinflammatory mechanisms in Alzheimer's disease. Front Immunol 11(21):456. https://doi.org/10.3389/FIMMU.2020.00456

Hemedan AA, Niarakis A, Schneider R, Ostaszewski M (2022) Boolean modelling as a logic-based dynamic approach in systems medicine. Comput Struct Biotechnol J 20:3161–3172. https://doi.org/10.1016/J.CSBJ.2022.06.035

Hieter P, Boguski M (1997) Functional genomics: it's all how you read it. Science 278(5338):601–602. https://doi.org/10.1126/SCIENCE.278.5338.601

Horgan RP, Kenny LC (2011) 'Omic' technologies: genomics, transcriptomics, proteomics and metabolomics. Obst Gynaecol 13(3):189–195. https://doi.org/10.1576/TOAG.13.3.189.27672

Housley WJ, Pitt D, Hafler DA (2015) Biomarkers in multiple sclerosis. Clin Immunol 161(1):51–58. https://doi.org/10.1016/J.CLIM.2015.06.015

Ideker T, Galitski T, Hood L (2001) A new approach to decoding life: systems biology. Annu Rev Genomics Hum Genet 2:343–372. https://doi.org/10.1146/ANNUREV.GENOM.2.1.343

Ienca M, Fins JJ, Jox RJ et al (2022) Towards a governance framework for brain data. Neuroethics 15(2):1–14. https://doi.org/10.1007/S12152-022-09498-8/FIGURES/2

Khodadadian A, Darzi S, Haghi-Daredeh S et al (2020) Genomics and transcriptomics: the powerful technologies in precision medicine. Int J Gen Med 13:627. https://doi.org/10.2147/IJGM.S249970

Kohl P, Crampin EJ, Quinn TA, Noble D (2010) Systems biology: an approach. Clin Pharmacol Ther 88(1):25–33. https://doi.org/10.1038/CLPT.2010.92

Kraljevic S, Stambrook PJ, Pavelic K (2004) Accelerating drug discovery. EMBO Rep 5(9):837. https://doi.org/10.1038/SJ.EMBOR.7400236

Kuo TC, Tian TF, Tseng YJ (2013) 3Omics: a web-based systems biology tool for analysis, integration and visualization of human transcriptomic, proteomic and metabolomic data. BMC Syst Biol 7(1):1–15. https://doi.org/10.1186/1752-0509-7-64/FIGURES/6

Larijani B, Salari P, Larijani B (2017) Ethical issues surrounding personalized medicine: a literature review. Acta Med Iran 55(3):209–217. https://acta.tums.ac.ir/index.php/acta/article/view/6215
Lee SJC, Nam E, Lee HJ, Savelieff MG, Lim MH (2017) Towards an understanding of amyloid-β oligomers: characterization, toxicity mechanisms, and inhibitors. Chem Soc Rev 46(2):310–323. https://doi.org/10.1039/C6CS00731G
Likić VA, McConville MJ, Lithgow T, Bacic A (2010) Systems biology: the next frontier for bioinformatics. Adv Bioinforma 2010:268925. https://doi.org/10.1155/2010/268925
Loscalzo J, Barabasi AL (2011) Systems biology and the future of medicine. Wiley Interdiscip Rev Syst Biol Med 3(6):619–627. https://doi.org/10.1002/WSBM.144
Lozano AM, Lipsman N, Bergman H et al (2019) Deep brain stimulation: current challenges and future directions. Nat Rev Neurol 15(3):148. https://doi.org/10.1038/S41582-018-0128-2
Maghuly F, Marzban G (2021) Editorial: omics technologies toward systems biology. Front Genet 12:756847. https://doi.org/10.3389/FGENE.2021.756847/BIBTEX
Maniatis S, Äijö T, Vickovic S et al (2019) Spatiotemporal dynamics of molecular pathology in amyotrophic lateral sclerosis. Science 364(6435):89–93. https://doi.org/10.1126/SCIENCE.AAV9776
Mayeux R (2004) Biomarkers: potential uses and limitations. NeuroRx 1(2):182. https://doi.org/10.1602/NEURORX.1.2.182
Messner CB, Demichev V, Wang Z et al (2023) Mass spectrometry-based high-throughput proteomics and its role in biomedical studies and systems biology. Proteomics 23(7-8):2200013. https://doi.org/10.1002/PMIC.202200013
Moore JB, Weeks ME (2011) Proteomics and systems biology: current and future applications in the nutritional sciences. Adv Nutr 2(4):355–364. https://doi.org/10.3945/AN.111.000554
Naylor S, Chen JY (2010) Unraveling human complexity and disease with systems biology and personalized medicine. Pers Med 7(3):275. https://doi.org/10.2217/PME.10.16
Nord AS, Pattabiraman K, Visel A, Rubenstein JLR (2015) Genomic perspectives of transcriptional regulation in forebrain development. Neuron 85(1):27–47. https://doi.org/10.1016/J.NEURON.2014.11.011
Noris García MA, Guerra F, Hurtado V et al (2022) Immune dysregulation in autism spectrum disorder: what do we know about it? Int J Mol Sci 23(6):3033. https://doi.org/10.3390/IJMS23063033
Palacios R, Goni J, Martinez-Forero I et al (2007) A network analysis of the human T-cell activation gene network identifies Jagged1 as a therapeutic target for autoimmune diseases. PLoS One 2(11):e1222. https://doi.org/10.1371/JOURNAL.PONE.0001222
Parikshak NN, Gandal MJ, Geschwind DH (2015) Systems biology and gene networks in neurodevelopmental and neurodegenerative disorders. Nat Rev Genet 16(8):441–458. https://doi.org/10.1038/nrg3934
Paul D, Sanap G, Shenoy S, Kalyane D, Kalia K, Tekade RK (2021) Artificial intelligence in drug discovery and development. Drug Discov Today 26(1):80. https://doi.org/10.1016/J.DRUDIS.2020.10.010
Perlman ZE, Slack MD, Feng Y, Mitchison TJ, Wu LF, Altschuler SJ (2004) Multidimensional drug profiling by automated microscopy. Science 306(5699):1194–1198. https://doi.org/10.1126/SCIENCE.1100709/SUPPL_FILE/PERLMAN.SOM.ZIP
Rinchetti P, Rizzuti M, Faravelli I, Corti S (2018) MicroRNA metabolism and dysregulation in amyotrophic lateral sclerosis. Mol Neurobiol 55(3):2617–2630. https://doi.org/10.1007/S12035-017-0537-Z
Rochfort S (2005) Metabolomics reviewed: a new "omics" platform technology for systems biology and implications for natural products research. J Nat Prod 68(12):1813–1820. https://doi.org/10.1021/NP050255W/ASSET/IMAGES/LARGE/NP050255WF00001.JPEG
Rose KM, Howell EL, Scheufele DA et al (2018) The values of synthetic biology: researcher views of their field and participation in public engagement. Bioscience 68(10):782–791. https://doi.org/10.1093/BIOSCI/BIY077

Savelieff MG, Noureldein MH, Feldman EL (2022) Systems biology to address unmet medical needs in neurological disorders. Methods Mol Biol 2486:247–276. https://doi.org/10.1007/978-1-0716-2265-0_13

Sordo Vieira L, Laubenbacher RC (2022) Computational models in systems biology: standards, dissemination, and best practices. Curr Opin Biotechnol 75:102702. https://doi.org/10.1016/J.COPBIO.2022.102702

Strianese O, Rizzo F, Ciccarelli M et al (2020) Precision and personalized medicine: how genomic approach improves the management of cardiovascular and neurodegenerative disease. Gene 11(7):1–24. https://doi.org/10.3390/GENES11070747

Toscano WA, Oehlke KP (2005) Systems biology: new approaches to old environmental health problems. Int J Environ Res Public Health 2(1):4–9. https://doi.org/10.3390/IJERPH2005010004

Wahbeh MH, Avramopoulos D (2021) Gene-environment interactions in schizophrenia: a literature review. Gene 12(12):1850. https://doi.org/10.3390/GENES12121850

Wang ZT, Tan CC, Tan L, Yu JT (2019) Systems biology and gene networks in Alzheimer's disease. Neurosci Biobehav Rev 96:31–44. https://doi.org/10.1016/J.NEUBIOREV.2018.11.007

Wang XX, Yu PC, Li J (2020) High-throughput metabolomics for identification of metabolic pathways and deciphering the effect mechanism of dioscin on rectal cancer from cell metabolic profiles coupled with chemometrics analysis. Front Pharmacol 11:487362. https://doi.org/10.3389/FPHAR.2020.00068/BIBTEX

Wood LB, Winslow AR, Strasser SD (2015) Systems biology of neurodegenerative diseases. Integr Biol 7(7):758–775. https://doi.org/10.1039/C5IB00031A

Wynn ML, Consul N, Merajver SD, Schnell S (2012) Logic-based models in systems biology: a predictive and parameter-free network analysis method. Integr Biol 4(11):1332–1337. https://doi.org/10.1039/C2IB20193C

Xue G, Chen C, Lu ZL, Dong Q (2010) Brain imaging techniques and their applications in decision-making research. Xin Li Xue Bao 42(1):120. https://doi.org/10.3724/SP.J.1041.2010.00120

Yan J, Risacher SL, Shen L, Saykin AJ (2018) Network approaches to systems biology analysis of complex disease: integrative methods for multi-omics data. Brief Bioinform 19(6):1370. https://doi.org/10.1093/BIB/BBX066

Yu Y, Wang X, Wang Q, Wang Q (2020) A review of computational modeling and deep brain stimulation: applications to Parkinson's disease. Appl Math Mech 41(12):1747. https://doi.org/10.1007/S10483-020-2689-9

Zhang Q, Bhattacharya S, Andersen ME, Conolly RB (2010) Computational systems biology and dose-response modeling in relation to new directions in toxicity testing. J Toxicol Environ Health B Crit Rev 13(2-4):253–276. https://doi.org/10.1080/10937404.2010.483943

Zhang B, Gaiteri C, Bodea LG et al (2013) Integrated systems approach identifies genetic nodes and networks in late-onset Alzheimer's disease. Cell 153(3):707–720. https://doi.org/10.1016/J.CELL.2013.03.030

Zou J, Zheng MW, Li G, Su ZG (2013) Advanced systems biology methods in drug discovery and translational biomedicine. Biomed Res Int 2013:742835. https://doi.org/10.1155/2013/742835

Systems Biology Approaches for Cancer Biology

21

Bandita Dutta, Dibyajit Lahiri, Moupriya Nag, Niladri Sarkar, Rina Rani Ray, and Debasmita Bhattacharya

Abstract

Cancers are complex heterogeneous diseases characterized by the unconditional and abnormal division of eukaryotic cells. When cell division remains unchecked within the body, they give rise to tissue masses known as tumors. This disease exhibits a high level of robustness to various therapeutic treatment inventions. The diagnosis and treatment of cancer is a key component of any overall cancer control plan. In the last few decades biological science has seen the emergence of new domain, aiming at solving the complexity of biological phenomena. The Systems-level understanding of biological science has become one of the essential biological analytical tools to detect powerful insights into human health and diseases. In the recent era of high scientific and medicinal research advancements, it's becoming difficult for scientists to integrate the disease in the conventional way. So, they are trying to shift their views to genomics, proteomics specifically the 'omics' science, computational biology and Systems. The combined omics information is leading to the proper profiling of health and diseases at global level. The system biology is mainly focused on the integrative nature of the evolutionary developed biological systems and the fundamental principle in

B. Dutta · N. Sarkar · R. R. Ray (✉)
Microbiology Research Laboratory, Department of Biotechnology, Maulana Abul Kalam Azad University of Technology, Kolkata, West Bengal, India

D. Lahiri · M. Nag
Department of Biotechnology, University of Engineering and Management, Kolkata, West Bengal, India

D. Bhattacharya (✉)
Department of Basic Science and Humanities, Institute of Engineering and Management, University of Engineering and Management, Kolkata, West Bengal, India
e-mail: debasmita.bhattacharya@iem.edu.in

S. Joshi et al. (eds.), *Systems Biology Approaches: Prevention, Diagnosis, and Understanding Mechanisms of Complex Diseases*,
https://doi.org/10.1007/978-981-99-9462-5_21

order to govern them. This technique focuses on engineering techniques to inscribe human diseases on biological perspective. The analysis of such analysis is illustrated to decipher complex pathological diseases such as prion disease, liver toxicity, diabetes and cancer.

This chapter is mainly focused on high-throughput technologies and approaches including diagnostic and therapeutics that help to understand the underlying biological processes involved in cancer biology. This study helps to improve in understanding the complex interaction that occurs between the normal cells and the onco-genes present in a human body.

Keywords

Cancer · Medicinal research · Omics · Computational biology · Therapeutics · System biology

21.1 Introduction

In the cancer domain system biology is directed to develop a comprehensive perspective in the development of cancer and their progression. System biology is aimed to develop a blueprint of cancer analogous deregulations along with its malignancy and phenotypes in multiple biological scales.

Thus system biology in cancer domain provides an aggregated view of cancer in several biological scale including genetics and epigenetics, signaling networks, mechanical properties, epidemiology, cellular behavior, histology and clinical manifestation. The histology of cancer cells is being explained according to the properties of a lower scale, that is, cell behavior.

Cancer System Biology (CSB) concedes individual disciples and data which draw a parallel province to study cancer biology systematically alone or in combinations. Cancer system biology comprises diverse domains including cancer biology along with engineering, biochemistry, bioinformatics, physics, mathematics and computer science that provide a novel outlook to encounter elemental challenges in the research field of cancer biology. The field of CSB is found as a dynamic, versatile and complex process where experimental methods form a bridge to the analytical approaches in order to buckle down consequent challenges.

21.2 Cellular Basis of Cancer Biology

Uncontrolled growth with indefinitely dividing ability and cell proliferation apart from cell normal growth control machinery leads to the development of cancer. Augmentation of genetic changes including activation of protooncogenes and their amelioration to oncogenes, desynchronization of tumor suppressor and DNA repair genes with the time enables the immortalization. Reprogramming of cellular energy metabolism and immune escape are the two major attribute of cancer biology which is possessed by the tumor microenvironment. The non-malignant stromal cells have ascendency to the tumor microenvironment that is perceived as the significant determinants in cancer biology. Tumor associated macrophages (TAM) along with fibroblasts and inflammatory cells have direct cell lineages through the secreted factors to the tumor cells and stimulate the tumor growth by inhibiting the metastasis and tumor therapy in their microenvironment.

21.3 Tumor Microenvironment and Their Characteristics

The malignancy of cancer cells is being regulated by their local environment comprising tumor cells, endothelial, nerve, fibroblast and immune cells and their reciprocity influencing their growth, invasion, regulatory signaling and metastasis.

Macrophages which arise from peripheral blood monocytes get circulated throughout the body after being derived from bone marrow precursors. Depending on the microenvironment, the phenotype of macrophages are divided into two types, that is, type 1 or M1 which are systemically activated and type 2 or M2 which are alternatively activated (Murray and Wynn 2011). Tumor associated macrophages (TAMs) are M2 activated cells which are influenced by cytokines including interleukin-4, interleukin-10 and many more. In addition, with a new classification, macrophages is referred to as pro-inflammatory or pro-angiogenic which helps in the development of solid tumor through a broad spectrum of array including chemokines, cytokines and inflammatory mediators (Lewis and Pollard 2006). Several clinical studies showed the dependency of malignancy with increased density of macrophages in tumor stroma.

Tumor infiltrating lymphocytes (TILs) are the lymphocytes which migrate within the tumor environment especially targeting the tumor cells in order to induce immune response. According to the cohorts studies of human tumors, TILs of varying frequency revealed their strong association in obstructing tumor growth (Swartz et al. 2012). Different T lymphocytes are associated with different functions to proliferate immune responses in tumor microenvironment.

- CD8+ cytotoxic T lymphocytes (CTLs) are associated with direct execution of tumor cells (Reissfelder et al. 2015).
- CD4+ T helper lymphocytes (Th) are recognized as heterogeneous cytokine-secreting T lymphocytes. Th1 or T helper type 1 lymphocytes helps in CTLs activation and Th2 or T helper type 2 lymphocytes mediate the stimulation of humoral immunity.
- Natural killer (NK) cells are incorporated in restricting the growth and destruction of tumor cells. Reduction of NK cell activity increases the cancer mortality in cancer patients (Gottschalk et al. 2010).

IFN-γ and TNF-α (tumor necrosis factor-α) stimulate the Th1- and CTL-mediated tumor infiltration and enhance the prophecy of cancer patients (Schreiber et al. 2011).

Endothelial cells (ECs) are the prime constituent of the most critical steps of tumor vascularization which enables cell growth and development. ECs harmonize the pathophysiologic processes in order to promote normal to prostatic tissues. The angiogenic vessels of ECs of tumor provide nutrients to the growing tumor cells (Chouaib et al. 2010). ECs also regulate the cell function and formation of metastasis by controlling leukocyte conscription. Several pro-angiogenic factors like VEGF are produced by tumor cells inside the tumor microenvironment which stimulate tumor cell mortality, angiogenesis and metastasis.

The most profuse cell, fibroblasts, are present in connective tissues and secrete extracellular matrix components in order to form a structural network. Cancer-Associated Fibroblasts (CAFs) are the most important factor in tumor microenvironment and abundantly present in tumor stroma. CAFs are associated with extracellular matrix protein production and their remodeling. CAFs provide oncogenic signals via secreting growth factors and enhance the tumor growth and angiogenesis with an elevated secretion of SDF-1/CXCL12 (Orimo et al. 2005). CAFs have a mutual relation to the tumor cells which enhance the crosstalk between the host cells and cancer cells.

21.4 Cancer Bioinformatics

It is anticipated that cancer bioinformatics has an important role in identifying and substantiating the biomarkers in order to diagnose and monitor the phenotype specific progression of diseases along with therapy-mediated response and enhancement of patients' health. Several genes, proteins, peptides or physicochemical variables are used to form single or multiple biomarkers which can be used to forward expression to functional indications by the formation of dynamic networks. Protein annotations, their interactions and signaling pathways are consolidated for the protein-protein interactions-mediated network biomarker. The dynamic network biomarker is able to diagnose the development of diseases at different stages with the time variation. This system corresponds to the patients-related clinical informatics including their history, clinical symptoms, diseases phenotype, biochemical analysis, therapies, pathologies, imaging profiles and many more (Wang 2011).

System biology in developing cancer medicine has a novel strategy which is focused towards **P4** medicine (predictive, preventive, personalized and participatory) (Hood and Friend 2011). According to Tian et al., high-throughput technologies are used in patients to generate virtual cloud of billions of data which includes one or more than one disease-mediated networks within the cell of diseased organs (Tian et al. 2012). The molecular network associated with the disease is able to detect the abnormality and function of P4 medicine at an early stage. Clinical bioinformatics have advancement in detecting development and progression of disease along with their clinical symptoms and signs via the combination of human cancer tissue-generated bioinformatics along with clinical signs and their measurement. In cancer bioinformatics global functions under omic model collectively incorporate discrete elements under a single mapping. Therapeutic strategy based on clinical perusal, pathology followed by biochemical analysis, imaging and therapies are materialized under cancer bioinformatics (Wang 2011).

Cancer bioinformatics is also able to predict the potency and accuracy of the medicine by measuring and speculating the medicine in order to provide gene and protein associated effective and safe therapeutic strategy. In the field of molecular, pathological, radiological and clinical domain data generated through epigenetics, proteomics, microarrays and next generation sequencing provide heterogenetic data that provides ontology-based solutions for the databases. Present study on genomic analysis of human hepatocellular carcinoma showed a genetic modification in liver cancer 11q13.3 gene augmentation enabling FGF19 gene activation that effectuates selective susceptibility in inhibiting FGF19 gene in mouse model and RNAi (Sawey et al. 2011). Both the system biology and cancer bioinformatics are assumed to enhance diagnosis, prevention and treatment in order to design the therapy. In addition to -omic data sets, biological sequences with genome evaluation through the bioinformatics and statistics and their classical approaches, three dimensional structures of protein are considered as the significant backbone of computational cancer research (Sylvia et al. 2006).

21.5 Integrating Network Modeling in Cancer System Biology

The development of metastatic cancer from a single cell with an oncogenic profile is a complex process. In most of the cancer profile cell proliferation along with apoptosis resistance and angiogenesis followed by metastasis are found familiar. Unregulated cell proliferation is analogous to cancer development. The ErbB-signaling network model is found to be important in downregulation of tumor-associated cell proliferation (Lazzara et al. 2009). This networking model is able to detect the dominant mechanism in cross-talk pathways. For example, the cross-talk mechanism between insulin signal and ErbB network through the mass action kinetic model signifies the role of GAB1 in synergic effects between insulin and epidermal growth factor-mediated treatment (Borisov et al. 2009). The novel mass action model is devised depending on intrinsic pathway, that is, stress-induced and mitochondrial pathway and extrinsic pathway, that is, activation of death receptor along with specific molecular network. Phosphorylation of p53 induces the ephemeral DNA damage that enables cell cycle arrest whereas perpetual DNA damage induces cell apoptosis. An equilibrium between cell proliferation and apoptosis plays an important role in developing a tumor. The developmental analysis of tumor at molecular level reveals the cell responses against pro-proliferative and pro-death stimuli (Gaudet et al. 2005). Epidermal growth factor and insulin within the pro-survival and pro-death autocrine loops are the fundamental elements which are associated with TNFα in HT-29 apoptosis. These molecular equilibria elucidate the impotency of cytotoxic ligands in killing cells in chemotherapeutics (Janes et al. 2006).

21.6 Molecular Interventions and Their Effects

Computational network model is based on the regulatory mechanism of identification of novel targets and effective treatment of the tumor. The mass-action kinetic model of ErbB1-3 is incorporated with the betacellulin and heregulin1-β ligand binding followed by dimerization of receptor, manifestation, breakdown, recycling and PI3K to phosphorylated AKT-mediated downstream signaling. At the phosphorylation stage of AKT, the dominant role of ErbB3 was measured through sensitivity analysis which indicated the decrease of phosphorylated AKT in ErbB3 level compared to ErbB1/2 level through the in silico monoclonal antibody testing. Depending on the prediction of this system biology model artificially originated human IgG2 monoclonal antibody MM-121 against ErbB3 that is efficient to block AKT phosphorylation (Schoeberl et al. 2009). It was observed that the ACHN xenograft-implemented mice showed slower tumor growth when it was treated with MM-121 irrespective of time. This model usually targets the kinase-dead ErbB3 over the mutated and overexpressed protein. The outrageous growth of the tumor cells is regulated by the variable receptors and ligand nature compared to exclusive oncogenic change.

It was observed that most of the system models predict the alteration of cell regulatory networks as an effect of therapeutic disorder in spite of predicting altercation and their consequences at the molecular level. Relational modeling methods include partial least square regression which is able to connect the cellular functional behavior in molecular-level network. According to Kumar et al., pharmacological inhibitors-mediated AKT, ERK and p38 pivotal kinase pathway modulations and alterations of migratory behavior of ErbB2 overexpressing mammary epithelial cell correspondence to the Heregulin and epidermal growth factor can be detected via this partial least square regression (Kumar et al. 2008). It was also perceived that combined five phosphorylation sites of four prime proteins are able to elucidate the consequences of kinase inhibitors on phenotypic behavior of cells under heterogeneous treatment conditions. Thus, collective signaling pathways are involved in signal to response correlation and interpret cell behavior based on genetic or therapeutic alterations.

21.7 Experimental Cancer-Based Model

Cancer is considered as one of the detrimental diseases in the world. However, with the improvement of therapy, the mortality rate of cancer is decreasing gradually. Several therapeutic solutions have been developed and their effectiveness are studied against cancer cell lines and 3D spheroids, patients-derived PDXs (in vivo xenografts) and PDOs (ex vivo organoids). With the advancement of the experimental model systems, the mechanism of cancer development starting from initiation, progression and lastly the metastatic cascade of cancer cells need to be understood (Dugger et al. 2018). Traditional model systems compared to advanced model systems became unsuccessful to enumerate the response against drugs in clinical trials.

Although the chemotherapeutic agents and targeted therapies showed positive results in preclinical trials, they failed in clinical trials as enumeration of disease progression in specific model systems is visionary (Dhandapani and Goldman 2017). Therefore, the cancer model system along with system biology against the therapeutic responses was designed and studied.

- Cell line-based system biology model: 2D monolayer system in tumor etiology and cancer therapy anticipate the understanding and provide the advantage in the cancer biology system after the inception of human cancer cell line in 1951. Although the 2D monolayer system in cancer biology is found to be amiable in long-term culture using economic materials and in gene manipulation through insertion and deletion (Masters 2000), there are limitations in mimicking 3D growth pattern of tumor. Cell-to-extracellular matrix (ECM) is one of the major components of cell system which is able to detect cell viability and cell extinction, metabolism of drug and tumor-mediated genes and protein expressions. 2D monolayer system is unable to utilize the oxygen and nutrients precisely compared to the 3D culture system which exactly mimic the tumor environment (Kapałczyńska et al. 2018; Pampaloni et al. 2007).
- In 1977, the 3D culture was first implemented by Hamburger and Salmon using soft agar solution (Hamburger and Salmon 1977). With the flow of time, various 3D culture methods were designed and documented. The 3D culture system has been categorized into three groups depending on the usage of materials: (1) cultivation into the non-adherent plates, (2) implanted into the matrigel medium and (3) seeded into the scaffold-based system. The 3D culture is grounded on the spheroid structure which enables the cancer cells in forming several layers and imitating the tumor attributes along with functional aspects of tissue-specific developing stages. Cardiomyocyte spheroids are one of the 3D spheroids that manifest heart-like rhythms. Hepatocyte spheroids are another 3D spheroids that is able to exhibit liver biochemical functions (Fukuda and Nakazawa 2005; Desroches et al. 2012).
- The 3D culture system exhibits the microenvironment through the communication between the cell-to-cell and cell-to-ECM which enable the signal transduction between the tumor tissues and helps in cell proliferation and development of cell morphology. It was observed that the 3D culture system is more capable in providing the structural and functional similarity with inclusive success rate in preclinical trials in comparison with 2D culture system (Schilsky 2010). Difference of sensitive responses of the chemotherapeutic agents in 2D and 3D culture systems of the same cell line was reported. The 2D and 3D model systems of HCT-116 cells and their respective sensitivities against the four anticancer agents 5-Fluorouracil, irinotecan, melphalan and oxaliplatin were distinguished which indicated that these inhibitors were more effective towards the 2D culture system compared to the 3D culture system due to the phenotypic differences and specific cellular interactions (Karlsson et al. 2012).
- Patient sample-based system biology model: Patient-derived preclinical models are substantiated via the direct transplantation of patient-derived tumor speci-

mens into nude mice, for example, patient-derived xenografts (PDXs) (Lai et al. 2017). It was observed that PDXs is able to mimic the anti-cancer drugs test mechanism and thus it is considered as the most optimistic model system for preclinical trials (Hidalgo et al. 2014). In addition, by predicting the drug responses and their effectiveness, PDXs is able to preserve the patient-acquired tumor heterogeneity, genotype and clinicopathological attributes. Although as preclinical model system PDXs has showed significant results but it also has some limitations. Bottleneck during engraftment of tumor samples into the host mouse in significant proportion is the major limitation. The prolonged engraftment also increases the PDXs maintenance costs. The probability of prevailing non-reproducible results due to not choosing the standardized engraftment methods for specific cancer cells is another drawback of this model system. The acquired drug resistance mechanism in metastatic colorectal cancer (mCRC) provides a new dimension over these limitations. Several studies have shown the usage of large PDX biobank to evaluate the resistance mechanism of drug molecules in mCRC. The anti-EGFP antibody blockage and their genomic topography along with functional ramification were studied in PDXs model of mCRC (Bertotti et al. 2015). Thus, PDXs is considered as the significant model system systemic analysis of cancer therapies.

- Another patient-derived ex vivo 3D structure of tumor, that is, patient-derived tumor organoids (PDOs) were obtained from patients having cancer and it is grown in an extracellular matrix. It was evident that the PDOs is able to predict the significant drug responses in cancer patients in clinical trials without damaging genomic heterogeneity of the original tumor (Sachs and Clevers 2014). According to the tumor histopathological features, it was observed that PDOs are able to sustain the morphological characteristics of the original tumor and also able to mimic the genetic characteristics of tumors even after long derived ex vivo culture. 3D cultures of PDOs model including bladder, breast, colorectal, gastrointestinal, glioblastoma, ovarian, pancreatic and prostate cancer anticipate the consequences of drug treatment and provide a different aspect in determining the advantages of customized medicine on the cancer patients. The entrenched PDO biobank comprising 20 primary CRC patients-derived tumors with conserved genetic heterogeneity and molecular epitopes. These PDOs from the PDO biobank showed the increased success rate in phases I and II clinical trials and Food and Drug Administration (FDA)-approved drugs showed the positive predicting value of 88% against PDOs indicating the potency of PDOs in cancer treatment model.

21.8 System Biology-Driven Precision Medicine for Cancer

Precision medicine is an emerging domain in cancer biology that is found to be efficient in enhancing health benefits (Fig. 21.1, Box 21.1). Data-driven science has opposed the clinical province as precision medicine became pertinent to the patients' needs. The molecular profile in terms of genomic and epigenomic alterations of each cancer tissue is different. The exploitation of genomic and systemic

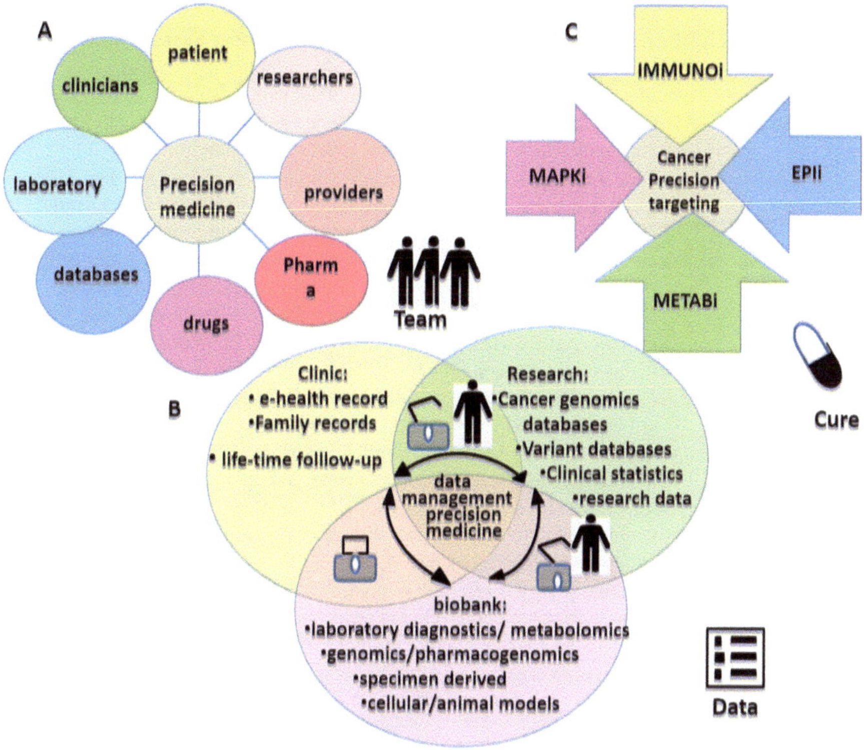

Fig. 21.1 Important unit in precision medicines and their applications in cancer. The framework of precision medicine depends on the collaboration of clinicians and scientists in a research group (**a**). Rapid and coherent interpretation of individual data (**b**). Target-based pharmacological strategies where the precision disease management is comprehended with target-based personalized treatment based on patients' genotype for proper cure (**c**)

apprehension for coherent cancer treatment sounds logical and certain steps towards the data-intensified disease management ameliorate survival. The comprehensive system-based treatment approaches in the oncology field have been challenged by the highly efficient omics data along with deep sequencing of patient cohorts. With the flow of time, the system biology has evolved in order to design prognostic and multiscale models which help in biomarker discovery and understanding biochemical networks in the genomic and non-genomic profiles including epigenomic, metabolic, proteomic and transcriptomic. Without the diagnosis of the genetic cause of disease expression, determination of the most suitable and effective treatment is impossible. Thus, the identification of disease causes become pivotal in system-based drug design and precision medicine-mediated disease management. It was also evident that system biology conciliated synergistic effects of multidimensional and complex data sets are robust compared to the entire summation of individual nodes. Cancer genome data is the biggest input in the complex puzzle and has unlocked an unusual window of opportunities in the system biology domain.

Box 21.1 Definition of Terminology Used in the Field of Cancer System Biology

Precision medicine	Medical model customized on individual patient. Accumulated research knowledge from biomarkers and patients genomic profiles help to accomplish significant medical predictions and determination
Big data	Data sets required for data management and processing. Advanced architecture and algorithms are required to manage the data sets and extract the value from it. Big data with a V-type attribute is able to analyze and corelate the trends and patterns via the network interactions
Genomics	Genomics is referred to the study of whole genes in human genome. Genomics is a storage of genes and mutations, both inherited and somatic that are associated to the development of an organism and also regulate homeostatic function or diseased states
Epigenomics	Chemical modifications without changing DNA sequence is able to affect gene activity. Epigenomics is defined as capturing epigenetic marks via genomic scales including DNA methylation and post-translation modifications of histone chromatin. In epigenetics, regulatory factors and reversible processes mainly affect gene expression, non-genomic inheritance, and phenotypes which are not associated with variation of DNA sequences
Systems biology	System biology is referred as quantification, modeling and visualization of network interactions. On the basis of designed model system, biology corelate the testable hypotheses and experimental validation
Genotyping	Genotyping is a determination process of the individual genetic makeup. This genotyping is associated with an entire genome or a clinical phenotype
SNP	Alteration of single nucleotide in DNA sequence of an individual and continuation of this alteration at a specific position in more than 1% of a population is referred as single-nucleotide polymorphism (SNP). Apart from genes SNP are associated on-annotated regions of DNA
SCNA	SCNAs is somatic copy number alterations, which is described as multiple deletions of chromosome arms of focal regions that evolve non-germline tissue like tumors. Whereas copy number variations (CNVs) indicate the changes of germline cells and development of organisms
Master regulator	Master regulator is hierarchy rate-limiting step. These master regulators are found at nodes of networks and regulate the metabolic flux or signaling pathways of cells
MAPK pathway	Group of signaling molecules that transmit a signal from a cell surface receptor into the cell nucleus to regulate a gene expression program that includes proliferation, mitosis, differentiation and cell survival. A mitogen-activated protein kinase, also known as a MAPK or MAP kinase, belongs to a class of kinases that are only capable of phosphorylating the amino acids serine, threonine and tyrosine. The next kinase is activated when one kinase is phosphorylated, acting as a switch in an amplifying cascade. Constitutively activated kinases can be produced as a result of oncogenic mutations, which can then lead to unchecked cell growth. Malignant melanoma exhibits dysregulation of a significant mechanism

Immunotherapy	It uses certain immune system components to combat conditions like cancer. Antibodies, vaccinations or immune system activation can all help with this. Cancer cells use immune evasion as a cloaking technique to suppress the expression of immune-recognizable surface markers, enabling unchecked survival. Immune detection and assault of cancer cells are improved as a result of T-cell immune checkpoint inhibitors, which mute immune system regulatory components
Neoantigen	Newly generated epitope which immune system has not yet identified. Mutations that are particular to a tumor might result in the emergence of new epitopes. The immune system can react to or repress tumor-reactive T-cell populations based on such epitopes, which significantly affects the efficacy of immunotherapies including T-cell checkpoint blockade and adoptive T-cell treatment
Oncometabolite	Tiny molecules that, via the disruption of metabolism and signaling, cause the development and spread of cancer. In cancer, certain enantiomers of hydroxy-keto acids or amino acids build up as a result of malfunctioning enzymes and alter the expression of genes across the genome by inhibiting regulatory proteins

21.9 Drug Reposition Through System Biology

The utilization of complete medical data sets is still challenging even after establishment of medical databases. System biology is able to provide latent solutions to overcome these challenges via evaluating the molecular interactions of individual biomolecules. The combined wet-lab experiments and their consecutive data in order to measure cellular components are utilized by the computational approaches to design the model that is able to analyze these multiple data sets of biological systems. The system biology framework implements computational approaches that obtain quantitative outcomes experimentally (Zou et al. 2013). System biology is able to interpret disease mechanisms along with classification of diseases, their diagnosis and respective therapy, prognostic or theranostic aspects of biomarkers. The system biology also enhances drug development efficiently to various diseases. An integrated method for predicting drug combinations, for example, network mining, synergy from gene expression and SynGeNet along with connectivity mapping and network analysis has significance in cancer treatment. The combination of tretinoin (retinoic acid receptor agonist) and vemurafenib (BRAF inhibitor) is effective in treating BRAF-mutant melanoma (Regan-Fendt et al. 2019). DrugPredict is a computational profile-based drug recycle approach which stipulates the use of indomethacin to treat epithelial ovarian cancer (Nagaraj et al. 2018). The FDA-approved non-oncological drugs have shown their chemo adjuvants potentials in clinical trials. Systems pharmacology is a domain where system biology is merged with pharmacology and able to speculate the molecular interactions of the drug to their targets. Using novel medicine in therapeutic approaches for effective treatment is an advanced strategy in systems pharmacology. System pharmacology function based on drug target network that predicts drug combinations, drug-target interactions and drug repositioning (Turanli and Arga 2017). The accumulation of drug-related experimental data and their integration and investigations are the outcome of advanced omics technologies which includes two predominant strategies signature-based (drug-induced transcriptomics research in particular) and network-based (computational approaches and integrations) (Iorio et al. 2013).

1. Signature-based approaches: Drug repository signature-based approach consists of cell- and target-based screening assays which usually compare the gene expression profile of diseased and drug-induced states in order to remodel the drug via signature revision for treating the disease of interest. In spite of being expensive, prolonged, trained robot dependent, signature-based approach is preferred. Signature-based approach is able to provide an intricate map on the disease and drug connections and also corroborate the preclinical and clinical data with low false-positive count (Shim and Liu 2014). Along with discovering the biological significance of new drug candidates, the signature-based approach is able to assess the computationally predicted drug candidates. The novel deep-learning approach is comprehended with genomic mutational fingerprint of human cancer cell lines and structural contour of 244 drugs, that is, Cancer Drug Response profile scan (CDR scan). Molecular docking interactions of 14 oncology and 23 non-oncology drugs and 1487 approved drugs showed their potency to repurpose the cancer treatment (Chang et al. 2018). With the advancement of technology and exponential development of omics data, diverse opportunities get opened up and computational novel methods enable the reposition of apprised drugs. It was observed that though the RNA interference transcriptome-based technique candidate therapeutic gene is able to target discrete subtypes of breast cancer. According to the candidate genes analysis, FDA-approved drugs are targeted by the 80 candidate proteins and the epilepsy target GABRP has the potency to stimulate the repositioning of the drug for treating breast cancer (Liu et al. 2017). Omics study is not confined to the human cell lines study; it is also studied on eukaryotic model organisms including nematode worm (*Caenorhabditis elegans*), yeast (*Saccharomyces cerevisiae*) and zebrafish (*Danio rerio*). Omics-based approaches are able to reveal the connections between the disease phenotype and drug reposition. GWAS (genomewide association studies) and PheWAS (Phenome-wide association studies) are the two harmonizing methods to find pioneering interrelation that links genetic variants to human diseases (Rastegar-Mojarad et al. 2015).
2. Network-based approaches: Network plays an important role in system biology in providing a profound framework in order to assimilate qualitative and quantitative biological attributes. For example, correlation of gene expressions or interactions of gene expressions. Molecular networks including co-expression networks, proteomic connections, transcriptional regulatory networks, signaling networks, metabolic networks, disease to disease, target and disease, drug to drug, target and disease, drug and gene, are all important in systems pharmacol-

ogy (Azuaje 2013). The identification of underlying molecular mechanisms and utilization of respective biomarkers for diagnosis of diseases, metabolic disorders and cancer are accomplished via network-based models (Calimlioglu et al. 2015). The network modeling the pharmacology domain is enlightening drug target, drug candidate, their interactions and their respective side-effects. The network modeling is the prime method in drug reposition, and in creating DGD (drug-gene-drug) triangle. The DGD node usually represents any one attribute amongst the drug, disease or gene and their consecutive relationships. A framework, which is consisting 4742 tumor genomes cancer mutation data and protein interactions network, specify that 62 predicted genes through in vitro studies and TFDP2 and AKT2 are tumor-inducing genes which are responsible for causing oncogene-negative lung adenocarcinomas (Horn et al. 2018). The proposed novel drug-target interactions pipeline (DTINet) utilizes low-dimensional features vectors and is able to amalgamate heterogeneous biological data on a large scale. DTINet also manifest the preventing potential of three drugs alendronate, chlorpropamide and telmisartan against cyclooxygenase proteins-mediated inflammatory diseases (Luo et al. 2017). The systemic, comprehensive and integrative network analysis of genomic and transcriptomic data apprehended the drug reposition through gene expression.

21.10 Examples of Repositioned Drugs Using System Biology

Drug repurposing is also known as therapeutic switching which is the current research interest. It usually refers to the use of traditionally used FDA-approved drug molecules which are very specific to a particular disease and are observed as novel medicine against different diseases or therapy. This repurposing leads to the development of the cancer biological research in new ways. Some of the widely acceptable FDA-approved drug molecules are Metformin, Aspirin, Digoxin, Itraconazole, Statins, Disulfiram. Based on computational system biological and pharmacological approaches, several studies revealed the target-specific mechanism of repurposed drug to prevent the tumor cell and malignancies (Table 21.1).

1. *Metformin*: In the field of cancer prevention and treatment, Metformin became a popular name as a hypoglycemic oral drug that is widely used to defeat cancer cells (Suissa and Azoulay 2014). The exact molecular mechanism of the Metformin has two different mechanisms. The direct mechanism includes the inhibition of the growth of tumor cells. There are various cell-signaling pathways, for example, insulin/INF-1, a very critically regulated network for cell proliferation. Mitogen-activated protein kinase (MAPKs) (Li et al. 2018) pathway in eukaryotes. Serine and Threonine protein kinase are mainly involved in regulation of multiple cellular activity by transducing the extracellular signals to the intracellular responses. The conservative transcription factor NF-kB pathway involves regulation of diverse inflammatory responses, cell growth and apoptosis, JNK/p38 pathway is very sensitive to intracellular or extracellular agents,

Table 21.1 The repurposing drugs and their methods of identification to detect for carcinogenic diseases are depicted here

Drug	Methods	Types of cancer
Metformin	DSPathNet	Breast, pancreas, prostate cancer
	SIGNOR database	Breast cancer
	SMiR-NBI	Breast cancer
Aspirin	SDTNBI	Various cancer types
	Molecular docking, immunoprecipitation, fluorescence assay	Colon cancer
	Investigation of gene-environment interactions at the genome scale	Colon cancer
Digoxin	Signaling network based method	Grp 3,4 medulloblastoma
	CMap	Myeloid leukemia
Proton pump inhibitors	Molecular docking	Various cancer types
Itraconazole	NBI	Breast cancer
Disulfiram	b-SDTNBI	Breast cancer

In silico and in vitro screening of the repurposing drugs have shown the impact effectiveness on cancer, hence they are potentially used as either anti-cancer drugs or directly as chemotherapeutics or combination with traditional cytotoxic agents

mutagens as well such as dietary agent, alcohol, radiation, insulin/IGF-1 pathway leads to auto phosphorylation causes abnormal cell proliferation (Sun et al. 2015). Metformin works on signaling pathways to reduce cellular activity in order to inhibit the various malignancies in lungs, colorectal, breast, prostate. Another route of mechanism of Metformin is an indirect way in which it typically binds in the glucose metabolic pathway to inhibit the peripheral insulin sensitivity and hepatic gluconeogenesis. Molecular dynamics is widely accepted study for any drug disease interaction strategy based on the quantitative changes in genetic expression associated with different genes in the body. Again, proteomic analysis provides the specific concept of drug-specialized cellular responses to the researchers as proteomic study helps in detecting the relationship between genotype and phenotype. SIGnaling Network Open Resource (SIGNOR) (Perfetto et al. 2016) is a signaling network of human breast cancer cells that mapped the transcriptome, proteome and phosphoproteome data sets for malignant cells (Sacco et al. 2016). Protein-protein interaction leads to identify the Metformin-mediated pathway mechanism while it is applied in the cancer cells. Twenty possible binding sites, including five kinases (the SGK1, EGFR, CDK7, MAPK14, and MAP2K2 genes), three genetic factors (TP53, AKT1, and PCNA genes) are observed (Hart et al. 2016) where metformin directly binds to them. This study gives an impact on drug target interaction of a cancer cell. It was identified that Metformin-specific binding novel pathway is centered around Myc and seven gene expression via the network CDKN1A, ESR1, MAX, MYC, PPARGC1A, STK11 and SP1 analysis of diabetes and cancer by constructing a network of signaling pathways exclusive to Metformin.

Advanced study showed that (Li et al. 2016) a miRNA network-based inference model (SMiR-NBI) stimulates the miRNA-mediated remodeling of drug-target interaction. Breast cancer cell line, associated with Metformin binding is the example of miRNA-mediated pharmacogenomic biomarkers in accordance with the experimental data of 11 miRNAs (e.g. miR-20a-5p, miR-27a-3p, miR-29a-3p, and miR-146a-5p).

2. *Aspirin*: Acetylsalicylic acid (ASA) or commonly known as Aspirin is a promising drug molecule to treat pain, inflammation, fever and thrombotic diseases. It notably plays an important role in Kawasaki disease, pericarditis and rheumatic fever and also has anti-cancer properties. Various types of malignancies including breast, colon, stomach, lungs, esophageal are being treated for decades by aspirin. At the molecular level, it works on metformin-specific signaling pathway network includes the NF-kB and STAT3 pathway, Wnt/-catenin pathway, mTOR pathway, and suppression of COX-1/COX-2 (Alfonso et al. 2014) involves the secretion of prostaglandin to ensure the safety of gastrointestinal mucosa and creating inflammation. Substructure-drug-target network-based inference (SDTNBI) method (Wu et al. 2018) is mainly based on the drug chemical substructure data with the drug gene disease network, that is, 29 distinct cancer kinds and 21 non-steroidal anti-inflammatory medications. According to systemic evolution, COX-2, AKR1C3, CA-9, CA-12, and CDK2 are the key targets of aspirin. Another observation through immunoprecipitation, fluorescence assay and Molecular docking showed that two allosteric sites of salicylic acid Asp145 and Lys33 interact with CDK2. Salicylic acid-mediated CDK2 binding at an allosteric site revealed a new way of target detection and mechanism of action of chemotherapeutics in cancer treatment. Furthermore, system biology revealed the association of Glutathione (GSH) metabolism with carcinomic risk as its key role to modulate the cell proliferation and DNA synthesis (Holmgren 1981) and repair mechanisms. The adverse effect of aspirin in interruption of tumor growth, inhibition of angiogenesis and negatively accelerating the metastasis spread was also detected (Chan et al. 2009). Recent study revealed that regular use of aspirin enhances the risk of colorectal cancer in accordance to genetic variation of chromosome no. 12 and 15 at highly expressive single nucleotide polymorphisms (SNPs) which are associated with DNA mismatch repair, cell cycle regulation with genetic susceptibility to cancer.
3. *Digoxin*: *Digitalis lanata*, an active producer of the drug digoxin, showed the anti-cancer properties. Generally, digoxin is a glycoside that is widely used in cardiac treatment including heart failure, heart blockage. In silico and meta-analysis of clinical evidence have shown that digoxin is potentially able to fight against malignancies including breast, lungs, melanoma, renal and, most importantly, leukemia. It is observed that digoxin functions as an inhibitor of Na+/k+-ATPase pumping process and regulate different downstream signaling pathways, for example, Src kinase, MAPK and Ca2+ signaling pathway alteration and HIF-1α inhibition (Lin et al. 2015). In silico and in vitro both the studies have shown the anti-cancer effect of digoxin on the human cancer cell. It was observed that non-small lung cancer cells A549 and H1299 are inhibited by cell

cycle arrest at G0/G1 and G2/M phase, respectively, while treated with digoxin. Phosphatidyl inositol-3-kinase/Akt (PI3K/AKT) pathway plays a crucial role in tumor cell proliferation and cell survival, metastasis as well. Digoxin is also found to inhibit the PI3K/AKT pathway to suppress cell proliferation. In the field of leukemia, digoxin became an important drug. According to Zhang XH et al., digoxin is highly applicable on leukemia cells as it has been found to inhibit the oncogenic AKT/mTOR signaling pathway of malignant cells. Furthermore, the combination of digoxin and the unfolded protein response (UPR)-signaling pathway are able to resist cell growth, cellular responses via G2/M phase cell cycle arrest and apoptosis of leukemia cells (Zhang et al. 2017).

4. *Proton pump inhibitors*: Proton pump inhibitors (PPIs) include various medicinal drugs that are helpful in lowering the acid formation in the stomach mainly for patients who are suffering from ulcers, gastrointestinal diseases, acid reflux (Strand et al. 2017), heartburn. The anti-cancer properties of PPIs have not been revealed extensively. The detection of particular agents that directly work on the specific molecular target instead of cytotoxic agents, having high toxicity is the main aim of recent research. Target-specific drug molecules are able to bind the malignant cells present in the body and follows the inhibitory mechanism on various metabolic pathways or cell-signaling pathways, that is, H+/K+-ATPase, vacuolar ATPase and organic cation transporter 2 (OCT2) (Ikemura et al. 2017) that are responsible for growth, proliferation and apoptosis in tumor cells. For example, T-lymphokine-activated killer cell-originated protein kinase (TOPK) and Ser/Thr protein kinase are found in the various cancer-mediated tissues. The highly expressive TOPK has been implicated with the tumor cell to regulate mitotic checkpoints, cell apoptosis and chronic inflammation mainly (Matsumoto et al. 2004). TOPK is also involved in the phosphorylation of H3 histone at Ser10 (Kwon et al. 2010). The FDA-approved medication database may be searched using the structure-based virtual ligand screening approach to find the target-related medicinal compounds. Pantoprazole is observed to have an inhibitory effectiveness against TOPK pathways. However, there were two TOPK inhibitor, HI-TOPK-032 (Kim et al. 2012) and OTS964 (Matsuo 2014) were identified in the year 2012 and 2014, respectively. Experimental studies using seven different prazoles have shown that ilaprazole, lansoprazole and omeprazole are three potent drugs that have inhibitory efficiency against TOPK (Zeng et al. 2016). The inhibition of de novo synthesis of fatty acids is another prospect of cancer treatment as intensify lipid biosynthesis and deregulation of lipogenesis enhances the survival of cancer and tumor cells and conversion of Acetyl-CoA and Malonyl-CoA into palmitate enzyme via enzyme fatty acid synthase (FASN) (Fako et al. 2015) stimulate the oncogenic cellular functions overexpression of cancer-associated cells.
5. *Itraconazole*: Itraconazole is highly effective oral steroid that is widely acceptable in treating fungal infection including aspergillosis, blastomycosis and histoplasmosis. Fungal cytochrome P450 enzyme lanosterol-14α-demethylase (LDM) (Perfect 2017) plays an important role in biosynthesis of sterols and becomes the target of the maximum antifungal agent for inhibition. Most impor-

tantly, Itraconazole is reported as an anti-cancer drug in the year 2007. Various in silico and in vitro analysis have shown the capability of this steroid to inhibit the formation of new blood vessels from existing vessels and help in wound healing. According to molecular analysis, the mechanism of the itraconazole is very specific. It works on the signaling pathway including AMPK, mTOR, Hedgehog that involves in the signal transmission from the cell membrane to the nucleus and Wnt/β-catenin, a highly conserved evolutionary pathway that basically maintains the homeostasis in cells that lead to develop cell growth, proliferation and apoptosis and cancer-related to inflammation (Liang et al. 2017). According to the preclinical and clinical studies, itraconazole has potency to defeat different malignancies (Chong et al. 2007). Using the NBI method, the binding of itraconazole and dipeptidyl peptidase-IV enzymes to the estrogen receptor has been established as both has the polypharmacological property. NBI refers to the network-based inference method to understand (Cheng et al. 2012) drug-target interaction. Synergistic of the itraconazole with another conventional drug molecule paclitaxel showed an effective treatment of cancer as a first-line chemotherapeutics agent.

6. *Disulfiram*: Disulfiram is a special type of organic disulfide that results from oxidative dimerization of N,N-diethyldithiocarbamic acid, that are widely famous for treating chronic alcoholism. It mainly inhibits the nuclear factor—kappa B (NF-κB) that is involved in the reformation of NF-κB protein in cytoplasm and drains the same from the nucleus of unstimulated cells, NPL4 and phosphoglycerate dehydrogenase (Spillier et al. 2019) through copper- and zinc-associated pathway. Several in vitro and in silico studies have showed the anti-cancer properties of disulfiram including breast, colon, uterine, lung cancer and many more.

21.11 System Biology for Targeted Cancer Therapy and Cancer Immunotherapy

The advancement of Human Genome Project (HGP) allows the human DNA sequencing which enables technological evolution in detecting profuse data sets. At the conjunction of personalized medicines with new technologies in drug development, the diagnosis, therapies and prevention of genetic diseases get an evolutionary precedence. The targeted therapy induces the amelioration of patients' survival and development of translational medical pursuit. It was manifested that all the domains of anti-cancer drug instigation incorporate drug design followed by drug reposition and cancer therapy based on computational network analysis and multi-omics data analysis. System biology plays an important role in developing an extensive amount of multi-omics data whose accuracy is further verified by the cancer researchers for effective anti-cancer medications.

The targeted therapy has limitations of low-response rate in clinical trials. In order to overcome these limitations, the cancer system biology is incorporated with mathematical and computational models to analyze patients' molecular data by identifying targets and their respective biomarkers. The recent study is focused on liver and colorectal cancers and their whole genome sequence in order to understand gene expression and their respective therapy. With an unbiased data analysis, it was observed that specific pathways and expression pattern-mediated tumor mutation is able to alter the drug response against cancer (Sadanandam et al. 2013). The principle regulatory nodes including oxidative phosphorylation, glycolysis and citric acid cycle of a major metabolic network act as favorable therapeutic targets. System biological approach along with gene expression data is required in extensive amounts against diseases having limited therapeutic alternatives.

Cancer is the uncontrollable cell multiplication that eventually spreads all over the body due to the production of new gene products via the joining of open reading frames of two genes which mediate chromosomal modifications and development of malignancy (Yaman et al. 2021). The T cells of the adaptive immune system induce negative selections by reacting against the antigens of cancer cells. The concept of "personalized medicine" enhances the multi-dimensional blueprint for patients' urgency. The genesis of personalized mutanome vaccines requires the rational vaccine design along with cancer-specific mutation of tumor, neoepitope prediction and individualized immunotherapies (Okada et al. 2020).

Neoepitopes are examples of individualized mutanome vaccines required for the immune system to treat cancer cells (Park et al. 2018). This treatment exhibits positive results against the various cancer treatments in clinical trials. NGS (next generation sequencing) technologies make the patients sample accessible to the researchers in order to investigate and identify the disease-associated genes and the most suitable therapy. Another personalized vaccination, that is, neoantigen-based work under multiple stages through the detection of mutations followed by prediction of potential epitopes. Immunogenicity of mutation-derived peptides is required to screen the mutation of NGS data and identify the neoantigens target and their consecutive biomarkers (Rahman et al. 2022).

21.12 Advantages of System Biology

System biology is the prime factor to connect the bioinformatics methods in the cancer domain based on their disease specificity and metabolism, cell communication through signaling and cell proliferation. System biology is an emerging field with the combination of bioinformatics, mathematics and several omics technologies. Thus, system biology has appeared as the critical aspect in diagnosis of diseases and their therapeutic responses for patients' cure.

- Novel methodologies: System biology has introduced novel tools in the cancer domain which includes Semantic Web technology in developing quantitative semantic models for understanding high-throughput clinical data. Corvus is a data warehouse that based on systemic biological knowledge provides the multiple omics data and their consecutive applications in SPARQL endpoint (Holford et al. 2012). Semantic models comprise genomic, epigenomic and transcriptomic data of melanoma specimens along with gene ontology data. The regulatory network of this model holds the information regarding transcription-binding factors and regulates the cellular functions and their responses against anti-cancer therapy. Multivariate assay is able to characterize error and apply centroid-based genomic predictors PAM50 to treat breast cancer and predict the information regarding variability. An integrative tool miRTrail is able to analyze the all-inclusive gene interactions and miRNAs and regulate the noninterference of pathogenic processes. This tool is able to integrate the information of 20,000 genes with 1000 miRNAs and generate 280,000 accepted interactions (Laczny et al. 2012). The regulatory interaction between the miRNAs and genes helps the computational mode to detect gene interactions at the target tumor, cell differentiations, patients' symptoms and their therapeutic responses. In system biology, medical imaging is another important factors in the field of clinical pathology, computerized tomography, ultrasonic, positron emission tomography and nuclear magnetic resonance imaging in order to get "early and accurate" cancer detection followed by their diagnosis.

- Novel approach of biomarkers: Identification and corroboration biomarkers is an important stage of system biology which enables early detection of disease phenotype and their progress and therapeutic response. The development of biomarkers from single to multiple markers, their functionality and contribution in developing dynamic network from a network provide a new dimension to the system biology. Network biomarker is based on protein-protein annotations, interactions and signaling pathway. Dynamic network biomarkers not only detect the disease progression, it is also able to correlate the clinical information of patients-related therapies, biochemical reports, imaging profiles and many more (Wang and Liotta 2011).
- Novel precision medicine: Emergence of P4 (predictive, preventive, personalized and participatory) medicines is a novel strategy in system cancer medicine. With the help of high-throughput technologies, billions of patients' data has been generated through the virtual cloud. The molecular network is able to predict the early stage abnormality and functional deformities of cancer cells and incorporate the P4 medicine. According to the recent study of genomic analysis of human hepatocellular carcinoma, an alteration of 11q13.3 amplification in liver cancer activate FGF19 (Sawey et al. 2011). The molecular network along with a wide range of biological and omics data sets are able to detect the genesis of tumors and help to design the diagnosis, treatment and prevention. Thus, system biology is an emerging approach in clinical research, medicine development and their applications and is beneficial in the healthcare domain of cancer biology.

21.13 Conclusions

With emergence of omics technologies and computational evolution, the study of cancer etiology and fundamental mechanisms get enhanced and discovery of clinical diagnostic techniques and prognostic biomarkers, therapeutic targets and development of therapeutic involvement get another dimension. In spite of having significant progress, several aspects remain unexplored in this domain. The ability of identifying only 5% of cancers and hidden cognition of metastasis are the reasons for fundamental gaps of being unexplored. These fundamental gaps lead to the inefficient development of designs for therapeutic interventions. In the last few decades, cancer is no longer considered as the genomic disease, rather it is appraised as cellular network-mediated diseases. The advancement of technologies enables the effective studies in cellular systems and disorders. Thus, computational skills, with mathematical model and omics data, are the pillars of workflow design and data analysis for significant therapy and disease controlling.

References

Alfonso L, Ai G, Spitale RC, Bhat GJ (2014) Molecular targets of aspirin and cancer prevention. Br J Cancer 111(1):61–67

Azuaje F (2013) Drug interaction networks: an introduction to translational and clinical applications. Cardiovasc Res 97(4):631–641. https://doi.org/10.1093/cvr/cvs289

Bertotti A, Papp E, Jones S, Adleff V, Anagnostou V, Lupo B, Sausen M, Phallen J, Hruban CA, Tokheim C et al (2015) The genomic landscape of response to EGFR blockade in colorectal cancer. Nature 526:14969. https://doi.org/10.1038/nature14969

Borisov N et al (2009) Systems-level interactions between insulin-EGF networks amplify mitogenic signaling. Mol Syst Biol 5:256

Calimlioglu B, Karagoz K, Sevimoglu T, Kilic E, Gov E, Arga KY (2015) Tissue-specific molecular biomarker signatures of type 2 diabetes: an integrative analysis of transcriptomics and protein–protein interaction data. Omi A J Integr Biol 19:563–573. https://doi.org/10.1089/omi.2015.0088

Chan AT, Ogino S, Fuchs CS, Meklin GP, Moeslein G, Olschwang et al (2009) Aspirin use and survival after diagnosis of colorectal cancer. JAMA 302:649–659. https://doi.org/10.1001/jama.2009.1112

Chang Y, Park H, Yang HJ et al (2018) Cancer drug response profile scan (CDRscan): a deep learning model that predicts drug effectiveness from cancer genomic signature. Sci Rep 8(1):8857. https://doi.org/10.1038/s41598-018-27214-6

Cheng F, Liu C, Jiang J et al (2012) Prediction of drug-target interactions and drug repositioning via network-based inference. PLoS Comput Biol 8(5):e1002503

Chong CR, Xu J, Lu J, Bhat S, Sullivan DJ Jr, Liu JO (2007) Inhibition of angio-genesis by the antifungal drug itraconazole. ACS Chem Biol 2(4):263–270

Chouaib S et al (2010) Endothelial cells as key determinants of the tumor microenvironment: interaction with tumor cells, extracellular matrix and immune killer cells. Crit Rev Immunol 30(6):529–545

Desroches BR, Zhang P, Choi BR, King ME, Maldonado AE, Li W, Rago A, Liu G, Nath N, Hartmann KM et al (2012) Functional scaffold-free 3-D cardiac microtissues: a novel model for the investigation of heart cells. Am J Physiol Heart Circ Physiol 302:H2031. https://doi.org/10.1152/ajpheart.00743.2011

Dhandapani M, Goldman A (2017) Preclinical cancer models and biomarkers for drug development: new technologies and emerging tools. J Mol Biomark Diagn 8:1000356. https://doi.org/10.4172/2155-9929.1000356

Dugger SA, Platt A, Goldstein DB (2018) Drug development in the era of precision medicine. Nat Rev Drug Discov 17:183–196. https://doi.org/10.1038/nrd.2017.226

Fako VE, Wu X, Pflug B, Liu JY, Zhang JT (2015) Repositioning proton pump inhibitors as anti-cancer drugs by targeting the thioesterase domain of human fatty acid synthase. J Med Chem 58(2):778–784
Fukuda J, Nakazawa K (2005) Orderly arrangement of hepatocyte spheroids on a microfabricated chip. Tissue Eng 11:1254–1262. https://doi.org/10.1089/ten.2005.11.1254
Gaudet S et al (2005) A compendium of signals and responses triggered by prodeath and prosurvival cytokines. Mol Cell Proteomics 4:1569–1590
Gottschalk A et al (2010) Review article: the role of the perioperative period in recurrence after cancer surgery. Anesth Analg 110(6):1636–1643
Hamburger AW, Salmon SE (1977) Primary bioassay of human tumor stem cells. Science 197:560061. https://doi.org/10.1126/science.560061
Hart S, Dider W, Han H, Xu Z, Zhao L (2016) Toward repurposing metformin as a precision anti-cancer therapy using structural systems pharmacology. Sci Rep 6:20441
Hidalgo M, Amant F, Biankin AV, Budinská E, Byrne AT, Caldas C, Clarke RB, de Jong S, Jonkers J, Maelandsmo GM et al (2014) Patient-derived xenograft models: an emerging platform for translational cancer research. Cancer Discov 4:998. https://doi.org/10.1158/2159-8290.CD-14-0001
Holford ME, McCusker JP, Cheung KH, Krauthammer M (2012) A semantic web framework to integrate cancer omics data with biological knowledge. BMC Bioinformatics 13(Suppl 1):S10
Holmgren A (1981) Regulation of ribonucleotide reductase. Curr Top Cell Regul 19:47–76
Hood L, Friend SH (2011) Predictive, personalized, preventive, participatory (P4) cancer medicine. Nat Rev Clin Oncol 8:184–187. https://doi.org/10.1038/nrclinonc.2010.227
Horn H, Lawrence MS, Chouinard CR et al (2018) NetSig: network-based discovery from cancer genomes. Nat Methods 15(1):61–66. https://doi.org/10.1038/nmeth.4514
Ikemura K, Hiramatsu S, Okuda M (2017) Drug repositioning of proton pump inhibitorsfor enhanced efficacy and safety of cancer chemotherapy. Front Pharmacol 8:911
Iorio F, Rittman T, Ge H, Menden M, Saez-Rodriguez J (2013) Transcriptional data: a new gateway to drug repositioning? Drug Discov Today 18:350–357. https://doi.org/10.1016/j.drudis.2012.07.014
Janes KA et al (2006) The response of human epithelial cells to TNF involves an inducible autocrine cascade. Cell 124:1225–1239
Kapałczyńska M, Kolenda T, Przybyła W, Zajączkowska M, Teresiak A, Filas V, Ibbs M, Bliźniak R, Łuczewski Ł, Lamperska K (2018) 2D and 3D cell cultures—a comparison of different types of cancer cell cultures. Arch Med Sci 14:63743. https://doi.org/10.5114/aoms.2016.63743
Karlsson H, Fryknäs M, Larsson R, Nygren P (2012) Loss of cancer drug activity in colon cancer HCT-116 cells during spheroid formation in a new 3-D spheroid cell culture system. Exp Cell Res 318:26. https://doi.org/10.1016/j.yexcr.2012.03.026
Kim DJ, Li Y, Reddy K (2012) Novel TOPK inhibitor HI-TOPK-032 effectively suppresses colon cancer growth. Cancer Res 72:3060–3068. https://doi.org/10.1158/0008-5472.can-11-3851
Kumar N et al (2008) Multipathway model enables prediction of kinase inhibitor cross-talk effects on migration of Her2-overexpressing mammary epithelial cells. Mol Pharmacol 73:1668–1678
Kwon HR, Lee KW, Dong Z, Lee KB, Oh SM (2010) Requirement of T-lymphokine-activated killer cell-originated protein kinase for TRAIL resistance of human HeLa cervical cancer cells. Biochem Biophys Res Commun 391:830–834. https://doi.org/10.1016/j.bbrc.2009.11.147
Laczny C, Leidinger P, Haas J, Ludwig N, Backes C, Gerasch A, Kaufmann M, Vogel B, Katus HA, Meder B et al (2012) miRTrail - a comprehensive webserver for analyzing gene and miRNA patterns to enhance the understanding of regulatory mechanisms in diseases. BMC Bioinformatics 13(1):36
Lai Y, Wei X, Lin S, Qin L, Cheng L, Li P (2017) Current status and perspectives of patient-derived xenograft models in cancer research. J Hematol Oncol 10:106. https://doi.org/10.1186/s13045-017-0470-7
Lazzara MJ et al (2009) Quantitative modeling perspectives on the ErbB system of cell regulatory processes. Exp Cell Res 315:717–725

Lewis CE, Pollard JW (2006) Distinct role of macrophages in different tumor microenvironments. Cancer Res 66(2):605–612

Li J, Lei K, Wu Z et al (2016) Network-based identification of microRNAs as potential pharmacogenomic biomarkers for anticancer drugs. Oncotarget 7(29):45584–45596

Li M, Li X, Zhang H, Lu Y (2018) Molecular mechanisms of metformin for diabetes and cancer treatment. Front Physiol 9:1039. https://doi.org/10.3389/fphys.2018.01039

Liang G, Liu M, Wang Q, Shen Y, Mei H, Li D, Liu W (2017) Itraconazole exerts its anti-melanoma effect by suppressing Hedgehog, Wnt, and PI3K/mTOR signaling pathways. Oncotarget 8(17):28510–28525

Lin SY, Chang HH, Lai YH et al (2015) Digoxin suppresses tumor malignancy through inhibiting multiple Src-related signaling pathways in non-small cell lung cancer. PLoS One 10(5):e0123305

Liu Y, Yin X, Zhong J et al (2017) Systematic identification and assessment of therapeutic targets for breast cancer based on genome-wide RNA interference transcriptomes. Gene 8(3):E86

Luo Y, Zhao X, Zhou J et al (2017) A network integration approach for drug target interaction prediction and computational drug repositioning from heterogeneous information. Nat Commun 8(1):573. https://doi.org/10.1038/s41467-017-00680-8

Masters JRW (2000) Human cancer cell lines: fact and fantasy. Nat Rev Mol Cell Biol 1:233–236. https://doi.org/10.1038/35043102

Matsumoto S, Abe Y, Fujibuchi T, Takeuchi T, Kito K, Ueda N, Shigemoto K, Gyo K (2004) Characterization of a MAPKK-like protein kinase PBK/TOPK. Biochem Biophys Res Commun 325:997–1004. https://doi.org/10.1016/j.bbrc.2004.10.133

Matsuo Y (2014) TOPK inhibitor induces complete tumor regression in xenograft models of human cancer through inhibition of cytokinesis. Sci Transl Med 6:259. https://doi.org/10.1126/scitranslmed.3010277

Murray PJ, Wynn TA (2011) Obstacles and opportunities for understanding macrophage polarization. J Leukoc Biol 89(4):557–563

Nagaraj AB, Wang QQ, Joseph P et al (2018) Using a novel computational drug repositioning approach (drug predict) to rapidly identify potent drug candidates for cancer treatment. Oncogene 37(3):403–414. https://doi.org/10.1038/onc.2017.328. Epub 2017 Oct 2

Okada M, Shimizu K, Iyoda T, Ueda S, Shinga J, Mochizuki Y et al (2020) PD-L1 expression affects neoantigen presentation. iScience 26(23):101238

Orimo A et al (2005) Stromal fibroblasts present in invasive human breast carcinomas promote tumor growth and angiogenesis through elevated SDF-1/CXCL12 secretion. Cell 121(3):335–348

Pampaloni F, Reynaud EG, Stelzer EHK (2007) The third dimension bridges the gap between cell culture and live tissue. Nat Rev Mol Cell Biol 8:839–845. https://doi.org/10.1038/nrm2236

Park YJ, Kuen DS, Chung Y (2018) Future prospects of immune checkpoint blockade in cancer: from response prediction to overcoming resistance. Exp Mol Med 22:50

Perfect JR (2017) The antifungal pipeline: a reality check. Nat Rev Drug Discov 16(9):603–616

Perfetto L, Briganti L, Calderone A et al (2016) SIGNOR: a database of causal relationships between biological entities. Nucleic Acids Res 44(D1):548–554

Rahman MM, Islam MR, Rahman F, Rahaman MS, Khan MS, Abrar S et al (2022) Emerging promise of computational techniques in anti-cancer research: at a glance. Bioengineering 25(9):335

Rastegar-Mojarad M, Ye Z, Kolesar JM, Hebbring SJ, Lin SM (2015) Opportunities for drug repositioning from phenome-wide association studies. Nat Biotechnol 33(4):342–345. https://doi.org/10.1038/nbt.3183

Regan-Fendt KE, Xu J, DiVincenzo M et al (2019) Synergy from gene expression and network mining (SynGeNet) method predicts synergistic drug combinations for diverse melanoma genomic subtypes. NPJ Syst Biol Appl 26(5):6. https://doi.org/10.1038/s41540-019-0085-4

Reissfelder C et al (2015) Tumor-specific cytotoxic T lymphocyte activity determines colorectal cancer patient prognosis. J Clin Invest 125(2):739–751

Sacco F, Silvestri A, Posca D et al (2016) Deep proteomics of breast cancer cells reveals that metformin rewires signaling networks away from a pro-growth state. Cell Syst 2(3):159–171. https://doi.org/10.1016/j.cels.2016.02.005. Epub 2016 Mar 3
Sachs N, Clevers H (2014) Organoid cultures for the analysis of cancer phenotypes. Curr Opin Genet Dev 24:68–73. https://doi.org/10.1016/j.gde.2013.11.012
Sadanandam A et al (2013) A colorectal cancer classification system that associates cellular phenotype and responses to therapy. Nat Med 19:619–625
Sawey ET, Chanrion M, Cai C, Wu G, Zhang J, Zender L, Zhao A, Busuttil RW, Yee H, Stein L et al (2011) Identification of a therapeutic strategy targeting amplified FGF19 in liver cancer by Oncogenomic screening. Cancer Cell 19(3):347–358. https://doi.org/10.1016/j.ccr.2011.01.040
Schilsky RL (2010) Personalized medicine in oncology: the future is now. Nat Rev Drug Discov 9:363–366. https://doi.org/10.1038/nrd3181
Schoeberl B et al (2009) Therapeutically targeting ErbB3: a key node in ligand-induced activation of the ErbB receptor-PI3K axis. Sci Signal 2:31
Schreiber RD, Old LJ, Smyth MJ (2011) Cancer immunoediting: integrating immunity's roles in cancer suppression and promotion. Science 331(6024):1565–1570
Shim JS, Liu JO (2014) Recent advances in drug repositioning for the discovery of new anticancer drugs. Int J Biol Sci 10:654–663. https://doi.org/10.7150/ijbs.9224
Spillier Q, Vertommen D, Ravez S et al (2019) Anti-alcohol abuse drug disulfiram inhibits human PHGDH via disruption of its active tetrameric form through a specific cysteine oxidation. Sci Rep 9(1):4737
Strand DS, Kim D, Peura DA (2017) 25 years of proton pump inhibitors: a comprehensive review. Gut Liver 11(1):27–37
Suissa S, Azoulay L (2014) Metformin and cancer: mounting evidence against an association. Diabetes Care 37(7):1786–1788. https://doi.org/10.2337/dc14-0500
Sun J, Zhao M, Jia P et al (2015) Deciphering signaling pathway networks to understand the molecular mechanisms of metformin action. PLoS Comput Biol 11:e1004202
Swartz MA et al (2012) Tumor microenvironment complexity: emerging roles in cancer therapy. Cancer Res 72(10):2473–2480
Sylvia N et al (2006) Cancer bioinformatics; from therapy design to treatment, vol 30. Wiley, Hoboken, p 287
Tian Q, Price ND, Hood L (2012) Systems cancer medicine: towards realization of predictive, preventive, personalized and participatory (P4) medicine. J Intern Med 271(2):111–121. https://doi.org/10.1111/j.1365-2796.2011.02498.x
Turanli B, Arga KY (2017) Systems biomedicine acts as a driver for the evolution of pharmacology. Ann Pharmacol Pharm 2:1–3
Wang XD (2011) Role of clinical bioinformatics in the development of network-based biomarkers. J Clin Bioinforma 1:28. https://doi.org/10.1186/2043-9113-1-28
Wang XD, Liotta L (2011) Clinical bioinformatics: a new emerging science. J Clin Bioinforma 1(1):1. https://doi.org/10.1186/2043-9113-1-1
Wu Z, Li W, Liu G, Tang Y (2018) Network-based methods for prediction of drug-target interactions. Front Pharmacol 9:1134
Yaman G, Çini N, Altuntaş İ, Erbaş O (2021) What does CRISPR technology provide to cancer treatments? JEB Med Sci 2:41–49
Zeng X, Liu L, Zheng M, Sun H, Xiao J, Lu T, Huang G, Chen P, Zhang J, Zhu F, Li H, Duan Q (2016) Pantoprazole, an FDA-approved proton-pump inhibitor, suppresses colorectal cancer growth by targeting T-cell-originated protein kinase. Oncotarget 7(16):22460–22473. https://doi.org/10.18632/oncotarget.7984. PMID: 26967058; PMCID: PMC5008373
Zhang XH, Wang XY, Zhou ZW, Bai H, Shi L, Yang YX, Zhou SF, Zhang XC (2017) The combination of digoxin and GSK2606414 exerts synergistic anticancer activity against leukemia in vitro and in vivo. Biofactors 43(6):812–820. https://doi.org/10.1002/biof.1380. Epub 2017 Aug 17. PMID: 28817203
Zou J, Zheng MW, Li G et al (2013) Advanced systems biology methods in drug discovery and translational biomedicine. Biomed Res Int 2013:742835. https://doi.org/10.1155/2013/742835

Prevention, Diagnosis, and Understanding Mechanisms of Metabolic Diseases Using Systems Biology

22

Bhavani P. V, B. Vinutha, Renuka B, and B. B. Borse

Abstract

Human metabolic diseases are typically network diseases. These diseases including cardiovascular diseases (CVDs), type 2 diabetes (T2D), obesity, diseases of kidney, liver, intestine, and neurological disorders to name a few. These metabolic diseases are noted as one of the most serious hazards to global health. Clearly, progress in multifactorial metabolic diseases requires an understanding of the control and regulation of the underlying metabolic networks. This could be achieved by application of systems biology as it offers powerful abstraction tools to simulate all known metabolic reactions in a cell, therefore providing a snapshot that is close to its observable phenotype. Collecting valuable information out of systems biology approach is a vital preventive measure, allowing us to remain one step ahead of this vicious condition. A systems biology approach to understand the metabolism in these organs will enable one to leverage recent advances in computational technology and methods to integrate diverse sets of data. Coupled with state-of-the-art machine learning methods, this will allow characterizing each patient's disease phenotype and delivering precision medicine solutions, therefore, leading to early diagnosis, preventative medicine,

B. P. V
Pentavalent Bio Sciences Private Limited, Bangalore, India

B. Vinutha
G. R Medical College, Hospital and Research Centre, Mangalore, India

R. B
Promic Svasthya Private Limited, Mysore, India

B. B. Borse (✉)
PPSFT/SFS Department, Central Food Technological Research Institute, Mysore, India
e-mail: borsebb@cftri.res.in

S. Joshi et al. (eds.), *Systems Biology Approaches: Prevention, Diagnosis, and Understanding Mechanisms of Complex Diseases*,
https://doi.org/10.1007/978-981-99-9462-5_22

reduced treatment, and management of the diseases. With this background, recent developments in the application of systems biology will be discussed with specific relevance to diagnosis, treatment, prevention, and management.

Keywords

System biology · Metabolic diseases · Multi-organ disease · Cardiovascular disease · Diabetes · Obesity · Nervous system

22.1 Introduction

Diseases that occur due to the disruption of normal metabolism in some organs, and or the whole body are considered metabolic diseases. Typically, these diseases are hereditary and sometimes occur due to environmental or lifestyle issues. Metabolic diseases are non-communicable and are also termed "metabolic disorders" and "non-communicable diseases." Examples of metabolic diseases include diabetes mellitus, hypothyroidism, hyperthyroidism, marasmus, kwashiorkor, albinism, sickle-cell anemia, obesity, cancer, osteoarthritis, psoriasis, liver diseases, kidney diseases, glucose intolerance, cardiac diastolic dysfunction, alcoholism, chronic respiratory disorders, cognitive disorders, neurological disorders such as Alzheimer's and many others.

The origin of metabolic diseases dates back to 1908 when Sir Archibald Edward Garrod (1857–1936) postulated the inheritability of inborn errors of metabolism. He conceptualized the 'chemical individuality' and gave evidence for the dynamic nature of metabolism by showing the involvement of metabolites in metabolic pathways (Scriver 2008). As per the World Health Organization (WHO) reports of 2022, non-communicable diseases (NCDs) are responsible for nearly three-quarters of deaths worldwide. Chief among the NCDs are cardiovascular diseases (heart disease and stroke), cancer, diabetes, chronic respiratory diseases, and mental health. Seventeen million people below the age of 70 years are known to die of NCDs every year. Around 86% of this population is from low- and middle-income countries (World Health Organization 2020, 2022). As per the global health estimates of 2019, 41 million deaths are caused due to NCD each year, and this is equivalent to 74% of the deaths. Many millions are affected by at least one NCD leading to a reduced quality of life and life expectancy (World Health Organization 2020).

Metabolic diseases are considered to be one of the greatest global health challenges of this century. Although, the incidence of these diseases is often preventable and death of affected people could be delayed or avoided by eliminating risk factors including unhealthy diet, physical activity, alcohol and tobacco consumption, hypertension, obesity, blood glucose levels, cholesterol levels, and air pollution (Hastings et al. 2006). Tobacco usage, one of the main causes of metabolic diseases, results in the death of one in every two users (tobacco chewers, active and passive smokers) (WHO 2022a). Alcohol consumption is known to cause metabolic diseases including liver cirrhosis, cardiovascular disorders, and cancers in organs including oropharynx, larynx, esophagus, liver, colon, rectum, and breast (WHO

2018; Bagnardi et al. 2015). Overeating is known to cause obesity, diabetes, stroke, heart disease, and cancer. Malnourishment and insufficient levels of micronutrients in the body are also known to cause metabolic diseases (WHO 2022b). As per the 2020 report by Institute for Health Metrics and Evaluation, 19% of deaths out of 8 million NCDs is caused by unhealthy diet (Global Burden of Disease Collaborative Network 2020). Though obesity by itself is a metabolic disease, people suffering from it are known to disease secondary metabolic diseases such as cardiovascular diseases, musculoskeletal disorders, diabetes, and even some kind of cancers (WHO 2021). Two percent of the deaths caused by NCDs worldwide are a result of physical inactivity, indicating that physical activity helps in the prevention and management of metabolic diseases. More than 80% of the adolescent population, one in every three women, and one in four men are not physically active (World Health Organization 2022). Hypertension is well established as a major risk factor responsible for cardiovascular diseases. In 2019, the World Bank country income classification reported that almost half of the population suffering from hypertension are unaware of their condition itself (World Health Organization 2022). Raise in cholesterol levels results in blockage in arteries in turn causing heart ailments and stroke (World Health Organization 2022). Similarly, diabetes is responsible for 20% of deaths caused by cardiovascular disorders (Global Burden of Disease Collaborative Network 2020).

In the past few decades, scientists have attempted to understand metabolic diseases using metabolomics. Metabolomics is a comprehensive and metric study of all metabolites in cells, tissues, organs, or a whole biological system. This appeared to be a potential approach because any variations in the regulation of metabolism are known to directly contribute to the diseases that are associated with dysregulated metabolism, like CVDs, diabetes, obesity, cancers, cognitive disorders, and several others (Newgard 2017). Metabolomics enables one to obtain the chemical phenotype of the subject of interest using highly sensitive techniques such as nuclear magnetic resonance (NMR) spectroscopy, mass spectrometry (MS), and statistical analysis of the spectra, which can further be used to deduce promising biomarkers of a given disease. These biomarkers have a unique potential of providing multiple and important information about the disease including onset, incidence, progression, and severity, along with associated phenotypic abnormalities. In other words, the metabolic phenotype of an organism obtained using metabolomics offers high-resolution information about various diseases (Dumas et al. 2014). In studies using this non-invasive and metric tool, that is, metabolomics, scientists have deduced the branched-chain amino acid-related metabolites associated with cardiometabolic diseases (Newgard et al. 2009). Wang et al. reported that the branched-chain amino acid-related metabolites are suitable as biomarkers to predict the incidence of type 2 diabetes and obesity (Wang et al. 2011a). Metabolomics has been successfully applied in the investigation of gestational diabetes and to predict outcomes in the offspring (Scholtens et al. 2014; Allalou et al. 2016). Similar studies have been adopted in establishing the positive correlation between metabolites and diabetes outcomes in obese children and adolescents (McCormack et al. 2013).

The dysregulation of metabolites is influenced by several factors including environment, behavior, and genetics. The environmental factors may include psychosocial and socioeconomic variations that would impact the behavior of an organism resulting in stress, in turn leading to hormonal and metabolite changes. The genetic predisposition to such metabolic diseases also plays a major role in incidence and severity in the offspring as these diseases are hereditary in nature (McCormack et al. 2013; Björntorp 1996). Besides the formerly mentioned factors, studies also support the role of gut microbiota and their role in the occurrence of metabolic disorders in animals and humans (Fernandes et al. 2006). The environmental, behavioral, microbial, and even metabolic factors influence the interaction between the genetic features and incidence of metabolic disease(s) in a given organism in a dynamic way throughout its life. However, it cannot be neglected that the metabolic profile of an organism also results from its genomic features too (Fernandes et al. 2006). The metabolic phenotype profile obtained using metabolomics is descriptively examined using high-throughput annotation, and simulation models followed by preclinical and clinical studies. The approach of using sophisticated annotation and simulation models is completely data-driven and requires computational and mathematical analysis, in other words, defined as *systems biology*. In recent years, systems biology technologies are employed along with metabolic phenotyping to analyze and interpret metabolic biomarkers to identify the genomic and metagenomic determinants of metabolism and disease associated. In this chapter, we have discussed the role and applications of systems biology in understanding the complexity associated with major metabolic diseases namely cardiovascular diseases, respiratory disorders, obesity, diabetes, liver diseases, renal diseases, and neurological disorders. We have also covered in length and depth how this approach can be used in the prevention, diagnosis, treatment, and management of these metabolic diseases.

22.2 Systems Biology in Health and Diseases

The advent of systems biology has enabled us to understand the biological processes involved in human physiology at intracellular and biomolecular strata and in turn obtain the phenotypes in health and diseased individuals. For this same reason, systems biology has a promising application in disease diagnosis and treatment using the complex data obtained on the biological, genetic, and physiological networks of the disease in study. Computational approach in combination with systems biological techniques can decode the disease dynamics. Systems biology uses combination of techniques including theory, computational modelling, technology development and experimental research. Use of computational modelling is the key approach that sets systems biology as a more advanced, and endurable technology to decipher confirmative information on a disease condition from complex and often dynamic datasets. In terms of metabolic diseases, disturbance in metabolic pathways is the primary cause for disease manifestation. And thus, metabolomics information is intensively studied and analyzed to explicitly understand the metabolic profile of the disease using computational modelling. The metabolic profiles are achieved using nuclear magnetic resonance (NMR) spectroscopy (Dumas et al. 2014), mass spectrometry (MS) techniques such as UHPLC-MS, UPLC-MS/MS, LC-MS/MS, LC-TOF/MS, LC-Q/TOF/MS, GC-MS, and others (Kordalewska and Markuszewski 2015).

22.3 Cardiovascular Diseases

Cardiovascular diseases occur due to disorders of the heart and blood vessels. Some of these diseases include coronary heart disease, cerebrovascular disease, peripheral arterial disease, rheumatic heart disease, congenital heart disease, deep vein thrombosis, and pulmonary embolism. Coronary heart disease, cerebrovascular disease, and peripheral arterial disease are the disease of blood vessels supplying blood to the heart muscle, brain, and arms and legs, respectively. Rheumatic heart disease occurs due to the damage in the heart muscles and valves caused by rheumatic fever. Congenital heart disease, as the name itself indicates, is a birth defect and it is caused due to deformities in the heart's development and functioning. Thrombosis and embolism are caused due to blood clots in leg veins which then move to the heart and lungs.

Cardiovascular diseases are the major metabolic diseases that are one of the major causes of death worldwide. The prevalence of CVDs is growing every year and a range of risk factors including genetic, ethnic, unhealthy diet, tobacco and alcohol consumption, environmental, physical inactivity, and socioeconomic parameters have a major impact on the increasing burden of these diseases. Exposure to these risk factors affects individuals by inducing a raise in pressure, glucose, and lipids levels in the blood, and obesity. As per WHO reports, 17.9 million deaths occur each year due to cardiovascular diseases. This death rate represents 32% of deaths worldwide, over 75% of them being in low- and middle-income countries. Thirty-eight percent of 17 million deaths under the age of 70 years in the year 2019, which is considered premature death, were caused by CVDs. Common reasons for CVD deaths in around 85% of the deaths are heart attack and stroke. Majority of the CVDs can be prevented by avoiding exposure to the risk factors and disease management is possible through early detection.

22.4 Understanding the Mechanism of CVDs Using Metabolomics

In systems biology, a metabolic disease phenotype is derived to understand the disease mechanism, using computational and mathematical models of data integrated from genomics, transcriptomics, proteomics, and metabolomics. All the data derived from these high-throughput "omics" approach, finally leads to the discovery of those metabolites of clinical importance with respect to the disease of interest. Thus, the application of metabolomics plays a pivotal role in discovering valuable clinical biomarkers of a given metabolic disease. However, unlike other common diseases, the application of metabolomics to cardiovascular diseases is not easy because of the disease complexity. The metabolic disease of the human cardiovascular system is dependent on the changes in hundreds of metabolites that occur in the body as a response to the exposed risk factors. Thus, the metabolites and the rapid changes caused in their physiochemical properties are considered the representative reporters of cardiovascular disease phenotypes in affected individuals. For instance, omics approach has established an evident link between iron and CVDs and furthermore, many studies have supported the role of ferroptosis, a form of regulated cell death which is dependent on iron, in the initiation and progression of several CVDs (Fang et al. 2022). Recent studies have reported that elevated levels of uric acid in serum, the end-product of purine metabolism, as a predictive marker for CVDs such as heart failure, coronary artery disease, and hypertension (Saito et al. 2021). Irrespective of the type, all CVDs display loss of proteostasis indicated by the accumulation of misfolded and/or unfolded proteins in the endoplasmic reticulum (ER) leading to ER stress (Ren et al. 2021).

22.5 Hypertension

Hypertension is another common CVDs, with elevated blood pressure as the primary symptom. Upon exploration of the pathomechanisms of hypertension in experimental rat model with salt-sensitivity, Wang and co-workers determined several folds of alteration in compounds of TCA cycle including citrate, isocitrate, succinate, fumarate, and cis-aconitate. This investigation emphasized the association between TCA cycle disorders and salt-sensitive hypertension (Wang et al. 2014). A cohort study in prehypertensive patients indicated evident alterations in levels of lysophosphatidylcholine in plasma and disturbance in amino acid metabolism (Kim et al. 2014). Even in young population of hypertensive men, similar results wherein levels of amino acids including glycine, lysine, cysteine were observed (Wang et al. 2015b). Studies in animal models have contributed in identifying role of several important metabolites in hypertension cases. Decrease in levels of citrate and α-ketoglutarate in urine is reported in spontaneously hypertensive rats (Akira et al. 2008) and then is connected to the impairments of TCA cycle. Contrarily, in another study increase in levels of succinate were noticed in older spontaneously hypertensive male rats (Fujiwara et al. 2005). Yet in another study, changes in blood samples of hypertensive rats showed decrease in serine, methionine, phenylalanine, ornithine but increase in lysine levels. Also increase in free fatty acids were reported in this study. The study emphasized on use of total ginsenosides, extracts of ginseng, as a promising therapeutic intervention to treat hypertension as there was significant reduction in blood pressure and free fatty acids levels in treated animals (Aa et al. 2010). Though studies on metabolomic profiling using human samples are limited, increased levels of free fatty acids in serum are reported (Liu et al. 2011). This change may have an effect of membrane fluidity with respect to ions causing raise in vascular muscle tone in turn leading to high blood pressure (Ordway et al. 1991). Metabolomics in blood pressure establishes a direct relation of inflammation to hypertension, as raise in levels of α-1 acid glycoprotein, an inflammatory protein, is observed in hypertensive patients (Boos and Lip 2006). A large-scale cohort study in United States of America, United Kingdom, China, and Japan, reported inverse association between blood pressure and excretion of formate, alanine and hippurate in urine based on their dietary habits (Holmes et al. 2008). Studies on metabolomics of essential hypertension have collectively reported pyruvate, valine, tryptophan, and lactic acid as biomarkers. Also, reports suggest that metabolic pathways of alanine, aspartate, and glutamate metabolism, arginine biosynthesis, and aminoacyl-tRNA biosynthesis are associated with metabolomics of essential hypertension. It is noteworthy that these biomarkers and metabolic pathways are essentially involved in insulin resistance, inflammation, and unregulated nitric oxide production, which all collectively contribute to induction of essential hypertension (Deng et al. 2021).

22.6 Heart Failure

Biologists have studied the metabolic changes during heart failure, as it is common in individuals suffering from CVDs. The human heart has high energy demand for its effective function. In order to produce this energy, it selects suitable substrates for metabolism and thus availability of the appropriate substrates in abundance is crucial for myocardial function (Taegtmeyer et al. 2004; Ingwall 2002). The common metabolic reactions during efficient heart functioning include oxidative phosphorylation, ATP production, and oxygen consumption. All these three metabolic reactions are found to be reduced in the heart during failure causing the deficit of ATPs for healthy functioning of the organ (Neubauer 2007). The heart losses the ability to select substrates for energy production. At early stages of heart failure, both glucose and fatty acid metabolism are enhanced. With the progression of the cardiac failure, carbohydrate metabolism is noted to surpass that of fatty acid, which untreated leads to chronic heart failure characterized by reduction of both carbohydrates and fatty acids for metabolism. Resultant low energy condition in the organ leads to a drop in the activity of creatine kinase and increase in anaerobic glycolysis. Metabolomic profiles in a heart failure condition shows upsurge in levels of lactate, ketone bodies, branch-chained amino acids, and L-C acylcarnitine (Müller et al. 2021). Low energy state during heart failure results in abnormalities in ion transport in turn affecting Krebs cycle and mitochondrial metabolism (Murgia et al. 2009; Kohlhaas and Maack 2010).

22.7 Stroke

In recent years the incidents of stroke are escalating in an alarming rate in developing countries. Metabolomics in stroke using plasma and urine samples have been studied in this connection. Reduction in levels of glutamine and methanol, and increased levels of pyruvate, formate, lactate, and glycolate were observed in plasma samples. On the other hand, levels of citrate, hippurate, and glycine were found to be decreased in urine samples (Taylor Kimberly et al. 2013). Plasma samples from humans and rats, and cerebrospinal fluid samples from rats affected with cardioembolic stroke showed decreased rate of leucine, isoleucine, and valine (Zhang et al. 2009). Findings of cohort studies evidenced higher levels of L-carnitine, and trimethylamine-N-oxide in plasma of stroke incidents (Koeth et al. 2013; Tang et al. 2013). Among individuals with high risk of stroke incidence exhibited lower levels of lysophosphatidylcholine, validating the metabolite as a potential biomarker for early diagnosis of the disease (Jové et al. 2015).

22.8 Atherosclerosis

Metabolomic profile of atherosclerosis tissues including plaque and intima using UHPLC-MS approach along with advanced bioinformatic models shed light on the alterations in levels of acylcarnitines in the tissue samples. In the same study, analysis of extensive dataset suggested phosphotidylethanoleamine-ceramides as potential biomarkers that can be used to derive information on pathogenesis of atherosclerosis (Vorkas et al. 2015). Studies using atherosclerosis-induced animal models indicated a relationship between changes in levels of fatty acids, lysophosphatidylcholines and bile acids and the disease development (Ming-Qian et al. 2012). Induction of atherosclerosis in hamsters suggested variation of taurocholic acid levels in plasma as a plausible biomarker that could be exploited in early diagnosis of the disease (Jove et al. 2013). Rizza and co-workers identified a total of 49 metabolites in sera of elderly patients as putative predictive biomarkers (Rizza et al. 2014). Recent studies on clinical diagnosis of coronary atherosclerosis have revealed metabolic disorders that could be identified using blood-based diagnostic tools (Müller et al. 2021). In case of atherosclerosis with increased risk for myocardial infarction and stroke, trimethylamine-N-oxide is reported as one of important predictors (Tang and Hazen 2014; Zhu et al. 2016). This metabolite is basically formed in liver by conversion of trimethylamine produced by the gut microbiome of the host, and then released into the blood stream (Hoyles et al. 2018). Gut microbes are known to produce trimethylamine by converting dietary carnitine and choline and if this process is intervened, is proven to aggravate atherosclerotic lesions in mouse model (Wang et al. 2015a). Increased levels of arginine and decreased levels of lysophosphatidylcholines in circulation are associated with increased risk for myocardial infarction (Park et al. 2015). So are the circulating phosphatidylcholines containing palmitic acid, ceramide, diacylglycerol or sphingomyelin (Muoio and Newgard 2008). A large population study wherein serum metabolic profiles of individuals with cardiovascular risk were recorded using NMR technique revealed in involvement of docosahexaenoic acid, tyrosine and glutamine in the disease development and thus were recommended as diagnostic markers of atherosclerosis (Würtz et al. 2012). In another similar study among atherosclerotic patients using GC-MS technique, metabolites including palmitate, stearate, and

1-monolinoleoylglycerol in serum were suggested as promising diagnostic biomarkers of atherosclerosis (Chen et al. 2010). Extensive genetic and epidemiologic studies with respect to atherosclerosis have encouraged testing and treating elevated levels of Lipoprotein(a) as it is proven to have a strong association with cardiovascular diseases globally (Kronenberg et al. 2022).

22.9 Coronary Heart Disease

Application of metabolomic profiling in case of coronary heart disease (CHD) on blood samples of CHD patients in comparison with healthy individuals revealed changes in dicarboxylacylcarnitines, acylcarnitines and fatty acids levels in disease development (Shah et al. 2012). Circulating branch-chained amino acids including leucine, isoleucine, valine, glutamate, glutamine, proline, and methionine are also indicative of future coronary artery disease (Yang et al. 2015). Groups also have attempted on understanding myocardial ischemia in this line since it is the main cause of CHD development. One such study demonstrated changes in citric acid cycle in inducible ischemia (Sabatine et al. 2005). Yet, another study revealed variations in purine metabolism in the diseased rat model of myocardial ischemia (Lva et al. 2010). Studies have indicated rise in circulating as well as myocardial lactate concentrations in case of ischemia and this is the result of disturbed glucose metabolism during ischemic period. Amino acid metabolism is also disturbed during ischemia and this is indicated by rise in glutamic oxaloacetic transaminase in serum, leading to increased levels of glutamate in ischemic tissue (Müller et al. 2021). Non-ischemic cardiomyopathy is another cause of CHDs. In such cases, glycolysis is increased in affected individuals and is uncoupled from glucose oxidation and reduced fatty acid oxidation (el Alaoui-Talibi et al. 1992). Metabolomic profiling of simulated hypertrophic cardiomyopathy, a type of non-ischemic cardiomyopathy, in mice model resulted upregulation of fatty acids, lysolipids, acylcarnitines, and purines in heart and ascorbate, heme and pyrimidines are found to be downregulated (Sansbury et al. 2014).

22.10 Pulmonary Heart Disease

Pulmonary heart disease is a severe condition of hypertension that occurs due to complications in hemodynamics. Explorative studies on pathomechanisms and metabolomics in case of pulmonary heart conditions in animal model and affected lung tissue samples of patients have revealed disturbance in the processes of glycolysis, oxidative phosphorylation, TCA cycle, metabolism, and oxidation of fatty acids (Bujak et al. 2014; Zhao et al. 2014).

22.11 Peripheral Arterial Disease

Patients with peripheral arterial disease are at a two-fold risk of death compared to other CVDs (Huang et al. 2013). Idiopathic and hereditary pulmonary arterial hypertension was found to alter concentrations of 62 metabolites in the plasma samples of the affected patients and 36 of them were established prognostic markers. Levels of tRNA-specific modified nucleosides, TCA cycle intermediates, glutamate, tryptophan, fatty acid acylcarnitines, and polyamine metabolites were increased and levels of steroids, sphingomyelins, and phosphatidylcholines were decreased in patients (Rhodes et al. 2017).

22.12 Obesity

According to the WHO, obesity is defined as an abnormal and excessive accumulation of fat in an individual causing to the impaired health. Common practice to classify an individual as obese or overweight is to calculate his/her body mass index (BMI). Again, as per WHO, BMI greater than or equal to 25 and 30 in adults is considered as overweight and obesity, respectively. Global estimates revealed that in the year 2016, 1.9 billion adults were overweight and over 650 million adults among them were categorized to be obese. This estimation included 39% of men and 40% of women under overweight category, summing up to 39% of global adult population. Under obesity category which was about 13% of the global adult population, 11% was represented by men and 15% by women population. Additional observation was that since 1975 the cases of obesity was tripled in 2016. Changes resulting from urbanization including increased intake of energy-dense food products, fat and sugar-enriched foods, sedentary lifestyle, changes in modes of transportation, and family history of the disease have caused in this drastic increase in obesity and overweight incidences. Obesity not only results in energy imbalance but also increases the risk factors for non-communicable diseases including CVDs, majorly heart failure and stroke, type 2 diabetes, non-alcoholic fatty live disease, chronic kidney disease, musculoskeletal disorders, cancers of endometrium, breasts, ovary, prostate, liver, kidney, gallbladder, and colon. Obesity in children increases the incidences of premature death, disability in adulthood, hypertension, CVDs, diabetes, fractures, and psychological disorders. Like any other non-communicable disease, obesity is also preventable by making lifestyle changes like limiting intake of processed foods, high calorie-based, fats and sugars-enriched foods, increasing intake of fruits, vegetables, legumes, whole grains, nuts, leafy vegetables, and developing habit of regular physical activity.

22.13 Metabolomics in Obesity

Recent study on metabolomic profiling of blood serum in young adults with obesity using proton NMR spectroscopy showed increased levels of lipids, glucose, N-acetyl glycoprotein, glutamate, alanine, lactate, branch chain amino acid, and 3-hydroxybutyrate. Besides, decreased levels of choline in the were recorded in serum samples from obese young adults. Serum metabolic profile in non-hyperlipidemic obese adults upon comparison with that of hyperlipidemic obese adults showed decreased levels of lipids, acetoacetate, N-acetyl glycoprotein, lactate, glutamate, and isoleucine, and increased levels of choline and glutamine. The study indicated glutamine, lactate, branch chain amino acids, acetoacetate, and 3-hydroxybutyrate are serum biomarkers of obesity. Serum triglyceride, acetoacetate, N-acetyl glycoprotein, glutamate, isoleucine, visceral fat, and unsaturated lipid were higher in hyperlipidemic obese individuals, indicating the said metabolites as biomarkers of hyperlipidemic obesity (Htun et al. 2021). Obesity in children is also rising globally and over 340 million cases of obesity were reported. In a recent metabolomic study wherein three assays (gas chromatography-mass spectrometry, hydrophilic interaction liquid chromatography coupled to mass spectrometry and lipidomics) were employed, higher concentrations of branch chain amino acids and lipids including triglycerides, phosphatidylcholines, and cholesteryl esters were found to be potential biomarkers to study pathomechanisms of pediatric obesity (Szczerbinski et al. 2022). These results were in alignment with research findings using mice model, wherein lipid metabolites were proven to be major contributing factors of obesity (Bao et al. 2021). Another study of metabolomics in combination with microarray approach on rat model of diet-induced obesity revealed that this metabolic disease is dependent on several environmental as well as genetic factors. A total of 23 genes, and 9 metabolites demonstrated significant relation with induction of obesity. The findings indicated an enhancement of glycolysis and disturbance in Krebs cycle in case of diet-induced obesity. The study in a nutshell identified metabolites as promising biomarkers for obesity treatment (Liu et al. 2021). All these studies indicate disturbance in metabolism of glycerophospholipids, and linolenic acids is associated with obesity. Studies also support the possibility of dysfunction in adipose tissue as one of the contributing factors to obesity, as it causes inflammatory cell infiltration in intra-abdominal fat (Blüher 2013). Elevated levels of inflammation markers including C-reactive protein, resistin, IL-6, and TNF-α in serum of obese patients are reported. The underlying mechanism of elevated inflammation markers indicates these inflammation-associated molecules, for instance, nuclear factor kappa B (NF-κB) as one of the therapeutic targets for obesity. Other molecular markers that are considered as possible targets for obesity treatment include sirtuin 1 (a metabolic regulator with critical role in inflammation and lipid accumulation), G protein-coupled receptors, peroxisome proliferator-activated receptors, programmed death-ligand 1, programmed cell death 1, toll-like receptors, adhesion molecules (intercellular adhesion molecule 1, vascular cell adhesion molecule 1), sterol regulatory element-binding transcription factor 1c signaling pathways and others (Yang et al. 2022).

22.14 Diabetes

Diabetes is a chronic metabolic disease wherein the body fails to produce insulin or use the insulin produced to regulate glucose levels in blood, causing an imbalance in levels of glucose in the body in turn leading to several health complications. Possible health complications that occur due to diabetes in patients include CVDs, kidney failure, loss of vision, delay in wound healing often leading to amputation of the affected region, nerve damage, and so on. Diabetes in pregnant women, also called gestational diabetes, increases the risk of fetal mortality and other complications. This disease is one of the major non-communicable diseases that has been targeted for action globally. Despite the best efforts by the world leaders, the disease continues to steadily increase over the past few decades, majorly in low- and middle-income countries than in high-income countries. In 2014, 422 million adults were found to be diabetes worldwide. In 2012, 1.5 million deaths were reported due to diabetes. Besides, 2.2 million deaths were caused in diabetic patients due to the other related complications such as CVDs, obesity, kidney failure and other diseases. Above 40% of deaths in diabetic patients is represented by adult population, aging below 70 years. In 2019, 436 people were affected with diabetes, projections of diabetes in adult population by 2045 is predicted to be 700 million. Influencing factors for occurrence of diabetes include family history, socioeconomic causes, environmental factors, ethnicity, obesity, overweight, physical inactivity, smoking, and dietary habits.

Basically, there are two types of diabetes: Type 1 and Type 2. In case of type 1 diabetes (T1D), the pancreas fails to produce sufficient insulin, a hormone required to regulate the blood glucose levels. Management of this type of diabetes is done by periodical administration of insulin injections and thus it is called insulin-dependent diabetes. Symptoms of type 1 diabetes include excessive thirst, frequent urination, constant hunger, fatigue, loss of weight and vision. Contrarily, type 2 diabetes (T2D) is a condition wherein the body cannot use the insulin produced by pancreas and majority of diabetes cases are type 2 condition. Type 2 diabetes symptoms are like type 1 diabetes but in a less intensity. Thus, the disease may sometimes go undiagnosed for a long period. Healthy individuals often display conditions of impaired glucose tolerance (IGT) and impaired fasting glycaemia (IFG) before developing type 2 diabetes and people with these two conditions are at high risk of heart attack and stroke. Unlike type 1, type 2 diabetes is preventable by implementation of practices to eat healthily, avoid smoking, exercise regularly, controlling blood pressure and cholesterol levels. Early diagnosis is recommended for management of diabetes with periodic testing and assessment of treatment course.

22.15 Metabolomics in Diabetes

The human metabolome database is known to contain >40,000 metabolites out of which 43 metabolites have been found to have association with type 2 diabetes. These metabolites have been encountered in the blood and urine samples. In case of diabetes, variation in the levels of free fatty acids, branched chain and aromatic amino acids, and carnitines have been recorded and thus considered as potential biomarkers with respect to dysglycemia. Dysregulation in β-oxidation due to the high concentrations or non-functionality of insulin is also witnessed in a metabolic phenotype of dysglycemia. Subjects with impaired glucose tolerance showed increased levels of free fatty acids and decreased levels of lysophosphatidylcholines in blood; increased levels of tryptophan, xanthine and decreased hippuric acid in urine. Insulin-sensitive subjects with normal blood glucose levels have a raise in concentrations of free fatty acids, glycine, acylcarnitines, and α-hydroxybutyrate. Blood of female obese subjects with insulin resistance had increased levels of branched chain amino acids, valine, and tyrosine and that of obese male subjects showed raise in tyrosine, alanine, and ketones. However, both groups had decreased levels of β-hydroxybutyrate and anhydroglucitol (Pallares-Méndez et al. 2016).

The list of major classes of metabolites that are explored commonly in diabetes research using NMR and MS techniques is exhaustive. To name a few metabolites of interest in diabetes research include acyl-carnitines, acyl-CoAs, acyl-glycerols, metabolism of amino acids, bile acids, cholesterol esters, fatty acids, phospholipids, glycolysis, lipid profiling, nucleotides, organic acids, oxidized amino acids and lipids, urea cycle, sterols, and uremic solutes (Sas et al. 2015). Aromatic amino acids such as phenylalanine and tyrosine are associated with increased risk of occurrence of T1D. Branched-chain amino acids are known to promote insulin activity and signaling in turn worsening the insulin resistance in patients with T1D. Changes in levels of plasma lipids including triglycerides, cholesterol, sphingolipids, glycerophospholipids, and phospholipids are indicative of development of T1D-associated autoimmunity. In comparison with non-diabetic individuals, levels of saturated lipids such as palmitic acid, myristic acid, and stearic acid are known to be higher in case of patients with T1D. Reduced levels of sphingomyelins and glycerophospholipids, and increase of lysophosphatidylethanolamine and lysophosphatidylcholine are reported in T1D cases. Through the approach of metabolomics, several promising biomarker metabolites of T2D for diagnosis and prognosis of the disease have been identified. Patients with T2D revealed raise in glucose, citric acid, kynurenic acid, pipecolic acid, glycine, and betaine in the urine metabolomic analysis, and were categorized as promising markers of the disease (Zhang et al. 2014). Five amino acids, leucine, isoleucine, valine, tyrosine, and phenylalanine, were proposed to be useful in prediction of the disease development and thus amino acid profiling could be helpful in risk assessment of diabetes incidence (Wang et al. 2011b). Results of another study emphasized that three metabolites *viz.* glycine, acetylcarnitine, and lysophosphatidylcholine are key biomarkers that may help in T2D prevention (Wang-Sattler et al. 2012). In a study of urinary metabolomics in T2D patients demonstrated alteration in nucleotide metabolism, indicating the role of N-methylnicotinamide and N-methyl-2-pyridone-5-carboxamide as potential markers to predict disease progression (Salek et al. 2007). Another independent biomarker for diabetes risk development is 2-aminoadipic acid, an inducer of insulin secretion (Sas et al. 2015). Reduced levels of glycline another diagnostic biomarker of T2D. A study to derive metabolic footprint of T2D revealed disturbance in key metabolic pathways linked to kidney function, lipid metabolism, and the interaction with the gut microflora (Suhre et al. 2010). With the progression of diabetes as indicated by glucose intolerance and insulin resistance significant changes in serum metabolome is recorded and hence metabolomic profiling of patient serum would help in assessment of disease progression and effective of treatment undertaken.

T2D is often known to cause diabetes associated kidney disease as well as CVDs resulting in increased death rates. Metabolomic profiles of blood and urine samples from patients with diabetic kidney disease (DKD) indicates potential biomarkers that provide disease information including disease mechanism, progression, and possible therapeutic targets. Increased levels of symmetric dimethylarginine in serum, increased metabolism of asymmetric dimethylarginine into citrulline and dimethylamine was observed in rapid kidney dysfunction in T2D cases. Upregulated tryptophan metabolism and higher levels of tryptophan or kynurenine have been recorded in patients with DKD (both T1D and T2D). Loss of activity of phenylalanine hydroxylase in case of DKD leads to impaired conversion of phenylalanine into tyrosine, causing raise in phenylalanine and reduced levels of tyrosine in patient's serum. Alteration of uracil levels, reduction of glycine, glycolic acid, 2-methylacetoacetate (an intermediate in metabolism of isoleucine), and elevated ratio of albumin-creatinine in urine samples of patients with T2D. Raise in metabolites of ceramide and sphingomyelin in plasma of patients with DKD is reported. Elevation in sphingomyelin levels is indicative of kidney disease progression in patients with T1D. Clinical trials using sodium-glucose cotransporter-2 inhibitors reduced the risk of DKD progression in patients with T2D. Asymmetric dimethylarginine is known to increase in T2D patients with CVDs also. Dysfunctional high-density lipoprotein cholesterol and an inverse association with the CVDs risk and associated complications, by causing alteration in cholesterol efflux, anti-oxidant, and anti-inflammatory capacity of the cardiovascular system. Altered levels of ethanolamine, tyrosine, tryptophan, glutamic acid, hydroxyproline, and 3-methylhistidine were predictive of CVDs. Role of amino acids and their altered levels in cases of diabetes, DKD and CVDs suggests the possibility of shared pathways and metabolic mechanisms in these diseases (Jin and Ma 2021).

Glucose intolerance is common in pregnancy which leads to diabetes, termed as "gestational pregnancy." This type of diabetes is associated with hypertension during pregnancy, increased rate of C-section delivery, reduced lactogenesis and difficulty in breastfeeding. Gestational diabetes increases the risk of future development of T2D, CVDs, macrosomia, and affects fetus by causing hypoglycemia, hyocalcemia, and jaundice. Metabolomics of gestational diabetes not only provides refined information on pathophysiology of the disease but also helps in early diagnosis and screening of the disease. Increase in fatty acids including linoleic acid, linolenic acid, arachidonic acid, eicosapentaenoic acid, and docoasapentaenoic acid during the third trimester of pregnancy has been reported. Metabolites of gluconeogenesis (alanine, hexitol, fructose, and lactate) and lipid metabolism (medium-chain fatty acids, acylcarnitines, glycerol, beta-hydroxybutyrate, and palmitoleic acid) are associated with glycemia in pregnant women. Variations in several amino acids levels in plasma are also recorded. Maternal diabetes has effects on the infant too. In a cohort study, gestational diabetes was found to induce hyperinsulinemia in fetus causing to lower cord artery serum glucose levels, along with alteration in metabolic pathways as indicated by defects in placental amino acid transportation.

22.16 Kidney Diseases

The kidney plays a critical role in maintaining homeostasis of the body by regulating acid-base balance, plasma volume and secretion of hormones. Imbalance in the renal metabolism and function lead to many kidney diseases causing loss of the body's homeostasis. A chronic kidney disease (CKD) is a condition of kidney damage due to the drop in the glomerular filtration rate to <60 mL/min/1.73 m^2 for three months or more. There are five stages of CKD namely (1) kidney damage, (2) kidney damage with glomerular filtration rate of 60–89 mL/min/1.73 m^2, (3) glomerular filtration rate of 30–49 mL/min/1.73 m^2, (4) glomerular filtration rate of 15–29 mL/min/1.73 m^2, and (5) end-stage renal failure with glomerular filtration rate of <15 mL/min/1.73 m^2. In general, CKD is a global public health problem, which is increasing due to diabetes and CVDs. Some of the other influencing factors of this disease include genetic, environmental, and health factors. If untreated, a CKD can progress to complete loss of kidney function resulting in end-stage renal disease (ESRD) which requires to put the affected patients under dialysis. Other causes for a patient to develop ESRD include diabetes, hypertension, glomerulonephritis, and tubulointerstitial injury. Currently, diagnosis of CKD is done by measuring the levels of serum creatinine, blood urea nitrogen (BUN), urine routine analysis, histopathology, and immunohistochemistry. Of all these tests, the most common test done is serum creatinine. However, this test has poor sensitivity and specificity because the concentration of this clinical biomarker varies based on age, gender, body weight, rate of metabolism, state of nutrition and hydration of the body. Also, the significant concentration changes of serum creatinine occur after the renal dysfunction and so this test is less than helpful in preventing kidney damage. Due to these and several other limitations with the current approaches of CKD diagnosis, need has arisen for promising markers for early diagnosis, therapy, prevention of disease occurrence and progression, and response to treatment. This need could be addressed by deriving the disease phenotype and the pathophysiology underlying the same. Advanced techniques such as genomics, proteomics, and metabolomics are explored collectively in this line seeking for potential markers of CKD.

22.17 Metabolomics in Kidney Diseases

CKD proteomics have resulted in encounter of novel potential biomarkers, of acute renal tubular injury, namely neutrophil gelatinase-associated lipocalin (NGAL), kidney injury molecule-1, and liver-type fatty acid binding protein from urine samples. Promising biomarkers for CKD from plasma include cystatin C, apolipoprotein A-IV, adiponectin, asymmetric dimethylarginine, fibroblast growth factor 23, natriuretic peptides, and so on. Unlike serum creatinine, NGAL is useful in early diagnosis of renal dysfunction. Rate of cystatin production during renal injury is independent of tissue type and muscle mass. Metabolomics research in CKD using animal models have reported changes of metabolites such as palmitic acid, phytosphingosine, chenodeoxychrolic acid, 12-hydroxy-3-oxocholadienic acid, 7-ketolithocholic acid, and lysophosphatidylethanolamine in fecal profile. These metabolites are considered as promising biomarkers for early diagnosis of chronic kidney failure. Studies have reported that uremic toxins such as indoxyl sulfate and p-cresyl sulfate, and polyunsaturated fatty acids as metabolites of CKD importance. Indoxyl sulfate is the first metabolite that shows up in case of CKD and thus it is considered as a reliable biomarker in understanding the disease progression, followed by phenyl sulfate, hippuric acid, and p-cresyl sulfate. Indoxyl sulfate, hippuric acid, p-cresyl sulfate, and 4-ethylphenyl sulfate are also indicators of host response to therapy. During the CKD condition, metabolic pathways including citrate cycle, and branched-chain amino acid metabolism are found to be disrupted. In case of polycystic kidney disease (PKD), energy metabolism and amino acid metabolism are found to be disturbed as reflected by the changes in the metabolites of these pathways. Kidney disease in diabetic patients is resulted by the upregulated pathways of lipid synthesis, and ketone body synthesis, and downregulation in pathways of TCA cycle or glycolysis. Concentrations of serum metabolites including phenylalanine, tryptophan, creatinine, kynurenine, and lysophosphatidylcholines are used as biomarkers for early chronic renal failure. Disturbance in triglyceride catabolism and β-oxidation in case of ESRD patients is observed leading to a decrease in lower-molecular-weight triacylglycerols and increase in intermediate-molecular-weight triacylglycerols in the plasma. Success of treatment with hemodialysis in ESRD patients was estimated by a promising metabolite biomarker, 1-methylinosine, in the plasma samples (Zhao 2013).

Some of the metabolic pathways identified to have a role in CKD include metabolism of steroid, tryglyceride, cinnamic acid, serine, arginine, alanine, aspartate, glutamate, histidine, methylamine, phenylalanine, tryptophan, taurine, hypotaurine, tyrosine, valine, purine, cysteine, bile acid, nucleotide, phospholipid, lipid, methionine cycle, TCA cycle, glycolysis, gluconeogenesis, glycolate pathway, fatty acid oxidation, and metabolism pathways in gut bacteria (Wang et al. 2019). In case of CKD and PKD, metabolomics analysis using urine, serum, plasma, and tissue samples revealed hippuric acid and uric acids are the markers for both kind of kidney diseases. Usually, uric acid is excreted in the urine in humans unlike other mammals where uric acid is oxidized to allantoin before its excretion in the urine. However, studies have reported excretion of allantoin in diseased conditions including CKD, PKD, diabetic nephropathy, and kidney transplantation. Since allantoin is a by-product of purine metabolism, role of this mechanism has been suggested in kidney diseases. Similarly, metabolites of tryptophan metabolism, including kynurenic acid, indoleacetic acid, and tryptophan are reported in cases of CKD, transplant, nephropathy, and kidney cancer (Abbiss et al. 2019). Potential biomarkers for the diabetic nephropathy include hexadecenoic acid, linolelaidic acid, linoleic acid, trans-4-hydoxy-L-proline, piperidine, 6-aminocaproic acid, azoxystrobin acid,

L-dihydroorotic acid, lysophosphotidylcholine, 6-methylmercaptopurine, and cuminaldehyde (Zhang et al. 2020). In case of metabolomics in aristolochic acid nephropathy, a tubulointerstitial nephritis that occurs in ESRD and urothelial malignancy, revealed lysophospholipids, chenodeoxycholic acid, lysophosphotidylethanolamine, and cholic acid as potential biomarkers for early and late diagnosis of the disease. Studies have confirmed improving levels of glucose 6-phosphate, glucose 1-phosphate, ratio of malate and asparate in plasma may lessen the CKD progression. Altered levels of metabolites such as glutamate, acetate, alanine, 3-hydroxybutyrate, pyruvate, acetoacetate, lactate, isoleucine, and valine are observed in case of renal fibrosis and thus these metabolites could be used in the disease diagnosis (Wang et al. 2019).

22.18 Liver Diseases

Major diseases of liver that effect hepatic function include cirrhosis, non-alcoholic fatty liver disease, alcoholic hepatitis, hepatitis due to viral infection (a communicable disease), and cancer. Global burden of liver diseases accounts to more than 2 million deaths every year, more than half of them due to cirrhosis and the rest due to hepatitis and hepatic cancer. There is a striking relation between liver diseases and influencing factors including geography, race, gender, ethnicity, genetics, and socioeconomic. Among the liver diseases, alcohol-associated liver disease (AALD) is the most common one worldwide. Annual alcohol consumption beyond permissible amount, 6.2 l/person, for a longer duration has injurious effect on liver. According to WHO, about 2 billion people worldwide consume alcohol and more than 75 million are affected with alcohol-use disorders. The mortality rate in case of cirrhosis due to alcoholism is 50%. Mortality in alcoholic hepatitis, an AALD, is also higher. Prevalence of non-alcoholic fatty liver disease (NAFLD) has also been globally increasing though the rate of mortality (7%) is lesser than AALD (36%). Reports reveal that a vast majority of deaths among NAFLD patients is due to issues not related to liver, but cardiovascular-related causes. As excessive alcohol consumption is precursive to AALD, co-existing diabetes and obesity are analogous risk factors with respect to NAFLD. Clinical conditions associated with NAFLD are distinguished into two distinctive forms: (1) NAFL including steatosis (fatty liver) with or without mild lobular inflammation and (2) non-alcoholic steato-hepatitis (NASH) with fibrosis, cirrhosis, and hepatocellular carcinoma. Infection of liver with viruses causes hepatitis leading to fatal health problems. Hepatitis virus includes five strains namely A, B, C, D, and E. These strains vary in geographical distribution, modes of transmission, and severity of illness. Of all, hepatitis B and C viruses lead to chronic hepatitis in large population, majorly in low- and middle-income countries, and the disease causes cirrhosis, and carcinoma of liver. According to WHO estimates, around 354 million cases of hepatitis B or C are recorded worldwide and for most of them, diagnosis and treatment is not available. Deaths due to hepatitis and occurrence of the disease can be prevented by vaccination programs, early diagnosis, treatment, and education campaigns. WHO aims to reduce new cases of hepatitis infections by 90% and deaths by 65% by 2030.

Hepatitis A is caused by hepatitis A virus and the virus can transmit through food, water, fecal-oral route, unhygienic sex, and direct contact with an infected person. Though the disease is not severe in many cases, it could rarely cause acute liver failure leading to death. Hepatitis caused by hepatitis B virus (HBV) is life-threatening, often caused due to cirrhosis and liver cancer developed as long-term complications. Other symptoms include acute illness, yellowing of skin and eyes due to jaundice, dark urine, abdominal pain, nausea, vomiting, and fatigue. Vaccination is 98–100% efficacious and thus preventable. Hepatitis C is caused by acute and chronic infection by hepatitis C virus. In case of acute infection, the virus is cleared within 6 months without any treatment. However, during chronic HCV infection, the patient is prone to develop cirrhosis and liver cancer within 20 years. Virus transmission occurs through unscreened blood transfusions, unsafe sex, unsafe injection practices and other unsafe health care practices. Upon treatment, recovery rate can be raised up to 95%. However, there is no effective vaccine against hepatitis C. Hepatitis D occurs only in patients already suffering with hepatitis B (super-infection). This is because the hepatitis D virus (HDV) requires HBV for its replication. Thus, this form of hepatitis occurs as a co-infection (infection with both B and D type) or super-infection. The infection rates of HDV are decreasing gradually worldwide, due to the successful HBV vaccination program. Though the disease is preventable by hepatitis B immunization, treatment success and recovery rates of hepatitis D are low in case of unvaccinated patients. Symptoms include fever, fatigue, anorexia, vomiting, nausea, jaundice, and fulminant hepatitis. Infection with hepatitis E virus causes inflammation of liver and the virus is transmitted via fecal-oral route, primarily via contaminated water. The cases of hepatitis E are high in South East Asian countries than in the rest of the world. Mortality rate as per 2015 WHO estimation was 3.3% and annual infection cases are estimated to be 20 million of with 3.3 million cases are symptomatic of hepatitis E. Symptoms include fever, anorexia, nausea, vomiting, abdominal pain, jaundice, and liver inflammation. In acute cases, infection results in liver failure leading to risk of death.

22.19 Metabolomics in Alcoholic-Associated Liver Disease (AALD)

Alcohol-associated liver disease may progress from fatty liver condition, that is, steatosis to steatohepatitis, liver cirrhosis and further developing into hepatic carcinoma. Excessive consumption of alcohol is one of the main causes of liver disease-associated morbidity and mortality. Howbeit, the pathogenesis of AALD is yet poorly understood, causing delay in diagnosis of disease progression as well as treatments. In recent decades, the role of gastrointestinal (GI) tract and the gut microbiota are related to the AALD pathogenesis and progression. Owing to the alcohol consumption, alterations in the equilibrium of the gut microbiome are noticed as demonstrated by the unwanted bacterial over growth, release of excessive bacteria by-products into the human system, in turn effecting the function and pathophysiology of GI tract. The pathophysiological conditions disturbed by alcohol include drastic changes in the concentrations of GI tract metabolites including bile acids, branched chain amino acids, and short chain fatty acids. These metabolic changes have critical contribution in the induction and progression of AALD. These findings have been deciphered using AALD models. Bacterial overgrowth has been observed in many of these models. This bacterial overgrowth has a direct impact on ethanol metabolism. In experimentally induced conditions, bacterial overgrowth causes enhanced endogenous and exogenous ethanol metabolism in turn causing raise in acetaldehyde concentrations in the intestinal lumen and the blood stream. Not only bacterial overgrowth, but passage of live bacteria from the GI tract to other sites like lymph node, liver, kidney, and blood stream was also observed in rats with hepatic inflammation. This bacterial translocation was observed in a few rat models, by 14 days of alcohol administration and in some no significant translocation was reported even after 2 weeks of alcohol administration. Not only the gut bacteria, but their endotoxins are also altered due to alcohol consumption. Under normal physiological conditions, the endotoxins produced by the gut bacteria are eliminated from body, but since their production is increased during alcohol intake, their level also increase in serum resulting in endotoxemia, a condition that is well encountered in patients in AALD. Endotoxemia has a direct effect on TNF-α levels, and inflammatory cytokines and chemokines are produced by hepatic Kupffer cells, resulting in neutrophils and monocytes infiltration into the liver. Other than endotoxins, bacterial biomolecules including DNA, peptidoglycan, flagellin and such are translocated to extraintestinal space and organs due to intestinal barrier dysfunction causing AALD progression. TLR9 activation due to the accumulation of bacterial DNA further sensitizes the liver to endotoxin-induced injury. Accumulation of bacterial DNA in the plasma of alcoholic cirrhosis patients is well reported. Alcohol consumption not only results in bacterial overgrowth but also induces dysbiosis, an imbalance in the intestinal bacterial composition affecting the GI tract. Overall take is that the ill effect of alcohol consumption of GI tract develops to damage liver and thus liver diseases are induced and continued alcohol consumption enables the disease progression (Zhong and Zhou 2014).

22.20 Metabolomics in Non-alcoholic Fatty Liver Disease (NAFLD)

NAFLD clinically present itself from simple steatosis to non-alcoholic steatohepatitis (NASH) with fibrosis or cirrhosis. Standard diagnosis of NASH involves liver biopsy and no precise non-invasive diagnostics are available. Liver lipidomics, amino acid, and bile acid metabolomics have been recently explored as potential diagnostic methodologies in the field of NAFLD in general, and NASH in particular. Numerous metabolomic biomarkers relevant to the progression of NASH have been unraveled. Metabolomic prediction models have identified aspartate aminotransferase (AST), insulin, and patatin-like phospholipase domain-containing-3

(PNPLA3) genotypes as promising diagnostic markers of NASH. Furthermore, MS-based lipids and metabolites including glutamate, glycine, isoleucine, lysophosphatidylcholine and phosphatidylethanolamine were also noted to be promising disease markers. Serum triglycerides are also found to be highly promising NASH diagnostic biomarkers based on a biopsy-proven cohort study. Alterations in bile acid have also been found to be characteristic of NAFLD and its severity. For instance, taurocholate in case of steatosis and hepatocyte ballooning, glycocholate for lobular, taurolithocholate for portal inflammation, and total bile acid ratio and conjugated cholate for fibrosis. Similarly, alterations in plasma amino acid also outlines the severity of NAFLD based on metabolomics prediction. For instance, increased plasma concentrations of branched amino acids and aromatic acids, and decrease in amino acids of glutathione synthesis are discovered in NAFLD patients (Kim 2021). Collectively, these findings that noninvasive assessment of NAFLD is a possibility upon even more intensive and accurate exploration in the field of metabolomics.

22.21 Metabolomics in Cirrhosis

Early diagnosis of cirrhosis is difficult with current approaches and novel biomarkers are being deciphered for specific and early detection of the disease using metabolomics research. Biological samples such as serum, plasma, urine, and liver tissue are subjected to analytical techniques including NMR, GC/MS, UPLC, LC-QTRAP, and others for metabolic profiling. Some of the metabolites of disease importance in case of liver cirrhosis identified with this approach include LDL, VLDL, aceone, glucose, lactate, choline, pyruvate, phenylalanine, lysine, glutamine, histidine, alanine, creatine, glutamate, isoleucine, leucine, acetoacetate, choline, low-density lipoproteins, unsaturated lipids, N-acetylglycoproteins, glycerol, hexadecanedioate, taurocholate, taurochenodeoxycholate, glycocholate, glycerophospholipids, steroid hormone, cholesterol, arachidonic acid, ratio of stearic acid to oleic acid, xanthine, and some more.

22.22 Metabolomics in Hepatocellular Carcinoma

Biological samples such as serum, plasma, urine, and liver tissue are subjected to analytical techniques including NMR, GC/MS, UPLC, HPLC, GC/TOF-MS, RPLC and others for metabolic profiling. Some of the metabolites of disease importance in case of hepatocellular carcinoma identified with this approach include LDL, VLDL, unsaturated lipids, acetate, pyruvate, glutamine, tyrosine, valine, phenylalanine, serine, glycine, aspartate, leucine, glutamate, alanine, threonine, lactate, inosine, glycerol, alpha-ketoglutarate, acetoacetate, xanthine, oxidative stress-associated metabolites, metabolites of TCA cycle, normetanephrine, cysteic acid, creatine, canavaninosuccinate, lysophosphatidylcholilne, N-acetyl glycoprotein, monounsaturated palmitic acid, myo-inositol, malate, linoleic acid, metabolites of glycolysis, attenuated mitochondrial oxidation, arachidonic acid synthesis, hippurate, citrate, dihydrouracil, and many more.

22.23 Neurological Diseases

Neurological diseases or disorders (NDs) are heterogeneous diseases that affect the autonomic, peripheral, and central nervous system of the human body. Non-communicable NDs include migraines, non-migraine headaches, multiple sclerosis, epilepsy, stroke, Alzheimer's disease, Parkinson's disorder, anxiety, depression, autism, and anorexia. In 2019, regionwide neurological conditions account for 533,172 deaths, 213,129 (40%) in men, and 320,043 (60%) in women. Nearly 32.9 deaths per 100,000 population, 33.1 deaths per 100,000 population in men and 32.2 deaths per 100,000 population in women was recorded. Highest age-standardized death rates due to NDs are recorded in the USA, Canada, Uruguay, Suriname, Honduras, Bolivia, and Cuba. Years of life lost due to premature mortality in case of NDs accounted for 7.5 million (3.5 million in men and 3.9 million in women). On the other hand, years lived with disability in 2019 due to NDs was accounted for 8.2 million (3.1 million in men and 5.1 million in women). However, a comprehensive understanding of pathogenesis of the NDs is still not available.

Systems biology is an ocean of approaches each developed based on specific concept. To name few of the concepts, it includes systems thinking (in the context that the systems are related), complexity theory (studying complex systems on generation of properties), non-linear dynamics (systems evolution with lack of linearity with time), control theory (behavior of dynamical systems), feedback control (part of system output signal is passed to the input), network (interconnected group with shared information), network analysis (topology, dynamics), gene regulatory networks (interaction of DNA segments with other cell components like RNA, protein, mRNA, interferons), pathways, pathway modularity (metabolic modules), pathway analysis, dynamic disease (includes variables of the organism like behavior, underlying pathogenesis, trauma, tumor), systems biomarker discovery, systems drug discovery, bottom-up approach (piecing together systems to create grander system), top-down approach (breaking down a system to obtain compositional subsystems), and several other (Villoslada et al. 2009).

On the platform of above-mentioned systems biology platform, several disciplines have been newly developed to study and understand neurological diseases. Some of them include computational neuroscience, neuroinformatics, neurophysiology and others. All these approaches have established a direct relationship between many of the neurological disorders and gut microbiota, termed "gut-brain axis." The analyses emphasize that the alterations in the gut and/or intestinal microbiota has impact on the metabolites of the nervous, immune, and endocrine systems. Several animal and human studies have implicated the direct effect of changes in specific metabolites that can be considered as promising biomarkers, on several NDs.

22.24 Metabolomics in Autism

Difficulty in social interaction, verbal and non-verbal communication and repetitive behaviors are the characteristics of autism or autism spectrum disorder (ASD). Studies have indicated an association of this disorder with intestinal microbiota. Intestinal inflammation and dysbiosis have a role in autism (Berding and Donovan 2016). Increase in PPA levels, an inflammatory biomarker, is observed in autistic rodent model. Administration of probiotics in such models not only demonstrated positive effect on gut health but also improved the communicative and sensory-motor behaviors. In case of human autistic subjects too, raise in PPA levels is observed. Also, increase in proinflammatory cytokines levels have been reported to have a strong association with poor social behavior and autistic disorder (Larroya-García et al. 2019). Metabolomic analysis using LC-MS/MS technique in ASD children revealed alterations in histidine and glutathione metabolism and aromatic amino acid biosynthesis pathway. Behavioral developments in autistic children demonstrated alterations in neurotransmitter metabolism. Methylation was also altered in aminoacyl-tRNA biosynthesis, and branched chain amino-acid and nicotinamide metabolism. Signature markers were identified in the energy metabolism (oxidative stress and inflammation) and aminoacyl-tRNA biosynthesis. Energy metabolism and L-glutamine also showed alterations in mitochondrial metabolism of ASD children (Brister et al. 2022).

22.25 Metabolomics in Anxiety

Anxiety disorder displays fear, exaggerated thinking, oppression, and depression in the affected individuals. Factors including psychology, genetics, biology, and chemistry of the patients are attributed to the etiology of anxiety disorders. Yet, the diagnosis is majorly symptoms-based. For better treatment of this disorder, personalized and guided medicines and processes are essential and this cannot be achieved by symptoms-based diagnosis. The underlying biological and genetic factors can be of help in this respect and metabolomics studies in this field are being conducted to discover promising signature biomarkers. Analytical methods such as GC-MS, LC-MS, LC-MS/MS, photometric and immunoassays are used to explore suitable biomarkers from biological samples such as plasma, urine, and brain. Some of the potential metabolites that are identified thus far, which can be used as biomarkers, include myoinositol, glutamate, tricarboxylic cycle-intermediates, dehydroascorbate, xylose, succinic acid, carnitine, acetylcarnitine, cytosine, glutamine, γ-glutamyl–glutamine, phosphatidyl-cholines, ceramides, cholesterol, adrenocorticotropic hormone, triglycerides, free fatty acids, and dehydro-epiandrosterone-sulfate (Humer et al. 2020).

22.26 Metabolomics in Alzhiemer's Disease

Alzheimer's disease (AD) is an aging-related neurodegenerative disease, causing loss of memory and cognitive functions in a progressive pattern. This is no cure for this disease. However, early diagnosis of the predisposing factors and possibility of the disease would help in taking necessary preventive treatments or lifestyle changes to delay the onset of the disease. Several metabolomic studies have been done to determine promising metabolites with potential of being employed as AD diagnostic biomarkers using plasma, brain, and cerebrospinal fluid (CSF) samples in human studies and animal studies. Here too, analytical techniques such as LC-MS, GC-MS, CE-TOF-MS, UHPLC-TOF-MS, UPLC-MS, UPLC-ESI-MS/MS, HPLC-MS/MS, and NMR spectroscopy are used. Several altered metabolites have been identified with potential of being further explored as promising biomarkers. Some of them include cholesterol, Desmosterol, Nonanoylglycine, Dopamine-quinone, Arachidic acid, cis-Vaccenic acid, Cerotic acid, Erucic acid, Linoleic acid, Asparagine, Asymmetric dimethyl-Arginine, Carnitine, Choline, Creatinine, Histidine, Methionine, N-acetyl-spermidine, Proline betaine, Tripeptide, Capric acid, Caprylic acid, Lauric acid, Myristic acid, Palmitic acid, Palmitoleic acid, Prostaglandin, Serotonin, Ceramides (Cers), Diacylglycerols (DAG), Polyunsaturated fatty acids (PUFAs), Urea, Uric acid, Valine, α-Ketoglutarate, progesterones, Pregnanolones, metabolites belonging to brain energetics, urea, amino acid, and glucose utilization metabolism (Yin et al. 2023).

22.27 Conclusions and Future Directions

Considering the diversified yet evident alterations in the metabolic pathways, the non-communicable metabolic diseases are of great importance to be explored in width and depth. Since the metabolic alterations occur before the onset of symptoms or the progression of the disease, they offer a window period for the physicians and patients to act on prevention or delay of the disease onset. Systems biology has a commendable and critical role in achieving this objective in this era. With the expanding knowledge base in the field of genetics, metabolic pathways, and epigenetics, systems biology has a vast data for exploration of potential biomarkers that can be employed in diagnosis, prevention, treatment, and management of diseases in general, and metabolic diseases in particular. Currently, all or majority of the biomarkers that have been identified are of benefit in early diagnosis of a given disease and very few biomarkers have been related of use in treatment. Future studies are required to further narrow down the identified pool of early diagnostic biomarkers to one or few highly reliable signature markers for each disease. Thus, narrowed down biomarkers need to be clinically validated on high sample numbers with critical importance to use easily accessible clinical samples such as blood, plasma, and urine for ease of diagnosis. Early diagnosis enables one to take necessary preventive actions and lifestyle changes to delay the disease onset, and manage the disease symptoms. However, more studies are to be undertaken to determine potential biomarkers specifically for treatment of metabolic diseases. With the ever-growing database on metabolic pathways and their inter-relation among health and diseases would be of great merit to achieve this objective in near future.

References

Aa JY, Wang GJ, Hao HP, Huang Q, Lu YH, Yan B et al (2010) Differential regulations of blood pressure and perturbed metabolism by total ginsenosides and conventional antihypertensive agents in spontaneously hypertensive rats. Acta Pharmacol Sin 31:930–937

Abbiss H, Maker GL, Trengove RD (2019) Metabolomics approaches for the diagnosis and understanding of kidney diseases. Meta 9(2):34

Akira K, Masu S, Imachi M, Mitome H, Hashimoto M, Hashimoto T (2008) 1H NMR-based metabonomic analysis of urine from young spontaneously hypertensive rats. J Pharm Biomed Anal 46:550–556

Allalou A, Nalla A, Prentice KJ, Liu Y, Zhang M, Dai FF, Ning X, Osborne LR, Cox BJ, Gunderson EP, Wheeler MB (2016) A predictive metabolic signature for the transition from gestational diabetes mellitus to type 2 diabetes. Diabetes 65(9):2529–2539

Bagnardi V, Rota M, Botteri E, Tramacere I, Islami F, Fedirko V et al (2015) Alcohol consumption and site-specific cancer risk: a comprehensive dose-response meta-analysis. Br J Cancer 112(3):580–593

Bao L, Yang C, Shi Z, Wang Z, Jiang D (2021) Analysis of serum metabolomics in obese mice induced by high-fat diet. Diabetes 14:4671

Berding K, Donovan SM (2016) Microbiome and nutrition in autism spectrum disorder: current knowledge and research needs. Nutr Rev 74(12):723–736

Björntorp P (1996) Behavior and metabolic disease. Int J Behav Med 3(4):285–302

Blüher M (2013) Adipose tissue dysfunction contributes to obesity related metabolic diseases. Best Pract Res Clin Endocrinol Metab 27(2):163–177

Boos CJ, Lip GY (2006) Is hypertension an inflammatory process? Curr Pharm Des 12:1623–1635

Brister D, Rose S, Delhey L, Tippett M, Jin Y, Gu H, Frye RE (2022) Metabolomic signatures of autism spectrum disorder. J Personal Med 12(10):1727

Bujak R, García-álvarez A, Rupérez FJ, Nuno-Ayala M, García A, Ruiz-Cabello J, Fuster V, Ibánez B, Barbas C (2014) Metabolomics reveals metabolite changes in acute pulmonary embolism. J Proteome Res 13:805–816

Chen X, Liu L, Palacios G, Gao J, Zhang N, Li G, Lu J, Song T, Zhang Y, Lv H (2010) Plasma metabolomics reveals biomarkers of the atherosclerosis. J Sep Sci 33:2776–2783

Deng Y, Huang C, Su J, Pan CW, Ke C (2021) Identification of biomarkers for essential hypertension based on metabolomics. Nutr Metab Cardiovasc Dis 31(2):382–395

Dumas ME, Kinross J, Nicholson JK (2014) Metabolic phenotyping and systems biology approaches to understanding metabolic syndrome and fatty liver disease. Gastroenterology 146(1):46–62

el Alaoui-Talibi Z, Landormy S, Loireau A et al (1992) Fatty acid oxidation and mechanical performance of volume-overloaded rat hearts. Am J Phys 262:1068–1074

Fang X, Ardehali H, Min J, Wang F (2022) The molecular and metabolic landscape of iron and ferroptosis in cardiovascular disease. Nat Rev Cardiol 4:1–7

Fernandes J, Saudubray JM, Van den Berghe G, Walter JH (eds) (2006) Inborn metabolic diseases: diagnosis and treatment. Springer, Cham

Fujiwara M, Arifuku K, Ando I, Nemoto T (2005) Pattern recognition analysis for classification of hypertensive model rats and diurnal variation using 1H-NMR spectroscopy of urine. Anal Sci 21:1259–1262

Global Burden of Disease Collaborative Network (2020) Global burden of disease study 2019. Results. Institute for Health Metrics and Evaluation, Seattle

Hastings G, McDermott L, Angus K, Stead M, Thomson S (2006) The extent, nature and effects of food promotion to children: a review of the evidence. World Health Organization, Geneva
Holmes E, Loo RL, Stamler J, Bictash M, Yap IK, Chan Q et al (2008) Human metabolic phenotype diversity and its association with diet and blood pressure. Nature 453:396–400
Hoyles L, Jiménez-Pranteda ML, Chilloux J et al (2018) Metabolic retroconversion of trimethylamine N-oxide and the gut microbiota. Microbiome 6:73
Htun KT, Pan J, Pasanta D, Tungjai M, Udomtanakunchai C, Chancharunee S, Kaewjaeng S, Kim HJ, Kaewkhao J, Kothan S (2021) Identification of metabolic phenotypes in young adults with obesity by 1H NMR metabolomics of blood serum. Life 11(6):574
Huang CC, McDermott MM, Liu K, Kuo CH, Wang SY, Tao H, Tseng YJ (2013) Plasma metabolomic profiles predict near-term death among individuals with lower extremity peripheral arterial disease. J Vasc Surg 58(4):989–996
Humer E, Pieh C, Probst T (2020) Metabolomic biomarkers in anxiety disorders. Int J Mol Sci 21(13):4784
Ingwall JS (2002) ATP and the heart. Springer, Cham
Jin Q, Ma RC (2021) Metabolomics in diabetes and diabetic complications: Insights from epidemiological studies. Cell 10(11):2832
Jove M, Ayala V, Ramirez-Núnez O, Serrano JCE, Cassanyé A, Arola L, Caimari A, del Bas JM, Crescenti A, Pamplona R, Portero-Otin M (2013) Lipidomic and metabolomic analyses reveal potential plasma biomarkers of early atheromatous plaque formation in hamsters. Cardiovasc Res 97:642–652
Jové M, Mauri-Capdevila G, Suárez I, Cambray S, Sanahuja J, Quílez A, Farré J, Benabdelhak I, Pamplona R, Portero-Otín M, Purroy F (2015) Metabolomics predicts stroke recurrence after transient ischemic attack. Neurology 84:36–45. https://doi.org/10.1212/WNL.0000000000001093
Kim HY (2021) Recent advances in nonalcoholic fatty liver disease metabolomics. Clin Mol Hepatol 27(4):553
Kim M, Jung S, Kim SY, Lee SH, Lee JH (2014) Prehypertension-associated elevation in circulating lysophosphatidlycholines, Lp-PLA2 activity, and oxidative stress. PLoS One 9(5):e96735
Koeth RA, Wang Z, Levison BS, Buffa JA, Org E, Sheehy BT, Britt EB, Fu X, Wu Y, Li L, Smith JD, DiDonato JA, Chen J, Li H, Wu GD, Lewis JD, Warrier M, Brown JM, Krauss RM, Tang WH, Bushman FD, Lusis AJ, Hazen SL (2013) Intestinal microbiota metabolism of L-carnitine, a nutrient in red meat, promotes atherosclerosis. Nat Med 19:576–585. https://doi.org/10.1038/nm.3145
Kohlhaas M, Maack C (2010) Adverse bioenergetic consequences of Na+-Ca2+ exchanger-mediated Ca2+ influx in cardiac myocytes. Circulation 122:2273–2280
Kordalewska M, Markuszewski MJ (2015) Metabolomics in cardiovascular diseases. J Pharm Biomed Anal 113:121–136
Kronenberg F, Mora S, Stroes ES, Ference BA, Arsenault BJ, Berglund L, Dweck MR, Koschinsky M, Lambert G, Mach F, McNeal CJ (2022) Lipoprotein (a) in atherosclerotic cardiovascular disease and aortic stenosis: a European Atherosclerosis Society consensus statement. Eur Heart J 43(39):3925–3946
Larroya-García A, Navas-Carrillo D, Orenes-Piñero E (2019) Impact of gut microbiota on neurological diseases: diet composition and novel treatments. Crit Rev Food Sci Nutr 59(19):3102–3116
Liu Y, Chen T, Qiu Y, Cheng Y, Cao Y, Zhao A, Jia W (2011) An ultrasonication-assisted extraction and derivatization protocol for GC/TOFMS-based metabolite profiling. Anal Bioanal Chem 400:1405–1417
Liu W, Xu J, Dai T, Shen G, Feng J (2021) An integrating strategy for serum metabolomics and microarray analysis to expand the understanding of diet-induced obesity. Anal Methods 13(28):3127–3135
Lva Y, Liua X, Yanb S, Lianga X, Yang Y, Dai W, Zhang W (2010) Metabolomic study of myocardial ischemia and intervention effects of compound Danshen Tablets in rats using

ultra-performance liquid chromatography/quadrupole time-of-flight mass spectrometry. J Pharmaceut Biomed 52:129–135

McCormack SE, Shaham O, McCarthy MA, Deik AA, Wang TJ, Gerszten RE, Clish CB, Mootha VK, Grinspoon SK, Fleischman A (2013) Circulating branched-chain amino acid concentrations are associated with obesity and future insulin resistance in children and adolescents. Pediatr Obes 8(1):52–61

Ming-Qian S, Jian-Xun L, Lan M, Jin C, Cheng-Ren L, Lei L, Jianxun R (2012) LC coupled with TOFMS for metabonomics study of mini-pigs with atherosclerosis. Chromatographia 75:491–497

Müller J, Bertsch T, Volke J, Schmid A, Klingbeil R, Metodiev Y, Karaca B, Kim SH, Lindner S, Schupp T, Kittel M (2021) Narrative review of metabolomics in cardiovascular disease. J Thorac Dis 13(4):2532

Muoio DM, Newgard CB (2008) Mechanisms of disease: molecular and metabolic mechanisms of insulin resistance and beta-cell failure in type 2 diabetes. Nat Rev Mol Cell Biol 9:193–205

Murgia M, Giorgi C, Pinton P et al (2009) Controlling metabolism and cell death: at the heart of mitochondrial calcium signalling. J Mol Cell Cardiol 46:781–788

Neubauer S (2007) The failing heart—an engine out of fuel. N Engl J Med 356(11):1140–1151

Newgard CB (2017) Metabolomics and metabolic diseases: where do we stand? Cell Metab 25(1):43–56

Newgard CB, An J, Bain JR, Muehlbauer MJ, Stevens RD, Lien LF, Haqq AM, Shah SH, Arlotto M, Slentz CA et al (2009) A branched chain amino acid-related metabolic signature that differentiates lean humans and contributes to insulin resistance. Cell Metab 9:311–326

Ordway RW, Singer JJ, Walsh JV Jr (1991) Direct regulation of ion channels by fatty acids. Trends Neurosci 14:96–100

Pallares-Méndez R, Aguilar-Salinas CA, Cruz-Bautista I, del Bosque-Plata L (2016) Metabolomics in diabetes, a review. Ann Med 48(1-2):89–102

Park JY, Lee SH, Shin MJ et al (2015) Alteration in metabolic signature and lipid metabolism in patients with angina pectoris and myocardial infarction. PLoS One 10:e0135228

Ren J, Bi Y, Sowers JR, Hetz C, Zhang Y (2021) Endoplasmic reticulum stress and unfolded protein response in cardiovascular diseases. Nat Rev Cardiol 18(7):499–521

Rhodes CJ, Ghataorhe P, Wharton J, Rue-Albrecht KC, Hadinnapola C, Watson G, Bleda M, Haimel M, Coghlan G, Corris PA, Howard LS (2017) Plasma metabolomics implicates modified transfer RNAs and altered bioenergetics in the outcomes of pulmonary arterial hypertension. Circulation 135(5):460–475

Rizza S, Copetti M, Rossi C, Cianfarani MA, Zucchelli M, Luzi A, Pecchioli C, Porzio O, Di Cola G, Urbani A, Pellegrini F, Federici M (2014) Metabolomics signature improves the prediction of cardiovascular events in elderly subjects. Atherosclerosis 232:260–264

Sabatine MS, Liu E, Morrow DA, Heller E, McCarroll R, Wiegand R, Berriz GF, Roth FP, Gerszten RE (2005) Metabolomic identification of novel biomarkers of myocardial ischemia. Circulation 112:3868–3875

Saito Y, Tanaka A, Node K, Kobayashi Y (2021) Uric acid and cardiovascular disease: a clinical review. J Cardiol 78(1):51–57

Salek RM, Maguire ML, Bentley E, Rubtsov DV, Hough T, Cheeseman M, Nunez D, Sweatman BC, Haselden JN, Cox RD, Connor SC (2007) A metabolomic comparison of urinary changes in type 2 diabetes in mouse, rat, and human. Physiol Genomics 29(2):99–108

Sansbury BE, DeMartino AM, Xie Z et al (2014) Metabolomic analysis of pressure-overloaded and infarcted mouse hearts. Circ Heart Fail 7:634–642

Sas KM, Karnovsky A, Michailidis G, Pennathur S (2015) Metabolomics and diabetes: analytical and computational approaches. Diabetes 64(3):718–732

Scholtens DM, Muehlbauer MJ, Daya NR, Stevens RD, Dyer AR, Lowe LP, Metzger BE, Newgard CB, Bain JR, Lowe WL Jr, HAPO Study Cooperative Research Group (2014) Metabolomics reveals broad-scale metabolic perturbations in hyperglycemic mothers during pregnancy. Diabetes Care 37(1):158–166

Scriver CR (2008) Garrod's Croonian Lectures (1908) and the charter 'Inborn Errors of Metabolism': albinism, alkaptonuria, cystinuria, and pentosuria at age 100 in 2008. J Inherit Metab Dis 31(5):580–598
Shah SH, Sun J-L, Stevens RD, Bain JR, Muehlbauer MJ, Pieper KS, Haynes C, Hauser ER, Kraus WE, Granger CB, Newgard CB, Califf RM, Newby LK (2012) Baseline metabolomic profiles predict cardiovascular events in patients at risk for coronary artery disease. Am Heart J 163:844–850
Suhre K, Meisinger C, Döring A, Altmaier E, Belcredi P, Gieger C, Chang D, Milburn MV, Gall WE, Weinberger KM, Mewes HW (2010) Metabolic footprint of diabetes: a multiplatform metabolomics study in an epidemiological setting. PLoS One 5(11):e13953
Szczerbinski L, Wojciechowska G, Olichwier A, Taylor MA, Puchta U, Konopka P, Paszko A, Citko A, Goscik J, Fiehn O, Fan S (2022) Untargeted metabolomics analysis of the serum metabolic signature of childhood obesity. Nutrients 14(1):214
Taegtmeyer H, Golfman L, Sharma S, Razeghi P, van Arsdall M (2004) Linking gene expression to function: metabolic flexibility in the normal and diseased heart. Ann N Y Acad Sci 1015(1):202–213
Tang WH, Hazen SL (2014) The contributory role of gut microbiota in cardiovascular disease. J Clin Invest 124:4204–4211
Tang WH, Wang Z, Levison BS, Koeth RA, Britt EB, Fu X, Wu Y, Hazen SL (2013) Intestinal microbial metabolism of phosphatidylcholine and cardiovascular risk. N Engl J Med 368:1575–1584. https://doi.org/10.1056/NEJMoa1109400
Taylor Kimberly W, Wang Y, Pham L, Furie KL, Gerszten RE (2013) Metabolite pro-filing identifies a branched chain amino acid signature in acute cardioembolic stroke. Stroke 44:1389–1395
Villoslada P, Steinman L, Baranzini SE (2009) Systems biology and its application to the understanding of neurological diseases. Ann Neurol 65(2):124–139
Vorkas PA, Shalhoub J, Isaac G, Want EJ, Nicholson JK, Holmes E, Davies AH (2015) Metabolic phenotyping of atherosclerotic plaques reveals latent associations between free cholesterol and ceramide metabolism in atherogenesis. J Proteome Res 14:1389–1399
Wang TJ, Larson MG, Vasan RS, Cheng S, Rhee EP, McCabe E, Lewis GD, Fox CS, Jacques PF, Fernandez C et al (2011a) Metabolite profiles and the risk of developing diabetes. Nat Med 17:448–453
Wang TJ, Larson MG, Vasan RS, Cheng S, Rhee EP, McCabe E, Lewis GD, Fox CS, Jacques PF, Fernandez C, O'Donnell CJ (2011b) Metabolite profiles and the risk of developing diabetes. Nat Med 17(4):448–453
Wang L, Hou E, Wang Z, Sun N, He L, Chen L, Liang M, Tian Z (2014) Analysis of metabolites in plasma reveals distinct metabolic features between Dahl salt-sensitive rats and consomic SS. 13BN rats. Biochem Biophys Res Commun 450(1):863–869
Wang Z, Roberts AB, Buffa JA et al (2015a) Non-lethal inhibition of gut microbial trimethylamine production for the treatment of atherosclerosis. Cell 163:1585–1595
Wang L, Hou E, Wang L, Wang Y, Yang L, Zheng X, Xie G, Sun Q, Liang M, Tian Z (2015b) Reconstruction and analysis of correlation networks based on GC–MS metabolomics data for young hypertensive men. Anal Chim Acta 854:95–105
Wang YN, Ma SX, Chen YY, Chen L, Liu BL, Liu QQ, Zhao YY (2019) Chronic kidney disease: biomarker diagnosis to therapeutic targets. Clin Chim Acta 499:54–63
Wang-Sattler R, Yu Z, Herder C, Messias AC, Floegel A, He Y, Heim K, Campillos M, Holzapfel C, Thorand B, Grallert H (2012) Novel biomarkers for pre-diabetes identified by metabolomics. Mol Syst Biol 8(1):615
WHO (2018) Global status report on alcohol and health. World Health Organization, Geneva
WHO (2021) Obesity and overweight. World Health Organization, Geneva
WHO (2022a) Tobacco. World Health Organization, Geneva
WHO (2022b) Diet. World Health Organization, Geneva
World Health Organization (2020) Global health estimates 2019: deaths by cause, age, sex, by country and by region, 2000–2019. World Health Organization, Geneva

World Health Organization (2022) Invisible numbers: the true extent of noncommunicable diseases and what to do about them. World Health Organization, Geneva

Würtz P, Raiko JR, Magnussen CG, Soininen P, Kangas AJ, Tynkkynen T, Thomson R, Laatikainen R, Savolainen MJ, Laurikka J, Kuukasjärvi P, Tarkka M, Karhunen PJ, Jula A, Viikari JS, Kähönen M, Lehtimäki T, Juonala M, Ala-Korpela M, Raitakari OT (2012) High-throughput quantification of circulating metabolites improves prediction of subclinical atherosclerosis. Eur Heart J 33:2307–2316

Yang RY, Wang SM, Sun L et al (2015) Association of branched-chain amino acids with coronary artery disease: a matched-pair case-control study. Nutr Metab Cardiovasc Dis 25:937–942

Yang M, Liu S, Zhang C (2022) The related metabolic diseases and treatments of obesity. Healthc Forum 10(9):1616

Yin C, Harms AC, Hankemeier T, Kindt A, de Lange EC (2023) Status of metabolomic measurement for insights in Alzheimer's disease progression—what is missing? Int J Mol Sci 24(5):4960

Zhang H-Y, Chen X, Hu P, Liang Q-L, Liang X-P, Wang Y-M, Luo G-A (2009) Metabolomic profiling of rat serum associated with isoproterenol-induced myocardial infarction using ultra-performance liquid chromatography/time-of-flight mass spectrometry and multivariate analysis. Talanta 79:254–259

Zhang AH, Qiu S, Xu HY, Sun H, Wang XJ (2014) Metabolomics in diabetes. Clin Chim Acta 429:106–110

Zhang H, Zuo JJ, Dong SS, Lan Y, Wu CW, Mao GY, Zheng C (2020) Identification of potential serum metabolic biomarkers of diabetic kidney disease: a widely targeted metabolomics study. J Diabetes Res 2020:3049098

Zhao YY (2013) Metabolomics in chronic kidney disease. Clin Chim Acta 422:59–69

Zhao Y, Peng J, Lu C, Hsin M, Mura M, Wu L, Chu L, Zamel R, Machuca T, Waddell T, Liu M, Keshavjee S, Granton J, de Perrot M (2014) Metabolomic heterogeneity of pulmonary arterial hypertension. PLoS ONE 9:e88727

Zhong W, Zhou Z (2014) Alterations of the gut microbiome and metabolome in alcoholic liver disease. World J Gastrointest Pathophysiol 5(4):514

Zhu W, Gregory JC, Org E et al (2016) Gut microbial metabolite TMAO enhances platelet hyperreactivity and thrombosis risk. Cell 165:111–124

MIX
Papier aus verantwortungsvollen Quellen
Paper from responsible sources
FSC® C105338

If you have any concerns about our products,
you can contact us on
ProductSafety@springernature.com

In case Publisher is established outside the EU,
the EU authorized representative is:
Springer Nature Customer Service Center GmbH
Europaplatz 3, 69115 Heidelberg, Germany

Printed by Libri Plureos GmbH
in Hamburg, Germany